HANDBOOK OF PRODUCT GRAPHS

SECOND EDITION

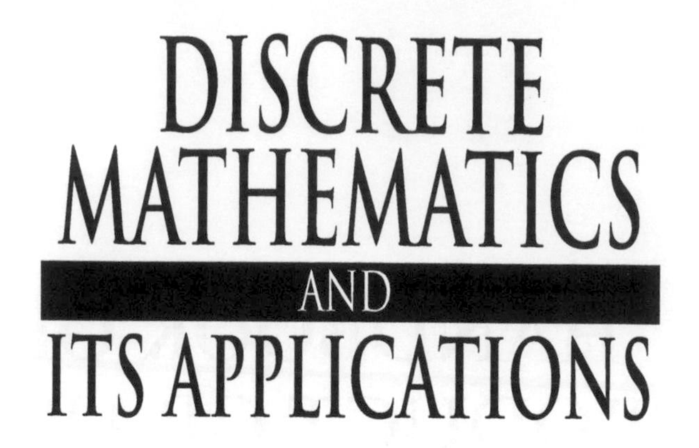

Series Editor
Kenneth H. Rosen, Ph.D.

R. B. J. T. Allenby and Alan Slomson, How to Count: An Introduction to Combinatorics, Third Edition

Juergen Bierbrauer, Introduction to Coding Theory

Donald Bindner and Martin Erickson, A Student's Guide to the Study, Practice, and Tools of Modern Mathematics

Francine Blanchet-Sadri, Algorithmic Combinatorics on Partial Words

Richard A. Brualdi and Dragoš Cvetković, A Combinatorial Approach to Matrix Theory and Its Applications

Kun-Mao Chao and Bang Ye Wu, Spanning Trees and Optimization Problems

Charalambos A. Charalambides, Enumerative Combinatorics

Gary Chartrand and Ping Zhang, Chromatic Graph Theory

Henri Cohen, Gerhard Frey, et al., Handbook of Elliptic and Hyperelliptic Curve Cryptography

Charles J. Colbourn and Jeffrey H. Dinitz, Handbook of Combinatorial Designs, Second Edition

Martin Erickson, Pearls of Discrete Mathematics

Martin Erickson and Anthony Vazzana, Introduction to Number Theory

Steven Furino, Ying Miao, and Jianxing Yin, Frames and Resolvable Designs: Uses, Constructions, and Existence

Mark S. Gockenbach, Finite-Dimensional Linear Algebra

Randy Goldberg and Lance Riek, A Practical Handbook of Speech Coders

Jacob E. Goodman and Joseph O'Rourke, Handbook of Discrete and Computational Geometry, Second Edition

Jonathan L. Gross, Combinatorial Methods with Computer Applications

Jonathan L. Gross and Jay Yellen, Graph Theory and Its Applications, Second Edition

Jonathan L. Gross and Jay Yellen, Handbook of Graph Theory

David S. Gunderson, Handbook of Mathematical Induction: Theory and Applications

Richard Hammack, Wilfried Imrich, and Sandi Klavžar, Handbook of Product Graphs, Second Edition

Darrel R. Hankerson, Greg A. Harris, and Peter D. Johnson, Introduction to Information Theory and Data Compression, Second Edition

Darel W. Hardy, Fred Richman, and Carol L. Walker, Applied Algebra: Codes, Ciphers, and Discrete Algorithms, Second Edition

Titles (continued)

Daryl D. Harms, Miroslav Kraetzl, Charles J. Colbourn, and John S. Devitt, Network Reliability: Experiments with a Symbolic Algebra Environment

Silvia Heubach and Toufik Mansour, Combinatorics of Compositions and Words

Leslie Hogben, Handbook of Linear Algebra

Derek F. Holt with Bettina Eick and Eamonn A. O'Brien, Handbook of Computational Group Theory

David M. Jackson and Terry I. Visentin, An Atlas of Smaller Maps in Orientable and Nonorientable Surfaces

Richard E. Klima, Neil P. Sigmon, and Ernest L. Stitzinger, Applications of Abstract Algebra with Maple™ and MATLAB®, Second Edition

Patrick Knupp and Kambiz Salari, Verification of Computer Codes in Computational Science and Engineering

William Kocay and Donald L. Kreher, Graphs, Algorithms, and Optimization

Donald L. Kreher and Douglas R. Stinson, Combinatorial Algorithms: Generation Enumeration and Search

Hang T. Lau, A Java Library of Graph Algorithms and Optimization

C. C. Lindner and C. A. Rodger, Design Theory, Second Edition

Nicholas A. Loehr, Bijective Combinatorics

Alasdair McAndrew, Introduction to Cryptography with Open-Source Software

Elliott Mendelson, Introduction to Mathematical Logic, Fifth Edition

Alfred J. Menezes, Paul C. van Oorschot, and Scott A. Vanstone, Handbook of Applied Cryptography

Richard A. Mollin, Advanced Number Theory with Applications

Richard A. Mollin, Algebraic Number Theory, Second Edition

Richard A. Mollin, Codes: The Guide to Secrecy from Ancient to Modern Times

Richard A. Mollin, Fundamental Number Theory with Applications, Second Edition

Richard A. Mollin, An Introduction to Cryptography, Second Edition

Richard A. Mollin, Quadratics

Richard A. Mollin, RSA and Public-Key Cryptography

Carlos J. Moreno and Samuel S. Wagstaff, Jr., Sums of Squares of Integers

Dingyi Pei, Authentication Codes and Combinatorial Designs

Kenneth H. Rosen, Handbook of Discrete and Combinatorial Mathematics

Douglas R. Shier and K.T. Wallenius, Applied Mathematical Modeling: A Multidisciplinary Approach

Alexander Stanoyevitch, Introduction to Cryptography with Mathematical Foundations and Computer Implementations

Jörn Steuding, Diophantine Analysis

Douglas R. Stinson, Cryptography: Theory and Practice, Third Edition

Roberto Togneri and Christopher J. deSilva, Fundamentals of Information Theory and Coding Design

W. D. Wallis, Introduction to Combinatorial Designs, Second Edition

W. D. Wallis and J. C. George, Introduction to Combinatorics

Lawrence C. Washington, Elliptic Curves: Number Theory and Cryptography, Second Edition

Titles (continued)

DISCRETE MATHEMATICS AND ITS APPLICATIONS
Series Editor KENNETH H. ROSEN

HANDBOOK OF PRODUCT GRAPHS

SECOND EDITION

Richard Hammack
Virginia Commonwealth University
Richmond, USA

Wilfried Imrich
Montanuniversität Leoben
Austria

Sandi Klavžar
University of Ljubljana and University of Maribor
Slovenia

CRC Press is an imprint of the
Taylor & Francis Group, an informa business
A CHAPMAN & HALL BOOK

CRC Press
Taylor & Francis Group
6000 Broken Sound Parkway NW, Suite 300
Boca Raton, FL 33487-2742

First issued in paperback 2016

Version Date: 20110801

ISBN 13: 978-1-138-19908-8 (pbk)
ISBN 13: 978-1-4398-1304-1 (hbk)

Library of Congress Cataloging-in-Publication Data

Hammack, Richard H.
Handbook of product graphs / Richard Hammack, Wilfried Imrich, Sandi Klavzar. -- 2nd ed.
p. cm. -- (Discrete mathematics and its applications)
Extensively revised, reorganized, updated, and expanded ed. of: Product graphs, structure, and recognition / Wilfried Imrich, Sandi Klavžar. 2011.
Includes bibliographical references and index.
ISBN 978-1-4398-1304-1 (hardcover : alk. paper)
I. Imrich, Wilfried, 1941- II. Klavžar, Sandi, 1962- III. Imrich, Wilfried, 1941- Product graphs, structure, and recognition. IV. Title. V. Series.

QA166.I47 2011
511'.52--dc23 2011026632

Visit the Taylor & Francis Web site at
http://www.taylorandfrancis.com

and the CRC Press Web site at
http://www.crcpress.com

Contents

Foreword

It is my pleasure to introduce you to the marvelous world of graph products, as presented by three experts in a hugely expanded and updated edition of the year 2000 classic by Imrich and Klavžar. This version, really a new book (thirty-three chapters, up from nine!), contains streamlined proofs, new applications, solutions to conjectures (such as Vizing's conjecture for chordal graphs), and new results in graph minors and flows. Every graph theorist, most combinatorialists, and many other mathematicians will want this volume in their collection.

Graphs are, of course, basic combinatorial structures; and products of structures are a fundamental construction in mathematics, for which theorems abound in set theory, category theory, and universal algebra. Thus, it is not surprising that good things happen when we take products of graphs. But the nature of those good things is very surprising indeed; many unique and new ideas emerge, taking both combinatorialists and algebraists by surprise.

For example, we expect many of the nice properties of products to be a result of a role played in some category-theoretic construct, but the authors' (and my own) favorite graph product, the Cartesian product, does not arise in that way. Nonetheless, it not only behaves beautifully, but carries metric space structure with it. Similarly, we might expect it to be trivial to determine how natural graph parameters like independence number and chromatic number behave in any product of graphs; but Hedetniemi's conjecture (Chapter 26), among other open problems, suggests that something quite deep is happening.

In graph theory, we often expect it to be algorithmically difficult to compute or even estimate most parameters or to recognize structure in a given graph. If we can't even get a decent approximation of the size of the largest clique in a graph, how can we expect to compute a more complex parameter? If we can't even tell when a graph is the covering graph of a partially ordered set, how can we expect to recognize products?

Consider the "windex" problem of Chapter 14. A server resides at a vertex of some fixed graph, and requested vertices arrive in a stream. At each arrival we pay a cost equal to the distance of the server to the request; between arrivals we can move the server, again at distance-cost, to a new vertex. How far in advance must we see the requests in order to play optimally?

This looks like a tough problem; indeed, it takes some study to see that the answer is "2" when the graph consists of two vertices connected by an edge. Yet, with the help of graph product recognition, we can compute this value in polynomial time for an arbitrary graph!

The authors have paid careful attention to algorithmic issues (indeed, many of the most attractive algorithms are products of their own research). Readers will find a gentle but incisive introduction to graph algorithms here, and a persuasive lesson on the insights to be gained by algorithmic analysis.

In sum—and product—Hammack, Imrich, and Klavžar have put together a world of elegant and useful results in a cogent, readable text. The old book was already a delight, and you will want the new one in an accessible place on your bookshelf.

PETER WINKLER *Hanover, New Hampshire*

Preface

EVERY branch of mathematics employs some notion of a product that enables the combination or decomposition of its elemental structures. In graph theory there are four main products, each with its own set of applications and theoretical interpretations.

The structure and applicability of these products are full of surprises. For example, large networks such as the Internet graph, with several hundred million hosts, can be efficiently modeled by subgraphs of powers of small graphs with respect to the direct product.[1] This is one of many examples of the dichotomy between the structure of products and that of their subgraphs, which is a main theme of the book.

A second theme is the design of efficient algorithms that recognize products and their subgraphs; a third, with the deepest results, is the relationship between graph parameters of the product and those of the factors.

The authors are fascinated by this rich, fertile field and wish to pass this enthusiasm to the reader. They have done their best to write a comprehensive overview of graph products.

Handbook

The first edition of this book has established itself as a standard reference on graph products. This second edition, extensively revised, reorganized, updated and expanded, can justly be called a handbook. It is a thorough introduction to the subject, with full proofs of most of the important results in the first four of its six parts. It is also an extensive survey of the field, with up-to-date research results and conjectures.

Contents

The book is organized into six parts. The first three parts present graph products in detail. They cover algebraic properties such as factorization and cancellation, and treat interesting, important classes of subgraphs. They contain many new, comprehensive proofs, and a wealth of new results. The fourth part pertains to algorithms, many of them new, described in such a way that they can be easily implemented. The fifth part focuses on graph invariants and the sixth on infinite, directed, and product-like graphs.

Part I prepares the reader for any one of Parts II, III, and V. Other dependencies will be described in the introduction to the corresponding parts, but we mention here that Part IV uses some material from Parts I, II, and III, while Part VI employs ideas from throughout the entire book.

Prerequisites

Part I is a concise introduction to graphs and their products. It is intended to make the book accessible to the nonspecialist, and requires few mathematical prerequisites. Part II and III build on Part I but are otherwise essentially self-contained. Part IV focuses on algorithms. For the reader's convenience it contains two chapters that introduce graph algorithms and do not depend on the rest of the book. This part thus requires only very modest previous knowledge of data structures and algorithms.

Parts V and VI pertain to many different areas of combinatorics and graph theory. While

[1] Leskovec, Chakrabarti, Kleinberg, Faloutsos, and Ghahramani (2010).

the statements of the theorems should be clear, their proofs, when given, require diverse prerequisites of varying depth.

To the instructor

The book is suitable as a text for a one- or two-semester course at the graduate level. Such a course would cover Part I, with the remaining material dictated by the interests of the instructor and students. We recommend Parts I, II, and III as a core. A course that emphasizes algorithms and complexity would include Part IV, whereas one that highlights invariants would cover Part V. However, we underscore that any sequence that respects the dependencies of the chapters could be the foundation of a meaningful course.

The text contains well over 300 exercises, almost all of which have hints or full solutions in the Appendix. In particular, Part V (on invariants) alone has 80 exercises; combined with the fact that its chapters are independent of each other, it is ideal for the instructor of an advanced course who wishes to cover selected topics.

To the student

This book, in particular Parts I through IV, is well-suited for self study. There are numerous exercises, with generous hints and solutions in the Appendix.

Although algorithms are an integral part of the book, and while the algorithmic material of Part IV can deepen one's understanding of product graphs, it should be mentioned that very little of the book depends directly on Part IV. The reader who chooses to ignore Part IV will find only several passages, in Part VI, that depend on the omitted material.

New material

We include numerous results and algorithms—some solving long-standing questions—that have appeared since the publication of the first edition. They include

- Cancellation results in Chapter 9
- A quadratic recognition algorithm for partial cubes in Chapter 11
- Results on the strong isometric dimension in Chapter 15
- Computing the Wiener index via canonical isometric embedding in Chapter 19
- Most of the connectivity results in Chapter 25
- A fractional version of Hedetniemi's conjecture in Chapter 26
- Results on the independence number of Cartesian powers of vertex-transitive graphs in Chapter 27
- Verification of Vizing's conjecture for chordal graphs in Chapter 28
- The majority of the results on minimum cycle bases in Chapter 29
- Numerous selected recent results (for instance, on complete minors and nowhere-zero flows) in Chapter 30

Applications

Products are often viewed as a convenient language with which to describe structures, but they are increasingly being applied in more substantial ways. Computer science is one of the many fields in which graph products are becoming commonplace. As one specific example, we mention load balancing for massively parallel computer architectures.

While it is not our intention to cover applications in depth, we nonetheless have reserved separate chapters for two of them, namely, Chapter 14 for the dynamic location problem and Chapter 19 for chemical graphs. The applications of median graphs in human genetics, and powers of direct products to model large networks are treated in Sections 12.5 and 30.5, respectively. We can give only short introductions to these topics, as they are vast enough to require books of their own.

Other applications motivate the study of approximate graph products in Chapter 33. One involves making the definition of traits of species more objective, a second aims for more effective methods of visualizing graphs, and a third involves applications in structural mechanics. A similar remark holds for the application of lattice dimension to graph visualization.

We hope this indicates the wealth and variety of applications. It is gratifying to watch the applications accumulate, and we sincerely hope that this book provides the impetus for further growth.

Free, dot, zig-zag, and replacement product

The free product of graphs has recently reappeared in the literature. We provide a concrete example in the section on median graphs with finite blocks. It is linked to the free product of groups in combinatorial group theory.

The dot product, the random dot product, and applications to the generation of random graphs with prescribed properties are briefly described in Section 30.5.

The zig-zag and the more transparent replacement product were introduced for the creation of expander graphs. They have similarities with products considered in this book and allow relatively simple proofs of expansion. We define them in Section 33.4.

Caveat

If you find errors or misleading formulations, please send a note to one of the authors. An errata, sample implementations of algorithms from Part IV, and other useful information will appear on the *Handbook of Product Graphs* website. For the actual address follow the links in the home pages of the authors.

Acknowledgments

We are indebted to many colleagues who helped us with this book by giving useful remarks on the first edition, reading parts of the present book and suggesting improvements, or providing information about the very recent developments.

For the second edition we especially thank Yaad Blum, Boštjan Brešar, Michalina Cienciała, Agelos Georgakopoulos, Marc Hellmuth, Gabi Imrich–Schwarz, Justyna Jabłońska-Gos, Aleksandra Jędrzejaszek, Janja Jerebic, Tomáš Kaiser, Donald Knuth, Tomáš Kupka, Gašper Mekiš, Terry Mills, P. Paulraja, Paweł Petecki, Iztok Peterin, Doug Rall, Heather Smith, Peter Stadler, Simon Špacapan, Marcin Wardyński, Ania Wasieczko, Christiaan van de Woestijne, David Wood, and Xuding Zhu.

Likewise, we wish to recall the support we received for the the first edition: Boštjan Brešar, Ross McConnell, Riste Škrekovski, Claude Tardif, and Xuding Zhu have read large parts of the manuscript and gave numerous invaluable comments. For helpful suggestions and ideas for additional material, we thank Bojan Mohar and Walter Wallis (one-factorizations), Pranava K. Jha (Hamiltonian decomposable graphs), Glenn Chappell and Eugen Mândrescu (perfect product graphs), Tomaž Pisanski (topological embeddings), and Derek Corneil (modular product). We are indebted to Martyn Mulder for his encouragement, and to Bojan Gorenec for designing the cover (of the first edition), as well as to Gabi Imrich–Schwarz for suggesting numerous corrections and improvements.

For this edition special thanks are due to CRC press. In particular, we appreciate the help and enthusiasm of David Grubbs and Bob Stern in the gestation period and beyond.

We wish to thank Karen Simon, the Project Editor, who guided us through critical stages during the final half year. Shashi Kumar deserves credit for invaluable technical support and Katy Smith for the preparation of the back cover and other marketing material. Express thanks are further due for the layout and design of the extraordinarily nice cover.

Last, but not least, we wish to acknowledge support by all those whose work in the background made the book possible.

We are also obliged to Kenneth H. Rosen for inclusion of our book into his series of *Discrete Mathematics and its Applications.*

Finally, we thank our wives — Micol Hammack, Gabi Imrich–Schwarz, and Maja Klavžar — and our children — Adriana and Sage Hammack; Daniel and Peter Imrich; Simon, Tamara, and Julija Klavžar — for their love and support while we worked on this project. Their patience and encouragement made this book possible. We gratefully dedicate it to them.

RICHARD HAMMACK — *Richmond, Virginia*
WILFRIED IMRICH — *Leoben, Austria*
SANDI KLAVŽAR — *Trzin, Slovenia*

Part I

A Brief Introduction to Graphs and Their Products

Introduction to Part I

PART I is a brief introduction to graph theory and product graphs, with an emphasis on ideas and constructions that are used throughout the book. The first two chapters review the elements of graph theory: graphs, subgraphs, homomorphisms, complete and bipartite graphs, paths and cycles, trees, planar graphs, automorphisms, and invariants.

Chapter 2 also treats symmetries of graphs and group actions on trees and hypercubes. We prove that a graph with transitive Abelian automorphism group has a spanning hypercube; this shows how the automorphism group, considered as a permutation group on the vertex set, can decisively determine the structure of a graph. The chapter culminates with the highly-useful No-Homomorphism Lemma.

Chapter 3 develops themes that are central to the remainder of the book: hypercubes, isometric subgraphs, median graphs, and retracts.

Our main objects of study appear in Chapter 4, where we introduce three fundamental graph products, namely the Cartesian product, the direct product, and the strong product. This is followed by an (optional) classification of certain associative products, providing an explanation of why the three products mentioned above are the most natural of all products. The classification also leads to a fourth product worthy of special attention, the lexicographic product, which is the only noncommutative product studied in this book.

Chapter 5 investigates the metric and connectedness properties of the four standard products—material that is of utmost importance in the remainder of the book.

We shall see that multiplication and disjoint union of graphs share many properties with ordinary multiplication and addition. Indeed, the direct product of two relations—and thus graphs—is first discerned in Whitehead and Russell (1912, p. 384), where the operations of arithmetic are extended to binary relations. The lexicographic product is due to Hausdorff (1914). Until the 1950s these concepts were studied mostly in algebraic settings; see Jónsson (1982) for a survey.

We remark that readers who are well-versed in graph theory may opt to skip Chapter 1, or just scan it to glean the notation. However, most readers will find new material in the remaining four chapters of this part.

Chapter 1

Graphs

We begin with a review of the main concept of this book: the idea of a graph. Here we are concerned with graphs, subgraphs, homomorphisms, complete and bipartite graphs, paths, cycles, and planar graphs. Because the great majority of the results in this book pertains to finite, undirected graphs without loops or multiple edges, we focus on this class of graphs. Occasionally directed graphs or graphs with loops or multiple edges are admitted in particular when they help to shorten proofs or improve the presentation.

1.1 Graphs and Subgraphs

By a *simple graph* G we mean a set $V(G)$ of *vertices*, together with a set $E(G)$ of unordered pairs $[u,v]$ of distinct vertices of G, the *edges* of G. We represent graphs graphically by drawing the vertices as nodes and the edges as line segments connecting the nodes, as in Figure 1.1. For an edge $e = [u,v]$ we call u and v its *endpoints* and say e *joins* them. Two vertices u and v are *adjacent* when they are joined by an edge. A vertex u adjacent to v is a *neighbor* of v. One says u and v are *incident* with the edge $[u,v]$; two edges are *incident* if they have a common endpoint. We often denote the edge $[u,v]$ as uv.

A graph is *finite* if its vertex set is finite, and the set of finite simple graphs is denoted by Γ. Included in Γ is the *empty graph* O, for which $V(O) = \emptyset$. Unless stated otherwise, the term *graph* will always mean a member of Γ. A graph is called *nontrivial* if $|V(G)| > 1$. $|V(G)|$ is also known as the *order* and $|E(G)|$ as the *size* of G.

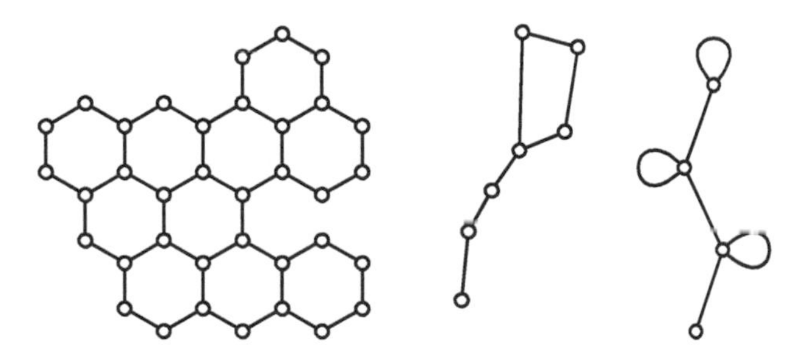

FIGURE 1.1 Simple graphs (left, center) and a graph with loops (right).

Occasionally we admit edges with identical endpoints, calling them *loops*; see Figure 1.1. The set of finite graphs in which loops are admitted is denoted as Γ_0. Note that $\Gamma \subset \Gamma_0$.

Given a vertex v in a graph G, the *neighborhood of* v is defined as

$$N(v) = \{u \mid uv \in E(G)\},$$

that is, the set consisting of all neighbors of v. Note that if there is a loop at v, then $v \in N(v)$. The set $N(v)$ is sometimes referred to as the *open neighborhood* of v, in order to distinguish it from the *closed neighborhood*

$$N[v] = N(v) \cup \{v\}$$

of v. If there is danger of ambiguity, we write $N_G(v)$ and $N_G[v]$ to emphasize the underlying graph.

Sometimes it is useful to assign directions to the edges of a graph. A *digraph* (or *directed graph*) D consists of vertices $V(D)$ together with a set of *arcs* $A(D) \subseteq V(D) \times V(D)$[1]. We visualize an arc (u, v) as an arrow pointing from u to v. Note that it is possible to have both $(u, v), (v, u) \in A(D)$, that is, we allow for the possibility that there is an arrow pointing from u to v and another from v to u.

An *oriented graph* is a digraph obtained from a graph by replacing each edge $[u, v]$ with exactly one arc (u, v) or (v, u).

We emphasize that for brevity we often write uv instead of $[u, v]$ or (u, v). However, whenever the possibility of confusion between the directed and undirected case arises, we revert to the notation (u, v) for ordered pairs of vertices and $[u, v]$ for unordered ones.

Many of the graphs that we investigate are subgraphs of other graphs. By a *subgraph* H of a graph G, we mean a graph H for which $V(H) \subseteq V(G)$ and $E(H) \subseteq E(G)$. We write $H \subseteq G$ to indicate that H is a subgraph of G.

If all pairs of vertices of a subgraph H of G that are adjacent in G are also adjacent in H, then H is an *induced subgraph*. Clearly, an induced subgraph of G is uniquely determined by its vertices. We write $\langle S \rangle$ or $\langle S \rangle_G$ for the subgraph of G induced by $S \subseteq V(G)$. Often we will consider so-called *vertex-deleted subgraphs*, namely the subgraphs of G induced by $V(G) \setminus \{v\}$, where $v \in V(G)$. Such a graph is denoted by $G - v$. Similarly, for $e \in E(G)$, the *edge-deleted subgraph* $G - e$ is defined by $V(G - e) = V(G)$ and $E(G - e) = E(G) \setminus \{e\}$. (See Figure 1.2.) In a similar spirit, if $X \subseteq V(G)$, then $G - X$ is the subgraph of G induced on $V(G) \setminus X$. If $Y \subseteq E(G)$, then $G - Y$ has vertex set $V(G)$ and edges $E(G) \setminus Y$.

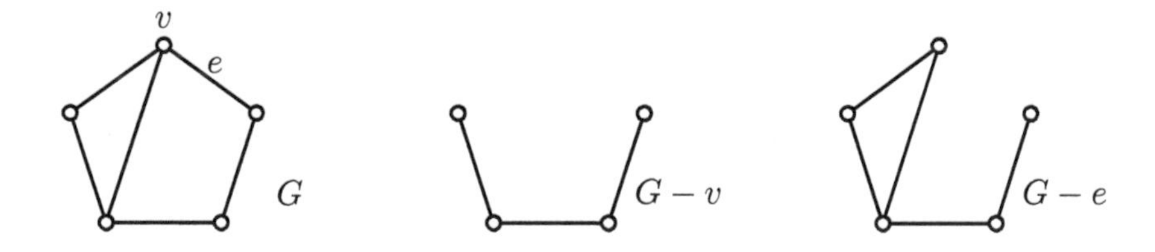

FIGURE 1.2 A graph G with vertex- and edge-deleted subgraphs.

A subgraph H of a graph G is a *spanning subgraph* if it has the same vertex set as G, that is, if $V(H) = V(G)$. Clearly, edge-deleted subgraphs are spanning subgraphs.

Our definitions imply that two graphs G and H are the same, in symbols $G = H$, when $V(G) = V(H)$ and $E(G) = E(H)$. Nonidentical graphs with the same structure are called

[1]The *Cartesian product* $U \times V$ of two sets U and V is defined as the set of all ordered pairs (u, v) with $u \in U$ and $v \in V$.

isomorphic. More precisely, two graphs G and H are called *isomorphic*, in symbols $G \cong H$, if there exists a bijection φ from $V(G)$ onto $V(H)$ that preserves adjacence and nonadjacence, in other words, a mapping for which $[\varphi(u), \varphi(v)] \in E(H)$ if and only if $[u, v] \in E(G)$. Such a mapping φ is called an *isomorphism* between G and H.

We will often be interested in homomorphic images of graphs. We say that φ is a *homomorphism* from a graph G into a graph H if it is an adjacency preserving mapping from $V(G)$ into $V(H)$, namely a mapping for which $[\varphi(u), \varphi(v)] \in E(H)$ whenever $[u, v] \in E(G)$. Note that φ induces a natural mapping from $E(G)$ into $E(H)$, which we also denote by φ. If φ is onto, both for vertices and edges, then H is a *homomorphic image* of G.

Let v be a vertex of a graph G. The *degree* $d(v)$ of a vertex v is defined as the number of vertices to which v is adjacent. Vertices of degree 0 are called *isolated.*

Theorem 1.1 *The sum of the degrees of the vertices of a graph is twice the number of its edges.*

Proof Let G be a graph. The sum $\sum_{u \in V(G)} d(u)$ can be regarded as a counting of the edges of G, where for each vertex u we count the number of edges incident with u and sum the result. This counts each edge exactly twice, once for each of its two endpoints. Thus the sum equals $2|E(G)|$. $\square$

Theorem 1.1 is often called "the first theorem of graph theory." It implies that the number of vertices of odd degree in a graph must be even.

A graph in which every vertex has degree r is *r-regular.* A 3-regular graph is also called *cubic.* By Theorem 1.1 every cubic graph has an even number of vertices.

A graph is the *complete graph* K_n if it is isomorphic to a graph on n vertices for which any two distinct vertices are adjacent. The graph K_1 is also called the *trivial graph,* K_2 an edge, and K_3 a *triangle.* Figure 1.3 depicts the graphs K_i for $3 \leq i \leq 5$. Note that K_n has $\binom{n}{2}$ edges.

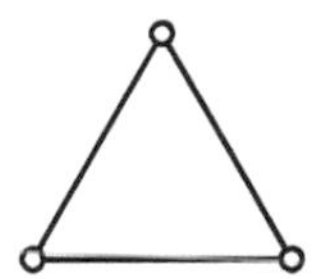
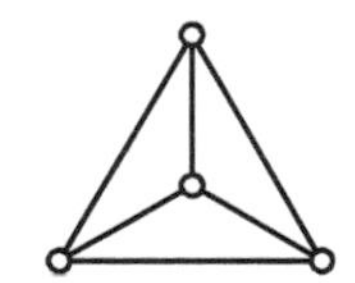
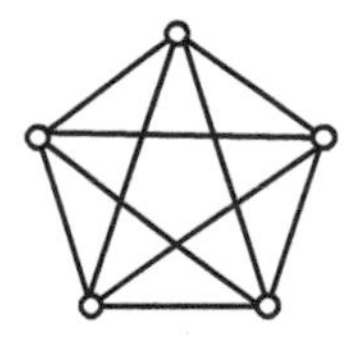

FIGURE 1.3 Complete graphs K_3, K_4, and K_5.

To indicate that G is a complete graph on n vertices, one usually writes $G = K_n$, and not $G \cong K_n$. Thus the equality sign for graphs can mean that the graphs on both sides are the same, namely have the same vertex- and edge-sets, or that the graph on the left side is a member of the class of graphs isomorphic to the graph on the right. In this sense, statements like $G \neq H$ and $G, H = K_n$ do not contradict each other.

A graph G is called *bipartite* if its vertex set can be represented as the union of two disjoint sets V_1 and V_2, such that every edge of G connects an element of V_1 with one of V_2. We call V_1, V_2 a *bipartition* of V. All classes of graphs investigated in Chapter 11 are bipartite.

If every element of V_1 is adjacent to every element of V_2, then G is called a *complete bipartite graph.* It is denoted by $K_{m,n}$, where m and n are the cardinalities of V_1 and V_2. The graph $K_{2,3}$ will play a special role in the characterization of median graphs, and $K_{3,3}$ is the well-known *utility graph.* These graphs are displayed in Figure 1.4.

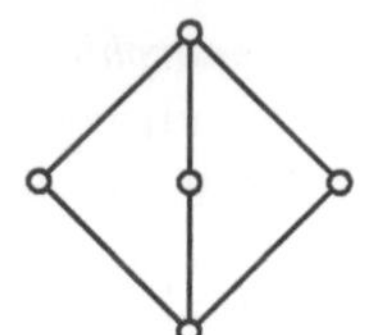
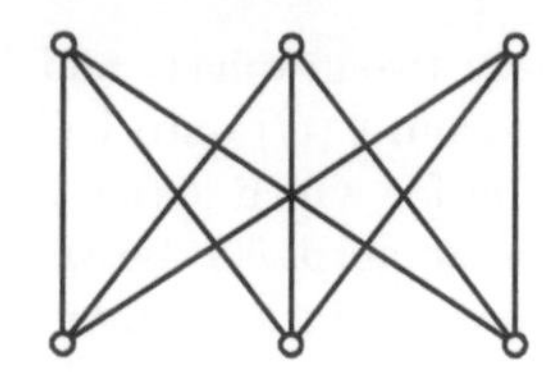

FIGURE 1.4 Graphs $K_{2,3}$ and $K_{3,3}$.

1.2 Paths and Cycles

The path P_n is the graph whose vertices are $1, 2, \ldots, n$, and for which two vertices are adjacent precisely if their difference is ± 1. A *path in a graph* G is a subgraph of G that is isomorphic to some P_n; in other words, a sequence of distinct vertices $v_1 v_2 \ldots v_n$ in G where $v_i v_j$ is an edge of G whenever $i - j = \pm 1$. The vertices v_1 and v_n are called the *endpoints* of the path. A path with endpoints u and v is called a *u, v-path.*

A *walk* in G is a sequence of (not necessarily distinct) vertices $v_1 v_2 \ldots v_n$, such that $v_i v_{i+1} \in E(G)$ for $i = 1, 2, \ldots, n-1$. We call such a walk a *v_1, v_n-walk.* If $v_1 = v_n$, we say it is a *closed walk.* Notice that a walk—unlike a path—is not a subgraph. (A walk contains information about order of traversal, etc.) However there is no harm in regarding a path as a walk joining its endpoints.

If P is a path or a walk, then its *length* $|P|$ is its number of edges. (A given edge of a walk is counted as many times as it appears on the walk.) Note that $|P_n| = n - 1$.

Proposition 1.2 *Every walk contains a path with the same endpoints.*

Proof We use induction on length. The assertion is true for all walks of length zero and one. Suppose that it holds for all walks of length at most n. Let $Q = v_1 v_2 \ldots v_{n+2}$ be a walk of length $n + 1$. If Q is a path, then we are done. Otherwise we have $v_i = v_j$, where $1 \leq i < j \leq n+2$. Removing the edges $v_i v_{i+1}, \ldots, v_{j-1} v_j$ and the vertices v_k with $i < k < j$ from Q, we obtain a walk Q' of length less than n with the same endpoints as Q. By the induction hypothesis it contains a path with the endpoints v_1 and v_{n+2}. □

A graph is *connected* if any two of its vertices are the endpoints of one of its paths. Thus complete graphs, complete bipartite graphs, and paths are connected. A graph is called *disconnected* if it is not connected.

A set X of vertices of a connected graph G is a *separating set* if the graph that we obtain from G by deleting the vertices of X is disconnected. A vertex x of a connected graph is a *cut vertex* if $X = \{x\}$ is a separating set. For disconnected graphs, it is often useful to consider the connected subgraphs that are maximal with respect to inclusion. They are uniquely determined and called *connected components* or simply *components.* If every component of a disconnected graph consists of a single vertex, we say the graph is *totally disconnected.* The totally disconnected graph on n vertices is denoted by D_n.

The *distance* $d_G(u, v)$ between vertices u and v of a graph G is the length of a shortest u, v-path. If no u, v-path exists, then $d_G(u, v) = \infty$. Sometimes we simply write $d(u, v)$ instead of $d_G(u, v)$.

In a connected graph G, the distance function d_G is an integer-valued *metric*, that is, an integer-valued function with the following properties:

1. $d_G(u, v) \geq 0$, equality holding if and only if $u = v$
2. $d_G(u, v) = d_G(v, u)$
3. $d_G(u, v) \leq d_G(u, w) + d_G(w, v)$

The *interval* $I(u, v)$ between two vertices u and v of a graph G is the set of vertices on shortest paths between u and v. Note that $I(u, v)$ contains u and v.

The *diameter* $\mathrm{diam}(G)$ of a connected graph G is the maximum distance between two vertices of G. Thus $\mathrm{diam}(K_n) = 1$ for $n \geq 2$ and $\mathrm{diam}(K_{m,n}) = 2$ if $m + n \geq 3$.

For graphs G and H, we let $G + H$ denote their *disjoint union*, that is, $V(G + H) = V(G) \cup V(H)$ and $E(G + H) = E(G) \cup E(H)$. (We assume here that $V(G) \cap V(H) = \emptyset$. Otherwise we first replace G and H by isomorphic copies of themselves with disjoint vertex sets.) For a positive integer n, we interpret nG as the disjoint union of n copies of G. Thus $D_n = nK_1$.

For an integer $n \geq 3$, the *cycle of length* n, or n-cycle for short, is the graph C_n whose vertices are $0, 1, 2, \ldots, n - 1$, and whose edges are the pairs $[i, i + 1]$, where the arithmetic is done modulo n. Clearly each C_n is connected. The cycle C_3 is also called a *triangle*, and C_4 is called a *square* or *2-cube*. (It is a special case of the r-cube Q_r, which is treated in Chapter 2.) A *cycle in a graph* is a subgraph that is isomorphic to some C_n.

We now prove an analogue to Proposition 1.2.

Proposition 1.3 *Every closed walk without subsequences of the form uvu contains a cycle.*

Proof The assertion is true for all closed walks of length 3. Suppose that it is true for all closed walks of length at most $n \geq 3$. Let $v_1 v_2 \ldots v_{n+1} v_1$ be a closed walk W of length $n+1$. If it is not a cycle, not all v_i can be distinct. Suppose that $v_i = v_j$, where $1 \leq i < j \leq n+1$. Because there is no subsequence of the form uvu, the walk $v_i v_{i+1} \ldots v_j$ has length at least 3. It cannot contain all edges of W, so its length is at most n, and thus it contains a cycle by the induction hypothesis. □

The following result characterizes bipartite graphs:

Proposition 1.4 *A graph is bipartite if and only if all of its cycles have even lengths.*

Proof Let G be bipartite and V_1 and V_2 be two disjoint sets into which $V(G)$ can be partitioned such that every edge has one endpoint in V_1 and one in V_2. Let $v_1 v_2 \ldots v_k v_1$ be a cycle and v_1 be an element of, say, V_1. Then all vertices with even subscripts are in V_2, and v_k can only be adjacent to v_1 if k is even.

For the converse, let G be a graph without odd cycles. We can assume that G is connected; otherwise, we consider the connected components of G separately. We choose a vertex $v \in V(G)$ and define V_1 as the set of all vertices of G that can be reached from v by walks of even lengths. All other vertices constitute V_2. They can only be reached by walks of odd lengths, and thus there can be no edge with both endpoints in V_2. Hence, if G is not bipartite, there must be an edge in G with both endpoints in V_1, which means that G contains a closed walk of odd length. Let $v_1 v_2 \ldots v_{2k+1} v_1$ be such a walk of minimum length. If there are indices $1 \leq i < j < 2k + 1$ for which $v_i = v_j$, then either $v_i v_{i+1} \ldots v_j$ or $v_j v_{j+1} \ldots v_{2k+1} v_1 \ldots v_i$ is a shorter closed walk of odd length. Hence, $v_1 v_2 \ldots v_{2k+1} v_1$ is an odd cycle, contrary to the assumption that G contains no odd cycles. □

As the vertices that can be reached by a walk of even length from v are exactly those that can be reached by a walk of even length from any vertex w in V_1, we infer that the sets V_1 and V_2 are independent of the choice of v. This leads to the following proposition.

Proposition 1.5 *The bipartition of the vertex set of a connected bipartite graph is unique.*

1.3 Trees and Forests

A graph without cycles is called *acyclic*. Clearly, acyclic graphs are bipartite. They are called *trees* when they are connected and *forests* in general. Hence, the connected components of forests are trees. Familiar examples of trees are organization charts, genealogies, and structural formulas for certain chemical compounds. Trees are used to analyze networks or structures and naturally arise in many areas of computer science. This simple concept is one of the most important ones in graph theory and its applications.

In this book trees are of interest not only in themselves but also because they are examples of median graphs and because many of the types of graphs that we will investigate have treelike structure. Moreover, they are a convenient tool in proofs and are relevant to the data structures that we will introduce later.

Here we restrict attention to two characterizations and the fact that every connected graph has a spanning tree.

Proposition 1.6 *A graph is a forest if and only if any two vertices are connected by at most one path.*

Proof Suppose that G is not a forest, that is, not acyclic. Let $v_1v_2\dots v_kv_1$ be a cycle of G. Then $v_1v_2\dots v_k$ and v_1v_k are two distinct paths between v_1 and v_k.

Conversely, let $P = u_1u_2\dots u_r$ and $Q = u_1v_2\dots v_su_r$ be two distinct paths from u_1 to u_r in G. Let i be the smallest index for which $u_{i+1} \neq v_{i+1}$ and $j > i$ the smallest index for which $v_j \in V(P)$, say $v_j = u_k$. Then $u_iu_{i+1}\dots u_kv_{j-1}v_{j-2}\dots v_{i+1}v_i$ is a cycle. □

Corollary 1.7 *A graph is a tree if and only if any two of its vertices are connected by exactly one path.*

Proposition 1.8 *Let G be a connected graph on $n \geq 1$ vertices and m edges. Then $n-1 \leq m$, with equality holding if and only if G is a tree.*

Proof We first show by induction that the number of edges in a tree on n vertices is $n-1$. This is true for the trivial graph. Suppose that it is true for all trees with fewer than n vertices. Let G be a tree on n vertices. Clearly, removal of any edge uv disconnects G because uv is the only u,v-path in G. The resulting graph $G-uv$ has exactly two components, one consisting of the vertices w of G for which the shortest path in G from w to u meets v and the other one consisting of the vertices for which this path does not meet v. Because both components are connected and acyclic, the assertion readily follows.

Next, suppose that G is a connected graph on n vertices with m edges. If it is not a tree, it must contain a cycle. Removal of an edge from this cycle yields another connected graph on n vertices with $m-1$ edges. We continue this process until all cycles are removed. Because the resulting graph is connected and acyclic on n vertices, it has $n-1$ edges and we conclude that m must have been at least that large.

Finally, let G be a connected graph with $n-1=m$. Suppose that it has a cycle C. Let e be an edge of C. Then $G-e$ is a connected graph on n vertices with only $n-2$ edges, which is not possible. □

Corollary 1.9 *Every nontrivial tree has at least two vertices of degree* 1.

Proof Let T be a tree. If it is nontrivial it must have at least $n \geq 2$ vertices. Suppose that x of these vertices are of degree 1. Then $2(n-x)+x$ is a lower bound for the sum of the degrees of T. Because the sum of the degrees of the vertices in a graph is twice the number of edges, we have $2n - x \leq 2|E(T)| = 2n - 2$ and therefore $x \geq 2$. □

The proof of Proposition 1.8 shows that every finite connected graph has an acyclic, connected spanning subgraph. Such a graph is called a *spanning tree*. We formulate this as a proposition.

Proposition 1.10 *Every finite connected graph has a spanning tree.*

For infinite graphs this result still holds, except that the simple induction proof must be replaced by transfinite induction or, equivalently, an application of Zorn's lemma. In fact, the existence of spanning trees in connected, infinite graphs is equivalent to the axiom of choice and could therefore be postulated as an axiom.

In Chapter 20 we will investigate when a graph can be represented as the union of k edge-disjoint spanning trees and as the union of edge-disjoint spanning forests. The minimum number of edge-disjoint spanning forests that partition the edge set of a graph G will play a role in bounding the running time of several algorithms. It is called the *arboricity* $a(G)$ of G.

1.4 Planar Graphs

A graph G is *planar* if it can be drawn in the plane such that no two edges cross. More rigorously, a simple graph is planar if it can be represented in the plane such that the edges correspond to simple Jordan curves, where each curve has two endpoints. The vertices correspond to these endpoints, and two curves are either disjoint or meet only at a common endpoint. Any such drawing is called a *plane drawing* of G, and a planar graph together with its plane drawing is a *plane graph.*

One can show that the study of finite planar graphs does not lead to topological difficulties and that one can even represent them such that the edges are straight line segments. For a proof of this result, which is due to Fáry (1948) and Wagner (1936), we refer to West (1996). Below we derive Euler's formula and some of its consequences, and discuss Kuratowski's characterization of planar graphs by forbidden subgraphs.

We begin with the observation that trees and cycles are planar graphs. Another example is the 3-cube Q_3, whose planar representation is shown in the left part of Figure 3.1.

Any plane drawing of a planar graph G divides the plane into regions that are called *faces.* For instance, the number of faces of a plane drawing of a tree is one, of a cycle two, and that of Q_3 from Figure 3.1 is six. The following result, known as *Euler's formula* for planar graphs, establishes a connection between the number of vertices, edges, and faces of any plane drawing of a connected planar graph.

Theorem 1.11 *Let G be a connected planar graph on n vertices and m edges, and let f be the number of faces of a plane drawing of G. Then*

$$n - m + f = 2\,.$$

Proof We first note that the number of faces of G is one if and only if G is a tree and that the result is true in this case, because $n - 1 = m$ for trees.

Thus suppose that $f > 1$. Clearly, G must have a cycle. Consider an edge e on this cycle. It is on the boundary of two faces. Since these two faces form a single new face when e is removed, the graph $G-e$ has the same number of vertices as G, but one edge and one face less. Now the result easily follows by induction on the number of faces. □

Thus no matter how we draw a planar graph in the plane, the number of faces is invariant.

Corollary 1.12 *A planar (simple) graph with at least* 3 *vertices has at most* $3n-6$ *edges.*

Proof Because every face must have at least three boundary edges and because every edge can be in at most 2 faces, we infer that $3f \leq 2m$. Hence $3(2+m-n) \leq 2m$, and therefore $m \leq 3n-6$. □

This corollary shows that planar graphs have few edges. It will be invoked later to prove that the arboricity of planar graphs is at most 3.

Corollary 1.13 *If a planar graph can be drawn in the plane such that every face is a* k*-cycle, then* $m = k(n-2)/(k-2)$.

Proof We have already noted that every edge of a cycle is in exactly two faces. Thus, $kf = 2m$, which immediately implies the corollary. □

This corollary is often used to prove the nonplanarity of graphs. We apply it to show that K_5 and $K_{3,3}$ are nonplanar. To see this, we first note that all faces of any planar drawing of K_5 must be triangles and that all faces in a plane drawing of $K_{3,3}$ must be 4-cycles. Now the observation that $|E(K_5)| = 10 \neq 9 = 3\,(|V(K_5)|-2)/(3-2)$ and $|E(K_{3,3})| = 9 \neq 8 = 4\,(|V(K_{3,3})|-2)/(4-2)$ shows that neither of these graphs can be planar.

It is easy to see that the graphs obtained from K_5 or $K_{3,3}$ by replacing some or all of their edges by paths of arbitrary lengths that have at most endpoints in common are also nonplanar. Such graphs are called *subdivisions*, that is, a graph H obtained in this way from a graph G is called a *subdivision* of G. Clearly, no graph containing a subdivision of K_5 or $K_{3,3}$ can be planar. Surprisingly, the converse also holds.

Theorem 1.14 *A graph is planar if and only if it contains no subdivision of* K_5 *or* $K_{3,3}$.

This characterization of planar graphs by forbidden subgraphs is due to Kuratowski (1930). A (relatively) short proof due to Thomassen that uses a reduction to 3-connected graphs can be found in West (1996). With this theorem, planarity becomes a purely combinatorial concept.

Exercises

1.1. Let G be an r-regular graph on n vertices. Determine $|E(G)|$.

1.2. Let e be an edge of a connected graph G. Show that $G-e$ has at most two components.

1.3. Let e be an edge of a connected graph G. Show that $G-e$ is connected if and only if e is an edge of a cycle of G.

1.4. Let G be a graph with n vertices. Prove that G is connected if its minimum vertex degree is at least $\frac{1}{2}(n-1)$.

1.5. Show that every homomorphism of an odd cycle into itself is a bijection.

1.6. For which integers n is there a cubic graph of order n?

1.7. Find a formula for the number of squares in $K_{m,n}$.

1.8. Determine the largest number of edges that a bipartite graph on n vertices can have.

1.9. Let T be a tree all of whose vertices are of degree 1 or 3. Determine the relationship between the number of vertices of degree 1 and the number of vertices of degree 3.

1.10. (Kel′mans, 1967) Let T be a tree with k vertices of degree 1. Let L be a set of $k-1$ of them. Show that for any $x, y \in V(T)$ there exists an element $v \in L$ such that $d(v,x) \neq d(v,y)$.

1.11. Prove that a graph is a forest if and only if all of its connected subgraphs are induced.

1.12. Show that the Petersen graph (see Figure 2.4 in Chapter 2) is not planar.

Chapter 2

Automorphisms and Invariants

In this chapter we introduce the automorphism group of a graph and investigate its action on trees and hypercubes. We also demonstrate how the structure of the automorphism group, considered a permutation group on the vertex set, can shed light on a graph's structure. In particular, we show that a connected graph with transitive Abelian automorphism group always has a spanning hypercube.

We also present several graph parameters: the chromatic number, the independence number, the clique number, and the domination number. These parameters are preserved under graph automorphisms and are therefore known as invariants. However, they are not preserved under graph homomorphisms, and this has some interesting consequences.

2.1 Automorphisms

An isomorphism of a graph G onto itself is called an *automorphism*. In other words, an automorphism of G is a permutation φ of $V(G)$ with the property that $[u, v]$ is an edge if and only if $[\varphi(u), \varphi(v)]$ is an edge.

For example, the mapping that interchanges the vertices of K_2 is an automorphism of K_2; for the cycle $C_n = v_1 v_2 \ldots v_n v_1$, the mapping $\varphi : v_i \mapsto v_{i+1}$, indices taken modulo n, is an automorphism. Moreover, any permutation of the vertex set of a complete or totally disconnected graph is an automorphism.

On the other hand, the *identity mapping* id : $V(G) \to V(G)$, defined by id : $u \mapsto u$ for all $u \in V(G)$, is the only automorphism of the graph of Figure 2.1. Such graphs are called *asymmetric*. Actually, this is the typical situation, as most finite graphs admit only the identity automorphism.

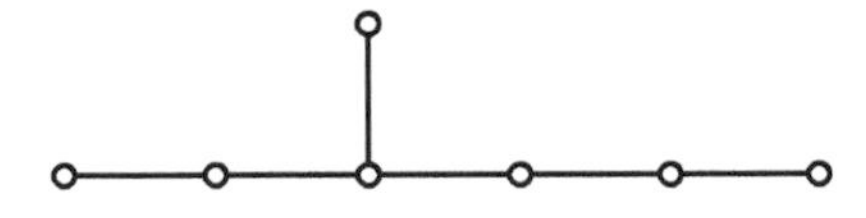

FIGURE 2.1 Asymmetric tree.

Nonetheless, many graphs are highly symmetric: for example, complete graphs and cycles, but also complete bipartite graphs, hypercubes, Hamming graphs, and many others described in this book. In this section we discuss some basic properties of graph automorphisms and continue with automorphisms of trees and hypercubes.

We use juxtaposition to denote the composition of mappings. Thus, for two mappings $\psi : V \to W$, $\varphi : U \to V$, the mapping $\psi\varphi : U \to W$ is defined by $\psi\varphi : u \mapsto \psi(\varphi(u))$. We use Greek or Latin lowercase letters for mappings. For automorphisms ψ and φ, the mapping $\psi\varphi$ is also called the product of ψ and φ. It is an automorphism. Moreover, the identity map is a unit, and the inverse φ^{-1} of a permutation φ is an automorphism if and only if φ is. Thus the set of automorphisms of a graph is closed under composition and inversion. We have therefore shown the following:

Proposition 2.1 *The automorphisms of a graph form a group.*

This group is the *automorphism group* of G and denoted by $\mathrm{Aut}(G)$. Sometimes it is simply called the group of G.

The automorphism group $\mathrm{Aut}(G)$ of a graph G is a subgroup of the group of all permutations of $V(G)$, the so-called symmetric group $\mathrm{Sym}(V(G))$.

As we already mentioned, $\mathrm{Aut}(K_n) = \mathrm{Sym}(V(K_n))$. The same holds for $\mathrm{Aut}(D_n)$. This is a special case of the fact that a graph and its complement have the same automorphism group. The *complement* $\overline{G}$ of a graph G is defined on $V(G)$ by setting

$$E(\overline{G}) = \{xy \mid x, y \in V(G), x \neq y, xy \notin E(G)\}.$$

In other words, $\overline{G}$ is obtained from G by making adjacent exactly those pairs of vertices that are nonadjacent in G. It readily follows that $\mathrm{Aut}(G) = \mathrm{Aut}(\overline{G})$ and $\mathrm{Aut}(D_n) = \mathrm{Aut}(K_n)$, as claimed above.

It is easy to see that not every permutation group on a set V can be realized as the automorphism group of a graph G with vertex set V. For example, one readily checks that the automorphism groups of the four simple graphs on three vertices, say v_1, v_2, v_3, are either equal to $\mathrm{Sym}(\{v_1, v_2, v_3\})$ or contain only one nonidentity element, which fixes one vertex and interchanges the other two. Hence, none of the automorphism groups of these graphs is a group of order 3 generated by a cycle of length three, say (v_1, v_2, v_3) or (v_1, v_3, v_2). (Note that "cycle" denotes the cycle of a permutation here, not a cycle of the underlying graph.)

This is a special case of the next proposition about graphs admitting doubly transitive groups of automorphisms, where a permutation group A on V is *doubly transitive* if there is a permutation $\varphi \in A$ to any two pairs u, v and x, y of distinct elements of V such that $x = \varphi(u)$ and $y = \varphi(v)$.

Proposition 2.2 *If the automorphism group of a graph G with at least one edge contains a doubly transitive subgroup, then G is complete.*

Proof Let G be a graph with a doubly transitive subgroup A of $\mathrm{Aut}(G)$ and uv be an edge of G. By assumption there is an automorphism $\varphi \in A$ to any pair $\{x, y\}$ of distinct elements that maps $\{u, v\}$ into $\{x, y\}$. Then xy is an edge too. Because the pair $\{x, y\}$ was arbitrarily chosen, G is complete. □

Clearly the automorphism group of a complete or totally disconnected graph G is $\mathrm{Sym}(V(G))$. Combining it with Proposition 2.2 we thus obtain Corollary 2.3:

Corollary 2.3 $\mathrm{Aut}(G) = \mathrm{Sym}(V(G))$ *if and only if G is complete or totally disconnected.*

The following theorem of Frucht (1938) for abstract groups contrasts the fact that not every permutation group on a set V can be realized as the automorphism group of a graph with vertex set V. We cite it without proof.

Theorem 2.4 *To every finite group A there exists a graph G for which* $\mathrm{Aut}(G) \cong A$.

As we have seen above, none of the automorphism groups of the graphs on three vertices are the cyclic permutation group of order 3 on these vertices. By Frucht's theorem there exists a graph whose automorphism group is isomorphic to the cyclic group of order 3. We depict such a graph in Figure 2.2 and leave it to the reader to verify that its group is as asserted. The only thing needed is the obvious fact that automorphisms preserve degrees and distances.

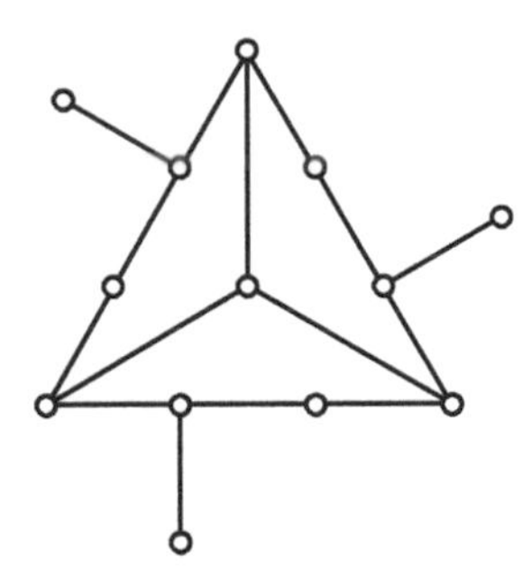

FIGURE 2.2 Graph with cyclic automorphism group.

The example of Figure 2.2 shows that a graph may contain vertices that are fixed by every automorphism. If a set of automorphisms maps a set of vertices or edges into itself, we say this set is *stabilized* by these automorphisms or *invariant* under these automorphisms. Clearly, every maximal set of vertices of equal degree in a graph is invariant under all automorphisms. In particular, this holds for vertices of degree 1, which are called *pendant vertices.*

Theorem 2.5 *Every tree T contains an edge or a vertex that is invariant under every automorphism of T.*

Proof We proceed by induction. The assertion is true for trees on one or two vertices. Suppose that it is true for every tree on at most $n-1$ vertices. Let T be a tree on n vertices. Every automorphism of T stabilizes the set of pendant vertices. We delete them and call the new tree S. By Corollary 1.9, the set of pendant edges is non-empty; thus S has fewer than n vertices and therefore an edge e or a vertex v that is stabilized by every automorphism of S. Because the restriction of every automorphism of T to S is an automorphism of S, the assertion of the theorem also holds for T. □

The *center* of a graph is the set of vertices u for which $\min_{u \in V(G)} \max_{v \in V(G)} d(u,v)$ is attained. In the case of trees, the center consists of one vertex or two adjacent ones, see Exercise 2.1. Clearly, the center is invariant under all automorphisms. It is the invariant element of Theorem 2.5.

Theorem 2.5 is a special case of the Fixed Cube Theorem 12.21 and of the more general "fixed box theorems" from Chapter 16. Because the fixed cubes of Theorem 12.21 are hypercubes, we conclude the section with a few remarks about the hypercube and its automorphisms.

The *hypercube* Q_r *of dimension* r is defined on the vectors $(v_1, v_2, \ldots, v_r)$ with $v_i \in \{0,1\}$. Two vertices are adjacent if the corresponding vectors differ in precisely one coordinate. In other words, two vertices $u = (u_1, u_2, \ldots, u_r)$, $v = (v_1, v_2, \ldots, v_r)$ are adjacent if there is an index j such that $u_j \neq v_j$ and $u_i = v_i$ for all $i \neq j$, $1 \leq i \leq r$. Figure 2.3

shows several hypercubes, where for simplicity we abbreviate the r-tuples $(u_1, u_2, \ldots, u_r)$ as $u_1u_2\ldots u_r$. We shall later see that Q_r is the Cartesian product of r copies of K_2.

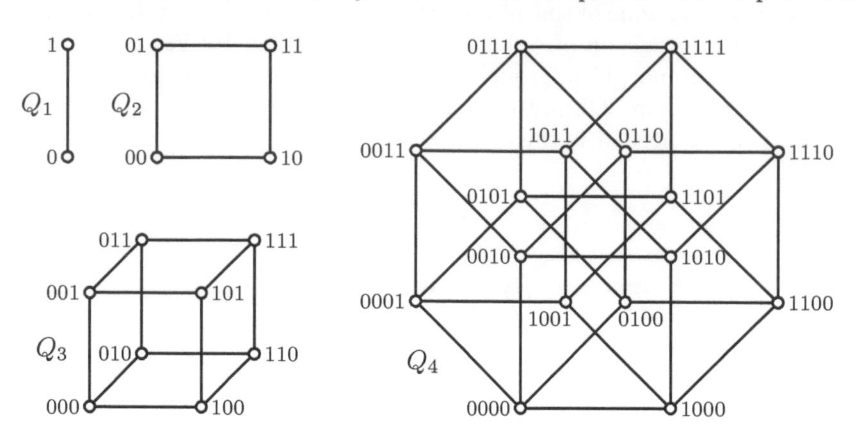

FIGURE 2.3 Some hypercubes.

Let $i < j$ be two integers between 1 and r. Then the mapping

$$\varphi_{i,j} : (v_1, v_2, \ldots, v_i, \ldots, v_j, \ldots, u_r) \mapsto (v_1, v_2, \ldots, v_j, \ldots, v_i, \ldots, v_r)$$

is an automorphism of Q_r. Furthermore, for any i, $1 \leq i \leq r$,

$$\psi_i : (v_1, v_2, \ldots, v_i, \ldots, v_r) \mapsto (v_1, v_2, \ldots, v_i + 1, \ldots, v_r),$$

where addition is modulo 2, is also an automorphism. Clearly, $\mathrm{id} = \psi_i^2$ for all i, and $\psi_i\psi_j = \psi_j\psi_i$ for all $1 \leq i, j \leq r$. Hence, the subgroup of $\mathrm{Aut}(Q_r)$ generated by the ψ_i is Abelian, and each nontrivial element has order 2. Such groups are known as *elementary Abelian 2-groups* and also called *Boolean.*

By Corollary 6.11, the $\varphi_{i,j}$ and the ψ_i generate $\mathrm{Aut}(Q_r)$, but this can also be shown directly, as in Exercises 2.5 and 2.6.

Furthermore, for any two vertices $u, v \in Q_r$ that differ in the coordinates $i_1, i_2, \ldots, i_j$, we clearly have $\psi_{i_1}\psi_{i_2}\cdots\psi_{i_j}(u) = v$. This leads to the concept of vertex-transitivity.

2.2 Vertex-Transitivity

This section demonstrates the strong interaction between a graph's structure and its automorphisms. We define vertex-transitivity and show that every graph whose automorphism group is transitive and Abelian contains a hypercube as a spanning subgraph.

We wish to note, however, that this is just one aspect of vertex-transitive graphs and that there exists a huge literature about them; see Babai (1995) and Godsil and Royle (2001). In this book they also play an important role in Chapter 27, that is, in the chapter that treats the independence number, and in the classification of infinite median graphs in Chapter 31.

The automorphism group of a graph G is *transitive* if there exists an automorphism φ to any pair u, v of vertices in G such that $\varphi(u) = v$. In this case, G is called *vertex-transitive.*

If one requires the existence of exactly one such automorphism to every pair u, v of vertices, then one says that $\mathrm{Aut}(G)$ is *regular* or *sharply transitive.*

For example, K_n and Q_r are vertex-transitive, but not sharply vertex-transitive unless they have only one or two vertices.

Lemma 2.6 *A transitive permutation group* Γ *on a set* V *is sharply transitive if and only if the stabilizer* $\Gamma_v = \{\varphi \in \Gamma \mid \varphi(v) = v\}$ *is trivial for every* $v \in V$.

Proof Suppose that $\alpha \in \Gamma_v$ for an arbitrarily chosen v. If Γ is sharply transitive, we infer from $\alpha(v) = v$ and $\mathrm{id}(v) = v$ that $\alpha = \mathrm{id}$. Hence, Γ_v consists only of the identity element. On the other hand, suppose that $|\Gamma_v| = 1$ for every $v \in V$. Let $\varphi(v) = u$ and $\psi(v) = u$ for a pair of vertices u, v. Then $\psi^{-1}\varphi(v) = v$, and thus $\psi^{-1}\varphi = \mathrm{id}$. Therefore, $\psi = \varphi$ and Γ is sharply transitive. □

We note that if the automorphism group of a graph is transitive but not sharply transitive, then it is possible that it has a subgroup whose action is sharply transitive. (For the smallest vertex-transitive graph that has no sharply transitive subgroup, see Exercise 2.10.)

Lemma 2.7 *If* Γ *is a transitive Abelian permutation group on a set* V, *then* Γ *is sharply transitive.*

Proof Let $\alpha \in \Gamma_v$. By Lemma 2.6 we have to show that $\alpha(u) = u$ for all $u \in V$. Let u be arbitrarily chosen. By the transitivity of Γ there exists a $\varphi \in \Gamma$ such that $\varphi(v) = u$. But then $\alpha(u) = \alpha\varphi(v) = \varphi\alpha(v) = \varphi(v) = u$. □

Before stating the next theorem—due to Imrich (1970)—we recall that groups in which every nontrivial element has order 2 are called elementary Abelian 2-groups.

Theorem 2.8 *Let* G *be a nontrivial graph with transitive Abelian automorphism group. Then* $\mathrm{Aut}(G)$ *is an elementary Abelian* 2*-group. If* G *has more than two vertices, then it is connected and contains a proper spanning hypercube.*

Proof Let G be a nontrivial graph with transitive Abelian group. Then $\mathrm{Aut}(G)$ is sharply transitive by Lemma 2.7. If G is disconnected, then it can only be D_2 by transitivity and sharp transitivity. All other graphs with transitive Abelian group are connected; in particular, all graphs on at least three vertices with transitive Abelian group are connected.

We first show that every element of $\mathrm{Aut}(G)$ has order 2. We can assume that G is connected, and we choose an arbitrary but fixed vertex v in G. Furthermore, we denote the unique automorphism that maps v to $x \in V(G)$ by φ_x. Because every automorphism maps v into some vertex, the φ_x constitute all automorphisms of G, so $|\mathrm{Aut}(G)| = |V(G)|$.

Let $x, y \in V(G)$. Because φ_x and φ_y are automorphisms, $[x, y]$ is an edge if and only if $\varphi_x^{-1}\varphi_y^{-1}[x, y] = \varphi_x^{-1}\varphi_y^{-1}[\varphi_x(v), \varphi_y(v)] = [\varphi_y^{-1}(v), \varphi_x^{-1}(v)]$ is also an edge. Let $\psi : V(G) \to V(G)$ be defined by $\psi : x \mapsto \varphi_x^{-1}(v)$. Then $[x, y]$ is an edge if and only if $[\psi(y), \psi(x)]$ is an edge. Because $[\psi(y), \psi(x)] = [\psi(x), \psi(y)]$ and ψ is bijective, ψ is an automorphism.

We wish to point out that we only needed the existence of a transitive Abelian subgroup of $\mathrm{Aut}(G)$ to show that $\psi : x \mapsto \varphi_x^{-1}(v)$ is an automorphism. Because we assume that $\mathrm{Aut}(G)$ itself is transitive and Abelian (and hence sharply transitive), we can say more.

Note that $\psi(v) = v$ and that $\mathrm{Aut}(G)$ is sharply transitive. Hence, $\psi = \mathrm{id}$. But then $\psi x = x$ for all $x \in V(G)$, which implies that

$$\varphi_x^{-1}(v) = \psi(x) = x = \varphi_x(v).$$

Then $\varphi_x^2(v) = v$, and thus $\varphi_x^2 = \mathrm{id}$ for all x.

We now introduce the notation α_u for φ_u if $u \in N(v)$. If $[x, y]$ is an edge, then $\varphi_x[x, y] =$

$[v, \varphi_x(y)] = [v, \varphi_x\varphi_y(v)]$ and $\varphi_x\varphi_y = \alpha_u$ for some $u \in N(v)$. Then $\alpha_u(x) = \varphi_x\varphi_y(x) = \varphi_y\varphi_x(x) = \varphi_y(v) = y$. Clearly, α_u is uniquely determined by $[x, y]$. Moreover we note that $\alpha_u(y) = x$, as all elements of Aut(G) have order 2. Thus, to every edge $[x, y]$ corresponds a unique $u \in N(v)$ with $\alpha_u(x) = y$ and $\alpha_u(y) = x$. We can therefore mark the edges with elements from $u \in N(v)$ and say $[x, y]$ has color u.

Let E_u be the set of edges of color u in G. Let $S \subseteq N(v)$ be a set of minimum cardinality such that H, defined by $V(H) = V(G)$ and

$$E(H) = \bigcup_{u \in S} E_u\,,$$

is connected. We claim that H is a hypercube.

First, we show that $\{\alpha_u \mid u \in S\}$ generates Aut(G). Let φ_x be an arbitrary automorphism of G, and P be a path $v_0v_1 \ldots v_p$ from $v = v_0$ to $x = v_p$ in H. Set $\alpha_{u_i}(v_i) = v_{i+1}$ for $i = 0, 1, \ldots, p-1$. Clearly, the u_i are in S and

$$\Big(\prod_{i=0}^{p-1} \alpha_{u_i}\Big) v_0 = v_p.$$

But then $\prod_{i=0}^{p-1} \alpha_{u_i} = \varphi_x$.

Second, $H - E_u$ is disconnected; otherwise $k-1$ color classes would suffice. This implies that $\{\alpha_u \mid u \in S\}$ is a minimal generating set. For a proof, suppose that this is not the case. Let $\alpha_u = \alpha_{u_1}\alpha_{u_2}\cdots\alpha_{u_j}$, where u and $u_1, u_2, \ldots, u_j$ are in S. Then $H - E_u$ is connected, as we can replace every edge of color u by a path consisting of edges of colors $u_1, u_2, \ldots, u_j$, but this contradicts the minimality of k.

This means that every element of Aut(G) can be represented as a product of elements of S. Because Aut(G) is Abelian, and because every nontrivial element has order 2, every group element corresponds to a subset of S and can be represented as a vector of length S with entries from $\{0, 1\}$. Because every automorphism of G is of the form φ_x, this also holds for $V(H)$. Clearly, any two vertices of H are connected if their vector representations differ in exactly one coordinate. As every vertex of H has $|S|$ neighbors, H has no other edges, so H is a hypercube.

Because every hypercube Q_r with $r > 1$ has nonregular group, we infer that H is a proper subgraph if G has more than two vertices. □

This theorem does not tell us much about the existence of graphs with transitive Boolean groups. Clearly, K_1 and K_2 are examples; but for larger graphs G, the theorem only asserts that they must contain a spanning hypercube Q_r and that the fixed point-free automorphisms of Q_r are also automorphisms of G. It is trivial to check that no such graphs exist for $r = 2$, less trivial for $r = 3$ (see Exercise 2.8), and more elaborate for $r = 4$. The existence of such graphs for $r > 5$ has been shown by Imrich (1970), and for $r = 5$ by Imrich and Watkins (1976).

Thus, finite graphs G with transitive Abelian automorphism group exist if and only if Aut$(G) = \mathbb{Z}_2^k$ for $k \neq 2, 3, 4$. This had already been asserted by McAndrew (1965). Proofs that the groups of such graphs must be elementary Abelian were obtained independently by Chao (1964) and Sabidussi (1964).

Likewise, for any infinite cardinal $\mathfrak{n}$, there exists a graph G on $\mathfrak{n}$ vertices and transitive Abelian automorphism group. Any such G must have a spanning hypercube (of dimension $\mathfrak{n}$) and Aut(G) must be elementary Abelian; see Imrich (1969b).

Nowitz and Watkins (1972a,b) call a graph G a *Graphical Regular Representation* or

GRR[1] of the group A if $\mathrm{Aut}(G)$ is sharply transitive and isomorphic to A. Hence, in their language, every elementary Abelian 2-group of order $2^k, k \neq 2, 3, 4$ has a GRR.

No other Abelian group and, as shown by Nowitz (1968), no non-Abelian group for which the mapping $a \mapsto a^{-1}$ is an automorphism, has a GRR. Beyond that, there exist only eight more exceptional groups, all of order at most 32, without a GRR. Any other group has a GRR; see Godsil (1981).

For more detailed information about GRRs see Watkins (2004), that is, Section 6.1 in Gross and Yellen (2004).

We conclude the section with a second look at the proof of Theorem 2.8. It reveals that the theorem also holds if $\mathrm{Aut}(G)$ contains a transitive, elementary Abelian 2-group as a subgroup.

Corollary 2.9 *Let G be a connected graph that admits a transitive, elementary Abelian 2-group of automorphisms. Then G has a spanning hypercube.*

2.3 Graph Invariants

An *n-coloring* of a graph G is a function f from $V(G)$ onto a set X of n elements such that $xy \in E(G)$ implies that $f(x) \neq f(y)$. The elements of X are called *colors*. Vertices with the same image (that is, vertices of the same color) are said to form a *color class*. The smallest number n for which an n-coloring exists is the *chromatic number* $\chi(G)$ of G.

Clearly, $\chi(K_n) = n$. It is also easily seen that a graph G with at least one edge is bipartite if and only if $\chi(G) = 2$.

The following result establishes a connection between the chromatic number and graph homomorphisms:

Proposition 2.10 *Let G be a graph. Then*

(i) *$\chi(G)$ is the smallest integer n for which there exists a homomorphism $G \to K_n$.*
(ii) *If there exists a homomorphism $G \to H$, then $\chi(G) \leq \chi(H)$.*

Proof (i) Let $V(K_n) = \{1, 2, \ldots, n\}$ and $f : V(G) \to \{1, 2, \ldots, n\}$ be an n-coloring of G. Define $g : V(G) \to V(K_n)$ by $g(u) = f(u)$ for any $u \in V(G)$. Then g is a homomorphism. Because every such homomorphism gives rise to an n-coloring of G with color classes $g^{-1}(i)$, the assertion follows.

(ii) Let g be a homomorphism from G into H and $\chi(H) = n$. By (i), there exists a homomorphism $h : H \to K_n$. Then hg is a homomorphism from G into K_n and by (i) $\chi(G) \leq n$. □

A graph G is called *χ-critical* if $\chi(G - v) < \chi(G)$ for every $v \in V(G)$. Complete graphs and odd cycles are χ-critical. It is well-known and easy to see that every nontrivial graph contains a χ-critical subgraph with the same chromatic number; see also Exercise 2.9.

A graph G is called *uniquely n-colorable* if any n-coloring of G determines the same partition of $V(G)$ into color classes. Trivial examples of uniquely colorable graphs are complete graphs and connected bipartite graphs. Using graph products, we demonstrate in Section

[1]In the theory of algebraic and topological groups, a *principal homogeneous space*, or *torsor*, for a group A is a set X on which A acts fixed point freely and transitively. Hence, GRRs are instances of principal homogeneous spaces.

26.4 that to every pair of natural numbers $n, s \geq 3$, there is a uniquely n-colorable graph without odd cycles shorter than s.

In our investigations of the chromatic number in Chapter 26, Kneser graphs play a special role. They are defined as follows: Let n and k be integers with $n \geq 2k$. Then the vertex set of the *Kneser graph* $K(n, k)$ consists of all k-subsets of $\{1, 2, \ldots, n\}$, two vertices being adjacent if they are disjoint. (Note that if we allowed $n < 2k$, then the corresponding graphs would have no edges.) Clearly, $K(n, 1) \cong K_n$. The Kneser graph $K(5, 2)$ is better known as *Petersen graph.* See Figure 2.4.

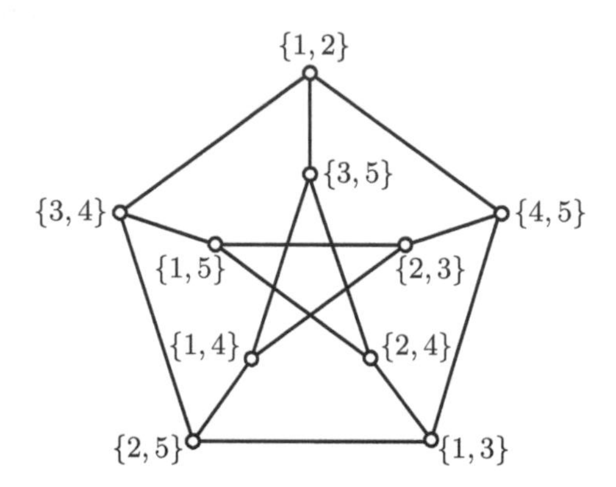

FIGURE 2.4 The Petersen graph as the Kneser graph $K(5, 2)$.

We already saw in Exercise 1.12 that the Petersen graph is nonplanar. It is known as a graph that is exceptional in many respects; for example, it is the smallest vertex-transitive graph that does not contain a subgroup of its automorphism group that is a sharply transitive subgroup on its vertex set. (See Exercise 2.10.)

A *clique* of a graph G is a maximal complete subgraph. By maximal we mean maximal with respect to inclusion. The size of a largest clique, that is, of a maximum complete subgraph, is called the *clique number* of G and denoted by $\omega(G)$. Clearly, $\chi(G) \geq \omega(G)$. The inequality can be strict, as the example of C_{2k+1} for $k \geq 2$ shows. (Evidently, $\omega(C_{2k+1}) = 2$ and $\chi(C_{2k+1}) = 3$ for $k \geq 2$.)

A set X of vertices of a graph G is called *independent* if no two distinct vertices of X are adjacent. The size of a largest independent set is called the *independence number* of G and denoted by $\alpha(G)$. For instance, $\alpha(K_n) = 1$ and $\alpha(C_{2k+1}) = k$. A set F of edges of G is called *independent* or a *matching* if no two (distinct) edges of F have a common endpoint. A matching that meets every vertex of G is called a *perfect matching* or a *1-factor.*

The chromatic number and the independence number of a graph are related by the following basic inequality.

Proposition 2.11 *For any graph G,*

$$\chi(G)\alpha(G) \geq |V(G)|.$$

Proof Let X be an arbitrary color class of a $\chi(G)$-coloring of G. Then X is an independent set of G and thus $|X| \leq \alpha(G)$. Hence in each color class we have at most $\alpha(G)$ vertices, and so $\chi(G) \geq |V(G)|/\alpha(G)$. □

A set D of vertices of a graph G is called *dominating* if every vertex $w \in V(G) \setminus D$ is adjacent to a vertex of D. The *domination number* $\gamma(G)$ of a graph G is the size of a smallest dominating set of G. A dominating set D with $|D| = \gamma(G)$ is called a *minimum dominating set*. For instance, $\gamma(K_n) = 1$ and $\gamma(C_n) = \lceil n/3 \rceil$.

The independence number of a graph and its domination number are related as follows:

Proposition 2.12 *For any graph G,*

$$\alpha(G) \geq \gamma(G).$$

Proof Let X be an independent set of G with $|X| = \alpha(G)$. Then any vertex of $V(G) \setminus X$ must be adjacent to at least one vertex of X, for otherwise X would not be a largest independent set. Thus X is a dominating set. □

2.4 The No-Homomorphism Lemma

Proposition 2.10 expresses the close connection between the chromatic number and homomorphisms. The next lemma establishes a connection between the independence number and homomorphisms. It is known as the *No-Homomorphism Lemma* because it can be used to show the nonexistence of homomorphisms.

Lemma 2.13 (No-Homomorphism Lemma) *Suppose there exists a homomorphism from a graph G to a vertex-transitive graph H. Then*

$$\frac{\alpha(G)}{|V(G)|} \geq \frac{\alpha(H)}{|V(H)|}.$$

Proof Let $\{I_1, I_2, \ldots, I_q\}$ be the maximum independent sets of H. Because H is vertex-transitive, each vertex of H belongs to the same number, say p, of the sets I_i. Therefore, $q\,\alpha(H) = p\,|V(H)|$, and so $p/q = \alpha(H)/|V(H)|$.

Now let φ be a homomorphism from G to H. Setting $J_i = \varphi^{-1}(I_i)$, and using the fact again that every vertex of H is in exactly p of the I_i, we obtain

$$\sum_{1 \leq i \leq q} |J_i| = \sum_{1 \leq i \leq q} |\varphi^{-1}(I_i)| = p \sum_{v \in V(H)} |\varphi^{-1}(v)| = p\,|V(G)|.$$

Note that this holds even if not every vertex v has a preimage.

Clearly, the J_i are independent, and thus $\alpha(G) \geq |J_i|$. Because there are q sets J_i, this implies that $q\,\alpha(G) \geq \sum_{1 \leq i \leq q} |J_i| = p\,|V(G)|$. Thus,

$$\frac{\alpha(G)}{|V(G)|} \geq \frac{p}{q} = \frac{\alpha(H)}{|V(H)|}.$$

□

The quotient $\alpha(G)/|V(G)|$ is also known as the *independence ratio* $i(G)$. With this notation, the lemma asserts that $i(G) \geq i(H)$ if there is a homomorphism from G to H, where H is vertex-transitive.

The lemma is due to Albertson and Collins (1985). They present several applications; for example, they use it to show that there is no homomorphism from the ith Cartesian power of the Petersen graph into the jth if $i > j$.

Another immediate, simple application shows that there is no homomorphism from C_{2k+1} to C_{2n+1} if $k < n$. For if such a homomorphism exists, then

$$\frac{\alpha(C_{2k+1})}{2k+1} \geq \frac{\alpha(C_{2n+1})}{2n+1}.$$

Because $\alpha(C_{2k+1}) = k$, this would imply that $k \geq n$.

See Exercise 2.11 for the nonexistence of homomorphisms from odd cycles to bipartite graphs. Moreover, Lemma 2.13 will be used in Exercise 27.6 for a proof of Theorem 27.13.

Our proof of Theorem 27.13 will use a probabilistic argument involving uniformly distributed sets of vertices in vertex-transitive graphs. For later reference we formulate a lemma about uniformly distributed sets of vertices. But first a definition. If a random variable x has any of n possible values that are equally probable, then one says that x is *discretely uniformly distributed.*

Lemma 2.14 *Let v be a vertex of a vertex-transitive graph G, and $\varphi_1, \varphi_2, \ldots, \varphi_k$ be uniformly distributed, randomly and independently chosen automorphisms. Then the vertices $\varphi_1(v), \varphi_2(v), \ldots, \varphi_k(v)$ are also uniformly distributed.*

Proof We first consider two automorphisms φ, ψ that map v into u. Then $\psi^{-1}\varphi(v) = v$, hence $\psi^{-1}\varphi$ is in the stabilizer $\mathrm{Aut}(G)_v$ of v. It follows that for fixed ψ, the map $\varphi \mapsto \psi^{-1}\varphi$ is a bijection from the set of automorphisms mapping v to u onto $\mathrm{Aut}(G)_v$.

Hence $|\mathrm{Aut}(G)_v|$ is the number of automorphisms that map v into u. This implies that $|\mathrm{Aut}(G)| = |V(G)| \cdot |\mathrm{Aut}(G)_v|$,[2] and thus the probability that an automorphism φ maps v into u is $|\mathrm{Aut}(G)_v|/|\mathrm{Aut}(G)| = 1/|V(G)|$. □

Simon Špacapan[3] extended the No-Homomorphism Lemma 2.13 and Theorem 27.13 on the independence number of direct powers of vertex transitive graphs to assertions about the k-independence number. Notice that the k-independence number $\alpha_k(G)$ of a graph G is the size of the largest k-colorable induced subgraph of G.

Both the No-Homomorphisms Lemma and Theorem 27.13 hold if $\alpha(G)$ is replaced by $\alpha_k(G)$.

Exercises

2.1. Show that the center of a tree consists of a single vertex or the endpoints of an edge.

2.2. Determine the number of automorphisms of P_n and C_n.

2.3. Let G be a connected graph, $\alpha \in \mathrm{Aut}(G)$, and $x \in V(G)$. Show that x is in the center of G if and only if $\alpha(x)$ is in the center.

2.4. Show that Q_r is connected.

2.5. Let φ be an automorphism of Q_r that fixes a vertex and all of its neighbors. Show that φ is the identity.

[2]This is a special case of the fact that the size of a group is the size of an orbit of an element v multiplied by the size of the stabilizer of v.

[3]Simon Špacapan, The k-independence number of direct products of graphs and Hedetniemi's conjecture, European Journal of Combinatorics, 32 (2011), 1377-1383

2.6. Show that the $\varphi_{i,j}$ and the ψ_i, as defined after the definition of Q_r, generate $\mathrm{Aut}(Q_r)$.

2.7. Show that Q_r has $2^r r!$ automorphisms.

2.8. Show that there is no graph on n vertices, $3 \leq n \leq 8$, with transitive, Abelian automorphism group.

2.9. Show that every nontrivial graph contains a χ-critical subgraph with the same chromatic number.

2.10. Show that the automorphism group of the Petersen graph $K(5,2)$ contains no subgroup that is sharply transitive on $K(5,2)$.

2.11. Show that there is no homomorphism from an odd cycle to a bipartite graph G.

2.12. Let $n(X,Y)$ denote the maximum number of vertices in an induced subgraph of X that is homomorphic to Y, that is, for which there exists a homomorphism $\varphi : X \to Y$. Given graphs G, H, K, show that

$$\frac{n(G,K)}{|V(G)|} \geq \frac{n(H,K)}{|V(H)|}$$

if H is vertex-transitive.

Chapter 3

Hypercubes and Isometric Subgraphs

We continue our investigation of hypercubes, which were introduced in Chapter 2. Hypercubes have numerous applications, for example in coding theory (where Gray codes are Hamilton cycles in hypercubes), computer architecture, mathematical chemistry, and phylogenetics. Hypercubes and their subgraphs are the main topic of interest not only in this chapter, but also in Chapters 18, 19, and 21. Here we introduce isometric subgraphs of hypercubes, known as partial cubes, and then continue with median graphs as retracts of hypercubes and generalizations of trees.

3.1 Hypercubes Are Sparse

As we will see in this chapter, hypercubes, also known as r-cubes, are the simplest class of Cartesian products. In Chapter 2 we defined the hypercube of dimension r to be the graph Q_r whose vertex set consists of all 0-1 vectors $(v_1, v_2, \ldots, v_r)$, where two vertices are adjacent if and only if they differ in precisely one coordinate. For brevity, we often abbreviate the vertex $(v_1, v_2, \ldots, v_r)$ as $v_1v_2\ldots v_r$.

Note that the vertices of Q_r can also be understood as characteristic vectors of subsets of an r-set. Specifically, any given vertex $v_1v_2\ldots v_r$ corresponds to the subset $\{i \mid v_i = 1\}$ of $\{1,2,\ldots,r\}$. Thus the vertices of Q_r can be labeled with the subsets of $\{1,2,\ldots,r\}$, where two subsets are adjacent provided that one is obtained from the other by deletion of a single element. Figure 3.1 shows two representations of Q_3. On the left, vertices are shown as 3-tuples; on the right as subsets of $\{1,2,3\}$. The characteristic vectors of these subsets are the 3-tuple representations of the corresponding vertices. For example, $\{2,3\}$ corresponds to 011, and $\emptyset$ to 000.

If two vertices of Q_r are adjacent, then one has an even number of 1's and the other has an odd number. It follows that Q_r is bipartite. Clearly, it has 2^r vertices. Note also that every vertex is adjacent to r other vertices. Hence the sum of the degrees in Q_r is $r2^r$, and because this is twice the number of the edges, we conclude that $|E(Q_r)| = r2^{r-1}$.

We show now that the distance between any two vertices $u = u_1u_2\ldots u_r$ and $v = v_1v_2\ldots v_r$ of an r-cube is the number of places in which they differ. To see this, let $i_1, i_2, \ldots, i_s$ be the places in which u and v differ, and for $k = 1,2,\ldots,s$, let u^k be formed from u by replacing $u_{i_1}, u_{i_2}, \ldots, u_{i_k}$ by $v_{i_1}, v_{i_2}, \ldots, v_{i_k}$. Then the sequence $u, u^1, u^2, \ldots, u^s = v$ is a u,v-path. This must be a shortest path because, as adjacent vertices differ in exactly one space, any u,v-path has length at least s. It is also clear that

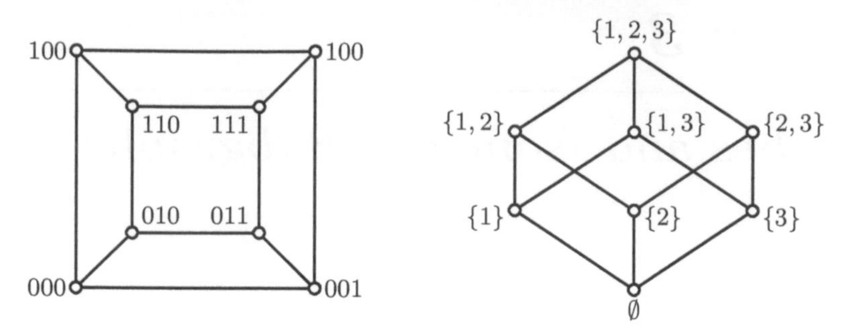

FIGURE 3.1 Representations of Q_3.

any shortest u, v-path can be obtained by switching the indices i_j in a prescribed order. (There are thus $s!$ different shortest paths from u to v.) Note that if $u_i = v_i$ for some i, then $x_i = u_i = v_i$ for every vertex x on a shortest u, v-path.

Let S be the set $\{i_1, i_2, \ldots, i_s\}$ of places in which u and v differ. Then the condition that $x_i = v_i$ for all $x \in I(u, v)$ whenever $i \notin S$ and that the places x_j can be arbitrarily assigned the values u_j or v_j for $j \in S$ imply that $I(u, v)$ induces an s-cube. In other words, the intervals of a hypercube induce the subcubes.

We have thus shown:

Proposition 3.1 *Let Q_r be a hypercube. Then*

(i) *Q_r is connected, bipartite, r-regular, and has diameter r.*
(ii) *$|V(Q_r)| = 2^r$ and $|E(Q_r)| = r2^{r-1}$.*
(iii) *For any pair of vertices $u, v \in V(Q_r)$, the subgraph induced by the interval $I(u, v)$ is a hypercube of dimension $d(u, v)$.*

Letting n denote the number of vertices of an r-cube and m its number of edges, we see that $m = \frac{1}{2}n \log_2 n$. This is quite small, considering the fact that the number of edges in a complete graph is $\binom{n}{2}$. In general we call a class of graphs *sparse* if m is proportional[1] to $n \log n$, so hypercubes are sparse. (Some authors require proportionality to n for sparseness; for our purposes, $n \log n$ suffices.) The next proposition shows that the sparseness property is inherited by the subgraphs of Q_r.

Lemma 3.2 (Density Lemma) *Let G be a subgraph of a hypercube. Then*

$$|E(G)| \leq \frac{1}{2}|V(G)| \cdot \log_2 |V(G)|,$$

equality holding if and only if G is a hypercube.

Proof The proof is by induction on the number n of vertices of G. The assertions are clearly true for $n \leq 2$. Let $|V(G)| \geq 3$. Because G is a subgraph of a hypercube Q_r, every vertex of G is an r-tuple over $\{0, 1\}$. We may assume without loss of generality that the first coordinates of the vertices of G are not all equal. Let G_1 be the subgraph of G induced by the vertices for which the first place of this r-tuple is 0 and G_2 the subgraph induced by the other vertices of G. Set $x = |V(G_1)|$ and $y = |V(G_2)|$, where the indexing is chosen so that

[1] More precisely, a class is sparse when $m = O(n \log n)$; see Chapter 17 for the definition of $O(f(n))$.

$x \geq y \geq 1$. Then, by the induction hypothesis, $|E(G_1)| \leq \frac{x}{2}\log_2 x$ and $|E(G_2)| \leq \frac{y}{2}\log_2 y$. As every vertex of G_2 has at most one neighbor in G_1, it thus is enough to show that

$$\frac{x}{2}\log_2 x + y + \frac{y}{2}\log_2 y \leq \frac{x+y}{2}\log_2(x+y).$$

We prove that this inequality actually holds for all real numbers $x \geq y \geq 1$. For $x = y$, both sides are equal. For $x > y$, we show strict inequality. It suffices to prove that $\frac{\partial}{\partial x}$ of the left side is strictly smaller than $\frac{\partial}{\partial x}$ of the right side, namely that

$$\frac{1}{2}\log_2 x + \frac{1}{2}\log_2 e < \frac{1}{2}\log_2(x+y) + \frac{1}{2}\log_2 e.$$

Because $y \geq 1$, this is indeed the case.

Equality can only occur, when $y = x$, when every vertex of G_2 has a neighbor in G_1, and when $|E(G_1)| = |E(G_2)| = \frac{x}{2}\log_2 x$. Then G_1 and G_2 are hypercubes by the induction hypothesis, and G is a hypercube too. □

This result has several important implications for algorithms. In particular, it implies that the arboricity of subgraphs of hypercubes is bounded by $\lceil \log_2 n \rceil$, that subgraphs of hypercubes cannot have more than $m \log_2 n$ squares, and that the squares can be found and listed in time proportional to $m \log n$. (The first assertion is part of Theorem 20.3, the second is part of Corollary 20.7, and the third a consequence of Proposition 20.6.) For a sparseness result related to the Density Lemma 3.2 for a larger class of graphs, see Graham (1970).

3.2 Isometric Subgraphs

For every subgraph H of a graph G, the inequality $d_H(u,v) \geq d_G(u,v)$ holds. If $d_H(u,v) = d_G(u,v)$ for all $u,v \in V(H)$, we say H is an *isometric* subgraph. More generally, if G and H are arbitrary graphs, then a mapping $f : V(G) \to V(H)$ is an *isometric embedding* if

$$d_H(f(u), f(v)) = d_G(u,v)$$

for any $u,v \in V(G)$. Note that an isometric embedding is necessarily injective.

Not every subgraph is isometric. For example, a path of length 3 in C_5 is not isometric, but paths of lengths 1 or 2 are. All isometric subgraphs are induced, but the converse is false. For example, the vertex-deleted subgraph in Figure 1.2 is induced but not isometric.

Proposition 3.3 *Let C be a shortest cycle or a shortest odd cycle of a graph G. Then C is isometric in G.*

Proof Let C_{2k+1} be a shortest odd cycle $v_1v_2\ldots v_{2k+1}$ of a graph G. In other words, we assume that there are no shorter cycles of odd length in G, but there may be shorter cycles of even length. If C_{2k+1} is not isometric, there must be two vertices, say v_1 and v_r, for which $d_G(v_1,v_r) < d_{C_{2k+1}}(v_1,v_r) = r-1$. Without loss of generality, we can assume that r is smallest possible. Clearly it is not larger than k. Then there must be an isometric path $P = v_1w_2\ldots w_sv_r$ of length less than $r-1$ that meets C only in v_1 and v_r. But then the cycles $v_1v_2\ldots v_rw_sw_{s-1}\ldots w_2$ and $v_1w_2\ldots w_sv_rv_{r+1}\ldots v_{2k}v_{2k+1}$ are both shorter than C. Because the sum $2s + (2k+1)$ of their lengths is odd, at least one of them must be odd, contradicting the minimality of C_{2k+1}.

Now, let C be a shortest cycle in G. If it is odd, there is nothing to show. If it is a cycle $v_1v_2\ldots v_{2k}$ of even length and not isometric, an argument analogous to the above yields two cycles $v_1v_2\ldots v_rw_sw_{s-1}\ldots w_2$ and $v_1w_2\ldots w_sv_rv_{r+1}\ldots v_{2k-1}v_{2k}$ that are both shorter than C, contradicting the minimality of C. □

The example of $K_4 - e$, that is, the graph obtained by deleting an edge from K_4, shows that shortest even cycles need not be isometric.

Isometric subgraphs of hypercubes are called *partial cubes*. They constitute a large class of graphs with many applications and includes, for example, benzenoid graphs, which will be treated in Chapter 19. Another subclass of partial cubes are median graphs, the main topic of the next section. For their characterization, the vertex-deleted subgraph $Q_3^- = Q_3 - v$, shown in Figure 3.2, plays an important role. It is a partial cube but not a median graph, as we shall see.

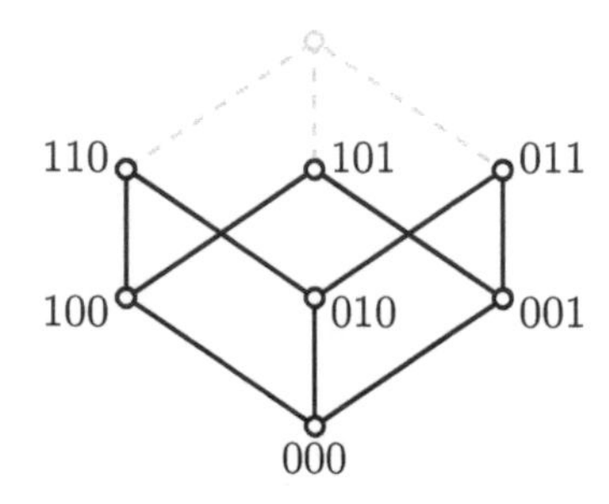

FIGURE 3.2 Q_3^- as an induced subgraph in Q_3.

3.3 Median Graphs

To any triple of vertices u, v, w in an arbitrary tree there exists a unique vertex z that lies on shortest paths between any two of them. This observation (Proposition 3.4) leads to the concepts of medians and median graphs.

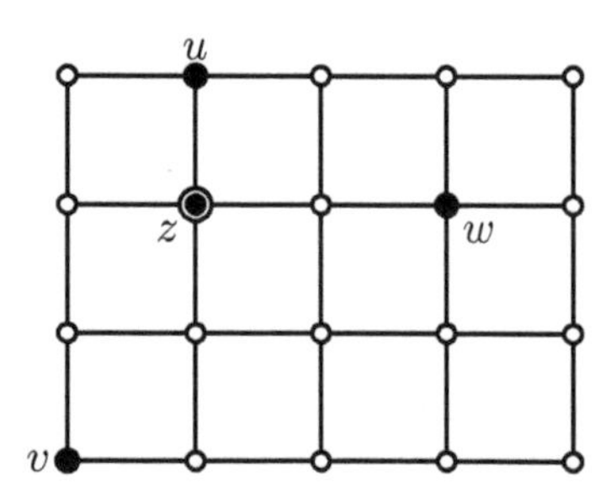

FIGURE 3.3 Median z of the triple u, v, and w.

A *median* of a triple of vertices u, v, w of a graph G is a vertex z that lies on a shortest u,v-path, on a shortest u,w-path, and on a shortest v,w-path. Note that z can be one of the vertices u, v, w. In Figure 3.3 the vertex z is a median of u, v, w. Alternatively, the medians of u, v, w can be defined as the vertices in $I(u, v) \cap I(u, w) \cap I(v, w)$.

A graph is a *median graph* if every triple of its vertices has a unique median, namely if

$$|\, I(u, v) \cap I(u, w) \cap I(v, w)\,| = 1$$

for every triple $u, v, w \in V(G)$. These graphs were introduced by Avann (1961), Nebeský (1971), and, independently, by Mulder (1978; 1980b; 1980a).

Neither $K_{2,3}$ nor Q_3^- is a median graph. The graph $K_{2,3}$ is not a median graph because its two vertices of degree 3 are medians of the other three vertices. Likewise, Q_3^- (Figure 3.2) is not a median graph because the triple 110, 101, 011 has no median. Thus, $K_{2,3}$ fails to be a median graph because it has too many medians, and Q_3^- because it does not have enough.

Proposition 3.4 *Trees are median graphs.*

Proof Let u, v, w, be vertices of a tree T and P, Q, R the unique paths in T from u to v, from u to w and from v to w, as depicted in Figure 3.4. Because these paths are unique, they are also shortest paths. The paths P and Q have the vertex u in common. Let z be the (uniquely defined) vertex on P and Q of largest distance from u. Then the subpaths P_{zv} of P from z to v and Q_{zw} of Q from z to w have only z in common. Therefore their union is a v, w-path, which is the path R. Notice that all vertices of R that are different from z are either not in P or not in Q. Hence, z is the only vertex that these three paths have in common and thus a median of u, v, w. Because P, Q, R are the only paths between u, v, and w, the vertex z is unique. □

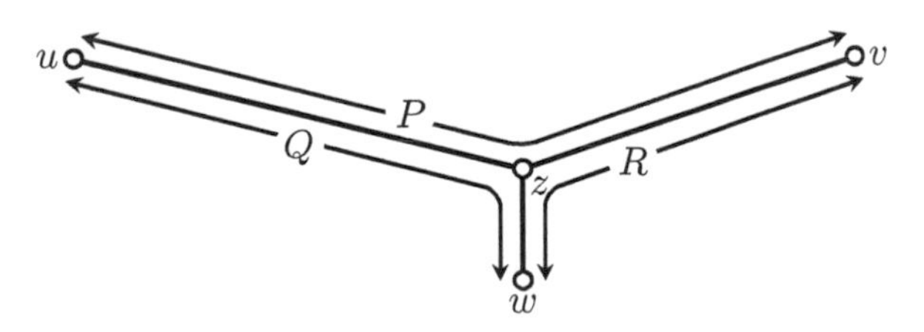

FIGURE 3.4 Median of u, v, w in a tree.

Lemma 3.5 *Let u_1, u_2, u_3 be vertices of a graph G. If they have a median z, then*

$$d(u_i, z) = \frac{1}{2}\left(d(u_i, u_j) + d(u_i, u_k) - d(u_j, u_k)\right),$$

where $\{i, j, k\} = \{1, 2, 3\}$.

Proof Because $d(u_i, u_j) = d(u_i, z) + d(z, u_j)$ for $1 \leq i, j \leq 3$ and $i \neq j$, the assertion can be proved by substitution of the appropriate identities on the right side of the equality. □

Proposition 3.6 *Graphs in which every triple has a median are bipartite.*

Proof Let C be a shortest odd cycle $v_1 v_2 \ldots v_{2k+1}$ of a graph satisfying the hypothesis of the proposition. By Proposition 3.3, it is isometric. Consider the vertices v_1, v_{k+1}, v_{k+2}. Because C is isometric, $d_G(v_1, v_{k+1}) = d_G(v_1, v_{k+2}) = k$. Thus, by Lemma 3.5, the distance of the median of v_1, v_{k+1}, v_{k+2} from v_1 is $\frac{1}{2}(2k - 1)$, which is impossible, as this is not an integer. □

Proposition 3.7 *Every hypercube is a median graph.*

Proof Consider a triple $u = u_1u_2 \ldots u_r$, $v = v_1v_2 \ldots v_r$, and $w = w_1w_2 \ldots w_r$ of vertices of Q_r. Form $z = z_1z_2 \ldots z_r$, where each z_i is defined by "majority rule," that is $z_i = 1$ if at least two of u_i, v_i, w_i are 1, and $z_i = 0$ otherwise. Then, given any two vertices of the triple, say u and v, differing in (say) k spaces, we have $d(u, v) = k = d(u, z) + d(z, v)$, so z is on a shortest u, v-path. Thus z is a median of u, v, and w.

On the other hand, if x is on a shortest path from u to v, then $u_i = v_i$ implies $x_i = v_i$, as noted on p. 28. Similar remarks hold if x is on a shortest u, w- or v, w-path. It follows that any median of u, v and w obeys majority rule, so the median is unique. □

3.4 Retracts

Median graphs can be characterized as retracts of hypercubes, a result that we will prove in Section 12.3 and that could also be deduced from Theorem 14.13. By a *retraction* φ of a graph G, we mean a homomorphism of G into itself with the property $\varphi^2(u) = \varphi(u)$ for all $u \in V(G)$. Notice that the condition $\varphi^2 = \varphi$ means φ restricts to the identity on its image. The homomorphic image H of G under φ is called a *retract*. Clearly, retracts are induced subgraphs.

A homomorphism of a graph into itself is also called an *endomorphism*. An endomorphism φ is *idempotent* if $\varphi^2 = \varphi$. We could therefore have defined retracts as *idempotent endomorphisms*.

A similar concept is the *weak retract*. It is based on weak homomorphisms, where a *weak homomorphism* $\varphi : G \to H$ is a map $\varphi : V(G) \to V(H)$ for which $uv \in E(G)$ implies $\varphi(u)\varphi(v) \in E(H)$ or $\varphi(u) = \varphi(v)$. A *weak retraction* is then an *idempotent weak homomorphism*. The image H of G under φ is called a *weak retract* of G. Clearly, every retract is a weak retract.

Every subgraph K_2 of a bipartite graph is a retract. Figure 3.5 shows further examples of retracts and weak retracts. The retractions are indicated by arrows; the corresponding retracts are induced by the subgraphs in the shaded regions.

We repeat that every retract is a weak retract. The converse need not be true. For instance, the path P_3 of G_2 of Figure 3.5 is not a retract of G_2 because a triangle cannot be mapped onto a path by a homomorphism.

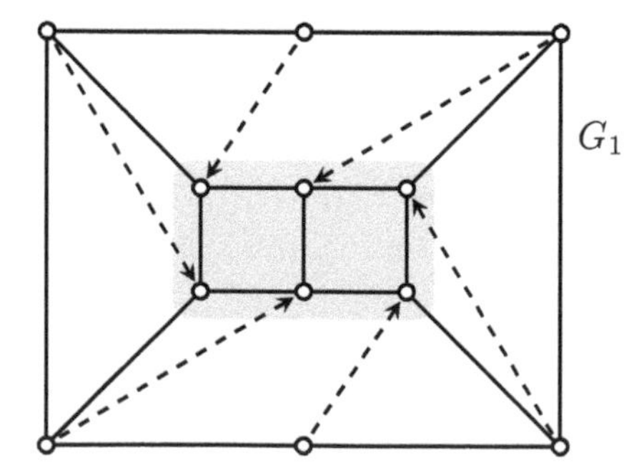

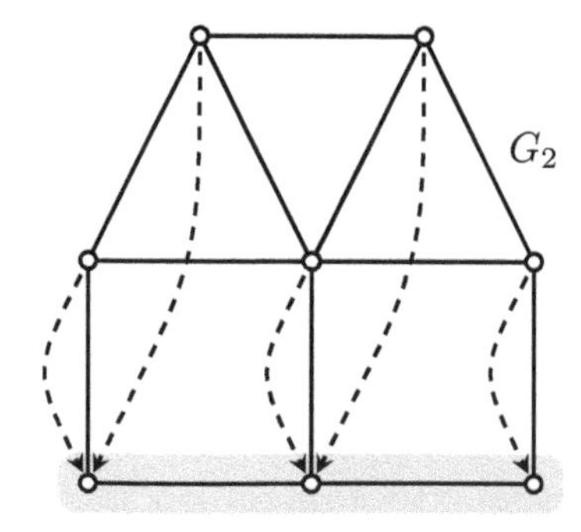

FIGURE 3.5 Retraction and a weak retraction.

A map $\varphi : V(G) \to V(H)$ for which $d_H(\varphi(u), \varphi(v)) \leq d_G(u, v)$ for any pair $u, v \in V(G)$ is called a *nonexpansive map*. This concept is equivalent to that of a weak homomorphism, as the next result shows.

Proposition 3.8 *A map is a weak homomorphism if and only if it is a nonexpansive map.*

Proof Clearly, a nonexpansive map is a weak homomorphism.

Suppose now that $\varphi : V(G) \to V(H)$ is a weak homomorphism. Let $u, v \in V(G)$ and P be a shortest u,v-path in G. Then $\varphi(P)$ contains a $\varphi(u), \varphi(v)$-walk in H. Consequently, $\varphi(P)$ contains a $\varphi(u), \varphi(v)$-path in H, and hence $d_H(\varphi(u), \varphi(v)) \leq d_G(u, v)$. □

Thus, weak retractions, and therefore also retractions, are nonexpansive maps. Because for a subgraph H of G we clearly have $d_G(u, v) \leq d_H(u, v)$ for any $u, v \in V(G)$, we infer the following result about (weak) retracts.

Corollary 3.9 *Weak retracts (and thus also retracts) are isometric subgraphs.*

Another basic and important property of retractions is that they preserve the chromatic number.

Proposition 3.10 *If H is a retract of G, then $\chi(H) = \chi(G)$.*

Proof Because H is a subgraph of G, we have $\chi(H) \leq \chi(G)$. Moreover, a retraction from G onto H is a homomorphism. Thus, $\chi(G) \leq \chi(H)$, by Proposition 2.10 (ii). □

This property does not hold for weak retracts, as can be seen from the example in Figure 3.5. While the depicted retract of G_2 is bipartite, we have $\chi(G_2) = 3$.

It requires some argument to show that median graphs are retracts of hypercubes, but the converse is easy. Below we show that the class of median graphs is closed under retractions. Because hypercubes are median graphs, this implies that retracts of hypercubes are median graphs too.

Proposition 3.11 *Retracts of median graphs are median graphs.*

Proof Let H be a retract of a median graph G and u, v, w be three vertices of H. Note that H is isometric by Corollary 3.9. Thus, every median of u, v, w in H is also a median in G. Hence the median is unique, if it exists.

Let z be the median of u, v, w in G. It lies on shortest u, v-, u, w- and v, w-paths P, Q, and R. Consider a retraction φ from G onto H and the images $\varphi(P), \varphi(Q)$, and $\varphi(R)$. These images connect vertices in H and must be shortest paths because H is isometric. But then $\varphi(z)$ is a median of u, v, w in H. □

Exercises

3.1. Let G be the graph obtained from $K_{4,4}$ by removal of an independent set of four edges. Show that $G \cong Q_3$.

3.2. Show that Q_r is not planar for any $r \geq 4$.

3.3. Show that for any $r \geq 3$, the vertex-deleted r-cube, $Q_r^- = Q_r - v$, is a partial cube but not a median graph.

3.4. For any $r \geq 2$, find an explicit isometric embedding of C_{2r} into Q_r.

3.5. Show that C_{2r} is not a retract of Q_r for $r \geq 3$.

3.6. Let H be a retract of G. Show that $\omega(H) = \omega(G)$ and that the shortest odd cycles of H and G have the same lengths.

3.7. Let G be a graph with $\chi(G) = n$. Show that K_n is a retract of G if and only if $\chi(G) = \omega(G)$.

3.8. Show that a χ-critical graph contains no proper retract.

3.9. A graph G is a *core* if no proper subgraph of G is a retract of G. Show that G is a core if and only if every homomorphism from G to itself is an automorphism of G.

3.10. Let f be a nonexpansive map of Q_r and x be the median of u, v, w. Show that f fixes x if it fixes u, v, and w.

3.11. For r-tuples a, b over the alphabet $\{0, 1\}$, let $a \vee b$ denote the maximum, taken coordinatewise, and $a \wedge b$ the minimum. For instance, $0011 \vee 0101 = 0111$ and $0011 \wedge 0101 = 0001$. Show that the median x of u, v, and w in Q_r is $(u \vee v) \wedge (u \vee w) \wedge (v \vee w) = (u \wedge v) \vee (u \wedge w) \vee (v \wedge w)$.

Chapter 4

Graph Products

We now introduce the primary object of interest in this book: the idea of a graph product. Broadly speaking, a graph product is a binary operation on Γ or Γ_0. However, under reasonable and natural restrictions (such as associativity), the number of different products is actually quite limited. The chapter begins with definitions of three main products that have been studied in the literature: the Cartesian product, the direct product, and the strong product. We then treat the issue of associativity (which allows for the easy extension of these products to arbitrarily many factors) and we examine the projections of products to their factors.

This is followed by a section that classifies all associative products, and justifies why the three main products are in a sense the most natural of all products. Along the way we will meet one additional product worthy of special attention, the so-called lexicographic product.

Although many products can be defined on Γ_0, for simplicity we assume unless noted otherwise that all graphs are in Γ, that is, they have no loops.

4.1 Three Fundamental Products

We now introduce three fundamental graph products: the *Cartesian product*, the *direct product*, and the *strong product*. These products, which have been widely investigated and have many significant applications, are the central theme of this book. In each case, the product of graphs G and H is another graph whose vertex set is the Cartesian product $V(G) \times V(H)$ of sets. However, each product has different rules for adjacencies.

The *Cartesian product* of G and H is a graph, denoted as $G \,\square\, H$, whose vertex set is $V(G) \times V(H)$. Two vertices (g,h) and (g',h') are adjacent precisely if $g = g'$ and $hh' \in E(H)$, or $gg' \in E(G)$ and $h = h'$. Thus,

$$\begin{aligned} V(G \,\square\, H) &= \{(g,h) \mid g \in V(G) \text{ and } h \in V(H)\}, \\ E(G \,\square\, H) &= \{(g,h)(g',h') \mid g = g',\ hh' \in E(H), \text{ or } gg' \in E(G),\ h = h'\}. \end{aligned}$$

The graphs G and H are called *factors* of the product $G \,\square\, H$. As an example, Figure 4.1 (left) shows the Cartesian product $P_4 \,\square\, P_3$. For clarity, P_4 and P_3 are displayed below and to the left of the product $P_4 \,\square\, P_3$.

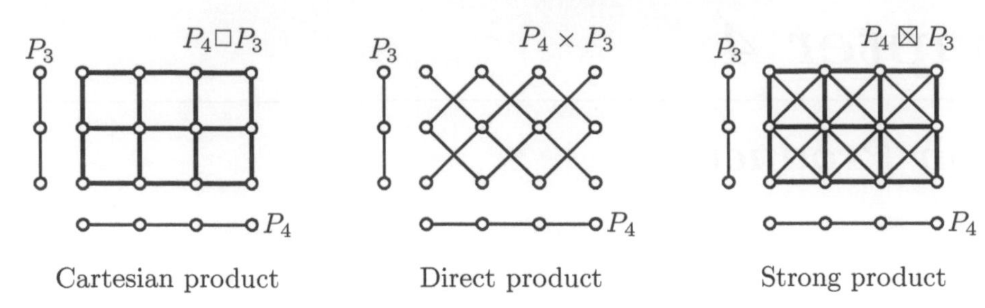

FIGURE 4.1 Examples of products.

The *direct product* of G and H is the graph, denoted as $G \times H$, whose vertex set is $V(G) \times V(H)$, and for which vertices (g, h) and (g', h') are adjacent precisely if $gg' \in E(G)$ and $hh' \in E(H)$. Thus,

$$\begin{aligned} V(G \times H) &= \{(g, h) \mid g \in V(G) \text{ and } h \in V(H)\}, \\ E(G \times H) &= \{(g, h)(g', h') \mid gg' \in E(G) \text{ and } hh' \in E(H)\}. \end{aligned}$$

Figure 4.1 (center) shows the direct product $P_4 \times P_3$. Notice that this particular example is disconnected. Other names for the direct product that have appeared in the literature are *tensor product*, *Kronecker product*, *cardinal product*, *relational product*, *cross product*, *conjunction*, *weak direct product*, *Cartesian product*, *product*, or *categorical product*. From the point of view of category theory, only *product* and *categorical product* are appropriate. For details, see p. 54.

Finally, the *strong product* of G and H is the graph denoted as $G \boxtimes H$, and defined by

$$\begin{aligned} V(G \boxtimes H) &= \{(g, h) \mid g \in V(G) \text{ and } h \in V(H)\}, \\ E(G \boxtimes H) &= E(G \square H) \cup E(G \times H). \end{aligned}$$

Occasionally one also encounters the names *strong direct product* or *symmetric composition* for the strong product. Note that $G \square H$ and $G \times H$ are subgraphs of $G \boxtimes H$. Figure 4.1 (right) shows the strong product $P_4 \boxtimes P_3$. For clarity, the edges of the subgraph $G \square H$ are drawn in bold.

Figure 4.2 shows other examples of our three fundamental products. Notice that in general a prism is the Cartesian product of a cycle by an edge, and the square lattice is the Cartesian product of a two-sided infinite path by itself. Also, note that $K_m \boxtimes K_n = K_{mn}$.

We can envision the edges of the Cartesian and strong products as being roughly aligned

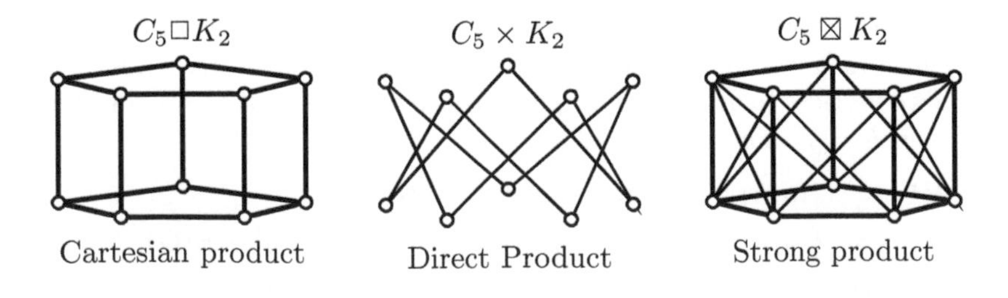

FIGURE 4.2 Examples of products.

with a Cartesian coordinate axis, and consequently we can often create natural and easily visualized drawings that highlight these products' structures. By contrast, the direct product can present challenges. For instance, note that the drawing of $C_5 \times K_2$ in Figure 4.2 is merely C_{10} twisted around on itself, and it would perhaps have a more natural representation as a ten-sided regular polygon. For another example, consider the direct product in Figure 4.3. As drawn on the left, the product $G \times P_3$ is not very appealing. Its structure is much more apparent when it is redrawn as on the right.[1]

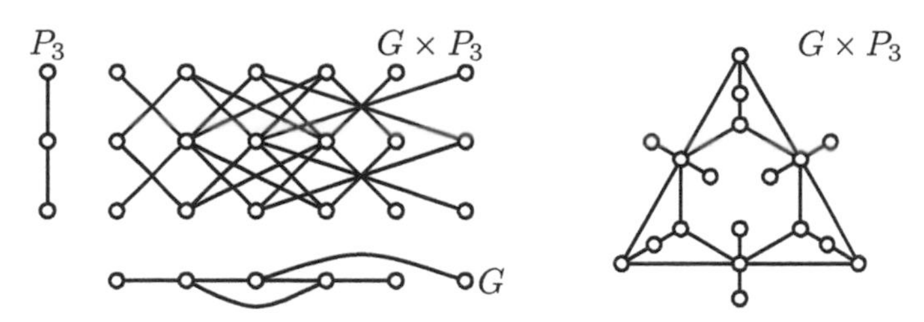

FIGURE 4.3 A direct product (left) and a more symmetric representation (right).

Finally, observe that $K_2 \,\square\, K_2 = C_4$, $K_2 \times K_2 = K_2 + K_2$, and $K_2 \boxtimes K_2 = K_4$, as shown in Figure 4.4. This explains the rationale behind the symbols for the three products: The square of K_2 produces either the shape $\square$, $\times$, or $\boxtimes$, depending on whether we use the Cartesian, direct, or strong product. The notation is due to Nešetřil (1981).

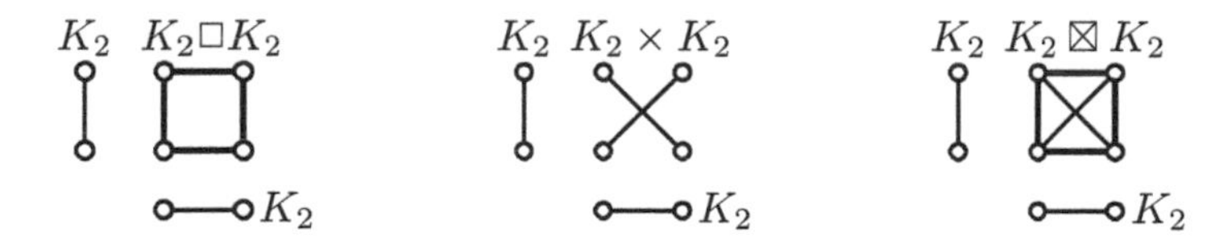

FIGURE 4.4 Rationale for the notation.

We will deduce various elementary properties of the three products in Chapter 5. For now we turn to the important issue of associativity.

4.2 Commutativity, Associativity, and Multiple Factors

It is immediate from the definitions of the three products that the map $(g, h) \mapsto (h, g)$ is an isomorphism from $G * H$ to $H * G$, where $*$ stands for any one of the three fundamental products. Thus, the three products are commutative in the sense that $G * H \cong H * G$ for all graphs G and H. Proving associativity needs a little more care.

Proposition 4.1 *The Cartesian, the direct, and the strong product are each associative. In particular, given graphs G_1, G_2, and G_3, the map $((x_1, x_2), x_3) \mapsto (x_1, (x_2, x_3))$ is an isomorphism $(G_1 * G_2) * G_3 \to G_1 * (G_2 * G_3)$, where $*$ stands for either the Cartesian, the direct, or the strong product.*

[1]This drawing was the logo for the Sixth Slovenian International Conference on Graph Theory in Bled, Slovenia, 2007.

Proof Let us first confirm this for the Cartesian product. Referring to the definition of the Cartesian product, note that $((x_1, x_2), x_3)((y_1, y_2), y_3) \in E((G_1 \square G_2) \square G_3)$ if and only if $x_i y_i \in E(G_i)$ for exactly one index $i \in \{1, 2, 3\}$, and $x_i = y_i$ for the other two indices. Similarly, the same conditions characterize $(x_1, (x_2, x_3))(y_1, (y_2, y_3)) \in E(G_1 \square (G_2 \square G_3))$. Thus, the map $((x_1, x_2), x_3) \mapsto (x_1, (x_2, x_3))$ is indeed an isomorphism, so the Cartesian product is associative.

According to the definition of the direct product, we have $((x_1, x_2), x_3)((y_1, y_2), y_3) \in E((G_1 \times G_2) \times G_3)$ if and only if $x_i y_i \in E(G_i)$ for each $i \in \{1, 2, 3\}$. Similarly, the same conditions characterize $(x_1, (x_2, x_3))(y_1, (y_2, y_3)) \in E(G_1 \times (G_2 \times G_3))$. It follows that the map $((x_1, x_2), x_3) \mapsto (x_1, (x_2, x_3))$ is an isomorphism from $(G_1 \times G_2) \times G_3$ to $G_1 \times (G_2 \times G_3)$, so the direct product is associative.

Turning now to the strong product, its definition gives $((x_1, x_2), x_3)((y_1, y_2), y_3) \in E((G_1 \boxtimes G_2) \boxtimes G_3)$ if and only if $x_i y_i \in E(G_i)$ or $x_i = y_i$ for each $i \in \{1, 2, 3\}$, and $x_i \neq y_i$ for at least one index i. Similarly, these same conditions characterize $((x_1, x_2), x_3)((y_1, y_2), y_3) \in E(G_1 \boxtimes (G_2 \boxtimes G_3))$. Thus, the strong product is associative. □

Associativity gives us license to omit parentheses when dealing with products with more than two factors. Indeed, Proposition 4.1 and its proof allow us to unambiguously define $G_1 \square G_2 \square G_3$ as the graph with vertex set $V(G_1) \times V(G_2) \times V(G_3)$, where two vertices (x_1, x_2, x_3) and (y_1, y_2, y_3) are adjacent if and only if $x_i y_i \in E(G_i)$ for some index i, and $x_j = y_j$ for $j \neq i$. Figure 4.5 is an example of a Cartesian product of three graphs, where, for brevity, the vertices (x_1, x_2, x_3) are written as $x_1 x_2 x_3$.

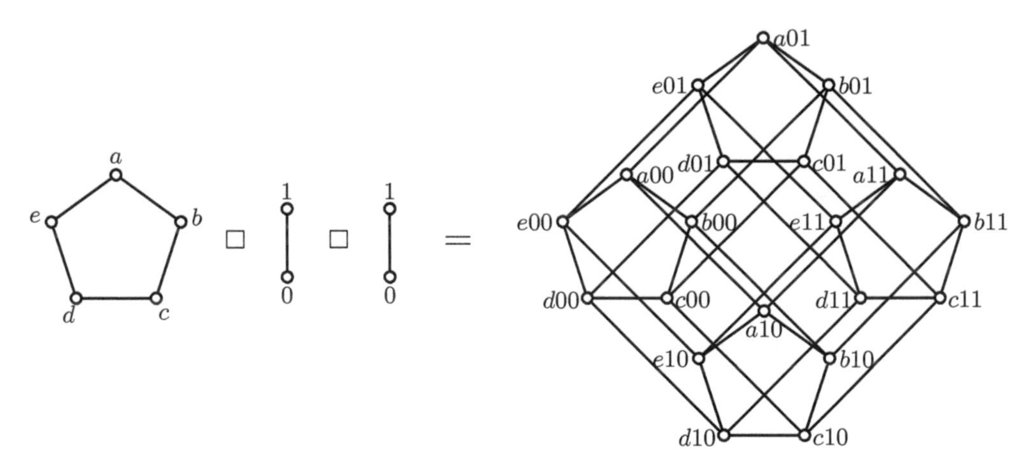

FIGURE 4.5 The product $C_5 \square K_2 \square K_2$.

In general, given graphs $G_1, G_2, \ldots, G_k$, then $G_1 \square G_2 \square \cdots \square G_k$ is the graph with vertex set $V(G_1) \times V(G_2) \times \cdots \times V(G_k)$, where two vertices $(x_1, x_2, \ldots, x_k)$ and $(y_1, y_2, \ldots, y_k)$ are adjacent if and only if $x_i y_i \in E(G_i)$ for some index i, and $x_j = y_j$ for $j \neq i$. We often use the notation $G_1 \square G_2 \square \cdots \square G_k = \square_{i=1}^{k} G_i$.

Generalizing the direct product to multiple factors, $G_1 \times G_2 \times \cdots \times G_k = \times_{i=1}^{k} G_i$ is the graph whose vertex set is $V(G_1) \times V(G_2) \times \cdots \times V(G_k)$, and for which vertices $(x_1, x_2, \ldots, x_k)$ and $(y_1, y_2, \ldots, y_k)$ are adjacent precisely if $x_i y_i \in E(G_i)$ for each index i.

Finally, $G_1 \boxtimes G_2 \boxtimes \cdots \boxtimes G_k = \boxtimes_{i=1}^{k} G_i$ has vertex set $V(G_1) \times V(G_2) \times \cdots \times V(G_k)$, and distinct vertices $(x_1, x_2, \ldots, x_k)$ and $(y_1, y_2, \ldots, y_k)$ are adjacent if and only if either $x_i y_i \in E(G)$ or $x_i = y_i$ for each $1 \leq i \leq k$. We note that in general $E(G_1 \boxtimes \cdots \boxtimes G_k) \neq E(G_1 \square \cdots \square G_k) \cup E(G_1 \times \cdots \times G_k)$, unless $k = 2$.

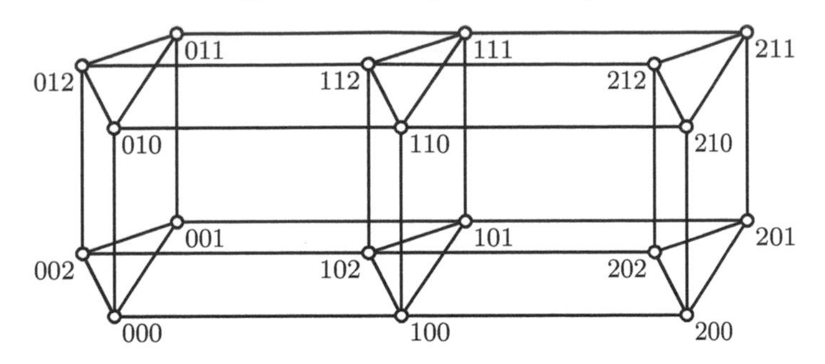

FIGURE 4.6 $P_3 \,\square\, K_2 \,\square\, K_3$.

Figures 4.6 through 4.8 show products with three factors P_3, K_2, and K_3, where the vertices of P_3 and K_3 are labeled by $0, 1, 2$, and those of K_2 by 0 and 1.

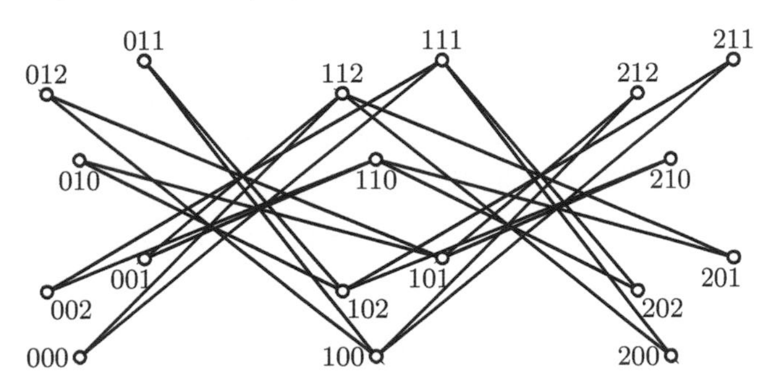

FIGURE 4.7 $P_3 \times K_2 \times K_3$.

As we mentioned before, Cartesian products of graphs are often relatively easy to visualize and draw. (Though of course the images can become increasingly complex as the number of factors grows.) By contrast, some care may be needed in interpreting the structure of a direct product of multiple factors. Notice that $P_3 \times K_2 \times K_3$ in Figure 4.7 has two components, each isomorphic to two hexagons joined at alternating vertices.

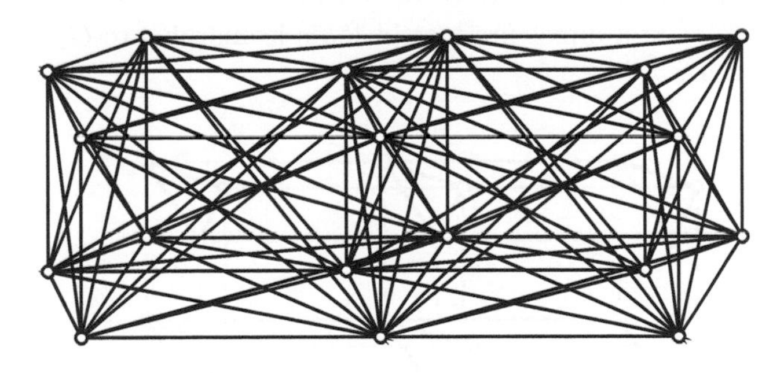

FIGURE 4.8 $P_3 \boxtimes K_2 \boxtimes K_3$.

The kth power of G with respect to the Cartesian product is denoted as $G^{\square,k}$, that is $G^{\square,k} = \square_{i=1}^{k} G$. Similarly, we denote kth powers of G with respect to the direct and strong products as $G^{\times,k}$ and $G^{\boxtimes,k}$.

4.3 Projections and Layers

Here we define the notion of projections from a product to its factors, and note that these maps respect adjacency in the sense that they are weak homomorphisms. We then introduce an important notion: the layers of a product.

Let $*$ represent either the Cartesian, the direct, or the strong product operation, and consider a product $G_1 * G_2 * \cdots * G_k$. For any index $1 \le i \le k$, there is a *projection map* $p_i : G_1 * G_2 * \cdots * G_k \to G_i$ defined as $p_i(x_1, x_2, \ldots, x_k) = x_i$. We call x_i the ith *coordinate* of the vertex $(x_1, x_2, \ldots, x_k)$. (Occasionally, when dealing with a product $G * H$, we may write the projections as p_G and p_H, and refer to the corresponding G- or H-coordinate.)

No matter which product $*$ represents, each projection p_i is a weak homomorphism. Indeed, the definitions imply that if $(x_1, x_2, \ldots, x_k)(y_1, y_2, \ldots, y_k)$ is an edge of $G_1 * G_2 * \cdots * G_k$, then either $x_i = y_i$ or $x_i y_i \in E(G_i)$ for each $1 \le i \le k$, so each p_i is a weak homomorphism. Even more is true for the direct product. Because $(x_1, x_2, \ldots, x_k)(y_1, y_2, \ldots, y_k)$ is an edge of $G_1 \times G_2 \times \cdots \times G_k$ if and only if $x_i y_i \in E(G_i)$ for each $1 \le i \le k$, each projection p_i is actually a homomorphism.

For Cartesian and strong products, the projections restrict to isomorphisms on certain subgraphs. Given a vertex $a = (a_1, a_2, \ldots, a_k)$ of the product $G = G_1 * G_2 * \cdots * G_k$, the *$G_i$-layer through a* is the induced subgraph

$$\begin{aligned} G_i^a &= \langle \{x \in V(G) \mid p_j(x) = a_j \text{ for } j \neq i\} \rangle \\ &= \langle \{(a_1, a_2, \ldots, x_i, \ldots, a_k) \mid x_i \in V(G_i)\} \rangle. \end{aligned}$$

Note that $G_i^a = G_i^b$ if and only if $p_j(a) = p_j(b)$ for each index $j \neq i$.

If $*$ is either the Cartesian or the strong product, then the restriction $p_i : G_i^a \to G_i$ is an isomorphism for each a and i. Figure 4.9 shows the Cartesian product $P_2 \,\square\, K_2 \,\square\, K_3$ and its P_2-, K_2-, and K_3-layers. Note that these layers are isomorphic to their respective factors. Gross and Yellen (2006) even use this idea in defining the Cartesian product as

$$G \,\square\, H = G \times V(H) \cup V(G) \times H,$$

emphasizing the H-layers through vertices of G, and G-layers through vertices of H.

While each G_i-layer of a Cartesian or a strong product is isomorphic to G_i, the layers of a direct product of graphs in Γ are totally disconnected. (However, in Section 5.3 we will see that a layer $G_i^{(a_1,\ldots,a_k)}$ of a direct product in Γ_0 is isomorphic to G_i provided that G_j has a loop at a_j for each $j \neq i$.)

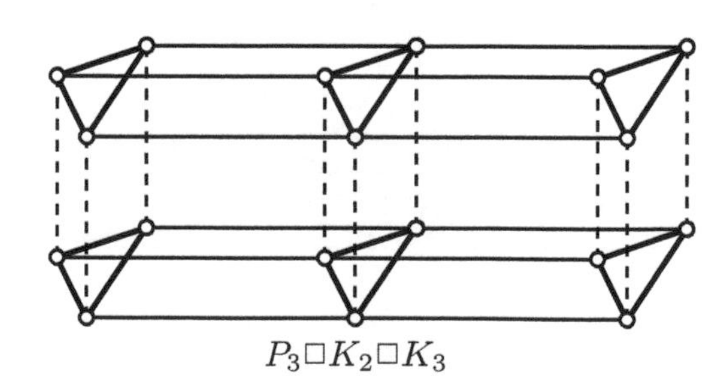

FIGURE 4.9 Product $P_3 \,\square\, K_2 \,\square\, K_3$, the P_3-layers (light), the K_2-layers (dashed), and the K_3-layers (bold).

4.4 Classification of Products

Are there other products in addition to the three main ones introduced in the previous section? The answer depends on what kinds of structures we admit as "graph products." To make the question tractable, we concentrate on binary operations on Γ that have the salient features of the three products introduced in the previous section. Specifically, we examine associative binary operations $*$ on Γ for which $V(G * H) = V(G) \times V(H)$ and for which the projections onto the factors are weak homomorphisms. In the first part of this section we show that the only "sufficiently interesting" products that meet these criteria are the Cartesian, direct, and strong products.

We will then relax the requirement that both projections be weak homomorphisms to obtain a fourth product of interest, called the *lexicographic product*. Finally, we classify all associative graph products for which $V(G*H) = V(G) \times V(H)$. We will discover that there are exactly twenty such products. Of these twenty products, only the Cartesian, direct, strong, and lexicographic products have the property that at least one projection is a weak homomorphism, and, in addition, are nontrivial enough to merit further study.

This section justifies the fact that this book mainly treats Cartesian, direct, strong, and lexicographic products. A reader who does not require such a justification may safely skip the remainder of this chapter.

We begin by enumerating the associative products having the property that projections to both factors are weak homomorphisms.

Products for which both projections are weak homomorphisms

We consider graph products $*$ for which $V(G * H) = V(G) \times V(H)$. Let us say that such a product $*$ is *associative* if the map $((g, h), k) \mapsto (g, (h, k))$ is an isomorphism from $(G*H)*K$ to $G * (H * K)$. We seek to classify all such products for which both projections $G * H \to G$ and $G * H \to H$ are weak homomorphisms.

But first, there is an additional criterion for a graph product that is so fundamental that it might almost be overlooked: The edge set of the product should be determined by some definite rule. The fact that the projections are weak homomorphisms means that $(g, h)(g', h') \in E(G*H)$ implies $gg' \in E(G)$ or $g = g'$, and $hh' \in E(H)$ or $h = h'$. Conversely, whether $(g, h)(g', h')$ is an edge of $G*H$ should be determined by some definite rule involving the incidence (or equality) of g and g', and h and h'. (For example, $(g, h)(g', h')$ is an edge of $G \times H$ if and only if $gg' \in E(G)$ and $hh' \in E(H)$.)

To formalize this, we use an *incidence function*: For any graph G, there is a function $\delta : V(G) \times V(G) \to \{\Delta, 1, 0\}$ defined as follows. (Here Δ is a previously undefined symbol.)

$$\delta(g, g') = \begin{cases} \Delta & \text{if } g = g', \\ 1 & \text{if } g \neq g' \text{ and } gg' \in E(G), \\ 0 & \text{if } g \neq g' \text{ and } gg' \notin E(G). \end{cases}$$

Thus, δ simply encodes the incidence relation of G. We require for any product $*$, that $\delta((g, h), (g', h'))$ be a function of $\delta(g, g')$ and $\delta(h, h')$. Such a function is simply a binary operation on the set $\{\Delta, 1, 0\}$, and with a slight bending of notation we write it as $\delta((g, h), (g', h')) = \delta(g, g') * \delta(h, h')$. For example, multiplication tables for this operation are shown below for the Cartesian, direct, and strong products.

$*$	Δ	1	0
Δ	Δ	1	0
1	1	0	0
0	0	0	0

Cartesian product

$*$	Δ	1	0
Δ	Δ	0	0
1	0	1	0
0	0	0	0

direct product

$*$	Δ	1	0
Δ	Δ	1	0
1	1	1	0
0	0	0	0

strong product

Looking at it this way, we can enumerate all possible products by simply filling in the tables in various ways. Fortunately there are some restrictions that make our work even easier. Note that in a table for any product, the symbol Δ can appear only in the upper-left corner. Also, observe that the projections onto the factors are weak homomorphisms if and only if the third row and column of the table consist entirely of 0's. This leaves only three entries to fill in (namely $\Delta * 1$, $1 * \Delta$, and $1 * 1$), giving potentially $2^3 = 8$ products, three of which we have already considered.

One possibility is to fill in all entries with 0's, as illustrated in Table (a) below. This leads to the "trivial product," $G * H = D_{|V(G)|} \,\square\, D_{|V(H)|}$, which is a totally disconnected graph. (In this section, we regard the graphs $D_{|V(G)|}$ and $K_{|V(G)|}$ as having vertex set $V(G)$.) Clearly, the trivial product is associative, and the projections to the factors are (vacuously) weak homomorphisms. We therefore have a new product, albeit not a very exciting one.

$*$	Δ	1	0
Δ	Δ	0	0
1	0	0	0
0	0	0	0

trivial product
$D_{|V(G)|} \,\square\, D_{|V(H)|}$

(a)

$*$	Δ	1	0
Δ	Δ	0	0
1	1	1	0
0	0	0	0

nonassociative
product

(b)

$*$	Δ	1	0
Δ	Δ	0	0
1	1	0	0
0	0	0	0

"uninteresting" product
$G \,\square\, D_{|V(H)|}$

(c)

Another possibility is to fill in the missing entries as in Table (b). However, this operation is not associative, as $(1 * \Delta) * 1 = 1 * 1 = 1 \neq 0 = 1 * 0 = 1 * (\Delta * 1)$. Likewise, the transpose of this table is not associative. It is easy to confirm (Exercise 4.15) that the corresponding graph products are not associative, so we do not consider them worthy of attention.

Table (c) is readily seen to be associative, so it leads to a new product. However, notice that $(g, h)(g', h')$ is an edge of this product if and only if $gg' \in E(G)$ and $h = h'$. Thus, $G * H = G \,\square\, D_{|V(H)|}$ merely consists of $|V(H)|$ copies of G. Similarly, the table's transpose leads to an associative product $G * H = D_{|V(G)|} \,\square\, H$. Although these are perfectly fine associative products, they are not particularly interesting, as they completely ignore all properties of one factor (other than its number of vertices).

We have now exhausted all eight possibilities, and the only sufficiently nontrivial cases are the Cartesian product, the direct product, and the strong product. We summarize our findings in Table 4.1 listing the associative products for which both projections are weak homomorphisms.

TABLE 4.1 Associative Products where Both Projections are Weak Homomorphisms.

$G \,\square\, H$	$G \times H$	$G \boxtimes H$	$D_{\|V(G)\|} \,\square\, D_{\|V(H)\|}$	$G \,\square\, D_{\|V(H)\|}$	$D_{\|V(G)\|} \,\square\, H$

Products for which only one projection is a weak homomorphism

Let us now turn our attention to associative products for which only one (say the first) projection is a weak homomorphism. A partial table for such a product looks as follows. The fact that projection to the first factor is a weak homomorphism is reflected in the bottom row of 0's. Because projection to the second factor is not a weak homomorphism, the third column is not all 0's, but we do not know a priori if one of its two missing entries is a 0.

$*$	Δ	1	0
Δ	Δ		
1			
0	0	0	0

Suppose it happens that the upper-right entry is 1, that is, $\Delta * 0 = 1$. Then for any $x \in \{\Delta, 1, 0\}$ we have $1 * x = (\Delta * 0) * x = \Delta * (0 * x) = \Delta * 0 = 1$, so the second row is all 1's. Also $\Delta * 1 = \Delta * (\Delta * 0) = \Delta^2 * 0 = 1$. Thus, we arrive at Table (c) below, which describes an associative operation.

On the other hand, suppose the upper-right entry is 0, that is, $\Delta * 0 = 0$. Then $1 * 0 = 1$, because the right-hand column must contain a 1. Also, $1 * x = (1 * 0) * x = 1 * (0 * x) = 1 * 0 = 1$, so the entire second row consists of 1's. Thus we arrive at either Table (a) or Table (b), depending on whether we make $\Delta * 1$ equal to 1 or 0. Both of these operations are readily seen to be associative, so they yield associative products.

$*$	Δ	1	0
Δ	Δ	1	0
1	1	1	1
0	0	0	0

$G \circ H$

(a)

$*$	Δ	1	0
Δ	Δ	0	0
1	1	1	1
0	0	0	0

$G \circ D_{|V(H)|}$

(b)

$*$	Δ	1	0
Δ	Δ	1	1
1	1	1	1
0	0	0	0

$G \circ K_{|V(H)|}$

(c)

The product corresponding to Table (a) is quite interesting. Here $(g, h)(g', h')$ is an edge of $G * H$ if and only if $gg' \in E(G)$, or $g = g'$ and $hh' \in E(H)$, so the structures of both G and H are indeed encoded in the product. This product is often called the lexicographic product in the literature. Its formal definition follows.

The *lexicographic product* of graphs G and H is the graph $G \circ H$ with

$$\begin{aligned} V(G \circ H) &= \{(g, h) \mid g \in V(G), h \in V(H)\}, \\ E(G \circ H) &= \{(g, h)(g', h') \mid gg' \in E(G), \text{ or } g = g' \text{ and } hh' \in E(H)\}. \end{aligned}$$

Figure 4.10 shows $P_3 \circ \overline{K_{2,2}}$ and $\overline{K_{2,2}} \circ P_3$. One of these products is connected and the other is not, so the lexicographic product—although associative—is not commutative. This is also evident by the noncommutative nature of Table (a) above.

The products described in Tables (b) and (c) above are easily seen to be $G * H = G \circ D_{|V(H)|}$ and $G * H = G \circ K_{|V(H)|}$, respectively. These are not particularly interesting, as they ignore the structure of the second factor.

Thus we have established that there are exactly three associative products for which projection to only the first factor is a weak homomorphism; namely, $G * H = G \circ H$, $G * H = G \circ D_{|V(H)|}$, and $G * H = G \circ K_{|V(H)|}$. Of course we could repeat our reasoning for products for which projection to only the *second* factor is a weak homomorphism. The arguments would be the same as above, but would involve the transposes of the respective

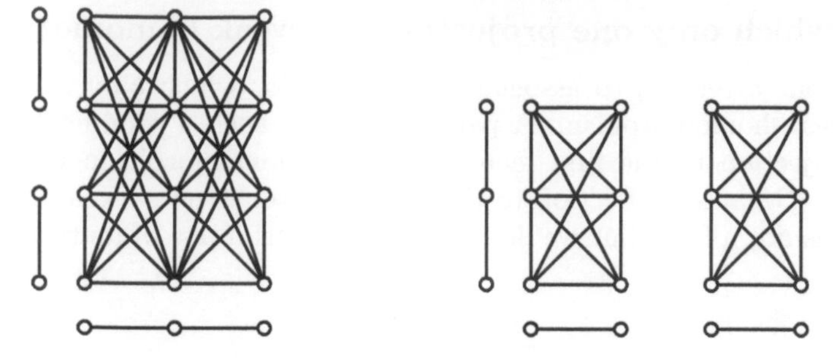

FIGURE 4.10 Lexicographic products $P_3 \circ \overline{K_{2,2}}$ and $\overline{K_{2,2}} \circ P_3$.

tables. Let us designate the products obtained this way as $G \circ^T H$, as well as $D_{|V(G)|} \circ^T H$ and $K_{|V(G)|} \circ^T H$, where $\circ^T$ could be called the *transposed lexicographic product.* We have now shown that there are only six associative products with the property that projection to one factor is a weak homomorphism but projection to the other factor is not. They are summarized in Table 4.2.

TABLE 4.2 Associative Products where Only One Projection is a Weak Homomorphism.

$G \circ H$	$G \circ D_{\|V(H)\|}$	$G \circ K_{\|V(H)\|}$	$G \circ^T H$	$D_{\|V(G)\|} \circ^T H$	$K_{\|V(G)\|} \circ^T H$

The above discussion establishes that there are essentially four associative graph products for which at least one projection is a weak homomorphism, and which employ the adjacency structure of both factors. They are $G \,\square\, H$, $G \times H$, $G \boxtimes H$, and $G \circ H$. (Products $\circ$ and $\circ^T$ are similar enough that we consider $\circ$ to the exclusion of $\circ^T$.)

All associative graph products

Let us proceed to classify *all* associative graph products. From the above considerations, we now need only examine those products for which neither projection is a weak homomorphism.

A primary tool in this investigation will be the idea of a complementary product. Given a graph product $*$, its *complementary product* $\overline{*}$ is the product defined as

$$G \,\overline{*}\, H = \overline{\overline{G} * \overline{H}}.$$

It is easy to verify (Exercise 4.16) that if a product is associative, then its complementary product is also associative. Further, any product is the complement of its complement, that is, $\overline{\overline{*}} = *$.

One checks that $\circ$ is its own complement, that is, $G \,\overline{\circ}\, H = \overline{\overline{G} \circ \overline{H}} = G \circ H$. Moreover, the complementary product of $G * H = G \circ D_{|V(H)|}$ is $G \,\overline{*}\, H = G \circ K_{|V(H)|}$. In fact, by Exercise 4.18, the complementary product of any product in Table 4.2 remains in Table 4.2.

By contrast, we claim that if $*$ is a product in Table 4.1, the complementary product $\overline{*}$ is such that neither of its projections are weak homomorphisms. To see this, take factors G and H and edges $gg' \in E(G)$ and $hh' \in E(\overline{H})$. Because the first projection of $*$ is a weak homomorphism, the pair $(g,h)(g',h')$ cannot be an edge of $\overline{G} * \overline{H}$, and hence it is an edge of $\overline{\overline{G} * \overline{H}} = G \,\overline{*}\, H$. But as $p_H(g,h)p_H(g',h') = hh' \notin E(H)$, the projection $G \,\overline{*}\, H \to H$ is not a weak homomorphism. Repeating this argument for edges $gg' \in E(\overline{G})$ and $hh' \in E(H)$, we see that the projection $G \overline{*} H \to G$ is also not a weak homomorphism.

It follows that if we form the complementary products of the six products in Table 4.1, then we get six wholly new and distinct associative products that do not appear in either of the Tables 4.1 or 4.2. So, thus far we have eighteen associative graph products: those in Tables 4.1 and 4.2, and the complements of those in Table 4.1.

If there is an associative product $*$ that is not one of the eighteen considered above, then both $*$ and its complement $\overline{*}$ are such that neither of their projections are weak homomorphisms. By Exercise 4.19, there are only two such products, which are complements of each other, and their incidence tables are as follows,

$\Diamond$	Δ	1	0
Δ	Δ	1	0
1	1	1	0
0	0	0	1

$\overline{\Diamond}$	Δ	1	0
Δ	Δ	1	0
1	1	0	1
0	0	1	0

The product corresponding to the table on the left is often called the *modular product*, and we will denote it by $\Diamond$. Note that the edge set of this product is

$$E(G \Diamond H) = E(G \,\Box\, H) \cup E(G \times H) \cup E(\overline{G} \times \overline{H}).$$

See Imrich (1972a) for further results on the modular product, including issues of prime factorization.

The following theorem summarizes the above discussion:

Theorem 4.2 *There are exactly twenty associative graph products.*

Six of these products (those from Table 4.1, including the Cartesian, direct, and strong products) have the property that projections to both factors are weak homomorphisms.

Another six (those from Table 4.2, including the lexicographic product) have the property that exactly one projections is a weak homomorphism.

An additional eight (the complementary products of those from Table 4.1, as well as the modular product and its complementary product) have the property that neither projection is a weak homomorphism in general.

This classification is based on Imrich and Izbicki (1975). Of course, there is no reason to restrict the investigation of associative products to simple graphs. The same, or at least similar, reasoning applies to directed graphs or even to graphs with multiple edges, both directed and undirected. For example, Imrich and Izbicki (1975) show that the lexicographic product (and its transpose) is the only associative product that is closed in the class of tournaments.[2]

A different approach yielding other products was followed by Pultr (1970, 1972) and Imrich and Pultr (1991). The latter classification is closer to the one here, because it restricts attention to products in the category of symmetric graphs defined on the Cartesian product of the vertex sets of the factors. It requires basic knowledge of category theory.

We conclude with a result about the complexity of finding cliques in graphs. The starting point is the observation that the weak modular product $G \nabla G$, as defined in Exercise 4.23, has a clique of size $|V(G)|$. Moreover, the product $G \nabla H$ of two graphs on n vertices has a clique of size n if and only if $G \cong H$ (Exercise 4.24). Kozen (1978) proved the following:

Theorem 4.3 *Let G and H be graphs of order n. The problem of finding a clique of order n in $G \nabla H$ is equivalent to the isomorphism problem, whereas the problem of determining whether $G \nabla H$ has a clique of size $n(1-\epsilon)$ is NP-complete.*

[2]A *tournament* is an oriented graph obtained from a complete graph by giving every edge a direction. By closure in the class of tournaments we mean that the product of two tournaments is a tournament.

Exercises

4.1. Draw pictures of $K_3 \,\square\, P_3$, $K_3 \times P_3$, $K_3 \boxtimes P_3$, and $K_3 \circ P_3$.

4.2. Find formulas for $|E(G \,\square\, H)|$, $|E(G \times H)|$, $|E(G \boxtimes H)|$, and $|E(G \circ H)|$.

4.3. Show that $K_{3,3}$ and $K_3 \,\square\, K_2$ are the only cubic graphs on six vertices.

4.4. Find a formula for the number of triangles in $K_n \times K_m$.

4.5. Show that $N_{G\times H}(g,h) = N_G(g) \times N_H(h)$ for any $(g,h) \in V(G \times H)$.

4.6. Show that $N_{G\boxtimes H}[(g,h)] = N_G[g] \times N_H[h]$ for any $(g,h) \in V(G \boxtimes H)$.

4.7. Verify that $K_m \boxtimes K_n = K_{mn}$.

4.8. Verify that $K_2 \circ D_n = K_{n,n}$.

4.9. Show that $\overline{K_3 \,\square\, K_3} \cong K_3 \,\square\, K_3$.

4.10. Show that for any $n \geq 3$ and any $m \geq 3$, $\overline{K_m \,\square\, K_n} = K_m \times K_n$.

4.11. Show that $\overline{K_3 \times K_3} \cong K_3 \times K_3$.

4.12. Show that $K_3 \times K_3 \cong K_3 \,\square\, K_3$.

4.13. Show that a graph product $*$ is commutative if and only if the corresponding binary operation $*$ on $\{\Delta, 0, 1\}$ is commutative.

4.14. Show that a graph product $*$ has a unit if and only if the corresponding binary operation $*$ on $\{\Delta, 0, 1\}$ has a unit.

4.15. Show that a graph product $*$ is associative if and only if the corresponding binary operation $*$ on $\{\Delta, 0, 1\}$ is associative.

$((g,h),k) \mapsto (g,(h,k))$ is an isomorphism, so the graph product $*$ is associative.

4.16. Verify that if a product is associative, then its complementary product is also associative. Also, any product $*$ is the complement of its complement, that is, $\overline{\overline{*}} = *$.

4.17. Suppose that $*$ is an associative graph product. Show that the multiplication table for the complementary product $\overline{*}$ (as a binary operation on $\{\Delta, 0, 1\}$) is obtained from the corresponding table for $*$ by interchanging the second and third rows, then interchanging the second and third columns, and finally changing all 1's to 0's and 0's to 1's.

4.18. Verify that the complementary product of any product in Table 4.2 is also in Table 4.2.

4.19. Suppose $*$ is an associative graph product for which neither the first nor the second projection is a weak homomorphism, and, in addition, neither the first nor the second projection of the complementary product $\overline{*}$ is a weak homomorphism. Show that $*$ and $\overline{*}$ are necessarily the modular product and its complementary product.

4.20. Show that the modular product is disconnected if and only if one factor is complete and the other disconnected, or if both factors have exactly two components, each complete.

4.21. Let G and H be nontrivial graphs. Show that $G \circ H \cong G \Diamond H$ if and only if H is complete.

4.22. Give an example of nonunique prime factorization of a graph with respect to the modular product.

4.23. Is the *weak modular product* ∇, defined by the following table, associative?

∇	Δ	1	0
Δ	Δ	0	0
1	0	1	0
0	0	0	1

4.24. Show that the weak modular product $G\nabla H$ (Exercise 4.23) of two graphs on n vertices has a clique of size n if and only if $G \cong H$.

4.25. Show that the only associative products closed in the class of tournaments (see the footnote on p. 45) are the lexicographic product and its transpose.

4.26. (Izbicki, private communication) Suppose that G and H have the same vertex set V. For two distinct vertices $u, v \in V$ we have the possibilities $uv \in E(G)$ or $uv \in E(\overline{G})$. Similarly, $uv \in E(H)$ or $uv \in E(\overline{H})$. In how many ways can we determine a "product" $G \cdot H$ with vertex set V taking recourse to these possibilities (and no others)? How many of these products are associative?

4.27. Extend the multiplication table for products of graphs in Γ to graphs in Γ_0.

4.28. Extend the multiplication table to products of directed graphs.

Chapter 5

The Four Standard Graph Products

Chapter 4 laid out the definitions of four standard graph products: the Cartesian product, the direct product, the strong product, and the lexicographic product. It also demonstrated that it is these four products alone—among all possible associative graph products of the type considered here—that involve the structure of both factors in a meaningful way, and have the additional property that at least one projection is a weak homomorphism. For this reason the four standard products are, by far, the most extensively studied, and have the widest range of applications. They are thus the primary topics of this book.

This chapter has four sections, one for each of the four standard products. Each is concerned mainly with the semiring structure over Γ or Γ_0, and the metric properties of distance, connectedness, and bipartiteness.

5.1 The Cartesian Product

The Cartesian product of graphs, introduced in Section 4.1, is a straightforward and natural construction. It has been widely investigated, has numerous interesting algebraic properties, and is in many respects the simplest graph product. Many classes of graphs considered in this book are Cartesian products, isometric subgraphs of Cartesian products, or retracts of Cartesian products. We now prepare for this by recalling the main results from Section 4.1, and by proving some fundamental results concerning distance and connectedness.

Recall that if $G_1, G_2, \ldots, G_k$ are graphs in Γ, then their *Cartesian product* is the graph

$$G_1 \,\Box\, G_2 \,\Box\, \cdots \,\Box\, G_k = \mathop{\Box}_{i=1}^{k} G_i$$

with vertex set $\{(x_1, x_2, \ldots, x_k) \mid x_i \in V(G_i)\}$, and for which two vertices $(x_1, x_2, \ldots, x_k)$, $(y_1, y_2, \ldots, y_k)$ are adjacent whenever $x_i y_i \in E(G_i)$ for exactly one index $1 \leq i \leq k$, and $x_j = y_j$ for each index $j \neq i$. Recall also that the kth power of G with respect to the Cartesian product is denoted as $G^{\Box,k}$, that is, $G^{\Box,k} = \Box_{i=1}^{k} G$.

In Section 4.2 we noted that the Cartesian product is commutative and associative in the sense that the maps $(x_1, x_2) \mapsto (x_2, x_1)$ and $((x_1, x_2), x_3) \mapsto (x_1, (x_2, x_3))$ are isomorphisms:

$$\begin{aligned} G_1 \,\Box\, G_2 &\cong G_2 \,\Box\, G_1, \\ (G_1 \,\Box\, G_2) \,\Box\, G_3 &\cong G_1 \,\Box\, (G_2 \,\Box\, G_3). \end{aligned}$$

It is immediate that the Cartesian product distributes over disjoint union:

$$G_1 \,\square\, (G_2 + G_3) \;=\; G_1 \,\square\, G_2 \;+\; G_1 \,\square\, G_3.$$

Moreover, the trivial graph K_1 is a unit with respect to the Cartesian product, that is,

$$K_1 \,\square\, G \cong G$$

for any simple graph G. Because Γ is a commutative monoid with respect to disjoint union, with the empty graph O as the neutral element, and because $O \,\square\, G = G \,\square\, O = O$, we conclude that Γ is a commutative semiring with unit K_1 under the operations $\square$ and $+$.

Also, by Section 4.3, each projection $p_i : G_1 \,\square\, G_2 \,\square\, \cdots \,\square\, G_k \to G_i$ is a weak homomorphism, that is, if $(x_1, x_2, \ldots, x_k)(y_1, y_2, \ldots, y_k)$ is an edge of $G_1 \,\square\, G_2 \,\square\, \cdots \,\square\, G_k$, then $x_i y_i \in E(G_i)$ or $x_i = y_i$ for each index i.

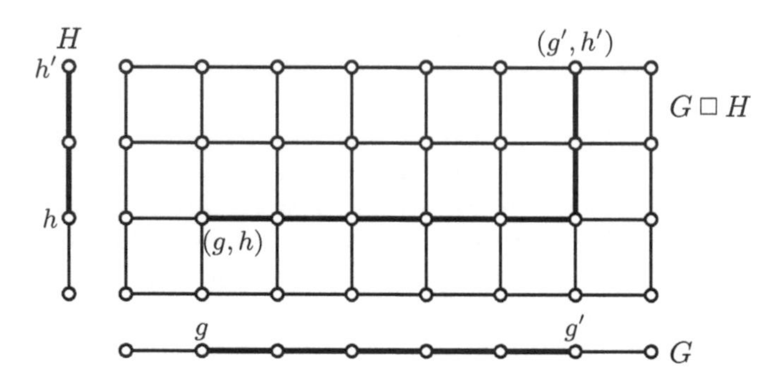

FIGURE 5.1 Illustration of the formula $d_{G \,\square\, H}((g,h),(g',h')) = d_G(g,g') + d_H(h,h')$.

We next turn our attention to distance in Cartesian products. Figure 5.1 suggests that if G and H are paths, then $d_{G \,\square\, H}((g,h),(g',h')) = d_G(g,g') + d_H(h,h')$. In fact, this is true for arbitrary G and H, according to the following proposition. In reading the proof, it may be useful to keep in mind the simple gestalt illustrated in Figure 5.1.

Proposition 5.1 *If (g,h) and (g',h') are vertices of a Cartesian product $G \,\square\, H$, then*

$$d_{G \,\square\, H}((g,h),(g',h')) = d_G(g,g') + d_H(h,h').$$

Proof First suppose that $d_G(g,g') = \infty$. Then G is a disjoint union $G = G_1 + G_2$ with $g \in V(G_1)$ and $g' \in V(G_2)$. Therefore, $G \,\square\, H = (G_1 + G_2) \,\square\, H = G_1 \,\square\, H + G_2 \,\square\, H$, with $(g,h) \in V(G_1 \,\square\, H)$ and $(g',h') \in V(G_2 \,\square\, H)$. Hence, $d_{G \,\square\, H}((g,h),(g',h')) = \infty$, and the proposition follows. By identical reasoning, the proposition follows if $d_H(h,h') = \infty$.

Thus we assume that both $d_G(g,g')$ and $d_H(h,h')$ are finite. Let $P = a_0 a_1 a_2 \ldots, a_{d_G(g,g')}$ be a path in G from $g = a_0$ to $g' = a_{d_G(g,g')}$. Let $Q = b_0 b_1 b_2 \ldots b_{d_H(h,h')}$ be a path in H from $h = b_0$ to $h' = b_{d_H(h,h')}$. This gives rise to two paths

$$\begin{aligned} P \times \{h\} &= (g,h)(a_1,h)(a_2,h)\ldots(g',h) \\ \{g'\} \times Q &= (g',h)(g',b_1)(g',b_2)\ldots(g',h') \end{aligned}$$

in $G \,\square\, H$ whose concatenation is a path of length $d_G(g,g') + d_H(h,h')$ from (g,h) to (g',h'). Hence, $d_{G \,\square\, H}((g,h),(g',h')) \leq d_G(g,g') + d_H(h,h')$.

Conversely, let R be a shortest path between (g,h) and (g',h'). Every edge of R is mapped into a single vertex by one of the projections p_G or p_H and into an edge by the other. This implies that

$$d_G(g,g') + d_H(h,h') \leq |E(p_G(R))| + |E(p_H(R))| = |E(R)| = d_{G \square H}((g,h),(g',h')),$$

and the proof is complete. □

Using associativity and applying Proposition 5.1 inductively, we immediately obtain an analogue for multiple factors (Exercise 5.10) and a corollary.

Corollary 5.2 (Distance Formula) *If $G = G_1 \square G_2 \square \cdots \square G_k$ and $x, y \in V(G)$, then*

$$d_G(x,y) = \sum_{i=1}^{k} d_{G_i}(p_i(x), p_i(y)).$$

Corollary 5.3 *A Cartesian product of graphs is connected if and only if every one of its factors is connected.*

5.2 The Strong Product

The strong product was introduced in Section 4.1. We now investigate its elementary properties in greater detail, with particular attention to distance and connectedness.

Recall that if $G_1, G_2, \ldots, G_k$ are graphs in Γ, then their *strong product* is the graph

$$G_1 \boxtimes G_2 \boxtimes \cdots \boxtimes G_k = \boxtimes_{i=1}^{k} G_i$$

with vertex set $\{(x_1, x_2, \ldots, x_k) \mid x_i \in V(G_i)\}$, and for which two distinct vertices $(x_1, x_2, \ldots, x_k)$ and $(y_1, y_2, \ldots, y_k)$ are adjacent provided that $x_i y_i \in E(G_i)$ or $x_i = y_i$ for each $1 \leq i \leq k$. As noted in Chapter 4, we use $G^{\boxtimes,k}$ for the kth power of G with respect to $\boxtimes$, that is, $G^{\boxtimes,k} = \boxtimes_{i=1}^{k} G$.

In Section 4.2 we noted that the strong product is commutative and associative in the sense that the maps $(x_1, x_2) \mapsto (x_2, x_1)$ and $((x_1, x_2), x_3) \mapsto (x_1, (x_2, x_3))$ are isomorphisms:[1]

$$\begin{aligned} G_1 \boxtimes G_2 &\cong G_2 \boxtimes G_1, \\ (G_1 \boxtimes G_2) \boxtimes G_3 &\cong G_1 \boxtimes (G_2 \boxtimes G_3). \end{aligned}$$

It is also immediate that the strong product distributes over disjoint union:

$$G_1 \boxtimes (G_2 + G_3) = G_1 \boxtimes G_2 + G_1 \boxtimes G_3.$$

Again, the trivial graph K_1 is a unit, that is,

$$K_1 \boxtimes G \cong G.$$

[1]The equations are the same as in the case of the Cartesian product; we include them for the sake of completeness.

Thus, as with the Cartesian case, Γ is a commutative semiring with unit K_1 under the operations $\boxtimes$ and $+$.

Recall from Section 4.3 that each projection $p_i : G_1 \boxtimes G_2 \boxtimes \cdots \boxtimes G_k \to G_i$ is a weak homomorphism. In the other direction, given a graph H and a collection of weak homomorphisms $\varphi_i : H \to G_i$, for $1 \le i \le k$, observe that the map $x \mapsto (\varphi_1(x), \varphi_2(x), \ldots, \varphi_k(x))$ is a weak homomorphism $H \to G_1 \boxtimes G_2 \boxtimes \cdots \boxtimes G_k$. From the two facts just mentioned, we see that every weak homomorphism $\varphi : H \to G_1 \boxtimes G_2 \boxtimes \cdots \boxtimes G_k$ necessarily has form $x \mapsto (\varphi_1(x), \varphi_2(x), \ldots, \varphi_k(x))$ for weak homomorphisms $\varphi_i : H \to G_i$. (Because $\varphi_i = p_i\varphi$ is a composition of weak homomorphisms.)

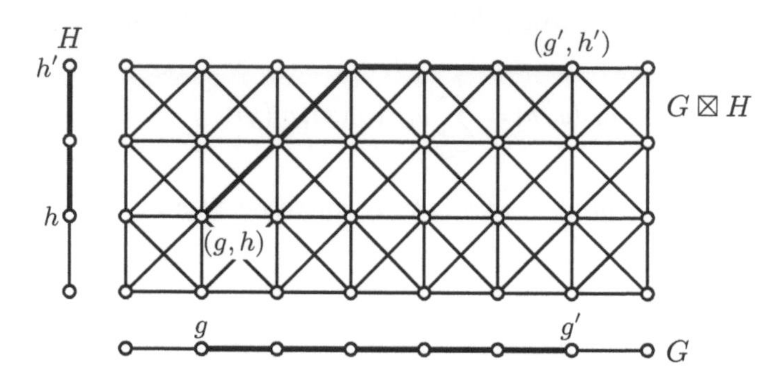

FIGURE 5.2 Illustration of the formula $d_{G\boxtimes H}((g,h),(g',h')) = \max\{d_G(g,g'), d_H(h,h')\}$.

We next turn our attention to distance in strong products. Figure 5.2 suggests that if G and H are paths, then $d_{G\boxtimes H}((g,h),(g',h')) = \max\{d_G(g,g'), d_H(h,h')\}$. In fact, this is true for arbitrary G and H, according to the following proposition, whose proof employs the idea suggested by Figure 5.2.

Proposition 5.4 *If (g,h) and (g',h') are vertices of a strong product $G \boxtimes H$, then*

$$d_{G\boxtimes H}\left((g,h),(g',h')\right) = \max\{d_G(g,g'), d_H(h,h')\}.$$

Proof First suppose that $d_G(g,g') = \infty$. Then the graph G is a disjoint union $G = G_1 + G_2$ with $g \in V(G_1)$ and $g' \in V(G_2)$. Therefore, $G \boxtimes H = (G_1 + G_2) \boxtimes H = G_1 \boxtimes H + G_2 \boxtimes H$, with $(g,h) \in V(G_1 \boxtimes H)$ and $(g',h') \in V(G_2 \boxtimes H)$. Hence, $d_{G\boxtimes H}((g,h),(g',h')) = \infty$, and the proposition follows. By identical reasoning, the proposition holds if $d_H(h,h') = \infty$.

Thus assume that both $d_G(g,g')$ and $d_H(h,h')$ are finite. Let $P = a_0a_1a_2\ldots, a_{d_G(g,g')}$ be a path in G from $g = a_0$ to $g' = a_{d_G(g,g')}$. Let $Q = b_0b_1b_2\ldots b_{d_H(h,h')}$ be a path in H from $h = b_0$ to $h' = b_{d_H(h,h')}$. By commutativity, we may assume $|P| \ge |Q|$. Consider the following two paths in $G \boxtimes H$:

$$\begin{aligned} Q' &= (g,h)(a_1,b_1)(a_2,b_2)\ldots(a_{d_H(h,h')},h'), \\ P' &= (a_{d_H(h,h')},h')(a_{d_H(h,h')+1},h')(a_{d_H(h,h')+2},h')\ldots(g',h'). \end{aligned}$$

The concatenation of Q' and P' is a path of length $|P| = \max\{d_G(g,g'), d_H(h,h')\}$ from (g,h) to (g',h'). Hence, $d_{G\boxtimes H}((g,h),(g',h')) \le \max\{d_G(g,g'), d_H(h,h')\}$.

Conversely, because the projection p_G is a weak homomorphism, it follows that

$$d_G(g,g') = d_G\left(p_G(g,h), p_G(g',h')\right) \le d_{G\boxtimes H}\left((g,h),(g',h')\right).$$

Also $d_H(h,h') \le d_{G\boxtimes H}((g,h),(g',h'))$, so $\max\{d_G(g,g'), d_H(h,h')\} \le d_{G\boxtimes H}((g,h),(g',h'))$, and the proof is complete. □

Applying Proposition 5.4 inductively, we immediately obtain an analogue for multiple factors, and a consequent corollary.

Corollary 5.5 (Distance Formula) *If $G = G_1 \boxtimes G_2 \boxtimes \cdots \boxtimes G_k$ and $x, y \in V(G)$, then*

$$d_G(x,y) = \max_{1\leq i\leq k} \{d_{G_i}(p_i(x), p_i(y))\}.$$

Corollary 5.6 *A strong product of graphs is connected if and only if every one of its factors is connected.*

5.3 The Direct Product

We introduced the direct product in Section 4.1. We now investigate its elementary properties in greater detail, deducing results analogous to those for the Cartesian and strong products in the previous two sections. Whereas the Cartesian and the strong products are usually regarded as operations on the class of simple graphs Γ, the most natural setting for the direct product (as we shall see) is the class of graphs Γ_0. Therefore, although we initially defined it as a product on Γ, we broaden our definition slightly, allowing it to apply to graphs in Γ_0:

If $G_1, G_2, \ldots, G_k$ are graphs in Γ_0, then their *direct product* is the graph

$$G_1 \times G_2 \times \cdots \times G_k = \bigtimes_{i=1}^{k} G_i$$

with vertex set $\{(x_1, x_2, \ldots, x_k) \mid x_i \in V(G_i)\}$, and for which vertices $(x_1, x_2, \ldots, x_k)$ and $(y_1, y_2, \ldots, y_k)$ are adjacent precisely if $x_iy_i \in E(G_i)$ for every $1 \leq i \leq k$. As noted earlier, the kth power of a graph G with respect to the direct product is denoted as $G^{\times,k}$.

Figure 5.3 shows two examples. Observe that $P_5 \times P_3$, displayed on the left, is disconnected. (For clarity, one component is drawn bold.) The example on the right illustrates several noteworthy facts, which follow immediately from the definitions: A product $G \times H$ has a loop at (g,h) if and only if both G and H have loops at g and h, respectively. Moreover, if G has no loop at g, then the H-layer $H^{(g,h)}$ is totally disconnected; whereas if G has a loop at g, then $H^{(g,h)}$ is isomorphic to H. (Analogous remarks hold for the G-layers.)

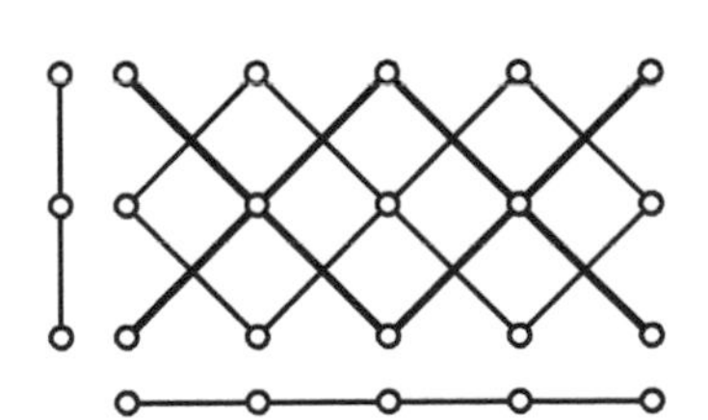

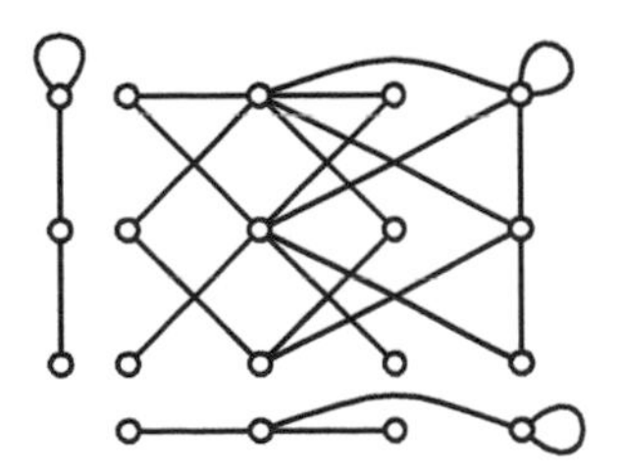

FIGURE 5.3 Two examples of direct products.

In Section 4.2 we proved that the direct product is commutative and associative. Although our reasoning was then restricted to graphs in Γ, a review of the proof reveals that it remains valid in the class Γ_0. Thus the maps $(x_1, x_2) \mapsto (x_2, x_1)$ and

$((x_1,x_2),x_3) \mapsto (x_1,(x_2,x_3))$ give rise to the following isomorphisms, where all factors belong to Γ_0.

$$\begin{aligned} G_1 \times G_2 &\cong G_2 \times G_1, \\ (G_1 \times G_2) \times G_3 &\cong G_1 \times (G_2 \times G_3). \end{aligned}$$

Clearly, the direct product distributes over the disjoint union:

$$G_1 \times (G_2 + G_3) \;=\; G_1 \times G_2 \;+\; G_1 \times G_3.$$

Although the trivial graph K_1 is a unit for both the Cartesian product and the strong product, this is decidedly not the case for the direct product. Indeed, $K_1 \times G$ is the completely disconnected graph on $|V(G)|$ vertices, so $K_1 \times G \ncong G$ in general. However, let $K_1^s \in \Gamma_0$ denote the graph with exactly one vertex, on which there is a loop. Observe that

$$K_1^s \times G \cong G$$

for any $G \in \Gamma_0$. Therefore, under the operations $\times$ and $+$, the set Γ_0 is a commutative semiring with unit K_1^s.

Let $G = G_1 \times G_2 \times \cdots \times G_k$. By simple rewording of the definitions, each projection $p_i : G \to G_i$ is a homomorphism. Furthermore, given a graph H and a collection of homomorphisms $\varphi_i : H \to G_i$, for $1 \le i \le k$, observe that the map $\varphi : x \mapsto (\varphi_1(x), \varphi_2(x), \ldots, \varphi_k(x))$ is a homomorphism $H \to G$. From the two facts just mentioned, we see that every homomorphism $\varphi : H \to G$ has the form $\varphi : x \mapsto (\varphi_1(x), \varphi_2(x), \ldots, \varphi_k(x))$, for homomorphisms $\varphi_i : H \to G_i$, where $\varphi_i = p_i\varphi$. Clearly φ is uniquely determined by the p_i and φ_i.

Notice that this property makes the direct product the product of graphs in the sense of category theory, justifying the name *categorical product*.[2]

The question of distance in direct products, although simple, is somewhat more subtle than for other products. Consider the distance between vertices (g,h) and (g',h') in $G \times H$. Take a walk $W : (g,h)(a_1,b_1)(a_2,b_2)\ldots(a_{n-1},b_{n-1})(g',h')$ of length n joining these vertices. Because the projections are homomorphisms, it follows that $p_G(W) : ga_1a_2\ldots a_{n-1}g'$ and $p_H(W) : hb_1b_2\ldots b_{n-1}h'$ are g,g'- and h,h'-walks of length n in G and H, respectively. Conversely, given walks $ga_1a_2\ldots a_{n-1}g'$ in G and $hb_1b_2\ldots, b_{n-1}h'$ in H, both of length n, we can construct a walk $(g,h)(a_1,b_1)(a_2,b_2)\ldots(a_{n-1},b_{n-1})(g',h')$ of length n in $G \times H$. We have thus proved the following proposition.

Proposition 5.7 *Suppose (g,h) and (g',h') are vertices of a direct product $G \times H$, and n is an integer for which G has a g,g'-walk of length n and H has an h,h'-walk of length n. Then $G \times H$ has a walk of length n from (g,h) to (g',h'). The smallest such n (if it exists) equals $d_{G\times H}((g,h),(g',h'))$. If no such n exists, then $d_{G\times H}((g,h),(g',h')) = \infty$.*

Figure 5.4 illustrates this proposition. The set of integers n for which the factor $G = C_9$ has a g,g'-walk of length n is $\{2,4,6,7,8,9,10,\ldots\}$. The set of integers n for which $H = P_4$ has an h,h'-walk of length n is $\{3,5,7,9,11,13,\ldots\}$. Because 7 is the smallest integer in both of these sets, we have $d_{G\times H}((g,h),(g',h')) = 7$. The figure shows a shortest path (of length 7) from (g,h) to (g',h'). Notice that this path projects to a g,g'-walk of length 7 in G and an h,h'-walk of length 7 in H.

By associativity, the previous proposition has an extension to arbitrarily many factors.

[2]Let C be a category and $\{X_i | i \in I\}$ a family of (not necessarily distinct) objects in C, $X \in C$ and $p_i : X \to X_i$ a collection of morphisms (called the canonical projections). Then X is the product of the X_i (with respect to the p_i) if they satisfy the following universal property: for any object Y and any collection of morphisms $\varphi_i : Y \to X_i$, there exists a unique morphism $\varphi : Y \to X$ such that $\varphi_i = p_i\varphi$ for all $i \in I$.

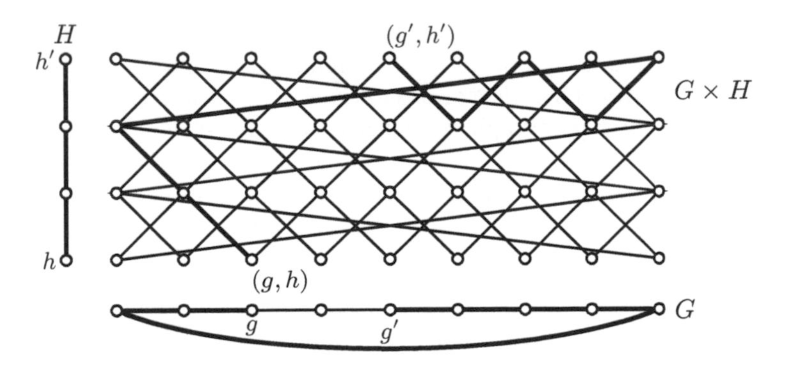

FIGURE 5.4 Illustration of Proposition 5.7.

Proposition 5.8 *Suppose x and y are vertices of $G = G_1 \times G_2 \times \cdots \times G_k$. Then*

$$d_G(x, y) = \min\{n \in \mathbb{N} \mid \text{ each factor } G_i \text{ has a walk of length } n \text{ from } p_i(x) \text{ to } p_i(y)\},$$

where it is understood that $d_G(x, y) = \infty$ if no such n exists.

For other formulations of distance in the direct product, see Kim (1991) and Lamprey, and Barnes (1974).

The connectedness properties of the direct product are much richer than those of the Cartesian and the strong product. Although all factors of a connected direct product must be connected, as one can see by projection into the factors, the converse is not true. For instance, suppose G and H are connected bipartite graphs with bipartitions $V(G) = G_0 \cup G_1$ and $V(H) = H_0 \cup H_1$. Take vertices $(g, h), (g', h') \in V(G \times H)$ for which $g, g' \in G_0$, $h \in H_0$ and $h' \in H_1$. Then any g, g'-walk in G has even length, while any h, h'-walk in H has odd length. Thus the lengths of a g, g'-walk and an h, h'-walk are never equal, and the Proposition 5.7 produces $d_{G \times H}((g, h), (g', h')) = \infty$. Consequently, the direct product of any two bipartite graphs is disconnected. The following theorem, first proved by Weichsel (1962), characterizes connectedness in direct products of two factors.

Theorem 5.9 (Weichsel's Theorem) *Suppose G and H are connected nontrivial graphs in Γ_0. If at least one of G or H has an odd cycle, then $G \times H$ is connected. If both G and H are bipartite, then $G \times H$ has exactly two components.*

Proof Suppose H has an odd cycle. Given two vertices (g, h) and (g', h') in $G \times H$, we wish to show that there is a walk from one to the other. By assumption, there is a nontrivial g, g'-walk P in G and a nontrivial h, h'-walk Q in H. If they have the same parity, we extend the shorter one to a walk of the same length as the longer one by traversing the last edge backward and forward as many times as necessary. Proposition 5.8 then guarantees a walk from (g, h) to (g', h').

If the parities are different, then we extend Q by walking from h to an odd cycle of H, traversing the odd cycle once, returning to h along the same route, and then traversing Q. Thus we obtain a new h, h'-walk Q' that has the same parity as P, and we proceed as in the previous paragraph. This completes the proof of the first statement.

For the second statement, suppose both G and H are connected and bipartite. We have already noted that $G \times H$ is disconnected. Now we show it has just two components. Take any vertex (g, h), and let (g', h') and (g'', h'') be vertices that are not in the same component as (g, h). It suffices to prove $d((g', h'), (g'', h'')) < \infty$. Take g', g- and h', h-walks P' and Q'

in G and H. These walks have opposite parity, for otherwise we would get a walk from (g,h) to (g',h'), as in the first paragraph of the proof. Similarly, we have g,g''- and h,h''-walks P'' and Q'' of opposite parity. Then the concatenations $P'+P''$ and $Q'+Q''$ have the same parity. Arguing as in the first paragraph, we get a walk from (g',h') to (g'',h''). □

Corollary 5.10 *A direct product of connected nontrivial graphs is connected if and only if at most one of the factors is bipartite. In fact, the product has* 2^{k-1} *components, where* k *is the number of bipartite factors.*

Finally, we point out a connection between the direct product of graphs and the Kronecker product of matrices. Recall that the *Kronecker product* of matrices U and V is the matrix $U \otimes V$ obtained by replacing each entry u_{ij} of U with the block $u_{ij}V$. The rows of $U \otimes V$ can be indexed by ordered pairs, so that (i,j) indexes the row corresponding to the jth row of V in the ith row block. Columns can be indexed similarly. Thus the entry of $U \otimes V$ in row (i,j) and column (k,ℓ) equals $u_{ik}v_{j\ell}$.

Suppose G and H have adjacency matrices[3] U and V, respectively, relative to vertex orderings $g_1, g_2, \ldots, g_m$ and $h_1, h_2, \ldots, h_n$, respectively. Then it is simple to verify that $G \times H$ has adjacency matrix $U \otimes V$ relative to the ordering $(g_1,h_1),(g_1,h_2),\ldots,(g_1,h_n)$, $(g_2,h_1),(g_2,h_2),\ldots,(g_2,h_n),\ldots,(g_m,h_1),(g_m,h_2),\ldots,(g_m,h_n)$ of its vertices.

This point of view can yield quick proofs. For example, suppose G is bipartite, so it has an adjacency matrix with block form $U = \left(\begin{smallmatrix} 0 & A \\ A^T & 0 \end{smallmatrix}\right)$, where the T indicates transpose. If H is another graph, with matrix V, then $G \times H$ has matrix $U \otimes V = \left(\begin{smallmatrix} 0 & A \\ A^T & 0 \end{smallmatrix}\right) \otimes V = \left(\begin{smallmatrix} 0 & A\otimes V \\ A^T\otimes V & 0 \end{smallmatrix}\right)$. The block form reveals that $G \times H$ is bipartite. Thus the direct product of a bipartite graph with an arbitrary graph is always bipartite.

5.4 The Lexicographic Product

We met the lexicographic product in Section 4.4. The lexicographic product of graphs G and H is the graph $G \circ H$ whose vertex set is $V(G) \times V(H)$, and for which $(g,h)(g',h')$ is an edge of $G \circ H$ precisely if $gg' \in E(G)$, or $g = g'$ and $hh' \in E(H)$.

This product was introduced as the *composition* of graphs by Harary (1959); see also Harary (1969). Although it essentially dates back to Hausdorff (1914), it was Harary's paper that initiated the investigations by graph theorists, independent of the work of C. C. Chang and Morel (1960) and C. C. Chang (1961). The lexicographic product is also known as graph *substitution*, a name that bears witness to the fact that $G \circ H$ can be obtained from G by substituting a copy H_g of H for every vertex g of G and then joining all vertices of H_g with all vertices of $H_{g'}$ if $gg' \in E(G)$. This is illustrated in Figure 5.5. In this figure, the copies of H in $G \circ H$ are indicated by dashed lines.

Figures 4.10 and 5.5 underscore that the lexicographic product is not commutative. We mentioned in passing that it is associative, but a proof is in order.

Proposition 5.11 *The lexicographic product is associative. In particular, the map* $((x_1,x_2),x_3) \mapsto (x_1,(x_2,x_3))$ *is an isomorphism from* $(G_1 \circ G_2) \circ G_3$ *to* $G_1 \circ (G_2 \circ G_3)$.

Proof From the definition, $((x_1,x_2),x_3)((y_1,y_2),y_3)$ is an edge of $(G_1 \circ G_2) \circ G_3$ precisely if one of the following three conditions holds: $x_1y_1 \in E(G_1)$, or $x_1 = y_1$ and $x_2y_2 \in E(G_2)$,

[3] The adjacency matrix and some of its fundamental properties are covered in Section 17.4.

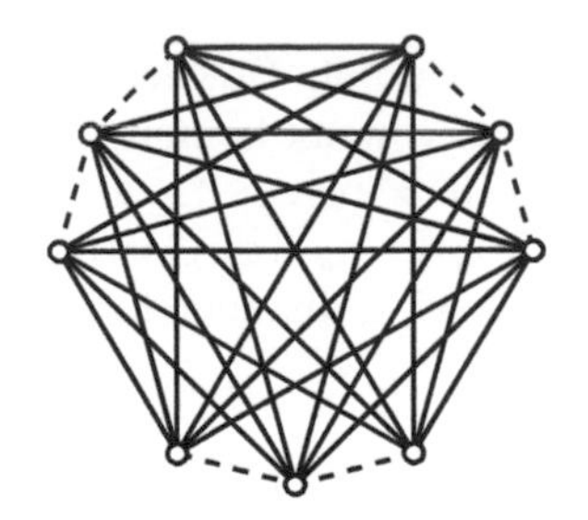

FIGURE 5.5 Lexicographic products $K_3 \circ P_3$ and $P_3 \circ K_3$.

or $x_1 = y_1$ and $x_2 = y_2$ and $x_3y_3 \in E(G_3)$. On the other hand we readily confirm that these same conditions characterize $(x_1, (x_2, x_3))(y_1, (y_2, y_3)) \in E(G_1 \circ (G_2 \circ G_3))$. □

Proposition 5.11 and its proof allow us to unambiguously extend the definition of the lexicographic product to more than two factors. Given graphs $G_1, G_2, \ldots, G_k$, we define $G_1 \circ G_2 \circ \cdots \circ G_k$ as follows: the vertex set is $V(G_1) \times V(G_2) \times \cdots \times V(G_k)$, and two vertices $(x_1, x_2, \ldots, x_k)$, $(y_1, y_2, \ldots, y_k)$ are adjacent if for some index $j \in \{1, 2, \ldots, k\}$ we have $x_jy_j \in E(G_j)$ and $x_i = y_i$ for each $1 \leq i < j$.

Notice that this definition is analogous to the lexicographic ordering of the Cartesian product of ordered sets $X_1, X_2, \ldots, X_k$, where $(x_1, x_2, \ldots, x_k) \prec (y_1, y_2, \ldots, y_k)$ provided that for some index $1 \leq j \leq k$ we have $x_j \prec y_j$, and $x_i = y_i$ for each $1 \leq i < j$. Replacing the X_i with graphs and $\prec$ with the adjacency relation, we get the lexicographic product; hence its name.

Let us summarize what we have seen of the algebraic properties of the lexicographic product. It is not commutative, but it is associative and it is easily seen to have K_1 as both a left and right unit:

$$\begin{aligned} (G_1 \circ G_2) \circ G_3 &\cong G_1 \circ (G_2 \circ G_3), \\ K_1 \circ G &\cong G, \\ G \circ K_1 &\cong G. \end{aligned}$$

It is also very easy to establish the following right-distributive rule, which holds for all graphs G, H, and K:

$$(G + H) \circ K = G \circ K + H \circ K\,.$$

However, there is no corresponding left-distributive rule: Consider that $K_2 \circ (K_1 + K_1) = C_4$, but $K_2 \circ K_1 + K_2 \circ K_1 = K_2 + K_2$. Nonetheless, algebraically speaking, Γ is a near-semiring with respect to the operations $+$ and $\circ$.

The breakdown of such fundamental properties makes the algebraic structure of the lexicographic product exceedingly rich, and there are many properties that have no analogues in the other products. For instance, we have

$$\overline{G \circ H} = \overline{G} \circ \overline{H},$$

as the reader is invited to verify. (In fact, we noted this equation in Section 4.4 when we remarked that the lexicographic product is its own complementary product.) We will meet this equation again, along with many new ones, in Chapter 10, where we investigate the deeper algebraic properties of the lexicographic product. For now we turn to distance and connectedness.

Observe that the first projection $p_1 : G_1 \circ G_2 \circ \cdots \circ G_k \to G_1$ is a weak homomorphism, though in general the projections to the other factors are not. However, by associativity, any projection $G_1 \circ G_2 \circ \cdots \circ G_k \to G_1 \circ G_2 \circ \cdots \circ G_i$ is a weak homomorphism for $i \leq k$. Also, it follows from the definitions that given vertex $a = (a_1, a_2, \ldots, a_k)$, the map p_i is an isomorphism from any G_i-layer $G_i^a = \langle\{(a_1, a_2, \ldots, x_i, \ldots a_k) \mid x_i \in V(G_i)\rangle$ to the factor G_i. These observations are used in the proof of the following distance formula. Recall that $d_G(g)$ denotes the degree of the vertex g of G.

Proposition 5.12 *Suppose (g, h) and (g', h') are two vertices of $G \circ H$. Then*

$$d_{G \circ H}((g,h),(g',h')) = \begin{cases} d_G(g,g') & \text{if } g \neq g', \\ d_H(h,h') & \text{if } g = g', \text{ and } d_G(g) = 0, \\ \min\{d_H(h,h'), 2\} & \text{if } g = g', \text{ and } d_G(g) \neq 0. \end{cases}$$

Proof Let (g, h) and (g', h') be as stated. Let M be the value of the right-hand side of the equation in the proposition.

First suppose $g \neq g'$. Because p_1 is a weak homomorphism, we have $d_{G \circ H}((g,h),(g',h')) \geq d_G(g,g') = M$. Conversely, let $P = ga_1a_2 \ldots g'$ be a shortest g, g'-path in G. Then $(g,h)(a_1,h')(a_2,h')(a_3,h') \ldots (g',h')$ is a path of length $d_G(g,g')$ in $G \circ H$; therefore $d_{G \circ H}((g,h),(g',h')) \leq d_G(g,g') = M$. Thus the proposition is true when $g \neq g'$.

Next suppose $g = g'$ and $d_G(g) = 0$. Because g is isolated, each component of the layer $H^{(g,h)}$ is a component of $G \circ H$. By virtue of the isomorphism $p_2 : H^{(g,h)} \to H$, we now have $M = d_H(h,h') = d_{G \circ H}((g,h),(g',h'))$.

Finally, if $g = g'$ and $d_G(g) \geq 0$, then there is an edge $gc \in E(G)$, and $(g,h)(c,h')(g,h')$ is a path of length 2 joining (g, h) to (g', h'). Combined with the isomorphism $p_2 : H^{(g,h)} \to H$, this yields $d_{G \circ H}((g,h),(g',h')) = \min\{d_H(h,h'), 2\} = M$. □

By associativity, we have the following immediate generalization to multiple factors.

Corollary 5.13 *Suppose $x = (x_1, x_2, \ldots, x_k)$ and $y = (y_1, y_2, \ldots, y_k)$ are distinct vertices of $G = G_1 \circ G_2 \circ \cdots \circ G_k$, and let i be the smallest index for which $x_i \neq y_i$. Then*

$$d_G(x,y) = \begin{cases} d_{G_i}(x_i, y_i) & \text{if } d_{G_\ell}(x_\ell) = 0 \text{ for each } 1 \leq \ell < i, \\ \min\{d_{G_i}(x_i, y_i),\ 2\} & \text{if } d_{G_\ell}(x_\ell) \neq 0 \text{ for some } 1 \leq \ell < i. \end{cases}$$

Finally, we get a characterization of connectedness.

Corollary 5.14 *A lexicographic product $G_1 \circ G_2 \circ \cdots \circ G_k$ of nontrivial graphs is connected if and only if G_1 is connected.*

Exercises

5.1. Show that $C_{2k+1} \times C_{2k+1} \cong C_{2k+1} \,\square\, C_{2k+1}$ for any $k \geq 1$. What happens if we replace odd cycles with even cycles?

5.2. Strengthen Proposition 5.3 as follows: $G \,\square\, H$ is connected and has no cut vertex if and only if G and H are connected.

5.3. Verify that the connected components of $K_{m,n} \times K_{m',n'}$ are $K_{mm',nn'}$ and $K_{mn',nm'}$.

5.4. A graph G is called *antipodal* if there exists a vertex v to any vertex $u \in V(G)$, such that $V(G) = I(u, v)$. Show that the Cartesian product of antipodal graphs is antipodal.

5.5. Let G be an isometric subgraph of a Cartesian product $\square_{i=1}^{k} G_i$, such that $p_i(G) = G_i$ for $i = 1, \ldots, k$. Show that the G_i are antipodal if G is antipodal. Find a counterexample to the converse.

5.6. Show that, given vertices $u, v \in V(G \,\square\, H)$, there is a unique vertex $x \in G^u$ such that $d(v, x) = \min\{d(v, y) \mid y \in V(G^u)\}$. Moreover, $p_G(v) = p_G(x)$.

5.7. (Behzad and Mahmoodian, 1969) Let G and H be connected graphs different from K_1 and K_2. Show that $G \,\square\, H$ is planar if and only if both factors are paths, or one is a path and the other a cycle.

5.8. (Behzad and Mahmoodian, 1969) A graph is *outerplanar* if it is planar and embeddable into the plane such that all vertices lie on the outer face of the embedding. Let G be an outerplanar graph. Show that $G \,\square\, K_2$ is planar.

5.9. (Jha and Slutzki, 1993) Show that the Cartesian product of two graphs is outerplanar if and only if one factor is a path and the other a K_2.

5.10. Prove by induction that if $G = G_1 \,\square\, G_2 \,\square \cdots \square\, G_k$ and $x, y \in V(G)$, then $d_G(x, y) = \sum_{i=1}^{k} d_{G_i}(p_i(x), p_i(y))$.

5.11. Show that the diameter of a strong product of k graphs is the maximum of the diameters of the factors.

5.12. (Abay-Asmerom, Hammack, and D. T. Taylor 2009) A *perfect r-code* of a graph G is a subset C of $V(G)$ such that each vertex of G is of distance at most r from exactly one vertex of C. Show that each $G_1, G_2, \ldots, G_n$ has a perfect r-code if and only if $G_1 \boxtimes G_2 \boxtimes \cdots \boxtimes G_n$ has a perfect r-code.

5.13. (Jha and Slutzki, 1993) Show that the strong product of two connected graphs is planar if and only if one of the following conditions is satisfied:

a. One factor is a tree and the other a K_2.
b. Both factors are P_3's.

5.14. Show that $\operatorname{diam}(G^{\times,k}) \leq 2\operatorname{diam}(G) + c$, where c is the length of a shortest odd cycle of G.

5.15. (Farzan and Waller, 1977) Let G be a connected graph. Show that $C_4 \times G$ is planar if and only if G is a tree.

5.16. (Farzan and Waller, 1977) Show that the direct product $G \times H$ of connected graphs G and H on at least five vertices is planar if and only if one of the following conditions is satisfied:

a. One factor is a path, and removal of pendant vertices from the other produces a path or a cycle.
b. One factor is a cycle, and removal of pendant vertices from the other produces a path.

5.17. This problem concerns the graphs illustrated in Figure 5.6.

(a) (Bottreau and Métivier, 1998) Let G be the graph obtained by subdividing two independent edges of $K_{3,3}$, as illustrated in Figure 5.6, left. Show that $G \times K_2$ is planar.

(b) (Beaudou, Dorbec, Gravier, and Jha, 2009) Let H be the graph obtained by subdividing two incident edges of K_5, as illustrated in Figure 5.6, right. Show that $H \times K_2$ is planar.

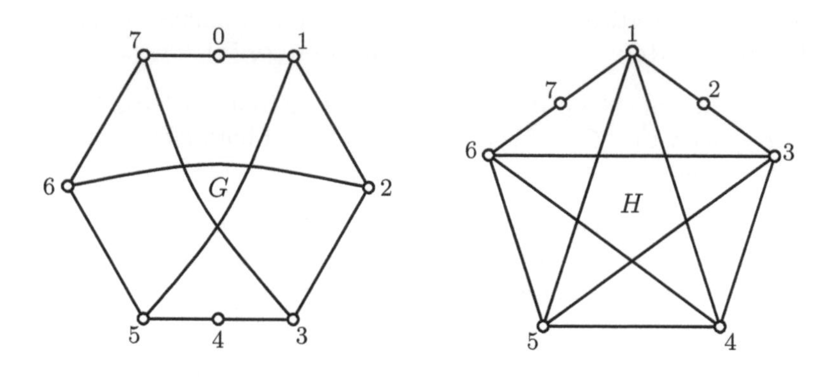

FIGURE 5.6 Subdivided $K_{3,3}$ and K_5.

Part II

Factorization and Cancellation

Introduction to Part II

GRAPH PRODUCTS, as we have seen, obey fundamental algebraic laws such as identity, associativity, and commutativity. It is natural to seek the extent of these algebraic connections. Do graphs factor uniquely into primes over a given product? Does cancellation hold? Part II answers these questions affirmatively, given certain mild restrictions such as connectedness and nonbipartiteness. As a dividend, we are able to describe the automorphism group of a graph in terms of the automorphisms of its prime factors.

We begin, in Chapter 6, with the Cartesian product. Uniqueness of prime factorization of connected graphs over this product was discovered in the early 1960s, independently by Sabidussi and Vizing. Our proof (which uses convexity arguments) leads immediately to automorphism and cancellation results.

Chapter 7 deduces analogous results for the strong product and follows the approach of Dörfler and Imrich (1970).

Chapter 8 proves that connected nonbipartite graphs factor uniquely into primes over the direct product. The primary tool is the so-called *Cartesian skeleton operation* $S : \Gamma_0 \to \Gamma$, which, under suitable conditions, satisfies $S(G \times H) = S(G) \,\square\, S(H)$. This reduces factorization over the direct product to the more manageable Cartesian product. Once developed, this theory easily describes the automorphism structure of direct product graphs and provides an alternative path to the results of Chapter 7.

The cancellation properties of the direct product are exceedingly rich, and Chapter 9 is devoted to characterizing the conditions under which cancellation in this product holds or fails. This also leads to a cancellation law for the strong product.

Chapter 10 treats the lexicographic product. Here, prime factorization is not unique but there is a strong and predictable connection between different prime factorizations of the same graph. We also deduce surprisingly strong cancellation results and characterize the conditions under which $G \circ H \cong H \circ G$.

Part II depends directly on the material presented in Part I. Within Part II, the material in Chapter 6 (Cartesian Product) is needed in both Chapter 7 (Strong Product) and Chapter 8 (Direct Product). In other respects the chapters are independent of one another, with the exception that insights gained in one chapter may facilitate (but are not essential to) the understanding of other chapters.

Chapter 6

Cartesian Product

The uniqueness of the prime factor decomposition of connected graphs with respect to the Cartesian product was first shown by Sabidussi (1960), and independently by Vizing (1963). This book presents two proofs, one here and the other in Chapter 23. The proof in this chapter invokes the convexity of layers and the fact that convex subgraphs are subproducts.

We show that the automorphism group of a graph is determined by the groups of its prime factors. We also characterize Cartesian products with transitive or Abelian automorphism groups and consider cancellation with respect to the Cartesian product. Finally we consider graphs that are nontrivial subgraphs of Cartesian products.

6.1 Prime Factor Decompositions

A graph is *prime* with respect to a given graph product if it is nontrivial and cannot be represented as the product of two nontrivial graphs. For the Cartesian product, this means that a nontrivial graph G is prime if $G = G_1 \,\square\, G_2$ implies that G_1 or G_2 is K_1. We show first that every graph has a prime factor decomposition with respect to the Cartesian product.

Proposition 6.1 *Every nontrivial graph G has a prime factor decomposition with respect to the Cartesian product. The number of prime factors is at most* $\log_2 |V(G)|$.

Proof Because the product of k nontrivial graphs has at least 2^k vertices, a graph G can have at most $\log_2 |V(G)|$ factors. Thus there is a presentation of G as a product $G_1 \,\square\, G_2 \,\square\, \cdots \,\square\, G_\ell$ with a maximal number of factors. Clearly, every factor is prime. □

Note that the same argument holds for all standard products.

As we will prove shortly, any connected graph factors uniquely into prime graphs with respect to the Cartesian product. In other words, if a connected graph G factors into primes as $G = G_1 \,\square\, G_2 \,\square\, \cdots \,\square\, G_k$ and $G = H_1 \,\square\, H_2 \,\square\, \cdots \,\square\, H_\ell$, then $k = \ell$ and the indices can be ordered so that $G_i \cong H_i$. However, for disconnected graphs, this need not be the case.

Theorem 6.2 *Prime factorization is not unique for the Cartesian product in the class of possibly disconnected simple graphs.*

Proof It is easy to see that

$$(K_1 + K_2 + K_2^{\Box,2}) \Box (K_1 + K_2^{\Box,3}) = (K_1 + K_2^{\Box,2} + K_2^{\Box,4}) \Box (K_1 + K_2).$$

We just need to show that the factors on the left- and right-hand sides are prime.

To see this, we observe that the number of components of a Cartesian product is the product of the numbers of components in the factors. Thus, if a graph consisting of two or three components is represented as a product of two graphs, one of these graphs must have one component and the other two or three. The graphs to be investigated are of the form $K_1 + A_1 + A_2$, where A_1 has as least two vertices, the same holding for A_2 if it is present. Its representation as a product of two factors must be of the form $B \Box (C_1 + C_2 + C_3)$. Hence

$$K_1 + A_1 + A_2 \cong B \Box C_1 + B \Box C_2 + B \Box C_3.$$

Then $K_1 \cong B \Box C_i$ for some i. Thus B is trivial, so $K_1 + A_1 + A_2$ is prime. □

The decomposition used in the proof of Theorem 6.2 is of the form

$$(1 + x + x^2)(1 + x^3) = (1 + x^2 + x^4)(1 + x).$$

This is an example of nonunique prime factorization in the subsemiring $\mathbb{Z}^+[x]$ of nonzero polynomials with positive coefficients in the polynomial ring $\mathbb{Z}[x]$ with integer coefficients over the indeterminate x. It seems to have been first exploited by Nakayama and Hashimoto (1950) for the construction of finite reflexive structures without unique prime factor decomposition in the ring of polynomials in an indeterminate x over the set of positive integers.

Notice that by Exercises 6.12 and 6.13 the prime factorization of every graph with fewer than six connected components is unique with respect to the Cartesian product. In fact, Exercise 6.13 implies that all polynomials in the subsemiring $\mathbb{Z}^+[x_1, x_2, x_3, \ldots]$ of $\mathbb{Z}[x_1, x_2, x_3, \ldots]$ that are the sum of four monomials (with coefficient 1) have unique prime factorizations in $\mathbb{Z}^+[x_1, x_2, x_3, \ldots]$.

For a more algebraic treatment that reduces the problem to a system of linear equations, which has a unique solution, and for further results about nonunique prime factorizations of disconnected graphs, see van de Woestijne (2011).

An important concept for all products are subproducts, which we call boxes: A *box* in a product $G = G_1 \Box \cdots \Box G_k$ is a subgraph of form $U_1 \Box \cdots \Box U_k$, where $U_i \subseteq G_i$ for each index i. In order to characterize boxes of Cartesian products, we first prove the following lemma:

Lemma 6.3 (Unique Square Lemma) *Let e and f be two incident edges of a Cartesian product $G_1 \Box G_2$ that are in different layers, that is, one in a G_1-layer and the other one in a G_2-layer. Then there exists exactly one square in $G_1 \Box G_2$ containing e and f. This square has no diagonals.*

Proof We may assume $e = uw = (u_1, u_2)(v_1, u_2)$ and $f = wv = (v_1, u_2)(v_1, v_2)$. In particular, this means $u_1 \neq v_1$ and $u_2 \neq v_2$. Suppose $z = (z_1, z_2)$ is adjacent to both u and v. As z is adjacent to $u = (u_1, u_2)$, we have $z_1 = u_1$ or $z_2 = u_2$. As z is adjacent to $v = (v_1, v_2)$, we have $z_1 = v_1$ or $z_2 = v_2$. These constraints force either $z = (v_1, u_2) = w$ or $z = (u_1, v_2)$. We now have a unique (and diagonal-free) square $(u_1, u_2)(v_1, u_2)(v_1, v_2)(u_1, v_2)$ containing e and f. □

Notice that this lemma also holds for arbitrarily many factors.

We say a subgraph W of a Cartesian product G has the *square property* if for any two adjacent edges e, f that are in different layers, the unique square of G that contains e and f is also in W.

Lemma 6.4 *A connected subgraph W of a Cartesian product is a box if and only if it has the square property.*

Proof Boxes have the square property by Lemma 6.3.

Suppose W is a connected subgraph of a Cartesian product G with the square property. It suffices to prove the lemma for $G = G_1 \Box G_2$. Let $a = (a_1, a_2)$ and $b = (b_1, b_2)$ be two vertices of W. We have to show that (a_1, b_2) and (b_1, a_2) are also in W. We may suppose that the vertices (a_1, a_2), (b_1, a_2), (a_1, b_2), and (b_1, b_2) are distinct. Because W is connected, there is a path P from a to b. Let us call an edge e of P a G_1-edge, if $p_2(e)$ consists only of one vertex, and a G_2-edge otherwise. By Lemma 6.3, we can replace every sequence e, f of two edges in P, where e is a G_1-edge and f a G_2-edge, by two edges e', f' in W, where e' is a G_2-edge and f' a G_1-edge. Thus we can assume that P consists of a sequence of G_1-edges followed by a sequence of G_2-edges, and that there also exists a path P' from a to b in W in which a sequence of G_2-edges is followed by a sequence of G-edges. But then the vertex (a_1, b_2) is on P and (b_1, a_2) on P', whence both are in W. □

For the uniqueness proof of the prime factor decomposition for connected graphs, we continue with a lemma about convex subgraphs. A subgraph $W \subseteq G$ is *convex* in G if every shortest G-path between vertices of W lies entirely in W. Notice that convex subgraphs of Cartesian products have the square property.

Lemma 6.5 *A subgraph W of $G = G_1 \Box \cdots \Box G_k$ is convex if and only if $W = U_1 \Box \cdots \Box U_k$, where each U_i is convex in G_i.*

Proof Suppose W is convex in G. Then it is connected and has the square property, and it is a box by Lemma 6.4. It follows that $W = p_1(W) \Box \cdots \Box p_k(W)$. We have to show that each $p_i(W)$ is convex. Fix i and take vertices a_i and b_i of $p_i(W)$. Let x_i be on a shortest a_i, b_i-path in G_i. We must show that x_i belongs to $p_i(W)$.

Choose vertices $a = (a_1, \ldots, a_k)$ and $b = (b_1, \ldots, b_k)$ of W with $p_i(a) = a_i$ and $p_i(b) = b_i$. Define $x = (x_1, \ldots, x_k)$ as follows. For each index $j \neq i$, let x_j be on a shortest a_j, b_j-path in G_j. Thus $d_{G_s}(a_s, b_s) = d_{G_s}(a_s, x_s) + d_{G_s}(x_s, b_s)$ for each $1 \leq s \leq k$. From this, Corollary 5.2 implies $d_G(a, b) = d_G(a, x) + d_G(x, b)$. It follows that x lies on a shortest a, b-path in G, so $x \in W$ by convexity of W. Hence $x_i = p_i(x) \in p_i(W)$.

The converse is reserved for Exercise 6.1. □

Notice that Lemma 6.5 implies that every layer $G_i^a = \{a_1\} \Box \cdots \Box G_i \Box \cdots \Box \{a_k\}$ in $G_1 \Box \cdots \Box G_k$ is a convex box.

We are now ready for the main result of this section.

Theorem 6.6 (Sabidussi-Vizing) *Every connected graph has a unique representation as a product of prime graphs, up to isomorphism and the order of the factors.*

Because we already know that every finite graph has a prime factorization, we only have to show that it is unique. To this end, the next lemma completes the proof of Theorem 6.6.

Lemma 6.7 *Let φ be an isomorphism between the connected graphs G and H that are representable as products $G = G_1 \Box \cdots \Box G_k$ and $H = H_1 \Box \cdots \Box H_\ell$ of prime graphs. Then $k = \ell$, and to any $a \in V(G)$ there is a permutation π of $\{1, 2, \ldots, k\}$ such that $\varphi(G_i^a) = H_{\pi(i)}^{\varphi(a)}$ for $1 \leq i \leq k$.*

Proof Fix $a = (a_1, \ldots, a_k)$, and say $\varphi(a) = b = (b_1, \ldots, b_\ell)$. As mentioned above, any G_i^a is convex in G, so its image $\varphi(G_i^a)$ is convex in H. Lemma 6.5 implies

$$(b_1, \ldots, b_\ell) \in \varphi(G_i^a) = U_1 \Box \cdots \Box U_\ell.$$

But $G_i \cong G_i^a \cong \varphi(G_i^a)$ is prime, so $U_i = \{b_i\}$ for all but one index, call it $\pi(i)$. In other words, $\varphi(G_i^a) \subseteq H_{\pi(i)}^{\varphi(a)}$. But then $G_i^a \subseteq \varphi^{-1}(H_{\pi(i)}^{\varphi(a)})$. Because $\varphi^{-1}(H_{\pi(i)}^{\varphi(a)})$ is convex, it is a box; and because it is prime, it must be contained in G_i^a. Therefore, $\varphi(G_i^a) = H_{\pi(i)}^{\varphi(a)}$.

We claim that the map $\pi : \{1, 2, \ldots, k\} \to \{1, 2, \ldots, \ell\}$ is injective. If $\pi(i) = \pi(j)$, then

$$\varphi(G_i^a) = H_{\pi(i)}^{\varphi(a)} = \varphi(G_j^a).$$

Because $H_{\pi(i)}^{\varphi(a)}$ is nontrivial (it is prime), it follows that G_i^a and G_j^a have a nontrivial intersection. This means $i = j$, so π is injective. Thus $k \leq \ell$.

Repeating the above argument for φ^{-1} gives $\ell \leq k$, so $k = \ell$ and π is a permutation. □

Figure 6.1 depicts a graph and its prime factors. We will also use it to illustrate the action of the automorphism group on a product of prime graphs; see Exercise 6.4.

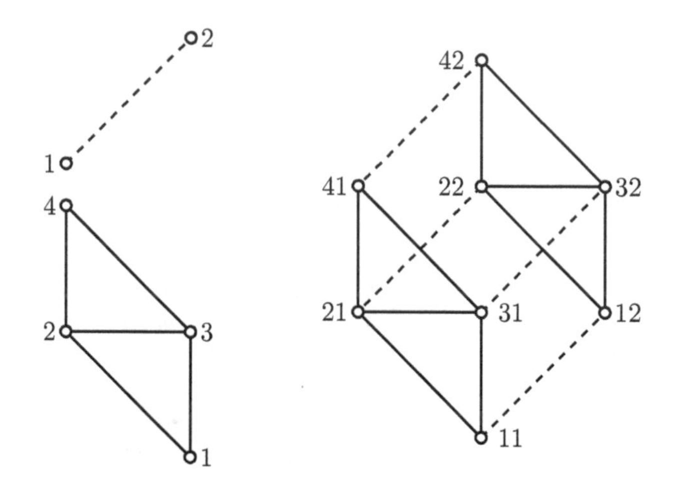

FIGURE 6.1 A graph and its prime factors.

We continue with a theorem that describes the structure of isomorphisms between connected Cartesian products.

Theorem 6.8 *Let G and H be isomorphic connected graphs with prime factorizations $G = G_1 \,\square \cdots \square\, G_k$ and $H = H_1 \,\square \cdots \square\, H_k$. Then for any isomorphism $\varphi : G \to H$, there is a permutation π of $\{1, 2, \ldots, k\}$ and isomorphisms $\varphi_i : G_{\pi(i)} \to H_i$ for which*

$$\varphi(x_1, x_2, \ldots, x_k) = \big(\varphi_1(x_{\pi(1)}), \varphi_2(x_{\pi(2)}), \ldots, \varphi_k(x_{\pi(k)})\big). \tag{6.1}$$

Proof Fix a vertex $a = (a_1, \ldots, a_k)$ of G. By Lemma 6.7, there is a permutation π of $\{1, 2, \ldots, k\}$ for which φ restricts to an isomorphism $G_i^a \to H_{\pi(i)}^{\varphi(a)}$ for each index i. Replacing π with π^{-1}, we can say that, for each i, φ restricts to an isomorphism

$$G_{\pi(i)}^a \to H_i^{\varphi(a)}.$$

To finish the proof, we will show that $p_i\varphi(x_1, \ldots, x_k)$ depends only on $x_{\pi(i)}$. Then we can just put $\varphi_i(x_{\pi(i)}) = p_i\varphi(x_1, \ldots, x_k)$, which yields Equation (6.1), and it is immediate that the φ_i are isomorphisms.

For any $x_{\pi(i)} \in V(G_{\pi(i)})$, consider the box $B[x_{\pi(i)}] = G_1 \square G_2 \square \cdots \square \{x_{\pi(i)}\} \square \cdots \square G_k$ whose $\pi(i)th$ factor is the single vertex $x_{\pi(i)}$. This box is convex, so by Lemma 6.5 its image $\varphi(B[x_{\pi(i)}])$ is a box in H.

Now, $B[x_{\pi(i)}] \cap G^a_{\pi(i)} = \{(a_1, a_2, \ldots, x_{\pi(i)}, \ldots, a_k)\}$. Thus the box $\varphi(B[x_{\pi(i)}])$ meets the box $\varphi(G^a_{\pi(i)}) = H_i^{\varphi(a)}$ at the single vertex $\varphi(a_1, a_2, \ldots, x_{\pi(i)}, \ldots, a_k)$. This means all vertices in $\varphi(B[x_{\pi(i)}])$ have the same ith coordinate $p_i\varphi(a_1, a_2, \ldots, x_{\pi(i)}, \ldots, a_k)$, so

$$p_i\big(\varphi(B[x_{\pi(i)}])\big) = p_i\varphi(a_1, a_2, \ldots, x_{\pi(i)}, \ldots, a_k).$$

Now, any $(x_1, \ldots, x_{\pi(i)}, \ldots, x_k) \in V(G)$ belongs to $B[x_{\pi(i)}]$. Thus $p_i\varphi(x_1, \ldots, x_{\pi(i)}, \ldots, x_k) = p_i\varphi(a_1, a_2, \ldots, x_{\pi(i)}, \ldots, a_k)$, which depends only on $x_{\pi(i)}$. □

Because we can replace any factor of a graph by an isomorphic copy without changing the structure of the product, we often relabel the vertices H_i such that the φ_i are the identity mapping. This yields a much more agreeable version of Equation (6.1). We formulate this observation as a corollary:

Corollary 6.9 *Suppose there is an isomorphism $\varphi : G_1 \square \cdots \square G_k \to H_1 \square \cdots \square H_k$, where each G_i and H_i is prime. Then the vertices of the H_i can be relabeled such that*

$$\varphi(x_1, x_2, \ldots, x_k) = (x_{\pi(1)}, x_{\pi(2)}, \ldots, x_{\pi(k)})$$

for some permutation π of $\{1, \ldots, k\}$.

6.2 Cartesian Product and Its Group

The automorphisms of a graph G are the isomorphisms of G to itself. Therefore, in the present development, the description of the automorphisms of the Cartesian product of connected prime graphs is an immediate corollary of Theorem 6.8. It is due to Imrich (1969a) and Miller (1970a).

Theorem 6.10 *Suppose φ is an automorphism of a connected graph G with prime factor decomposition $G = G_1 \square G_2 \square \cdots \square G_k$. Then there is a permutation π of $\{1, 2, \ldots, k\}$ and isomorphisms $\varphi_i : G_{\pi(i)} \to G_i$ for which*

$$\varphi(x_1, x_2, \ldots, x_k) = \big(\varphi_1(x_{\pi(1)}), \varphi_2(x_{\pi(2)}), \ldots, \varphi_k(x_{\pi(k)})\big).$$

We consider two special cases now:

1. The permutation π is the identity. Then every φ_i is an automorphism of G_i. We say φ is *generated by automorphisms of the factors G_i*. If all factors are pairwise nonisomorphic, these automorphisms already generate the full automorphism group of G.

2. At least two prime factors G_r and G_s are isomorphic. Let π be the transposition $(r\,s)$, and φ_r, φ_s be a pair of isomorphisms from G_r onto G_s, respectively from G_s onto G_r. Furthermore, for indices i other than r or s, let φ_i be the identity on $V(G_i)$. Then the map φ corresponding to π, φ_r, φ_s and the φ_i for $i \neq r, s$ is an automorphism. We call it a *transposition of two isomorphic prime factors* of G.

Corollary 6.11 *The automorphism group of a connected graph with prime factor decomposition $G_1 \square G_2 \square \cdots \square G_k$ is generated by automorphisms and transpositions of the prime factors.*

Because only isomorphic prime factors can be transposed, we obtain a corollary for relatively prime graphs, where two graphs are called *relatively prime* if there exists no nontrivial graph that is a factor of both of them.

Corollary 6.12 *Let G be the Cartesian product $G_1 \square G_2 \square \cdots \square G_k$ of connected, relatively prime graphs. Then every automorphism φ of G preserves the layer structure of G with respect to the given product decomposition and can be written in the form*

$$\varphi(x_1, x_2, \ldots, x_k) = \big(\varphi_1(x_1), \varphi_2(x_2), \ldots, \varphi_k(x_k)\big),$$

where the φ_i are automorphisms of G_i. In this case, $\mathrm{Aut}(G)$ is the direct product of the automorphism groups of the factors.

These results imply a simple theorem, which helps us visualize the structure of the automorphism group of a product of prime graphs.

Theorem 6.13 *The automorphism group of the Cartesian product of connected prime graphs is isomorphic to the automorphism group of the disjoint union of the factors.*

Proof Let $G_1, G_2, \ldots, G_k$ be the connected components of a graph G. Assume each of these components is prime. Then an automorphism φ of G_i yields an automorphism of G by applying φ on G_i and fixing all vertices of the other components. In addition, if components G_i and G_j are isomorphic, interchanging G_i with G_j and fixing all other vertices also gives an automorphism of G. Every other automorphism of G is generated by automorphisms of these two types. Hence, the structure of the automorphism group of G is the same as that of the automorphism group of the corresponding Cartesian product. □

In other words, the automorphism group is the direct product of the wreath products on the sets of pairwise isomorphic factors.

Distinguishing number of Cartesian products

Albertson and Collins (1996) introduced the *distinguishing number* $D(G)$ of a graph G as the smallest number of labels that can destroy all of G's nontrivial automorphisms. More precisely, $D(G)$ is the least integer d such that G has a labeling with d labels that is preserved only by the trivial automorphism. This concept has received a lot of attention; for some recent developments, see Arvind, Cheng, and Devanur (2008) and references therein. Here we list what is known about the distinguishing number of Cartesian product graphs.

Based on the work of Bogstad and Cowen (2004), Albertson (2005), and Klavžar and Zhu (2007), the final result about the distinguishing number of Cartesian graph powers was established by Imrich and Klavžar (2006):

Theorem 6.14 *If $k \geq 2$, then $D(G^{\square,k}) = 2$ for all nontrivial, connected graphs $G \neq K_2, K_3$. Furthermore, $D(K_n^{\square,k}) = 2$ if $n = 2, 3$ and $n + k \geq 6$.*

Hence, all but three Cartesian powers can be distinguished with two labels, the exceptions being $K_2^{\square,2}$, $K_2^{\square,3}$, and $K_3^{\square,2}$, each of which has distinguishing number 3. This situation is typical; usually most graphs in a given family can be distinguished by two labels. Not surprisingly, Theorem 6.10 is an indispensable tool in the proof of Theorem 6.14 and related results.

As $\mathrm{Aut}(K_k \,\square\, K_n)$ is very rich, it takes some effort to determine $D(K_k \,\square\, K_n)$. The next result was obtained independently by M. J. Fisher and Garth (2008) and by Imrich, Jerebic, and Klavžar (2008):

Theorem 6.15 *Let k, n, d be integers with $2 \leq d$, $k < n$ and $(d-1)^k < n \leq d^k$. Then*

$$D(K_k \,\square\, K_n) = \begin{cases} d & \text{if } n \leq d^k - \lceil \log_d k \rceil - 1\,, \\ d+1 & \text{if } n \geq d^k - \lceil \log_d k \rceil + 1\,. \end{cases}$$

If $n = d^k - \lceil \log_d k \rceil$, then $D(K_k \,\square\, K_n)$ is either d or $d+1$. It can be computed recursively in $O(\log^* n)$ time, where $\log^*$ denotes the iterated logarithm.

For a more detailed treatment (including proofs) of the distinguishing number of Cartesian products, see Chapter 17 of the book by Imrich, Klavžar, and Rall (2008).

6.3 Transitive Group Action on Products

We now apply the above results to Cartesian products with transitive and sharply transitive automorphism groups. In particular, we show that a Cartesian product, connected or not, has transitive automorphism group if and only if every factor has transitive automorphism group. Moreover, we prove that connected or disconnected graphs with transitive automorphism groups have unique prime factor decompositions with respect to the Cartesian product.

Proposition 6.16 *A Cartesian product of connected graphs has transitive automorphism group if and only if every factor has transitive automorphism group.*

Proof Suppose that $G = G_1 \,\square\, G_2 \,\square\, \cdots \,\square\, G_k$. Let $v = (v_1, v_2, \ldots, v_k)$, $u = (u_1, u_2, \ldots, u_k)$ be arbitrary vertices of G. If all G_i have transitive group, there are automorphisms $\varphi_i \in \mathrm{Aut}(G)$ with $\varphi_i(v_i) = u_i$, so the automorphism $\varphi = (\varphi_1, \varphi_2, \ldots, \varphi_k)$ of G maps v to u. Thus $\mathrm{Aut}(G)$ is transitive if all $\mathrm{Aut}(G_i)$ are transitive.

For the converse, let G be a connected graph with transitive automorphism group. Suppose that G is the Cartesian product $G_1 \,\square\, G_2 \,\square\, \cdots \,\square\, G_k$ of relatively prime graphs and that we are given vertices $v_i, u_i \in V(G_i)$ for every $i \in \{1, 2, \ldots, k\}$. Because G has transitive group, there is an automorphism φ of G that maps $(v_1, v_2, \ldots, v_k)$ into $(u_1, u_2, \ldots, u_k)$. By Corollary 6.12, φ can be represented in the form

$$\varphi(v_1, v_2, \ldots, v_k) = (\varphi_1(v_1), \varphi_2(v_2), \ldots, \varphi_k(v_k)),$$

where the φ_i are automorphisms of G_i. Hence $\varphi_i(v_i) = u_i$ for each i. In other words, each G_i has transitive group.

If some of the prime factors of G are isomorphic, we can collect them together into relatively prime collections of factors. Hence it remains to show that a connected graph has transitive group if a Cartesian power of it has transitive group. Thus let $G = \square_{i=1}^{k} G_i$, where all G_i are isomorphic. There is no loss of generality in assuming that all the G_i are, in fact, equal, so $G = G_1 \,\square\, G_1 \,\square\, \cdots \,\square\, G_1$. For arbitrary vertices $a, b \in V(G_1)$, let φ be the automorphism of G with $\varphi(a, a, a, \ldots, a) = (b, a, a, \ldots, a)$. By Theorem 6.10, there are isomorphisms (in this case, automorphisms) $\varphi_i : G_1 \to G_1$ for which

$$(\varphi_1(a), \varphi_2(a), \ldots, \varphi_k(a)) = (b, a, a, \ldots, a).$$

Thus the automorphism φ_1 of G_1 maps a to b. □

This result does not hold for infinite graphs. By the results of Chapter 31, a connected infinite graph can have transitive automorphism group even if all of its prime divisors with respect to the Cartesian product have trivial automorphism groups; see Exercise 31.15. In the case of finite graphs, however, we have the following theorem:

Theorem 6.17 *A Cartesian product has transitive automorphism group if and only if every factor has transitive automorphism group.*

Proof If every factor has transitive group, we argue as in the first paragraph of the proof of Proposition 6.16 to conclude that the product has transitive group.

Conversely, Proposition 6.16 takes care of the connected case. To complete the proof, we need only show that transitivity of $\mathrm{Aut}(G \,\Box\, H)$ implies the transitivity of $\mathrm{Aut}(G)$ and $\mathrm{Aut}(H)$ for disconnected $G \,\Box\, H$. Denote the components of G by $X_1, \ldots, X_g$ and those of H by $Y_1, \ldots, Y_h$. Clearly, all components of $G \,\Box\, H$ must have transitive automorphism group, and they are all isomorphic. Because they are of the form $X_i \,\Box\, Y_j$, all components of G and H must have transitive group, by Proposition 6.16. We have to show that any two X_i are isomorphic, and likewise for any two Y_j.

Without loss of generality, we can assume that G is disconnected. It suffices to show that X_1 and X_2 are isomorphic, namely that any prime factor of X_1 is also a prime factor of X_2 and has the same multiplicity. Let $X_1 = P^r \,\Box\, U$, $X_2 = P^s \,\Box\, W$, and $Y_1 = P^t \,\Box\, Z$, where P is not a divisor of U, W, or Z. Because $X_1 \,\Box\, Y_1 = P^{r+t} \,\Box\, U \,\Box\, Z$ and $X_2 \,\Box\, Y_1 = P^{s+t} \,\Box\, W \,\Box\, Z$ are isomorphic, we conclude that $r + t = s + t$, and hence that $r = s$. □

Corollary 6.18 *Every graph with transitive automorphism group has unique prime factor decomposition with respect to the Cartesian product.*

Proof It suffices to prove the theorem for disconnected G. Suppose that $X_1, \ldots, X_r$ are the components of G and that $P_1 \,\Box\, P_2 \,\Box\, \cdots \,\Box\, P_k$ is the prime factor decomposition of X_1.

We first show that every disconnected prime factor must be totally disconnected. For, if $\sum_{i=1}^{r} Y_i$ is a factor, then any two Y_i, Y_j must be isomorphic by Theorem 6.17. Hence $\sum_{i=1}^{r} Y_i = rY_1 = D_r \,\Box\, Y_1$. Because $r \neq 1$, this can only be prime if $Y_1 = K_1$. Therefore the product of all disconnected prime factors is a totally disconnected graph D_s for some s, and because of the unique prime factorization of s, the prime factors are uniquely determined. From $D_s \,\Box\, Y \cong G$ follows $Y \cong X_1$, and this determines the connected prime factors. □

Corollary 6.19 *The automorphism group of a connected graph G with prime factor decomposition $H_1 \,\Box\, H_2 \,\Box\, \cdots \,\Box\, H_k$ is sharply transitive if and only if the prime factors are pairwise nonisomorphic and have sharply transitive automorphism groups.*

Proof The proof follows from the above considerations and the observation that $\mathrm{Aut}(H \,\Box\, H)$ cannot be regular if H is nontrivial because transpositions of the two factors have fixed points but are not the identity. □

Similarly one proves that this corollary remains true if the term "sharply transitive" is replaced by "fixed point-free."

Corollary 6.20 *The automorphism group of connected graph G with prime factor decomposition $H_1 \,\Box\, H_2 \,\Box\, \cdots \,\Box\, H_k$ is Abelian if and only if the following conditions are satisfied:*

(i) *Every prime factor has Abelian automorphism group.*
(ii) *Prime factors with nontrivial groups are pairwise nonisomorphic.*
(iii) *There are no three pairwise isomorphic factors with trivial automorphism group.*

Proof Note that $H \square H \square H$ has non-Abelian group for nontrivial H because the transpositions of the factors generate the symmetric group on three elements, which is not Abelian.

If φ is a nontrivial automorphism of H mapping a into b, we consider $H \square H$, denote the transposition of the two factors by α and the mapping $(u, v) \mapsto (\varphi(u), v)$ by β. Then $\alpha\beta(a, a) = (b, a)$, but $\beta\alpha(a, a) = (a, b)$, so $\mathrm{Aut}(H \square H)$ is not Abelian. □

Proposition 6.16 and Corollaries 6.19 and 6.20 are from Imrich (1969a), and Theorem 6.17 with Corollary 6.18 is from Imrich (1972a).

6.4 Cancellation

If G, H, and K are connected graphs, and $G \square K \cong H \square K$, then Theorem 6.6 immediately guarantees that $G \cong H$, that is, the common factor can be cancelled from the product. In fact, this cancellation property holds even if the assumption of connectedness is removed. The following proof is based on the approach of Fernández, Leighton, and López-Presa (2007). Recall that O denotes the empty graph.

Theorem 6.21 *Suppose $G, H, K \in \Gamma$ and $K \neq O$. If $G \square K \cong H \square K$, then $G \cong H$.*

Proof The idea is to embed Γ in the ring $\mathcal{R} = \mathbb{Z}[x_1, x_2, x_3, \ldots]$ of polynomials in countably many indeterminates, and then transfer cancellation in $\mathcal{R}$ back to Γ.

Let $G_1, G_2, G_3, \ldots$ be a list of all connected prime graphs in Γ. Define a map $\varphi : \Gamma \to \mathcal{R}$ as follows. Any connected nontrivial graph G factors uniquely as $G = G_{i_1}^{\square, j_1} \square G_{i_2}^{\square, j_2} \square \cdots \square G_{i_k}^{\square, j_k}$, where the G_{i_s} are prime and pairwise nonisomorphic, and we set $\varphi(G) = x_{i_1}^{j_1} x_{i_2}^{j_2} \cdots x_{i_k}^{j_k}$. Thus, in particular, $\varphi(G_i) = x_i$ for $i \geq 1$, and we further define $\varphi(K_1) = 1$ and $\varphi(O) = 0$. Thus φ is well-defined on the set of connected graphs. Finally, any arbitrary G can be decomposed uniquely as a disjoint union $G = H_1 + H_2 + \cdots + H_k$ of connected components, and we set $\varphi(G) = \sum_{i=1}^{k} \varphi(H_i)$.

The map φ is well-defined and is a bijection of Γ onto the set polynomials in $\mathbb{Z}[x_1, x_2, x_3, \ldots]$ with nonnegative coefficients. Moreover, the distributive and commutative properties of the Cartesian product yield $\varphi(G \square H) = \varphi(G)\varphi(H)$ for all $G, H \in \Gamma$.

Now, if $G \square K \cong H \square K$, we have $\varphi(G \square K) = \varphi(H \square K)$, so $\varphi(G)\varphi(K) = \varphi(H)\varphi(K)$. If $K \neq O$, then $\varphi(K) \neq 0$, and $\varphi(G) = \varphi(H)$ by cancellation in the integral domain $\mathcal{R}$. Finally, $G \cong H$ because φ is injective. □

Certain cancellation-type results also hold for graph homomorphisms. Suppose G is a connected graph with odd girth at least $2k + 1$. (Recall that the *odd girth* of a graph is the length of its shortest odd cycle.) The graph G is called *strongly* $(2k+1)$*-angulated* if for all pairs $u, v \in V(G)$ there is a sequence of $(2k + 1)$-cycles $C_1, \ldots C_t$ such that $u \in C_1$, $v \in C_t$, and C_i and C_{i+1} share at least one edge for every $1 \leq i < t$. The following is a typical result from Che, Collins, and Tardif (2008):

Theorem 6.22 *Let G be a strongly $(2k + 1)$-angulated graph with a vertex that is fixed by every endomorphism of G. Suppose that H is a strongly $(2k + 1)$-angulated graph and T a* *graph with the odd girth at least $2k+1$. If there exists a homomorphism $G \square H \to G \square T$, then there is also a homomorphism $H \to T$.*

The key insight of the proof of this theorem (as well as for other theorems in the paper) is that, roughly speaking, strongly $(2k+1)$-angulatedness assures that homomorphisms preserve layers of Cartesian products.

6.5 S-Prime Graphs

Given a graph product $*$, it is natural to ask which graphs are nontrivial subgraphs of $*$-products. As we will see, the answer is not difficult for the direct product, the strong product, and the lexicographic product. This section suggests that the question for the Cartesian product is quite intriguing.

If G and H are graphs on at least two vertices, then a subgraph X of $G * H$ is called *nontrivial* if each of $p_G(X)$ and $p_H(X)$ has at least two vertices. A graph X is called $*$-*S-prime*[1] if it cannot be represented as a nontrivial subgraph of a $*$-product graph. Graphs that are not $*$-S-prime are called $*$-*S-composite.* Clearly, $*$-S-prime graphs are prime with respect to $*$.

Sabidussi (1975) proved that complete graphs and complete graphs with an edge removed are the only $\times$-S-prime graphs. (He named these graphs subdirectly irreducible, see p. 104.) Lamprey and Barnes (1981) followed by noting that the only $\boxtimes$-S-prime graphs and $\circ$-S-prime graphs are K_1, $K_1 + K_1$, and K_2. (Exercise 6.14.) Hence it remains to consider $\square$-S-prime graphs; for simplicity we will call them S-prime henceforward. Likewise, $\square$-S-composite graphs will be called S-composite.

This section describes four different approaches to S-prime graphs (equivalently S-composite graphs), the first one being rather trivial.

Plottings

A *plotting* of a graph is a drawing of it on the plane such that the endpoints of each edge either have the same abscissa or the same ordinate. Note that a standard coordinate drawing of a product graph is a plotting. Thus, clearly, a graph is S-prime if and only if in any of its possible plottings, all the vertices are either plotted on the same horizontal or on the same vertical line.

Basic S-prime graphs

For a nontrivial characterization of S-prime graphs, Lamprey and Barnes (1995) introduced *basic S-prime graphs* as S-prime graphs on at least three vertices that contain no **proper basic** S-prime subgraphs. They proved that a graph on at least three vertices is either a basic S-prime graph or can be constructed from such graphs by two special rules.

This characterization prompts us to list all basic S-prime graphs. However, the task seems to be quite difficult. Lamprey and Barnes (1995) noted that K_3 and $K_{2,3}$ are the only basic S-prime graphs on at most six vertices and provided two more sporadic examples, one of them shown on the left of Figure 6.2.

Klavžar, Lipovec, and Petkovšek (2002) proved that basic S-prime graphs are precisely

---

[1] S stands for subgraph.

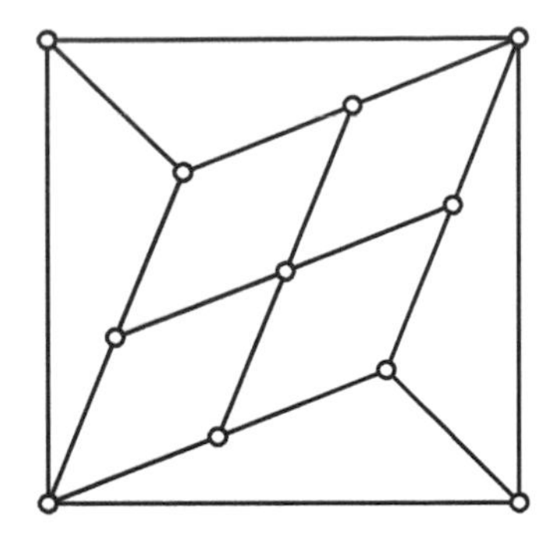

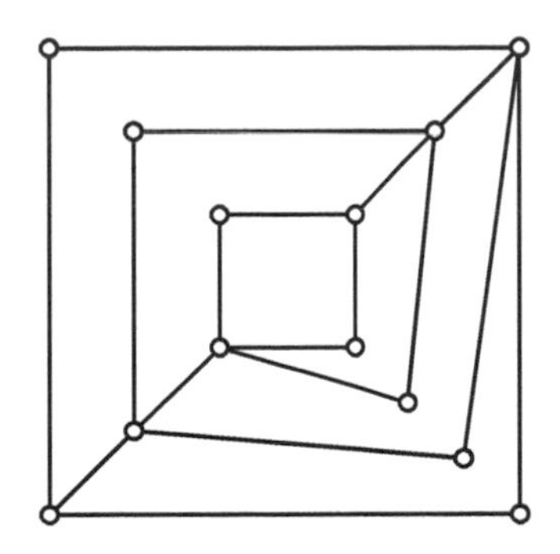

FIGURE 6.2 Basic S-prime graphs.

the graphs on at least three vertices that contain no **proper** S-prime subgraph on at least three vertices. Using this simplified definition, an infinite family of basic S-prime graphs was constructed. The line of research was continued by Brešar (2004), who gave several characterizations of basic S-prime graphs. This in particular enabled him to construct several infinite families of basic S-prime graphs. (The graph on the right of Figure 6.2 is an example from such a series.) Moreover, there exist basic S-prime graphs of arbitrary diameter. (The one in the figure has diameter 4.) Brešar also noticed that the graph on the left of Figure 6.2 is the only known basic S-prime graph that is 3-connected; he posed the problem of whether there are more such graphs, in particular whether the number of such graphs is finite.

The above investigations lend the impression that the variety of basic S-prime graphs is probably too complicated to allow a clear classification.

Vertex labelings

Let X be a graph and $c : V(X) \to \{1, 2, \ldots, k\}$ a surjective mapping. A nontrivial path P of X is *well-colored* if $c(u) \neq c(v)$ holds for any consecutive vertices u and v of P. The mapping c is a *path k-coloring* if $c(u) \neq c(v)$ holds for any well-colored u, v-path of X. Using this concept, Klavžar et al. (2002) characterized S-composite graphs as follows:

Theorem 6.23 *Let X be a connected graph on at least three vertices. Then X is S-composite if and only if there exists a path k-coloring of X with $2 \leq k \leq |V(X)| - 1$.*

Proof Let X be S-composite. Then X is a subgraph of a Cartesian product $G \,\square\, H$ such that X intersects at least two G-layers and at least two H-layers. Let $V(G) = \{g_1, \ldots, g_k\}$ and $V(H) = \{h_1, \ldots, h_{|V(H)|}\}$. Consider X as a subgraph of $G \,\square\, H$ and for a vertex $(g_i, h_j) \in V(X)$ set $c(g_i, h_j) = i$. Then $c : V(X) \to \{1, \ldots, k\}$. An arbitrary well-colored path P of X (with respect to the labeling c) must necessarily lie within a fixed G-layer. But then the endvertices of P have different labels. So c is a path k-coloring of X. Clearly, $2 \leq k \leq |V(X)| - 1$.

For the converse, assume that c is a path k-coloring of X with $2 \leq k \leq |V(X)| - 1$. Remove from X all edges uv such that $c(u) = c(v)$ to obtain a spanning subgraph X'. Let $C_1, \ldots, C_t$ be the connected components of X'. As $k \leq |V(X)| - 1$, there are vertices u and v of X with $c(u) = c(v)$. Then u and v belong to different connected components of X' for otherwise a well-colored u, v-path would exist. It follows that $t \geq 2$.

We claim that X is a subgraph of $K_k \square K_t$. Let $V(K_n) = \{1, \ldots, n\}$ and let $g : V(X) \to V(K_t)$ be the natural contraction: for $u \in C_i$, set $g(u) = i$. Then define

$$f : V(X) \to V(K_k \square K_t)$$

with

$$f(u) = (c(u), g(u)).$$

The claim follows after verifying that f is injective and that it maps edges to edges. The details are not difficult and are left to the reader.

Because the image of c consists of at least two elements, X lies in at least two K_t-layers. The same argument applied to the image of g implies that X lies in at least two K_k-layers. Hence, X is S-composite. □

Hellmuth, Gringmann, and Stadler (2011) obtained a (technical) characterization of path k-colorings in the class of Cartesian products of S-prime graphs. Combining their characterization with Theorem 6.23, they deduced the following appealing result, where a *diagonalized Cartesian product* is a graph obtained from a Cartesian product graph by adding an edge between two of its diametrical vertices.

Theorem 6.24 *A diagonalized Cartesian product of S-prime graphs is S-prime.*

Edge labelings

Klavžar and Peterin (2005) showed that S-composite graphs can also be characterized in terms of edge labelings:

Theorem 6.25 *Let X be a connected graph. Then X is S-composite if and only if $E(X)$ can be labeled with two labels such that on any induced cycle of X on which both labels appear, the labels change at least three times while passing the cycle.*

One direction of the proof of Theorem 6.25 is easy. First observe (Exercise 6.15) that it suffices to consider nontrivial subgraphs of products of complete graphs. Then, if X is a subgraph of $K_k \,\Box\, K_t$, label its edges that project to K_k with 1 and the remaining edges with 2. The other direction uses techniques similar to those that will be introduced for the canonical isometric embedding in Chapter 13.

In the above paper, characterizations via edge labelings are also obtained for induced and isometric subgraphs of Cartesian products of complete graphs.

Exercises

6.1. Prove the converse of Lemma 6.5.

6.2. Let G be a connected graph with a vertex that is contained in no square of G. Show that G is prime with respect to the Cartesian product. Find an example of a graph demonstrating that the converse is not true in general.

6.3. (Imrich, 1972b) Show that the complement of a Cartesian product of a graph on at least ten vertices is prime.

6.4. Show that the automorphism group of the graph in Figure 6.1 is $\mathbb{Z}_2^3$.

6.5. Give an example of two nonisomorphic connected graphs G and H for which $\mathrm{Aut}(G \,\Box\, H)$ is not the direct product $\mathrm{Aut}(G) \times \mathrm{Aut}(H)$.

6.6. Let G be a connected graph on at least ten vertices with automorphism group A. Construct a graph whose group is the direct product $A \times A$.

6.7. Let n, k, and m be integers with $k \leq n$. Prove that there exists a homomorphism $C_{2k+1} \,\square\, C_{2n+1} \to C_{2m+1}$ if and only if $k = m$.

6.8. (Baron, 1968) For a connected graph G, the *tree graph* of G is defined on the set of spanning trees of G. Two spanning trees S, T of G are adjacent in the tree graph if there are edges $e \in E(S)$ and $f \in E(T)$, such that $T = (S \setminus e) \cup f$. Also, a subgraph of G is a *block of* G if it is a maximal subgraph without cut vertices or edges whose removal increases the number of components.

Show that the tree graph of a graph G is the Cartesian product of the tree graphs of the blocks of G.

6.9. Let G be a cycle. Show that the tree graph of G (as defined in Exercise 6.8) is prime.

6.10. (Nowakowski and Rival, 1988) Let G be a connected graph such that each edge of G is in a triangle and H be a connected triangle-free graph. Show that any retract of $G \,\square\, H$ is a box.

6.11. Show that every graph G with a prime number of connected components is prime if and only if the connected components of G have no common nontrivial factor.

6.12. Show that the prime factorization of every graph with a prime number of connected components is unique with respect to the Cartesian product.

6.13. Show that the prime factorization of every graph G with at most five connected components is unique with respect to the Cartesian product.

6.14. Show that K_1, $K_1 + K_1$, and K_2 are the only $\boxtimes$-S-prime graphs the only $\circ$-S-prime graphs.

6.15. Show that a graph is S-composite if and only if it is a nontrivial subgraph of the Cartesian product of two complete graphs.

Chapter 7

Strong Product

This chapter treats prime factorizations and automorphisms of the strong product of graphs. As in the Cartesian case, prime factorization is unique for connected graphs, but not necessarily for disconnected ones. Again, there is a strong connection between the automorphism group of a graph and the automorphism groups of its prime factors. For so-called S-thin graphs, the relationship is the same as in the case of the Cartesian product.

The main theorems in this chapter are special cases of results about the direct product that will be proved in Chapter 8. Nonetheless, we prove them directly here. The proofs are shorter and provide a different perspective.

7.1 Basic Properties and S-Thin Graphs

We defined the strong product in Section 4.1 and studied its distance and connectedness properties in Section 5.2. We begin this section with remarks about the existence and non-uniqueness of prime factorizations. Then we motivate the introduction of S-thin graphs and derive several of their basic properties.

Recall that every nontrivial graph has a prime factorization with respect to any of the standard products, because a graph on n vertices cannot have more than $\log_2 n$ nontrivial factors, and so any factorization into a product of nontrivial graphs with a maximal number of factors is a prime factorization; the proof is the same as that of Proposition 6.1.

Prime factorization need not be unique for disconnected graphs. To see this, consider

$$(K_1 + K_2 + K_2^{\boxtimes,2}) \boxtimes (K_1 + K_2^{\boxtimes,3}) = (K_1 + K_2^{\boxtimes,2} + K_2^{\boxtimes,4}) \boxtimes (K_1 + K_2)\,.$$

By arguments similar to those for the Cartesian product (as in the proof of Theorem 6.2), we see that the factors on the left- and right-hand sides are prime.

Despite the many similarities between the Cartesian and the strong product there are several important differences. The fact that the diameter of a strong product is the maximum of the diameters of the factors and not the sum of the diameters, as in the case of the Cartesian product, is the first remarkable difference. As we saw in Exercise 4.6, closed neighborhoods in strong products are boxes, that is,

$$N_{G\boxtimes H}[(g,h)] = N_G[g] \times N_H[h]\,, \tag{7.1}$$

unlike the case of the Cartesian product. We will see that this implies that coordinatization in strong products may not be unique, and this will motivate the introduction of S-thin graphs.

For another difference between the strong product and the Cartesian product, we note that layers in a strong product need not be convex. (See Figure 7.1.) Because the convexity of layers was essential in our approach to the unique prime factorization of connected graphs over the Cartesian product, we must follow a different approach here. Nonetheless, layers in strong products are isomorphic to the factors. By the Distance Formula (Corollary 5.5), they are also isometric.

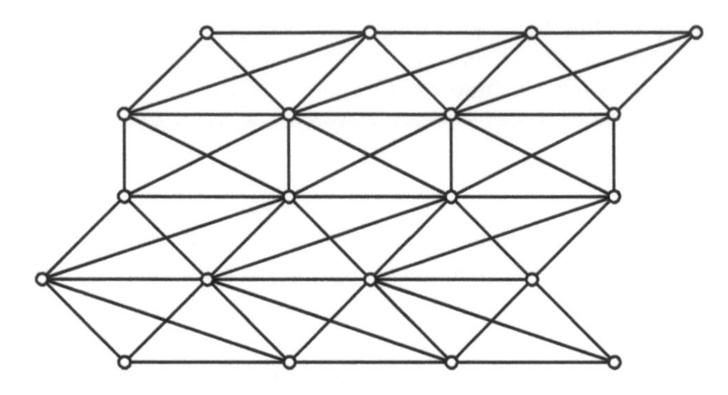

FIGURE 7.1 $P_4 \boxtimes P_5$.

As an example for nonunique coordinatization consider $K_4 = K_2 \boxtimes K_2$. Figure 7.2 shows the three different ways of representing K_4 as a nontrivial product. They correspond to the three ways of choosing a pair of independent edges in K_4.

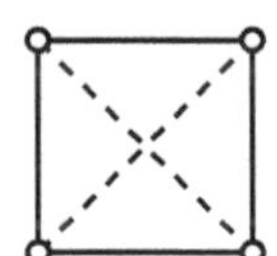
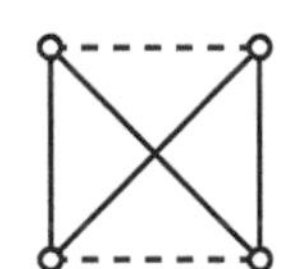
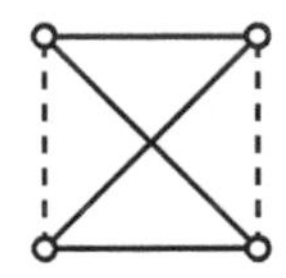

FIGURE 7.2 Squares in $K_4 = K_2 \boxtimes K_2$.

For a more general example, consider $K_{mn} = K_m \boxtimes K_n$. Here any two partitions $U_1, U_2, \ldots, U_n$ and $V_1, V_2, \ldots, V_m$ of $V(K_{mn})$ that satisfy $|U_i \cap V_j| = 1$ for $1 \leq i \leq n$ and $1 \leq j \leq m$ can be considered as the vertex sets of the K_m- and K_n-layers of $K_m \boxtimes K_n$.

Let us call the edges of $G = G_1 \boxtimes G_2 \boxtimes \cdots \boxtimes G_k$ that differ in exactly one coordinate *Cartesian*, and the others *non-Cartesian*. In other words, the Cartesian edges correspond to the edges of $\square_{i=1}^{k} G_i \subseteq G_1 \boxtimes G_2 \boxtimes \cdots \boxtimes G_k$. By the above examples, the set of Cartesian edges depends on the factorization.

But under certain conditions, the set of Cartesian edges of a product is intrinsic, that is, preserved under automorphisms of the graph. This is the case for $P_4 \boxtimes P_5$ in Figure 7.1. We will treat this in more detail in Section 7.2. There the set of Cartesian edges plays an important role in decomposition algorithms of graphs with respect to the strong product.

The above examples illustrate the general fact that if two vertices of a graph have the same closed neighborhood, then their transposition is an automorphism. Such an automorphism need not respect the layers, and that is why layers cannot be uniquely coordinatized. This motivates a definition. We say vertices x, y of a graph G are in the *relation* S, written xSy, provided that $N[x] = N[y]$. (We write S_G if there is a chance of ambiguity.) The next result (whose proof is Exercise 7.1) summarizes some essential features of this relation.

Proposition 7.1 *If $G \in \Gamma$, then S is an equivalence relation on $V(G)$. The equivalence classes induce complete subgraphs of G. Given classes U and V, either all vertices of U are adjacent to all vertices of V, or all vertices of U are nonadjacent to all vertices of V.*

Given a vertex x of G, we denote the S-equivalence class containing x as

$$[x] = \{x' \in V(G) \,|\, N_G[x'] = N_G[x]\}.$$

We define the quotient G/S in the usual way. Specifically,

$$V(G/S) = \{[x] \mid x \in V(G)\},$$

and distinct classes $[x]$ and $[y]$ are adjacent if $x'y' \in E(G)$ for some $x' \in [x]$ and $y' \in [y]$. Observe that Proposition 7.1 implies $[x][y] \in E(G/S)$ if and only if $xy \in E(G)$ and $[x] \neq [y]$.

We say G is *S-thin* if $G/S = G$, that is, if its S-equivalence classes are single vertices. Note that G/S is S-thin for any graph G. The next lemma shows that the S-classes of a product $G \boxtimes H$ are precisely the sets $U \times V$, where U is an S-class of G and V is an S-class of H. One significant consequence of the lemma is that a graph is S-thin if and only if all of its factors with respect to the strong product are S-thin.

Lemma 7.2 *If G and H are graphs, then $V((G\boxtimes H)/S) = \{[x]\times[y] \mid x \in V(G), y \in V(H)\}$. Moreover, the map $[(x,y)] \mapsto ([x],[y])$ is an isomorphism $(G \boxtimes H)/S \to G/S \boxtimes H/S$.*

Proof To see that $V((G \boxtimes H)/S) = \{[x] \times [y] \mid x \in V(G), y \in V(H)\}$, just observe

$$\begin{aligned}
(x', y') \in [(x, y)] &\iff N_{G\boxtimes H}[(x', y')] = N_{G\boxtimes H}[(x, y)] \\
&\iff N_G[x'] \times N_H[y'] = N_G[x] \times N_H[y] \\
&\iff N_G[x'] = N_G[x] \text{ and } N_H[y'] = N_H[y] \\
&\iff x' \in [x] \text{ and } y' \in [y] \\
&\iff (x', y') \in [x] \times [y].
\end{aligned}$$

Therefore, $[(x, y)] = [x] \times [y]$, and the assertion follows.

By Proposition 7.1 and the remarks that follow, we have $[(x,y)][(x',y')] \in E((G\boxtimes H)/S)$, if and only if $(x,y)(x',y') \in E(G \boxtimes H)$ and $[(x,y)] \neq [(x',y')]$, if and only if $(x,y)(x',y') \in E(G \boxtimes H)$ and $[x] \neq [x']$ or $[y] \neq [y']$, if and only if $([x],[y])([x'],[y']) \in E(G/S \boxtimes H/S)$. □

7.2 Cliques and the Extraction of Complete Factors

The principal idea followed in this chapter for the investigation the factorizations of graphs over the strong product is a careful examination of the mapping of cliques by isomorphisms between two composite graphs. We begin with the following observation.

Lemma 7.3 *Let G and H be graphs and Q a clique of $G \boxtimes H$. Then $Q = p_G(Q) \boxtimes p_H(Q)$, where $p_G(Q)$ and $p_H(Q)$ are cliques of G and H, respectively.*

Proof Clearly, $p_G(Q)$ and $p_H(Q)$ are complete subgraphs of G and H, respectively. (See Figure 7.3.) Thus there is a clique Q_G of G with $p_G(Q) \subseteq Q_G$ and a clique Q_H of H with $p_H(Q) \subseteq Q_H$. Because $Q \subseteq Q_G \boxtimes Q_H$ and since Q is a maximal complete subgraph, we infer that $Q = Q_G \boxtimes Q_H$. Moreover, $Q_G = p_G(Q)$ and $Q_H = p_H(Q)$. □

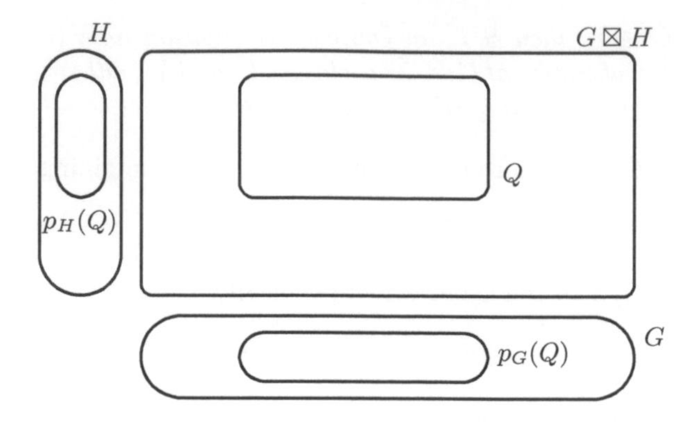

FIGURE 7.3 Clique of a strong product.

Lemma 7.4 *Let G be a graph, Q a clique of G and $v \in V(G)$. Furthermore, let Q_v denote $V(Q)$ if $v \in V(Q)$ and $V(G) \setminus V(Q)$ otherwise. Then the S-equivalence class $[v]$ is*

$$[v] = \bigcap_{Q \in \mathfrak{Q}} Q_v \,,$$

where $\mathfrak{Q}$ denotes the set of all cliques of G.

In other words, the S-equivalence class of a vertex v is the intersection of all cliques that contain v with the complements of all cliques that do not contain v.

Proof If u is a vertex with $v\,S\,u$, then u and v are clearly contained in the same cliques. So $Q_v = Q_u$ for all cliques Q, and $[v] \subseteq \bigcap_{Q\in\mathfrak{Q}} Q_v$.

On the other hand, let $u \in \bigcap_{Q\in\mathfrak{Q}} Q_v$. If u is different from v, it must be adjacent to v; otherwise, there is a clique containing v but not u and $u \notin \bigcap_{Q\in\mathfrak{Q}} Q_v$. This is also the case if there exists a vertex w that is adjacent to only one vertex of the pair u, v. Thus, $v\,S\,u$ and $\bigcap_{Q\in\mathfrak{Q}} Q_v \subseteq [v]$. □

For an S-thin graph G, Lemma 7.4 implies that $\bigcap_{Q\in\mathfrak{Q}} Q_v$ is the one-element set $\{v\}$.

Lemma 7.5 *If $G \cong K_k \boxtimes H$, then k divides $|U|$ for any $U \in V(G/S)$. Conversely, if k divides $|U|$ for all $U \in V(G/S)$, then there is a graph H for which $G \cong K_k \boxtimes H$.*

Proof If $G \cong K_k \boxtimes H$, then Lemma 7.2 implies that $V(G/S) = \{V(K_k) \times U \mid U \in V(H)\}$.

Conversely, suppose G is such that k divides $|U|$ for all $U \in V(G/S)$. Fix a labeling of the vertices of each U by ordered pairs (i, U_j), so that

$$U = \left\{(i, U_j) \mid 1 \le i \le k,\ 1 \le j \le \frac{|U|}{k}\right\}.$$

Then $G = K_k \boxtimes H$, where $V(K_k) = \{1, 2, \ldots, k\}$ and H is the graph with vertices $V(H) = \{U_j \mid U \in V(G/S), 1 \le j \le |U|/k\}$, and $E(H) = \{U_jV_\ell \mid U = V \text{ or } UV \in E(G/S)\}$. □

Before continuing, we note the following simple but useful proposition on isomorphisms. The proof makes repeated use of the definitions and Proposition 7.2.

Proposition 7.6 *Two graphs G and H are isomorphic if and only if*

(i) *There is an isomorphism* $\pi : G/S \to H/S$, *and*
(ii) $|U| = |\pi(U)|$, *for all* $U \in V(G/S)$.

Proof Suppose $\varphi : G \to H$ is an isomorphism. Then $\varphi([x]) = [\varphi(x)]$, so $|[x]| = |[\varphi(x)]|$. Define $\pi : G/S \to H/S$ by $\pi([x]) = [\varphi(x)]$. One checks easily that π satisfies (i) and (ii).

Conversely, suppose (i) and (ii) hold. For $[x] \in V(G/S)$, let $\varphi_{[x]} : [x] \to \pi([x])$ be a bijection. Then $\varphi : V(G) \to V(H)$, defined by $\varphi|_{[x]} = \varphi_{[x]}$, is an isomorphism. □

Lemma 7.7 *For any integer* $k \geq 1$, *if* $K_k \boxtimes G \cong K_k \boxtimes H$, *then* $G \cong H$.

Proof Suppose there is an isomorphism $\varphi : K_k \boxtimes G \to K_k \boxtimes H$. We just need to produce an isomorphism $\pi : G/S \to H/S$ satisfying condition (ii) of Proposition 7.6. Recall that the S-classes of $K_k \boxtimes G$ have form $V(K_k) \times U$, for $U \in V(G/S)$, and the S-classes of $K_k \boxtimes H$ have form $V(K_k) \times V$, for $V \in V(H/S)$. Thus φ sends any S-class $V(K_k) \times U$ to an S-class $V(K_k) \times \pi(U)$. Clearly, $\pi : G/S \to H/S$ is a bijection and $|U| = |\pi(U)|$, as both cardinalities equal $|V(K_k) \times U|/k$. It is straightforward to check that π is an isomorphism. □

In Section 10.2 we will see that Lemma 7.7 is a special case of a cancellation theorem for the lexicographic product, that is, of Theorem 10.8.

It will be convenient to say a graph G *divides* a graph H with respect to a given graph product if G is a factor of H with respect to this product. If not explicitly stated, it should be clear from the context which product is meant.

Lemma 7.8 *If* $K_m \boxtimes G \cong K_n \boxtimes H$ *and neither* G *nor* H *is divisible by* K_k *for any* $k > 1$, *then* $m = n$ *and* $G \cong H$.

Proof As G is not divisible by any K_k for $k > 1$, Lemma 7.5 implies that the numbers in the set $\{|U| \mid U \in G/S\}$ are relatively prime. By the same reasoning, the numbers in $\{|V| \mid V \in H/S\}$ are relatively prime. But because $K_m \boxtimes G \cong K_n \boxtimes H$ and the S-classes of $K_m \boxtimes G$ have form $V(K_m) \times U$ for $U \in G/S$ (and similarly for $K_m \boxtimes H$), we get

$$\{m|U| \mid U \in G/S\} = \{n|V| \mid U \in H/S\}.$$

It follows that $m = n$. An application of Lemma 7.7 completes the proof. □

Before proving unique prime factorization for connected graphs, we note that the fundamental theorem of arithmetic combined with the equation $K_m \boxtimes K_n = K_{mn}$ implies that complete graphs have unique prime factorizations. Indeed, if an integer p has prime factorization $p = p_1 p_2 \cdots p_k$, then K_p factors uniquely into primes as $K_p = K_{p_1} \boxtimes K_{p_2} \boxtimes \cdots \boxtimes K_{p_k}$.

7.3 Unique Prime Factorization for Connected Graphs

In the case of Cartesian products, we have seen that automorphisms preserve layers with respect to prime factors of a connected graph. The situation for the strong product is different. It will be convenient to make the layers thicker and to define so-called towers.

Let G be a graph and H' a subgraph of H. The *G-tower over* H' is the subproduct $G \boxtimes H'$ of $G \boxtimes H$. More generally, if $G = G_1 \boxtimes G_2 \boxtimes \cdots \boxtimes G_n$, we define the G_k-tower in G as the product of G_k by a subgraph of $G_1 \boxtimes G_2 \boxtimes \cdots \boxtimes G_{k-1} \boxtimes G_{k+1} \boxtimes \cdots \boxtimes G_n$.

Lemma 7.9 *Let* $\varphi : G \boxtimes H \to G' \boxtimes H'$ *be an isomorphism and* T *be a* G*-tower over a clique* Q *of* H. *If* G *and* H *are connected, then* $\varphi(T)$ *is a box of* $G' \boxtimes H'$.

Proof The Distance Formula (Corollary 5.5) implies that T is an isometric subgraph of $G \boxtimes H$, so $\varphi(T)$ is an isometric subgraph of $G' \boxtimes H'$. This means that $\varphi(T)$ is an induced subgraph.

It is straightforward to check that $\varphi(T)$ is a subgraph of $p_{G'}(\varphi(T)) \boxtimes p_{H'}(\varphi(T))$. We need to show $\varphi(T) = p_{G'}(\varphi(T)) \boxtimes p_{H'}(\varphi(T))$. As $\varphi(T)$ is induced, we just need to show that each vertex of $B = p_{G'}(\varphi(T)) \boxtimes p_{H'}(\varphi(T))$ belongs to $\varphi(T)$.

Thus let (g, h) be a vertex of B. Then $\varphi(T)$ must have vertices of form (g, h') and (g', h). We will use induction on the distance from (g, h') to (g', h) to show that $(g, h) \in \varphi(T)$.

First suppose the distance is 1. Then the edge $\varphi^{-1}(g, h')\varphi^{-1}(g', h)$ of T lies in some clique $Q'' \boxtimes Q$ in T, where Q'' is a clique in G. Thus $\varphi(Q'' \boxtimes Q) \subseteq \varphi(T)$ is a clique in $G' \boxtimes H'$. By Lemma 7.3, $\varphi(Q'' \boxtimes Q)$ is a box in $G' \boxtimes H'$, and because this box contains the edge $(g, h')(g', h)$, it also contains (g, h). Thus $(g, h) \in \varphi(T)$.

Now suppose $d((g, h'), (g', h)) > 1$. Because $\varphi(T)$ is connected and isometric, it has a shortest path $(g, h')(x, y) \dots (g', h)$. The induction hypothesis applied to paths $(g, h')(x, y)$ and $(x, y) \dots (g', h)$ yields $(g, y), (x, h) \in \varphi(T)$. Corollary 5.5 gives $d((g, y), (x, h)) < d((g, h'), (g', h))$, and the induction hypothesis once more gives $(g, h) \in \varphi(T)$. □

Lemma 7.10 *Let $G \boxtimes H$ be connected and $\varphi : G \boxtimes H \to G' \boxtimes H'$ an isomorphism. If G is prime and not complete, then either all images of G-towers over cliques of H lie in G'-towers over cliques of H' or all lie in H'-towers over cliques of G'.*

Proof Let T be a G-tower over a clique K_k of H. We first show that $\varphi(T)$ is contained in a G'-tower over a clique of H' or in an H'-tower over a clique of G'.

By Lemma 7.9, we have $\varphi(T) = A \boxtimes B$, where $A \subseteq G'$ and $B \subseteq H'$. Moreover, by Lemma 7.8, we can uniquely represent A and B in the form

$$A \cong K_n \boxtimes A' \quad \text{and} \quad B \cong K_m \boxtimes B',$$

where A' and B' are not divisible by a K_p for any $p > 1$. Hence

$$A \boxtimes B \cong K_k \boxtimes G \cong (K_n \boxtimes K_m) \boxtimes (A' \boxtimes B').$$

By Lemma 7.5, the greatest common divisor of the cardinalities of the vertices of A'/S and of the cardinalities of the vertices of B'/S is 1. By Lemma 7.2, the same holds for the cardinalities of the vertices of $(A' \boxtimes B')/S$. Invoking Lemma 7.5 again, we see that $A' \boxtimes B'$ is not divisible by a K_p for any $p > 1$. But then Lemma 7.8 implies that $G \cong A' \boxtimes B'$. Because G is prime, A' or B' must be the one vertex graph, and hence A or B complete.

It remains to show that G-towers are either all mapped into G'-towers or all mapped into H'-towers. Assume that this is not the case. Then, because $G \boxtimes H$ is connected, there are G-towers T_1 over Q_1 and T_2 over Q_2 such that $\varphi(T_1)$ lies in an H'-tower over Q'_1, $\varphi(T_2)$ in a G'-tower over Q'_2, and $\varphi(T_1) \cap \varphi(T_2) \neq \emptyset$. Note that $\varphi(T_1) \cap \varphi(T_2)$ is contained in $Q'_1 \boxtimes Q'_2$ and thus complete. However, the preimage of $\varphi(T_1) \cap \varphi(T_2)$ is $T_1 \cap T_2 = G \boxtimes (Q_1 \cap Q_2)$. But G is not complete by assumption, so $T_1 \cap T_2$ cannot be complete either, a contradiction. □

Lemma 7.11 *Under the assumptions of Lemma 7.10, suppose that the images of G-towers over cliques of H lie in G'-towers over cliques of H'. Then G divides G'.*

Proof Let T be a G'-tower over a clique Q of H'. Then the preimages of the cliques of T lie in G-towers $T_1, T_2, \dots, T_s$ over cliques $Q_1, Q_2, \dots, Q_s$ of H. By Lemma 7.10,

$$T = \varphi(T_1) \cup \varphi(T_2) \cup \dots \cup \varphi(T_s).$$

Let $R = Q_1 \cup Q_2 \cup \dots \cup Q_s$. Then

$$G' \boxtimes Q = T \cong G \boxtimes R.$$

As G is prime, Lemma 7.5 implies that $R \cong Q \boxtimes G''$ for some G'', so $G' \boxtimes Q \cong G \boxtimes (Q \boxtimes G'') \cong (G \boxtimes G'') \boxtimes Q$. By Lemma 7.7, we can cancel Q, which completes the proof. □

Lemma 7.12 *Suppose φ is an isomorphism from the connected graph $G \boxtimes H$ to $G' = G_1 \boxtimes G_2 \boxtimes \cdots \boxtimes G_k$. Suppose also that $G_1, G_2, \ldots, G_k$ are prime and that G is prime but not complete. Then for some G_i, the images of G-towers over cliques of H are G_i-towers over cliques of the product $G_1 \boxtimes \cdots \boxtimes G_{i-1} \boxtimes G_{i+1} \boxtimes \cdots \boxtimes G_k$.*

Proof Let T be a G-tower over a clique Q of H, and suppose that Q_1 and Q_2 are cliques of T with nonempty intersection. Then $\varphi(Q_1)$ and $\varphi(Q_2)$ are cliques of G'. By Lemma 7.3, their projections to the factors G_i are cliques. Clearly, as $Q_1 \neq Q_2$, there exists an index i with $p_{G_i}(\varphi(Q_1)) \neq p_{G_i}(\varphi(Q_2))$. Set $H' = G_1 \boxtimes \cdots \boxtimes G_{i-1} \boxtimes G_{i+1} \boxtimes \cdots \boxtimes G_k$. Then $G' = G_i \boxtimes H'$.

We claim that $p_{G_j}(\varphi(Q_1)) = p_{G_j}(\varphi(Q_2))$ for any $j \neq i$. Suppose that this is not the case. Then $p_{H'}(\varphi(Q_1)) \neq p_{H'}(\varphi(Q_2))$, and because $G' = H' \boxtimes G_i$, the cliques $\varphi(Q_1)$ and $\varphi(Q_2)$ belong to different G_i-towers and different H'-towers, in contradiction to Lemma 7.10. This proves the claim and that $\varphi(T)$ lies in a G_i-tower. Moreover, because G_i is prime, $\varphi(T)$ is a G_i-tower over a clique of H' by Lemma 7.11. An application of Lemma 7.10 completes the proof. □

Lemma 7.13 *Under the assumptions of Lemma 7.12, there is an i such that $G \cong G_i$ and $H \cong G_1 \boxtimes \cdots \boxtimes G_{i-1} \boxtimes G_{i+1} \boxtimes \cdots \boxtimes G_k$. In other words, if $G \boxtimes H \cong G \boxtimes H'$, and G is prime but not complete, then $H \cong H'$.*

Proof By Lemma 7.12, there exists a G_i such that the images of G-towers over cliques of H are G_i-towers over cliques of

$$H' = G_1 \boxtimes \cdots \boxtimes G_{i-1} \boxtimes G_{i+1} \boxtimes \cdots \boxtimes G_k.$$

By Lemma 7.11, the graph G divides G_i and as G is prime, $G \cong G_i$.

Let B_j be a vertex of H/S. Let $\mathcal{Q}_H$ denote the set of cliques in H. Then $B_j = \bigcap_{Q \in \mathcal{Q}_H} Q_v$ for any $v \in B_j$, and it is not difficult to see that the vertices of H'/S are

$$B'_j = p_{H'}\left(\bigcap_{Q \in \mathcal{Q}_H} \varphi(Q_v \times V(G))\right).$$

Moreover, the assignment $B_j \mapsto B'_j$ satisfies conditions (i) and (ii) of Proposition 7.6, so $H \cong H'$. □

Theorem 7.14 *Every connected graph has unique prime factor decomposition over the strong product.*

Proof Suppose a graph G has prime factor decompositions $G_1 \boxtimes \cdots \boxtimes G_k$ and $G'_1 \boxtimes \cdots \boxtimes G'_\ell$. We may assume that $G_{r+1}, \ldots, G_k$ and $G'_{s+1}, \ldots, G'_\ell$ are complete, and the other factors are not complete. Hence $G_1 \boxtimes \cdots \boxtimes G_r$ and $G'_1 \boxtimes \cdots \boxtimes G'_s$ are not divisible be a nontrivial complete graph, and thus, by Lemma 7.8,

$$G_1 \boxtimes \cdots \boxtimes G_r \cong G'_1 \boxtimes \cdots \boxtimes G'_s$$

and

$$G_{r+1} \boxtimes \cdots \boxtimes G_k \cong G'_{s+1} \boxtimes \cdots \boxtimes G'_\ell.$$

As prime factorization is unique for complete graphs, the graphs $G_{r+1}, \ldots, G_k$ coincide with the graphs $G'_{s+1}, \ldots, G'_\ell$. Finally, Lemma 7.13 implies that the $G_1, \ldots, G_r$ coincide with the graphs $G'_1, \ldots, G'_s$. □

This proof of the unique prime factor decomposition of connected graphs with respect to the strong product is modeled after the proof presented by Dörfler and Imrich (1970). At about the same time, McKenzie (1971) published refinement theorems for product representations of infinite relational structures.[1] McKenzie's theorems imply the results of this chapter, as well as many other results.

7.4 Automorphisms

Every automorphism of a graph G induces an automorphism of G/S. These automorphisms form a subgroup, say $\mathrm{Aut}(G)/S$, of $\mathrm{Aut}(G/S)$. Clearly, we have

$$|B| = |\varphi(B)|$$

for every $\varphi \in \mathrm{Aut}(G)/S$ and every $B \in G/S$.

Conversely, every automorphism $\varphi \in \mathrm{Aut}(G/S)$ with this property is in $\mathrm{Aut}(G)/S$. Evidently the elements of $\mathrm{Aut}(G)/S$ are exactly those automorphisms of G/S that preserve the cardinalities of the equivalence classes of S in G.

Because of this close relationship between $\mathrm{Aut}(G)$ and $\mathrm{Aut}(G/S)$, we will mainly be concerned with the description of $\mathrm{Aut}(G/S)$ in the sequel; that is to say, we will mainly be concerned with groups of S-thin structures.

By Corollary 7.2, a graph is S-thin if and only if all its factors are S-thin. We show that the relationship between the automorphism group of a connected S-thin graph and the groups of its prime factors with respect to the strong product is the same as that in the case of the Cartesian product.

To see this, it clearly suffices to show that automorphisms of strong products of S-thin prime graphs preserve layers.

Theorem 7.15 *Let $G = G_1 \boxtimes G_2 \boxtimes \cdots \boxtimes G_k$ be the product of connected, S-thin prime graphs and φ an automorphism of G. Then there exists a G_i to every G_j, $1 \leq i,j \leq k$, such that φ maps every G_j-layer into a G_i-layer.*

Proof By Lemma 7.12, there exists a G_i such that φ maps every G_j-tower over cliques of $H = G_1 \boxtimes \cdots G_{j-1} \boxtimes G_{j+1} \boxtimes \cdots \boxtimes G_k$ into G_i-towers over cliques of $H' = G_1 \boxtimes \cdots G_{i-1} \boxtimes G_{i+1} \boxtimes \cdots \boxtimes G_k$.

By Lemma 7.4, every vertex $v \in V(H)$ is the unique element of the intersection $\bigcap_{Q \in \mathcal{Q}} Q_v$. Let G_j^a be the G_j-layer with $p_H(a) = v$. Then G_j^a is the G_j-tower over v and can be represented in the form $\bigcap_{Q \in \mathcal{Q}} Q_v \boxtimes G_j$. Its image under φ is the intersection of G_i-towers over cliques or complements of cliques in H'; that is, it is a G_i-tower over a subset of H'. This subset must be a one-element set because G_i is prime; it must be a G_i-layer. □

We are now in a position to prove the analogue to Theorem 6.10.

Theorem 7.16 *Suppose φ is an automorphism of a connected S-thin graph G with prime*

[1] See Section 31.6.

factor decomposition $G = G_1 \boxtimes G_2 \boxtimes \cdots \boxtimes G_k$. *Then there exists a permutation* π *of* $\{1, 2, \ldots, k\}$ *and isomorphisms* $\varphi_i : G_{\pi(i)} \to G_i$ *for which*

$$\varphi(x_1, x_2, \ldots, x_k) = \big(\varphi_1(x_{\pi(1)}), \varphi_2(x_{\pi(2)}), \ldots, \varphi_k(x_{\pi(k)})\big).$$

Proof By Theorem 7.15, there is a G_i to every G_j, $1 \leq i, j \leq k$, such that φ maps every G_j-layer into a G_i-layer. Set $j = \pi(i)$. Clearly, π is a permutation of the index set $\{1, 2, \ldots, k\}$.

Because every Cartesian edge is in some G_i-layer, the automorphism φ maps the set of Cartesian edges of G onto itself. In other words, φ maps

$$\mathop{\square}_{i=1}^{k} G_i \subseteq \mathop{\boxtimes}_{i=1}^{k} G_i$$

into itself. Because it preserves adjacency and because every G_i-layer is induced, φ induces an automorphism of $\square_{i=1}^{k} G_i$ that preserves the layer structure. Hence, we can complete the proof with the same arguments that were used in the proof of Theorem 6.10. □

As in the case of the Cartesian product, this immediately implies that all automorphisms of a strong product of connected, S-thin prime graphs are generated by automorphisms and transpositions of the prime factors. Clearly, this is not the case for a nontrivial strong product of graphs if at least one factor is not S-thin.

Corollary 7.17 *The automorphism group* Aut(G) *of a connected graph* G *with nontrivial prime factor decomposition* $G_1 \boxtimes G_2 \boxtimes \cdots \boxtimes G_k$ *is generated by automorphisms and transpositions of the prime factors if and only if* G *is* S*-thin.*

As in the Cartesian case, the automorphism group of a strong product of connected, S-thin prime graphs is the automorphism group of the disjoint union of its prime factors. The proof is identical to the proof of Theorem 6.13.

Theorem 7.18 *The automorphism group of the strong product of connected,* S*-thin prime graphs is isomorphic to the automorphism group of the disjoint union of the factors.*

We continue with several other results about the automorphism groups of strong products that are similar to the ones in the Cartesian case. First, consider Theorem 6.17, which states that a Cartesian product has transitive group if and only if every factor has transitive group. It immediately carries over to S-thin strong products. Because a graph G has transitive group if and only if all equivalence classes of S_G have the same cardinality and if G/S has transitive group, we even have the following result:

Theorem 7.19 *A strong product has transitive automorphism group if and only if every factor has transitive automorphism group.*

Similar to the case of the Cartesian product, we also infer the following corollary:

Corollary 7.20 *The automorphism group of a connected graph* G *with prime factor decomposition* $G_1 \boxtimes G_2 \boxtimes \cdots \boxtimes G_k$ *is regular if and only if the prime factors are pairwise nonisomorphic and have regular automorphism groups.*

Again one can show that this corollary remains true if the term "regular" is replaced by "fixed point-free."

For graphs with Abelian group, we have the following analogue to Corollary 6.20:

Corollary 7.21 *The automorphism group of a connected graph G with prime factor decomposition $G_1 \boxtimes G_2 \boxtimes \cdots \boxtimes G_k$ is Abelian if and only if the following conditions are satisfied:*

(i) *Every prime factor has Abelian automorphism group.*
(ii) *If G is S-thin, then the prime factors with nontrivial groups are pairwise nonisomorphic, and there are no three pairwise isomorphic factors with trivial group.*
(iii) *If G is not S-thin, then G has exactly one prime factor that is not S-thin. All the other factors are pairwise nonisomorphic and have trivial groups.*

The proofs are straightforward and left to the reader.

Exercises

7.1. Prove Proposition 7.1.

7.2. Show that P_k is S-thin for $k \neq 2$.

7.3. Give an example of a graph G for which $\mathrm{Aut}(G/S)$ is different from $(\mathrm{Aut}(G))/S$.

7.4. Determine the number of automorphisms of $P_n \boxtimes P_m$, and compare it to the number of automorphisms of $P_n \,\Box\, P_m$.

7.5. Show that every graph with transitive group has unique prime factor decomposition with respect to the strong product.

7.6. Show that every connected graph G that is decomposable with respect to the strong product is prime with respect to the Cartesian one (and vice versa).

7.7. Show that G can be prime with respect to the strong product even if G/S is composite.

7.8. Prove Corollary 7.21.

Chapter 8

Direct Product

We defined the direct product of graphs in Chapter 4 and deduced some of its elementary properties in Section 5.3. Now we explore the problem of prime factorization over this product. The chapter culminates with a proof that connected nonbipartite graphs in Γ_0 factor uniquely into primes in Γ_0. However, it is necessary to develop some preliminary ideas. We first define a relation R that is analogous to the relation S from Chapter 7. We then define a so-called *Cartesian skeleton operation* $S : \Gamma_0 \to \Gamma$ that satisfies $S(G \times H) = S(G) \,\Box\, S(H)$ for R-thin graphs. This allows us to exploit unique factorization over $\Box$ to prove unique factorization of connected nonbipartite thin graphs over $\times$. Subsequently we extend this result to graphs that are not R-thin. Prime factorization leads naturally to a classification of the automorphisms of direct products.

We close the chapter by showing how these results yield alternative proofs of prime factorization over the strong product.

8.1 Nonuniqueness of Prime Factorization

Recall that the loop K_1^s is a unit for the direct product in the sense that $K_1^s \times G \cong G$ for every graph $G \in \Gamma_0$. A graph G is *prime* with respect to the direct product if it has more than one vertex and $G \cong G_1 \times G_2$ implies that either G_1 or G_2 equals K_1^s.

An expression $G \cong G_1 \times G_2 \times \cdots \times G_k$, with each G_i prime, is called a *prime factorization* of G. By the same argument used for the Cartesian product, we see that every nontrivial graph has a prime factorization over the direct product. We will prove that this factorization is unique for connected nonbipartite graphs. But to set the stage—and also to convey the complexity of the issue—we first examine how uniqueness may fail.

Theorem 8.1 *Prime factorization with respect to the direct product is not unique in*

(i) *The class of graphs with loops at each vertex;*
(ii) *The class of connected graphs in Γ;*
(iii) *The class of connected graphs in Γ_0.*

Proof For the first assertion, let K_p^s denote the graph obtained from K_p by adding a loop at each vertex. Recall that $G^{\times,n}$ denotes the nth direct power of G, and consider the equation

$$\left(K_1^s + K_2^s + (K_2^s)^{\times,2}\right) \times \left(K_1^s + (K_2^s)^{\times,3}\right) = \left(K_1^s + (K_2^s)^{\times,2} + (K_2^s)^{\times,4}\right) \times \left(K_1^s + K_2^s\right).$$

Equality holds because both sides equal $K_1^s + K_2^s + (K_2^s)^{\times,2} + (K_2^s)^{\times,3} + (K_2^s)^{\times,4} + (K_2^s)^{\times,5}$. The factors involved are prime, by Exercise 8.3.

For (ii), the following equation shows that connected graphs may have nonunique factorizations in Γ:

Here the graph $N \times \Delta$ on the left further factors into three terms in Γ_0. Applying the associative property and re-multiplying produces the graph $K_2 \times A$ on the right. Now, N is prime in Γ, as it could only factor nontrivially as a product of two graphs on two vertices, yet $N \neq K_2 \times K_2$. Turning to the other factors, Δ and K_2 are clearly prime. The graph A on the right is prime in Γ because it has six vertices, so it could only factor as $A = K_2 \times G$ for some graph G with three vertices. But then A would be bipartite, a contradiction. Thus $N \times \Delta$ and $K_2 \times A$ are two different prime factorizations (in Γ) of the same graph.

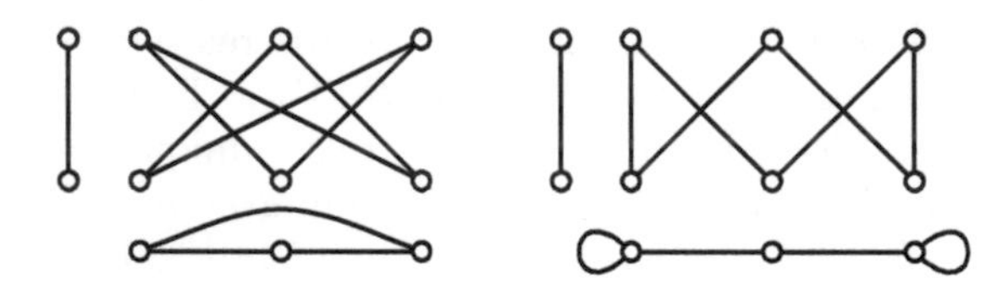

FIGURE 8.1 Two prime factorizations of C_6 in Γ_0.

For (iii), Figure 8.1 shows two different prime factorizations of C_6 in Γ_0. Further, Figure 8.2 shows a graph with two prime factorizations with different numbers of factors. $\square$

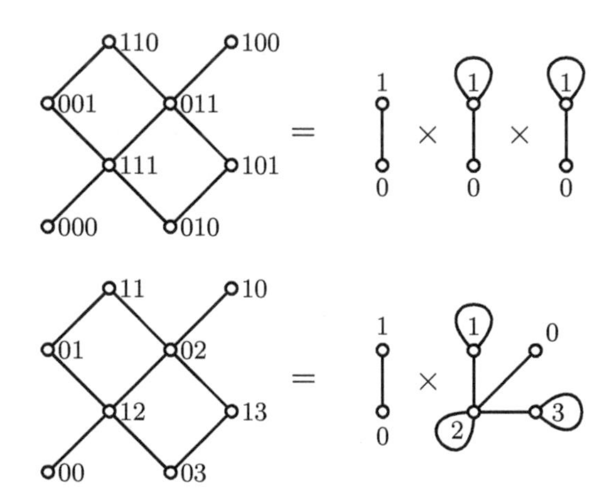

FIGURE 8.2 Two prime factorizations with a different number of prime factors.

8.2 R-Thin Graphs

Having noted the failure of prime factorization over the direct product, we now prepare for the proof that connected nonbipartite graphs in Γ_0 decompose uniquely into prime factors in Γ_0. One issue that arises here and elsewhere is the notion of so-called R-thinness.

To motivate this topic, recall that the layers of the prime factorization of a connected graph over the Cartesian product are uniquely determined. This is not so with the direct product, as is illustrated in Figure 8.3. The figure shows a graph G that has different coordinatizations with respect to a prime factorization. In one case, the sets $\{1, a, 2\}$ and $\{3, b, 4\}$ are layers, and in the other case, sets $\{1, b, 2\}$ and $\{3, a, 4\}$ are layers.

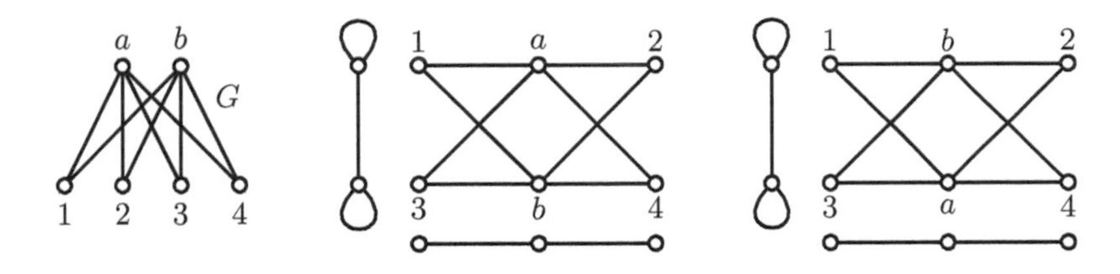

FIGURE 8.3 Layers of a prime factorization of a graph G are not unique.

A closer look at the figure reveals the source of the problem. Vertices a and b have the same neighborhood $\{1, 2, 3, 4\}$, so their positions in the factorization can be interchanged.

Evidently, then, the existence of vertices with identical neighborhoods complicates the discussion of prime factorizations over the direct product. To overcome this difficulty, we introduce a relation R on the vertices of a graph. Two vertices x and x' of a graph G are in *relation* R, written xRx', precisely if $N_G(x) = N_G(x')$. (For clarity, we may occasionally write R_G for R.) It is a simple matter (Exercise 8.4) to show that R is an equivalence relation on $V(G)$. (For example, the equivalence classes for G in Figure 8.3 are $\{a, b\}$ and $\{1, 2, 3, 4\}$.) We easily see that the subgraph induced on an R-equivalence class is either totally disconnected or is a complete graph with loops at each vertex.

We denote the quotient (in Γ_0) of G by its R-equivalence classes as G/R. For example, $G/R = K_2$ for the graph G in Figure 8.3. Also, if G is a complete graph with loops at each vertex, then $G/R = K_1^s$, while $K_n/R = K_n$ and $D_n/R = K_1$.

A graph is called R-*thin* if all of its R-equivalence classes contain just one vertex. In this case, $G/R \cong G$. The reader should check that G/R is R-thin for any $G \in \Gamma_0$ (Exercise 8.5).

Given $x \in V(G)$, let $[x] = \{x' \in V(G) \mid N_G(x') = N_G(x)\}$ denote the R-equivalence class containing x. Because the relation R is defined entirely in terms of adjacencies, it is clear that given an isomorphism $\varphi : G \to H$, we have xR_Gy if and only if $\varphi(x)R_H\varphi(y)$. Thus φ maps equivalence classes of R_G to equivalence classes of R_H, and, in particular, $\varphi([x]) = [\varphi(x)]$.

We now collect a number of relevant results concerning quotients G/R.

Lemma 8.2 *Suppose $G \in \Gamma_0$. Then $xy \in E(G)$ if and only if $[x][y] \in E(G/R)$.*

Proof If $xy \in E(G)$, then $[x][y] \in E(G/R)$ by the definition of G/R. Conversely, let $[x][y] \in E(G/R)$, so there is an edge $x'y' \in E(G)$ with $x' \in [x]$ and $y' \in [y]$. Thus $N(x') = N(x)$ and $N(y') = N(y)$. Because $x'y' \in E(G)$, we have $x' \in N(y') = N(y)$, hence $y \in N(x') = N(x)$, so $xy \in E(G)$. □

Proposition 8.3 *For any isomorphism $\varphi : G \to H$, there is a corresponding isomorphism $\widetilde{\varphi} : G/R \to H/R$, defined as $\widetilde{\varphi}([x]) = [\varphi(x)]$. Moreover, $\widetilde{\varphi}^{-1} = \widetilde{\varphi^{-1}}$.*

Proof As the isomorphism $\varphi : G \to H$ maps R-classes onto R-classes, the map $\widetilde{\varphi}$ is well-defined and bijective. Using Lemma 8.2, we have $[x][y] \in E(G/R)$ if and only if $xy \in E(G)$, if and only if $\varphi(x)\varphi(y) \in E(H)$, if and only if $[\varphi(x)][\varphi(y)] \in E(H/R)$, if and only if $\widetilde{\varphi}([x])\widetilde{\varphi}([y]) \in E(H/R)$.

To see that $\widetilde{\varphi}^{-1} = \widetilde{\varphi^{-1}}$, observe $\widetilde{\varphi^{-1}}\widetilde{\varphi}([x]) = \widetilde{\varphi^{-1}}([\varphi(x)]) = [\varphi^{-1}\varphi(x)] = [x]$, and likewise $\widetilde{\varphi}\widetilde{\varphi^{-1}}([x]) = [x]$. □

In general, the converse of Proposition 8.3 is false: if $G/R \cong H/R$, then it is not necessarily the case that $G \cong H$. (For example, $P_2/R \cong P_3/R$.) However, by adding an additional constraint, we do get the following characterization.

Proposition 8.4 *Suppose $G, H \in \Gamma_0$. Then $G \cong H$ if and only if $G/R \cong H/R$ and there is an isomorphism $\widetilde{\varphi} : G/R \to H/R$ with $|X| = |\widetilde{\varphi}(X)|$ for each $X \in V(G/R)$. In fact, given an isomorphism $\widetilde{\varphi} : G/R \to H/R$, any map $\varphi : V(G) \to V(H)$ that restricts to a bijection $\varphi : X \to \widetilde{\varphi}(X)$ for every $X \in G/R$ is an isomorphism from G to H.*

Proof Given an isomorphism $\varphi : G \to H$, let $\widetilde{\varphi} : G/R \to H/R$ be the isomorphism from Proposition 8.3. Then for each $[x] \in V(G/R)$, we have $|[x]| = |[\varphi(x)]| = |\widetilde{\varphi}([x])|$.

Conversely, suppose there is an isomorphism $\widetilde{\varphi} : G/R \to H/R$ with $|X| = |\widetilde{\varphi}(X)|$ for each $X \in V(G/R)$. Let $\varphi : G \to H$ be any map that carries each X bijectively onto $\widetilde{\varphi}(X)$, so $\varphi([x]) = \widetilde{\varphi}([x])$. Then φ is bijective, and it is an isomorphism: $xy \in E(G)$ if and only if $[x][y] \in E(G/R)$, if and only if $\widetilde{\varphi}([x])\widetilde{\varphi}([y]) \in E(H/R)$, if and only if $\varphi([x])\varphi([y]) \in E(H/R)$, if and only if $\varphi(x)\varphi(y) \in E(H)$. (Lemma 8.2 was used in the first and last equivalence.) □

The next proof (and many that follow) uses the fact $N_{G\times H}((x,y)) = N_G(x) \times N_H(y)$, which was proved in Exercise 4.5.

Proposition 8.5 *If graphs G and H in Γ_0 have no isolated vertices, then $V\big((G\times H)/R\big) = \{X \times Y \mid X \in V(G/R), Y \in V(H/R)\}$. In particular, $[(x,y)] = [x] \times [y]$. Furthermore, $(G \times H)/R \cong G/R \times H/R$, and $[(x,y)] \mapsto ([x],[y])$ is an isomorphism.*

Proof Consider an arbitrary vertex $[(x,y)]$ of $(G \times H)/R$, and note the following:

$$\begin{aligned}
(x',y') \in [(x,y)] &\iff N_{G\times H}((x',y')) = N_{G\times H}((x,y)) \\
&\iff N_G(x') \times N_H(y') = N_G(x) \times N_H(y) \\
&\iff N_G(x') = N_G(x) \text{ and } N_H(y') = N_H(y) \\
&\iff x' \in [x] \text{ and } y' \in [y] \\
&\iff (x',y') \in [x] \times [y]\,.
\end{aligned}$$

(The third equivalence uses the fact that there are no isolated vertices.) Thus $[(x,y)] = [x] \times [y]$. To finish the proof, we show that$[(x,y)] \mapsto ([x],[y])$ is an isomorphism. Using Lemma 8.2,

$$\begin{aligned}
[(x,y)][(x',y')] \in E\big((G\times H)/R\big) &\iff (x,y)(x',y') \in E(G \times H) \\
&\iff xx' \in E(G) \text{ and } yy' \in E(H) \\
&\iff [x][x'] \in E(G/R) \text{ and } [y][y'] \in E(H/R) \\
&\iff ([x],[y])([x'],[y']) \in E(G/R \times H/R).
\end{aligned}$$

The proof is now complete. □

We make two important remarks. First, the equation $(G \times H)/R \cong G/R \times H/R$ from Proposition 8.5 implies that a direct product is R-thin if and only if each factor is R-thin.

Second, the three above propositions, taken together, imply that any isomorphism $\varphi : G_1 \times \cdots \times G_k \to H_1 \times \cdots \times H_\ell$ of form

$$\varphi(x_1, \ldots, x_k) = \big(\varphi_1(x_1, \ldots, x_k), \ldots, \varphi_\ell(x_1, \ldots, x_k)\big)$$

induces an isomorphism $\widetilde{\varphi} : G_1/R \times \cdots \times G_k/R \longrightarrow H_1/R \times \cdots \times H_\ell/R$ of form

$$\begin{aligned} \widetilde{\varphi}([x_1], \ldots, [x_k]) &= (\widetilde{\varphi}_1([x_1], \ldots, [x_k])], \ldots, \widetilde{\varphi}_\ell([x_1], \ldots [x_k])) \\ &= (\,[\varphi_1(x_1, \ldots, x_k)]\,,\ \ldots\ ,\ [\varphi_\ell(x_1, \ldots x_k)]\,), \end{aligned}$$

having the property that φ maps the R-class $[x_1] \times \cdots \times [x_k]$ of $G_1 \times \cdots \times G_k$ bijectively to the R-class $\widetilde{\varphi}_1([x_1], \ldots, [x_k])] \times \cdots \times \widetilde{\varphi}_\ell([x_1], \ldots, [x_k])$ of $H_1 \times \cdots \times H_\ell$.

Proposition 8.5 shows that if G factors as $G = A \times B$, then G/R factors as $G/R = A/R \times B/R$. But conversely, a factorization $G/R = A \times B$ does not necessarily correspond to a factorization of G. However, the following result describes conditions under which $G/R = A \times B$ yields a factorization $G = A' \times B'$ with $A \cong A'/R$ and $B \cong B'/R$.

Proposition 8.6 *Suppose $G \in \Gamma_0$ has no isolated vertices, and $G/R \cong A \times B$. Then each vertex $(x, y) \in V(A \times B)$ labels a unique R-class of G, and we denote its cardinality as $|(x, y)|$. If there are functions $\alpha : V(A) \to \mathbb{N}$ and $\beta : V(B) \to \mathbb{N}$ with $|(x, y)| = \alpha(x) \cdot \beta(y)$, then there are graphs A' and B' for which $G \cong A' \times B'$. Further, there are isomorphisms $\widehat{\alpha} : A \to A'/R$ and $\widehat{\beta} : B \to B'/R$ with $|\widehat{\alpha}(x)| = \alpha(x)$ and $|\widehat{\beta}(y)| = \beta(y)$.*

Proof Suppose $G/R \cong A \times B$, and let α and β be as stated. Notice that A and B are R-thin, by Proposition 8.5.

Define A' as follows. Take a family $\{U_x \mid x \in V(A)\}$ of disjoint sets such that $|U_x| = \alpha(x)$ for each $x \in V(A)$. Put $V(A') = \bigcup_{x \in V(A)} U_x$, and say that a vertex in U_x is adjacent to a vertex in U_y if and only if $xy \in E(A)$. (Thus U_x induces a $K^s_{\alpha(x)}$ in A' if xx is a loop in A; otherwise U_x induces a $D_{\alpha(x)}$.) As A is R-thin, it is easy to verify that the R-classes of A' are precisely the sets U_x, and $\widehat{\alpha} : x \mapsto U_x$ is an isomorphism $A \to A'/R$ with $|\widehat{\alpha}(x)| = \alpha(x)$. Construct B' and $\widehat{\beta}$ similarly, that is, $V(B') = \bigcup_{y \in V(B)} V_y$ and $\widehat{\beta} : y \mapsto V_y$ is an isomorphism $B \to B'/R$.

Consider the following composition of isomorphisms, the second given by Proposition 8.5:

$$\begin{array}{ccccc} G/R = A \times B & \longrightarrow & A'/R \times B'/R & \longrightarrow & (A' \times B')/R, \\ (x, y) & \longmapsto & (U_x, V_y) & \longmapsto & U_x \times V_y\,. \end{array}$$

This is an isomorphism $G/R \to (A' \times B')/R$, where each $(x, y) \in G/R$ maps to the R-class $U_x \times U_y$, and $|(x, y)| = \alpha(x) \cdot \beta(y) = |U_x \times V_y|$. Then $G \cong A' \times B'$, by Proposition 8.4. □

For the next result, recall that K^s_p denotes the graph K_p with loops added to each vertex.

Corollary 8.7 *A graph $G \in \Gamma_0$ factors as $G = A \times K^s_p$ if and only if p divides the order of each R-class of G.*

Proof If $G = A \times K^s_p$, then every R-class of G has form $X \times V(K^s_p)$, where X is an R-class of A, by Proposition 8.5. Therefore p divides the order of each R-class of G.

Suppose p divides the order of each R-class of G. Put $V(K^s_1) = \{1\}$. Observe that $G/R = G/R \times K^s_1$, where $X \in G/R$ is identified with the pair $(X, 1)$. Define $\alpha : V(G/R) \to \mathbb{N}$ as $\alpha(X) = |X|/p$, and $\beta : V(K^s_1) \to \mathbb{N}$ as $\beta(1) = p$. Then $|(X, 1)| = \alpha(X) \cdot \beta(1)$, so Proposition 8.6 yields $G = A' \times B'$. The isomorphism $\widehat{\beta} : K^s_1 \to B'/R$ with $|\widehat{\beta}(1)| = p$ gives $B' = K^s_p$. □

8.3 The Cartesian Skeleton

This section defines the Cartesian skeleton $S(G)$ of an arbitrary graph G in Γ_0 as a certain graph having the same vertex set as G. Following the reasoning of Hammack and Imrich (2009), we prove that $S(G)$ is connected provided G is connected and nonbipartite, and show that $S(G \times H) = S(G) \,\Box\, S(H)$ for R-thin graphs. Ultimately this will allow us to transfer the unique prime factorization of connected graphs in $(\Gamma, \Box)$ to connected nonbipartite graphs in $(\Gamma_0, \times)$. Our point of departure is the Boolean square.

The *Boolean square* of a graph G is the graph G^s with $V(G^s) = V(G)$ and $E(G^s) = \{xy \mid N_G(x) \cap N_G(y) \neq \emptyset\}$. Thus, xy is an edge of G^s if and only if G has an x, y-walk of length two. For example, if $G = K_p$, then $G^s = K_p^s$ is K_p with a loop added to each vertex. We note in passing that the adjacency matrix of G^s is the Boolean second power of the adjacency matrix of G; that is, if G has adjacency matrix A, then the matrix of G^s is obtained from A^2 by replacing each nonzero entry by 1.

Observe that if G has an x, y-walk W of even length, then G^s has an x, y-walk of length $|W|/2$ on alternate vertices of W. It follows that G^s is connected if G is connected and has an odd cycle. (The presence of an odd cycle guarantees an even walk between any two vertices of G.) On the other hand, if G is connected and bipartite, then G^s has exactly two components and their respective vertex sets are the two partite sets of G.

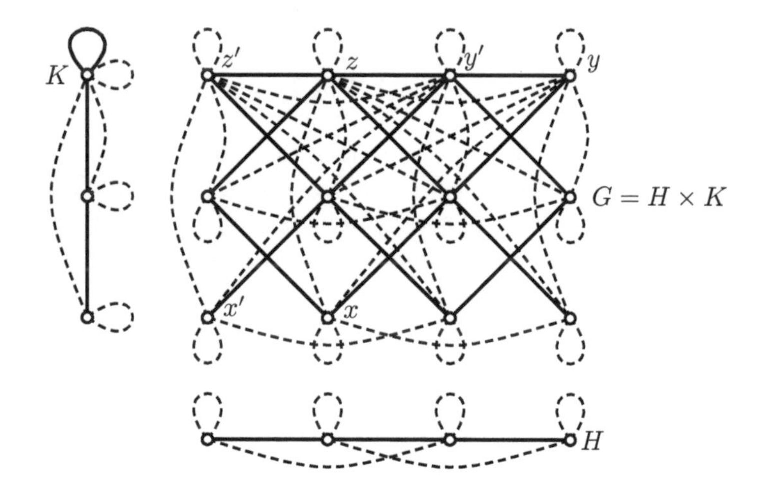

FIGURE 8.4 Graphs H, K, and $G = H \times K$ (solid) and their Boolean squares (dashed).

Figure 8.4 shows three graphs H, K, and $G = H \times K$ together with their Boolean squares H^s, K^s, and $(H \times K)^s$. Notice that $(H \times K)^s = H^s \times K^s$. This is in fact a general principle.

Lemma 8.8 *If $G_1, G_2, \ldots G_k$ are graphs, then $(G_1 \times G_2 \times \cdots \times G_k)^s = G_1^s \times G_2^s \times \cdots \times G_k^s$.*

Proof Observe $(x_1, \ldots, x_k)(y_1, \ldots, y_k) \in E\left((G_1 \times \cdots \times G_k)^s\right)$ if and only if there is a walk of length 2 joining $(x_1, \ldots, x_k)$ and $(y_1, \ldots, y_k)$, if and only if each G_i has a x_i, y_i-walk of length 2, if and only if $x_i y_i \in E(G_i^s)$ for each $1 \leq i \leq k$, if and only if $(x_1, \ldots, x_k)(y_1, \ldots, y_k)$ is an edge of $G_1^s \times \cdots \times G_k^s$. □

We now explain how to construct the Cartesian skeleton $S(G)$ of a graph G by removing strategic edges from G^s.

Given a factorization $G = H \times K$, we say that an edge $(h,k)(h',k')$ of the Boolean square G^s is *Cartesian* relative to the factorization if either $h = h'$ and $k \neq k'$, or $h \neq h'$ and $k = k'$. For example, in Figure 8.4, edges xz and zy of G^s are Cartesian, but edges xy and yy of G^s are not. We now identify two intrinsic criteria for non-loop edges of G^s that tell us if they may fail to be Cartesian relative to some factoring of G.

(i) In Figure 8.4 the edge xy of G^s is not Cartesian. For this edge, there is a $z \in V(G)$ with $N_G(x) \cap N_G(y) \subset N_G(x) \cap N_G(z)$ and $N_G(x) \cap N_G(y) \subset N_G(y) \cap N_G(z)$.

(ii) In Figure 8.4 the edge $x'y'$ of G^s is not Cartesian. For this edge, there is a $z' \in V(G)$ with $N_G(x') \subset N_G(z') \subset N_G(y')$.

Our aim is to remove from G^s all edges that meet one of these criteria. Now, these criteria are somewhat dependent on one another. For instance, $N_G(x) \subset N_G(z) \subset N_G(y)$ implies $N_G(y) \cap N_G(x) \subset N_G(y) \cap N_G(z)$. Also, $N_G(y) \subset N_G(z) \subset N_G(x)$ implies $N_G(x) \cap N_G(y) \subset N_G(x) \cap N_G(z)$. This allows us to pack the above criteria into the following definition.

Definition 8.1 *An edge xy of the Boolean square G^s is* dispensable *if it is a loop, or if there exists some $z \in V(G)$ for which both of the following statements hold:*

(1) $N_G(x) \cap N_G(y) \subset N_G(x) \cap N_G(z)$ *or* $N_G(x) \subset N_G(z) \subset N_G(y)$,

(2) $N_G(y) \cap N_G(x) \subset N_G(y) \cap N_G(z)$ *or* $N_G(y) \subset N_G(z) \subset N_G(x)$.

Note that the above statements (1) and (2) are symmetric in x and y. Also note the next remark, which follows from the paragraph preceding the definition. It will be used often.

Remark 8.1 *An edge xy is dispensable if and only if there is a z with $N(x) \subset N(z) \subset N(y)$, or $N(y) \subset N(z) \subset N(x)$, or $N(x) \cap N(y) \subset N(x) \cap N(z)$ and $N(y) \cap N(x) \subset N(y) \cap N(z)$.*

Now we come to the main definition of this section.

Definition 8.2 *The* Cartesian skeleton *$S(G)$ of a graph G is the spanning subgraph of the Boolean square G^s obtained by removing all dispensable edges from G^s.*

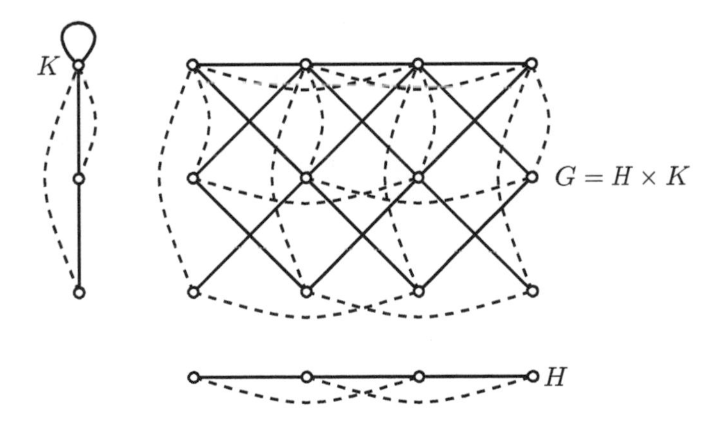

FIGURE 8.5 Graphs H, K, and $H \times K$ (solid) and their Cartesian skeletons (dashed).

Figure 8.5 is the same as Figure 8.4, except all dispensable edges of H^s, K^s, and $(H \times K)^s$ are deleted. Thus the remaining dashed edges are $S(H), S(K)$, and $S(H \times K)$. Note that although $S(G)$ was defined without regard to the factorization $G = H \times K$, we nonetheless have $S(H \times K) = S(H) \square S(K)$. The following lemma and proposition show this always holds for R-thin graphs. The proofs make frequent use of the fact that for $G = H \times K$,

$$N_G((h,k)) \cap N_G((h',k')) \;=\; (N_H(h) \cap N_H(h')) \times (N_K(k) \cap N_K(k')),$$

which follows from $N_G((h,k)) = N_H(h) \times N_K(k)$ and simple set theory.

Lemma 8.9 *If G is an R-thin graph with an arbitrary factorization $G = H \times K$, then every edge of $S(G)$ is Cartesian with respect to this factorization.*

Proof Suppose, for the sake of contradiction, that an edge $(h,k)(h',k') \in S(G)$ is not Cartesian. Because $S(G)$ has no loops, we have $h \neq h'$ and $k \neq k'$. Observe:

$$\begin{aligned} N_G((h,k)) \cap N_G((h',k')) &= (N_H(h) \cap N_H(h')) \times (N_K(k) \cap N_K(k')) \\ &\subseteq N_H(h) \times (N_H(k) \cap N_H(k')) \\ &= N_G((h,k)) \cap N_G((h,k')), \end{aligned}$$

$$\begin{aligned} N_G((h',k')) \cap N_G((h,k)) &= (N_H(h') \cap N_H(h)) \times (N_K(k') \cap N_K(k)) \\ &\subseteq (N_H(h') \cap N_H(h)) \times N_K(k') \\ &= N_G((h',k')) \cap N_G((h,k')). \end{aligned}$$

Now, because $(h,h')(k,k')$ is not dispensable, at least one of the above inclusions is equality and we get $N_H(h) \cap N_H(h') = N_H(h)$ in the first case or $N_K(k) \cap N_K(k') = N_K(k')$ in the second. From this, $N_H(h) \subseteq N_H(h')$ or $N_K(k') \subseteq N_K(k)$, and by R-thinness,

$$N_H(h) \subset N_H(h') \quad \text{or} \quad N_K(k') \subset N_K(k). \tag{8.1}$$

Repeating the above argument but interchanging h with h', and k with k', we get

$$N_H(h') \subset N_H(h) \quad \text{or} \quad N_K(k) \subset N_K(k'). \tag{8.2}$$

From inclusions (8.1) and (8.2), we see that $N_H(h) \subset N_H(h')$ and $N_K(k) \subset N_K(k')$, or $N_K(k') \subset N_K(k)$ and $N_H(h') \subset N_H(h)$. The first case gives

$$N_H(h) \times N_K(k) \subset N_H(h) \times N_K(k') \subset N_H(h') \times N_K(k'),$$

that is, $N_G((h,k)) \subset N_G((h,k')) \subset N_G((h',k'))$, so $(h,k)(h',k')$ is dispensable, a contradiction. Similarly, the second case yields $N_G((h',k')) \subset N_G((h,k')) \subset N_G((h,k))$. □

Proposition 8.10 *If H, K are R-thin graphs without isolated vertices, then $S(H \times K) = S(H) \square S(K)$.*

Proof First we show $S(H \times K) \subseteq S(H) \square S(K)$. By Lemma 8.9, all edges of $S(H \times K)$ are Cartesian, so we just need to show that $(h,k)(h',k) \in S(H \times K)$ implies $hh' \in S(H)$. (The same argument will work for edges of form $(h,k)(h,k')$.) Suppose for the sake of contradiction that $(h,k)(h',k) \in S(H \times K)$, but $hh' \notin S(H)$. Thus hh' is dispensable in H^s, so there is a z' in $V(H)$ for which both of the following conditions hold:

$$\begin{array}{lll} N_H(h) \cap N_H(h') \subset N_H(h) \cap N_H(z') & \text{or} & N_H(h) \subset N_H(z') \subset N_H(h') \\ N_H(h') \cap N_H(h) \subset N_H(h') \cap N_H(z') & \text{or} & N_H(h') \subset N_H(z') \subset N_H(h). \end{array} \tag{8.3}$$

Because $N_K(k) \neq \emptyset$ (there are no isolated vertices), we can multiply each neighborhood $N_H(u)$ in (8.3) by $N_K(k)$ on the right and still preserve the proper inclusions. Then the fact $N_H(u) \times N_K(k) = N_{H\times K}((u,k))$ yields the dispensability conditions (1) and (2), where $x = (h,k)$ and $y = (h',k)$ and $z = (z',k)$. Thus $(h,k)(h',k) \notin S(H \times K)$, a contradiction.

Now we show $S(H) \,\square\, S(K) \subseteq S(H \times K)$. Take an edge in $S(H) \,\square\, S(K)$, say $(h,k)(h',k)$ with $hh' \in S(H)$. We must show that $(h,k)(h',k)$ is not dispensable in $(H \times K)^s$. Suppose it was. Then there would be a vertex $z = (z', z'')$ in $H \times K$ such that the dispensability conditions (1) and (2) hold for $x = (h,k), y = (h',k)$, and $z = (z', z'')$. The various cases are considered below. Each leads to a contradiction.

Suppose $N_G(x) \subset N_G(z) \subset N_G(y)$. This means

$$N_H(h) \times N_K(k) \subset N_H(z') \times N_K(z'') \subset N_H(h') \times N_K(k),$$

so $N_K(z'') = N_K(k)$. Then the fact that $N_K(k) \neq \emptyset$ permits cancellation of the common factor $N_K(k)$, so $N_H(h) \subset N_H(z') \subset N_H(h')$, and hh' is dispensable. We reach the same contradiction if $N_G(y) \subset N_G(z) \subset N_G(x)$.

Finally, suppose there is a $z = (z', z'')$ for which both $N_G(x) \cap N_G(y) \subset N_G(x) \cap N_G(z)$ and $N_G(y) \cap N_G(x) \subset N_G(y) \cap N_G(z)$. Rewrite this as

$$\begin{array}{rcl} N_G((h,k)) \cap N_G((h',k)) & \subset & N_G((h,k)) \cap N_G((z',z'')) \\ N_G((h',k)) \cap N_G((h,k)) & \subset & N_G((h',k)) \cap N_G((z',z'')), \end{array}$$

which is the same as

$$\begin{array}{rcl} \big(N_H(h) \cap N_H(h')\big) \times N_K(k) & \subset & \big(N_H(h) \cap N_H(z')\big) \times \big(N_K(k) \cap N_K(z'')\big) \\ \big(N_H(h') \cap N_H(h)\big) \times N_K(k) & \subset & \big(N_H(h') \cap N_H(z')\big) \times \big(N_K(k) \cap N_K(z'')\big). \end{array}$$

Thus $N_K(k) \subseteq N_K(k) \cap N_K(z'')$, so $N_K(k) = N_K(k) \cap N_K(z'')$, whence

$$\begin{array}{rcl} N_H(h) \cap N_H(h') & \subset & N_H(h) \cap N_H(z') \\ N_H(h') \cap N_H(h) & \subset & N_H(h') \cap N_H(z'). \end{array}$$

Thus hh' is dispensable, a contradiction. □

Because $S(G)$ is defined entirely in terms of the adjacency structure of G, we have the following immediate consequence of Definition 8.2.

Proposition 8.11 *Any isomorphism $\varphi : G \to H$, as a map $V(G) \to V(H)$, is also an isomorphism $\varphi : S(G) \to S(H)$.*

We next consider connectivity of $S(G)$. The following lemma is needed.

Lemma 8.12 *Suppose G has no isolated vertices. If $x, y \in V(G)$ and $N(x) \subset N(y)$, then G^s has an x,y-path consisting of nondispensable edges.*

Proof Consider the following maximal chain of neighborhoods between $N(x)$ and $N(y)$, ordered by proper inclusion. (It is possible that $y_1 = y$.)

$$N(x) \subset N(y_1) \subset N(y_2) \subset N(y_3) \subset \cdots \subset N(y_k) \subset N(y). \tag{8.4}$$

We claim that xy_1 is a nondispensable edge of G^s. Certainly $N(x) \subset N(y_1)$ implies $xy_1 \in E(G^s)$, as $N(x) \neq \emptyset$. Also, there is no z with $N(x) \cap N(y_1) \subset N(x) \cap N(z)$, for if there were, the condition $N(x) \subset N(y_1)$ would yield $N(x) \subset N(x) \cap N(z)$, which is impossible. As the chain is maximal, there is no z for which $N(x) \subset N(z) \subset N(y_1)$. Clearly, $N(y_1) \subset N(z) \subset N(x)$ is impossible too. Thus xy_1 is a nondispensable edge of G^s.

The exact same argument shows that each y_iy_{i+1} is a nondispensable edge of G^s, as is y_ky. Thus we have the required path $xy_1y_2 \ldots y_ky$. □

Proposition 8.13 *Suppose a graph G is connected.*

(i) *If G has an odd cycle, then $S(G)$ is connected.*
(ii) *If G is bipartite, then $S(G)$ has two components whose respective vertex sets are the two partite sets of G.*

Proof Observe that the statement is true if $S(G)$ is replaced by G^s. As $S(G)$ is just G^s with the dispensable edges removed, we need only prove that for any edge $xy \in E(G^s)$, there is an x,y-path in G^s consisting of nondispensable edges.

For each edge $xy \in E(G^s)$, define the integer

$$k_{xy} = \max\{\, |N(u) \cap N(v)| - |N(x) \cap N(y)| \;\mid\; u, v \in V(G), u \neq v\}.$$

Notice $k_{xy} \geq 0$. (Put $u = x$ and $v = y$.) If $k_{xy} = 0$, then the definition of k_{xy} implies that there is no z for which $N(x) \cap N(y) \subset N(x) \cap N(z)$ or $N(y) \cap N(x) \subset N(y) \cap N(z)$. Then $N(x) \subset N(z) \subset N(y)$ is also impossible, as it implies $N(y) \cap N(x) \subset N(y) \cap N(z)$. Therefore xy is not dispensable if $k_{xy} = 0$.

Take $N > 0$, and assume that whenever G^s has an edge xy with $k_{xy} < N$, there is a x,y-path in G^s composed of nondispensable edges. Now suppose xy is dispensable and $k_{xy} = N$. If $N(x) \subset N(y)$ or $N(y) \subset N(x)$, then we are done, by Lemma 8.12, so assume $N(x) \not\subset N(y)$ and $N(y) \not\subset N(x)$. As xy is dispensable, there is a vertex z with

$$N(x) \cap N(y) \subset N(x) \cap N(z) \qquad \text{and} \qquad N(y) \cap N(x) \subset N(y) \cap N(z).$$

This implies $N(x) \cap N(z) \neq \emptyset \neq N(y) \cap N(z)$, so $xz, yz \in E(G^s)$. But it also means

$$|N(u) \cap N(v)| - |N(x) \cap N(z)| \;<\; |N(u) \cap N(v)| - |N(x) \cap N(y)|$$

for all u, v, so $k_{xz} < k_{xy}$. Similarly, $k_{zy} < k_{xy}$. The induction hypothesis gurantees there are x,z- and z,y-paths of nondispensable edges in G^s. □

8.4 Factoring Connected, Nonbipartite, R-Thin Graphs

In this section we prove that every connected, nonbipartite, R-thin graph has a unique prime factorization over the direct product. The result hinges on the following lemma, the proof of which uses the Cartesian skeleton to transfer questions of direct-product factorization to Cartesian-product factorization. Nonbipartiteness becomes a crucial property here, as it is necessary for connectedness of the skeleton.

Lemma 8.14 *Consider any isomorphism $\varphi : G_1 \times \cdots \times G_k \to H_1 \times \cdots \times H_\ell$, where $\varphi(x_1, \ldots, x_k) = \big(\varphi_1(x_1, \ldots, x_k), \varphi_2(x_1, \ldots, x_k), \ldots, \varphi_\ell(x_1, \ldots, x_k)\big)$, and all the factors are connected, non-bipartite, and R-thin. If a factor G_i is prime, then exactly one of the functions $\varphi_1, \varphi_2, \ldots, \varphi_\ell$ depends on x_i.*

Proof By commutativity and associativity, it suffices to prove the lemma for the case $k = \ell = 2$, and with G_1 prime. Thus take an isomorphism $\varphi : G_1 \times G_2 \to H_1 \times H_2$, where $\varphi(x_1, x_2) = \big(\varphi_1(x_1, x_2), \varphi_2(x_1, x_2)\big)$.

We will prove the lemma by showing that if it is not the case that exactly one of φ_1 and φ_2 depends on x_1, then G_1 is not prime.

Certainly if neither φ_1 nor φ_2 depends on x_1, then the fact that φ is bijective means that $|V(G_1)| = 1$, so G_1 is not prime. Thus assume that both φ_1 and φ_2 depend on x_1. This means each of G_1, H_1, and H_2 has more than one vertex. If G_2 had only one vertex, then $G_1 \cong H_1 \times H_2$, and G_1 would not be prime. Thus each factor G_1, G_2, H_1, and H_2 has more than one vertex. Taking skeletons, and applying Proposition 8.11, we see that φ is also an isomorphism $\varphi : S(G_1 \times G_1) \to S(H_1 \times H_2)$. Because all factors are R-thin, Proposition 8.10 applies, and we have an isomorphism

$$\varphi : S(G_1) \,\square\, S(G_2) \to S(H_1) \,\square\, S(H_2). \tag{8.5}$$

Keep in mind that φ is simultaneously an isomorphism $\varphi : G_1 \times G_2 \to H_1 \times H_2$ and an isomorphism $\varphi : S(G_1) \,\square\, S(G_2) \to S(H_1) \,\square\, S(H_2)$. Because each of G_1, G_2, H_1, and H_2 is connected and nonbipartite, each factor $S(G_1), S(G_2), S(H_1)$, and $S(H_2)$ is connected, by Proposition 8.13. Consider prime factorizations

$$\begin{array}{ll} S(G_1) = A_1 \,\square\, A_2 \,\square\, \cdots \,\square\, A_k, & S(H_1) = C_1 \,\square\, C_2 \,\square\, \cdots \,\square\, C_\ell, \\ S(G_2) = B_1 \,\square\, B_2 \,\square\, \cdots \,\square\, B_m, & S(H_2) = D_1 \,\square\, D_2 \,\square\, \cdots \,\square\, D_n. \end{array}$$

Our isomorphism (8.5) becomes

$$\varphi : (A_1 \,\square\, \cdots \,\square\, A_k) \,\square\, (B_1 \,\square\, \cdots \,\square\, B_m) \to (C_1 \,\square\, \cdots \,\square\, C_\ell) \,\square\, (D_1 \,\square\, \cdots \,\square\, D_n). \tag{8.6}$$

Corollary 6.9 applies here. In fact, in using it, we may order the factors A_i and B_i and relabel the vertices of the C_i and D_i so that the isomorphism (8.6) has form

$$\begin{array}{l} \varphi : (A_1 \,\square\, \cdots \,\square\, A_k) \,\square\, (B_1 \,\square\, \cdots \,\square\, B_m) \to \\ \quad \big(A_1 \,\square\, \cdots \,\square\, A_s \,\square\, B_1 \,\square\, \cdots \,\square\, B_t\big) \,\square\, \big(A_{s+1} \,\square\, \cdots \,\square\, A_k \,\square\, B_{t+1} \,\square\, \cdots \,\square\, B_m\big), \end{array} \tag{8.7}$$

for some $0 < s < k$ and $0 \le t \le m$, and where

$$\varphi((a_1, \ldots, a_k), (b_1, \ldots, b_m)) = ((a_1, \ldots, a_s, b_1, \ldots, b_t), (a_{s+1}, \ldots, a_k, b_{t+1}, \ldots, b_m)).$$

If $t = 0$, we have $\varphi((a_1, \ldots, a_k), (b_1, \ldots, b_m)) = ((a_1, \ldots, a_s), (a_{s+1}, \ldots, a_k, b_1, \ldots, b_m))$; and if $t = m$, then $\varphi((a_1, \ldots, a_k), (b_1, \ldots, b_m)) = ((a_1, \ldots, a_s, b_1, \ldots, b_m), (a_{s+1}, \ldots, a_k))$. But our assumption that both φ_1 and φ_2 depend on $x_1 \in V(G_1)$ forces $0 < s < k$.

We have now labeled the vertices of G_1 with $V(A_1 \,\square\, \cdots \,\square\, A_k)$, and those of G_2 with $V(B_1 \,\square\, \cdots \,\square\, B_m)$. We have labeled vertices of H_1 with $V(A_1 \,\square\, \cdots \,\square\, A_s \,\square\, B_1 \,\square\, \cdots \,\square\, B_t)$, and those of H_2 with $V(A_{s+1} \,\square\, \cdots \,\square\, A_k \,\square\, B_{t+1} \,\square\, \cdots \,\square\, B_m)$. To tame the notation, we denote a vertex $(a_1, \ldots, a_s, a_{s+1}, \ldots, a_k) \in V(G_1)$ as (x, y), where $x = (a_1, \ldots, a_s)$ and $y = (a_{s+1}, \ldots, a_k)$. Similarly, any $(b_1, \ldots, b_t, b_{t+1}, \ldots, b_m) \in V(G_2)$ is denoted (u, v), where $u = (b_1, \ldots, b_t)$ and $v = (b_{t+1}, \ldots, b_m)$. With this convention we regard vertices of H_1 and H_2 as (x, u) and (y, v), respectively, and we have

$$\varphi((x, y), (u, v)) = ((x, u), (y, v)).$$

Remember that this is the same isomorphism $\varphi : G_1 \times G_2 \to H_1 \times H_2$ that we began the proof with; all we have done is relabeled the vertices of the factors to put φ into a more convenient form.

Now we display a nontrivial factorization $G_1 = S \times S'$. Define S and S' as follows:

$$\begin{array}{rcl} V(S) & = & \{x \mid ((x, y), (u, v)) \in V(G_1 \times G_2)\}, \\ E(S) & = & \{xx' \mid ((x, y), (u, v))((x', y'), (u', v')) \in E(G_1 \times G_2)\}, \end{array}$$

$$\begin{array}{rcl} V(S') & = & \{y : ((x, y), (u, v)) \in V(G_1 \times G_2)\}, \\ E(S') & = & \{yy' \mid ((x, y), (u, v))((x', y'), (u', v')) \in E(G_1 \times G_2)\}. \end{array}$$

We claim $G_1 = S \times S'$, that is $(x,y)(x',y') \in E(G_1)$ if and only if $(x,y)(x',y') \in E(S \times S')$. Certainly if $(x,y)(x',y') \in E(G_1)$, then there is an edge

$$\big((x,y),(u,v)\big)\big((x',y'),(u',v')\big) \in E(G_1 \times G_2).$$

The definitions of S and S' then imply $(x,y)(x',y') \in E(S \times S')$.

Conversely, suppose $(x,y)(x',y') \in E(S \times S')$. Then $xx' \in E(S)$ and $yy' \in E(S')$. By definition of S and S', this means $G_1 \times G_2$ has edges

$$\big((x,y''),(u,v)\big)\big((x',y'''),(u',v')\big) \text{ and } \big((x'',y),(u'',v'')\big)\big((x''',y'),(u''',v''')\big).$$

Applying isomorphism φ, we see that $H_1 \times H_2$ has edges

$$\big((x,u),(y'',v)\big)\big((x',u'),(y''',v')\big) \text{ and } \big((x'',u''),(y,v'')\big)\big((x''',u'''),(y',v''')\big).$$

Consequently, $(x,u)(x',u') \in E(H_1)$ and $(y,v'')(y',v''') \in E(H_2)$. Thus $H_1 \times H_2$ has an edge $\big((x,u),(y,v'')\big)\big((x',u'),(y',v''')\big)$. Applying φ^{-1} to this, we get

$$\big((x,y),(u,v'')\big)\big((x',y'),(u,v''')\big) \in E(G_1 \times G_2),$$

hence $(x,y)(x',y') \in E(G_1)$. Thus $G_1 = S \times S'$, and the lemma is proved. □

We now easily obtain a result that implies unique prime factorization for connected, nonbipartite, R-thin graphs.

Theorem 8.15 *Consider any isomorphism $\varphi : G_1 \times G_2 \times \cdots \times G_k \to H_1 \times H_2 \times \cdots \times H_\ell$, where all the factors G_i and H_i are connected, nonbipartite, R-thin, and prime. Then $k = \ell$, and there is a permutation π of $\{1,2,\cdots,k\}$ and isomorphisms $\varphi_i : G_{\pi(i)} \to H_i$ for which $\varphi(x_1,x_2,\ldots,x_k) = \big(\varphi_1(x_{\pi(1)}),\varphi_2(x_{\pi(2)}),\ldots,\varphi_k(x_{\pi(k)})\big)$.*

Proof Assume the hypothesis. Note that Lemma 8.14 implies that for each $i = 1,2,\ldots,k$, exactly one φ_j depends on x_i. But no φ_j is constant, because φ is surjective and each H_i has more than one vertex (it is prime). It follows that $k \geq \ell$. The same argument applied to φ^{-1} gives $\ell \geq k$, therefore $k = \ell$.

In summary, each φ_j depends on exactly one x_i, call it $x_{\pi(j)}$. The result follows. □

8.5 Factoring Connected, Nonbipartite Graphs

We are now ready to prove that connected nonbipartite graphs in Γ_0 have unique prime factorizations, a result due to McKenzie (1971). The complete graphs K_p^s play a role here, and it is helpful to keep in mind that $K_p^s/R = K_1^s$. Also note that $K_p^s \times K_q^s \cong K_{pq}^s$, so that if p has prime factorization $p = p_1p_2\ldots p_k$, then we have a unique prime factorization

$$K_p^s = K_{p_1}^s \times K_{p_2}^s \times \cdots \times K_{p_k}^s.$$

The next lemma is analogous to Lemma 8.14, and is the key ingredient of our main theorem. Before stating it, we recall (see the remarks following Proposition 8.5) that any isomorphism $\varphi : G_1 \times \cdots \times G_k \to H_1 \times \cdots \times H_\ell$ induces a corresponding isomorphism $\widetilde{\varphi} : G_1/R \times \cdots \times G_k/R \to H_1/R \times \cdots \times H_\ell/R$, where

$$\begin{aligned}\widetilde{\varphi}([x_1],\ldots,[x_k]) &= (\widetilde{\varphi}_1([x_1],\ldots,[x_k]),\ldots,\widetilde{\varphi}_\ell([x_1],\ldots,[x_k])) \\ &= \big([\varphi_1(x_1,\ldots,x_k)],\ldots,[\varphi_\ell(x_1,\ldots,x_k)]\big),\end{aligned}$$

and φ maps the R-class $[x_1] \times \cdots \times [x_k]$ of $G_1 \times \cdots \times G_k$ bijectively to the R-class $\widetilde{\varphi}_1([x_1],\ldots,[x_k]) \times \cdots \times \widetilde{\varphi}_\ell([x_1],\ldots,[x_k])$ of $H_1 \times \cdots \times H_\ell$.

Lemma 8.16 *Suppose there is an isomorphism $\varphi : G_1 \times \cdots \times G_k \to H_1 \times \cdots \times H_\ell$, where $\varphi(x_1, x_2, \ldots, x_k) = (\varphi_1(x_1, \ldots, x_k), \ldots, \varphi_\ell(x_1, \ldots, x_k))$, and all factors are connected and have odd cycles. Let $\widetilde{\varphi} : G_1/R \times \cdots \times G_k/R \to H_1/R \times \ldots \times H_\ell/R$ be the induced isomorphism*

$$\widetilde{\varphi}([x_1], \ldots, [x_k]) = (\widetilde{\varphi}_1([x_1], \ldots, [x_k]), \ldots, \widetilde{\varphi}_\ell([x_1], \ldots, [x_k])).$$

If some G_i is prime, then at most one of the functions $\widetilde{\varphi}_j$ depends on $[x_i]$.

Proof By commutativity and associativity, we need only prove this for the case where G_1 is prime and $k = \ell = 2$. Let's write $\varphi : G_1 \times G_2 \to H_1 \times H_2$, as $\varphi(x, y) = (\varphi_1(x, y), \varphi_2(x, y))$, so $\widetilde{\varphi} : G_1/R \times G_2/R \to H_1/R \times H_2/R$ becomes $\widetilde{\varphi}([x], [y]) = (\widetilde{\varphi}_1([x], [y]), \widetilde{\varphi}_2([x], [y]))$.

If $G_1 = K_p^s$, then $|V(G_1/R)| = 1$, so neither $\widetilde{\varphi}_1$ nor $\widetilde{\varphi}_2$ depends on $[x]$, and we are done. Likewise, if either H_1 or H_2 is a K_p^s, then one of $\widetilde{\varphi}_1$ or $\widetilde{\varphi}_2$ is necessarily constant, and we are done. Thus, for the rest of the proof we assume that none of G_1, H_1, H_2 is isomorphic to a K_p^s, so each quotient G_1/R, H_1/R, and H_2/R has more than one vertex. Certainly each is R-thin. Moreover, each is connected and nonbipartite because G_1, H_1, and H_2 are.

Assume that both $\widetilde{\varphi}_1$ and $\widetilde{\varphi}_2$ depend on $[x]$. In what follows, we show that G_1 is not prime, a contradiction that proves the lemma.

Certainly if both $\widetilde{\varphi}_1$ and $\widetilde{\varphi}_2$ depend on $[x]$, then Lemma 8.14 implies that G_1/R is not prime. Take a prime factorization $G_1/R = A_1 \times \cdots \times A_n$. This gives a labeling $[x] = (a_1, \ldots, a_n)$ of R-classes of G_1 with vertices of $A_1 \times \cdots \times A_n$. Now $\widetilde{\varphi}$ is an isomorphism

$$\widetilde{\varphi} : A_1 \times \cdots \times A_n \times G_2/R \to H_1/R \times H_2/R.$$

Lemma 8.14 implies that, for each $i = 1, 2, \ldots, n$, exactly one of $\widetilde{\varphi}_1$ and $\widetilde{\varphi}_2$ depends on a_i. Order the factors of $G_1/R = A_1 \times \cdots \times A_n$ so that $\widetilde{\varphi}_1$ depends on $a_1, a_2, \ldots, a_s$, but not on $a_{s+1}, \ldots, a_n$; and $\widetilde{\varphi}_2$ depends on $a_{s+1}, \ldots, a_n$, but not on $a_1, \ldots, a_s$. Then we have

$$\widetilde{\varphi}(a_1, \ldots, a_s, a_{s+1}, \ldots, a_n, [y]) = (\widetilde{\varphi}_1(a_1, \ldots, a_s, [y]), \widetilde{\varphi}_2(a_{s+1}, \ldots, a_n, [y])).$$

Now, $(a_1, \ldots, a_n) = [x] \in V(G_1/R)$ is an R-class of G_1, so it is meaningful to speak of its cardinality $|(a_1, \ldots, a_n)|$. At this point we have $G_1/R = (A_1 \times \cdots \times A_s) \times (A_{s+1} \times \cdots \times A_n)$. According to Proposition 8.6, graph G_1 will have a nontrivial factoring (i.e., be nonprime) provided we can produce functions $\alpha : V(A_1 \times \cdots \times A_s) \to \mathbb{N}$ and $\beta : V(A_{s+1} \times \cdots \times A_n) \to \mathbb{N}$ for which $|(a_1, \ldots, a_s, a_{s+1}, \ldots, a_n)| = \alpha(a_1, \ldots, a_s) \cdot \beta(a_{s+1}, \ldots, a_n)$. The remainder of the proof is a construction of such functions.

Fix an R-class $[y_0] \in V(G_2/R)$. Observe that the isomorphism $\varphi : G_1 \times G_2 \to H_1 \times H_2$ carries any R-class $(a_1, \ldots, a_n) \times [y_0]$ of $G_1 \times G_2$ bijectively to the R-class $\widetilde{\varphi}_1(a_1, \ldots, a_s, [y_0]) \times \widetilde{\varphi}_2(a_{s+1}, \ldots, a_n, [y_0])$ of $H_1 \times H_2$. Therefore

$$\begin{aligned} |(a_1, \ldots, a_s, a_{s+1}, \ldots, a_n)| &= |\widetilde{\varphi}_1(a_1, \ldots, a_s, [y_0])| \cdot \frac{|\widetilde{\varphi}_2(a_{s+1}, \ldots, a_n, [y_0])|}{|[y_0]|} \\ &= |\widetilde{\varphi}_1(a_1, \ldots, a_s, [y_0])| \cdot \frac{N(a_{s+1}, \ldots, a_n)}{D(a_{s+1}, \ldots, a_n)}, \end{aligned}$$

where $N(a_{s+1}, \ldots, a_n)$ and $D(a_{s+1}, \ldots, a_n)$ are the numerator and denominator of the fully reduced fraction $\frac{|\widetilde{\varphi}_2(a_{s+1}, \ldots, a_n, [y_0])|}{|[y_0]|}$. The above equation shows that for every $(a_1, \ldots, a_n)$, the integer $D(a_{s+1}, \ldots, a_n)$ divides $|\widetilde{\varphi}_1(a_1, \ldots, a_s, [y_0])|$. Let d be the least common multiple of the $D(a_{s+1}, \ldots, a_n)$, taken over all $(a_1, \ldots, a_n)$. Then d divides $|\widetilde{\varphi}_1(a_1, \ldots, a_s, [y_0])|$, so

$$\begin{aligned} |(a_1, \ldots, a_s, a_{s+1}, \ldots, a_n)| &= \frac{|\widetilde{\varphi}_1(a_1, \ldots, a_s, [y_0])|}{d} \cdot d\frac{N(a_{s+1}, \ldots, a_n)}{D(a_{s+1}, \ldots, a_n)} \\ &= \alpha(a_1, \ldots, a_s,) \cdot \beta(a_{s+1}, \ldots, a_n) \end{aligned}$$

is a product of integer functions. Thus G_1 has a nontrivial factoring and is not prime. □

Theorem 8.17 *Any connected nonbipartite graph in Γ_0 with more than one vertex has a unique factorization into primes in Γ_0.*

Proof Suppose $G \in \Gamma_0$ is a connected nonbipartite graph with more than one vertex, and let $G \cong G_1 \times G_2 \times \cdots \times G_k$ and $G \cong H_1 \times H_2 \times \cdots \times H_\ell$ be two prime factorizations. We will show that $k = \ell$ and the indices can be ordered so that $G_i \cong H_i$.

By Weichsel's theorem 5.9, each factor G_i and H_j is connected and has an odd cycle. Take an isomorphism $\varphi : G_1 \times G_2 \times \cdots \times G_k \to H_1 \times H_2 \times \cdots \times H_\ell$, where $\varphi(x_1, \ldots, x_k) = (\varphi_1(x_1, \ldots, x_k), \ldots, \varphi_\ell(x_1, \ldots, x_k))$. Consider the induced isomorphism

$$\widetilde{\varphi} : G_1/R \times \cdots \times G_k/R \to H_1/R \times \cdots \times H_\ell/R,$$

where $\widetilde{\varphi}([x_1], \ldots, [x_k]) = (\widetilde{\varphi}_1([x_1], \ldots, [x_k]), \ldots, \widetilde{\varphi}_\ell([x_1], \ldots, [x_k]))$. By Lemma 8.16, for each $i = 1, 2, \ldots, k$, at most one $\widetilde{\varphi}_j$ depends on $[x_i]$. Using the fact that $\widetilde{\varphi}^{-1} = \widetilde{\varphi^{-1}}$, and applying Lemma 8.16 to φ^{-1}, we see that any $\widetilde{\varphi}_j$ depends on *at most one* $[x_i]$. We may thus assume that the factors G_i and H_i have been ordered so that

$$\widetilde{\varphi}([x_1], \ldots, [x_m], [x_{m+1}], \ldots, [x_k]) = (\widetilde{\varphi}_1([x_1]), \ldots, \widetilde{\varphi}_m([x_m]), [y_{m+1}], [y_{m+2}], \ldots, [y_\ell]).$$

Here the index m has the property that $1 \le i \le m$ if and only if exactly one $\widetilde{\varphi}_j$ depends on $[x_i]$. Also $m < i \le k$ if and only if no $\widetilde{\varphi}_j$ depends on $[x_i]$. We allow the possibility that $m = 0$, and in this case no $\widetilde{\varphi}_j$ depends on any $[x_i]$, so $\widetilde{\varphi}$ is constant. At the other extreme, if $m = k$, then for each $[x_i]$ there is exactly one $\widetilde{\varphi_j}$ (namely $\widetilde{\varphi}_i$) that depends on $[x_i]$.

Consider the indices i with $m < i \le k$, for which no $\widetilde{\varphi}_j$ depends on $[x_i]$. As $\widetilde{\varphi}$ is bijective, it must be that G_i/R has just one vertex $[x_i]$. Thus $G_i = K^s_{p_i}$ for some prime number p_i whenever $m < i \le k$. Similarly, for each j with $m < j \le \ell$, the coordinates $[y_j]$ do not depend on any $[x_i]$. Again, as $\widetilde{\varphi}$ is bijective, H_j/R has just one vertex $[y_j]$. Thus for $m < j \le \ell$, we have $H_j = K^s_{q_j}$ for some prime q_j.

But if $1 \le i \le m$, then $[x_i]$ and $\widetilde{\varphi}([x_i])$ may vary, so G_i/R and H_i/R have more than one vertex, so neither G_i nor H_i is a K^s_p. Note further that because $\widetilde{\varphi}$ is an isomorphism, each map $\widetilde{\varphi}_i : G_i/R \to H_i/R$ is an isomorphism, for $1 \le i \le m$.

Now φ carries the R-class $[x_1] \times [x_2] \times \cdots \times [x_k]$ of $G_1 \times \cdots \times G_k$ bijectively to the R-class $\widetilde{\varphi}_1([x_1]) \times \cdots \times \widetilde{\varphi}_m([x_m]) \times [y_{m+1}] \times \cdots \times [y_\ell]$ of $H_1 \times \cdots \times H_\ell$. Therefore

$$\prod_{i=1}^{k} |[x_i]| = \left(\prod_{i=1}^{m} |\widetilde{\varphi}_i([x_i])| \right) \cdot \left(\prod_{i=m+1}^{\ell} |[y_i]| \right). \tag{8.8}$$

For any $1 \le r \le m$, this yields

$$\frac{|[x_r]|}{|\widetilde{\varphi}_r([x_r])|} = \prod_{i=1}^{r-1} \frac{|\widetilde{\varphi}_i([x_i])|}{|[x_i]|} \cdot \prod_{i=r+1}^{m} \frac{|\widetilde{\varphi}_i([x_i])|}{|[x_i]|} \cdot \prod_{i=m+1}^{\ell} |[y_i]|.$$

This expression is constant, because the only variable that appears on the left is $[x_r]$, while $[x_r]$ does not appear on the right. Hence

$$\frac{|[x_r]|}{|\widetilde{\varphi}_r([x_r])|} = \frac{a}{b}, \qquad \text{so} \qquad |[x_r]| = a\frac{|\widetilde{\varphi}_r([x_r])|}{b}$$

for some reduced fraction a/b. This means a divides $|[x_r]|$ and b divides $|\widetilde{\varphi}_r([x_r])|$ for any $[x_r] \in V(G_r/R)$. Because G_r is prime and not a K^s_p, it must be that $a = 1$. (See Corollary 8.7.) As $\widetilde{\varphi}_r : G_r/R \to H_r/R$ is surjective, the same logic applied to H_r gives $b = 1$. Thus

$|[x_r]| = |\widetilde{\varphi}_r([x_r])|$ for each $[x_r] \in G_r/R$, so there is an isomorphism $\varphi_r : G_r \to H_r$ by Proposition 8.4.

Applying $|[x_r]| = |\widetilde{\varphi}_r([x_r])|$ for $1 \leq r \leq m$ to Equation (8.8) yields

$$\prod_{i=m+1}^{k} |[x_i]| = \prod_{i=m+1}^{\ell} |[y_i]|.$$

Recall that $|[x_i]| = p_i$ and $|[y_i]| = q_i$ are constant prime numbers, so the fundamental theorem of arithmetic gives $k = \ell$, and the p_i are the same as the q_i, up to order. Consequently for $m < i \leq k$, the factors $G_i = K_{p_i}^s$ are the same as $H_i = K_{q_i}^s$, up to order. As the previous paragraph has $G_i \cong H_i$ for $1 \leq i \leq m$, the proof is complete. □

8.6 Automorphisms

We now use prime factorization to describe the automorphism groups of direct products. Proposition 8.15 gives a perfect companion to Theorems 6.10 and 7.16.

Theorem 8.18 *Suppose φ is an automorphism of a connected nonbipartite R-thin graph G that has a prime factorization $G = G_1 \times G_2 \times \cdots \times G_k$. Then there exists a permutation π of $\{1, 2, \ldots, k\}$, together with isomorphisms $\varphi_i : G_{\pi(i)} \to G_i$, such that*

$$\varphi(x_1, x_2, \ldots, x_k) = (\varphi_1(x_{\pi(1)}), \varphi_2(x_{\pi(2)}), \ldots, \varphi_k(x_{\pi(k)})).$$

Thus Aut(G) *is generated by the automorphisms of the prime factors and transpositions of isomorphic factors. Consequently,* Aut(G) *is isomorphic to the automorphism group of the disjoint union of the prime factors of G.*

Suppose a connected nonbipartite graph G has prime factorization $G = G_1 \times \cdots \times G_k$. By the results of Section 8.2, any $\varphi \in \text{Aut}(G)$ induces an automorphism $\widetilde{\varphi} \in \text{Aut}(G/R)$. Conversely, any $\widetilde{\varphi} \in \text{Aut}(G/R)$ can be so induced by an automorphism φ of G if and only if $|X| = |\widetilde{\varphi}(X)|$ for every $X \in G/R$, and in this case φ can be any map carrying each X bijectively to $\widetilde{\varphi}(X)$. Because a graph has transitive group if and only if all its R-classes have the same cardinality and G/R has transitive group, we get the following result, which is entirely parallel to Theorem 7.19.

Theorem 8.19 *A direct product has transitive automorphism group if and only if each factor has a transitive automorphism group.*

Likewise Corollaries 7.20 and 7.21 (which are stated for the strong product) hold for the direct product as well. One has only to replace the phrases "connected graph G" and "S-thin" with "connected nonbipartite graph G in Γ_0" and "R-thin."

The results on the automorphism group of the direct product are from Dörfler (1974) and generalize the ones of Imrich (1969a) for the Cartesian product.

Dörfler (1974) characterized graphs with *primitive automorphism group*, namely graphs with transitive groups that leave only trivial partitions of the vertex set invariant. A graph G has primitive automorphism group if G is either a K_k^s or the direct power of a prime graph on at least three vertices with primitive group. This generalizes the corresponding result of Imrich (1969a) for the Cartesian product.

A graph G that has an irredundant[1] isomorphic embedding into a direct product (of graphs in Γ) is called a *subdirect product* by Sabidussi (1975). A graph G is then called *subdirectly irreducible* if any representation of G as a subdirect product contains at least one factor that is isomorphic to G. Sabidussi shows, among other results, that complete graphs and complete graphs with one edge missing are the only subdirectly irreducible graphs (recall that we already mentioned this result on p. 74, where subdirectly irreducible graphs were referred to as $\times$-S-prime graphs). Sabidussi also showed that R-thin graphs have subdirect product representations in which all factors are complete.

8.7 Applications to the Strong Product

There is a close relationship between the strong product of graphs in Γ and the direct product of graphs with loops at each vertex. We now examine that relationship and explore its potential for deducing properties of the strong product.

Given a graph G in Γ, let $\mathcal{L}(G) \in \Gamma_0$ be the graph obtained by adding a loop to each vertex of G. Clearly, for all $G, H \in \Gamma$, we have $G \cong H$ if and only if $\mathcal{L}(G) \cong \mathcal{L}(H)$. Moreover, we have the following easy consequence of the definition.

Lemma 8.20 *If* $G_1, \ldots, G_k \in \Gamma$, *then* $\mathcal{L}(G_1 \boxtimes \cdots \boxtimes G_k) = \mathcal{L}(G_1) \times \cdots \times \mathcal{L}(G_k)$.

Proof This follows from the following chain of equivalences:

$$
\begin{aligned}
& (x_1, \ldots, x_k)(y_1, \ldots, y_k) \in E(\,\mathcal{L}(G_1 \boxtimes \cdots \boxtimes G_k)\,) \\
\Longleftrightarrow\ & (x_1, \ldots, x_k) = (y_1, \ldots, y_k) \ \text{ or } \ (x_1, \ldots, x_k)(y_1, \ldots y_k) \in E(G_1 \boxtimes \cdots \boxtimes G_k) \\
\Longleftrightarrow\ & x_i = y_i \ \text{ or } \ x_i y_i \in E(G_i) \ \text{ for each } \ 1 \le i \le k \\
\Longleftrightarrow\ & x_i y_i \in E(\mathcal{L}(G_i)) \ \text{ for each } \ 1 \le i \le k \\
\Longleftrightarrow\ & (x_1, \ldots, x_k)(y_1, \ldots, y_k) \in \mathcal{L}(G_1) \times \cdots \times \mathcal{L}(G_k)\,.
\end{aligned}
$$

The proof is now complete. □

Using unique factorization of connected nonbipartite graphs over the direct product, we now have an easy alternate proof of Theorem 7.14.

Theorem 7.14 *Every connected graph in* Γ *has a unique prime factor decomposition with respect to the strong product.*

Proof Let G be a connected graph in Γ. Then $\mathcal{L}(G)$ is connected and nonbipartite, so it has a unique prime factorization over the direct product. Observe that because $\mathcal{L}(G)$ has a loop at each vertex, each of its prime factors must also have loops at all of their vertices. Thus each prime factor has form $\mathcal{L}(A_i)$ for some $A_i \in \Gamma$, so the prime factorization can be written as

$$\mathcal{L}(G) = \mathcal{L}(A_1) \times \mathcal{L}(A_2) \times \cdots \times \mathcal{L}(A_n)\,, \tag{8.9}$$

where the factors $\mathcal{L}(A_i)$ (and hence also each A_i) are uniquely determined by G.

Now consider any prime factorization

$$G = G_1 \boxtimes G_2 \boxtimes \cdots \boxtimes G_k \tag{8.10}$$

[1]An irredundant embedding has no unused factors or vertices. Compare the definition of ireducible isometric embeddings on p. 162.

over the strong product. From this, Lemma 8.20 yields

$$\mathcal{L}(G) = \mathcal{L}(G_1) \times \mathcal{L}(G_2) \times \cdots \times \mathcal{L}(G_k). \tag{8.11}$$

Observe that each $\mathcal{L}(G_i)$ is prime over $\times$. Indeed, any factoring of it must have form $\mathcal{L}(G_i) = \mathcal{L}(C) \times \mathcal{L}(C')$ for graphs $C, C' \in \Gamma$, and the lemma gives $\mathcal{L}(G_i) = \mathcal{L}(C \boxtimes C')$. Hence $G_i \cong C \boxtimes C'$ and primeness of G_i implies one of C or C' is K_1, and therefore one of the factors $\mathcal{L}(C)$ or $\mathcal{L}(C')$ is $\mathcal{L}(K_1)$. Thus $\mathcal{L}(G_i)$ is prime.

Comparing prime factorizations (8.9) and (8.11), and applying Theorem 8.17, we get $n = k$, and we may assume the ordering is such that $\mathcal{L}(G_i) \cong \mathcal{L}(A_i)$ for each $1 \le i \le n$. Consequently, $G_i \cong A_i$ for each $1 \le i \le n$. But, as was noted above, the A_i are uniquely determined by G, so the factorization (8.10) is unique. □

Suppose x and y are vertices of a graph $G \in \Gamma$. Observe that xSy in G if and only if xRy in $\mathcal{L}(G)$. Therefore a graph $G \in \Gamma$ is S-thin if and only if $\mathcal{L}(G)$ is R-thin. Combining this with Lemma 8.20, we see that any statement about the strong product of S-thin graphs in Γ can be translated to a corresponding statement about the direct product of R-thin graphs (with loops at each of their vertices) in Γ_0. As an example of this strategy, Lemma 8.20 and Theorem 8.15 yield the following companion to Theorem 8.15.

Theorem 8.21 *Consider any isomorphism $\varphi : G_1 \boxtimes \cdots \boxtimes G_k \to H_1 \boxtimes \cdots \boxtimes H_k$, where all the factors G_i and H_i are connected, nonbipartite, S-thin, and prime. Then there is a permutation π of $\{1, 2, \ldots, k\}$ and isomorphisms $\varphi_i : G_{\pi(i)} \to H_i$ for which $\varphi(x_1, x_2, \ldots, x_k) = (\varphi_1(x_{\pi(1)}), \varphi_2(x_{\pi(2)}), \ldots, \varphi_k(x_{\pi(k)}))$.*

Exercises

8.1. Let $G = G_1 \times G_2 \times \cdots \times G_k$ be the direct product of graphs in Γ_0. Show that G has no loops if and only if at least one of the G_i, $1 \le i \le k$, has no loops.

8.2. Recall that K_p^s denotes the complete graph K_p with loops added to each vertex. Show that $K_p^s \times K_q^s \cong K_{pq}^s$.

8.3. Show that the factors of the equation $(K_1^s + K_2^s + (K_2^s)^{\times,2}) \times (K_1^s + (K_2^s)^{\times,3}) = (K_1^s + (K_2^s)^{\times,2} + (K_2^s)^{\times,4}) \times (K_1^s + K_2^s)$ from Theorem 8.1 are indeed prime with respect to the direct product.

8.4. Given a graph G, prove that the relation R is an equivalence relation on $V(G)$.

8.5. Given a graph G, prove that the quotient G/R is R-thin.

8.6. Prove that a simple graph G is connected and nonbipartite if and only if the Boolean square G^s is connected and nonbipartite.

8.7. Given a graph G, let $\mathcal{N}(G)$ denote the graph obtained from G by removing all its loops. Find an example of a connected, nonbipartite, R-thin graph G for which $\mathcal{N}(G^s)$ is not S-thin.

8.8. Show that the cocktail-party graph $K_{3\times 2}$ (that is, the graph obtained from K_6 by removing a perfect matching) is the unique 4-regular graph on six vertices.

8.9. Demonstrate that $K_2 \times K_{3\times 2}$ and $K_3 \times C_4$ are isomorphic.

8.10. Prove that the graph H in Figure 8.6 has a direct product factorization in the class of simple graphs Γ.

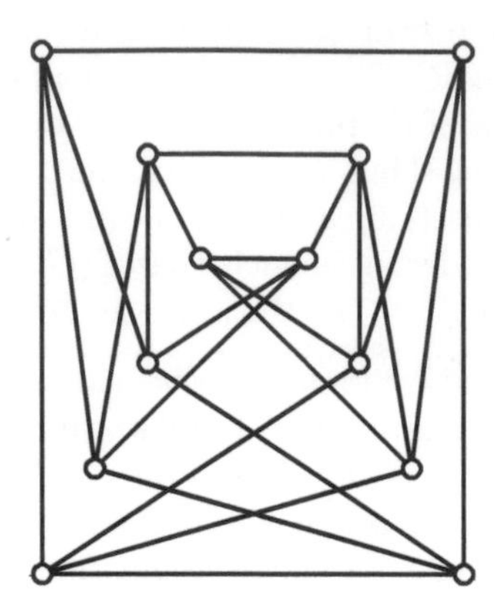

FIGURE 8.6 Graph H.

8.11. Prove that if one of G or H is bipartite, then $G \times H$ is bipartite.

8.12. Let G be a connected bipartite graph, so $G \times G$ is bipartite and has exactly two components. Show that at least one component of $G \times G$ admits an involution (i.e., an automorphism of order 2) that interchanges its partite sets.

8.13. Prove that if both G and H are nonbipartite, then $G \times H$ is nonbipartite.

8.14. Show that if G is bipartite, then $G \times K_2$ is isomorphic to the disjoint union $G + G$.

8.15. For any $n \geq 1$, describe the connected components of the direct product

$$\underbrace{K_2 \times K_2 \times \cdots \times K_2}_{n \text{ times}} \times K_4 .$$

8.16. Show that a simple connected graph G of odd order and with $\Delta(G) \leq 3$ is prime with respect to the direct product (in Γ as well as in Γ_0).

8.17. Show that C_{2n+1}, $n \geq 1$, is prime with respect to the direct product.

8.18. This problem concerns graphs H (possibly with loops) satisfying $K_2 \times H \cong C_{2p}$, where $p \geq 2$. Show that if p is even, then there is exactly one such graph H. Show that if p is odd, there are exactly two such graphs H.

8.19. Which paths are prime/composite with respect to the direct product? (That is, determine and verify a statement for paths P_n, as it is given for cycles in Exercises 8.17 and 8.18.)

8.20. This problem concerns graphs H (possibly with loops) satisfying $K_2 \times H \cong Q_3$. Find three such graphs H.

8.21. Let K be the connected graph on two vertices each with a loop. Verify that $C_4 = K_2 \times K$ and $K_{3\times 2} = K_3 \times K$, where $K_{3\times 2}$ is the cocktail-party graph. Based on these facts, give another solution of Exercise 8.9.

8.22. (Miller, 1968) Let G and H be nontrivial, connected graphs. Show that $G \times H \cong G \,\Box\, H$ if and only if $G \cong H \cong C_{2k+1}$.

Chapter 9

Cancellation

Given a graph product $*$, it is natural to seek the conditions under which $A * C \cong B * C$ implies $A \cong B$. We call this the *cancellation problem* for the product. For the Cartesian product, the solution is straightforward. Theorem 6.21 says $A \,\square\, C \cong B \,\square\, C$ implies $A \cong B$, provided that C is not the empty graph. Because of this, we say that *cancellation holds* for the Cartesian product.

We now consider this problem for the strong and direct products. As we will see, cancellation holds for the strong product but can fail for the direct. In fact, the situation for the direct product is remarkably rich, and much of this chapter is devoted to determining the exact conditions under which cancellation holds or fails. More generally, given graphs A and C, we enumerate and compute all graphs B for which $A \times C \cong B \times C$.

9.1 Cancellation for the Strong Product

We begin with a highly productive approach instigated by Lovász (1971). It was originally used to establish cancellation properties for the direct product, but we adapt it here for the strong product, leaving the direct product for the subsequent section.

Given graphs X and A, let $\hom(X, A)$ denote the number of homomorphisms from X to A. Similarly, $\hom_w(X, A)$ is the number of weak homomorphisms from X to A. The next proposition follows from the fact (Exercise 9.2) that the projection operations are homomorphisms for the direct product and weak homomorphisms for the strong product.

Proposition 9.1 *Suppose* $X \in \Gamma_0$.

(i) *If* $A, C \in \Gamma_0$, *then* $\hom(X, A \times C) = \hom(X, A) \cdot \hom(X, C)$.

(ii) *If* $A, C \in \Gamma$, *then* $\hom_w(X, A \boxtimes C) = \hom_w(X, A) \cdot \hom_w(X, C)$.

In essence, we will use these equations to reduce questions about cancellation of graphs to cancellation in the ring $\mathbb{Z}$. The central ingredient is the fact—first proved by Lovász (1971)—that $A \cong B$ if $\hom(X, A) = \hom(X, B)$ for all X. Before presenting the proof, some preliminary remarks are in order. The development that follows is based on that of Hell and Nešetřil (2004). (The only substantial difference is that we work with graphs, rather than digraphs.)

Given a partition Ω of the vertices of a graph $X \in \Gamma_0$, recall that the *quotient* X/Ω has vertex set Ω, and $U, V \in \Omega$ are adjacent provided that X has an edge uv with $u \in U$ and

$v \in V$. Notice that the map $X \to X/\Omega$ sending u to the element $U \in \Omega$ that contains u is a homomorphism.

Now let $\mathrm{inj}(X, A)$ be the number of injective homomorphisms $X \to A$.

Lemma 9.2 *Suppose $X, A \in \Gamma_0$, and let $\mathcal{P}$ be the set of all partitions of $V(X)$. Then*

$$\hom(X, A) = \sum_{\Omega \in \mathcal{P}} \mathrm{inj}(X/\Omega, A).$$

Proof Let $\mathrm{Hom}(X, A)$ be the set of all homomorphisms from X to A, and put

$$\Upsilon = \{(\Omega, f) \mid \Omega \in \mathcal{P},\ f \in \mathrm{inj}(X/\Omega, A)\},$$

so $|\Upsilon| = \sum_{\Omega \in \mathcal{P}} \mathrm{inj}(X/\Omega, A)$. It suffices to produce a bijection $\mathrm{Hom}(X, A) \to \Upsilon$. Now, for $f \in \mathrm{Hom}(X, A)$, let $f \mapsto (\Omega, f^*)$, where $\Omega = \{f^{-1}(a) \mid a \in V(A)\} \in \mathcal{P}$, and $f^* : X/\Omega \to A$ is defined as $f^*(U) = f(u)$, for $u \in U$. It is easy to check that this is an injective map to Υ. For surjectivity, if $(\Omega, f^*) \in \Upsilon$, then the composition $X \to X/\Omega \xrightarrow{f^*} A$ maps to (Ω, f^*). □

Theorem 9.3 *If $A, B \in \Gamma_0$ and $\hom(X, A) = \hom(X, B)$ for every $X \in \Gamma_0$, then $A \cong B$.*

Proof Suppose $\hom(X, A) = \hom(X, B)$ for every graph X. Our strategy is to show that this implies $\mathrm{inj}(X, A) = \mathrm{inj}(X, B)$ for every X. Then the theorem will follow because we get $\mathrm{inj}(B, A) = \mathrm{inj}(B, B) > 0$ and $\mathrm{inj}(A, B) = \mathrm{inj}(A, A) > 0$, so there are injective homomorphisms $A \to B$ and $B \to A$, whence $A \cong B$.

We use induction on $|X|$ to show $\mathrm{inj}(X, A) = \mathrm{inj}(X, B)$ for all X. If $|X| = 1$, then

$$\mathrm{inj}(X, A) = \hom(X, A) = \hom(X, B) = \mathrm{inj}(X, B).$$

If $|X| > 1$, then Lemma 9.2 applied to the equation $\hom(X, A) = \hom(X, B)$ produces

$$\sum_{\Omega \in \mathcal{P}} \mathrm{inj}(X/\Omega, A) = \sum_{\Omega \in \mathcal{P}} \mathrm{inj}(X/\Omega, B).$$

Let T be the trivial partition of $V(X)$ consisting of $|X|$ singleton sets. Then $X/T = X$ and the above equation becomes

$$\mathrm{inj}(X, A) + \sum_{\Omega \in \mathcal{P} - T} \mathrm{inj}(X/\Omega, A) \quad = \quad \mathrm{inj}(X, B) + \sum_{\Omega \in \mathcal{P} - T} \mathrm{inj}(X/\Omega, B).$$

The sums are equal by the inductive hypothesis, hence $\mathrm{inj}(X, A) = \mathrm{inj}(X, B)$. □

Theorem 9.3 has an easy analogue for weak homomorphisms.

Theorem 9.4 *If $A, B \in \Gamma$ and $\hom_w(X, A) = \hom_w(X, B)$ for every $X \in \Gamma_0$, then $A \cong B$.*

Proof For $G \in \Gamma$, let $\mathcal{L}(G)$ be G with loops added to each vertex. Observe that $\hom_w(X, G) = \hom(X, \mathcal{L}(G))$. Now, if $\hom_w(X, A) = \hom_w(X, B)$ for every graph $X \in \Gamma_0$, then also $\hom(X, \mathcal{L}(A)) = \hom(X, \mathcal{L}(B))$. By Theorem 9.3 we get $\mathcal{L}(A) \cong \mathcal{L}(B)$, whence $A \cong B$. □

Now we can prove cancellation for the strong product.

Theorem 9.5 *If $A, B, C \in \Gamma$ and C is nonempty, then $A \boxtimes C \cong B \boxtimes C$ implies $A \cong B$.*

Proof Suppose $A \boxtimes C \cong B \boxtimes C$. Then for any $X \in \Gamma_0$, we have $\hom_w(X, A \boxtimes C) = \hom_w(X, B \boxtimes C)$, and Proposition 9.1 yields

$$\hom_w(X, A) \cdot \hom_w(X, C) = \hom_w(X, B) \cdot \hom_w(X, C).$$

But $\hom_w(X, C) \neq 0$ because any constant map $X \to C$ is a weak homomorphism. Therefore $\hom_w(X, A) = \hom_w(X, B)$, and Theorem 9.4 produces $A \cong B$. □

9.2 Cancellation for the Direct Product

In general, cancellation for the direct product fails dramatically. If C is any bipartite graph, then there are always non-isomorphic graphs A and B for which $A \times C \cong B \times C$. Indeed, just take $A = K_2$ and $B = 2K_1^s$ (two loops). Then $A \times C \cong 2C \cong B \times C$.

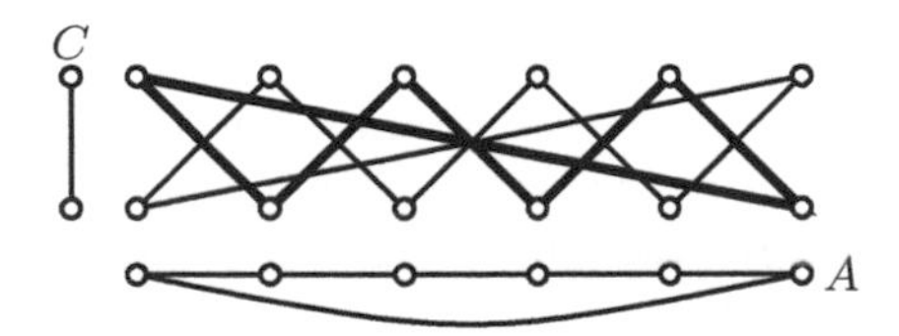

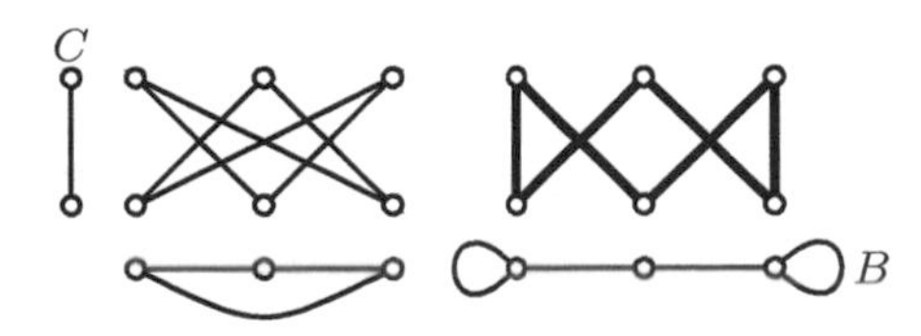

FIGURE 9.1 Failure of cancellation: $A \times C \cong B \times C$ but $A \ncong B$.

Figure 9.1 shows another example. Here $A \times C \cong 2C_6 \cong B \times C$, but $A \ncong B$. Despite such failures, the machinery built in the previous section yields several easy cancellation results. In what follows, all graphs are in Γ_0; all proofs are due to Lovász (1971).

Proposition 9.6 *If $A \times C \cong B \times C$, and C has a loop, then $A \cong B$.*

Proof Suppose $A \times C \cong B \times C$. Let X be any graph. Applying Proposition 9.1, we get

$$\hom(X, A) \cdot \hom(X, C) = \hom(X, B) \cdot \hom(X, C).$$

But $\hom(X, C) \neq 0$, as the constant map sending X to a vertex with a loop is a homomorphism. Thus $\hom(X, A) = \hom(X, B)$, and Theorem 9.3 gives $A \cong B$. □

Proposition 9.7 *If $A \times C \cong B \times C$, and there are homomorphisms $A \to C$ and $B \to C$, then $A \cong B$.*

Proof As before, $A \times C \cong B \times C$ implies $\hom(X, A) \cdot \hom(X, C) = \hom(X, B) \cdot \hom(X, C)$. If $\hom(X, C) \neq 0$, then $\hom(X, A) = \hom(X, B)$. On the other hand, if $\hom(X, C) = 0$, then it must be that $\hom(X, A) = 0$, for otherwise there is a homomorphism $X \to A \to C$. Similarly we argue $\hom(X, B) = 0$. In any case $\hom(X, A) = \hom(X, B)$, so $A \cong B$. □

Corollary 9.8 *If $A \times C \cong B \times C$, and A, B and C are bipartite, then $A \cong B$ if C has at least one edge.*

Proposition 9.9 *If $A \times C \cong B \times C$, and there is a homomorphism $D \to C$ then $A \times D \cong B \times D$.*

The proof of Proposition 9.9 is similar to that of Proposition 9.7 and is left as an exercise. One significant consequence of it is that if $A \times C \cong B \times C$ and C has at least one edge, then $A \times K_2 \cong B \times K_2$. This is useful, as it can reduce questions about cancellation to the simpler case involving a common factor of K_2.

In addition to the above results, Lovász (1971) also proved the following generalization of Proposition 9.6. Combined with the remarks at the beginning of this section, it tells us that in general $A \times C \cong B \times C$ implies $A \cong B$ if and only if C is not bipartite.

Theorem 9.10 *If $A \times C \cong B \times C$, and C has an odd cycle, then $A \cong B$.*

Lovász obtains this as a consequence of the following theorem. Its proof involves a theory of so-called k-partite structures. To date no simple graph-theoretic proof is known.

Theorem 9.11 *If $A, B, C \in \Gamma_0$ and $A \times C \cong B \times C$, then there is an isomorphism $A \times C \to B \times C$ with $(a, c) \mapsto (\varphi(a, c), c)$.*

9.3 Anti-Automorphisms and Factorials

Theorem 9.10 does not completely resolve the question of when C can be cancelled from $A\times C \cong B\times C$. Although it does imply that cancellation can fail if and only if C is bipartite, it does not address what properties of A (or B) might guarantee that cancellation holds. For example, if $A = K_1^s$, then surely $A \times C \cong B \times C$ implies $A \cong B$, whether or not C is bipartite. We might reasonably ask what other graphs A have this property.

This section answers that question. Given a graph A and a bipartite graph C, we classify those graphs B for which $A\times C \cong B\times C$. This leads to exact conditions on A that guarantee that $A \times C \cong B \times C$ implies $A \cong B$. Our methods involve two new ideas: the notion of an *anti-automorphism* of a graph and a *factorial* operation on graphs. The development is based on Hammack (2009), but we here use a slightly improved definition of a factorial.

An *anti-automorphism* of a graph A is a bijection $\alpha : V(A) \to V(A)$ with the property that $xy \in E(A)$ if and only if $\alpha(x)\alpha^{-1}(y) \in E(A)$ for all pairs $x, y \in V(A)$. The set of all anti-automorphisms of A is denoted $\mathrm{Ant}(A)$.

In general $\mathrm{Ant}(A)$ is not a group, though it contains the identity and is closed with respect to taking inverses. Notice that any automorphism of order 2 is an anti-automorphism. Figure 9.2 (left) shows an anti-automorphism of order 4.

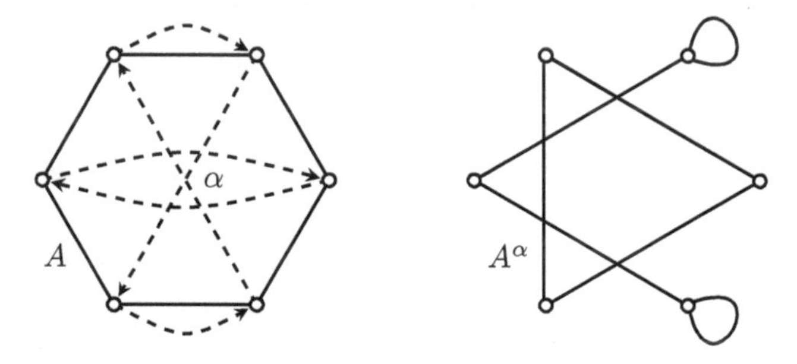

FIGURE 9.2 Left: a graph A and an $\alpha \in \mathrm{Ant}(A)$. Right: the graph A^α.

The following construction is of key importance. Given an anti-automorphism α of a graph A, we define a graph A^α as $V(A^\alpha) = V(A)$ and $E(A^\alpha) = \{x\,\alpha(y) \mid xy \in E(A)\}$. Figure 9.2 shows an example of graphs A and A^α.

We take care to point out that the statement $xy \in E(A) \Leftrightarrow x\alpha(y) \in E(A^\alpha)$ is true, and it follows not just from the definition of A^α, but also from the fact that α is an anti-automorphism. This is summarized in the following result, which will be used frequently and without further comment.

Proposition 9.12 *If $\alpha \in \mathrm{Ant}(A)$, then $xy \in E(A)$ if and only if $x\alpha(y) \in E(A^\alpha)$.*

Proof Certainly if $xy \in E(A)$, then $x\alpha(y) \in E(A^\alpha)$ by the definition of A^α. Conversely, suppose $x\alpha(y) \in E(A^\alpha)$. By the definition of A^α, this means that either $xy \in E(A)$ or $\alpha^{-1}(x)\alpha(y) \in E(A)$. In the second case, the fact that α is an anti-automorphism ensures that $xy \in E(A)$. □

Comparing Figures 9.1 and 9.2, we see that the graph A is the six-cycle in both figures, and $B = A^\alpha$ and $A \times C \cong A^\alpha \times C$. This illustrates a general principle.

Proposition 9.13 *Let A and B be graphs. If C is a bipartite graph that has at least one edge, then $A \times C \cong B \times C$ if and only if $B \cong A^\alpha$ for some $\alpha \in \mathrm{Ant}(A)$.*

Proof Suppose $A \times C \cong B \times C$. We will construct an anti-automorphism α of A for which $A^\alpha \cong B$. As C has an edge, there is a homomorphism $K_2 \to C$, and therefore Proposition 9.9 implies $A \times K_2 \cong B \times K_2$. By Theorem 9.11, there is an isomorphism $A \times K_2 \to B \times K_2$ of form $(x, c) \mapsto (\varphi(x, c), c)$. (Actually, this can easily be shown directly for the present simple situation of $C = K_2$.) Put $V(K_2) = \{0, 1\}$ and define maps $\lambda, \mu : V(A) \to V(B)$ as follows:

$$\begin{aligned} \lambda(x) &= \varphi(x, 0), \\ \mu(x) &= \varphi(x, 1). \end{aligned}$$

Because $(x, c) \mapsto (\varphi(x, c), c)$ is an isomorphism, it follows readily that λ and μ are bijective. We now show that the composition $\lambda^{-1}\mu$ is an anti-automorphism. Observe that

$$\begin{aligned} xy \in E(A) &\iff (x, 0)(y, 1) \in E(A \times K_2) \\ &\iff (\varphi(x, 0), 0)(\varphi(y, 1), 1) \in E(B \times K_2) \\ &\iff (\lambda(x), 0)(\mu(y), 1) \in E(B \times K_2) \\ &\iff \lambda(x)\mu(y) \in E(B). \end{aligned}$$

Thus we have

$$xy \in E(A) \iff \lambda(x)\mu(y) \in E(B), \tag{9.1}$$

and applying this again gives

$$\begin{aligned} xy \in E(A) &\iff \lambda(x)\mu(y) \in E(B) \\ &\iff \mu^{-1}\lambda(x)\lambda^{-1}\mu(y) \in E(A) \\ &\iff (\lambda^{-1}\mu)^{-1}(x)\, \lambda^{-1}\mu(y) \in E(A). \end{aligned}$$

This means $\lambda^{-1}\mu \in \text{Ant}(A)$. Set $\alpha = \lambda^{-1}\mu$. Notice that $\lambda : A^\alpha \to B$ is an isomorphism, as follows. By definition, any edge of A^α, has form $x\,\alpha(y) = x\,\lambda^{-1}\mu(y)$ for some $xy \in V(A)$. Taking λ of both endpoints produces $\lambda(x)\mu(y)$, which by (9.1) is an edge of B. On the other hand, if $uv \in E(B)$, then $\lambda^{-1}(u)\mu^{-1}(v) \in E(A)$, so $\lambda^{-1}(u)\alpha\mu^{-1}(v) \in E(A^\alpha)$, which reduces to $\lambda^{-1}(u)\lambda^{-1}(v) \in E(A^\alpha)$. Therefore $B \cong A^\alpha$.

Conversely, it suffices to prove that $A \times C \cong A^\alpha \times C$ for any bipartite graph C and $\alpha \in \text{Ant}(A)$. Let C_0 and C_1 be a bipartition of C, and define a map $\Theta : A \times C \to A^\alpha \times C$ as

$$\Theta(a, c) = \begin{cases} (a, c) & \text{if } c \in C_0, \\ (\alpha(a), c) & \text{if } c \in C_1. \end{cases}$$

This is clearly bijective. Suppose $(x, c)(y, c') \in E(A \times C)$. We may assume $c \in C_0$ and $c' \in C_1$. Then $\Theta(x, c)\Theta(y, c') = (x, c)(\alpha(y), c') \in E(A^\alpha \times C)$. In the other direction, any edge of $A^\alpha \times C$ must be either of form $(x, c)(\alpha(y), c')$ or $(\alpha(x), c)(y, c')$, where in each case $c \in C_0$, $c' \in C_1$ and $xy \in E(A)$. In the first case, $(x, c)(\alpha(y), c')$ is the image under Θ of the edge $(x, c)(y, c')$ of $A \times C$. In the second case, $(\alpha(x), c)(y, c')$ is the image under Θ of $(\alpha(x), c)(\alpha^{-1}(y), c')$, which is an edge of $A \times C$ because α is an anti-automorphism. □

Proposition 9.13 implies that the set $\text{Ant}(A)$ in some sense parameterizes the graphs B for which $A \times C \cong B \times C$. For any $\alpha \in \text{Ant}(A)$, the graph $B = A^\alpha$ satisfies $A \times C \cong B \times C$. Conversely for any B with $A \times C \cong B \times C$, there is some $\alpha \in \text{Ant}(A)$ for which $B \cong A^\alpha$. However, this correspondence need not be injective. There can exist distinct anti-automorphisms α and β for which $A^\alpha \cong A^\beta$. For example, if $A = K_3$, there are three distinct transpositions α_1, α_2, and α_3 that interchange two vertices and fix the third. Each is an anti-automorphism, and $A^{\alpha_1} \cong A^{\alpha_2} \cong A^{\alpha_3}$ is the path of length 2 with loops at each end. As a tool for sorting out which anti-automorphisms yield isomorphic graphs, we introduce the notion of a graph factorial.

The *factorial* of a graph A is the graph $A!$ whose vertices are the permutations of $V(A)$. Permutations λ and μ are adjacent in $A!$ exactly when $xy \in E(A) \iff \lambda(x)\mu(y) \in E(A)$ for all pairs $x, y \in V(A)$. We denote an edge joining the vertices λ and μ as $(\lambda)(\mu)$ in order to avoid confusion with composition.

Replacing xy in this definition with $\lambda^{-1}(x)\mu^{-1}(y)$, we see that $(\lambda)(\mu) \in E(A!)$ if and only if $(\lambda^{-1})(\mu^{-1}) \in E(A!)$, so $\mu \to \mu^{-1}$ is an involution of $A!$. Note that α is an anti-automorphism of A if and only if $(\alpha)(\alpha^{-1}) \in E(A!)$, so $\mathrm{Ant}(A)$ consists of the permutations of $V(A)$ that are joined to their inverses by an edge of $A!$. We remark also that there is a loop at a vertex μ of $A!$ if and only if μ is an automorphism of A.

As an example of a graph factorial, consider the complete graph K_p^s with loops at each vertex. Any pair of permutations of $V(K_p^s)$ must be adjacent in $K_p^s!$, so $K_p^s! \cong K_{p!}^s$. Consequently,

$$K_p^s! \cong K_p^s \times K_{p-1}^s \times K_{p-2}^s \times \cdots \times K_3^s \times K_2^s,$$

which explains our choice of the word "factorial" for this construction.

For another example, consider $K_p!$. Because every permutation of $V(K_p)$ is an automorphism of K_p, it follows that $K_p!$ has a loop at each of its $p!$ vertices. Moreover, given an edge $(\lambda)(\mu) \in E(K_p!)$ we must have $\lambda = \mu$, for otherwise there is an $x \in V(K_p)$ with $\lambda(x)\mu(x) \in E(K_p)$, forcing a loop at x. Therefore every edge of the factorial is a loop, so

$$K_p! = p!K_1. \tag{9.2}$$

Given a graph A, we define a relation $\simeq$ on $\mathrm{Ant}(A)$ by declaring $\alpha \simeq \beta$ if $\alpha = \lambda\beta\mu^{-1}$ for some $(\lambda)(\mu) \in E(A!)$. It is not hard to verify that this is an equivalence relation (Exercise 9.5).

Proposition 9.14 *If $\alpha, \beta \in \mathrm{Ant}(A)$, then $\alpha \simeq \beta$ if and only if $A^\alpha \cong A^\beta$.*

Proof Suppose $\alpha \simeq \beta$, so $\alpha = \lambda\beta\mu^{-1}$ for some $(\lambda)(\mu) \in E(A!)$. Then $\alpha\mu = \lambda\beta$ and

$$\begin{aligned} xy \in E(A) &\iff \lambda(x)\mu(y) \in E(A) \\ &\iff \lambda(x)\alpha\mu(y) \in E(A^\alpha) \\ &\iff \lambda(x)\lambda\beta(y) \in E(A^\alpha). \end{aligned}$$

Now, the edges of A^β are precisely the pairs $x\beta(y)$ for $xy \in E(A)$, and the above equivalences show that $\lambda(x)\lambda\beta(y) \in E(A^\alpha)$. Thus λ is a homomorphism from A^β to A^α. Further, observe that any edge $x\alpha(y)$ of A^α is the image under λ of some edge of A^β: Since $x\alpha(y) \in E(A^\alpha)$, we have $xy \in E(A)$, so $\lambda^{-1}(x)\mu^{-1}(y) \in E(A)$, and hence $\lambda^{-1}(x)\beta\mu^{-1}(y) \in E(A^\beta)$. Then λ sends this edge to $x\,\lambda\beta\mu^{-1}(y) = x\alpha(y)$. Therefore $\lambda : A^\beta \to A^\alpha$ is an isomorphism.

Conversely, let there be an isomorphism $\lambda : A^\beta \to A^\alpha$. Note $\alpha = (\lambda)\beta(\alpha^{-1}\lambda\beta)^{-1}$. We just need to show that $(\lambda)(\alpha^{-1}\lambda\beta) \in E(A!)$, and this involves showing that $xy \in E(A)$ if and only if $\lambda(x)\alpha^{-1}\lambda\beta(y) \in E(A)$. Now,

$$\begin{aligned} xy \in E(A) &\iff x\beta(y) \in E(A^\beta) \\ &\iff \lambda(x)\lambda\beta(y) \in E(A^\alpha) \\ &\iff \lambda(x)\alpha^{-1}\lambda\beta(y) \in E(A). \end{aligned}$$

If $\alpha^{-1}\lambda(x)\lambda\beta(y) \in E(A)$, the anti-automorphism property of α gives $\lambda(a)\alpha^{-1}\lambda\beta(y) \in E(A)$. □

For each $\alpha \in \mathrm{Ant}(A)$, let $[\alpha]$ denote the $\simeq$-equivalence class containing α. Propositions 9.13 and 9.14 imply the following:

Theorem 9.15 *Let A be a graph and C be a bipartite graph with at least one edge. If the $\simeq$-equivalence classes of* $\mathrm{Ant}(A)$ *are* $\{[\alpha_1],[\alpha_2],\ldots,[\alpha_k]\}$*, then the isomorphism classes of the graphs B for which $A \times C \cong B \times C$ are precisely those in* $\{A^{\alpha_1}, A^{\alpha_2}, \ldots, A^{\alpha_k}\}$.

We illustrate this theorem with an example. Let $A = K_4$, with vertex set $\{0,1,2,3\}$. We now compute all of the graphs B for which $A \times C \cong B \times C$, for an arbitrary bipartite graph C. According to Theorem 9.15, we must first compute $\mathrm{Ant}(A)$.

For this, notice that any permutation of $\{0,1,2,3\}$ satisfying $\alpha^2 = \mathrm{id}$ is either the identity or an automorphism of A or order 2, and therefore belongs to $\mathrm{Ant}(A)$. Conversely, consider any $\alpha \in \mathrm{Ant}(A)$. Given any $x \in V(A) = V(K_4)$, because $xx \notin E(K_4)$, we must have $\alpha(x)\alpha^{-1}(x) \notin E(K_4)$. Hence $\alpha(x) = \alpha^{-1}(x)$, so $\alpha^2 = \mathrm{id}$. Thus

$$\mathrm{Ant}(A) = \{\alpha \in \mathrm{Aut}(K_4) \mid \alpha^2 = \mathrm{id}\},$$

that is, $\mathrm{Ant}(A)$ consists of all the involutions of $A = K_4$.

Next we must compute the $\simeq$-equivalence classes of $\mathrm{Ant}(A)$. By Equation (9.2), the edges of $A!$ are just the loops $(\lambda)(\lambda)$, where λ is a permutation of $V(A)$. Thus $\alpha \simeq \beta$ provided that $\alpha = \lambda\beta\lambda^{-1}$, that is, if α and β are conjugate by a permutation of $V(A)$. There are only three conjugacy classes of involutions of $\{0,1,2,3\}$, namely [id], [(02)], and [(02)(13)].

Thus Theorem 9.15 asserts that there are exactly three graphs B with $A \times C \cong B \times C$, namely $B = A^{\mathrm{id}}$, $B = A^{(02)}$ and $B = A^{(02)(13)}$. These are illustrated in Figure 9.3. (As a quick check, the reader may verify that $A^{\mathrm{id}} \times K_2 \cong A^{(02)} \times K_2 \cong A^{(02)(13)} \times K_2 \cong Q_3$.)

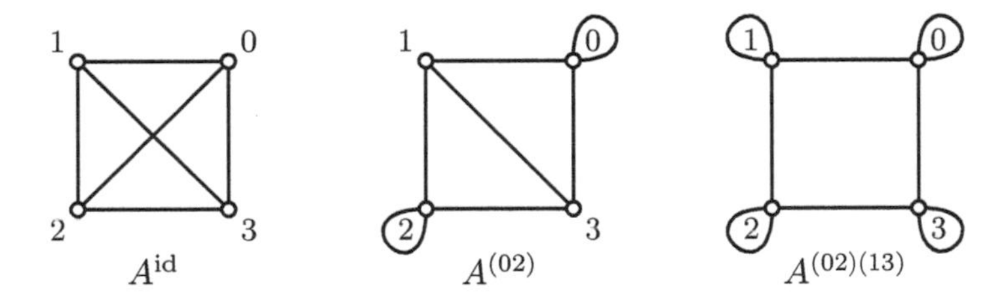

FIGURE 9.3 The three graphs B for which $A \times C \cong B \times C$.

We close with a summary of the consequences of Theorems 9.10 and 9.15. The simultaneous conditions $A \times C \cong B \times C$ and $A \ncong B$ are possible if and only if C is bipartite and A (hence also B) has more than one $\simeq$-equivalence class. Therefore we have the following:

Theorem 9.16 *Cancellation of $A \times C \cong B \times C$ is guaranteed to hold if and only if C has an odd cycle or A (hence also B) has exactly one $\simeq$-equivalence class.*

For other approaches to cancellation, see Abay-Asmerom, Hammack, Larson, and Taylor (2010); Hammack (2008), Fernández, Leighton, and López-Presa (2007); and Imrich, Klavžar, and Rall (2007).

9.4 Graph Exponentiation

Graph exponentiation is such a natural concept that we feel obliged to state at least the definitions.

For graphs G and H, the *exponential* or *direct power* G^H is a graph whose vertices are the functions $V(H) \to V(G)$. Two functions f and g are adjacent provided that $f(u)g(v) \in E(G)$ whenever $uv \in E(H)$. (Observe that the factorial $A!$ is a subgraph of A^A.)

We see immediately that $G^{K_1^s} \cong G$. Verifying the following identities is not much harder:

$$A^{(B+C)} = A^B \times A^C, \tag{9.3}$$
$$(A \times B)^C = A^C \times B^C, \tag{9.4}$$
$$A^{B \times C} = (A^B)^C. \tag{9.5}$$

Much more difficult is the question under which conditions the usual cancellation laws for bases and exponents hold:

$$A^G \cong A^H \quad \text{implies} \quad G \cong H,$$
$$A^G \cong B^G \quad \text{implies} \quad A \cong B,$$

or conditions that ensure that two direct powers

$$A^G \cong B^H$$

have common refinements

$$A \cong C^X, B \cong C^Y, G \cong Y \times Z, H \cong X \times Z.$$

For partially ordered sets, the definition of graph exponentiation dates back to Birkhoff; see, for example, Birkhoff (1940). There exists a sizable literature devoted to the solution of the above problems. For a survey, see Jónsson (1982). In general, however, graph exponentiation for simple graphs seems to be little explored.

Exercises

9.1. Find four different graphs B for which $C_{10} \times C_8 \cong B \times C_8$.

9.2. Prove Proposition 9.1.

9.3. Prove Proposition 9.9.

9.4. Prove that for any graph A, each nontrivial component of $A!$ either is K_p^s for some p or is a complete bipartite graph.

9.5. Given a graph A, we define a relation $\simeq$ on $\text{Ant}(A)$ by declaring $\alpha \simeq \beta$ if $\alpha = \lambda\beta\mu^{-1}$ for some edge (possibly a loop) $(\lambda)(\mu) \in E(A!)$. Prove that this is an equivalence relation.

9.6. (Hammack, 2008) Suppose A and C are bipartite. Show that there is a graph B that is not isomorphic to A, for which $A \times C \cong B \times C$, if A has a connected component with an automorphism of order 2 that reverses its bipartite sets.

9.7. Prove the validity of the laws (9.3) through (9.5) for exponents.

9.8. Suppose A and C are bipartite. Show that $A \times C \cong B \times C$ implies $A \cong B$ if and only if B is bipartite.

9.9. Suppose A is a connected bipartite graph with an automorphism α of order 2 that reverses its bipartite sets. Show that A^α is not bipartite.

Chapter 10

Lexicographic Product

We introduced the lexicographic product in Section 4.4 and examined its metric properties in Section 5.4. We now address deeper questions such as prime factorization and cancellation. We prove that all prime factorizations of a given graph G have the same number of factors and that there exists a canonical way of transforming any prime factorization into any other one. We also show the existence of a large class of graphs with unique prime factor decomposition with respect to the lexicographic product.

As in the case of the other products, a nontrivial lexicographic product has transitive automorphism group if and only if all factors have transitive groups. However, the automorphism group of a nontrivial lexicographic product cannot be regular. This motivates the study of lexicographic products that admit a regular subgroup of the automorphism group. In this vein we show that every graph with transitive automorphism group is a homomorphic image of a nontrivial lexicographic product that admits a regular group of automorphisms.

As shown by Sabidussi (1958), graphs that admit a regular group of automorphisms are Cayley graphs. Vertex transitive graphs can then be characterized as retracts of Cayley graphs.

10.1 Basic Properties

This section is concerned with distributive properties of the lexicographic product with respect to the disjoint union, the join of graphs, and basic cancellation laws. These results, together with properties of so-called externally related sets, provide the background for the investigation of the structure of prime factor decompositions of graphs with respect to the lexicographic product in Section 10.4.

Recall that the lexicographic product $G \circ H$ of two graphs G and H is defined on $V(G \circ H) = V(G) \times V(H)$, two vertices $(u, x), (v, y)$ of $G \circ H$ being adjacent whenever $uv \in E(G)$, or $u = v$ and $xy \in E(H)$. See Figures 4.10 and 5.5.

By Corollary 5.14, the product $G \circ H$ of nontrivial graphs is connected if and only if G is. Thus $G \circ H \not\cong H \circ G$ whenever one of the factors is disconnected and the other one connected and nontrivial, as in Figure 4.10. Figure 5.5 shows that commutativity can fail even when both factors are connected.

The lexicographic product is associative (Proposition 5.11) and has K_1 as a unit. With respect to the disjoint union, we have the distributive law

$$(A+B)\circ C = A\circ C + B\circ C.$$

With respect to taking complements, we note that

$$\overline{G\circ H} = \overline{G}\circ\overline{H}\,.$$

Because $\overline{\overline{G}} = G$, this implies that

$$G\circ H = \overline{\overline{G}\circ\overline{H}}\,.$$

Because of this, we say the lexicographic product is *self-complementary.* (That is—in the parlance of Section 4.4—it is its own complementary product.)

The *join* $G\oplus H$ of G and H is defined by

$$G\oplus H = \overline{\overline{G}+\overline{H}}\,.$$

From the right-distributive law

$$(A+B)\circ C = A\circ C + B\circ C$$

for the disjoint union, we obtain

$$(A\oplus B)\circ C = A\circ C\oplus B\circ C$$

for the join by self-complementarity. We also have the left-distributive law

$$K_n\circ(A\oplus B) = K_n\circ A\oplus K_n\circ B.$$

Taking complements, we deduce a second left-distributive law

$$D_n\circ(A+B) = D_n\circ A + D_n\circ B.$$

Although the lexicographic product is not commutative in general, $G\circ H\cong H\circ G$ if both G and H are complete or if both are totally disconnected. There are no other graphs that commute with complete or totally disconnected graphs.

Proposition 10.1 *Let $n\geq 2$ and G be a graph. Then*

(i) *$G\circ K_n\cong K_n\circ G$ if and only if G is complete, and*
(ii) *$G\circ D_n\cong D_n\circ G$ if and only if G is totally disconnected.*

Proof We first prove (ii). We have already seen that $G\circ D_n\cong D_n\circ G$ if G has no edges. Assume now that $G\circ D_n\cong D_n\circ G$. Evidently,

$$|E(G)|\cdot n^2 = |E(G\circ D_n)| = |E(D_n\circ G)| = n\cdot|E(G)|.$$

Hence $|E(G)|(n^2-n)=0$. For $n\geq 2$, this is only possible if $|E(G)|=0$.

To prove (i), we recall that $\overline{G\circ H} = \overline{G}\circ\overline{H}$. Hence G and K_n commute if and only if $\overline{G}$ and $D_n = \overline{K_n}$ commute. For $n\geq 2$ this is the case if and only if $\overline{G}$ has no edges, or equivalently if and only if G is complete. □

By associativity, two graphs commute if they are both (lexicographic) powers of one and the same graph. Thus two graphs G and H commute in each of the following trivial cases, where, just as for the other products, we will write $G^{\circ,k}$ to denote the kth power of G with respect to the lexicographic product:

1. G and H are both complete.
2. G and H are both totally disconnected.
3. There exists a graph K and integers $n\geq 1$, $m\geq 1$ such that $G = K^{\circ,n}$ and $H = K^{\circ,m}$.

These three cases are the only ones, but some argument is required to prove it; see Theorem 10.9.

10.2 Self-Complementarity and Cancellation Properties

Every graph has a prime factor decomposition with respect to the lexicographic product. In Section 10.4 we show that it need not be unique, not even for connected graphs. However, we will also prove that there is a canonical way of obtaining all prime factor decompositions from a given one and that the number of prime factors is preserved. The proof makes repeated use of self-complementarity and of distributive properties.

Here we investigate these properties and prove a cancellation law that is of interest in its own right.

Recall that the connected components of a graph are the largest indecomposable induced subgraphs of a graph with respect to the disjoint union. One could say that they are prime with respect to the disjoint union and that the representation of a graph as a disjoint union of indecomposable graphs is unique.

By complementarity, an analogous statement holds for the join. Here the indecomposable elements of a graph G are precisely the complements of the components of $\overline{G}$. We call them the *join-components* of G.

Clearly, a graph G is indecomposable with respect to the disjoint union and the join if and only if both G and its complement $\overline{G}$ are connected.

Components and join-components are special cases of so-called externally related subgraphs of a graph. Let G be a graph. We call a subset A of $V(G)$ and the subgraph $\langle A \rangle$ of G *externally related*[1] if every vertex $x \in V(G) \setminus A$ that is adjacent to at least one vertex in A is adjacent to all vertices of A. Components and join-components are externally related.

Lemma 10.2 *Let $\langle A \rangle$ and $\langle B \rangle$ be externally related subgraphs of a graph G with nonempty intersection. Then $\langle A \cup B \rangle$ is also externally related. If $\langle A \rangle$ and $\langle B \rangle$ are either both complete or both totally disconnected, then $\langle A \cup B \rangle$ is also either complete or totally disconnected.*

Proof Clear. □

Special cases of externally related subgraphs of a lexicographic product $G \circ H$ are the H-layers. If $\langle A \rangle$ is an externally related subgraph of G, then $\langle A \rangle \circ H$ is externally related in $G \circ H$.

Lemma 10.3 *The projections $p_G(X)$ and $p_H(X)$ of an externally related subgraph X of $G \circ H$ onto the factors G and H are externally related in G and H, respectively.*

Proof The first assertion is obvious. For the second it is useful to recall that contrary to the fact that the projection p_G of X into G is a weak homomorphism, p_H is in general neither a homomorphism nor a weak homomorphism. Let $X_H = p_H(X)$ and $a_2 \in V(H) \setminus X_H$ be adjacent to $b_2 \in X_H$. Let b be a vertex of X with $p_H(b) = b_2$, namely $b = (b_1, b_2)$. The vertex $a = (b_1, a_2)$ cannot be in X, and so $ab \in E(G \circ H)$ by the definition of the lexicographic product. If c_2 is any other vertex of X_H, let $c = (c_1, c_2) \in V(X)$. Clearly, $ac \in E(G \circ H)$. Thus, if $c_1 = b_1$, we have $a_2c_2 \in E(H)$. If not, $b_1c_1 \in E(G)$, and therefore $b = (b_1, b_2)$ is adjacent to $d = (c_1, a_2) \notin V(X)$. But then $c \in V(X)$ is also adjacent to d, and hence $c_2a_2 \in E(H)$. □

[1]Externally related sets have been introduced for various reasons by many authors, and the terminology is far from uniform. They are also known as *autonomous sets, closed sets, stable sets, intervals*, and *partitive sets*, to name just a few. Nowadays they are frequently called *modules*; see the book of Brandstädt, Le, and Spinrad (1999).

Lemma 10.4 *Let X be an externally related subgraph of $G \circ H$, where $|V(X)| \leq |V(H)|$. Then $X_G = p_G(X)$ is either complete or totally disconnected. Furthermore, if $|X_G| > 1$, then every join-component of X is a join-component of an H-layer of $G \circ H$, or every component of X is a component of an H-layer of $G \circ H$.*

Proof The first assertion trivially holds if $|X_G| = 1$.

Thus let $|X_G| > 1$. Suppose that X is connected. We wish to show that any two vertices $a, b \in V(X_G)$ are adjacent. Let $x \in V(X)$ be a vertex with $p_G(x) = a$. Because X is connected, there must be a neighbor y of x in $X - H^x$. This vertex is adjacent to every vertex in H^x and, in particular, to every vertex in $H^x - X$. Because $|V(X)| \leq |V(H)|$, the graph $H^x - X$ is nonempty. Let u be a vertex in $H^x - X$. As X is externally related, u is adjacent to every vertex in X, including the vertex $z \in V(X)$ with $p_G(z) = b$. But then $ab \in E(X_G)$.

If X is not connected, then $\overline{X}$ is connected and externally related in $\overline{G \circ H} = \overline{G} \circ \overline{H}$ by the above argument, and $\overline{X_{\overline{G}}}$ is complete. Thus $X_{\overline{G}}$ is totally disconnected.

The argument also shows that every join-component of X must be completely contained in an H-layer. Suppose that C is such a join-component and that $C \subseteq X \cap H^x$. We can then use the above notation and complete the proof with the observation that y is adjacent to every element w in $H^x \setminus X$, which implies that w must be adjacent to every element of $H^x \setminus X \supseteq C$. □

Corollary 10.5 *Let X be an externally related subgraph of $G \circ H$, where $|V(X)| \leq |V(H)|$. If both X and $\overline{X}$ are connected, then X is contained in an H-layer of $G \circ H$.*

Proof If X and $\overline{X}$ are connected, then X is indecomposable with respect to disjoint union and join. □

Proposition 10.6 *Let $X \circ Y \cong A \circ B$ and $|V(Y)| \leq |V(B)|$; then A has a right divisor R of cardinality* $\mathrm{lcm}(|V(Y)|, |V(B)|)/|V(B)|$. *$R$ is complete if Y is connected; otherwise, R is totally disconnected.*

Proof By complementarity, it suffices to prove the proposition for connected Y.

Let φ be an isomorphism from $X \circ Y$ onto $A \circ B$. Then the images φY^x of the Y-layers Y^x of $X \circ Y$ are externally related subgraphs of $A \circ B$ satisfying the conditions of Lemma 10.4.

Let two vertices a, b of A be in relation ρ if there exists a φY^x such that $a, b \in p_G(\varphi Y^x)$. Furthermore, let $A_1, A_2, \ldots, A_k$ be the subgraphs of A induced by the equivalence classes with respect to ρ^*. By Lemma 10.4 and repeated applications of Lemmas 10.3 and 10.2, one sees that the A_i are complete and externally related. Thus between any two different A_i-s, there are either all edges or none. Hence the complete graph on $\gcd(|V(A_1)|, |V(A_2)|, \ldots, |V(A_k)|)$ vertices is a right divisor of A, and the observation that $\mathrm{lcm}(|V(Y)|, |V(B)|)/|V(B)|$ divides every $|V(A_i)|$ completes the proof. □

Proposition 10.7 *Let $X \circ Y \cong A \circ B$ and $|V(Y)|$ be a divisor of $|V(B)|$. Then Y is a right divisor of B.*

Proof Let φ be an isomorphism from $X \circ Y$ onto $A \circ B$. If Y and $\overline{Y}$ are both connected, then every B-layer of $A \circ B$ is a union of images of Y-layers of $X \circ Y$ because of Lemma 10.4 and Corollary 10.5. But then the assertion of the proposition clearly holds.

Thus either Y or $\overline{Y}$ is disconnected. It suffices to consider the case in which Y is disconnected. We can also assume that φ does not map every Y-layer of $X \circ Y$ into a B-layer of $A \circ B$; otherwise, there is nothing to show. But then, by Lemma 10.4, every component of $\varphi(Y)$ is in a B-layer and thus isomorphic to a component of B. Furthermore, because φ is onto, every B-layer B^c of $A \circ B$ is the union of images of Y-layers of $X \circ Y$ that are completely contained in B^c and of components of images of Y-layers, that is, of images of components of Y-layers of $X \circ Y$. We can thus represent Y in the form

$$Y = \sum_{1 \le i \le k} a_i Y_i ,$$

where the Y_i are the nonisomorphic components of Y and the a_i are their multiplicities in Y. Let b_i be the multiplicity of Y_i in B. Observe that $B = H + \sum_{1 \le i \le k} b_i Y_i$, where H is induced on the union of the images of Y-layers contained in B^c. Then, because each Y-layer is externally related, it follows that $H = G \circ Y$ for the graph G for which $V(G)$ is the set of images of Y-layers contained in B^c and two vertices in G are adjacent if and only if the two Y layers are adjacent. Therefore,

$$B = G \circ Y + \sum_{1 \le i \le k} b_i Y_i .$$

Then $|V(X)| \cdot a_i = |V(A)| \cdot (|V(G)| \cdot a_i + b_i)$ for every $i, 1 \le i \le k$. Because $|V(Y)|$ is a divisor of $|V(B)|$, we conclude that $|V(A)|$ divides $|V(X)|$. For $h = (|V(X)| - |V(A)| \cdot |V(G)|)/|V(A)|$, we thus have $h \cdot a_i = b_i$. It follows that

$$B \cong G \circ Y + D_h \circ Y = (G + D_h) \circ Y ,$$

so Y is indeed a right divisor of B. □

Theorem 10.8 *Let $X \circ Y \cong A \circ B$ and $|V(Y)| = |V(B)|$. Then $Y \cong B$ and $X \cong A$.*

Proof By Proposition 10.7, one only has to show that $X \cong A$. We leave the details to the reader. □

Note that this *cancellation* property implies that $G \cong H$ if $G^{\circ,k} \cong H^{\circ,k}$ for some integer $k > 1$.

10.3 Commutativity

Recall that there are three trivial cases in which two graphs A and B commute with respect to the lexicographic product: A and B are complete, A and B are totally disconnected, or A and B are powers of one and the same graph. Moreover, we have shown that if A is complete (or totally disconnected) and if A and B commute, then B must be complete (or totally disconnected). As the following theorem shows, the proof of which follows the presentation of Imrich (1969c), there are no other cases.

Theorem 10.9 *Let A and B be two nontrivial graphs that commute with respect to the lexicographic product. Then they are either both complete, both totally disconnected, or both powers of one and the same graph C.*

Proof By the self-complementarity of the lexicographic product, we can assume that $A \circ B$ and $B \circ A$ are both connected and, hence, also A and B. We can choose the notation such that $|V(A)| \le |V(B)|$.

Let φ be an isomorphism of $B \circ A$ onto $A \circ B$.

Case 1. If the image φA^x of every A-layer A^x in $B \circ A$ is completely contained in a B-layer of $A \circ B$, then A is a right divisor of B. In this case there exists a graph C such that $B \cong C \circ A$.

Case 2. Otherwise, there is a vertex y such that φA^y meets at least two B-layers of $A \circ B$. Because A^y is externally related in $B \circ A$, its image under φ must be externally related in $A \circ B$. Now an application of Lemma 10.4 to $X = \varphi A^y$ shows that every join-component of A is isomorphic to one of B.

In Case 1, $A \circ (C \circ A) \cong (C \circ A) \circ A$ and thus $(A \circ C) \circ A \cong (C \circ A) \circ A$. By Theorem 10.8, $A \circ C \cong C \circ A$. If $C = K_1$, the assertion of the theorem is true; and if $C \neq K_1$, the theorem holds for A and B if it holds for A and C. This is because if A and C are complete, then B must be complete; and if A and C are powers of a graph G, then also B is such. We can repeat this process until either K_1 comes up as a factor or until Case 2 occurs.

We are thus left with the case where $A \circ B \cong B \circ A$, $|V(A)| \leq |V(B)|$, and where every join-component of A is isomorphic to a join-component of B. We consider three subcases:

Subcase 2.1 Every join-component of A is isomorphic to K_1. In this case, A is complete. By Proposition 10.1 (i), B must also be complete.

Subcase 2.2 No join-component of A is isomorphic to K_1. Let us denote the join-components of A by $A_1, A_2, \ldots, A_r$ and the join-components of B by $B_1, B_2, \ldots, B_s$. Because A contains no join-component isomorphic to K_1, the join-components of $A \circ B$ are $A_i \circ B$ for $i = 1, \ldots, r$.

Furthermore, the join-components of $B \circ A$ are $B_j \circ A$ for the B_j with $|V(B_j)| \neq 1$, and $B_j \circ A_i, i = 1, \ldots, r$, for the B_j with $|V(B_j)| = 1$.

Let A_1 be a join-component with minimum cardinality in A, and B_1 a join-component of minimum cardinality in B.

If $|V(B_1)| = 1$, we have $|V(A_1)| \cdot |V(B)| = |V(A_1)|$. This is only possible for $|V(B)| = 1$, but then A and B commute.

Otherwise, $|V(A_1)| \cdot |V(B)| = |V(B_1)| \cdot |V(A)|$. Because every join-component of A is isomorphic to one of B, we have $|V(A_1)| \geq |V(B_1)|$. Because we also assume $|V(B)| \geq |V(A)|$, equality must hold in all cases. But then $|V(A)| = |V(B)|$ and $A \cong B$ by Theorem 10.8.

Subcase 2.3 A contains join-components that are isomorphic to K_1 and join-components that are not. We show that A is a divisor of B, that is, we reduce this case to Case 1. Slightly changing the notation, we now denote the nonisomorphic join-components of A by

$$A_0, A_1, A_2, \ldots, A_r$$

and the nonisomorphic join-components of B by

$$B_0, B_1, B_2, \ldots, B_s.$$

For the multiplicities we introduce the notation $a_0, a_1, \ldots, a_r$ for the join-components of A and $b_0, b_1, \ldots, b_s$ for the join-components of B. Furthermore we let $A_0 \cong B_0 \cong K_1$ and set $a = a_0, b = b_0$. We also recall that every join-component of A is isomorphic to one of B. Then $A \circ B$ has the (pairwise nonisomorphic) join-components

$$K_1, B_1, B_2, \ldots, B_s, A_1 \circ B, \ldots, A_r \circ B,$$

and $B \circ A$ has the (pairwise nonisomorphic) join-components

$$K_1, A_1, A_2, \ldots, A_r, B_1 \circ A, \ldots, B_s \circ A.$$

We show first that every join-component B_i is of the form $A_j \circ A^{\circ,p}$. As usual, we adopt the convention $A^{\circ,0} \cong K_1$. Let B_i be any join-component of B. It is a join-component of $A \circ B$ and therefore isomorphic to a join-component of $B \circ A$. Thus $B_i \cong A_j$ or $B_i \cong B_{i_1} \circ A$ with $|V(B_i)| > |V(B_{i_1})|$. For B_{i_1}, we analogously have $B_{i_1} \cong A_{j_1}$ or $B_{i_1} \cong B_{i_2} \circ A$. As long as B_{i_k} is not isomorphic to A_{j_k}, we can continue this process. However, because of $|V(B_i)| > |V(B_{i_1})| > |V(B_{i_2})| > \cdots$, this process cannot continue indefinitely, which yields the desired result.

Therefore the join-components of B are K_1 with multiplicity b and products of the form $A_i \circ A^{\circ,j}, j = 0, \ldots, k_i$ for $i = 1, \ldots, r$. Let $b_{i,j}$ be their multiplicities. We thus obtain the following table for the multiplicities of the join-components of $A \circ B$ and $B \circ A$:

	Join-Component		Multiplicity
$A \circ B$	K_1		$a \cdot b$
	$A_i \circ A^{\circ,j}$	$0 \le j \le k_i, 1 \le i \le r$	$a \cdot b_{i,j}$
	$A_i \circ B$	$1 \le i \le r$	a_i
$B \circ A$	K_1		$b \cdot a$
	A_i	$1 \le i \le r$	$b \cdot a_i$
	$A_i \circ A^{\circ,j}$	$1 \le j \le k_i + 1, 1 \le i \le r$	$b_{i,j-1}$

It is not hard to see that the join-components of $A \circ B$, respectively of $B \circ A$, as listed above, are pairwise nonisomorphic. From $A \circ B \cong B \circ A$ we infer that $b \cdot a_i = a \cdot b_{i,0}$ and $b_{i,j-1} = a \cdot b_{i,j}$ for $1 \le j \le k_i$. Thus

$$b \cdot a_i = a^{k_i+1} \cdot b_{i,k_i} \quad \text{for} \quad 1 \le i \le r. \tag{10.1}$$

Moreover, the join-components $A_i \circ A^{\circ,k_i+1}$ of $B \circ A$ can only be isomorphic to the $A_j \circ B$ of $A \circ B$. From this we conclude that there must be a permutation π of the first r natural numbers with $A_i \circ A^{\circ,k_i+1} \cong A_{\pi i} \circ B$. Therefore $b_{i,k_i} = a_{\pi i}$, which implies that

$$b \cdot a_i = a_{\pi i} \cdot a^{k_i+1} \tag{10.2}$$

by (10.1). Hence

$$b^2 \cdot a_i = b \cdot a_{\pi i} \cdot a^{k_i+1} = a_{\pi^2 i} \cdot a^p,$$

where $p = (k_{\pi i+1}) + (k_i + 1)$. Now it is easy to see that there is an integer q to every integer $d > 0$ with

$$b^d \cdot a_i = a_{\pi^d i} \cdot a^q.$$

If d is the order of π, we have $\pi^d i = i$ and $b^d = a^q$. Thus a is a divisor of b, or vice versa. If $b|a$, we infer from (10.2) that

$$a_i = a_{\pi i} \cdot a^{k_i} \cdot \frac{a}{b}.$$

This, in turn, implies that $a_i \ge a_{\pi i}$, and hence

$$a_{\pi i} \ge a_{\pi^2 i} \ge \cdots \ge a_{\pi^{d-1} i} \ge a_{\pi^d i} = a_i,$$

namely, $a_i = a_{\pi i}$. With (10.2) we thus have $a|b$. From this and $b \cdot a_i = a \cdot b_{i,0}$, we finally obtain

$$a_i \cdot \frac{b}{a} = b_{i,0}.$$

Because the join-components K_1 and A_i of B have multiplicities b and $b_{i,0}$ and because the others are of the form $A_i \circ A^{\circ,j}$, where $j \ge 1$, this implies that A is a right divisor of B. □

10.4 Factorizations and Nonuniqueness

In this section we show that all prime factorizations of a graph with respect to the lexicographic product have the same number of factors and that they can be transformed into each other by sequences of so-called transpositions of complete or totally disconnected factors. This also yields a large class of graphs with unique prime factorizations (Theorem 10.11 and Corollary 10.12).

We begin with the observation that prime factorizations of connected graphs with respect to the lexicographic product need not be unique. To see this, consider

$$(K_2 \circ D_2 \oplus K_3) \circ K_2 \cong K_2 \circ (D_2 \circ K_2 \oplus K_3).$$

That both sides are equal can be seen from the distributive laws for the lexicographic product with respect to the join of graphs. That all factors are prime is clear, because they have either seven or two vertices. Finally, the observation that the first factor on the left side has nineteen edges and the second one on the right only seventeen shows that they cannot be isomorphic.

The key to our investigations is the following lemma.

Lemma 10.10 *Let G be a graph without isolated vertices and m, n natural numbers. Then $D_n \circ G + D_m$ is prime with respect to the lexicographic product if and only if $G \circ D_n + D_m$ is prime.*

If G has no trivial join-components, then $K_n \circ G \oplus K_m$ is prime with respect to the lexicographic product if and only if $G \circ K_n \oplus K_m$ is prime.

Proof It suffices to prove the first assertion. Because both $D_n \circ G + D_m$ and $G \circ D_n + D_m$ are decomposable if $\gcd(m, n) > 1$ (Exercise 10.1), we can assume that m and n are relatively prime. Moreover, the assertion is trivially true when G is totally disconnected. Suppose that $D_n \circ G + D_m$ is not prime. Let

$$D_n \circ G + D_m = \left(\sum_{i=0}^{a} a_i A_i\right) \circ \left(\sum_{j=0}^{b} b_j B_j\right)$$

be a decomposition, where A_i and B_j denote connected graphs and $a_i A_i$ the disjoint union of a_i copies of A_i. Furthermore let $A_0 \cong B_0 \cong K_1$. Then

$$D_n \circ G + D_m = a_0 b_0 (A_0 \circ B_0) + a_0 \sum_{j>0} b_j B_j + \sum_{i>0} \left(a_i A_i \circ \sum_j b_j B_j\right).$$

We note that the B_j and the $A_i \circ \sum b_j B_j$ are pairwise nonisomorphic and that n divides the multiplicity of every nontrivial component of $D_n \circ G + D_m$. Therefore n also divides $a_0 b_j$ and a_i for $i, j > 0$.

Because $m = a_0 b_0$ and $\gcd(m, n) = 1$, we conclude that $\gcd(n, a_0) = 1$. Because n divides $a_0 b_j$, it must also divide b_j. Therefore we can represent $D_n \circ G + D_m$ in the form $(D_n \circ A + a_0 K_1) \circ (D_n \circ B + b_0 K_1)$. For $D_n \circ G$, this implies that

$$D_n \circ G = a_0 D_n \circ B + D_n \circ A \circ (D_n \circ B + b_0 K_1)$$

and

$$G = a_0 B + A \circ (D_n \circ B + b_0 K_1).$$

Because of the distributive laws for the lexicographic product, substitution of this expression for G in $G \circ D_n + D_m$ yields

$$G \circ D_n + D_m = (A \circ D_n + a_0 K_1) \circ (B \circ D_n + b_0 K_1).$$

Hence $G \circ D_n + D_m$ is not prime.

Let us assume now that $G \circ D_n + D_m$ is not prime. Analogous to the above, we can represent $G \circ D_n + D_m$ in the form

$$a_0 b_0 (A_0 \circ B_0) + a_0 \sum_{j>0} b_j B_j + \sum_{i>0} \left(a_i A_i \circ \sum_j b_j B_j \right).$$

As before, $a_0 b_0 = m$. Together with $\gcd(m,n) = 1$, this implies that $\gcd(b_0, n) = 1$. Because every B_j is isomorphic to a component of $G \circ D_n$, we also infer that D_n is a right divisor of every B_j. Thus we can represent $\sum b_j B_j$ in the form $B \circ D_n + b_0 K_1$, where B has only nontrivial components.

Because every $A_i \circ \sum b_j B_j$ is isomorphic to a component of $G \circ D_n$, there must be a component W of G with

$$W \circ D_n \cong A_i \circ (B \circ D_n + b_0 K_1).$$

By Proposition 10.6, the totally disconnected graph D with

$$|V(D)| = \frac{\operatorname{lcm}(|V(D_n)|, |V(B \circ D_n + b_0 K_1)|)}{|V(B \circ D_n + b_0 K_1)|} = \frac{\operatorname{lcm}(n, n|V(B)| + b_0)}{n|V(B)| + b_0}$$

is a right divisor of A_i. Because $|V(D)| = n/\gcd(n, n|V(B)| + b_0) = n/\gcd(n, b_0) = n$, we can represent $\sum a_i A_i$ in the form $A \circ D_n + a_0 K_1$, where A has only nontrivial components. Now the proof can be completed like that of the first part. □

If $Q \circ D_p$ is a prime factorization of a graph G and if Q can be represented in the form $D_p \circ H + D_m$, where H has only nontrivial components, then $D_p \circ R$, where $R = H \circ D_p + D_m$, is another prime factorization of G. We say these two representations arise from each other by a *transposition of D_p*. If H is empty, then D_p and D_m are simply interchanged. Analogously, we define a *transposition of K_p*.

Note that our example for the nonunique prime factor decomposition of graphs with respect to the lexicographic product is a special case of the transposition of complete factors. The proof of the following result is adapted from Imrich (1972a), respectively Dörfler and Imrich (1972) for directed graphs. However, the results are already due to C. C. Chang (1961), who investigated factorizations of finite relations.

Theorem 10.11 *Any prime factor decomposition of a graph can be transformed into any other one by transpositions of totally disconnected or complete factors.*

Proof The proof is by induction with respect to the number n of vertices of G. For $n = 2$, the assertion of the theorem is evidently true.

Let it be true for all graphs with less than n vertices and G be a graph on n vertices. We consider two prime factor decompositions of G; denote the first factor in the first decomposition by P and the first factor in the second one by Q. Then

$$P \circ R \cong Q \circ S \cong G.$$

We can choose the notation such that $|V(R)| \leq |V(S)|$. Note that R and S will not be prime in general.

If $|V(R)|$ divides $|V(S)|$, we infer from Proposition 10.7 that R is a right divisor of S, namely that there exists a graph T with $S = T \circ R$. But then an application of Theorem 10.8 shows that $P \cong Q \circ T$. Because P is prime, T must be trivial and $P \cong Q$ and $R \cong S$. Now the validity of the assertion of the theorem follows from the induction hypothesis because $|V(R)| < |V(G)|$.

If $|V(R)|$ is not a divisor of $|V(S)|$, then Proposition 10.6 implies that Q has a nontrivial right divisor T that is complete or totally disconnected. Because Q is prime, it must be equal to this divisor T. Without loss of generality, we can assume that Q is the totally disconnected graph D_q, where q is prime.

Let $P = \sum p_i P_i$, where the P_i are the components of P and the p_i their multiplicities. We assume that $P_0 \cong K_1$ and admit that $p_0 = 0$. Analogously, we represent R by $\sum r_j R_j$ and S by $\sum s_k S_k$. By $P \circ R \cong D_q \circ S$, we have

$$\sum_{i>0} p_i(P_i \circ R) + p_0 \sum r_j R_j \cong q \sum s_k S_k.$$

If $p_0 = 0$, then every component of S has R as a right divisor, and therefore also S. Then $P \cong Q$, as above. Thus let $p_0 \neq 0$. We note that every p_i, for $i > 0$, must be divisible by q. But then $P = \sum p_i P_i$ can be represented in the form $D_q \circ X + p_0 K_1$, where X has no nontrivial components. Because P is prime, we have $\gcd(q, p_0) = 1$. But q must also divide all $p_0 r_j$, and therefore all r_j. Then D_q is a left divisor of R. Let $R \cong D_q \circ U$. From

$$(P \circ D_q) \circ U \cong D_q \circ S$$

we now infer by Proposition 10.7 that U is a right divisor of S. This means that there is a V with $S \cong V \circ U$. By Theorem 10.8, we thus have

$$P \circ D_q \cong D_q \circ V.$$

P has exactly p_0 isolated vertices, thus also V, and we can represent V in the form $Y + p_0 K_1$, where Y has only nontrivial components. Then

$$(D_q \circ X + p_0 K_1) \circ D_q \cong D_q \circ (Y + p_0 K_1)$$

implies that $D_q \circ X \circ D_q \cong D_q \circ Y$, and hence $X \circ D_q \cong Y$.

Now the observation that $P \cong D_q \circ X + p_0 K_1$ and $V \cong Y + p_0 K_1 \cong X \circ D_q + p_0 K_1$ shows that $P \circ D_q$ and $D_q \circ V$ arise from each other by transposition of D_q.

Moreover, let us recall that $Q \cong D_q$, $R \cong D_q \circ U$ and $S \cong V \circ U$, namely

$$G \cong P \circ D_q \circ U \cong D_q \circ V \circ U.$$

By the induction hypothesis, every prime factor decomposition of R can be transformed into $D_q \circ U$ by transpositions of complete or totally disconnected factors. A transposition of D_q transforms $P \circ D_q \circ U$ into $D_q \circ V \circ U$. Again by the induction hypothesis, we can transform every prime factor decomposition of $S = V \circ U$ into any other one by transpositions of complete or totally disconnected factors. This proves the theorem. □

Corollary 10.12 *All prime factor decompositions of a graph G with respect to the lexicographic product have the same number of factors.*

If there is a prime factorization of G without complete or totally disconnected factors, then G has unique prime factor decomposition.

If there is a prime factorization of G in which only complete factors have trivial join-components and only totally disconnected factors have trivial components, then G has unique prime factor decomposition.

10.5 Automorphisms

We derive a simple condition under which the automorphism group of a lexicographic product is the wreath product of the groups of the factors. Then we continue with lexicographic products with transitive groups. In particular, we consider products that allow regular group actions. This motivates the introduction of Cayley graphs and some of their properties, in particular, with respect to the lexicographic product.

The wreath product

Let $G \circ H$ be a lexicographic product, β an automorphism of H, and (g,h) a vertex of $G \circ H$. Then the permutation of $V(G \circ H)$ that maps (g,h) into $(g,\beta h)$ and is the identity elsewhere clearly is in $\mathrm{Aut}(G \circ H)$. Also, if $\alpha \in \mathrm{Aut}(G)$, then the mapping $(g,h) \mapsto (\alpha g, h)$ is an automorphism of $G \circ H$.

The group generated by such elements is known as the *wreath product* $\mathrm{Aut}(G) \circ \mathrm{Aut}(H)$. Evidently all its elements can be written in the form

$$(g,h) \mapsto (\alpha g, \beta_{\alpha g} h),$$

where α is an automorphism of G and where the $\beta_{\alpha g}$ are automorphisms of H.

As the example of $K_2 \circ K_2$ shows, $\mathrm{Aut}(G) \circ \mathrm{Aut}(H)$ can be a proper subgroup of $\mathrm{Aut}(G \circ H)$. The next theorem describes when it is equal to $\mathrm{Aut}(G \circ H)$.

For the statement of the theorem, we use the relations S and R that were defined in Chapters 7 and 8. For the reader who skipped those chapters, it suffices to know that R_G is nontrivial if and only if there exists an externally related set $\{u,v\}$ of two nonadjacent vertices, whereas S_G is nontrivial if and only if there exists an externally related set $\{u,v\}$ of two adjacent vertices.

Theorem 10.13 *Let $G \circ H$ be the lexicographic product of simple nontrivial graphs. Then* $\mathrm{Aut}(G \circ H) = Aut(G) \circ Aut(H)$ *if and only if H is connected in case R_G is nontrivial and $\overline{H}$ is connected in case S_G is nontrivial.*

Proof Note that the conditions of the theorem are not exclusive. Thus, it is possible that both R_G and S_G are nontrivial, just as both H and $\overline{H}$ can be connected. Nevertheless, by the self-complementarity of the lexicographic product, it suffices to prove the first assertion.

Let u, v be two vertices of G with uR_Gv and α be the automorphism of G that interchanges u and v. Suppose that U is a nontrivial component of H. Define $\varphi : V(G \circ H) \to V(G \circ H)$ by

$$\varphi(g,h) = \begin{cases} (g,h) & \text{if } h \in V(U) \text{ and} \\ (\alpha g, h) & \text{otherwise}. \end{cases}$$

Then φ is an automorphism of $G \circ H$ that is not in $\mathrm{Aut}(G) \circ \mathrm{Aut}(H)$.

On the other hand, let ψ be an automorphism of $G \circ H$ that is not in $\mathrm{Aut}(G) \circ \mathrm{Aut}(H)$. Then there must be an H-layer, say $H^{(g,h)}$, that is not mapped into an H-layer of $G \circ H$. Then $|p_G(\psi H^{(g,h)})| \geq 2$. Because $H^{(g,h)}$ is externally related, we can invoke Lemma 10.4. Thus $p_G(\psi H^{(g,h)})$ is either complete or totally disconnected. Because H is connected, $p_G(\psi H^{(g,h)})$ is complete; and by Lemma 10.3, it is also externally related. But then S_G is nontrivial, contrary to assumption. □

The remainder of this section is concerned with graphs with transitive and regular automorphism groups. To follow the proofs, it suffices to know the definitions, which we briefly recall: A graph G has transitive automorphism group if there exists an automorphism φ to any pair u, v of vertices in G such that $\varphi u = v$. If one requires that there exists only one such automorphism to every pair of vertices, then G has regular automorphism group.

As in the case of the other products considered so far, a lexicographic product has transitive automorphism group if and only if every factor has transitive group:

Theorem 10.14 *A lexicographic product $G \circ H$ has transitive automorphism group if and only if G and H have transitive automorphism groups.*

Proof Because $\mathrm{Aut}(G) \circ \mathrm{Aut}(H)$ is contained in $\mathrm{Aut}(G \circ H)$, the product clearly has transitive group if the factors do. If $\mathrm{Aut}(G) \circ \mathrm{Aut}(H) = \mathrm{Aut}(G \circ H)$, the converse also holds.

Thus the case remains in which these groups are not equal. By Theorem 10.13, this is only possible if either R_G or S_G is nontrivial. Because the lexicographic product is self-complementarity and because S_G is nontrivial exactly if $R_{\overline{G}}$ is nontrivial, we can assume that R_G is nontrivial. In this case, H must be disconnected. We have to show that both G and H have transitive group if $G \circ H$ has.

We show first that $\mathrm{Aut}(G \circ H)$ maps components of the H-layers of $G \circ H$ into components of H-layers. To see this, let $\varphi \in \mathrm{Aut}(G \circ H)$ and $H^{(g,h)}$ be arbitrarily chosen. Set $X = \varphi H^{(g,h)}$. If $|p_G(X)| = 1$, then the assertion is true. Otherwise, we infer from Lemma 10.4 that every component of X is one of an appropriate H-layer of the product.

One calls these components blocks and says they form a *system of imprimitivity* with respect to the automorphism group of $G \circ H$. Clearly, $\mathrm{Aut}(G \circ H)$ can only be transitive if the blocks are pairwise isomorphic and have transitive groups. Thus, H has transitive group.

We continue as in the proof of Proposition 10.6. We say two vertices a, b of G are in relation ρ if there exists a φH^x such that $a, b \in p_G(\varphi H^x)$, where $\varphi \in \mathrm{Aut}(G \circ H)$ and $x \in V(G \circ H)$. Furthermore, let $G_1, G_2, \ldots, G_k$ be the subgraphs of G induced by the equivalence classes with respect to ρ^*. By Lemma 10.4 and repeated applications of Lemmas 10.3 and 10.2, the G_i are totally disconnected and externally related. Thus, between any two different G_i's, there are either all edges or none. Because $\mathrm{Aut}(G \circ H)$ acts transitively on the subproducts $G_i \circ H$, it induces a transitive group acting on the G_i. Because these are totally disconnected and externally related, $\mathrm{Aut}(G)$ is transitive. □

Theorem 10.13 about automorphisms of lexicographic products is due to Sabidussi (1959). The fundamental characterizations of graphs admitting regular or transitive automorphism groups as laid down in Theorems 10.15 and 10.16 are also due to Sabidussi (1958, 1964).

Evidently the automorphism group of a nontrivial lexicographic product cannot be regular. This contrasts the situation for other products. However, lexicographic products may admit regular subgroups of their automorphism groups. To prepare for the characterization of this situation, we define Cayley graphs.

Cayley graphs

Given a group A and a subset S of $A \setminus \{1\}$, where 1 denotes the unit element of A, we define the *Cayley graph* $\Gamma(A, S)$ on A by setting $V(\Gamma(A, S)) = A$ and

$$E(\Gamma(A, S)) = \{[a, as] \mid a \in A, s \in S\}.$$

The element 1 is excluded from S to avoid loops. Moreover, because $[a, as] = [as, (as)s^{-1}]$, the Cayley graphs $\Gamma(A, S)$ and $\Gamma(A, S \cup S^{-1})$, where $S^{-1} = \{s^{-1} \mid s \in S\}$, are the same.

Furthermore it is easy to see that $\Gamma(A,S)$ is connected if and only if every element of A can be written as a product of elements in S, namely if S generates A.

We show now that there exists a homomorphism $\lambda : A \to \mathrm{Aut}(\Gamma(A,S))$ such that $\lambda(A)$ acts regularly on $\Gamma(A,S)$. To see this, define $\lambda : a \mapsto \lambda_a$ by

$$\lambda_a(x) = ax \text{ for } x \in V(\Gamma(A,S)).$$

Because $\lambda_a[x, xs] = [ax, (ax)s]$, the mappings λ_a are indeed automorphisms of $\Gamma(A,S)$. Moreover, for $a, b \in A$,

$$\lambda_b(\lambda_a(x)) = \lambda_b(ax) = b(ax) = (ba)x = \lambda_{ba}(x).$$

This means that λ is a homomorphism. We say *A acts on $\Gamma(A,S)$ by left multiplication.* Clearly, this action is transitive, because for any pair $a, b \in A$, we have $\lambda_{ba^{-1}}a = b$. The action is also regular, as $\lambda_x a = b$ and $\lambda_y a = b$ imply that $xa = b = ya$, from which we infer that $x = y$ and $\lambda_x = \lambda_y$.

Theorem 10.15 *A graph G is isomorphic to a Cayley graph $\Gamma(A,S)$ if and only if* $\mathrm{Aut}(G)$ *contains a subgroup A_0 that is regular as a permutation group on $V(G)$. In this case, $A = A_0$.*

Proof By the above, it suffices to show that $G \cong \Gamma(A_0, S)$ for some $S \subseteq A_0$ if $\mathrm{Aut}(G)$ contains a regular subgroup A_0. Let v_0 be a fixed vertex of G. By assumption, there is a unique element $a_v \in A_0$ to every $v \in V(G)$ such that $a_v v_0 = v$. For $N(v_0) = \{v_1, v_2, \ldots, v_k\}$, set $S = \cup_{i=1}^{k} a_{v_i}^{\pm 1}$. We wish to show that the mapping $a_v \mapsto v$ is an isomorphism from $\Gamma(A_0, S)$ onto G.

Clearly, the mapping is a bijection. Consider an a_v and a pair $\{x, xa\}$, where $x, a \in A_0$. Then $a_v(\{x, xa\}) = a_v(a_{x^{-1}}\{1, a\})$ is in $E(G)$ if and only if $a \in S$, namely if and only if $\{x, xa\}$ is in $E(\Gamma(A_0, S))$. □

Note that this implies that the graphs with transitive Abelian group of Chapter 6 are Cayley graphs.

Theorem 10.16 *Let G be a graph with transitive automorphism group and $n = |\mathrm{Aut}(G)|\,/\,|V(G)|$. Then $G \circ D_n$ admits a group of automorphisms $A \subseteq \mathrm{Aut}(G \circ D_n)$ that acts regularly on $G \circ D_n$ and is isomorphic to* $\mathrm{Aut}(G)$.

Proof Set $A = \mathrm{Aut}(G)$ and let v_0 be an arbitrary but fixed vertex of G, where A_{v_0} is the stabilizer of v_0 in A. Note that all stabilizers are of the form $\varphi A_{v_0} \varphi^{-1}$ because $\varphi A_{v_0} \varphi^{-1} = A_{\varphi v_0}$ for any $\varphi \in A$. Of course this implies that $|A| = |A_{v_0}| \cdot |V(G)|$, namely $n = |A_{v_0}|$.

Let A_{v_0} be the vertex set of D_n. We wish to define a regular group action of A on $G \circ D_n$. To this end we choose an element a_v in A for every $v \in V(G) - v_0$ such that $a_v(v_0) = v$. Let L denote the set of these elements. For the subset N of the elements a_v for which v is a neighbor of v_0, we consider the Cayley graph $\Gamma(A, NA_{v_0})$ and show that it is isomorphic to $G \circ D_n$.

Let $(v, a) \in V(G \circ D_n)$ and define $\psi : V(G \circ D_n) \to V(\Gamma(A, NA_{v_0})) = A$ by $\psi(v, a) = a_v a$. This mapping is onto, because every $b \in A$ can be represented in the form $a_{b(v_0)}((a_{b(v_0)})^{-1} b)$, where $(a_{b(v_0)})^{-1} b$ clearly is in A_{v_0}. Moreover, if $a_v a = a_w a'$ for $a, a' \in A_{v_0}$, then

$$v = a_v(v_0) = a_v(a v_0) = a_v a(v_0) = a_w a'(v_0) = a_w(a'(v_0)) = a_w(v_0) = w\,.$$

Then $a_v = a_w$; and from $a_v a = a_w a'$ and the cancellation property of group multiplication, we get $a = a'$. So ψ is a bijection.

Because the number of edges in $G \circ D_n$ and $\Gamma(A, NA_{v_0})$ is the same, it suffices to show that ψ maps edges to edges to complete the proof. Let $[(v,a),(w,b)]$ be an edge of $G \circ D_n$. Because D_n has no edges, $[v,w] \in E(G)$. Then $a_v^{-1}w$ is a neighbor of v_0. Moreover, $a^{-1}a_v^{-1}a_w b(v_0)$ is also a neighbor of v_0, say u. Then $a_u \in N$ and $a_u^{-1}a^{-1}a_v^{-1}a_w b$ stabilizes v_0. It is therefore equal to an element $c \in A_{v_0}$. Thus $a_w b = a_v a a_u c$ and $\psi(v,a) = a_v a$ is adjacent to $\psi(w,b) = a_w b$ in $\Gamma(A, NA_{v_0})$. □

Because the projection p_G of $G \circ D_n$ onto G is a homomorphism, every graph with transitive group is a homomorphic image of a Cayley graph. In fact, because G is isomorphic to a G-layer of $G \circ D_n$, every graph with transitive group is a retract of a Cayley graph. We formulate this as a corollary:

Corollary 10.17 *Every graph with transitive automorphism group is a retract of a Cayley graph.*

Naturally the question arises whether all graphs with transitive group are already Cayley graphs themselves. As so often, the Petersen graph is a counterexample, and taking Cartesian products with Cayley graphs that are prime with respect to the Cartesian product, one can obtain arbitrarily large counterexamples; see Sabidussi (1964).

We close the chapter on the lexicographic product with three remarks.

- The lexicographic product has numerous applications; some of them are mentioned in Chapter 26. We also refer to Jónsson (1982) for a survey of the role of the lexicographic product in the arithmetic of ordinal numbers and the study of total order.

- An important generalization of the lexicographic product is the *X-join*. It was introduced by Sabidussi (1961) as the graph formed from a given graph G by replacing every vertex v of G by a graph H_v and joining the vertices of H_v with those of H_u whenever $uv \in E(G)$. Note that the H_v need not be mutually isomorphic. We denote it by $G[H_{v_1}, ..., H_{v_n}]$. For some of its properties and a generalization of Theorem 10.13, see Hemminger (1968). For other results, see Habib and Maurer (1979) or Moehring and Radermacher (1984) for applications.

- Another product that has similarities with the lexicographic product is the replacement product. It was introduced for the construction of good expander graphs. We give a short account of it in Section 33.4.

Exercises

10.1. Let m and n be integers with $\gcd(m,n) > 1$. Show that both $D_n \circ G + D_m$ and $G \circ D_n + D_m$ are decomposable with respect to the lexicographic product.

10.2. (Knauer, 1987) Let G be a graph. Show that for any n, the lexicographic products $K_n \circ G$ and $C_{2n+1} \circ G$ have no proper retracts if and only if G has no proper retract.

10.3. Recall that a graph G is a core if no proper subgraph of G is a retract of G. A retract H of G is called a *core of* G if H is a core. Show that every graph G has a core and that any two cores of G are isomorphic.

10.4. By Exercise 10.3, we can speak about **the** core of a graph (as an abstract graph). Suppose that G and H are graphs such that there exists a homomorphism $G \to H$ and a homomorphism $H \to G$. Then show that G and H have the same core.

10.5. (Hahn and Tardif, 1997) Show that for connected graphs G and H, the core of $G \circ H$ can be represented as the lexicographic product $G' \circ H'$, where G' is a subgraph of G and H' the core of H.

10.6. Show that cycles and complete graphs are Cayley graphs.

10.7. Let $G = \Gamma(A, S)$ and $G' = \Gamma(A', S')$ be Cayley graphs. Show that $G \times G'$ is isomorphic to $\Gamma(A \times A', S \times S')$.

10.8. Show that hypercubes are Cayley graphs.

10.9. Let $G = \Gamma(A, S)$ and $G' = \Gamma(A', S')$ be Cayley graphs. Show that $G \circ G'$ is isomorphic to $\Gamma(A \times A', (\{1\} \times S') \cup (S \times A'))$.

10.10. Let G be a graph with transitive automorphism group and n the cardinality of a vertex-stabilizer of G. Show that $G \circ K_n$ is a Cayley graph.

Part III

Isometric Embeddings

Introduction to Part III

MOST GRAPHS are prime. Consequently, the characteristic structure of graph products makes them special, interesting, and rare. We can obtain a much wider class by considering the graphs that can be isometrically (and nontrivially) embedded into graph products. This is the theme of Part III.

The first three chapters investigate isometric subgraphs and retracts of hypercubes, and show how graphs can be canonically embedded into Cartesian products. The fourth chapter studies isometric subgraphs and weak retracts of Hamming graphs, and uses the results to solve the so-called dynamic location problem.

The fifth chapter develops analogous ideas for the strong product. It shows that every graph embeds isometrically into a strong product of paths, and we introduce the strong dimension of a graph as the minimum number of factors required for such an embedding.

Along the way we derive numerous fixed-cube results. For example, we show that any median graph contains a hypercube that is fixed by all automorphisms of the graph. Such results typically hold for graphs that are isometric subgraphs of hypercubes, Hamming graphs, or strong product of paths. The last chapter generalizes these ideas to arbitrary Cartesian products of graphs. The fixed-box theorems of Feder (1995), Tardif (1997), together with Theorem 16.25 of Feder (2006) about fixed points of several nonexpansive mappings, are the most general results of this type.

The Djoković-Winkler relation Θ is our primary tool. This relation on the edge set of a graph is indispensable in the first three chapters of this part. It will also be used for an independent proof of the unique prime factorization of connected graphs with respect to the Cartesian product (Theorems 23.2 and 23.4) and will provide the basis for a straightforward factorization algorithm for Cartesian products (Algorithm 23.1).

Other highlights of Part III include Mulder's convex expansion theorem (Theorem 12.8), Bandelt's characterization of median graphs as retracts of hypercubes (Theorem 12.18), and the canonical isometric embedding of graphs by Graham and Winkler (Theorem 13.2).

Part III assumes knowledge of Part I, but does not require any of the results in Part II. Part III begins with Chapter 11 on partial cubes and the relation Θ. Chapter 12 (Median Graphs) and Chapter 13 (Canonical Isometric Embedding) both build on Chapter 11, but are independent of each other. Chapter 14 (Dynamic Location Problem) uses Chapters 11 through 13. Chapter 15 is entirely independent of the other chapters in Part III, but Chapter 16 uses ideas from Chapters 11 through 15.

Chapter 11

The Relation Θ and Partial Cubes

The relation Θ plays an important role in the structural characterization of isometric and convex subgraphs of hypercubes. It is the basis of the so-called canonical isometric embedding of graphs into Cartesian products and, together with the relation τ (defined later), allows a characterization of prime factorizations of connected graphs with respect to the Cartesian product. This characterization is valid in both the finite and the infinite case, and leads to another proof of the unique prime factorization property.

These structural characterizations are the basis of numerous algorithms, among them recognition algorithms for partial cubes, Cartesian products, and an algorithm for the computation of the Wiener index for benzenoid graphs.

The first two sections of this chapter are concerned with the definition of Θ, its basic properties, and applications to characterizations of convex subgraphs and of partial cubes.

The third section demonstrates the astonishing richness of partial cubes, even when restricted to the cubic case. It also exhibits their close connection with geometric structures. Finally we discuss graphs that are scale embeddable into hypercubes, a broad generalization of partial cubes.

11.1 Definition and Basic Properties of Θ

The relation Θ was introduced by Djoković (1973), but the definition given here is due to Winkler (1984). We refer to it as the Djoković-Winkler relation. In this section we derive its basic properties and use it to characterize convex subgraphs of bipartite graphs.

Unlike the relations R and S, which are relations on vertices, Θ is a relation on edges. Two edges $e = ab$ and $f = xy$ in a graph are in relation Θ, in symbols $e\Theta f$, if

$$d(a,x) + d(b,y) \neq d(a,y) + d(b,x)\,.$$

See Figure 11.1, where $e\Theta f$ if the dashed and dotted distances have unequal sums.

We remark that if e and f are in different components, then they are not in relation Θ, because $d(a,x) + d(b,y) = d(a,y) + d(b,x) = \infty$.

The relation Θ is reflexive and symmetric, but need not be transitive. We denote its transitive closure, that is, the smallest transitive relation containing Θ, by Θ^*. If $G = C_{2n}$ is a cycle of even length, then Θ consists of all pairs of antipodal edges. Hence, Θ^* has n equivalence classes and $\Theta = \Theta^*$ in this case. On the other hand, any edge of an odd cycle is in relation Θ with its two antipodal edges. In this case, Θ^* has only one equivalence class.

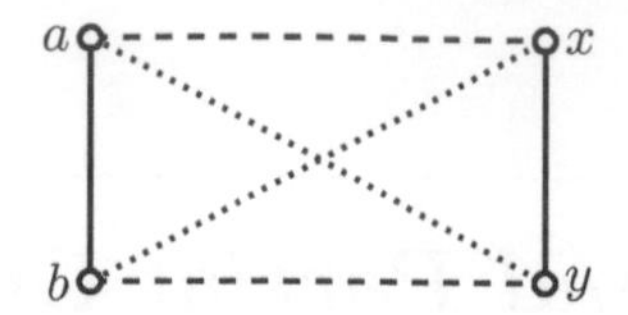

FIGURE 11.1 Definition of Θ.

Lemma 11.1 *No two distinct edges on a shortest path in a graph are in relation* Θ.

Proof Let e and f be two edges of a shortest path P in a graph. We can choose the notation such that $P = u_0u_1 \dots u_m$ and $e = u_iu_{i+1}$, $f = u_ju_{j+1}$, where $i < j$. Then

$$\begin{aligned} d(u_i, u_j) + d(u_{i+1}, u_{j+1}) &= \big(d(u_{i+1}, u_j) + 1\big) + \big(d(u_i, u_{j+1}) - 1\big) \\ &= d(u_{i+1}, u_j) + d(u_i, u_{j+1}), \end{aligned}$$

which means that e is not in relation Θ with f. □

For a tree on n vertices, we infer from Lemma 11.1 that Θ^* has $n-1$ equivalence classes, each consisting of a single edge. Lemma 11.1 also implies that two adjacent edges are in relation Θ if and only if they lie in a common triangle. For bipartite graphs, this means that incident edges cannot be in relation Θ.

Lemma 11.2 *Suppose* $e = ab$ *and* $f = xy$ *are two edges of a bipartite graph, and* $e\Theta f$. *Then the notation can be chosen such that*

$$d(a, x) = d(b, y) = d(a, y) - 1 = d(b, x) - 1.$$

Proof Clearly, $d(a, x) \neq d(a, y)$, for otherwise the bipartite graph would contain a closed walk of odd length (containing a, x and y). As x and y are adjacent, their distances to other vertices can differ by at most one. Choose the notation so that $d(a, y) = d(a, x) + 1$.

By the same argument, $d(b, x) \neq d(b, y)$. If $d(b, y) = d(b, x) + 1$, then $d(a, x) + d(b, y) = d(a, y) + d(b, x)$, contrary to $e\Theta f$. Thus

$$\begin{aligned} d(b, y) &= d(b, x) - 1 \\ &\leq d(b, a) + d(a, x) - 1 \\ &= d(a, x) \\ &= d(a, y) - 1 \\ &\leq d(a, b) + d(b, y) - 1 \\ &= d(b, y). \end{aligned}$$

Hence, equality holds everywhere. □

For ordered pairs $p = (a, b)$ and $q = (x, y)$ of vertices of a graph, we set

$$\mu(p, q) = d(a, y) - d(a, x) - d(b, y) + d(b, x).$$

Clearly, $\mu(p, q)$ changes its sign when the orientation of either p or q is reversed, but remains unaltered when both orientations are changed. Thus, if we view edges $e = ab$ and $f = xy$ as ordered pairs with arbitrary orientations, then $e\Theta f$ if and only if $\mu(e, f) \neq 0$.

Lemma 11.3 *Suppose that a walk* P *connects the endpoints of an edge* e *but does not contain it. Then* P *contains an edge* f *with* $e\Theta f$. *If it is the only edge of* P *with this property, then it is not incident with* e.

Proof Let the path $u_0u_1 \dots u_m$ connect the endpoints of $e = u_0u_m$. Set $e_i = u_{i-1}u_i$ for $i = 1, 2, \dots, m$. Considering e and e_i as ordered pairs (u_0, u_m) and (u_{i-1}, u_i), we set

$$s = \sum_{i=1}^{m} \mu(e, e_i).$$

By the definition of μ, we have $s = d(u_m, u_0) + d(u_0, u_m) = 2$. This means that at least one of the summands $\mu(e, e_i)$ is nonzero. But then $e\Theta e_i$. The observation that $|\mu(e, e_i)| \le 1$ for edges incident with e completes the proof. □

Suppose we are given a connected graph G with a spanning tree T. Then Lemma 11.3 implies that every edge of G is in relation Θ with some edge of T. Consequently, Θ^* has at most $|V(G)| - 1$ equivalence classes.

Lemma 11.4 *Let F be the union of one or more equivalence classes of Θ^* and P be a path whose edges are in F. Then every edge of any shortest path connecting the endpoints of P is also in F.*

Proof Let Q be such a shortest path. We may, without loss of generality, assume that Q has no other common vertex with P than the endpoints. No pair of edges of Q is in relation Θ by Lemma 11.1, but every edge of Q must be in relation Θ with some edge of P by Lemma 11.3. Thus, every edge of Q is also in F. □

Recall (p. 67) that a subgraph is convex if it contains every shortest path joining any two of its vertices. We can thus reformulate Lemma 11.4 as follows:

Lemma 11.5 *Let F be the union of one or more equivalence classes of Θ^* and H be the subgraph of G spanned by the edges in F. Then every connected component of H is convex.*

Figure 11.2 illustrates Lemma 11.5. It shows a graph G with five Θ^*-equivalence classes E_1, E_2, E_3, E_4, E_5. Part (a) shows the graph and a representative e_i of every class E_i, (c) shows the graph $(V(G), E_1)$, (d) shows the graph $(V(G), E_2)$, and (b) the spanning subgraph of G with edge set $E_1 \cup E_2$. In each case, all components are convex in G.

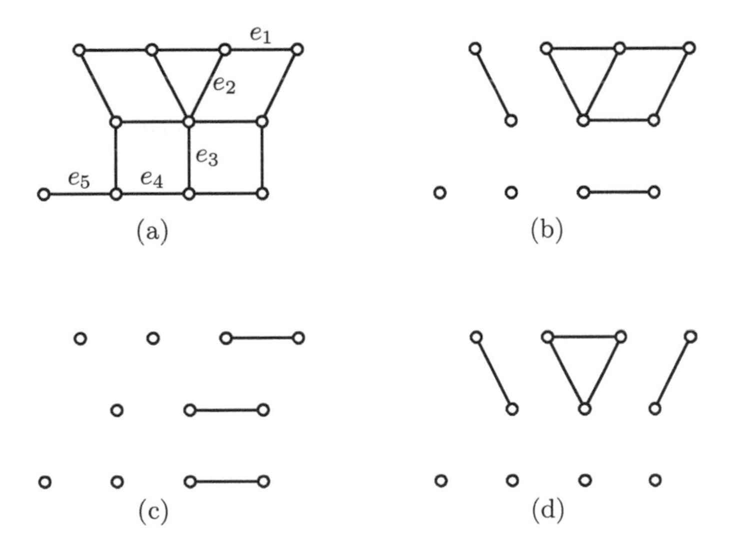

FIGURE 11.2 Graph G and components of $(V(G), E_1 \cup E_2)$.

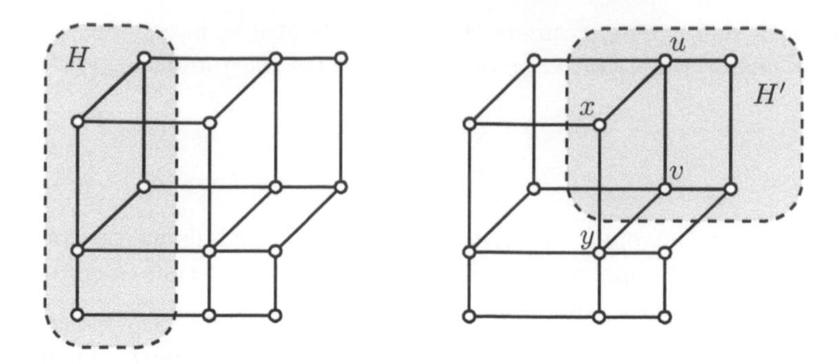

FIGURE 11.3 Convex and nonconvex subgraphs.

For bipartite graphs, we obtain a stronger result. First, we need a definition. If H is a subgraph of G, then ∂H is the set of all edges xy of G with $x \in V(H)$ and $y \notin V(H)$.

Lemma 11.6 (Convexity lemma) *An induced connected subgraph H of a bipartite graph G is convex if and only if no edge of ∂H is in relation Θ with any edge of H.*

Proof Suppose that H is convex and there are edges $ab \in E(H)$ and $xy \in E(\partial H)$ that are in relation Θ. Say that $x \in V(H)$ and $y \notin V(H)$. By Lemma 11.2 we can choose the notation such that xy is on a shortest a, x-path, in contradiction to the convexity of H.

Conversely, suppose that H is an induced connected subgraph of G, and no edge of ∂H is in relation Θ to an edge in H. Let a, b be two vertices of H, P a shortest a, b-path in G, and Q an arbitrary a, b-path in H. If P is not in H, then $|P| > 1$ (because H is induced), so P has an edge e in ∂H. By Lemma 11.1, the edge e is not in relation Θ with any other edge of P. However, by Lemma 11.3, it must be in relation Θ to an edge in $(P \cup Q) - e$ and thus in relation Θ to an edge in $Q \subseteq H$, contrary to assumption. □

Lemma 11.6 is from Imrich and Klavžar (1998). It is interesting in its own right and will be used later as the basis of a simple algorithm for the recognition of median graphs.

To illustrate the lemma, consider the subgraphs H and H' in Figure 11.3. For any edge e of ∂H and any edge f of H, we can easily find a shortest path containing e and f. By Lemma 11.1, e and f are not in relation Θ; hence H is convex by the Convexity lemma. On the other hand, the edge xy of $\partial H'$ is not in relation Θ with the edge uv of H', whence H' is not convex.

11.2 Characterizations of Partial Cubes

In this section we give several characterizations of partial cubes. (Recall that a partial cube is a graph that can be realized as an isometric subgraph of a hypercube.) We also show that convexity of intervals of a given bipartite graph does not guarantee that it is a partial cube.

For an edge ab of a graph G, let W_{ab} be the set of vertices of G that are closer to a than to b. In symbols,

$$W_{ab} = \{ w \mid w \in V(G),\ d_G(w, a) < d_G(w, b) \}.$$

If G is bipartite, then the two sets W_{ab} and W_{ba} partition $V(G)$.

Proposition 11.7 *Let $e = ab$ be an edge of a connected bipartite graph G and*

$$F_{ab} = \{ f \mid f \in E(G),\, e\Theta f \}.$$

Then $G - F_{ab}$ has exactly two connected components, namely $\langle W_{ab} \rangle$ and $\langle W_{ba} \rangle$. Moreover, if $w \in W_{ab}$, then every shortest a, w-path is completely contained in $\langle W_{ab} \rangle$.

Proof By Lemma 11.3, the vertices a and b belong to different connected components of $G - F_{ab}$. We claim that $G - F_{ab}$ has exactly two connected components.

Let w be an arbitrary vertex of W_{ab}, and P a shortest a, w-path in G. Because $w \in W_{ab}$ and G is bipartite, we infer that $ba \cup P$ is a shortest b, w-path. Lemma 11.1 implies that no edge of P is in relation Θ with ab. Thus, P belongs to $G - F_{ab}$, so w is in the component of $G - F_{ab}$ containing a. Because w was arbitrary, this component contains all vertices of W_{ab}. Analogously, all vertices of W_{ba} belong to the component of $G - F_{ab}$ containing b.

Finally, we must show that these components are induced subgraphs. Suppose an edge wx of G has both endpoints in W_{ab}. By bipartiteness, this is the terminal edge of some shortest path P originating at a. As in the previous paragraph, $ba \cup P$ is a shortest path; hence, ab is not in relation Θ with wx, so wx is an edge of $G - F_{ab}$. Thus, the component with vertex set W_{ab} is indeed $\langle W_{ab} \rangle$. The reasoning is identical for the other component. □

The following theorem puts forth two fundamental characterizations of partial cubes:

Theorem 11.8 *For a connected graph G, the following statements are equivalent:*

(i) *G is a partial cube.*
(ii) *G is bipartite, and $\langle W_{ab} \rangle$ and $\langle W_{ba} \rangle$ are convex subgraphs of G for all $ab \in E(G)$.*
(iii) *G is bipartite and $\Theta^* = \Theta$.*

Proof (i) $\Rightarrow$ (ii). Let G be a partial cube and let α be an isometric embedding of G into a hypercube. Thus, G is bipartite. Consider an edge ab of G and assume, without loss of generality, that the first coordinate of $\alpha(a)$ is 0 and the first coordinate of $\alpha(b)$ is 1. Then, for any vertex w of W_{ab}, the first coordinate of $\alpha(w)$ must be 0. It follows that any shortest path between two vertices of W_{ab} lies completely in W_{ab}. Hence, W_{ab} induces a convex subgraph of G. Clearly this is also true of W_{ba}.

(ii) $\Rightarrow$ (iii). Let $uv \,\Theta\, ab$ and $ab \,\Theta\, xy$. We have to show that $uv \,\Theta\, xy$. By Lemma 11.7, we may assume $u, x \in W_{ab}$ and $v, y \in W_{ba}$. Because G is bipartite, $d(u, x) \neq d(u, y)$. As W_{ab} is convex and $u, x \in W_{ab}$, we infer that $d(u, x) = d(u, y) - 1$. The same argument applied to W_{ba} yields $d(v, y) = d(v, x) - 1$. Thus, $d(u, x) + d(v, y) \neq d(u, y) + d(v, x)$, so $uv \,\Theta\, xy$.

(iii) $\Rightarrow$ (i). Let G be bipartite and $\Theta^* = \Theta$. Let $e_1 = x_1y_1$, $e_2 = x_2y_2$, …, $e_k = x_ky_k$ be representatives of the equivalence classes of Θ^*. Define an embedding $\alpha : V(G) \to V(Q_k) = \{0, 1\}^k$ as follows: For $v \in V(G)$ and $i = 1, 2, \ldots, k$, let the ith coordinate of $\alpha(v)$ be 0 if $v \in W_{x_iy_i}$ and 1 if $v \in W_{y_ix_i}$. We claim that α is an isometric embedding.

Let $uv \in E(G)$ and assume that uv belongs to the equivalence class of the edge e_i. By Proposition 11.7, $\alpha(u)$ and $\alpha(v)$ differ in the ith coordinate. Furthermore, if $j \neq i$, then the pair uv, x_jy_j is not in relation Θ, and therefore u and v must be either both in $W_{x_jy_j}$ or both in $W_{y_jx_j}$ by Proposition 11.7. Thus, $\alpha(u)$ and $\alpha(v)$ have the same jth coordinate. Consequently, α maps edges to edges.

Furthermore, if P is a shortest u, v-path, then by Lemma 11.1, $\alpha(u)$ and $\alpha(v)$ differ in just as many coordinates as P has edges, but this is the distance $d_G(u, v)$. □

The characterization (ii) is due to Djoković (1973), and (iii) to Winkler (1984).

Several other characterizations of partial cubes are known. Roth and Winkler (1986) characterized them as the bipartite graphs G whose distance matrix has exactly one positive eigenvalue. Chepoi (1988) proved (Exercise 11.7) that partial cubes are precisely the graphs that can be obtained from K_1 by a sequence of expansions; see also Chepoi (1994). The following characterization, first explicitly stated in Imrich and Klavžar (1993), will be used in the recognition algorithm for partial cubes—Algorithm 18.2.

Theorem 11.9 *Let G be a connected bipartite graph and $E_1, E_2, \ldots, E_k$ the equivalence classes of $E(G)$ with respect to Θ^*. Then G is a partial cube if and only if every $G - E_i$ has exactly two connected components.*

Proof Suppose that G is a partial cube. Then G is bipartite, $\Theta = \Theta^*$, and every $G - E_i$ has exactly two components by Proposition 11.7.

Conversely, suppose that G is a connected bipartite graph and that every $G - E_i$ has exactly two connected components. It suffices to show that $\Theta = \Theta^*$. Let E_i be arbitrarily chosen and H_1, H_2 be the two components of $G - E_i$. Because H_1 and H_2 are both convex by Lemma 11.5, we infer that every edge in E_i has one endpoint in H_1, the other one in H_2, and that any two edges $e, f \in E_i$ must be disjoint. Take arbitrary edges $e = ab$, $f = xy$ in E_i, where $a, x \in V(H_1)$ and $b, y \in V(H_2)$. Because G is bipartite, $d(a,x) \neq d(a,y)$. Hence, by the convexity of H_1, $d(a,y) = d(a,x) + 1$. Similarly, $d(b,x) = d(b,y) + 1$. Therefore,

$$d(a,x) + d(b,y) \neq d(a,y) + d(b,x)$$

and $e\Theta f$. Thus, any two edges of the Θ^* class E_i are in relation Θ. As E_i was arbitrarily chosen, $\Theta = \Theta^*$ and G is a partial cube. □

Chepoi and Tardif (1994, personal communication) asked whether partial cubes can also be characterized as bipartite graphs with convex intervals. Considering subdivisions of wheel graphs, Brešar and Klavžar (2002) showed that this is not the case. The smallest example they constructed is shown in Figure 11.4. It is a subdivision of K_4.

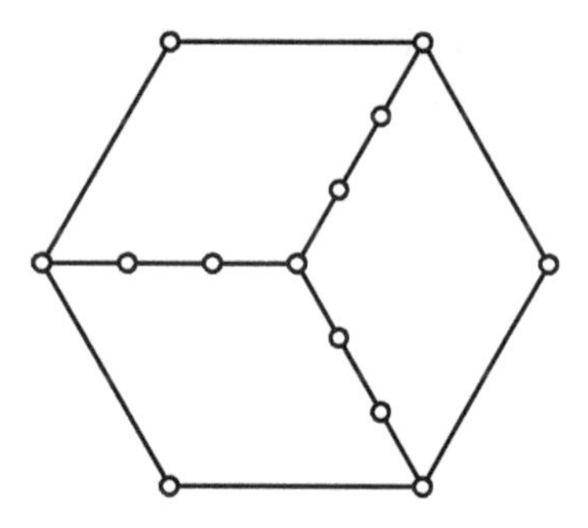

FIGURE 11.4 Bipartite graph with convex intervals that is not a partial cube.

On the positive side, Brešar and Klavžar (2002) proved that a bipartite graph with convex intervals that is not a partial cube always contains a subdivision of K_4.

11.3 Cubic Partial Cubes

We end the chapter with a brief look at (finite) cubic partial cubes. The only cubic median graph is the 3-cube Q_3 but the variety of cubic partial cubes that are not median graphs is astonishingly rich. Despite many efforts, the problem of their classification is still widely open. In particular, only one nonplanar cubic partial cube is known; it is the generalized Petersen graph $P(10, 3)$, also known as the Desargues graph.

Another, albeit trivial, infinite family of partial cubes are even prisms, that is, Cartesian products $C_{2k} \,\square\, K_2$ (cf. Exercise 11.9).

Eppstein (2006), who considered simplicial arrangements of lines in the real projective plane (finite sets of lines such that each of the regions is a triangle), proved:

Theorem 11.10 *To any simplicial line arrangement A in the projective plane, there exists a cubic partial cube with twice as many vertices as the number of triangles in A.*

Three infinite families of simplicial line arrangements are known; see Grünbaum (1972). One of them leads to the even prisms, the other two yield two new infinite families of cubic partial cubes. In addition, ninety-one examples of simplicial line arrangements that do not belong to any of the infinite families are known, see Grünbaum (1972), each of them leading to an additional example of a cubic partial cube.

Moreover, the same approach can be used on pseudoline arrangements. (A pseudoline arrangement is a collection of curves in the plane that are topologically equivalent to lines, where any two curves cross in a single point. Such an arrangement is simplicial provided that each of its regions is bounded by sides belonging to three curves.) Again, a partial cube can be associated to any pseudoline arrangement; and if the arrangement is simplicial, then the corresponding partial cube is cubic. Grünbaum (1972) mentions that there are seven infinite families of simplicial pseudoline arrangements, so there exist seven additional infinite families of cubic partial cubes.

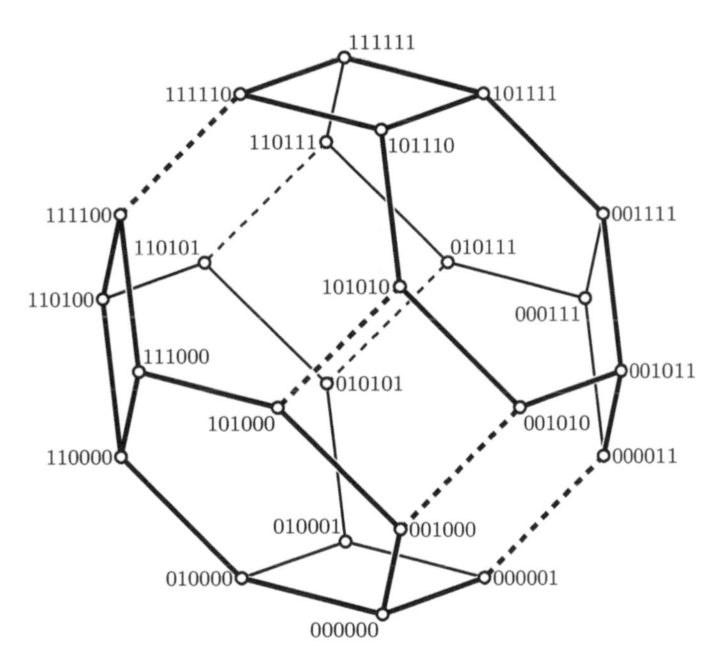

FIGURE 11.5 A cubic partial cube on twenty-four vertices, embedded in Q_6. The Θ-class of edges with endpoints differing in the fifth coordinate is shown dashed.

The partial cubes considered above are relatively large. In fact, with computer assistance, Bonnington, Klavžar, and Lipovec (2003) showed that for up to thirty vertices, there are only three nontrivial cubic partial cubes, namely the Desargues graph, the graph on twenty-four vertices from Figure 11.5, and one additional example on thirty vertices. The graph on twenty-four vertices appears in many different contexts. (See, for instance, Gedeonova (1990).) It is known as the *permutahedron* Π_3; cf. Ziegler (1995).

Finally, tribes of cubic partial cubes were introduced by Klavžar and Shpectorov (2007) as the smallest class of graphs that contains given cubic partial cubes as well as all of their cubic expansions and cubic contractions. We do not go into details here and only present three graphs from these investigation in Figure 11.6.

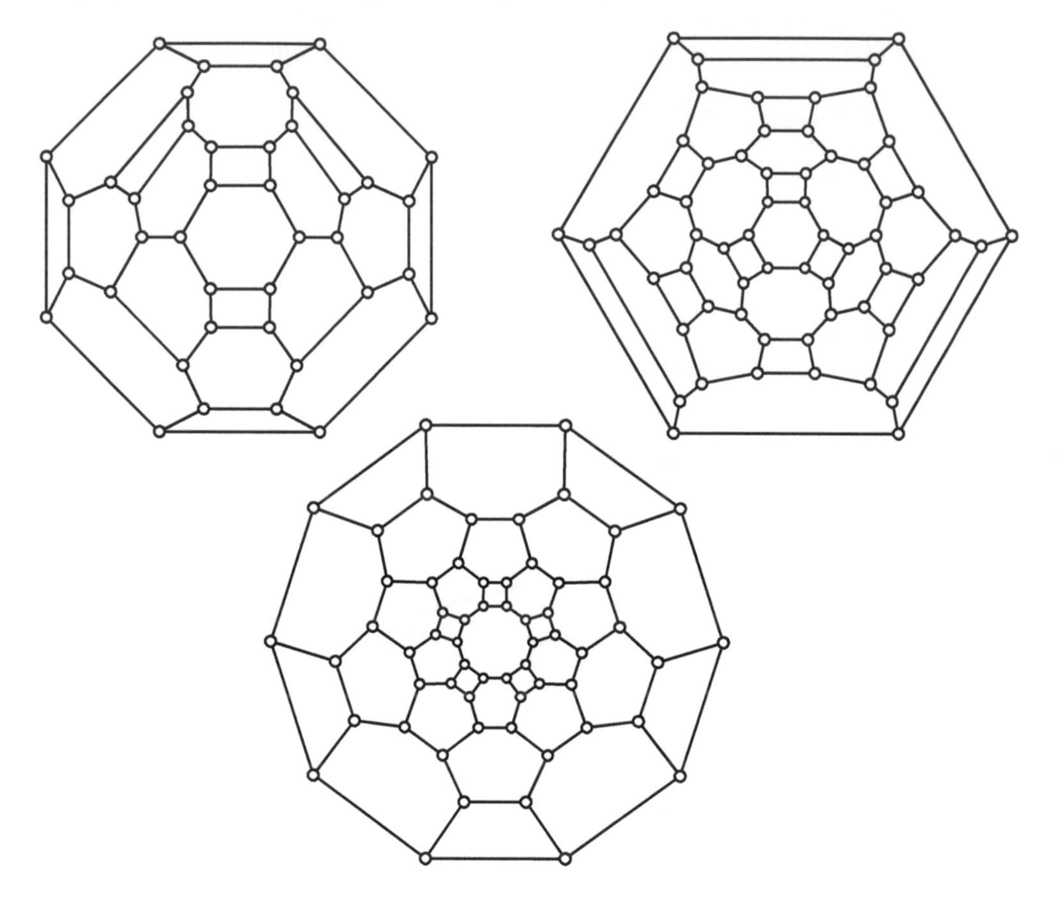

FIGURE 11.6 Three cubic partial cubes.

11.4 Scale Embeddings into Hypercubes

In this section we briefly describe two far-reaching generalizations of partial cubes: ℓ_1-graphs and hypermetric graphs. From our point of view, their central property is that they can be characterized as isometric subgraphs of Cartesian product graphs. For additional information on their rich theory, see the book by Deza and Laurent (1997).

Recall that points of the metric space ℓ_1 are real sequences $\{a_n\}$ with $\sum_{n=0}^{\infty} |a_n| < \infty$, and the distance between points $\mathbf{a} = \{a_n\}$ and $\mathbf{b} = \{b_n\}$ is $d_1(\mathbf{a}, \mathbf{b}) = \sum_{n=0}^{\infty} |a_n - b_n|$. A graph G is called an *ℓ_1-graph* if the metric space $(V(G), d_G)$ is isomorphic to a subspace of ℓ_1, that is, if there exists a mapping $f : V(G) \to \ell_1$ with $d_1(f(u), f(v)) = d_G(u, v)$ for all $u, v \in V(G)$.

The following notions are crucial for the characterization of ℓ_1-graphs. For a graph G, a mapping $\beta : V(G) \to Q_d$ is a *scale λ-embedding* (or an embedding with scale λ) if

$$d_{Q_n}(\beta(u), \beta(v)) = \lambda d_G(u, v)$$

for all vertices u and v of G.

Assouad and Deza (1980) proved the following fundamental result about finite ℓ_1-graphs:

Theorem 11.11 *A graph is an ℓ_1-graph if and only if it admits a scale λ-embedding in Q_d for some λ and d.*

Clearly, partial cubes are precisely the ℓ_1-graphs embeddable with scale 1. The large variety of ℓ_1-graphs that are not partial cubes includes complete graphs, the Petersen graph and its complement, and the five regular polyhedra, just to mention a few. Further, the Cartesian product of two ℓ_1-graphs is again ℓ_1.

To characterize ℓ_1-graphs as isometric subgraphs of Cartesian products, we need to introduce two families of graphs. The *cocktail-party graph* $K_{n\times 2}$ is obtained from K_{2n} by deleting the edges of a perfect matching. The half-cube graph $\frac{1}{2}Q_d$ consists of an (arbitrary) partite set of Q_d, where two vertices of $\frac{1}{2}Q_d$ are adjacent if they are at distance 2 in Q_d. Shpectorov (1993) proved:

Theorem 11.12 *A graph is an ℓ_1-graph if and only if it is an isometric subgraph of the Cartesian product of complete graphs, half-cubes, and cocktail-party graphs.*

Because K_n is a subgraph $K_{n\times 2}$, Theorem 11.12 can be formulated without a reference to complete graphs. We nevertheless use the present formulation because it immediately implies that partial Hamming graphs (which are introduced in Chapter 14) are a proper subclass of ℓ_1-graphs.

Shpectorov (1993) also designed an $O(mn)$ algorithm for recognizing ℓ_1-graphs, where n and m are the number of vertices and edges of the graph considered.

Theorem 11.12 was discovered independently by Deza and Grishukhin (1993) in a more general setting involving hypermetric graphs (for definition, see their paper or the book Deza and Laurent (1997)). They proved that hypermetric graphs can be characterized as isometric subgraphs of the Cartesian products of complete graphs, half-cubes, cocktail-party graphs, and copies of the so-called *Gosset graph* G_{56}.

To conclude the chapter, we add that Roth and Winkler (1986) proved that in the bipartite case, the metric hierarchy collapses to partial cubes. More precisely, a bipartite graph is a partial cube if and only if it is an ℓ_1-graph, if and only if it is a hypermetric graph.

Exercises

11.1. Let G be a connected graph, and let e be an edge of G such that $G-e$ is disconnected. Show that if f and f' are edges from different connected components of $G-e$, then f and f' are not in relation Θ.

11.2. Show that Lemma 11.6 is false if we remove the assumption that H is induced.

11.3. Show that a connected graph G is a partial cube if and only if every block of G is a partial cube.

11.4. Show that a connected, bipartite graph in which every edge is contained in at most one cycle is a partial cube.

11.5. Let H be a partial cube and F_{ab} a Θ-class of H. Let G be the graph obtained from H by contraction of every edge in F_{ab} to a single vertex and replacement of all double edges that may arise by single ones. (One says that G is obtained from H by *contraction*.) Show that G is a partial cube.

11.6. Let G be a graph and V_1, V_2 subsets of $V(G)$ with the following properties: $V_1 \cap V_2 \neq \emptyset$, $V_1 \cup V_2 = V(G)$, $\langle V_1 \rangle$ and $\langle V_2 \rangle$ are isometric in G, and there are no edges between $V_1 \setminus V_2$ and $V_2 \setminus V_1$. Then the *expansion* of G (with respect to V_1 and V_2) is the graph H that is obtained from G by the following operations:

(i) Replacement of each vertex $v \in V_1 \cap V_2$ by vertices v_1 and v_2 and the edge $v_1 v_2$.

(ii) Insertion of edges from v_1, resp. v_2, to all neighbors of v in $V_1 \setminus V_2$, resp. $V_2 \setminus V_1$.

(iii) Replacement of every edge $vu \in \langle V_1 \cap V_2 \rangle$ by the edges $v_1 u_1$ and $v_2 u_2$.

Show that H is a partial cube.

11.7. (Chepoi, 1988) Show that a graph is a partial cube if and only if it can be obtained from K_1 by a sequence of expansions.

11.8. Verify that the graph of Figure 11.5 is a partial cube.

11.9. Show that if G and H are partial cubes, then $G \,\Box\, H$ is a partial cube as well.

11.10. If G is a cubic partial cube, then either $G = C_{2n} \,\Box\, K_2$ for some $n \geq 2$ or G is prime with respect to the Cartesian product.

11.11. Show that partial cubes have convex intervals.

11.12. (Eppstein, 2006; Ovchinnikov, 2008) Let $\mathcal{A}$ be a finite set of hyperplanes in $\mathbb{R}^d$. Let $G_{\mathcal{A}}$ be the graph whose vertices are the regions of the arrangement. Two vertices are adjacent if the corresponding regions meet along a $(d-1)$-dimensional face of the arrangement. Prove that $G_{\mathcal{A}}$ is a partial cube.

11.13. Show that for any $n \geq 2$, the graph obtained from K_n by subdividing each of its edges exactly once is a partial cube.

11.14. (Wilkeit, 1990) Call a graph 5-*gonal* if

$$\sum_{i<j} d(x_i, x_j) + \sum_{i<j} d(y_i, y_j) \;\leq\; \sum_{i,j} d(x_i, y_j)$$

holds for any $\{x_1, x_2, y_1, y_2, y_3\} \subseteq V(G)$. Show that in a 5-gonal graph G, $\langle W_{uv} \rangle$ is convex for any edge uv of G.

11.15. (Avis, 1981) Show that a bipartite graph is a partial cube if and only if it is 5-gonal.

11.16. Show that a graph G embeddable into Q_d with scale 2 is an isometric subgraph of the half-cube $\frac{1}{2}Q_d$.

Chapter 12

Median Graphs

Median graphs were defined in Chapter 3 as graphs for which every triple of vertices has a unique median. We deduced some of their basic properties, namely that they are bipartite, and that trees, hypercubes, and Cartesian products of median graphs are median graphs. We now aim for deeper insight into the structure of these graphs.

In this chapter we derive a fundamental characterization of median graphs—Mulder's convex expansion theorem. Along the way we obtain new properties, including the fact that median graphs are partial cubes. Then we apply the expansion theorem to prove an inequality that relates the numbers of vertices, edges, and Θ-classes of a median graph, and we generalize this inequality to partial cubes. In the second part of the chapter we characterize median graphs as retracts of hypercubes, prove a fixed cube theorem, and relate an interesting application of median graph networks to human genetics.

12.1 Mulder's Convex Expansion

This section begins with several new properties of median graphs and ends with Mulder's convex expansion theorem. This most useful structure theorem provides the basis for advanced recognition algorithms.

Although their definition does not indicate any connection with r-cubes, median graphs are not only embeddable into r-cubes, they are also isometrically embeddable. To show this, we decompose a graph by removing all edges in relation Θ to a given edge. We proceed as in Proposition 11.7, but with the additional assumption that the graph is a median graph. First we extend our notation. Let $e = ab$ be an edge of a connected graph G. In addition to the previously defined sets

$$\begin{aligned} W_{ab} &= \{w \mid w \in V(G),\, d(w,a) < d(w,b)\}, \\ W_{ba} &= \{w \mid w \in V(G),\, d(w,b) < d(w,a)\}, \\ F_{ab} &= \{f \mid f \in E(G),\, e\Theta f\}, \end{aligned}$$

we introduce the sets

$$U_{ab} = \{u \mid u \in W_{ab},\, u \text{ is adjacent to a vertex in } W_{ba}\},$$
$$U_{ba} = \{u \mid u \in W_{ba},\, u \text{ is adjacent to a vertex in } W_{ab}\}.$$

(See Figure 12.1.) For clarity, we may sometimes write these sets as U_{ab}^G and W_{ab}^G, etc.

If G is bipartite, we have $F_{ab} = \{uv \in E(G) \mid u \in W_{ab},\ v \in W_{ba}\}$ by Proposition 11.7.

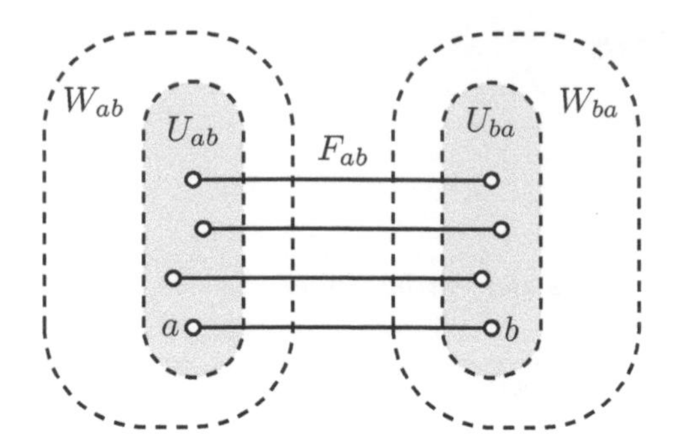

FIGURE 12.1 Fundamental sets in a median graph.

Lemma 12.1 *If ab is an edge of a median graph, then F_{ab} is a matching that induces an isomorphism between $\langle U_{ab}\rangle$ and $\langle U_{ba}\rangle$.*

Proof Let $uv, xy \in F_{ab}$, with $u, x \in U_{ab}$ and $v, y \in U_{ba}$. The result will follow if we can show that $d(u,x) = d(v,y)$. Suppose that this is not the case; say $d(u,x) < d(v,y)$.

Let c be the median of a, u, x, and d the median of b, v, y. Then $c \neq d$, as Proposition 11.7 implies $c \in W_{ab}$ and $d \in W_{ba}$. Notice that d is also the median of a, v, y. We will arrive at a contradiction by showing that c equals the median d of a, v, y.

Now, c is on a shortest u, x-path P, and d is on a shortest v, y-path Q. By assumption, $|P| < |Q|$. By bipartiteness, the v, y-path $P' = vu \cup P \cup xy$ has the same length as Q. Thus c is on the shortest v, y path P'.

Clearly, c is on a shortest a, v-path with final edge uv. Similarly, c is on a shortest a, y-path. Because c is also on the shortest v, y-path P', it is the median of a, v, y. □

To show that median graphs are partial cubes, the following two lemmas will be useful:

Lemma 12.2 *If ab is an edge of a median graph, then each shortest path from a to any $u \in U_{ab}$ is completely in $\langle U_{ab}\rangle$.*

Proof Let P be a shortest a, u-path that is not contained in U_{ab}. By Proposition 11.7, it lies completely in $\langle W_{ab}\rangle$. Let $w \in V(P)$ be the vertex closest to u that is not in U_{ab}. We can assume w is adjacent to u. Let v be the neighbor of u in U_{ba}.

Clearly, $d(b,w) = d(a,u) = d(b,v)$ and $d(w,v) = 2$. Let c be the median of b, v, w. Lemma 3.5 yields $d(w,c) = \frac{1}{2}\big(d(b,w) + d(v,w) - d(b,v)\big) = 1$, so w is adjacent to c.

Now, the median c is on a shortest path from b to v, so $c \in W_{ba}$, by Proposition 11.7. But w is not adjacent to any vertex in W_{ba}. □

Lemma 12.3 *For any edge ab of a median graph, $\langle U_{ab}\rangle$ and $\langle U_{ba}\rangle$ are isometric subgraphs.*

Proof Let $u, x \in U_{ab}$. The median c of a, u, x must be on shortest paths P from a to u and Q from a to x. By Lemma 12.2, these paths are in U_{ab}. On the other hand, the part of P from u to c, together with the part of Q from c to x, is a shortest u, x-path in U_{ab}. □

Proposition 12.4 *Median graphs are partial cubes.*

Proof Suppose $uv\Theta ab$ and $ab\Theta xy$ in a median graph. The preceding lemmas yield $uv\Theta xy$. Thus $\Theta = \Theta^*$ for median graphs, and the assertion follows by Theorem 11.8 (iii). □

The graph Q_3^- (Figure 3.2) is the standard example of a partial cube that is not a median graph. Other examples include even cycles of length 6 or greater.

Recall that there are two ways that a bipartite graph may fail to be a median graph: It may have too many medians or not enough. In the case of partial cubes, which are isometric subgraphs of r-cubes, only the latter applies, because medians in r-cubes are unique.

The next property of median graphs requires a new concept. We say a subgraph $H \subseteq G$ is *gated* in G if for every $v \in V(G)$ there is an $x \in V(H)$ that is on a shortest v, u-path for every $u \in V(H)$. Note that such an x must be unique. It is called the *gate* of v in H.

Suppose that H is a gated subgraph of G. Let u, v be vertices of H and x be a vertex on a shortest u, v-path P. Let y be the gate of x in H. If $x \neq y$, then P is not a shortest path. Thus $x = y \in V(H)$, which means that H is convex. Thus gated subgraphs are convex.

For median graphs, the converse holds as well. To show this, we introduce the distance of a vertex from a subgraph. Let H be a subgraph of a connected graph G and $v \in V(G)$. Then the distance $d(v, H)$ of v from H is defined as $\min\{d(v, x) \mid x \in V(H)\}$.

Lemma 12.5 *A subgraph of a median graph is convex if and only if it is gated.*

Proof We have already seen that gated subgraphs are convex. For the converse, let G be a median graph and H be a convex subgraph of G. Given a vertex $v \in V(G)$, fix a vertex $x \in V(H)$ that is closest to v. Given any $u \in V(H)$, we claim that x lies on a shortest u, v-path. Let y be the median of u, x and v, so y lies on a shortest u, v-path. We complete the proof by showing $y = x$. Because $u, x \in V(H)$, it follows by convexity that $y \in V(H)$, and thus $d(v, y) \geq d(v, x)$ by choice of x. Now we have $d(v, x) = d(v, y) + d(y, x) \geq d(v, x) + d(y, x)$, which implies $d(y, x) = 0$, so $x = y$. □

We now prove a characterization of median graphs due to Mulder (1978, 1980a). We first formulate his theorem in a form that is most suitable for the algorithms that we present later. Then we state the theorem in its original form. First a lemma.

Lemma 12.6 *For any edge ab of a median graph, $\langle W_{ab} \rangle$ is a median graph and $\langle U_{ab} \rangle$ is a convex subgraph of $\langle W_{ab} \rangle$.*

Proof Notice that W_{ab} is convex, by Proposition 12.4 and Theorem 11.8. Then the first assertion follows from the fact that a convex subgraph of a median graph is a median graph.

Suppose that $\langle U_{ab} \rangle$ is not convex. By the Convexity Lemma 11.6, there is an edge xy in $\langle W_{ab} \rangle$ with $x \in U_{ab}$ and $y \notin U_{ab}$ that is in relation Θ to an edge uv of $\langle U_{ab} \rangle$. By Lemma 11.2, we can assume that $d(x, u) = d(y, v)$. Let v' and x' be the neighbors of v and x in $\langle U_{ba} \rangle$. Clearly, $d(v, x) = d(v', x')$ and $d(x', y) = 2$. As in the proof of Lemma 12.2, it follows that v', x', and y have no median. □

Theorem 12.7 *Let ab be an edge of a connected, bipartite graph G. Then G is a median graph if and only if the following three conditions are satisfied:*

(i) *F_{ab} is a matching defining an isomorphism between $\langle U_{ab} \rangle$ and $\langle U_{ba} \rangle$.*
(ii) *$\langle U_{ab} \rangle$ is convex in $\langle W_{ab} \rangle$, and $\langle U_{ba} \rangle$ is convex in $\langle W_{ba} \rangle$.*
(iii) *$\langle W_{ab} \rangle$ and $\langle W_{ba} \rangle$ are median graphs.*

Proof If G is a median graph, then (i)–(iii) hold by Lemmas 12.1 and 12.6.

Conversely, say (i)–(iii) hold. In particular, $\langle U_{ab} \rangle$ is gated in $\langle W_{ab} \rangle$ by Proposition 12.5. We make a preliminary claim about paths in G.

Fix a vertex v in W_{ab}. Let $x \in U_{ab}$ be the gate of v in U_{ab}, and $xx' \in F_{ab}$. We claim that for any $z \in W_{ba}$, there is a shortest v,z-path that includes the edge xx'. To see this, note that condition (i) implies that any shortest v,z-path has exactly one edge uu' in F_{ab}. Moreover, by convexity of U_{ab}, such a path is a concatenation $P \cup Q \cup uu' \cup R$ of paths, where $P \subseteq \langle W_{ab}\rangle - E(\langle U_{ab}\rangle)$, $Q \subseteq \langle U_{ab}\rangle$, and $R \subseteq \langle W_{ba}\rangle$. (Possibly P or R is trivial.) We can assume $P \cup Q$ passes through the gate x, and x is the initial vertex of Q. By (i), the v,z-path $P \cup Q \cup uu' \cup R$ can be replaced by the shortest v,z-path $P \cup xx' \cup Q' \cup R$, where Q' is the image of Q under the isomorphism induced by F_{ab}. This proves the claim.

Note also that (using the same notation as in the previous paragraph) the concatenation of $P \cup xx'$ with a shortest path in $\langle W_{ba}\rangle$ is still a shortest path.

Now we can prove the theorem. Take $v, y, z \in V(G)$. We must show that these have a unique median in G. If they are all in W_{ab} (or W_{ba}), then there is nothing to prove. Thus say $v \in W_{ab}$ and $y, z \in W_{ba}$. The above remarks imply c is a median of v, y, z if and only if it is a median of $x', y, z \in W_{ba}$. Because W_{ba} is a median graph, c exists and is unique. □

To prepare for the original formulation of Theorem 12.7, we first define the concept of an expansion of a graph G; cf. Exercise 11.6.

Suppose $V(G) = V_1 \cup V_2$, where $V_1 \cap V_2 \neq \emptyset$, each $\langle V_i\rangle$ is an isometric subgraphs of G, and no edge of G joins $V_1 \setminus V_2$ to $V_2 \setminus V_1$. An *expansion* of G with respect to V_1 and V_2 is a graph H obtained from G by the following steps. (See Figure 12.2.)

(i) Replace each $v \in V_1 \cap V_2$ by vertices v_1, v_2, and insert the edge v_1v_2.
(ii) Insert edges between v_1 and all neighbors of v in $V_1 \setminus V_2$;
insert edges between v_2 and all neighbors of v in $V_2 \setminus V_1$.
(iii) Insert the edges v_1u_1 and v_2u_2 if $v, u \in V_1 \cap V_2$ are adjacent in G.

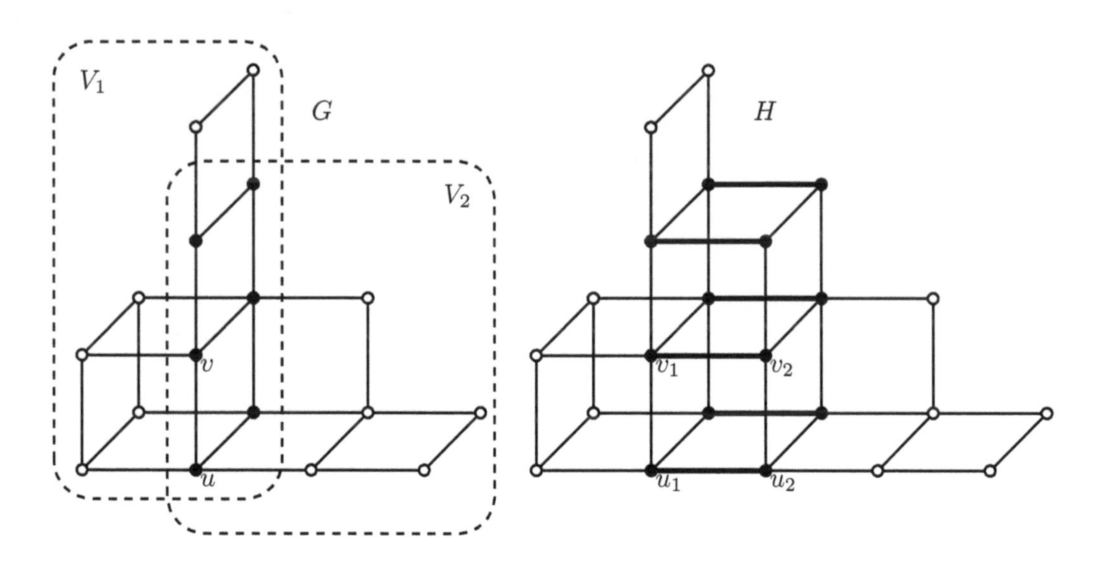

FIGURE 12.2 Expansion H of G.

An expansion is *connected* if $\langle V_1 \cap V_2\rangle$ is a connected subgraph of G. We also say that the graph H is a connected expansion of G with respect to $\langle V_1 \cap V_2\rangle$. *Isometric* and *convex* expansions are defined analogously.

A *contraction* is the inverse operation to an expansion, namely G is a contraction of H. We say that H can be obtained from G by an *expansion procedure* if it can be obtained from G by a sequence of expansions. This leads to another characterization of median graphs:

Theorem 12.8 (Mulder's Convex Expansion Theorem) *A graph is a median graph if and only if it can be obtained from the one-vertex graph by a convex expansion procedure.*

Proof We use induction to show that every graph obtained from K_1 by a convex expansion procedure is median. Clearly, K_1 is median. Suppose H is obtained from a median graph G by a convex expansion with respect to V_1 and V_2. Then $V(H)$ is the disjoint union of V_1 and V_2. Let F be the set of edges of H joining V_1 to V_2. It is easy to check that the sets F, V_1, and V_2 fulfill conditions (i) through (iii) of Theorem 12.7, where these sets correspond to F_{ab}, W_{ab}, and W_{ba}, respectively. Hence H is median by Theorem 12.7.

Conversely, we verify that a contraction of a median graph is median (Exercise 12.7). □

In this section we have proved two characterizations of median graphs. Dozens of other characterizations are known; see Klavžar and Mulder (1999) for a survey.

12.2 Inequalities for Median Graphs and Partial Cubes

A graph is called *cube-free* if it does not contain the 3-cube as an induced subgraph. For cube-free median graphs, Theorem 12.8 reduces to the following:

Corollary 12.9 *A graph is a cube-free median graph if and only if it can be obtained from K_1 by an expansion procedure in which each expansion is taken with respect to a convex tree.*

Proof Suppose that a graph G can be obtained from K_1 by an expansion procedure in which every expansion step is taken with respect to a convex tree. Then G is a median graph by Theorem 12.8. Moreover, as the expansion steps are taken with respect to trees, G is cube-free by a simple induction argument.

Conversely, let G be a cube-free median graph. Then it can be obtained from K_1 by a convex expansion procedure. Assume that an expansion step was taken with respect to a graph containing a cycle. Let H be a graph obtained in this step of the expansion procedure, and let H be an expansion with respect to G_1', G_2'. By our assumption, $G_0' = \langle G_1' \cap G_2' \rangle$ contains a cycle. Let C_n, $n \geq 4$, be a shortest cycle in G_0'. Clearly, C_n is an even isometric cycle. Let u, v, and w be three consecutive vertices of C, and let x be the antipodal vertex of v on C. Then the median y of u, w, and x together with u, v, and w form a C_4 in G_0'. Hence, after the expansion step is performed, we find a 3-cube in H and thus also in G. □

We are now prepared for an inequality for median graphs, due to Klavžar, Mulder, and Škrekovski (1998).

Theorem 12.10 *Let G be a median graph with n vertices, m edges, and k equivalence classes with respect to the relation Θ. Then*

$$2n - m - k \leq 2\,.$$

Moreover equality holds if and only if G is cube-free.

Proof We prove the inequality by induction on the number of vertices using Theorem 12.8. The inequality reduces to $2 \leq 2$ if $G = K_1$. So assume that G is the expansion of the median graph G' with respect to its convex subgraphs G_1', G_2'. By induction we have $2n' - m' - k' \leq 2$, where k', n', m' are the parameters of G'. Let t be the number of vertices in $G_0' = G_1' \cap G_2'$.

Then G'_0, being connected, has at least $t-1$ edges. Thus $n = n' + t$, $m \geq m' + 2t - 1$, and $k = k' + 1$. So

$$\begin{aligned} 2n - m - k &\leq 2(n' + t) - (m' + 2t - 1) - (k' + 1) \\ &= 2n' - m' - k' \\ &\leq 2\,. \end{aligned}$$

Clearly, the equality $2n - m - k = 2$ holds if and only if all of the expansions taken on the way from K_1 to G are taken with respect to two isometric subgraphs whose intersection is a tree. By Corollary 12.9, this is equivalent to G being cube-free. □

Corollary 12.11 *A planar embedding of a cube-free median graph with n vertices and k Θ-classes has $n - k$ regions.*

Proof By Euler's formula (Theorem 1.11), $n - m + f = 2$. Now apply Theorem 12.10. □

For a median graph without Q_4 as a subgraph, the following theorem can be proved:

Theorem 12.12 *Suppose a median graph G contains no subgraph isomorphic to Q_4. If G has n vertices, m edges, k Θ-classes, and h subgraphs isomorphic to Q_3, then*

$$2n - m - k + h = 2\,.$$

Because Q_4 is not planar, no planar graph has a 4-cube as a subgraph (Exercise 3.2). Thus Theorem 12.12 combined with Euler's formula yields a corollary:

Corollary 12.13 *If a planar embedding of a median graph has n vertices, k Θ-classes, h subgraphs Q_3, and f regions, then*

$$f = n - k + h\,.$$

One can prove Theorem 12.12 on almost the same lines as Theorem 12.10, except that the following result, which is of independent interest, is needed:

Proposition 12.14 *Any cube-free median graph with n vertices and m edges has exactly $m - n + 1$ squares.*

Proof We use induction on the number of expansion steps. The result is trivial for K_1.

Let G be a cube-free median graph obtained by an expansion of a cube-free median graph H with respect to H'. By Corollary 12.9, H' is a tree. Say H' has k vertices, so it has $k - 1$ edges. By the induction hypothesis, H has $m' - n' + 1$ squares, where m' and n' are its numbers of edges and vertices, respectively.

Now G contains all squares of H and an additional square for every edge of H'. Hence the number of squares of G is

$$m' - n' + 1 + (k - 1) = m' - n' + k\,.$$

On the other hand,

$$m - n + 1 = (m' + k + (k - 1)) - (n' + k) + 1 = m' - n' + k\,,$$

which proves the proposition. □

Despite the maxim "it is easier to generalize than to specialize," it took quite some time until Theorem 12.10 was generalized from median graphs to partial cubes. Before stating the generalization due to Klavžar and Shpectorov (2011), we need to define the *convex excess* of a graph.

For a graph G, let $\mathcal{C}(G) = \{C \mid C \text{ is a convex cycle of } G\}$. The *convex excess* of G is

$$ce(G) = \sum_{C \in \mathcal{C}(G)} \frac{|C| - 4}{2}.$$

Note that for bipartite graphs, $ce(G) = 0$ if and only if G contains no convex cycle of length 6 or more.

Let F be a Θ-class of a partial cube G. Then the *F-zone graph*, Z_F, is the graph with $V(Z_F) = F$, where f and f' are adjacent in Z_F if they belong to a common convex cycle of G. Now we can state the generalization of Theorem 12.10.

Theorem 12.15 *For a partial cube G with n vertices, m edges, and k Θ-classes,*

$$2n - m - k - ce(G) \leq 2\,. \tag{12.1}$$

Equality holds if and only all zone graphs of G are trees.

The insight that led to the proof of Theorem 12.15 is that, in an expansion step, convex cycles lift to convex cycles, but isometric cycles need not lift to isometric cycles. This perhaps explains the ten-year gap between Theorem 12.10 and Theorem 12.15.

12.3 Median Graphs as Retracts

In this section we characterize median graphs as retracts of a hypercubes. This result, which we mentioned in Chapter 3, was discovered by Bandelt (1984). Here we follow the approach of Mulder (1990), using the convex expansion theorem.

But first, a new concept is required. Let uv be an edge of a median graph G for which $U_{uv} = W_{uv}$. Then $\langle W_{uv} \rangle$ is called a *peripheral subgraph* of G. Note that in a tree, every vertex of degree 1 is a peripheral subgraph, and that in the r-cube, all $(r-1)$-cubes are peripheral. Figure 12.3 shows a more general example.

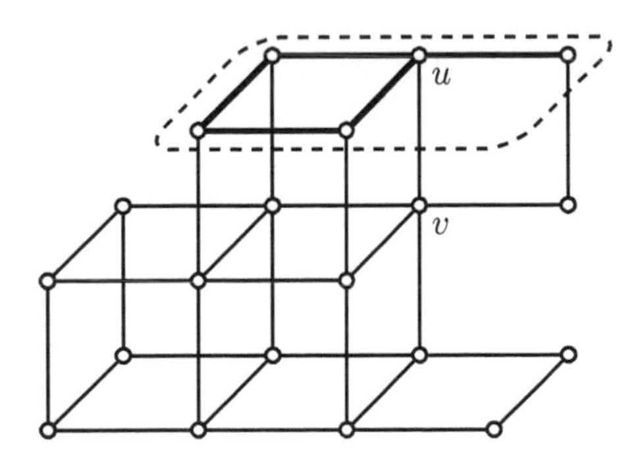

FIGURE 12.3 Peripheral subgraph of a median graph.

Lemma 12.16 *Every median graph has a peripheral subgraph.*

Proof Let ab be an arbitrary edge of a median graph G, and assume that U_{ab} is properly contained in W_{ab}. Let xy be an edge with $y \in U_{ab}$ and $x \in W_{ab} \setminus U_{ab}$. Now, $\langle U_{ab} \rangle$ is convex by Lemma 12.6; hence the Convexity Lemma 11.6 implies that no edge of $\langle U_{ab} \rangle$ is in relation Θ to xy. Therefore all edges of F_{xy} lie in $\langle W_{ab} \rangle$, and so W_{xy} is properly contained in W_{ab}. If W_{xy} induces a peripheral subgraph, we are done; otherwise, we repeat the argument until we arrive at an edge uv with $U_{uv} = W_{uv}$. □

Actually the proof of the above lemma shows a little more: If uv is an edge of a median graph, then both W_{uv} and W_{vu} contain peripheral subgraphs. In this sense, median graphs are treelike.

Our characterization of median graphs as retracts of hypercubes requires the following lemma.

Lemma 12.17 *If H is a convex subgraph of a median graph G, then there exists a non-expansive map $f : V(G) \to V(G)$ (i.e., a weak homomorphism) satisfying*

(i) *If $v \in V(H)$, then $f(v) = v$.*
(ii) *If $v \notin V(H)$, then $f(v)$ is adjacent to v, and it is on a shortest path from v to H.*

Proof We use induction on $|V(G)|$. The assertion is trivial if G has only one vertex. Suppose G has $n > 1$ vertices, and assume the assertion is true for all median graphs with fewer than n vertices.

If $H = G$, then the identity function f has the desired properties. Therefore assume H is a proper (convex) subgraph of G, so there is an edge xy with $x \in V(H)$ and $y \in V(G) \setminus V(H)$. By the Convexity Lemma 11.6, xy is not in relation Θ to any edge of H, so $H \subseteq G - F_{xy} = \langle W_{xy} \rangle + \langle W_{yx} \rangle$. Then H is a subgraph of $\langle W_{xy} \rangle$, and, as noted previously, $\langle W_{yx} \rangle$ contains a peripheral subgraph $\langle U_{ab} \rangle$ of G.

Now $G' = G - U_{ab} = \langle W_{ba} \rangle$ is a median graph (by Theorem 12.7) with fewer vertices than G, and H is a convex subgraph of G'. By the induction hypothesis, there is a map $f : V(G') \to V(G')$ satisfying the stated properties. We can extend f to $V(G)$ as follows: If $v \in U_{ab}$, let $vv' \in F_{ab}$, and put $f(v) = v'$.

It is straightforward to verify that this extended map is a weak homomorphism, and it clearly satisfies (i) and the first part of (ii). To complete the proof, note that if $v \in U_{ab}$, then the fact that F_{ab} induces an isomorphism $\langle U_{ab} \rangle \to \langle U_{ba} \rangle$ means that any shortest path from v to H can be rerouted as a shortest path from v to H that passes through $v' = f(v)$. □

Lemma 12.17 was also found by Tardif (1996). The map f in our version of the lemma is called a *mooring* there. For the proof, Tardif invoked Theorem 12.18. (See Exercise 12.13.)

Theorem 12.18 *A graph G is a median graph if and only if G is a retract of a hypercube.*

Proof As r-cubes are median graphs, Proposition 3.11 implies that every retract of an r-cube is a median graph.

Conversely, suppose G is a median graph. Then it is a partial cube, by Proposition 12.4. In other words, it is an isometric subgraph of a hypercube.

We will show that if an isometric subgraph G of a hypercube is a median graph, then G is a retract of the hypercube. We use induction on the number of Θ-classes of G. If G has one Θ-class, then $G = K_2$, and the assertion is trivial. Assume G has more than one Θ-class, and that the theorem is true for all median graphs with fewer Θ-classes than G.

Say that G is an isometric subgraph of Q_r. By Lemma 12.16, G has a Θ-class F_{uv} with $U_{uv} = W_{uv}$. Figure 12.4 is a schematic representation of this situation. Here the set

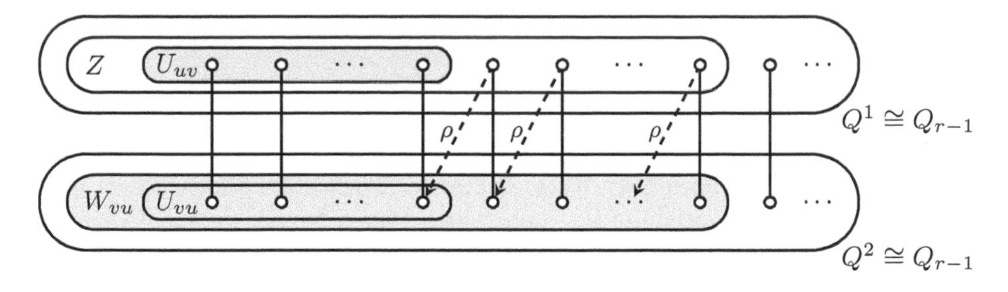

FIGURE 12.4 Sets from the proof of Theorem 12.18.

$U_{uv} = W_{uv}$ is in a copy of an $(r-1)$-cube $Q^1 \subseteq Q_r$, and W_{vu} lies in a copy of an $(r-1)$-cube Q^2. The edges of Q_r joining Q^1 to Q^2 induce an isomorphism $\mu : Q^1 \to Q^2$. Then $Z = \mu^{-1}(W_{vu})$ induces an isomorphic copy of $\langle W_{vu}\rangle$ in Q^1.

Now, Theorem 12.7 asserts that $\langle W_{vu}\rangle$ is a median graph. Also, because $\langle W_{vu}\rangle$ is convex in G, and G is isometric in Q_r, it is clear that $\langle W_{vu}\rangle$ is isometric in Q^2. Thus the induction hypothesis guarantees a retraction $r_2 : Q^2 \to \langle W_{vu}\rangle$. Let $r_1 = \mu^{-1} r_2 \mu$ be the corresponding retraction $Q^1 \to \langle Z\rangle$. It is straightforward to verify that the map $r : Q_r \to \langle Z \cup W_{vu}\rangle$ defined as

$$r(x) = \begin{cases} r_1(x) & \text{if } x \in V(Q^1)\,, \\ r_2(x) & \text{if } x \in V(Q^2) \end{cases}$$

is a retraction of Q_r to $\langle Z \cup W_{vu}\rangle$.

We complete the proof by constructing a retraction $\rho : \langle Z \cup W_{vu}\rangle \to G$, for then the composition ρr is the desired retraction of Q_r to G. Construct ρ as follows: By Lemma 12.6, $\langle U_{vu}\rangle$ is convex in $\langle W_{vu}\rangle$; hence Lemma 12.17 yields a map $f : W_{vu} \to W_{vu}$ that is the identity on U_{vu} and which sends vertices of $W_{vu} \setminus U_{vu}$ to adjacent vertices closer to U_{vu}. Define $\rho : \langle Z \cup W_{vu}\rangle \to G$ as

$$\rho(x) = \begin{cases} f\mu(x) & \text{if } x \in Z \setminus U_{uv}\,, \\ x & \text{otherwise}\,. \end{cases}$$

By construction, this is the identity on G. To show it is a retraction, we just need to show that it is a homomorphism. The straightforward details are left to the reader. □

In Exercise 3.11 we claimed that the median of three vertices u, v, and w of a hypercube can be denoted by the Boolean espression $(u \vee v) \wedge (u \vee w) \wedge (v \vee w)$. Feder (1995) showed that median graphs can similarly be characterized by Boolean equations. For example, the median graph consisting of the vertices 000, 100, 010, 110, and 011 in Q_3 can be described as the set of all 3-tuples $x_1x_2x_3$ over $\{0,1\}$ for which the expression $(\bar{x}_1 \vee \bar{x}_3) \wedge (x_2 \vee \bar{x}_3)$ has value 1. (Here $\bar{x}_i = 1$ if $x_i = 0$, and $\bar{x}_i = 0$ if $x_i = 1$.) This expression is a conjunction of clauses consisting of disjunctions of two elements. More generally, it is an instance of a conjunctive normal form with two literals per clause. The problem of satisfying such forms is known as the 2-satisfiability problem, or 2SAT. Feder showed that a set of vertices of a hypercube induces a median graph if and only if it is the set of solutions of a 2SAT instance with no equivalent variables. (Two nontrivial variables x_i and x_j are equivalent if either $x_i = x_j$ or $x_i = \bar{x}_j$ in all solutions.) This leads to an alternative proof of Theorem 12.18; see Theorem 3.38 of Feder (1995).

12.4 A Fixed Cube Theorem

In this section we present three proofs that every median graph G has a subcube that is stabilized by every automorphism of G. This generalizes Theorem 2.5 for trees. It is also a special case and probably the origin of the more general fixed-box theorems of Chapter 16. It deserves special attention, and we present different approaches that can, in turn, lead to different fixed cubes.

For the first proof we begin with the so-called Helly property of convex subgraphs of median graphs.

Theorem 12.19 *Let $\mathcal{H}$ be a set of convex subgraphs of a median graph G, where any two elements of $\mathcal{H}$ have nonempty intersection. Then $\bigcap_{H\in\mathcal{H}} H \neq \emptyset$.*

Proof We first show that any three subgraphs in $\mathcal{H}$ have nonempty intersection. Let A, B, C be three such subgraphs. Consider vertices u, v, and w from $A \cap B$, $B \cap C$, and $C \cap A$, resectively. Because G is a median graph, u, v, w have a unique median, say z. By convexity, z belongs to $A \cap B \cap C$.

Let the theorem be true for any collection of fewer than n convex subgraphs, and suppose that $\mathcal{H}$ consists of n pairwise intersecting subgraphs. Replace any two of them, say A and B, by $A \cap B$. The new family $(\mathcal{H} \setminus \{A, B\}) \cup \{A \cap B\}$ has fewer than n subgraphs and, by the above, still satisfies the pairwise intersection property. Its nonempty intersection is the nonempty intersection of $\mathcal{H}$. □

Lemma 12.20 *For a median graph G, the following are equivalent:*

(i) *G is a hypercube.*
(ii) *For any edge uv of G, W_{uv} has $|V(G)|/2$ vertices.*
(iii) *For any edge uv of G, $W_{uv} = U_{uv}$.*

Proof Clearly, (i) implies (ii).

To show that (iii) follows from (ii), assume that for some edge uv of G we have $W_{uv} \neq U_{uv}$. Let w be a vertex of $W_{uv} \setminus U_{uv}$. We may, without loss of generality, assume that w is adjacent to a vertex x in U_{uv}. By the Convexity Lemma, wx is not in relation Θ to any of the edges of $\langle U_{uv}\rangle$. Therefore the set of edges F_{wx} is completely contained in W_{uv}, and thus W_{wx} cannot have $|V(G)|/2$ vertices because $|W_{uv}| = |V(G)|/2$.

Assume now that (iii) holds for G. Let ab be an edge of G and H be the contraction of G with respect to F_{ab}. Then H satisfies condition (iii), and by induction on the number of Θ-classes, H is a hypercube, say Q_d. But then G is Q_{d+1}. □

Theorem 12.21 (Fixed cube theorem) *Every median graph G contains a subcube that is invariant under all automorphisms of G.*

First proof We note that theorem is true for the trivial median graph and proceed by induction on the number of vertices. Suppose that G is a median graph on n vertices and that the theorem is true for median graphs on fewer than n vertices. We may assume that G is not a hypercube; otherwise, the entire graph is an invariant hypercube.

By Lemma 12.20, there is a set W_{uv} with $|W_{uv}| > n/2$. The image of a W_{uv} under an automorphism ϕ is $W_{\phi(u)\phi(v)}$. It has the same size as W_{uv}. We say that the collection of all W_{uv} of this size is invariant under automorphisms of G. Because any two such W_{uv} have nonempty intersection, the intersection of all of them is nonempty by Theorem 12.19. It induces a median graph H on fewer than n vertices and is invariant under all automorphisms of G. Thus any subcube of H invariant under all elements of $\mathrm{Aut}(H)$ is also invariant under all elements of $\mathrm{Aut}(G)$. □

The second proof is modeled closely after that of Theorem 2.5 and uses Lemma 12.16, which asserts that every median graph has a peripheral subgraph.

Second proof Suppose that every subgraph $\langle W_{uv}\rangle$ of G is peripheral. Then $V(G) = U_{uv} \cup U_{vu}$ for every edge uv of G. Because the sets U_{uv} and U_{vu} are disjoint and have the same number of vertices, we infer that $|U_{uv}| = |V(G)|/2$ for every edge uv of G. Moreover, if all W_{uv} are peripheral, $|W_{uv}| = |V(G)|/2$ for $uv \in E(G)$. By Lemma 12.20, G is a hypercube and of course stabilized by every automorphism of G.

If not every $\langle W_{uv}\rangle$ of G is peripheral, we proceed by induction. We only have to note that every automorphism of G preserves the collection of peripheral subgraphs and the (nonempty) subgraph of G obtained by removing all peripheral $\langle W_{uv}\rangle$ from G. □

For another approach that also yields invariant subcubes, we introduce a new concept. The *distance center* of a connected graph G is defined as the set of vertices u of G that minimize $\sum_{v\in V(G)} d_G(u,v)$. For example, let T be the path of length 2 with two pendant edges attached to one endpoint. Then the distance center of T consists of the vertex of degree 3 in T. It is different from the center of T, which consists of the endpoints of the interior edge of T, that is, of the vertex of degree 2 and the vertex of degree 3 in T.

If G is a subgraph of H, then then distance center $C(G,H)$ of G in H is defined as

$$\{u \in V(H) \mid \sum_{v\in V(G)} d_H(u,v) \text{ is minimal}\}\,.$$

$C(G,H)$ need not be contained in G. For example, consider an isometric subgraph C_6 of Q_3. The $Q_3 = C(C_6, Q_3)$. However, the situation changes when G is a retract of H.

Lemma 12.22 *If G is a retract of H, then the distance center $C(G,H)$ of G in H is a subgraph of G.*

Proof Let $f : H \to G$ be a retraction and $u \in C(G,H)$. Because retractions are nonexpansive,

$$\sum_{v\in V(G)} d_H(f(u),v) \leq \sum_{v\in V(G)} d_H(u,v).$$

If u is not in $V(G)$, then $w = f(u) \in V(G)$, and so $d_H(f(u),w) = 0 < d_H(u,w)$. This means that the preceding inequality is strict, which is not possible. Thus $C(G,H) \subseteq V(G)$. □

The following lemma is due to Nieminen (1987). The proof presented here is adapted from Feder (1995), who also developed algorithms for minimal fixed cubes.

Lemma 12.23 *If G is a retract of Q_r, then the distance center $C(G,Q_r)$ is a subcube of G.*

Proof By Lemma 12.22, $C(G,Q_r) \subseteq V(G)$. Furthermore

$$\sum_{v\in V(G)} d_{Q_r}(u,v) = \sum_{i=1}^{r} \sum_{v\in V(G)} d_i(u_i,v_i),$$

where d_i denotes the distance in the ith factor K_2 of Q_r and u_i, v_i the ith coordinates of u, v in Q_r. Let C_i be the set of v_i that minimize $\sum_{v\in V(G)} d_i(u_i,v_i)$. Then $u \in C(G,Q_r)$ if and only if $u_i \in C_i$ for all i. This is equivalent to the assertion that $C(G,Q_r)$ induces a subcube of Q_r. □

Theorem 12.24 *Let G be a retract of Q_r. Then $C(G, Q_r)$ is a subcube of G that is invariant under all automorphisms of G.*

Proof Let G be a median graph represented as a retract $f(Q_r)$. By Lemma 12.23, $C(G, Q_r)$ induces a subcube of G. Because distances are preserved by automorphisms, $C(G, Q_r)$ is invariant under all automorphisms of G. □

An even stronger result is the following:

Theorem 12.25 *Every nonexpansive map from a median graph to itself has a fixed cube.*

Proof We first show that the statement holds for hypercubes. Let f be a nonexpansive map of Q_r. Clearly, $f^{j+1}(Q_r) \subseteq f^j(Q_r)$ for all natural numbers j. Because Q_r is finite, there is a k with $f^{k+1}(Q_r) = f^k(Q_r)$. Let $G = f^k(Q_r)$. Then the restriction of f to G is an automorphism of G. For $u \in V(G)$, let $n(u)$ be the least integer for which $f^{n(u)}(u) = u$, and let ℓ be the least common multiple of $\{n(u) \mid u \in V(G)\}$ that is greater than or equal to k. Then $f^\ell : Q_r \to G$ is a retraction. By Lemma 12.23, the distance center of G is the fixed cube.

For the general case, let G be a median graph represented as $g(Q_r)$, where g is a retraction, and h a nonexpansive map of G. Then hg is a nonexpansive map of Q_r for which the preceding argument applies. □

Theorem 12.25 is due to Bandelt and van de Vel (1987). Their results can also be presented in the framework of metric spaces. This is the reason for our use of the terminology "nonexpansive map" rather than "weak homomorphism."

Similar results hold for retracts of *Hamming graphs*, that is, Cartesian products of complete graphs, and general Cartesian products. They will be treated as "fixed box theorems" in Chapter 16.

12.5 Median Networks in Human Genetics

Median graphs have been applied to problems in human genetics, though a detailed treatment of this topic would be beyond the scope of this book. We thus take only a quick glance at the main ideas. See Bandelt (2006) for a brief overview on evolutionary networks in general and median networks in particular.

The first use of median networks in human genetics goes back to Bandelt, Forster, Sykes, and Richards (1995). The central object of their study was the mitochondrial DNA, or mtDNA for short. Realizing that traditional tree-building methods were unsatisfactory when analyzing human mtDNA, they developed an approach that distinguishes between unresolvable and resolvable character conflicts. They attained this goal with the following model.

In the first phase, the mtDNA of certain number of individuals from a given population is examined and converted to binary data. (This is possible because the great majority of observed changes are the so-called transitions A $\leftrightarrow$ G or C $\leftrightarrow$ T.) This preprocessing yields a set of binary vectors of a fixed length, say d, which are regarded as vertices of Q_d. The corresponding *median network* is the smallest median subgraph of Q_d containing the data vectors. Equivalently, this network is obtained by adding medians to triples of the original vectors, and continuing this process until no new vertices are added.

In their seminal paper, Bandelt et al. (1995) show that all the so-called most parsimonious trees[1] for the given vectors can be realized in the corresponding median network. (This is perhaps not surprising, as median graphs have a treelike structure, as we noted earlier.) They also give an efficient procedure for constructing median networks, as well as a reduction procedure to be used when median networks become too large. This median network approach is then applied to several concrete data sets. For example, Figure 12.5, which is adapted from their paper, is the median network for a set of twenty-eight Frisians.

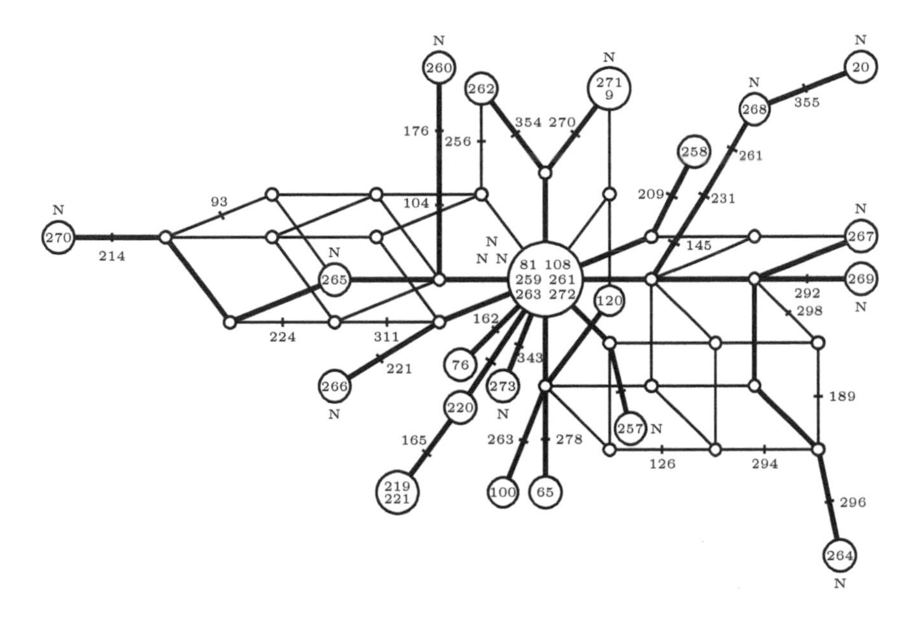

FIGURE 12.5 A median network.

We remark that the displayed network encodes more information than the mere graph structure. For instance, the areas of circles representing "haplotyles" are proportional to the number of individuals with that haplotype, while the small circles are hypothetical haplotypes. The letter N stands for individuals from North Frisian islands and the bold lines indicate the unique most parsimonious tree.

The median network approach led to numerous developments. We list a selection of important achievements in chronological order.

- See Richards et al. (1996) for another early application of median networks to human genetics. This influential paper studied the mtDNA of 821 individuals from Europe and the Middle East and concluded that ancestors of the great majority of modern, extant lineages entered Europe during the Upper Paleolithic. This research used median networks as the basis for the phylogenetic analysis.
- A significant problem with the mtDNA data in publications and databanks is that they can contain serious artifacts. Bandelt, Quintana-Murci, Salas, and Macaulay (2002) considered one type of such errors—the so-called phantom mutations—and designed a procedure for filtering out highly mutable sites. Their analysis made use of the sequence $(f_d)_{d \geq 0}$, where f_d is the number of d-cubes of the median network.
- Bandelt, Salas, and Lutz-Bonengel (2004) describe four main errors that occur in

[1] Roughly speaking, phylogenetic trees requiring the fewest evolutionary changes.

sequencing mtDNA for forensic purposes and give specific published data that are not appropriate for such purposes. They propose a solution, and give small, specifically selected databases that can be represented by full median networks.

- Kong et al. (2008) propose the technique of recombination diagrams in order to deal with errors that occur in virtually every mtDNA database.
- Using the median network method, Bandelt et al. (2009) contrast some early reports of complete mtDNA sequences to more recent total mtDNA data in studies of various mitochondrial diseases. They give diagrams displaying incomplete reading of sequences.

To close, we mention that quasi-median networks show promise in situations where the data do not allow binary representation. (Chapters 14 and 22 treat quasi-median graphs.)

Exercises

12.1. Find all retracts of the 3-cube.

12.2. (Klavžar and Škrekovski, 2000) Show that if e is an edge of a graph G that lies in at least two squares, then not both G and $G - e$ can be median graphs.

12.3. (Klavžar and Škrekovski, 2000) Show that a connected subgraph of the Cartesian product of two paths with n vertices and m edges is a median graph if and only if it contains $m - n + 1$ squares.

12.4. (Klavžar and Škrekovski, 2000) Let G be a plane median graph with n vertices and m edges that is a subgraph of the Cartesian product of two paths. Show that the length of its outer face is $4n - 2m - 4$. This is twice the number of its Θ-classes.

12.5. (Isbell, 1980) Show that a median graph G is a hypercube if and only if $V(G) = W_{ab} \cup W_{uv}$ for any disjoint sets W_{ab} and W_{uv}.

12.6. (Bandelt and Mulder, 1983) Show that every interval of a median graph induces a median graph.

12.7. Complete the proof of Theorem 12.8 by showing that a contraction of a median graph is median.

12.8. (Mulder, 1980b) Show that a graph is an n-cube if and only if it is an n-regular median graph.

12.9. Give an example of a partial cube that does not have a peripheral subgraph.

12.10. (Soltan and Chepoĭ, 1987; Škrekovski, 2001) Let q_r denote the number of subgraphs of a median graph isomorphic to Q_r and k be the number of its Θ-classes. Show that

$$\sum_{i \geq 0} (-1)^i q_i = 1 \quad \text{and} \quad k = -\sum_{i \geq 0} (-1)^i i q_i \,.$$

12.11. Deduce Theorem 12.10 as a consequence of Theorem 12.15.

12.12. Show that any median subgraph of a median graph G is a retract of G.

12.13. Prove Lemma 12.17 with the aid of Exercise 12.12.

Chapter 13

The Canonical Isometric Embedding

The main theme of this chapter is the canonical isometric embedding, which embeds any graph isometrically into a Cartesian product. We begin by defining the map, proving that it is an isometry, and deducing several of its properties. Then we take a closer look at the role of the relation Θ in Cartesian products. Among others things, we prove that the only isometric irredundant embeddings into Cartesian products of complete graphs are the canonical isometric embeddings. We close with a description of the automorphisms of canonically embedded graphs.

13.1 The Embedding and Its Properties

Suppose Π a partition of the vertices of a graph G. The *quotient graph* G/Π is a graph with vertex set Π, and for which distinct classes $C_1, C_2 \in \Pi$ are adjacent if some vertex in C_1 is adjacent to a vertex of C_2.

The canonical embedding of a connected graph G is defined as follows: Let the Θ^*-classes of G be $E_1, E_2, \ldots, E_k$, and for each index i, put $G_i = G - E_i$. Let Π_i be the partition of $V(G_i)$ whose classes are the vertices of the connected components of G_i. Let $\alpha_i : G \to G/\Pi_i$ be the map sending any v to the component of G_i that contains it. The *canonical embedding*

$$\alpha : G \to G/\Pi_1 \,\Box\, G/\Pi_2 \,\Box\, \cdots \,\Box\, G/\Pi_k$$

is defined by

$$\alpha(v) = \big(\alpha_1(v), \alpha_2(v), \ldots, \alpha_k(v)\big)\,.$$

Figure 13.1 shows an example. Here Θ^* has three equivalence classes, giving rise to quotients $K_4 - e$, K_2, and K_2. Thus we have an embedding $\alpha : G \to (K_4 - e) \,\Box\, K_2 \,\Box\, K_2$. Observe that this embedding is indeed isometric.

In Section 11.1 we defined a function μ on ordered pairs of vertices $p = (a, b)$ and $q = (x, y)$ by

$$\mu(p, q) = d(a, y) - d(a, x) - d(b, y) + d(b, x)\,.$$

Recall that μ is symmetric; that is $\mu(p, q) = \mu(q, p)$. By arbitrarily orienting edges e and f, we see that $e\Theta f$ if and only if $\mu(e, f) \neq 0$. Furthermore, for sets (or multisets) A, B of ordered pairs of vertices, we define

$$\mu(A, B) = \sum_{p\in A} \sum_{q\in B} \mu(p, q)\,.$$

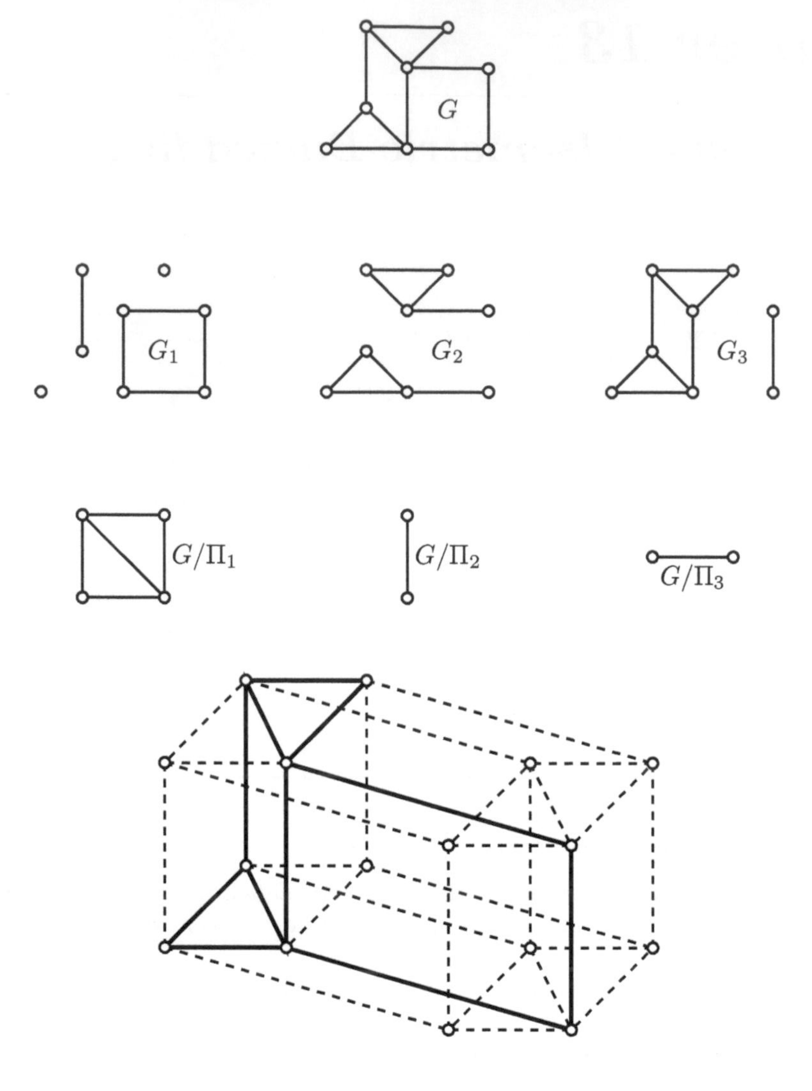

FIGURE 13.1 Canonical embedding of G.

We are interested in the case where B is a path $Q = y_0y_1 \ldots y_k$ with edges oriented as (y_{i-1}, y_i) for $i = 1, 2, \ldots, k$. If $p = (y_0, y_k)$ and q is an ordered pair of vertices, then it is easily seen that $\mu(q, Q) = \mu(q, p)$. (Lemma 11.3 is a special case.) Thus

$$\mu(A, Q) = \mu(A, p)$$

for any set (or multiset) A of ordered pairs of vertices. Also $\mu(Q, A) = \mu(p, A)$, by symmetry of μ.

The following lemma is the key for the proof of the most important property of the canonical embedding—the fact that it is an isometry.

Lemma 13.1 *If P is a shortest u, v-path in G and Q is an arbitrary u, v-path, then*

$$|E(P) \cap E_i| \leq |E(Q) \cap E_i|$$

for any Θ^*-class E_i.

Proof Orient P and Q so that each edge is directed toward the endpoint v. Set $p = (u, v)$. Let us abbreviate $E(P) \cap E_i$ as $P \cap E_i$, etc. In the following computations we agree that in expressions such as $P \cap E_i$, each edge from E_i that intersects P is given the same orientation as the edge of P that it coincides with. Then

$$\begin{aligned}\mu(p, P \cap E_i) &= \mu(Q, P \cap E_i) \\ &= \mu(Q \cap E_i, P \cap E_i) \\ &= \mu(Q \cap E_i, P) \\ &= \mu(Q \cap E_i, p)\,.\end{aligned}$$

To see this, we note that the first and the last equality hold by the above observation. The other two equalities follow from the fact that if e and f are arbitrarily directed edges of G, then $\mu(e, f) = 0$ whenever e and f are not in the relation Θ.

As P is a shortest path, the definition of μ yields $\mu(p, f) = 2$ for any directed edge f of P. Hence

$$\mu(p, P \cap B) = 2|P \cap B|$$

for any set (or multiset) B of directed pairs. Therefore

$$2|P \cap E_i| = \mu(p, P \cap E_i) = \mu(Q \cap E_i, p) \leq 2|Q \cap E_i|\,,$$

where the inequality follows from the observation that $|\mu((x, y), (x', y'))| \leq 2$ if at least one of xy and $x'y'$ is an edge. □

We have now prepared all prerequisites for the main theorem of this section.

Theorem 13.2 (Graham-Winkler) *The canonical embedding is an isometry.*

Proof Let $uv \in E(G)$ and assume that $uv \in E_i$. From Lemma 11.3 we infer that $\alpha_i(u)\alpha_i(v) \in E(G/\Pi_i)$. Moreover, if $j \neq i$, then $\alpha_j(u) = \alpha_j(v)$. Thus α maps edges onto edges. Lemma 11.3 also implies that α is injective.

Suppose α is not an isometry. Then there is a shortest path P joining vertices x and y in G, and an even shorter shortest path R from $\alpha(x)$ to $\alpha(y)$ in $\square_{i=1}^{k} G/\Pi_i$. Now, R can be written as

$$\alpha(x) = (C_{1,1}, C_{2,1}, \ldots, C_{k,1})(C_{1,2}, C_{2,2}, \ldots, C_{k,2}) \ldots (C_{1,\ell}, C_{2,\ell}, \ldots, C_{k,\ell}) = \alpha(y)\,,$$

where each $C_{i,j}$ is a component of $G - E_i$, and x is in each $C_{i,1}$, and y is in each $C_{i,\ell}$.

For each $1 \leq i \leq k$, let F_i be the set of edges in $\square_{i=1}^{k} G/\Pi_i$ whose endpoints differ only in the ith coordinate. Then

$$\sum_{i=1}^{k} |P \cap E_i| = |P| > |R| = \sum_{i=1}^{k} |R \cap F_i|,$$

so there is an index i for which $|P \cap E_i| > |R \cap F_i|$. Consider the sequence $C_{i,1}, C_{i,2}, \ldots, C_{i,\ell}$ of components of $G - E_i$. For any two successive terms $C_{i,j}, C_{i,j+1}$ in this sequence, either $C_{i,j} = C_{i,j+1}$ or there is an edge $e_j \in E_i$ joining $C_{i,j}$ to $C_{i,j+1}$, and there is a total of $r = |R \cap F_i|$ such edges. List them in order of traversal as $e_{j_1}, e_{j_2}, \ldots e_{j_r}$. These edges can be extended by edges in the components $C_{i,j}$ to form a path Q in G joining x to y with $|Q \cap E_i| = |R \cap F_i| < |P \cap E_i|$, in contradiction to Lemma 3.1. □

The isometry α possesses several other properties that are collected in the following theorem. First two definitions.

We call an isometric embedding $\beta : G \to \square_{i=1}^{m} H_i$ *irredundant* if $|V(H_i)| \geq 2$ for $i = 1, \ldots, m$, and if every vertex $h \in \bigcup_{i=1}^{m} V(H_i)$ occurs as a coordinate in the image of some $g \in V(G)$. (This means that an irredundant embedding has no unused factors or vertices.) The image of $\beta(G)$ in $\square_{i=1}^{m} H_i$ is called an *irredundant subgraph.*

A graph G is called *irreducible* if any irredundant isometric embedding $\beta : G \to \square_{i=1}^{m} H_i$ is trivial, namely $m = 1$ and $G = H_1$. Note that an irreducible graph is necessarily prime, but not conversely. (The path on three vertices is a counterexample.) The next theorem is by Graham and Winkler (1985).

Theorem 13.3 *If α is the canonical embedding of a connected graph G, then*

(i) *α is irredundant.*
(ii) *α has the most factors among all irredundant isometric embeddings of G.*
(iii) *Each quotient graph G/Π_i is irreducible.*
(iv) *α is unique among the embeddings satisfying* (ii).

Proof Consider (i). If uv is an edge of an equivalence class E_i, then, by Lemma 13.1, u and v belong to different connected components of G_i. Thus G/Π_i has at least two vertices. The second condition for irredundancy follows from the fact that each α_i is surjective.

For (ii), let $\beta : G \to \square_{i=1}^{m} H_i$ be any irredundant isometric embedding. Furthermore, let e and f be two edges of $\beta(G)$ that are in layers with respect to different factors H_i, H_j. We wish to show first that e and f are not in the relation Θ in this case.[1] For $m = 2$ this is obvious by Corollary 5.1. Thus e, f are not in the relation $\Theta_{H_i \,\square\, H_j}$. By the Distance Formula (Corollary 5.2), $H_i \,\square\, H_j$ is an isometric subgraph of $\square_{i=1}^{m} H_i$, hence $e\Theta f$ cannot hold either. This means that all edges of an equivalence class E_i must be in layers with respect to the same factor, say H_i. Hence α has the largest possible number of factors.

Statement (iii) follows from (ii), because the reducibility of a quotient graph G/Π_i would lead to an irredundant isometric embedding of G with a larger number of factors than α.

Statement (iv) follows by the same argument as the one used for (ii). □

Corollary 13.4 *Let G be a connected graph on n vertices. Suppose α embeds G into k factors. Then G is a tree if and only if $k = n - 1$.*

Proof We already know (see the remark after Lemma 11.3) that α embeds in at most $n-1$ factors. Furthermore, if G is a tree, no two edges of G are in the relation Θ, by Lemma 11.1. Thus $k = n - 1$.

Conversely, if G has a cycle, then at least two edges, say e and f, are in the relation Θ. Consider a spanning tree T of G containing the edges e and f. By Lemma 11.3, every edge of $E(G) \setminus E(T)$ is in relation Θ to an edge of T. Thus $k < n - 1$. □

13.2 The Relation Θ and the Cartesian Product

If $G \cong \square_{i=1}^{k} G_i$, then we say $\square_{i=1}^{k} G_i$ is a *product representation* of G. With respect to this representation, we introduce a *product coloring* $c : E(G) \to \{1, \ldots, k\}$ as follows: For $uv \in E(G)$ we set $c(uv) = i$ if u and v differ in coordinate i. Thus all edges in any layer G_i^a have color i. Note that c is not an edge-coloring in the usual sense, because incident edges may have the same color. Figure 13.2 shows an example of a product coloring.

[1]This fact also follows from the more general Lemma 13.5.

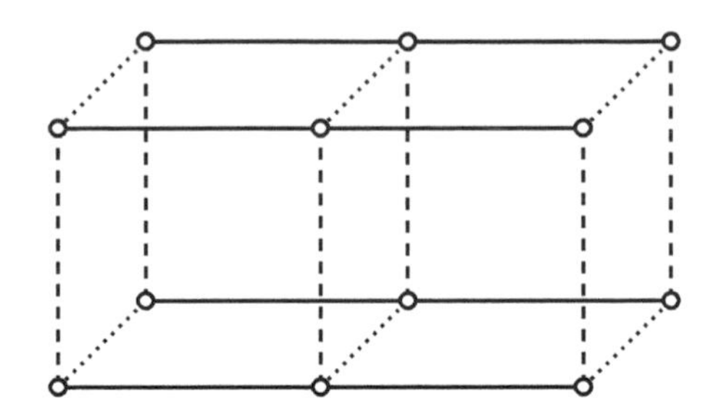

FIGURE 13.2 Product coloring of $G = P_3 \,\square\, K_2 \,\square\, K_2$.

Lemma 13.5 *Suppose* $G = \square_{i=1}^{k} G_i$ *and* $e, f \in E(G)$.

(i) *If* $c(e) = c(f) = i$ *and* $p_i(e) = p_i(f)$, *then* $e\Theta f$.

(ii) *If* $e\Theta f$, *then* $c(e) = c(f)$.

Proof Put $e = uv = (u_1, \ldots, u_k)(v_1, \ldots, v_k)$ and $f = xy = (x_1, \ldots, x_k)(y_1, \ldots, y_k)$.

(i) Because $c(e) = c(f) = i$, we have $v = (u_1, \ldots, v_i, \ldots, u_k)$ and $y = (x_1, \ldots, y_i, \ldots, x_k)$. In view of $p_i(e) = p_i(f)$, we may assume, without loss of generality, that $u_i = x_i$ and $v_i = y_i$. By the Distance Formula (Corollary 5.2), $d(u, x) < d(u, y)$ and $d(v, y) < d(v, x)$. Thus $e\Theta f$.

(ii) Suppose $c(e) \neq c(f)$. Without loss of generality, say $c(e) = i < j = c(f)$. Then

$$\begin{aligned} u &= (u_1, \ldots, u_i, \ldots, u_j, \ldots, u_k), \quad & x &= (x_1, \ldots, x_i, \ldots, x_j, \ldots, x_k), \\ v &= (u_1, \ldots, v_i, \ldots, u_j, \ldots, u_k), \quad & y &= (x_1, \ldots, x_i, \ldots, y_j, \ldots, x_k). \end{aligned}$$

By the Distance Formula, both $d_G(u, x) + d_G(v, y)$ and $d_G(u, y) + d_G(v, x)$ are equal to

$$d_{G_i}(u_i, x_i) + d_{G_j}(u_j, x_j) + d_{G_i}(v_i, x_i) + d_{G_j}(u_j, y_j) + 2 \sum_{1 \leq \ell \leq k}^{\ell \neq i,j} d_{G_\ell}(u_\ell, x_\ell).$$

Therefore $e\Theta f$ does not hold. □

Notice that $e\Theta f$ need not imply $p_i(e) = p_i(f)$, even if $c(e) = c(f)$. An example are edges e and f in different K_3-layers of $K_3 \,\square\, K_2$ with $p_1(e) \neq p_1(f)$.

The fact that $e\Theta f$ implies $c(e) = c(f)$ means that Θ is finer than the product relation c. Because c is an equivalence relation, the transitive closure Θ^* of Θ is also contained in c. $K_2 \,\square\, P_3$ shows that Θ^* can be strictly finer than c.

Let $G = \square_{i=1}^{k} G_i$ be the Cartesian product of connected graphs. Then every set of edges of color i is an equivalence class of c and thus a union of equivalence classes of Θ^*. Hence the connected components of the subgraph of G spanned by the edges of color i are convex by Lemma 11.5. As these connected components are the G_i-layers of G, we have just reproved the fact that layers in Cartesian products are convex, which we already know from Lemma 6.5.

As an application of Lemma 13.5 we consider embeddings into products of triangles. We know that partial cubes can be characterized as bipartite graphs with transitive Θ (Theorem 11.8). What happens if the bipartiteness condition is dropped? Following the approach of Winkler (1984), Lemma 13.5 and the canonical embedding combine to answer this question.

Corollary 13.6 *Let G be a connected graph. Then G is isometrically embeddable into a Cartesian product of triangles if and only if* $\Theta = \Theta^*$.

Proof Let G be a connected graph and $uv \in E(G)$. Recall that W_{uv} denotes the set of vertices closer to u than to v. Let ${}_uW_v$ be the set of vertices of G at equal distance from u and v. Clearly, W_{uv}, W_{vu}, and ${}_uW_v$ form a partition of $V(G)$. Note that $uv\Theta xy$ if and only if xy is an edge between two different parts of this partition.

Let $\Theta = \Theta^*$ and E_i be the equivalence class of Θ containing uv. Then G/Π_i is K_3 if ${}_uW_v$ is nonempty, and K_2 otherwise. Thus G canonically embeds into a Cartesian product of complete graphs on two or three vertices. It is easy to extend this embedding to an isometric embedding into a product of triangles.

Conversely, let β be an isometric embedding of G into a Cartesian product of triangles and c be the corresponding product coloring. Let $e = uv$ and $f = xy$ be two edges of G. We claim that $e\Theta f$ if and only if $c(e) = c(f)$.

If $e\Theta f$, then $c(e) = c(f)$ by Lemma 13.5 (i). Conversely, let $c(e) = c(f) = i$. Because we have an embedding into a product of triangles, we may assume without loss of generality that $p_i(u) = p_i(x)$. Then $d(u,x) < d(u,y)$ and $d(v,y) \leq d(v,x)$. It follows that $e\Theta f$, which proves the claim. Thus Θ is equal to the equivalence relation c. □

Another application of Lemma 13.5 (and Theorem 13.3), although not related to products of triangles, is the following:

Corollary 13.7 *Let $\beta : G \to H_1 \square \cdots \square H_m$ be an isometric irredundant embedding of a graph G into a product of complete graphs H_i. Then this embedding is the canonical isometric embedding.*

Proof Let $H = H_1 \square H_2 \square \cdots \square H_m$ and U and V be two H_i-layers. We claim that $p_i(\beta(G) \cap U) \subseteq p_i(\beta(G) \cap V)$, or vice versa. If this is not the case, there are vertices $u, u' \in U$ and $v, v' \in V$ such that

$$u \in \beta(G) \cap U,\ u' \notin \beta(G) \cap U,\ v \in \beta(G) \cap V,\ v' \notin \beta(G) \cap V\,,$$

where $p_i(u) = p_i(v')$, $p_i(u') = p_i(v)$. See Figure 13.3.

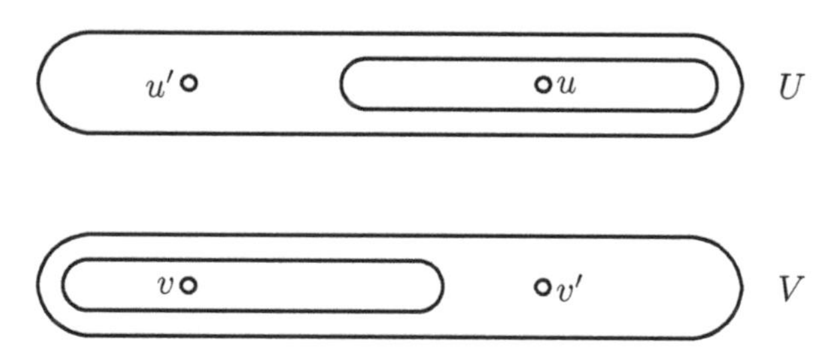

FIGURE 13.3 Layers U and V.

Suppose that the distance between U and V in H is k. Then, because $\beta(G)$ is isometric in H, we have $k+1 = d_H(u,v) = d_{\beta(G)}(u,v)$, which is only possible if $u' \in \beta(G)$ or $v' \in \beta(G)$. This proves the claim.

Let the edges e and e' in $\beta(G)$ belong to layers U and V, respectively. If $p_i(\beta(G) \cap U) \subseteq p_i(\beta(G) \cap V)$, then, by Lemma 13.5 (i), there is an edge e'' in $\beta(G) \cap V$ with $e\Theta e''$. Because $\beta(G) \cap V$ is complete, we have $e''\Theta^* e'$, hence $e\Theta^* e'$.

We have thus proved that any two edges e, e' in $\beta(G)$ are in relation Θ^* if they have the same color with respect to the product coloring of H. In other words, all edges of the same color lie in a common equivalence class of Θ^*. On the other hand, no two edges of different colors are in a common equivalence class by Lemma 13.5 (ii). Hence the number of colors is equal to the number of equivalence classes of Θ^*. Because, by Theorem 13.3 (iv), α is unique, we conclude that $\beta = \alpha$. □

13.3 Automorphisms of Canonical Embeddings

In this section we follow the approach of Imrich (1989) to show that the canonical isometric embedding of a graph G into a Cartesian product $G^* = G_1^* \,\square\, \cdots \,\square\, G_k^*$ establishes a strong connection between G and G^*.

We begin by showing that automorphisms of G give rise to automorphisms of G^*. Because any $\varphi \in \mathrm{Aut}(G)$ preserves distances between vertices, it follows that $e\Theta f$ if and only if $\varphi(e)\Theta\varphi(f)$. This means that φ permutes the equivalence classes $E_1, E_2, \ldots, E_k$ of Θ^*. Let π be the permutation of $\{1, 2, \ldots, k\}$ with $\varphi(E_{\pi(i)}) = E_i$.

Recall that the G_i^* are formed by contractions of the connected components of $G_i = (V(G), E(G) \setminus E_i)$ to single vertices. By the above, $\varphi(G_{\pi(i)}) = \big(V(G), E(G) \setminus \varphi(E_{\pi(i)})\big) = \big(V(G), E(G) \setminus E_i\big) = G_i$. In particular, the image $\varphi(C)$ of a connected component C of $G_{\pi(i)}$ is a component of G_i. Moreover, if an edge e joins two components C, C' of $G_{\pi(i)}$, then $\varphi(e)$ joins $\varphi(C)$ to $\varphi(C')$. We infer that φ induces an isomorphism $\psi_{\pi(i)} : G^*_{\pi(i)} \to G_i^*$.

For the contraction α_i of G_i to G_i^*, we have

$$\alpha_i \,\varphi(v) = \psi_{\pi(i)}\, \alpha_{\pi(i)}(v).$$

In coordinate representation this reads as

$$\begin{aligned} \alpha\,\varphi(v) &= \big(\alpha_1\varphi(v),\, \alpha_2\varphi(v),\, \ldots, \alpha_k\varphi(v)\big) \\ &= \big(\psi_{\pi(1)}\alpha_{\pi(1)}(v),\;\; \psi_{\pi(2)}\alpha_{\pi(2)}(v),\; \ldots,\; \psi_{\pi(k)}\alpha_{\pi(k)}(v)\big)\,. \end{aligned}$$

Let φ^* denote the automorphism of G^* that sends the vertex $(v_1, v_2, \ldots, v_k)$ of G^* to $\big(\psi_{\pi(1)}(v_{\pi(1)}),\, \psi_{\pi(2)}(v_{\pi(2)}),\, \ldots, \psi_{\pi(k)}(v_{\pi(k)})\big)$. Then

$$\alpha\,\varphi(v) = \varphi^*\,\alpha(v)$$

and

$$\alpha(G) = \alpha\varphi(G) = \varphi^*\alpha(G).$$

We have thus proved the following theorem:

Theorem 13.8 *Let α be the canonical isometric embedding of a graph G into a Cartesian product $G^* = G_1^* \,\square\, \cdots \,\square\, G_k^*$. Then every automorphism of $\alpha(G)$ is induced by an automorphism of G^*.*

Despite this theorem, it is in general not easy to deduce much about the group of a canonically embedded graph G from the group of G^*. For example, P_3 canonically embeds into $K_2 \,\square\, K_2$. Both factors have a transitive group, but not P_3. It works better the other way.

Theorem 13.9 *Let G be a connected graph with transitive automorphism group and α the canonical isometric embedding of G into the Cartesian product $G^* = G_1^* \,\square\, \cdots \,\square\, G_k^*$. Then* $\mathrm{Aut}(G^*)$ *is transitive.*

Proof It suffices to show that each G_i^* has a transitive group. Suppose that this is not the case. Without loss of generality, we can assume that $\mathrm{Aut}(G_1^*)$ is intransitive. Let V_1 be the orbit of a vertex $v_1 \in V(G_1^*)$, that is,

$$V_1 = \{\varphi(v_1) \,|\, \varphi \in \mathrm{Aut}(G_1^*)\,\}.$$

Because $\mathrm{Aut}(G_1^*)$ is not transitive, the set $V_2 = V(G_1^*) \setminus V_1$ is not empty.

If all the other G_i^* are nonisomorphic to G_1^*, we choose two vertices v, w in $\alpha(G)$ for which the first coordinate of v in $G_1^* \,\square\, \cdots \,\square\, G_k^*$, say v_1, is in V_1, whereas the first coordinate of w is in V_2. Such vertices exist because G is canonically embedded into G^*. Now, by Theorem 13.3, the G_i^* are irreducible, hence also prime. We infer from Theorem 6.10 that $\mathrm{Aut}(G^*)$ is not transitive.

Thus suppose that $G_1^* \cong G_2^* \cong \cdots \cong G_l^*$, where $2 \leq l \leq k$, and that no other factor of G^* is isomorphic to G_1^*. We can further assume that $G_1^* = G_2^* = \cdots = G_l^*$. Thus, for any $v \in G^*$, the ith coordinate, say v_i, of v, $1 \leq i \leq l$, is either in V_1 or in V_2. Set

$$n(v) = \big|\{v_1, v_2, \ldots, v_l\} \cap V_1\,\big|.$$

Clearly, $n(v) = n\,\varphi(v)$ for any automorphism φ of G^*.

As before, we choose vertices $v, w \in V(\alpha(G))$ for which $v_1 \in V_1$ and $w_1 \in V_2$. Let P be a path in $\alpha(G)$ from v to w. Every edge connects vertices that differ in exactly one coordinate. Thus there must be an edge, say $v'w'$, in P for which $v_1' \in V_1$, but $w_1' \in V_2$. Then $n(v') \neq n(w')$, and there is no automorphism of G^*, and thus of $\alpha(G)$, that maps v' into w'. But then G is not transitive. □

As an example, consider the six-cycle C_6. It has transitive group and canonically embeds into Q_3, whose every factor also has transitive group.

Exercises

13.1. Find the canonical isometric embedding of $P_3 \,\square\, P_3$.

13.2. Show that a graph G is irreducible if and only if G has a single Θ^*-class.

13.3. Prove that an irreducible graph has no cut vertex.

13.4. Find an irreducible graph with no cut vertex in which every edge is contained in an odd cycle.

13.5. (Graham, 1988) In the usual random graph model, an edge between two vertices of a random graph is selected with probability one half. Based on this model, prove that almost all graphs are irreducible. More precisely, prove that the number of irreducible graphs on n vertices divided by the total number of graphs on n vertices tends to 1 as $n \to \infty$.

13.6. Embed C_6 isometrically into Q_3. Show directly that every automorphism of C_6 is induced by an automorphism of Q_3.

Chapter 14

A Dynamic Location Problem

This chapter develops a graph invariant called windex, introduced by Chung, Graham, and Saks (1987, 1989) in the context of dynamic location theory. It is closely connected to Cartesian products of complete graphs. These graphs, known as Hamming graphs, are treated in the first section, while the second section introduces the dynamic location problem and the corresponding invariant windex. Then quasi-median graphs are introduced as a natural generalization of median graphs. The chapter culminates with a proof that graphs with finite windex coincide with quasi-median graphs.

14.1 Hamming Graphs

Hypercubes can be generalized as follows. Take integers $k_i \geq 2$ for $i = 1, 2, \ldots, r$. Form a graph whose vertices are the r-tuples $b_1 b_2 \ldots b_r$ with $b_i \in \{0, 1, \ldots, k_i - 1\}$. Two tuples are adjacent if they differ in precisely one place. Such a graph is called a *Hamming graph*. Clearly, a graph is a Hamming graph if and only if it is a Cartesian product of nontrivial complete graphs (Exercise 14.1). For example, Figure 14.1 shows the Hamming graph $K_4 \,\Box\, K_3 \,\Box\, K_2$.

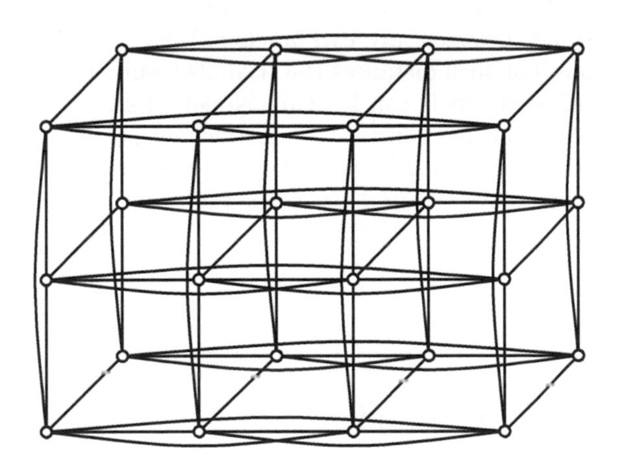

FIGURE 14.1 Hamming graph $K_4 \,\Box\, K_3 \,\Box\, K_2$.

Recall from Proposition 3.1 that the interval $I(u,v)$ between vertices u and v of a hypercube induces a $d(u,v)$-cube. This observation generalizes directly to Hamming graphs.

Proposition 14.1 *Let u and v be vertices at distance r in a Hamming graph. Then $I(u,v)$ induces a hypercube of dimension r.*

The proof of this proposition follows the lines of the analogous proof for hypercubes. We leave the details to the reader (Exercise 14.2).

Just as we defined partial cubes as isometric subgraphs of hypercubes, we define *partial Hamming graphs* as isometric subgraphs of Hamming graphs. The following is an equivalent definition of partial Hamming graphs: Let Σ be a finite alphabet. Given two r-tuples w_1 and w_2 of elements from Σ, their *Hamming distance* $H(w_1,w_2)$ is the number of positions in which they differ. A graph G is a partial Hamming graph if and only if each vertex $v \in V(G)$ can be labeled by a tuple $w(v)$ of fixed length such that $H(w(u),w(v)) = d_G(u,v)$ for all $u,v \in V(G)$. Such a labeling is a *Hamming labeling*.

Let us apply this concept to a communication network represented by a graph. Information between vertices is transmitted in packets, where many of these packets may be on the network simultaneously. A particular packet cannot be transmitted until a corresponding path is clear. Local routing realizes such a system by assigning an address to each vertex and using only local information to route each packet to its destination. Graham and Pollak (1971, 1972) suggested an addressing system in which the distance between two vertices is the Hamming distance between their addresses. Then, if a packet has distance k from its destination, it checks the addresses of neighboring vertices and moves to a vertex of distance $k-1$. Continuing this process, the packet reaches its destination via a shortest path. Such a scheme is possible if the graph corresponding to the network is a partial Hamming graph.

Chepoi (1988) obtained several interesting characterizations of partial Hamming graphs. One of them is an expansion theorem for which the expansion theorem for partial cubes mentioned in Section 11.2 is a special case. Another result asserts that a graph is a partial Hamming graph if and only if for any edge uv the sets W_{uv}, W_{vu}, ${}_uW_v$, and their complements induce convex subgraphs. (For the definition of ${}_uW_v$, see the proof of Corollary 13.6.) Wilkeit (1990) gives five other characterizations of partial Hamming graphs.

Not all graphs are partial Hamming graphs, that is, not all graphs can be embedded isometrically into Hamming graphs. To circumvent this difficulty, Graham and Pollak (1971) devised a method of labeling the vertices of a graph over the alphabet $\{0, 1, *\}$, where distances from $*$ to 0 and to 1 are defined as zero. (The symbol $*$ is the so-called "don't care" symbol.) The distance between any two labels is then the sum of the distances between respective positions. For such distances the triangle inequality need not hold, and the distance between two different labels can be zero. Nonetheless, they conjectured that for graphs on n vertices, one can always construct such labelings isometrically representing the distance of the respective graphs using labels of length at most $n-1$. This conjecture was known as the *squashed cube conjecture* and was settled affirmatively by Winkler (1983); see Chapter 9 of van Lint and Wilson (1992) for details.

14.2 Graphs with Finite Windex

We now describe the central problem of this chapter—the dynamic location problem and the associated invariant windex. We will show that Hamming graphs and their weak retracts have finite windex, thus paving the way for the rest of the chapter.

Suppose that a system is represented by a connected graph G, and a common resource is located at some vertex s. This vertex is called the *state* of the system, and it can be changed to any other vertex. However, the cost of moving the resource from state s to state s' is equal to $d(s, s')$. A sequence of requests is then made, each consisting of a vertex r at which some processing must be done. The cost of this processing is $d(s, r)$, where s is the current state of the system.

Let $s_0 = r_0$ be the *initial state* of the system, and $r_1 r_2 \dots r_l$ be a *request sequence.* When the system is in state s_{i-1}, it moves to state s_i and processes request r_i. (Possibly, $s_i = s_{i-1}$.) This procedure is repeated indefinitely.

Our goal is to find a *state sequence* $s_1 s_2 \dots s_l$ such that the *cost*

$$\sum_{i=1}^{l}(d(s_{i-1}, s_i) + d(s_i, r_i))$$

of the state sequence with respect to the request sequence is as small as possible. We call this the *dynamic location problem.* If the request sequence is known in advance, the problem can be solved by a standard dynamic programming approach. However, we wish to solve the problem in real time, because in general we have only incomplete knowledge of future requests. In fact, we will assume that for some integer k, the system knows the next k requests.

An algorithm for the dynamic location problem *works within window* k if the choice of each s_i depends only on the r_j with $i \leq j < i + k$. The *window index*, or *windex* of a graph G, denoted $WX(G)$, is the smallest k for which there exists an optimal algorithm for the dynamic location problem that works within window k. If no such k exists, then $WX(G) = \infty$.

We note that $WX(G) \geq 2$ for any graph with at least one edge. Indeed, let $uv \in E(G)$ and assume that $s_0 = u$. Suppose that we can "see" only the first request, which is v. If we set $s_1 = v$, then the request sequence $vuuuu\dots$ shows that our strategy is not optimal, for the state sequence $uuuu\dots$ has a lower cost. On the other hand, if we set $s_1 = u$ (or any other vertex), then the request sequence $vvvv\dots$ tells the same. Hence no algorithm can work within window 1.

In fact, the windex may even be infinite:

Lemma 14.2 *If* $G = K_4 - e$, *then* $WX(G) = \infty$.

Proof Let $V(K_4 - e) = \{u, v, w, z\}$ with nonadjacent u and z. Let $s_0 = v$, and consider the request sequence

$$\sigma = wuzuz\dots uz\,,$$

where the two-element subsequence uz is repeated k times. Because $d(u, z) = 2$ and $s_0 = v$, the cost of any state sequence for σ is at least $2k + 1$.

Assume that the request following σ is v. Then the cost of the state sequence consisting solely of v's is $2k+1$. Moreover, if $s_1 \neq v$, then we will either change the state of the system at least twice or end up with a state different from v. In any case, the cost is at least $2k+2$. It follows that $s_1 = v$ is optimal.

If, on the other hand, the request following σ is w, then we infer by a similar argument that $s_1 = w$ is optimal.

Because k can be arbitrarily large, no algorithm works within a finite window. □

Contrary to $K_4 - e$, complete graphs have finite windex.

Lemma 14.3 *For any* $n \geq 2$, $WX(K_n) = n$.

Proof Let $V(K_n) = \{u_0, u_1, \ldots, u_{n-1}\}$. Consider the request sequence $\sigma = u_0 u_1 \ldots u_{n-1}$, where u_0 is the initial state. If the first request after σ is u_0, an optimal state sequence must begin (and continue) with u_0, because then the cost is $n-1$ while in any other case it is at least n. Analogously we see that an optimal state sequence must begin with u_1 if the first request after σ is u_1. This proves that $WX(K_n) > n-1$.

To prove that $WX(K_n) \leq n$, consider the following strategy. Let s_i be the current state, and let $r_{i+1} r_{i+2} \ldots r_{i+n}$ be the next n requests. Let r be the first repeated vertex in the sequence $s_i r_{i+1} r_{i+2} \ldots r_{i+n}$ (e.g., in the sequence $u_1 u_3 u_2 u_3 u_2 u_1$, such a vertex is u_3). Put

$$s_{i+1} = \begin{cases} r_{i+1} & \text{if } r = r_{i+1}\,, \\ s_i & \text{otherwise}\,. \end{cases}$$

Clearly, this algorithm works within window n. Moreover, it is not difficult to show by induction on the length of the request sequence that the algorithm produces an optimal state sequence. □

We have just seen that complete graphs have finite windex. To obtain a much larger class of graphs with this property, two additional observations are helpful.

Proposition 14.4 *If H is a weak retract of G, then $WX(H) \leq WX(G)$.*

Proof Let $f : G \to H$ be a weak retraction and $\sigma = r_0 r_1 \ldots r_n$ a request sequence to H. Then σ is also a request sequence to G. Let $s_1 s_2 \ldots s_n$ be an optimal state sequence in G. We claim that $f(s_1) f(s_2) \ldots f(s_n)$ is an optimal state sequence in H for the request sequence σ. Let $s'_1 s'_2 \ldots s'_n$ be an optimal state sequence in H. Observe that

$$\begin{aligned} \sum_{i=1}^{n} \left(d_H(s'_{i-1}, s'_i) + d_H(s'_i, r_i)\right) &\geq \sum_{i=1}^{n} \left(d_G(s'_{i-1}, s'_i) + d_G(s'_i, r_i)\right) \\ &\geq \sum_{i=1}^{n} \left(d_G(s_{i-1}, s_i) + d_G(s_i, r_i)\right) \\ &\geq \sum_{i=1}^{n} \left(d_H(f(s_{i-1}), f(s_i)) + d_H(f(s_i), r_i)\right). \end{aligned}$$

(The third inequality follows because weak homomorphisms are nonexpansive by Proposition 3.8.) We conclude that $f(s_1) f(s_2) \ldots f(s_n)$ is indeed an optimal state sequence. But this means that, if $WX(G) = k < \infty$, we obtain a window k algorithm for H using a window k algorithm for G and converting the output $s_1 s_2 s_2 \ldots$ to the state sequence $f(s_1) f(s_2) f(s_3) \ldots$. □

Proposition 14.4 cannot be extended to isometric subgraphs. For example, C_6 is an isometric subgraph of the 3-cube Q_3, but $WX(C_6) = \infty$ (Exercise 14.5) and $WX(Q_3) = 2$ (Exercise 14.10).

We have now arrived at a point where the window index and Cartesian products meet.

Proposition 14.5 $WX(G \,\square\, H) = \max\{WX(G), WX(H)\}$ *for all graphs G and H.*

Proof Let $\sigma = (a_0, x_0)(a_1, x_1) \ldots (a_n, x_n)$ be a request sequence to $G \,\square\, H$ and $\tau = (b_1, y_1)(b_2, y_2) \ldots (b_n, y_n)$ be a state sequence. Then the Distance Formula (Corollary 5.2) implies that τ is a minimal state sequence for σ if and only if $b_1 b_2 \ldots b_n$ is a minimal state sequence for $a_0 a_1 \ldots a_n$ in G and $y_1 y_2 \ldots y_n$ a minimal state sequence for $x_0 x_1 \ldots x_n$ in H. Hence a window k algorithm in $G \,\square\, H$ induces window k algorithms for G and H, and window k_1 and k_2 algorithms for G and H produce a window $\max\{k_1, k_2\}$ algorithm for $G \square H$. $\square$

The generalization of Proposition 14.5 to finitely many factors is obvious. Note also that Proposition 14.5, together with Lemma 14.3, immediately implies that $WX(Q_n) = 2$ for $n \geq 1$.

From Propositions 14.4 and 14.5 and Lemma 14.3, we infer the main result of this section.

Theorem 14.6 *If G is a weak retract of a Hamming graph, then $WX(G)$ is finite.*

We conclude this section with a necessary condition for a graph to be of finite windex. This requires two definitions.

A graph has the *triangle property* if, for any edge uv and any vertex w with $d(u, w) = d(v, w) = k \geq 2$, there exists a common neighbor x of u and v with $d(x, w) = k - 1$. A graph has the *quadrangle property* if, for any vertices u, v, w and z, where $d(u, w) = d(v, w) = k = d(z, w) - 1$ and z is a common neighbor of u and v, there exists a common neighbor x of u and v with $d(x, w) = k - 1$. See Figure 14.2 for schematic representations of these properties.

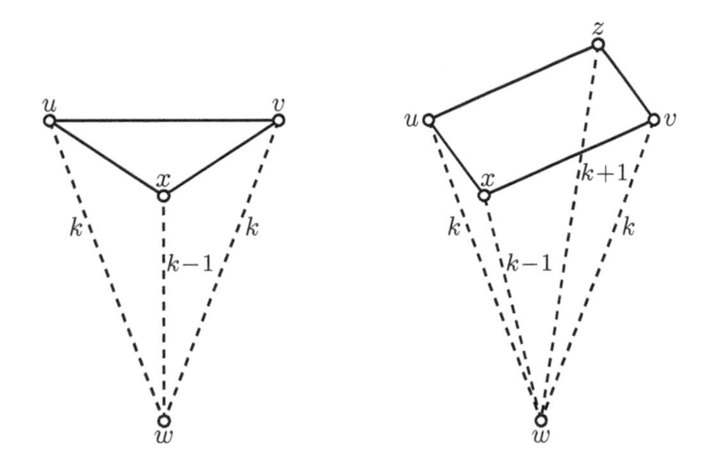

FIGURE 14.2 Triangle property (left) and quadrangle property (right).

A (connected) graph is *weakly modular* if it possesses the triangle and the quadrangle property.

Proposition 14.7 *Let G be a connected graph with finite windex. Then G is weakly modular and contains no $K_4 - e$ or $K_{2,3}$ as an induced subgraph.*

Proof Suppose that $K_4 - e$ is an induced subgraph of G. Then it is a weak retract of G. Indeed, map all vertices of $V(G)\backslash V(K_4 - e)$ to a vertex of $K_4 - e$ of degree 3. By Proposition 14.4, we have $WX(K_4 - e) \leq WX(G) < \infty$, in contradiction to Lemma 14.2.

In the remainder of the proof, we show that if G contains $K_{2,3}$ as an induced subgraph or if G is not weakly modular, then $WX(G) = \infty$. For each case we list an arbitrarily long request sequence σ such that the selection of an initial state depends on a request after σ. We will not provide all details; they can be found in Chung, Graham, and Saks (1989).

Assume that $K_{2,3}$ is an induced subgraph of G, and let $\{u, v, w\} \cup \{x, y\}$ be the bipartition of $V(K_{2,3})$. Let $\sigma = (uvw)^j$, where $3j > WX(G)$. It is not difficult to see that if the first request after σ is x, then x must be selected as the first state vertex and, if the first request after σ is y, this vertex must be the first state vertex.

Suppose that G does not satisfy the quadrangle property. Then there are vertices u, v, w, $z \in V(G)$ with $d(u,w) = d(v,w) = m = d(z,w) - 1$, where z is a common neighbor of u and v, $d(u,v) = 2$, and there is no common neighbor x of u and v with $d(x,w) = m-1$. Let $\sigma = uzw(vwu)^j$, where $3j > WX(G)$. Now, if the first request after σ is u, then u must be selected as the first state vertex; and if the first request after σ is v, then the vertex z must be the first state vertex.

Finally, for the triangle property where $uv \in E(G)$ and $d(u,w) = d(v,w)$, consider $\sigma = uwv(wu)^j$ with $2j > WX(G)$. □

14.3 Quasi-Median Graphs and Generalizations

Median graphs form a proper subclass of partial cubes, a class of graphs that in turn naturally generalizes to partial Hamming graphs. What could be a natural generalization of median graphs?

To answer this question, consider a median graph G isometrically embedded in Q_n. Let $u = u_1u_2\ldots u_n$, $v = v_1v_2\ldots v_n$, and $w = w_1w_2\ldots w_n$ be arbitrary vertices of G. Then the median $x = x_1x_2\ldots x_n$ of the triple u, v, w is obtained by majority rule in every coordinate; see the proof of Proposition 3.7. More precisely, we set x_i equal to the element that appears at least twice among u_i, v_i, w_i. Call a subgraph G of a hypercube *median-closed* if for any three vertices u, v, and w of G, the vertex obtained from them by majority rule belongs to G as well. This leads to the following observation.

Proposition 14.8 *A graph is a median graph if and only if it is a median-closed partial cube.*

Mulder (1978, 1980a) noticed that the conditions of Proposition 14.8 can be weakened:

Corollary 14.9 *A graph is a median graph if and only if it is a median-closed induced subgraph of a hypercube.*

Proof By Proposition 14.8, it remains to prove that a median-closed induced subgraph G of a hypercube H is an isometric subgraph.

Suppose not. Let u and v be vertices of G such that $k = d_G(u,v) > d_H(u,v)$, and assume that k is as small as possible. Because G is an induced subgraph, we have $k \geq 3$. Note also that $d_H(u,v) \leq k-2$ because H is bipartite.

Let P be a shortest u,v-path in G and w be the vertex of P of distance 2 from v. Let x be the vertex obtained by majority rule from u, v, w. Because G is induced and median-closed, x belongs to G and is adjacent to v and w. Furthermore, because of the way we selected k, we infer that $d_G(u,x) = d_H(u,x) = d_H(u,v) - 1 \leq k-3$, which implies that $d_G(u,v) \leq k-2$, a contradiction. □

To generalize median graphs, we extend the concept of median-closed subgraphs of hypercubes to arbitrary Cartesian products as follows: Let $G = G_1 \,\square\, G_2 \,\square\, \cdots \,\square\, G_n$. For vertices $u, v, w \in V(G)$, the the *imprint* $\mathrm{imp}(u,v;w)$ of u and v on w is the vertex x of G defined by

$$x_i = \begin{cases} u_i & \text{if } u_i = v_i\,, \\ w_i & \text{otherwise}\,. \end{cases}$$

Note that if $G_i = K_2$, then x_i is obtained from u_i, v_i, w_i by majority rule. Consequently, if $G = Q_n$, then $\text{imp}(u, v; w)$ is just the median of u, v, w. Therefore the imprint function generalizes the notion of a median from hypercubes to Hamming graphs.

In addition, we call a subgraph H of G *imprint-closed* if for any vertices u, v, and w of H, the imprint $\text{imp}(u, v; w)$ also belongs to H.

In view of Corollary 14.9, we now generalize median graphs as follows: A connected graph G is a *quasi-median graph* if G is an imprint-closed induced subgraph of a Hamming graph. We first observe:

Proposition 14.10 *A graph is a median graph if and only if it is a bipartite quasi-median graph.*

Proof Suppose G is a median graph. Then it is bipartite by Proposition 3.6, and by Corollary 14.9 it is a median-closed induced subgraph of a hypercube. Because for hypercubes $\text{imp}(u, v; w)$ equals the median of u, v, and w, G is an imprint-closed induced subgraph of a Hamming graph.

Conversely, suppose G is a bipartite quasi-median graph, which means it is a connected, bipartite, imprint-closed, induced subgraph of a Hamming graph $G_1 \Box G_2 \Box \cdots \Box G_k$. It suffices to show that the projection of G onto any factor contains at most two vertices. (For then G is an imprint-closed induced subgraph of a hypercube, hence also a median-closed induced subgraph, and therefore a median graph by Corollary 14.9.) We need only show that the projection to the first factor has fewer than three vertices, for the same argument will work for any factor. Suppose to the contrary that $\pi_1(G)$ has more than two vertices, and represent $G_1 \Box G_2 \Box \cdots \Box G_k$ as $G_1 \Box H$, where $H = G_2 \Box \cdots \Box G_k$. Because G is connected, it must have an edge joining some H-layer $\pi_1^{-1}(a_1)$ to another H-layer $\pi_1^{-1}(a_2)$, where $a_1, a_2 \in \pi_1(G)$. This edge must have form $(a_1, x)(a_2, x)$ for some $x \in V(H)$. Let $a_3 \in \pi_1(G) \setminus \{a_1, a_2\}$ and choose a vertex $(a_3, y) \in V(G)$. Note that

$$\text{imp}\big((a_1, x), (a_2, x); (a_3, y)\big) = (a_3, x)\,.$$

Because G is imprint-closed, it follows that $(a_3, x) \in V(G)$. But then the vertices (a_1, x), (a_2, x), and (a_3, x) induce a triangle, contradicting bipartiteness. □

The concept of quasi-median graphs is due to Mulder (1980a), who introduced them as follows: Let u_1, u_2, u_3 be a triple of vertices of a graph G. A *quasi-median* of u_1, u_2, u_3 is a triple x_1, x_2, x_3 such that for any distinct i and j,

(i) $d(u_i, u_j) = d(u_i, x_i) + d(x_i, x_j) + d(x_j, u_j)$,
(ii) $d(x_i, x_j) = k$, and
(iii) k is minimal with respect to (i) and (ii).

The integer k is called the *size* of the quasi-median. A quasi-median of size 0 is a median of the triple u_1, u_2, u_3. See Figure 14.3 for an example of size 3. Mulder defined quasi-median graphs as graphs satisfying the following conditions.

(i) Any triple of vertices in G has a unique quasi-median,
(ii) G does not contain $K_4 - e$ as an induced subgraph, and
(iii) The smallest convex subgraph of G that contains an isometric C_6 is Q_3.

Numerous additional characterizations of quasi-median graphs are known; see the survey by Bandelt, Mulder, and Wilkeit (1994). Among these characterizations we select the following, because it nicely connects various concepts defined in this book.

Recall that a subgraph H of G is gated in G if there exists a vertex $x \in V(H)$ to every $v \in V(G)$ such that x lies on a shortest v, u-path for every $u \in V(H)$. Let H_1 and H_2 be gated subgraphs of a graph G such that $V(H_1) \cup V(H_2) = V(G)$, $V(H_1) \cap V(H_2) \neq \emptyset$, and no edges join $H_1 \setminus H_2$ to $H_2 \setminus H_1$. Then G is called the *gated amalgamation* of H_1 and H_2.

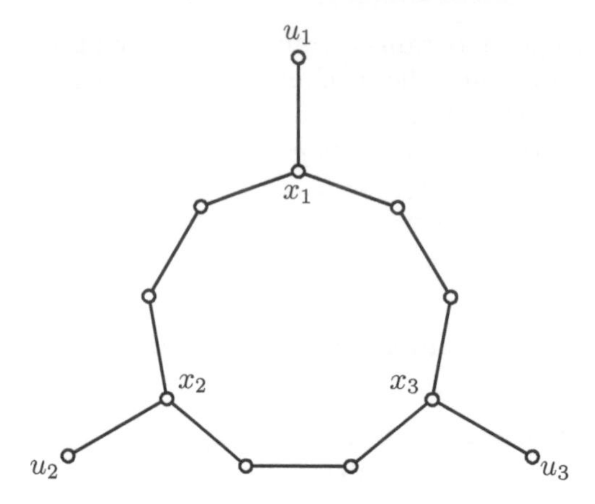

FIGURE 14.3 The triple u_1, u_2, u_3 has a quasi-median x_1, x_2, x_2 of size 3.

Theorem 14.11 *A connected graph is a quasi-median graph if and only if it can be obtained from Hamming graphs by a sequence of gated amalgamations.*

In Chapter 12 we showed that every nonexpansive map of a median graph has a fixed cube. Clearly, a subcube of a hypercube is a subproduct; thus every nonexpansive map of a median graph G contains a fixed box. For quasi-median graphs, a result analogous to Theorem 12.25 holds.

Theorem 14.12 *Every nonexpansive map of a quasi-median graph has a fixed box, namely a fixed Hamming graph.*

This theorem is due to Chastand (1992); see also Chastand and Polat (1996). The proof is identical to that of Theorem 12.25.

In addition to quasi-median graphs, there are numerous other important nonbipartite generalizations of median graphs. We briefly mention some of them.

Pseudo-median graphs A graph is called *pseudo-median* if every triple of vertices either has a unique median or there exists a unique triangle that is contained in the union of any three shortest paths joining the vertices of the triple. Clearly, a graph is a median graph if and only if it is a pseudo-median triangle-free graph. Unfortunately, the Cartesian product of two pseudo-median graphs need not be pseudo-median, but the Cartesian product of a median graph and a pseudo-median graph is pseudo-median. The main result of Bandelt and Mulder (1991) asserts that a graph is pseudo-median if and only if it can be obtained from graphs of the form $H \,\Box\, Q_n$ by a sequence of gated amalgamations, where H belongs to one of three relatively simple families of graphs. Bandelt and Mulder (1988) proved that a graph G is a regular pseudo-median graph if and only if G is either $Q_n \,\Box\, K_m$ or $Q_n \,\Box\, K_{m\times 2}$, where $K_{m\times 2}$ is the *cocktail-party graph*, that is, the graph obtained from K_{2m} by removing a set of m independent edges. This generalizes Exercise 12.8. The same paper extends Theorem 12.21 to pseudo-median graphs as follows: If $f : G \to G$ is a nonexpansive map of a pseudo-median graph, then G has a regular pseudo-median subgraph that is invariant under f.

Weakly median graphs A *weakly median graph* is a weakly modular graph in which every triple of vertices has a unique quasi-median. The latter condition can also be expressed by saying that they do not contain any two vertices with an unconnected triple of common neighbors that can in turn be described by a list of four forbidden induced subgraphs. Weakly median graphs contain quasi-median graphs as a proper subclass (cf. Theorem 14.13). Moreover, they also generalize pseudo-median graphs and are, contrary to pseudo-median graphs, closed under Cartesian product.

Weakly median graphs were first studied in Chepoi (1989). Later, Bandelt and Chepoi (2000) proved that they are precisely the graphs that can be obtained by successive applications of gated amalgamations from Cartesian products of the following graphs: K_2's, 5-wheels, induced subgraphs of cocktail-party graphs $K_{n\times 2}$ ($n \geq 3$) that contain either K_4 or an induced 4-wheel, and 2-connected bridged graphs not containing K_4 or the complete multipartite graph $K_{1,1,3}$ as an induced subgraph. Here a graph is called *bridged* if it does not contain any isometric cycle of length greater than 3. They also deduced that weakly median graphs are ℓ_1-graphs. (See p. 143 for the definition of ℓ_1-graphs.) Brešar (2003) proved that regular weakly median graphs are precisely Cartesian products of complete graphs and cocktail-party graphs. For further developments on weakly median graphs, see Polat (2004), Chastand and Polat (2006), Bandelt and Chepoi (2008a), and the survey by Bandelt and Chepoi (2008b).

Fiber-complemented graphs Chastand (2001, 2003) introduced a class of graphs that is even more general than the class of weakly median graphs, hence his graphs include quasi-median graphs, pseudo-median graphs, and weakly median graphs. For a graph G, a gated set A of G, and $x \in A$, let $k_A^{-1}(x)$ be the set of all vertices of G whose gate in A is x. A graph G is *fiber-complemented* if for any gated set A of G and any vertex $x \in A$, the set $k_A^{-1}(x)$ is gated. Chastand (2001) proved that fiber-complemented graphs can be characterized as the graphs obtained by successive applications of gated amalgamations from Cartesian products of graphs that are not decomposable with respect to these two graph operations. However, contrary to the weakly median case, the list of graphs from which all fiber-complemented graphs can be generated using these two operations is not known. Further results on fiber-complemented graphs were obtained in Brešar (2003), Bandelt and Chepoi (2007), and Brešar and Tepeh Horvat (2008); see also the survey in Bandelt and Chepoi (2008b).

Cage-amalgamation graphs These graphs were introduced by Brešar and Tepeh Horvat (2009) as a common generalization of median graphs and chordal graphs.[1] A graph is a *cage-amalgamation graph* if it can be obtained by successive applications of gated amalgamations from Cartesian products of 2-connected chordal graphs and K_2's. They give several characterizations of cage-amalgamation graphs and prove a couple of equalities for these graphs in the spirit of Exercise 12.10.

Brešar, Chalopin, Chepoi, Kovše, Labourel, and Vaxès (2010) proved that cage-amalgamation graphs are precisely the weak retracts of Cartesian products of chordal graphs. In addition, these graphs can be further characterized as weakly modular graphs that do not contain $K_{2,3}$, k-wheels, and k-wheels minus one spoke ($k \geq 4$) as induced subgraphs. This characterization was conjectured in Brešar and Tepeh Horvat (2009).

[1]Recall that a graph is *chordal* if each of its cycles on at least four vertices has a chord, that is, an edge joining two nonconsecutive vertices of the cycle.

14.4 Graphs with Finite Windex Are Quasi-Median Graphs

We are now ready for the central theorem of this chapter. Its proof uses several key ideas, including the canonical isometric embedding.

Theorem 14.13 *For a connected graph G, the following conditions are equivalent:*

(i) *G is a quasi-median graph.*
(ii) *G is a weak retract of a Hamming graph.*
(iii) *G has finite windex.*
(iv) *G is weakly modular and contains no $K_4 - e$ or $K_{2,3}$ as an induced subgraph.*

We first point out two important consequences of the theorem.

By Bandelt's Theorem 12.18 a graph is a median graph if and only if it is retract of a hypercube. Since every retract is a weak retract, the class of median graphs is contained in the class of weak retract of hypercubes. By condition (ii) of Theorem 14.13 every weak retract of a hypercube is a quasi-median graph, and as subraphs of hypercubes they are bipartite. Now Proposition 14.10 says that a quasi-median graph is bipartite if and only if it is a median graph. We conclude that every weak retract of a hypercube is a retract. We thus have the following corollary:

Corollary 14.14 *A weak endomorphism of a hypercube is a weak retract if and only if it is a retract.*

For the second consequence notice that weak retracts are isometric subgraphs by Corollary 3.9. By condition (ii) of Theorem 14.13 (ii) we thus infer:

Corollary 14.15 *Quasi-median graphs are partial Hamming graphs.*

The converse of Corollary 14.15 is false as we already know from the bipartite case.

The proof of Theorem 14.13 is not easy. The implication (ii) $\Rightarrow$ (iii) is Theorem 14.6 while the implication (iii) $\Rightarrow$ (iv) is Proposition 14.7. It remains to prove the implications (i) $\Rightarrow$ (ii) and (iv) $\Rightarrow$ (i).

Proof of Theorem 14.13, Step (i) $\Rightarrow$ (ii) We first extend our notation. Let $u = (u_1, \ldots, u_n)$ be a vertex of $G = G_1 \,\square\, G_2 \,\square\, \cdots \,\square\, G_n$ and $I = \{i_1, \ldots, i_j\}$ be a subset of $\{1, \ldots, n\}$. Then we set $p_I(u) = (u_{i_1}, \ldots, u_{i_j})$. For $X \subseteq V(G)$, let $p_I(X) = \{p_I(x) \mid x \in X\}$. Furthermore we define

$$S_i(u) = \{(w_1, \ldots, w_n) \in V(G) \mid w_i = u_i\},$$

and if $v = (v_1, \ldots, v_n)$ is another vertex of G and $i \neq j$, then

$$L(u, i; v, j) = S_i(u) \cup S_j(v)\,.$$

The set $L(u, i; v, j)$ is called an *L-set* and $\{i, j\}$ is its *support*. Figure 14.4 is a schematic representation of an L-set in a Cartesian product of three factors.

Assume now that H is a quasi-median graph; that is, H is a connected, imprint-closed subgraph of a Hamming graph $G = G_1 \,\square\, \cdots \,\square\, G_n$. (Clearly, every G_i is complete.) We can assume without loss of generality that H is proper and irredundant in G.

Lemma 14.16 *The set $V(H)$ is an intersection of L-sets of G. If $V(H) \subseteq L(u, i; v, j)$, then $L(u, i; v, j)$ is unique among all L-sets that contain $V(H)$ and have support $\{i, j\}$.*

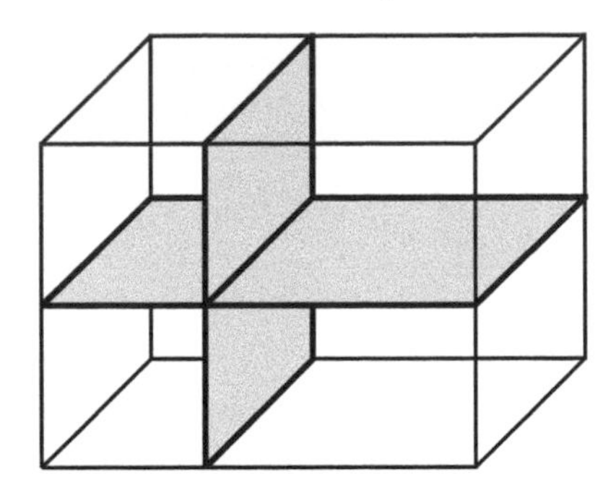

FIGURE 14.4 An L-set.

Proof Let $u \in V(G)\backslash V(H)$. To prove that $V(H)$ is an intersection of L-sets of G, it suffices to show that there exists an L-set of G that contains H but not u. Let $I \subseteq \{1, \ldots, n\}$ be a minimal set such that $p_I(u) \notin p_I(H)$. Because H is irredundant in G, $|I| \geq 2$.

Let $|I| \geq 3$ and $i_1, i_2, i_3 \in I$. By the choice of I, there are vertices x, y, and z of H with

$$\begin{aligned} p_{I\backslash\{i_1\}}(x) &= p_{I\backslash\{i_1\}}(u), \\ p_{I\backslash\{i_2\}}(y) &= p_{I\backslash\{i_2\}}(u), \\ p_{I\backslash\{i_3\}}(z) &= p_{I\backslash\{i_3\}}(u). \end{aligned}$$

Because H is imprint-closed, the vertex $w = \text{imp}(x, y; z)$ belongs to H. However, $p_I(w) = p_I(u)$, contrary to the choice of I.

It remains to consider the case $|I| = 2$. We may assume that $I = \{1, 2\}$. By irredundancy, there exist $v, w \in V(H)$ with $v_1 = u_1$ and $w_2 = u_2$ (clearly, $v_2 \neq u_2$ and $w_1 \neq u_1$), and

$$\begin{aligned} u &= (u_1, u_2, u_3, \ldots, u_n), \\ v &= (u_1, v_2, v_3, \ldots, v_n), \\ w &= (w_1, u_2, w_3, \ldots, w_n). \end{aligned}$$

Note that $u \notin L(w, 1; v, 2)$. We will show that $V(H) \subseteq L(w, 1; v, 2)$.

Define a bipartite graph B as follows: Set $V(B) = V(G_1) \cup V(G_2)$, and for $a \in V(G_1)$, $b \in V(G_2)$, let ab be an edge of B if there is a vertex $x \in V(H)$ such that $p_I(x) = (a, b)$. Because H is connected and irredundant, B is connected.

Let x be a vertex of H. By the definition of B, we have $x_1x_2 \in E(B)$. Assume now that x_1 has another neighbor in $V(G_2)$. Then there is a vertex $y \in V(H)$ with $x_1 = y_1$ and $x_2 \neq y_2$. Let z be any vertex of H. Then $\text{imp}(x, y; z)$ is a vertex of H and $p_I(\text{imp}(x, y; z)) = (x_1, z_2)$. It follows that x_1 is adjacent in B to all vertices of G_2. We conclude that x_1 is adjacent in B to either all vertices of G_2 or to a single vertex. If it is adjacent to a single vertex, then the connectedness of B implies that this neighbor of x_1 is adjacent to all vertices of $V(G_1)$.

In view of $v_1v_2 \in E(B)$, $w_1w_2 \in E(B)$, and $v_1w_2 = u_1u_2 \notin E(B)$, every edge of B must be incident with w_1 or v_2. Therefore $V(H) \subseteq S_1(w)$ or $V(H) \subseteq S_2(v)$ and we have proved that $V(H) \subseteq L(w, 1; v, 2)$.

For the second assertion of the lemma, assume that $V(H) \subseteq L(u, i; v, j)$. Because H is an irredundant subgraph, there are vertices $x, y \in V(H)$ with $x_i \neq u_i$, and $y_j \neq v_j$. As $V(H) \subseteq L(u, i; v, j)$, we also have $x_j = v_j$ and $y_i = u_i$. Consider an arbitrary path P in H between x and y (recall that H is connected), and let z be the first vertex of P with $z_i = u_i$. For any vertex w of P between x and z, we must have $w_j = v_j$; thus also $z_j = v_j$. We conclude that the only L-set with support $\{i, j\}$ containing x, y, z is $L(u, i; v, j)$. □

By Lemma 14.16, it suffices to prove that intersections of L-sets are weak retracts of G. Appropriate weak retractions will be composed from the mappings

$$f^{(u,i;v,j)} : V(G) \to L(u,i;v,j)$$

defined as

$$f^{(u,i;v,j)}(x) = \begin{cases} x & \text{if } x_i = u_i \text{ or } x_j = v_j\,, \\ (x_1, \ldots, u_i, \ldots, v_j, \ldots, x_n) & \text{otherwise}. \end{cases}$$

Lemma 14.17 *The map $f^{(u,i;v,j)}$ is a weak retraction.*

Proof Let xy be an edge of G and k an index for which $x_k \neq y_k$. If $k \neq i, j$, then $f^{(u,i;v,j)}(x)$ and $f^{(u,i;v,j)}(y)$ differ precisely in position k. On the other hand, if $k = i$ (or $k = j$), we see that $f^{(u,i;v,j)}(x)$ and $f^{(u,i;v,j)}(y)$ either differ in position k or not at all. □

The maps $f^{(u,i;v,j)}$ are called *elementary retractions*. Suppose that $V(H)$ is contained in an L-set with support $\{i, j\}$. Then, by Lemma 14.16, this L-set is unique, and we may denote it by L_{ij}. Let $f^{i,j}$ be the corresponding elementary retraction.

Lemma 14.18 *For any vertex x of G, there is a sequence of elementary retractions $f^{i,j}$ that maps x to H.*

Proof Let Γ be the graph with vertex set $\{1, \ldots, n\}$, where i is adjacent to j if $V(H)$ is contained in L_{ij}. We say that x satisfies the edge ij if $x \in L_{ij}$. If $L_{ij} = L(u,i;v,j)$, then we set $\ell_i^j = u_i$ and $\ell_j^i = v_j$.

We proceed by induction on the number of unsatisfied edges of Γ. If there is no such edge, then $x \in V(H)$, and the conclusion is trivial. So let there be at least one unsatisfied edge.

Define a relation $\to$ on $V(\Gamma)$ as follows: If ij is an edge of Γ, $x_i = \ell_i^j$, and $x_j \neq \ell_j^i$, then we set $i \to j$. Suppose that $i \to j$ and $j \to k$. Then $x_i = \ell_i^j$, $x_j \neq \ell_j^i$, $x_j = \ell_j^k$, and $x_k \neq \ell_k^j$. Because $x \in L_{ij}$, $x \in L_{jk}$, and $\ell_j^k \neq \ell_j^i$, we must have $x_i = \ell_i^j$ or $x_k = \ell_k^j$. Thus $x \in L_{ik}$. Furthermore, because $x_k \neq \ell_k^j$, we have $i \to k$. It follows that $\to$ is a transitive acyclic orientation. Moreover, it is clearly antisymmetric, whence $\to$ is a partial order.

Let ij be an arbitrary unsatisfied edge. Call an element of $\{1, \ldots, n\}$ a sink if no edge is directed away from it. Because $\to$ is a partial order, there is a sink h for which either $h = i$ or $i \to h$. Then hj is an unsatisfied edge of Γ. Likewise we see that we have a sink k if $k = j$ or $j \to k$ and hk is an unsatisfied edge of Γ. Now consider the elementary retraction $f^{h,k}$, and note that $f^{h,k}(x) \in L_{hk}$. Moreover, because h and k are sinks, we also have $f^{h,k}(x) \in L_{ij}$ whenever $x \in L_{ij}$. It follows that $f^{h,k}(x)$ has fewer unsatisfied edges, which completes the induction. □

To complete the proof of (i) $\Rightarrow$ (ii), we make repeated use of Lemma 14.18. Let x be any vertex of $V(G) \setminus V(H)$. Then the composition φ of the sequence of weak retractions (which is a weak retraction) of Lemma 14.18 maps x to H and fixes H. If there are vertices left in $V(G) \setminus V(H)$ not mapped into H by φ, then we continue by induction.

Proof of Theorem 14.13, Step (iv) $\Rightarrow$ (i) Let G be weakly modular, and assume that it contains no $K_4 - e$ or $K_{2,3}$ as an induced subgraph. The key step to prove that G is a quasi-median graph is to show that G is a partial Hamming graph. It suffices to show that the quotient graphs G/Π_i of the canonical embedding α are complete. We first construct the equivalence classes of Θ^*; then the result readily follows.

Recall that a subgraph H of G is gated in G if for every $v \in V(G)$, there exists a vertex $x \in V(H)$ such that, for every $u \in V(H)$, x lies on a shortest u,v-path. If such a vertex exists, it is unique and called the gate of v in H. We denote it by $k_H(v)$.

Let $\mathfrak{Q}$ be the set of all cliques of G. Because G contains no $K_4 - e$, each edge of G belongs to a unique member of $\mathfrak{Q}$. Another basic property of members of $\mathfrak{Q}$ is gatedness.

Lemma 14.19 *Every clique of G is gated.*

Proof Let $Q \in \mathfrak{Q}$, and $w \in V(G) \setminus V(Q)$. Assume that $d(w,Q) = d(w,u) = d(w,v)$ for $u,v \in V(Q)$. By the triangle property, there is a common neighbor x of u and v and, because G has no $K_4 - e$ as an induced subgraph, $x \in V(Q)$. But then $d(x,w) = d(u,w) - 1$, a contradiction. □

Let $Q, Q' \in \mathfrak{Q}$. Then Q and Q' are called *parallel* if the vertices of Q and Q' can be labeled $\{u_1,\ldots,u_k\}$ and $\{v_1,\ldots,v_k\}$, respectively, such that

$$d(u_i,v_j) = \begin{cases} d(Q,Q') & \text{if } i = j\,, \\ d(Q,Q') + 1 & \text{otherwise}\,. \end{cases}$$

Note that parallel cliques have equal sizes.

The cliques Q and Q' are called *opposite* if there exist unique vertices $u \in V(Q)$ and $u' \in V(Q')$ such that $d(u,u') = d(Q,Q')$, and $d(u,v') = d(v,u') = d(Q,Q') + 1$ as well as $d(v,v') = d(Q,Q') + 2$ for $v \in V(Q) - u$ and $v' \in V(Q') - u'$. Figure 14.5 shows examples of parallel and opposite cliques.

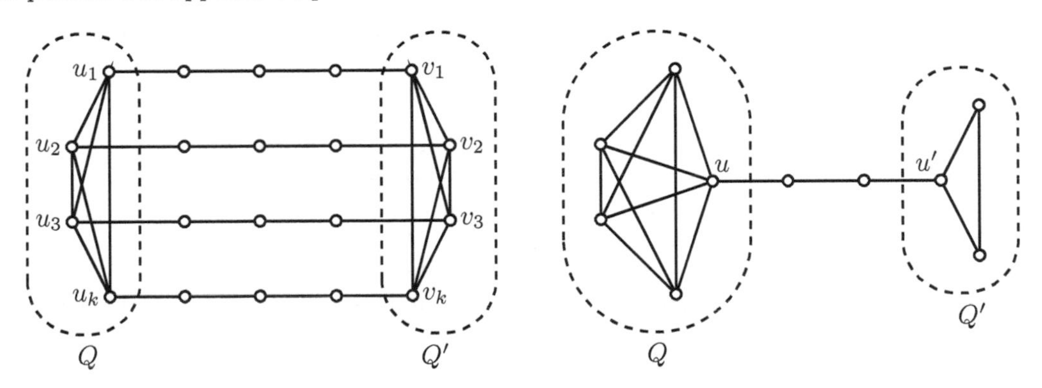

FIGURE 14.5 Parallel cliques (left) and opposite cliques (right).

Lemma 14.20 *Two cliques Q and Q′ of G are either parallel or opposite.*

Proof Let $X = \{u \in V(Q) \mid d(u,Q') = d(Q,Q')\}$ and $X' \subseteq V(Q')$ be defined analogously. By Lemma 14.19, the gate $k_{Q'}(u)$ lies in X' for any $u \in X$ and the gate $k_Q(u')$ in X for any $u' \in X'$. Moreover, the maps $k_{Q'}$ and k_Q are inverses of each other. Therefore, if $X = Q$ and $X' = Q'$, then Q and Q' are parallel. Suppose now that $v \in V(Q) \setminus X$. Then $d(v,Q') = 1 + d(u,k_{Q'}(u))$ for any $u \in X$. Because the gate $k_{Q'}(u)$ is unique, $|X| = 1$. Analogously one shows $|X'| = 1$. Thus, using Lemma 14.19, one proves that Q and Q' are opposite in this case. □

To indicate that Q and Q' are parallel cliques of G, we introduce the notation $Q|Q'$. We will show that $|$ is an equivalence relation on $\mathfrak{Q}$ and that the Θ^*-classes are determined by the equivalence classes of $|$. We first need the following lemma:

Lemma 14.21 *Let $Q_0, Q_m \in \mathbf{Q}$ with $d(Q_0, Q_m) = m$ and $Q_0|Q_m$. Then there exist cliques $Q_1, \ldots, Q_{m-1}$ such that $Q_i|Q_{i+1}$ and $d(Q_i, Q_{i+1}) = 1$ for $i = 0, \ldots, m-1$.*

Proof For $m = 1$, the statement is trivial. Thus let $m > 1$. Set $V(Q_0) = \{u_1, \ldots, u_k\}$ and $V(Q_m) = \{v_1, \ldots, v_k\}$, where $d(u_i, v_i) = m$. Let w_1 be the neighbor of v_1 on a shortest path from v_1 to u_1. Because $Q_0|Q_m$, the vertices w_1, v_2, u_2, and v_1 satisfy the conditions of the quadrangle property. Hence there is a common neighbor w_2 of w_1 and v_2 with $d(w_2, u_2) = m - 1$. Let Q_{m-1} be the clique containing the edge w_1w_2. Note that $v_1 \notin V(Q_{m-1})$, because G contains no induced $K_4 - e$. Furthermore, for $j > 1$, we have $v_j \notin V(Q_{m-1})$ because $d(w_1, v_j) = 2$. It follows that $Q_{m-1} \cap Q_m = \emptyset$, and therefore $d(Q_{m-1}, Q_m) = 1$. As $d(w_1, v_1) = d(w_2, v_2) = d(Q_{m-1}, Q_m)$, Lemma 14.20 implies that $Q_{m-1}|Q_m$. On the other hand, $d(Q_0, Q_{m-1}) = m - 1 = d(u_1, w_1) = d(u_2, w_2)$. Hence, invoking Lemma 14.20 again, $Q_0|Q_{m-1}$. Induction completes the argument. □

Lemma 14.22 *The relation $|$ is an equivalence relation on $\mathbf{Q}$.*

Proof The relation $|$ is clearly reflexive and symmetric. Assume that it is not transitive; then there are cliques Q, Q', and Q'' with $Q|Q'$, $Q'|Q''$ but not $Q|Q''$. Let $d(Q, Q') = m$ and $Q' = Q_0$, Q_1, $\ldots$, $Q_m = Q$ be a sequence of cliques as in Lemma 14.21 and j be the largest index for which $Q_j|Q''$. By our assumptions, $0 \leq j < m$. Set $p = d(Q_j, Q'')$, $V(Q_j) = \{u_1, \ldots, u_k\}$, $V(Q_{j+1}) = \{v_1, \ldots, v_k\}$, and $V(Q'') = \{w_1, \ldots, w_k\}$, where $d(u_i, v_i) = 1$ and $d(u_i, w_i) = p$ for $i = 1, \ldots, k$.

Case 1. $d(Q_{j+1}, Q'') = p + 1$.
In this case we have $d(v_i, Q'') = d(Q_{j+1}, Q'')$ for any i. Hence $Q_{j+1}|Q''$, in contradiction to the choice of j.

Case 2. $d(Q_{j+1}, Q'') = p - 1$.
Then $d(v_i, w_s) = p - 1$ for some $v_i \in V(Q_{j+1})$ and $w_s \in V(Q'')$. Because $d(u_i, w_s) \leq p$, Lemma 14.19 implies that $s = i$. Assume without loss of generality that $i = 1$. Then the vertices v_1, u_2, w_2, and u_1 satisfy the conditions of the quadrangle property. Hence there is a common neighbor v of v_1, and u_2 with $d(v, w_2) = p - 1$. Then u_1, v_2, and v are adjacent to both v_1 and u_2. If $v \notin V(Q_{j+1})$ we would have an induced $K_{2,3}$. Therefore $v \in V(Q_{j+1})$, and thus $v = v_2$. We conclude that $d(v_2, w_2) = p - 1$, and by Lemma 14.20, we infer that $Q_{j+1}|Q''$, a contradiction.

Case 3. $d(Q_{j+1}, Q'') = p$.
This case is treated similarly to the previous one and left to the reader. □

Let $[Q]$ denote the equivalence class with respect to $|$ containing a clique Q and $E_{[Q]}$ be the set of edges induced by the cliques of $[Q]$, namely

$$E_{[Q]} = \bigcup_{Q' \in [Q]} E(Q').$$

Lemma 14.23 *The Θ^*-classes coincide with the sets $E_{[Q]}$.*

Proof Let $e, f \in E_{[Q]}$ and $e \in E(Q')$, $f \in E(Q'')$. Because any two edges of a triangle are in relation Θ, any two edges of Q' and any two edges of Q'' are in relation Θ^*. Moreover, because $Q'|Q''$, the edge $e = u_1u_2$ is in relation Θ to v_1v_2, where $v_1 = k_{Q'}(u_1)$ and $v_2 = k_{Q'}(u_2)$. Thus $e\Theta^* f$.

For the converse, note first that if edges $e = uv$ and $e' = u'v'$ belong to opposite cliques, then $d(u, u') + d(v, v') = d(u, v') + d(v, u')$; that is, e and f are not in relation Θ. Therefore, if $e\Theta^* f$, then e and f must necessarily belong to the same set $E_{[Q]}$. □

Lemma 14.24 *Let $Q|Q'$ and $u \in V(Q)$. Then there is a shortest path from u to $k_{Q'}(u)$ consisting only of edges in $E(G) \setminus E_{[Q]}$.*

Proof Let $d(Q, Q') = m$. By Lemma 14.21, there exists a sequence of parallel cliques $Q = Q_0, Q_1, \ldots, Q_m = Q'$. For $i = 1, \ldots, m$, let $u_i = k_{Q_i}(u_{i-1})$. Then the sequence $u_0 = u, u_1, \ldots, u_m = k_{Q'}(u)$ defines a shortest path from u to $k_{Q'}(u)$. Furthermore, because $Q_i \in [Q]$ and because the members of $[Q]$ are disjoint, the clique containing the edge $u_{i-1}u_i$ is not a member of $[Q]$. □

We are now ready to complete the proof that G is a partial Hamming graph. As already mentioned, it suffices to show that the quotient graphs G/Π_i of α are complete. In other words, if H', H'' is an arbitrary pair of connected components of a graph $(V(G), E(G)\setminus E_{[Q]})$, then it is enough to show that there is an edge of $E_{[Q]}$ with one endpoint in H' and the other in H''. Because G is connected, there are vertices $v' \in V(H')$ and $v'' \in V(H'')$ such that $v' \in V(Q')$ and $v'' \in V(Q'')$ for $Q', Q'' \in [Q]$. By Lemma 14.24, there is a path from v' to $k_{Q''}(v')$ consisting only of edges from $E(G) \setminus E_{[Q]}$. The observation that $k_{Q''}(v') \in V(H')$ and $k_{Q''}(v')$ is in $E_{[Q]}$ adjacent to v'' proves the claim.

To complete the proof of step (iv) $\Rightarrow$ (i), let G be embedded isometrically into a Hamming graph and let u, v, $w \in V(G)$. We wish to show that $\text{imp}(u, v; w) \in V(G)$.

Let $I = \{i \mid u_i \neq v_i\}$, $I_u = \{i \in I \mid u_i = w_i\}$, $I_v = \{i \in I \mid v_i = w_i\}$, and $I' = I\setminus(I_u \cup I_v)$. Then $I = I_u \cup I_v \cup I'$. If $I = I_u$, then $u = \text{imp}(u, v; w)$; and if $I = I_v$, then $v = \text{imp}(u, v; w)$. We may therefore assume that $I' \neq \emptyset$.

Let $|I| = |I'| = 1$. Then $uv \in E(G)$ and $d(u, w) = d(v, w) = m$. If $m = 1$, then $\text{imp}(u, v; w) = w$; and if $m > 1$, then $\text{imp}(u, v; w)$ is the vertex adjacent to u and v and of distance $m - 1$ from w. Because G satisfies the triangle property, $\text{imp}(u, v; w) \in V(G)$.

Let $I = \{i, j\}$. Because G is isometric and $d(u, v) = 2$, there is a vertex $x \in V(G)$ adjacent to u and v. We may, without loss of generality, assume that $x_i = u_i$ and $x_j = v_j$. Let $u' = \text{imp}(u, x; w)$ and $v' = \text{imp}(v, x; w)$. Because x differs from both u and v in exactly one coordinate, we infer that $u', v' \in V(G)$. If $u_i = w_i$, then $\text{imp}(u, v; w) = u'$; and if $v_j = w_j$, then $\text{imp}(u, v; w) = v'$.

For $|I| > 2$, we proceed by induction. Let P be a shortest u, v-path, x the neighbor of u, and y the neighbor of v on P. By the induction hypothesis, $u' = \text{imp}(u, y; w) \in V(G)$ and $v' = \text{imp}(v, x; w) \in V(G)$. Note also that $d(u', v') = 2$. Hence, because $\text{imp}(u, v; w) = \text{imp}(u', v'; w)$, we conclude that $\text{imp}(u, v; w) \in V(G)$.

Exercises

14.1. Show that a graph is a Hamming graph if and only if it is a Cartesian product of complete graphs.

14.2. Show that every interval of a Hamming graph induces a hypercube.

14.3. Show that every interval of a partial Hamming graph is bipartite.

14.4. Show that median graphs satisfiy the quadrangle property. Is the same true for partial cubes?

14.5. Show directly that $WX(C_6) = \infty$.

14.6. A subgraph is called Δ-closed if, provided it contains an edge of a triangle, it contains the whole triangle. Show that a subgraph of a weakly modular graph is gated if and only if it is convex and Δ-closed.

14.7. (Bandelt, Mulder, and Wilkeit, 1994) Show that a graph G is a Hamming graph if and only if G is a quasi-median graph that contains no convex P_3.

14.8. Show that the graph in Figure 11.4 is a quasi-median graph.

14.9. Let G be a quasi-median graph and let x_1, x_2, x_3 be the quasi-median of u_1, u_2, u_3. Express the quasi-median in terms of the imprint function.

14.10. (Chung, Graham, and Saks, 1987) Show that a connected graph G is a median graph if and only if $WX(G) = 2$.

Chapter 15

Isometries in Strong Products and Product Dimensions

The first four chapters of Part III dealt with isometric subgraphs of Cartesian products. We now turn our attention to the strong product. We show that any connected graph isometrically embeds into the strong products of paths, a result that leads naturally to the notion of the strong isometric dimension of a graph. Then we consider special isometric subgraphs of strong products—their retracts—and characterize the weak retracts of strong products of paths as Helly graphs. We close the chapter with a brief overview of other notions of graph dimension that are analogous to the strong isometric dimension.

15.1 Strong Isometric Dimension

Theorem 13.2 asserts that any connected graph isometrically embeds into a Cartesian product. Unfortunately, these embeddings are quite often trivial (cf. Exercise 13.5) because they reduce to the identity map. In contrast to this, we now show that every graph is isometrically embeddable into a strong product of paths. Hence, except for paths, any such embedding is nontrivial.

As usual, we label the vertices of the path P_m consecutively as $0, 1, 2, \ldots, m-1$.

Let G be an isometric subgraph of a strong product of paths. By the Distance Formula for the strong product (Corollary 5.5), the distance between two vertices of G is the maximum absolute value of the differences of their corresponding coordinates. For instance, Figure 15.1 describes an isometric embedding of the Petersen graph into $P_3^{\boxtimes,5}$.

The following general embedding result goes back all the way to Schönberg (1938).

Theorem 15.1 *Any connected graph on n vertices can be isometrically embedded into a strong product of n paths.*

Proof Let G be a connected graph with vertices $\{v_1, \ldots, v_n\}$. Let $e(v_i)$ be the *eccentricity* of v_i, that is, the maximum distance from v_i to any other vertex of G. For each $1 \le i, j \le n$, set $\beta_i(v_j) = d_G(v_i, v_j)$. We define

$$\beta : V(G) \to H = \boxtimes_{i=1}^{n} P_{e(v_i)+1}$$

as $\beta(v_j) = (\beta_1(v_j), \beta_2(v_j), \ldots, \beta_n(v_j))$. We claim that β is an isometry. Note first that $\beta_i(v_j) \leq e(v_i)$, so β indeed maps $V(G)$ to $V(H)$. Let v_k and v_ℓ be two vertices of G. Because $\beta_k(v_k) = d_G(v_k, v_k) = 0$ and $\beta_k(v_\ell) = d_G(v_k, v_\ell)$, the Distance Formula for strong products (Corollary 5.5) implies that

$$d_H(\beta(v_k), \beta(v_\ell)) \geq d_G(v_k, v_\ell).$$

Given a v_i, we may assume without loss of generality that $d_G(v_i, v_k) \geq d_G(v_i, v_\ell)$. Then

$$|\beta_i(v_k) - \beta_i(v_\ell)| = d_G(v_i, v_k) - d_G(v_i, v_\ell) \leq d_G(v_k, v_\ell).$$

By Corollary 5.5 we thus conclude that $d_H(\beta(v_k), \beta(v_\ell)) = d_G(v_k, v_\ell)$. □

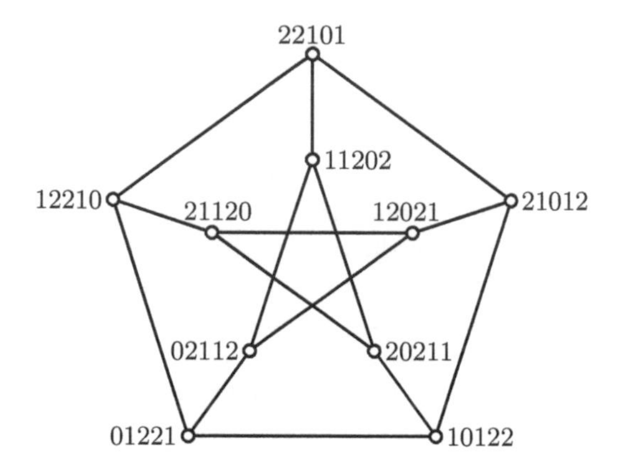

FIGURE 15.1 Isometric embedding of the Petersen graph into $P_3^{\boxtimes,5}$.

In view of Theorem 15.1, we define the *strong isometric dimension* of a connected graph G, denoted $\text{sdim}(G)$, as the smallest integer k such that G isometrically embeds into the strong product of k paths. Theorem 15.1 can be rephrased as follows.

Corollary 15.2 *If G is a connected graph, then* $\text{sdim}(G) \leq |V(G)|$.

A closer look at the proof of Theorem 15.1 reveals that the result still holds if the last coordinate β_n of β is deleted. Hence, $\text{sdim}(G) \leq |V(G)| - 1$. Fitzpatrick and Nowakowski (2000) observed that the bound can be further improved as follows. (The proof is reserved for Exercise 15.1.)

Corollary 15.3 *If G is a connected graph, then* $\text{sdim}(G) \leq |V(G)| - \text{diam}(G)$.

The strong dimension is generally difficult to compute, but some bounds and exact values are known. We collect several of these in the next theorem.

Theorem 15.4

(i) *For $n \geq 2$,* $\text{sdim}(K_n) = \lceil \log_2 n \rceil$.
(ii) *For $m + n \geq 3$,* $\text{sdim}(K_{m,n}) = \lceil \log_2 m \rceil + \lceil \log_2 n \rceil$.
(iii) *For $n \geq 4$,* $\text{sdim}(C_n) = \lceil n/2 \rceil$.
(iv) *For $n \geq 1$,* $\text{sdim}(Q_n) = 2^{n-1}$.
(v) *For a tree T with k leaves,* $\lceil \log_2 k \rceil \leq \text{sdim}(T) \leq 2\lceil \log_2 k \rceil$.
(vi) *The Petersen graph has strong dimension* 5.

(vii) *For* $k \geq 1$ *and* $\binom{k}{\lfloor k/2 \rfloor} < n \leq \binom{k+1}{\lfloor (k+1)/2 \rfloor}$, $\mathrm{sdim}(K_2 \,\square\, K_n) = k+1$.

Theorem 15.4 (i) is left for Exercise 15.2, (ii) is due to Dewdney (1977), while items (iii) through (v) were proved by Fitzpatrick and Nowakowski (2000). Equation (vi) is from Jerebic and Klavžar (2006), while (vii) is from Fronček, Jerebic, Klavžar, and Kovář (2007).

To prove the lower bound for trees, Fitzpatrick and Nowakowski showed that the contraction of an interior edge of a tree does not increase the dimension. Their proof of the upper bound is more involved and they ask if the true dimension is always the lower bound.

For the lower bound of (vii), see Exercise 15.3. To prove the upper bound, two additional tools were used. One is Sperner's celebrated theorem. The other is the following result, first obtained by Dewdney (1977) in a different context. The theorem deals with coverings by complete bipartite graphs, where we view $K_1 = K_{1,0}$ as a complete bipartite graph.

Theorem 15.5 *If* G *is a graph with* $\mathrm{diam}(G) = 2$, *then* $\mathrm{sdim}(G)$ *equals the smallest* r *for which the edges of* $\overline{G}$ *can be covered with complete bipartite subgraphs* $B_1, \ldots, B_r$ *of* $\overline{G}$, *such that for any* $xy \in E(G)$, *there exists a* B_i *with* $x \in V(B_i)$ *and* $y \notin V(B_i)$.

Proof Assume $\mathrm{sdim}(G) = r$ and let $\beta = (\beta_1, \ldots, \beta_r)$ be an isometric embedding of G into the strong product of r paths. Because $\mathrm{diam}(G) = 2$, we may assume (Exercise 15.4) that $\beta : G \to P_3^{\boxtimes, r}$. Set $V(P_3) = \{0, 1, 2\}$ and for $i = 1, \ldots, r$, let

$$X_i = \{u \in V(G) \mid \beta_i(u) = 0\} \quad \text{and} \quad Y_i = \{u \in V(G) \mid \beta_i(u) = 2\}\,.$$

Suppose $x \in X_i$ and $y \in Y_i$. Because β is an isometry, we infer that $d_G(x, y) = 2$, and hence $xy \in E(\overline{G})$. Therefore $X_i \cup Y_i$ forms the vertex set of a complete bipartite subgraph B_i of $\overline{G}$. Moreover, any edge of $\overline{G}$ is covered by at least one of these subgraphs. Finally, if $xy \in E(G)$, then there is an index i such that $|\beta_i(x) - \beta_i(y)| = 1$, say $\beta_i(x) = 0$ and $\beta_i(y) = 1$. Then x lies in B_i and y does not. If $\beta_i(x) = 1$ and $\beta_i(y) = 2$, then y lies in B_i and x does not, etc.

Conversely, assume that the edges of $\overline{G}$ can be covered with r complete bipartite graphs B_i with bipartitions X_i, Y_i, $1 \leq i \leq r$, such that for any edge xy of G, there is an i with $x \in V(B_i)$ and $y \notin V(B_i)$. Define $\beta : G \to P_3^{\boxtimes, r}$ as

$$\beta_i(x) = \begin{cases} 0 & \text{if } x \in X_i\,, \\ 2 & \text{if } x \in Y_i\,, \\ 1 & \text{otherwise}\,. \end{cases}$$

We next verify that β is a homomorphism. Let $xy \in E(G)$. Consider an index $1 \leq j \leq r$. If both x and y belong to B_j, then because $xy \notin E(\overline{G})$, we have $\beta_j(x) = \beta_j(y) \in \{0, 2\}$. If neither x nor y belongs to B_j, then $\beta_j(x) = \beta_j(y) = 1$. If $x \in B_j$, but $y \notin B_j$, then the definition of β gives $|\beta_i(x) - \beta_i(y)| = 1$. Because there does exist an index i with $x \in B_i$ and $y \notin B_i$, it follows that $d(\beta(x), \beta(y)) = 1$. Hence β maps edges to edges.

Assume $d_G(x, y) = 2$. Then $xy \in E(\overline{G})$ and so xy is covered with at least one B_i. Hence $|\beta_i(x) - \beta_i(y)| = 2$. We infer $d(\beta(x), \beta(y)) = 2$, so β is an isometry. □

Figure 15.2 illustrates Theorem 15.5 and its proof. In this example, $\overline{G}$ is covered with two complete bipartite graphs, $K_{1,3}$ and $K_{2,2}$, with bipartitions $\{z\}, \{u, v, w\}$ and $\{x, u\}, \{y, w\}$, respectively. The embedding into $P_3 \boxtimes P_3$ is shown on the right-hand side of the figure.

The condition concerning edges in Theorem 15.5 is rather technical but cannot be omitted in general. To see this, consider $G = K_4 - e$, which has strong dimension 2. Now, $\overline{G}$ is an edge and two isolated vertices, so its edge(s) can be covered with a single $B_1 = K_{1,1}$. However, the edge of G connecting vertices of degree 3 does not fulfill the condition of the theorem. We need to include one of the vertices of degree 3 as a second $B_2 = K_{1,0} = K_1$.

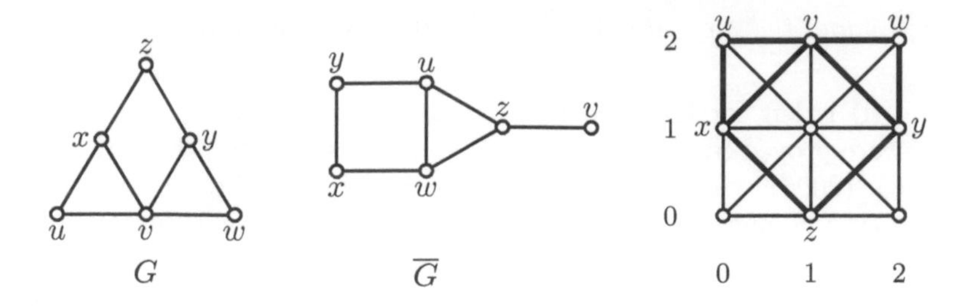

FIGURE 15.2 Graph G and its isometric embedding into $P_3 \boxtimes P_3$.

But a careful modification of the proof of Theorem 15.5 yields a result with simpler hypotheses:

Corollary 15.6 *Suppose G is a graph of diameter 2, for which any edge is contained in an induced path on three vertices. Then* $\text{sdim}(G)$ *equals the smallest r such that the edges of $\overline{G}$ can be covered with r complete bipartite subgraphs.*

Corollary 15.6 applies to the Petersen graph. We already know from Figure 15.1 that $\text{sdim}(P) \leq 5$. Hence to prove Theorem 15.4 (vi), one needs to argue that the edges of the Petersen graph cannot be covered with fewer than five complete bipartite graphs.

We close the section with an application of the strong isometric dimension to an interesting two-player game called *Cops and Robbers*, which is played on undirected graphs. Player 1 chooses an initial set of vertices, which represent the locations of cops. Then Player 2 chooses a vertex for a robber. Subsequently, they take turns alternately. Player 1 moves a subset of the cops along the edges of G to adjacent vertices, and Player 2 responds by either keeping the robber at the present vertex or moving it to an adjacent vertex. This is the so-called passive variant of the game. In the active variant, both the robber and a nonempty subset of the cops must move at their turns. Both players know each other's positions throughout, and we assume that they always play their optimal strategy. The *cop-number* of a graph G is the smallest number of cops such that in a finite number of moves, at least one of them occupies the same vertex as the robber. For the passive variant, this number is denoted by $c(G)$ and for the active variant by $c'(G)$.

Fitzpatrick and Nowakowski (2001) prove the following relationship with the strong isometric dimension.

Theorem 15.7 *If G is a graph with* $\text{sdim}(G) = 2$*, then* $c(G) \leq 2$.

For the Cartesian product, Tošić (1986) proved subadditivity:

Theorem 15.8 $c(G \,\square\, H) \leq c(G) + c(H)$.

Neufeld and Nowakowski (1998) improved this bound in several special cases. Maamoun and Meyniel (1987) determined the (passive) cop-number for Cartesian products of trees, and Neufeld and Nowakowski (1998) extended their investigation by considering Cartesian products of trees and cycles. The cops and robbers game was played on infinite graphs by Bonato, Hahn, and Tardif (2010). To construct largest possible classes of infinite graphs with finite cop number, they used the so-called weak strong product. (See p. 417 for the details.)

For the active version, we have the following theorem, which was anticipated by Tošić (1986). It was proved by Neufeld and Nowakowski (1998), who also considered the direct and the strong product.

Theorem 15.9 *If $n \geq 3$, then $c'(Q_n) = \lceil n/2 \rceil$.*

There is also a tandem version of the cops and robbers game: the cops move in pairs such that they are at distance at most one from each other after every move. This game was studied on Cartesian, direct, and strong products by Clarke and Nowakowski (2005).

15.2 Retracts of Strong Products

Note that a path of length 2 in $K_2 \,\square\, K_2$ is a retract, so retracts of Cartesian products need not be boxes. In this section we show that the situation is different for the strong product. In addition, we characterize weak retracts of strong products of paths in two ways, and show that they always contain a complete subgraph invariant under all automorphisms. First a lemma.

Lemma 15.10 *If R is a retract of $G \boxtimes H$ and $(a_1, a_2)(b_1, b_2)$ is a non-Cartesian edge of $G \boxtimes H$ in R, then (a_1, b_2) and (b_1, a_2) are also in R.*

Proof Let $\varphi : V(G \boxtimes H) \to V(R)$ be a retraction, and suppose $(a_1, a_2)(b_1, b_2)$ is a non-Cartesian edge in R. Then $a_1 \neq b_1$ and $a_2 \neq b_2$, so the vertices $\{(a_1, a_2), (b_1, b_2), (a_1, b_2), (b_1, a_2)\}$ induce a K_4 in $G \boxtimes H$. By symmetry, it is enough to prove that $(b_1, a_2) \in V(R)$. Suppose to the contrary that $(b_1, a_2) \notin V(R)$. Set $(c_1, c_2) = \varphi(b_1, a_2)$. Clearly, $(c_1, c_2) \neq (b_1, a_2)$. Because (c_1, c_2) is adjacent to (b_1, b_2), we have $b_1 = c_1$ or $b_1 c_1 \in E(G)$. Similarly, because (c_1, c_2) is adjacent to (a_1, a_2), we infer that $a_2 = c_2$ or $a_2 c_2 \in E(H)$. Thus $(c_1, c_2) = \varphi(b_1, a_2)$ is adjacent to (b_1, a_2), a contradiction. □

The next theorem is from Imrich and Klavžar (1992), who also prove that if G and H are triangle-free, then retracts of $G \boxtimes H$ are precisely products of retracts of the factors.

Theorem 15.11 *Let G and H be connected graphs and R a retract of $G \boxtimes H$. Then $R = G' \boxtimes H'$, where G' and H' are weak retracts of G and H, respectively.*

Proof Lemma 15.10 implies that the Cartesian edges of R induce a connected graph with the square property in $G \,\square\, H$. By Lemma 6.4, the Cartesian edges of R are a box in $G \,\square\, H$. Because convex subgraphs are induced, $R = G' \boxtimes H'$, where G' and H' are subgraphs of G and H.

To show that G' and H' are weak retracts, let $\varphi : G \boxtimes H \to G' \boxtimes H'$ be a retraction and $b \in V(H')$. Define a map $\varphi' : V(G) \to V(G')$ by $\varphi'(x) = p_1 \varphi(x, b)$. If $xy \in E(G)$, then $\varphi(x, b)$ is adjacent to $\varphi(y, b)$; hence either $\varphi'(x) = \varphi'(y)$ or $\varphi'(x)\varphi'(y) \in E(G')$. Finally, for $x \in V(G')$, we have $\varphi(x, b) = (x, b)$. Thus $\varphi'(x) = x$, and φ is a retraction. Use the same argument for H'. □

That the weak retracts G' and H' of Theorem 15.11 need not be retracts of G and H will be demonstrated later in Theorem 26.14.

Notice however that Theorem 15.11 implies that retracts of strong products of paths are boxes, that is, again strong products of paths. But in general we gain more structure by considering weak retracts. For this, the following concepts are needed.

We say a family $\mathcal{F}$ of sets has the *Helly property* if any finite collection of pairwise nondisjoint sets from $\mathcal{F}$ has nonempty intersection. We have already seen in Theorem 12.19 that the family of convex subgraphs of a median graph has the Helly property.

For the definition of a Helly graph, we first extend the concept of a neighborhood. The *ball of radius r and center u* is defined to be

$$N_r(u) = \{v \in V(G) \,|\, d(u,v) \leq r\}.$$

A *Helly graph* is a graph in which the family of balls has the Helly property. To show that Helly graphs can be characterized as weak retracts of strong products of paths, the following lemma is useful.

Lemma 15.12 *Suppose H is a connected graph, $W \subseteq V(H)$, and G is a Helly graph. Then any nonexpansive map $f : W \to V(G)$ can be extended to a nonexpansive map $\widetilde{f} : V(H) \to V(G)$ that coincides with f on W.*

Proof Consider a nonexpansive map $f : W \to V(G)$. We will enlarge the domain of f vertex-by-vertex. Let $w \in V(H) \setminus W$, and consider the family

$$\mathcal{B} = \big\{N_{d_H(u,w)}\big(f(u)\big) \mid u \in W\big\}$$

of balls in G. Let u and v be arbitrary vertices of W. Then

$$d_G(f(u), f(v)) \leq d_H(u,v) \leq d_H(u,w) + d_H(w,v).$$

Thus a shortest path in G from $f(u)$ to $f(v)$ has a vertex that is within a distance of $d_H(u,w)$ of $f(u)$ and $d_H(w,v)$ of $f(v)$. Therefore

$$N_{d_H(u,w)}\big(f(u)\big) \cap N_{d_H(v,w)}\big(f(v)\big) \neq \emptyset\,.$$

As G is a Helly graph, $\mathcal{B}$ has nonempty intersection. Let $\widetilde{w}$ be a vertex from the intersection. Extend f to $W \cup \{w\}$ by setting $f(w) = \widetilde{w}$. Clearly, this extended f remains nonexpansive. Repeat this procedure until the domain of f is all of $V(H)$. □

Corollary 15.13 *If a Helly graph G is an isometric subgraph of a graph H, then G is a weak retract of H.*

Proof Let $W = V(G) \subseteq V(H)$. The identity map $\mathrm{id}_W : W \to V(G)$ is nonexpansive, and by Lemma 15.12, it can be extended to a nonexpansive map $\widetilde{\mathrm{id}}_W : H \to G$. This is a weak retraction. □

We are now ready for the following theorem by Nowakowski and Rival (1983) and Quilliot (1983). In particular, Lemma 15.12 is from Quilliot (1983); see also Quilliot (1985b).

Theorem 15.14 *For a graph G, the following statements are equivalent:*

(i) *G is a weak retract of a strong product of paths.*
(ii) *G is a Helly graph.*
(iii) *G is a weak retract of any graph of which it is an isometric subgraph.*

Proof (i) $\Leftrightarrow$ (ii). It is immediate that paths are Helly graphs. From this it follows that the strong product of paths is a Helly graph as well. (Use the fact that balls in a strong product are products of balls in the factors.) Now let H be a strong product of paths, and suppose $f : H \to G$ is a weak retraction. Consider a family

$$N^G_{r_1}(u_1), N^G_{r_2}(u_2), \ldots, N^G_{r_k}(u_k)$$

of pairwise intersecting balls in G and the corresponding family

$$N^H_{r_1}(u_1), N^H_{r_2}(u_2), \ldots, N^H_{r_k}(u_k)$$

in H. Because $N^G_{r_i}(u_i) \subseteq N^H_{r_i}(u_i)$, the balls $N^H_{r_i}(u_i)$ are pairwise intersecting. Then, because H is a Helly graph, $\bigcap_{i=1}^k N^H_{r_i}(u_i) \neq \emptyset$. For any x in this intersection, we thus infer that

$$d_G(u_i, f(x)) \leq d_H(u_i, x) \leq r_i.$$

Hence

$$f(x) \in \bigcap_{i=1}^{k} N^G_{r_i}(u_i).$$

This shows that weak retracts of strong products of paths are Helly graphs.

For the converse, we isometrically embed the Helly graph G into a strong product H of paths (which is possible by Theorem 15.1) and apply Corollary 15.13.

(ii) $\Leftrightarrow$ (iii). Assume first that G is a weak retract of every graph into which it is isometrically embedded. Consider a family of its pairwise intersecting balls $\mathcal{B} = \{N_{r_1}(u_1), \ldots, N_{r_k}(u_k)\}$. Then $d(u_i, u_j) \leq r_i + r_j$. Form a new graph H by adding an extra vertex x to G and, for each $i = 1, \ldots, k$, a path of length r_i connecting x with u_i. In view of $d(u_i, u_j) \leq r_i + r_j$, we infer that G is an isometric subgraph of H. By assumption, there exists a weak retraction $f : H \to G$. Finally, because f is nonexpansive, $f(x)$ belongs to all elements of $\mathcal{B}$.

The converse follows by Corollary 15.13. □

For additional characterizations of Helly graphs, see Hell and Rival (1987), Bandelt and Pesch (1989), Bandelt and Prisner (1991), and the book by Pesch (1988). The equivalence between (i) and (iii) was proved by Hell (1972) in the following similar context: A graph is a retract of the direct product of paths if and only if it is a retract of any bipartite graph in which it is an isometric subgraph. For more information, see Bandelt, Dählmann, and Schütte (1987) and Bandelt, Farber, and Hell (1993). These problems were considered in the more general context of graphs, metric spaces, and partially ordered sets by Jawhari, Misane, and Pouzet (1986). We also add that Bandelt and Pesch (1989) presented two recognition algorithms for Helly graphs, one of complexity $O(n^4)$ and the other of complexity $O(mn^2)$.

We conclude the section with an appealing result on Helly graphs due to Quilliot (1985a). It is yet another generalization of the fact that the automorphisms of a tree stabilize its center.

Theorem 15.15 (Fixed Simplex Theorem) *Each Helly graph G has a complete subgraph that is invariant under all automorphisms of G.*

Proof Let G be a Helly graph. If $\mathrm{diam}(G) = 1$, then G is a complete graph, and we are done. Therefore assume that $d = \mathrm{diam}(G) > 1$. Set $k = \lceil d/2 \rceil$ and consider the family of all balls $N^G_k(u)$. Any two of them have nonempty intersection, and therefore the intersection

$$S = \bigcap_{u \in V(G)} N^G_k(u)$$

is nonempty. It is also invariant under automorphisms of G. Let $G' = \langle S \rangle$. The diameter d' of G' is smaller than d, because for any $x, y \in V(G')$, we have $x \in N_k(y)$, so $d_G(x, y) \leq k < d$.

If we can show that G' is a Helly graph, the theorem follows by induction on the diameter.

We show first that G' is isometric in G, that is, we prove $d_{G'}(x, y) = d_G(x, y)$ for any $x, y \in V(G')$. We use induction of $r = d_G(x, y)$. Certainly this is true if $r = 1$ because G' is induced; fix an $r > 1$ and assume it is true for smaller values of r. Now suppose $d_{G'}(x, y) = r$. Certainly

$$N^G_1(x) \cap N^G_{r-1}(y) \neq \emptyset.$$

But also, because x and y are within distance k of any vertex z of G, we have

$$N_1^G(x) \cap N_k^G(z) \neq \emptyset \quad \text{and} \quad N_{r-1}^G(y) \cap N_k^G(z) \neq \emptyset$$

for all $z \in V(G')$. Thus the family $\{N_1^G(x), N_{r-1}^G(y)\} \cup \{N_k^G(z) \mid z \in V(G)\}$ has pairwise nonempty intersections. This combined with the fact that G is a Helly graph guarantees a vertex $z \in N_1^G(x) \cap N_{r-1}^G(y) \cap S$. This vertex belongs to G', and we have $d_G(x,z) \leq 1$ and $d_G(z,y) \leq r-1$. Then, by induction, $d_G(x,z) = d_{G'}(x,z)$ and $d_G(z,y) = d_{G'}(z,y)$. Therefore $d_{G'}(x,y) \leq d_{G'}(x,z) + d_{G'}(z,y) = d_G(x,z) + d_G(z,y) \leq r = d_G(x,y)$. But $d_{G'}(x,y) \leq d_G(x,y)$ implies $d_{G'}(x,y) = d_G(x,y)$, so G' is isometric.

Because G' is isometric, every ball $N_r^{G'}(x)$ can be represented in the form

$$N_r^{G'}(x) = N_r^G(x) \cap \bigcap_{z \in V(G)} N_k^G(z).$$

With this representation and the fact that $N_r^{G'}(x) \subseteq N_r^G(x)$, it is easy to see that the family of balls in G' satisfies the Helly property. □

For a survey on (convexity and fixed-point properties in) Helly graphs, see Polat (2001).

15.3 Other Product Graph Dimensions

Graphs can be embedded into different products and as different kinds of subgraphs: as subgraphs, as induced subgraphs, or as isometric subgraphs. Provided that a general embedding theorem can be proved, one can consider the corresponding graph dimension, just as we did for the strong isometric dimension. In this short section we give a brief overview of several dimensions involving the Cartesian product and the direct product. For more information on such dimensions, including complexity issues, we refer to Nešetřil and Rödl (1985) and Nešetřil (1981).

Cartesian product dimensions

An obvious possibility for a (Cartesian) isometric dimension would be to define it as the number of factors of its canonical isometric embedding. Instead, the term *isometric dimension* of a graph G, $\mathrm{idim}(G)$, is reserved for the bipartite case: it is the smallest (and by Theorem 13.3 (ii) also the largest) integer d for which G embeds isometrically and irredundantly into the d-dimensional cube. If there is no such d, we set $\mathrm{idim}(G) = \infty$.

Clearly, $\mathrm{idim}(G) < \infty$ if and only if G is a partial cube and is in that case equal to the number of Θ-classes of G.

The *lattice dimension* of a graph G, $\mathrm{ldim}(G)$, is the smallest integer d such that G embeds isometrically into the d-dimensional integer lattice. Again, if there is no such d, we set $\mathrm{ldim}(G) = \infty$. Notice that the two-sided infinite path has lattice dimension 1. For finite graphs, the definition is equivalent to saying that d is the smallest integer (if it exists) such G is isometric in $P_n^{\Box,d}$ for some n. In Exercise 15.6 the reader is invited to prove the following:

Proposition 15.16 *For a finite graph G, $\mathrm{ldim}(G) < \infty$ if and only if G is a partial cube.*

Determining the lattice dimension of a partial cube is generally hard. However, Eppstein (2005) designed an $O(|E(G)|^2)$ algorithm for this task. The main idea is to reduce the determination of the lattice dimension of a given graph to the computation of the size of a maximum matching in an associated graph (the so-called semi-cube graph). Imrich and Kovše (2009) followed with a linear algorithm that isometrically embeds a given tree T into an integer lattice of dimension $\mathrm{ldim}(T)$.

Using a poset-based approach Cheng (2011) computes the lattice dimension of median graphs G and embeds them in $O(|V(G)|\,\mathrm{ldim}(G) + (\mathrm{ldim}(G))^{2.5})$ time. This can be considerably better than $O(|E(G)|^2)$. Cheng also computes the isometric dimension $\mathrm{idim}(G)$, that is, the number of Θ-classes, of a median graph in $O(|V(G)| + |E(G)|)$ time. The Θ-classes themselves are not determined however. In other words, the algorithm does not provide an embedding.

One of the motivations for embedding graphs into lattices comes from graph drawing. Compare Eppstein (2004) for applications in this direction.

Direct product dimensions

The *direct dimension* $\mathrm{ddim}(G)$ is the smallest number n for which G is an **induced** subgraph of the direct product of n complete graphs. This concept was introduced by Nešetřil and Rödl (1978).

Lovász, Nešetřil, and Pultr (1980) studied the direct dimension of a graph extensively and proved that, with the exception of some very special cases, $\mathrm{ddim}(G) \leq |V(G)| - 2$. They also computed the dimension for several classes of graphs. This problem has also been investigated by Křivka (1981a), who showed that $\mathrm{ddim}(Q_n) = n - 1$ for $n \geq 3$. The direct dimension of the disjoint union of graphs was investigated by Křivka (1981b, 1985) and Alles (1985). The problem of how the addition of an edge to a graph affects its direct dimensions was considered by Kříž (1984).

Eaton and Rödl (1996) obtained a very strong result. Extending a result of Alon (1986), which asserts that there exists a constant c to every positive integer d such that

$$\mathrm{ddim}(G) \leq c(d+1)^2 \log n + 1$$

for any graph G on n vertices with maximum degree $\Delta(G) \leq d$, they proved the following.

Theorem 15.17 *There exists a constant c to every positive integer d such that* $\mathrm{ddim}(G) \leq cd \log n$ *for every graph G on n vertices with maximum degree $\Delta(G) \leq d$.*

We close by mentioning that the *bipartite dimension* of a bipartite graph G is defined as the smallest number k such that G is an **induced** subgraph of the direct product of k copies of P_3; see Poljak and Pultr (1981) and Poljak, Rödl, and Pultr (1983).

Exercises

15.1. Show that $\mathrm{sdim}(G) \leq |V(G)| - \mathrm{diam}(G)$ for any connected graph G.

15.2. Show that $\mathrm{sdim}(K_n) = \lceil \log_2 n \rceil$ for $n \geq 2$.

15.3. Let $k \geq 1$ and $n \leq \binom{k}{\lfloor k/2 \rfloor}$. Then show that $\mathrm{sdim}(K_2 \,\square\, K_n) \leq k$.

15.4. Let G be a graph with $\text{diam}(G) = d$. Show that if G isometrically embeds into the strong product of n paths, then G also isometrically embeds into $P_{d+1}^{\boxtimes,n}$.

15.5. Determine $\text{idim}(P_n \,\square\, P_m)$.

15.6. Show that a finite graph has finite lattice dimension if and only if it is a partial cube.

Chapter 16

Fixed Box Theorems

This chapter is concerned with fixed box theorems. We saw that trees have two types of subgraphs that are stabilized by all automorphisms: the center and the distance center. This holds for median graphs in general. Theorem 12.21 states that every median graph (i.e., retract of a hypercube) contains a subcube that is stabilized by all automorphisms. One type of subcube is obtained by successive removal of vertices at the periphery, the other type is the distance center.

In Chapter 14 we also saw that every quasi-median graph (i.e., weak retract of a Hamming graph) contains a fixed Hamming graph, and from Chapter 15 we know that every Helly graph (i.e., weak retract of a strong product of paths) contains an invariant complete subgraph.

In all these cases, a weak retract of a Cartesian product of complete graphs or of a strong product of paths is considered, and the existence of an invariant subgraph is asserted. In the Cartesian case, the invariant subgraph is a box, and the theorems are called called *fixed box theorems*. The main aim of this chapter are extensions to weak retracts of Cartesian products of arbitrary simple graphs. They are due to Feder (1995) and Tardif (1997).

The chapter ends with a result of Feder (2006) about the computation of fixed points of several nonexpansive mappings. A complete characteristic of families $\mathcal{G}$ of graphs G is given, such that the fixed points of several nonexpansive mappings of $\Box_{G\in\mathcal{G}} G$ can be computed in polynomial time. When the conditions are not satisfied, very simple examples with two retractions give NP-completeness for the computation of a common fixed point.

16.1 Gated Subgraphs and Median Functions

To prepare for the first fixed box theorem, we begin with a generalization of the concept of a weak retraction. We continue with several new results about gated sets. In particular, we show that the family of gated sets in a graph has the Helly property. We also extend the concept of a median to that of a median function. Moreover, we introduce the concept of median-compatible relations, so-called tolerances.

Let H be a subgraph of G. A map $f : G \to H$ is called *weakly nonexpansive* if $d(f(u), v)) \leq d(u, v)$ holds for all vertices $u \in V(G)$ and $v \in V(H)$. Clearly, a weak retraction to H is a weakly nonexpansive map. Indeed, if f is a weak retraction, then $d(f(u), v)) = d(f(u), f(v)) \leq d(u, v)$.

Together with the concept of the distance center (p. 155) we obtain the following result, the proof of which is identical to the proof of Lemma 12.23.

Theorem 16.1 *Let $G = \Box_{i=1}^{k} G_i$ and let $f : G \to H$ be a weakly nonexpansive map. Then the distance center of H in G is a box. In particular, this holds if H is a weak retract of G.*

Despite its easy proof, this theorem is quite powerful. Clearly, Theorems 14.12 and 12.24 are immediate consequences. Various other theorems, such as Theorem 12.25, also easily follow.

The question whether there exist fixed boxes that are retracts is much more difficult. In particular, we have not yet shown that the fixed boxes of Theorems 14.12 and 12.24 are weak retracts. Different approaches to the solution of this and related problems are due to Feder and Tardif but neither approach is easy. We will follow the approach of Tardif because it is more along the line of this book.

Let H be a gated subgraph of a graph G. Recall that that gated subgraphs are convex. We denote the gate of v in H by $k_H(v)$, and we show next that k_H is a weak retraction.

Lemma 16.2 *Let H be a gated subgraph of a graph G. Then k_H is a weak retraction.*

Proof Clearly, k_H is idempotent. To show that it is nonexpansive, consider two vertices u, v. Suppose that $d(u,v) < d(k_H(u), k_H(v))$, and let the notation be chosen such that $d(u, k_H(u)) \leq d(v, k_H(v))$. Then

$$\begin{aligned} d(v, k_H(u)) &\leq d(v,u) + d(u, k_H(u)) \\ &< d(k_H(v), k_H(u)) + d(v, k_H(v)) \\ &= d(v, k_H(v)) + d(k_H(v), k_H(u)) \\ &= d(v, k_H(u)), \end{aligned}$$

which is impossible. □

Lemma 16.3 *Let H_1, H_2 be gated subgraphs of a graph G with $H_1 \cap H_2 \neq \emptyset$. Then $k_{H_1}(H_2) = k_{H_2}(H_1) = H_1 \cap H_2$ is gated, and $k_{H_1 \cap H_2} = k_{H_1} k_{H_2} = k_{H_2} k_{H_1}$.*

Proof Let v be a vertex in $H_1 \cap H_2$ and x an arbitrary vertex of H_2. Because there is a shortest x,v-path via $k_{H_1}(x)$ and because H_2 is convex, the gate $k_{H_1}(x)$ must also be in H_2. Thus $k_{H_1}(H_2) \subseteq H_2$. Because every $k_{H_1}(x)$ is in H_1 and because $k_{H_1}(y) = y$ for every y of $H_1 \cap H_2$, we infer that $k_{H_1}(H_2) = H_1 \cap H_2$. The assertion $k_{H_2}(H_1) = H_1 \cap H_2$ follows by interchanging the role of H_1 and H_2.

Consider an arbitrary vertex z and a vertex w in $H_1 \cap H_2$ of minimum distance from z. There is a shortest z,w-path via $k_{H_1}(z)$, and this path meets $H_1 \cap H_2$ in $k_{H_2} k_{H_1}(z)$. Thus w must be the unique vertex $k_{H_2} k_{H_1}(z)$. By the same argument, $w = k_{H_1} k_{H_2}(z)$, whence $k_{H_1} k_{H_2} = k_{H_2} k_{H_1}$. Because H_2 is gated, every vertex of $H_1 \cap H_2$ can be reached from z by a shortest path via $k_{H_2}(z)$. Furthermore, every such path contains $w = k_{H_1} k_{H_2}(z)$. This implies that $H_1 \cap H_2$ is gated too; the gate of z being w. Thus $k_{H_1 \cap H_2} = k_{H_1} k_{H_2}$. □

Because of this result, there exists a smallest gated set containing a given set S. We denote it $\ll S \gg$. It is the intersection of all gated sets containing S.

Corollary 16.4 *The family of gated sets in a graph has the Helly property.*

Proof We show first that the assertion holds for families of three sets and continue by induction.

Let H_1, H_2, H_3 be gated sets with pairwise nonempty intersection and let $u \in H_1 \cap H_2$. Then $k_{H_3}(u) \in H_3 \cap H_1$ and $k_{H_3}(u) \in H_3 \cap H_2$ because of Lemma 16.3. Then $k_{H_3}(u) \in H_1 \cap H_2 \cap H_3$ and H_1, H_2, H_3 have nonempty intersection.

Suppose now that the assertion holds for $k \geq 3$ sets, and let $H_1, \ldots, H_{k+1}$ be a family of

$k+1$ sets with pairwise nonempty intersections. Consider $H'_1, \ldots, H'_k$, where $H'_i = H_i \cap H_{k+1}$, $i = 1, \ldots, k$. Then $H'_i \cap H'_j = H_i \cap H_j \cap H_{k+1} \neq \emptyset$ for $1 \leq i, j \leq k$. Thus the family of sets $H_1, H_2, \ldots, H_{k+1}$ also has nonempty intersection. □

We now define a *median function* by setting $\mathrm{m}(x, y; z) = k_{\ll x,y \gg}(z)$. In the case of median graphs, it yields the median of x, y, z, and for quasi-median graphs, it coincides with the imprint function.

Lemma 16.5 *Let m be the median function of a graph G. Then*

(i) $\mathrm{m}(x, y; m(x, y; z)) = \mathrm{m}(x, y; z)$,
$\mathrm{m}(x, y; z) = \mathrm{m}(y, x; z)$,
$\mathrm{m}(x, x; y) = \mathrm{m}(x, y; x) = x$.

(ii) *If* $z \in I(x, y)$, *then* $\mathrm{m}(x, y; z) = z$.

(iii) *A subgraph H of G is gated if and only if* $\mathrm{m}(x, y; z) \in H$ *for each* $x, y \in V(H)$ *and* $z \in V(G)$.

Proof Parts (i), (ii), and the if part of (iii) are easily seen to be true. Thus suppose that H is a subgraph of G for which $\mathrm{m}(x, y; z) \in V(H)$ for $x, y \in V(H)$ and $z \in V(G)$. We have to show that H is gated; that is, we have to show for arbitrary $z \in V(G)$ that there exists a y in H such that $y \in I(x, z)$ for any $x \in V(H)$. Let y be a vertex in H of minimal distance from z. By assumption, $\mathrm{m}(x, y; z) \in V(H)$ for any $x \in V(H)$ and $d(\mathrm{m}(x, y; z), z) \leq d(y, z)$. Thus $y = \mathrm{m}(x, y; z) \in I(x, z)$. □

We have seen in the previous chapters that median graphs and quasi-median graphs can be characterized by closure properties under majority rule and the imprint function. We are therefore interested in subgraphs of Cartesian products that are closed under the median function and, given a subgraph with this property, wish to find other subgraphs with the same property. We thus define a relation on the vertex set of a graph that is compatible with the median function.

Let x_i, y_i, $i = 1, 2, 3$, be arbitrary vertices of G. We say a binary, reflexive, and symmetric relation α is a *tolerance* on $V(G)$ if

$$\mathrm{m}(x_1, x_2; x_3) \,\alpha\, \mathrm{m}(y_1, y_2; y_3)$$

whenever $x_i \,\alpha\, y_i$ for $i = 1, 2, 3$.

For $S \subseteq V(G)$, set $\alpha(S) = \{x \in V(G) \mid x \,\alpha\, y \text{ for some } y \in S\}$. For $S = \{s\}$, we simply write $\alpha(s)$. Furthermore, we call a subset B of $V(G)$ a *block of* α if $B = \bigcap_{b \in B} \alpha(b)$. In other words, the blocks of α are the maximal subsets of $V(G)$ any two elements of which are in relation α.

For induced subgraphs H of G we shall abuse the above notation and write $\alpha(H)$ for the subgraph $\langle \alpha(H) \rangle$ induced by $\alpha(H)$.

Lemma 16.6 *Let α be a tolerance on a graph G and H a gated subgraph of G. Then $\alpha(H)$ is also gated, and for $u \in V(G)$, $u \in \alpha(H)$ if and only if $k_H(u) \,\alpha\, u$.*

Proof Consider vertices $x, z \in V(\alpha(H))$ and $u, v \in V(H)$ for which $u \,\alpha\, x$ and $v \,\alpha\, z$ hold. Then for any $w \in V(G)$, $\mathrm{m}(u, v; w) \,\alpha\, \mathrm{m}(x, z; w)$. Now, $\mathrm{m}(u, v; w) \in H$ by Lemma 16.5 (iii) and thus $\mathrm{m}(x, z, w) \in \alpha(H)$. But then, again by Lemma 16.5 (iii), $\alpha(H)$ is gated.

This also implies that $\alpha(u)$ is convex for any $u \in V(G)$. Then $k_H(u) \in \alpha(u)$ if $\alpha(u) \cap H \neq \emptyset$. Finally, if $k_H(u) \in \alpha(u)$, then clearly $\alpha(u) \cap H \neq \emptyset$. □

Corollary 16.7 *Let α be a tolerance on a graph G. Then (the subgraphs induced by) the blocks of α are gated.*

Because the one-vertex subgraphs of G are gated, all $\alpha^i(u)$ are gated for any $u \in V(G)$. Here $\alpha^i(u)$ is defined by $\alpha(\alpha^{i-1}(u))$ for $i > 1$. We also set $\alpha^0(u) = u$, namely $\alpha^0 = \mathrm{id}$. It is easy to see that $\alpha^i \subseteq \alpha^{i+1}$. It is known that the transitive closure α^* is $\bigcup_{i\geq 0} \alpha^i$.

Lemma 16.8 *Let α be a tolerance on a connected graph G. Then $\alpha^* = V(G) \times V(G)$ if and only if $E(G) \subseteq \alpha$.*

Proof Suppose that $\alpha^* = V(G) \times V(G)$, and let uv be an edge of G. We have to show that $v\,\alpha\,u$ holds. Because $u \notin \alpha^0 v$ and because G is finite, there is a largest k with $u \notin \alpha^k(v)$. Set $H = \alpha^k(v)$. Clearly, $u \in \alpha(H)$. By Lemma 16.6, H is gated and $k_H(u)\,\alpha\,u$. Because $v = k_H(u)$, we conclude that $v\,\alpha\,u$.

The converse immediately follows from the connectedness of G. □

Let α be a tolerance on a graph G. Then the *block graph* $\mathcal{B}(G,\alpha)$ is defined on the blocks of α, two blocks being adjacent if they have nonempty intersection.

Lemma 16.9 *Suppose α is a tolerance on a graph G. Let A, B be blocks of α, and $r > 0$. Then the following statements are equivalent:*

(i) $B \subseteq \alpha^r(A)$.
(ii) $B \cap \alpha^{r-1}(A) \neq \emptyset$.
(iii) $B \in N_r(A)$.

Proof (i) $\Rightarrow$ (ii). Assume that $B \subseteq \alpha^r(A)$ and B is disjoint from $C = \alpha^{r-1}(A)$. Choose $u \in B$ and set $v = k_C(u)$. For any $w \in B$ we have $w \in \alpha^r(A) = \alpha(C)$. Therefore $w\,\alpha\,k_C(w)$ by Lemma 16.6. Setting $w' = k_C(w)$, we thus have $w = \mathrm{m}(w,w;v)\ \alpha\ \mathrm{m}(u,w';v) = v$, because $v = k_C(u) \in I(u,w')$. But this is not possible, as B is a block of α and $v \notin B$.

(ii) $\Rightarrow$ (iii). For an arbitrarily chosen element $a_{r-1} \in B \cap \alpha^{r-1}(A)$ for which $a_{r-1}\,\alpha\,a_{r-2}$, there exists a sequence of elements $a_{r-2}\,\alpha\,a_{r-3}\,\alpha \ldots \alpha\,a_1\,\alpha\,a_0$, where $a_i \in \alpha^i(A)$. For $i = 1,\ldots,r-1$, let C_i denote the smallest block containing $\{a_{i-1}, a_i\}$. Then $AC_1C_2\ldots C_{r-1}B$ is a path of length r in $\mathcal{B}(G,\alpha)$ from A to B.

(iii) $\Rightarrow$ (i). If $B \in N_r(A)$, then there exists a path

$$A = C_0C_1C_2\ldots C_{r-1}C_r = B$$

from A to B in $\mathcal{B}(G,\alpha)$. Let $u_0, u_1, u_2, \ldots, u_{r-1}$ be chosen such that $u_i \in C_i \cap C_{i+1}$. Then $u_0\,\alpha\,u_1\,\alpha\,u_2\ \alpha\ \ldots\alpha\,u_{r-1}$ and $u_{r-1}\,\alpha\,u$ for any $u \in B$. Hence $u \in \alpha^r(A)$. □

Corollary 16.10 *Let α be a tolerance on a connected graph G. Then $\mathcal{B}(G,\alpha)$ is connected if and only if $E(G) \subseteq \alpha$.*

Proof By the above, $\mathcal{B}(G,\alpha)$ is connected if and only if every pair of elements of G is in relation α^*. By Lemma 16.8, this is the case if and only if $E(G) \subseteq \alpha$. □

Proposition 16.11 *Let α be a tolerance on a graph G. Then each connected component of $\mathcal{B}(G,\alpha)$ is a Helly graph.*

Proof For an index set I, let $\{N_{r_i}(A_i)\}_{i\in I}$ be a family of pairwise intersecting balls in $\mathcal{B}(G,\alpha)$ and $i,j \in I$. Then there exists a block $B_{i,j}$ of α with $B_{i,j} \in N_{r_i}(A_i) \cap N_{r_j}(A_j)$. By Lemma 16.9, $B_{i,j} \cap \alpha^{r_i-1}(A_i) \neq \emptyset$ and $B_{i,j} \cap \alpha^{r_j-1}(A_j) \neq \emptyset$. Thus there exist elements $u_{i,j} \in \alpha^{r_i-1}(A_i)$ and $u_{j,i} \in \alpha^{r_j-1}(A_j)$ that are in relation α. Again by Lemma 16.9, $B_{i,j} \subseteq \alpha^{r_i}(A_i) \cap \alpha^{r_j}(A_j) \neq \emptyset$. This means that the family $\{\alpha^{r_i}(A_i)\}_{i\in I}$ consists of pairwise intersecting gated sets of G. By Corollary 16.4, this family has nonempty intersection. Let u be an element of this intersection and $u_i = k_{\alpha^{r_i-1}(A_i)}(u)$ for $i \in I$. By Lemma 16.6, $u\,\alpha\,u_i$.

Now, let $j \in I$. Then $\mathrm{m}(u, u_{i,j}; u_i) = u_i$, because $u_i \in I(u, u_{i,j})$. Also, $\mathrm{m}(u_j, u_{j,i}; u) = u_j$ because $\ll u_j, u_{j,i} \gg \subseteq \alpha^{r_j - 1}(A_j)$. Finally, $\mathrm{m}(u_j, u_{j,i}; u) = k_{\ll u_j, u_{j,i} \gg}(u)$.

We therefore conclude that $u_i = \mathrm{m}(u, u_{i,j}; u_i)\, \alpha\, \mathrm{m}(u_j, u_{j,i}; u) = u_j$ for any pair $i, j \in I$ of indices. This means that $\{u_i\}_{i \in I}$ is contained in a block B of α and that $B \cap \alpha^{r_i - 1}(A_i) \neq \emptyset$ for $i \in I$. By Lemma 16.9, $B \in \bigcap_{i \in I} N_{r_i}(A_i)$. □

16.2 A Fixed Box Theorem for Median Function-Closed Graphs

We now derive fixed box theorems for subgraphs of Cartesian products that are closed under the median function. Special cases include the invariance of the distance center of a tree under all automorphisms and Theorem 12.24. The main result of this section is Theorem 16.17, due to Imrich (2000, first edition of this book).

Lemma 16.12 *Let $G = \square_{i=1}^{k} G_i$ and H_i be a gated subgraph of G_i, $i = 1, \ldots, k$. Then $H = \square_{i=1}^{k} H_i$ is a gated subgraph of G with $k_H = k_{H_1} \times k_{H_2} \times \ldots \times k_{H_k} = \prod_{1 \leq i \leq k} k_{H_i}$.*

The proof follows from the Distance Formula (Corollary 5.2) and is omitted.

Lemma 16.13 *Let G be an irredundant isometric subgraph of $\square_{i=1}^{k} G_i$, and let H be a gated subgraph of G. Then for every $i = 1, \ldots, k$, the projections $H_i = p_i(H)$ are gated in G_i, $k_{H_i} p_i = p_i k_H$ and $H = \left(\square_{i=1}^{k} H_i\right) \cap G$. Also, $B = \square_{i=1}^{k} H_i$ is gated in G, and for $u \in G$, $k_B(u) = k_H(u)$.*

Proof For $u \in V(G)$, $v = k_H(u)$, and $w \in V(H)$, we have $d(u, w) = d(u, k_H(u)) + d(k_H(u), w)$. By the Distance Formula and the triangle inequality, this implies $d_{G_i}(u_i, w_i) = d_{G_i}(u_i, v_i) + d_{G_i}(v_i, w_i)$ for all $i = 1, \ldots, k$. Because $p_i(G) = G_i$, we infer that $H_i = p_i(H)$ is gated in G_i and that $k_{H_i} u_i = p_i k_H(u)$. By Lemma 16.12, $B = \square_{i=1}^{k} H_i$ is gated, and for $u \in V(G)$, $k_{H_i} u_i = p_i k_B(u)$; therefore $k_B = k_{H_i}$. □

Corollary 16.14 *Let $G = \square_{i=1}^{k} G_i$ and m be the median function on G. Then m is the product of the median functions m_i of the factors.*

Proof If $u, v, w \in V(G)$, then $\ll u, v \gg \subseteq \prod_{i=1}^{k} \ll u_i, v_i \gg$ and $\prod_{i=1}^{k} \ll u_i, v_i \gg \subseteq \ll u, v \gg$, by Lemma 16.13. Thus we infer that $\ll u, v \gg = \prod_{i=1}^{k} \ll u_i, v_i \gg$ and $p_i(\mathrm{m}(u, v; w)) = \mathrm{m}_i(u_i, v_i; w_i)$. □

Let G be the Cartesian product $G_1 \square G_2$ and H a subproduct of G. Suppose that (u_1, u_2), (v_1, v_2) are vertices of H. Then the vertices (u_1, v_2), (v_1, u_2) are also in H. Clearly, this property characterizes subproducts. For the characterization of subproducts that are retracts, we need a condition that somehow captures the property that subcubes of hypercubes are closed under the operation of taking medians. The problem is that $\mathrm{m}_H(u, v; w)$ can be different from $\mathrm{m}_G(u, v; w)$, because $\ll u, v \gg_H$ may be different from $\ll u, v \gg_G$. We call a subgraph H of G *median function-closed* if the median function of H is identical with the restriction of the median function of G to H.

Proposition 16.15 *Let H be a median function-closed subgraph of a Cartesian product $G_1 \square G_2$. Then the relation $\alpha(G_1, G_2)$ defined on H by setting $(u_1, u_2)\alpha(v_1, v_2)$ if $(u_1, v_2) \in V(H)$ and $(v_1, u_2) \in V(H)$ is a tolerance on H, and $E(H) \subseteq \alpha(G_1, G_2)$.*

Proof By assumption, the median function on H is the restriction of the median function m of G to H, which in turn is the product $\mathrm{m}_1 \times \mathrm{m}_2$ of the median functions m_1 and m_2 of G_1 and G_2, respectively. Now consider $u_1, \ldots, u_6 \in V(H)$ with $u_i \,\alpha\, u_{i+3}$ for $i = 1, 2, 3$ and let $u_i = (s_i, t_i)$ for $i = 1, \ldots, 6$. Then

$$\mathrm{m}(u_1, u_2; u_3) = ((\mathrm{m}_1(s_1, s_2; s_3), \mathrm{m}_2(t_1, t_2; t_3))$$

and

$$\mathrm{m}(u_4, u_5; u_6) = (\mathrm{m}_1(s_4, s_5; s_6), \mathrm{m}_2(t_4, t_5; t_6)).$$

By the definition of α, we further have

$$(\mathrm{m}_1(s_1, s_2; s_3), \mathrm{m}_2(t_4, t_5; t_6)) = \mathrm{m}((s_1, t_4), (s_2, t_5); (s_3, t_6)) \in V(H),$$
$$(\mathrm{m}_1(s_4, s_5; s_6), \mathrm{m}_2(t_1, t_2; t_3)) = \mathrm{m}((s_4, t_1), (s_5, t_2); (s_6, t_3)) \in V(H).$$

Thus $\mathrm{m}(u_1, u_2; u_3) \,\alpha\, \mathrm{m}(u_4, u_5; u_6)$ and α is a tolerance.

It remains to show that $E(H) \subseteq \alpha$. Let uv be in $E(H)$. Then $p_1 u = p_1 v$ or $p_2 u = p_2 v$, and $u \,\alpha\, v$ by the definition of α. □

Before stating and proving a fixed box theorem for median-closed subgraphs of a Cartesian product, we need a lemma that extends the Fixed Simplex Theorem 15.15. First a few remarks.

Let m be the median function on a graph G. Because the definition of the median function is invariant under automorphisms, we have $\varphi\mathrm{m}(u, v; w) = \mathrm{m}(\varphi(u), \varphi(v); \varphi(w))$. Moreover, if α is a tolerance, then $\varphi\alpha$, defined by $u \,\varphi\alpha\, v \equiv \varphi u \,\alpha\, \varphi v$, is also a tolerance on G. Also, $E(G) \subseteq \alpha$ if and only if $E(G) \subseteq \varphi\alpha$. Let β be the least tolerance on G containing $E(G)$. Then β is invariant under all automorphisms of G, and every automorphism permutes the blocks of G and induces an automorphism of $\mathcal{B}(G, \beta)$.

Lemma 16.16 *Let G be a graph and β the least tolerance of G that contains $E(G)$. Then there exists a gated subgraph H of G on which β is transitive and which is invariant under every automorphism of G.*

Proof Note that $\mathcal{B}(G, \beta)$ is a Helly graph by Corollary 16.10 and Proposition 16.11. By the Fixed Simplex Theorem, there exists a set $\{B_1, B_2, \ldots, B_k\}$ of pairwise intersecting blocks of β that is invariant under all automorphisms of G. By Corollary 16.7, these blocks induce gated subgraphs and have nonempty intersection H by Lemma 16.4. □

Theorem 16.17 (Fixed Box Theorem for Median Function-Closed Subgraphs) *Let H be a median function-closed subgraph of a Cartesian product $G = \Box_{i=1}^{k} G_i$. Then there exists a box $S \subseteq H$ that is a weak retract of H and that is invariant under every automorphism of H.*

Proof Let β be the least tolerance containing $E(H)$ and let S be the gated set given by Lemma 16.16. Any two elements of S are in relation β, and S is invariant under all automorphisms of H. Because S is gated, it is a weak retract. It remains to show that $S = \prod_{i=1}^{k} p_i(S)$. Let $v \in \prod_{i=1}^{k} p_i(S)$. Then there are vertices $u_1, u_2, \ldots, u_k$ in S with $p_i(v) = p_i(u_i)$. We now recursively define vertices $v_1, v_2, \ldots, v_k \in S$, beginning with $v_1 = u_1$. Suppose that v_{i-1} has already been defined. To define v_i, we consider the product tolerance $\alpha = \alpha(G_i, \prod_{j=1, j \neq i}^{k} G_j)$ of Proposition 16.15. Because $\beta \subseteq \alpha$, and as any two elements of S are in relation β, we infer $v_{i-1} \,\alpha\, u_i$. By the definition of α, there exists a $v_i \in S$ for which $p_i(v_i) = p_i(u_i)$ and $p_j(v_i) = p_j(v_{i-1})$ for all $j \neq i, 1 \leq i \leq k$. Then $v_i \in S$, because S is convex. Thus the $v_1, v_2, \ldots, v_k$ are well defined and $v_k \in S$. □

Note that the subgraph S is gated, isometric, and induced. In Section 14.3, quasi-median graphs were defined as imprint-closed induced subgraphs of Hamming graphs. Because the median function generalizes the imprint function, we immediately deduce the following generalization of Theorem 14.12.

Theorem 16.18 *Every quasi-median graph G contains a weak retract that is a Hamming graph and which is invariant under all automorphism of G.*

Of course, this also strengthens Theorem 12.21. In general, retracts of a product are a proper subset of the class of graphs satisfying the conditions of Theorem 16.17. For example, consider $G = K_2 \,\Box\, C_5$ and $H = K_2^v \cup C_5^v$ for $v \in V(G)$. Note that H satisfies the conditions of Theorem 16.17 but is not a weak retract.

Brešar (2002) continued the study of median-function closed subgraphs of Cartesian products. He introduced *absolute C-median graphs* as graphs G such that whenever G is an isometric subgraph of a Cartesian product graph, G is a median-function closed subgraph of it. Absolute C-median graphs form a large class of graphs that includes numerous median-lime graphs, as for instance the pseudo-median graphs introduced at the end of Section 14.3. To decide whether a graph is an absolute C-median graph, it suffices to look at the canonical isometric embedding:

Theorem 16.19 *Let G be a graph and $\alpha : G \to \Box_{i=1}^{k} G/\Pi_i$ the canonical isometric embedding. Then G is as absolute C-median graph if and only if $\alpha(G)$ is a median-function closed subgraph of $\Box_{i=1}^{k} G/\Pi_i$.*

Brešar also conjectured that absolute C-median graphs can be, roughly speaking, obtained by a sequence of gated amalgamations along gated boxes. The truth of the conjecture would lead to an alternative, more transparent proof of the Fixed Box Theorem for Median Function-Closed Subgraphs.

16.3 Feder-Tardif's Fixed Box Theorems

We now prove the two original fixed box theorems for graphs, both of which are due to Feder (1995) and Tardif (1997). They are Theorem 16.22, which states that every retract of a Cartesian product has a fixed box, and Theorem 16.23, which asserts that every endomorphism of a Cartesian product stabilizes a fixed box.

We wish to show that retracts of Cartesian products are median function-closed under a mild additional assumption. We begin with the intersection of gated subgraphs and retracts.

Lemma 16.20 *Let R be a weak retract of a graph G. Then the intersection of every gated subgraph H of G with R is gated in R.*

Proof Let $\varphi : G \to R$ be a weak retraction. Then $\varphi(I(u, w)) \subseteq I(u, w)$ for $u, w \in V(R)$ by Proposition 3.8.

Suppose that $u \in V(R)$ and v is a vertex from $H \cap R$. Then $\varphi(I(u, v)) \subseteq I(u, v)$. Therefore $(\varphi k_H)^{i+1}(u) \in I((\varphi k_H)^i(u), v) \subseteq I(u, v)$ for all $i \geq 0$. By the finiteness of G, there exists an i for which $(\varphi k_H)^{i+1}(u) = (\varphi k_H)^i(u)$. Because retractions are nonexpansive, we infer $d((\varphi k_H)^i(u), v) \leq d(k_H(\varphi k_H)^i(u), v) \leq d((\varphi k_H)^{i+1}(u), v) = d((\varphi k_H)^i(u), v)$. But then $(\varphi k_H)^i(u) \in V(R \cap H)$. Because $(\varphi k_H)^i(u) \in I(u, v)$, we conclude that $(\varphi k_H)^i(u) = k_H(u)$. $\square$

Theorem 16.21 *Let R be an irredundant weak retract of a Cartesian product $G = \square_{i=1}^{k} G_i$, and m the median function on G. Then $m(x, y; z) \in V(R)$ for all $x, y, z \in V(R)$.*

Proof Let $u, v, w \in V(R)$ and H be the smallest gated subgraph of R containing u and w. Because of Lemmas 16.12 and 16.13, $B = \prod_{i=1}^{k} p_i H$ is gated in G, $H = B \cap R$, and $k_H(v) = k_B(v)$. To prove $\mathrm{m}(u, v; w) = k_B(v)$, it suffices to show that B is the smallest gated subgraph of G containing u and w. Let C be this subgraph. Then $C \cap R$ is a gated subgraph of R by Lemma 16.20, whence $H \subseteq C \subseteq B$. Therefore $p_i(H) \subseteq p_i(C) \subseteq p_i(B) = p_i(H)$ for $i = 1, \ldots, k$. Because $C = \prod_{i=1}^{k} p_i(C)$, we conclude $C = B$. □

This is not quite enough for an application of Proposition 16.15 in the proof of Theorem 16.17, because we need median closure without the additional condition that $p_i(R) = G_i$ for every $i = 1, 2, \ldots, k$. But this is easy to achieve; we simply consider $\prod_{i=1}^{k} p_i(R)$. Clearly, R is also a weak retract with respect to this product, and thus median function-closed with respect to the median function of $\prod_{i=1}^{k} p_i(R)$. We therefore have the following theorem:

Theorem 16.22 (Feder-Tardif's Fixed Box Theorem I) *Let R be a weak retract of a Cartesian product G. Then there exists a box $S \subseteq R$ that is a weak retract of R and has the property that $\varphi(S) = S$ for each automorphism φ of R.*

Actually, Feder's theorem is more general, because it holds for any subgraph R of G that is the image of a weakly nonexpansive map of G.

We know from the proof of Theorem 12.25 that the images of iterates of endomorphisms of finite graphs stabilize after a finite number of steps. Repeating this argument, we obtain:

Theorem 16.23 (Feder-Tardif's Fixed Box Theorem II) *Let G be a Cartesian product and $\varphi : G \to G$ be a nonexpansive map. Then there exists a box $S \subseteq G$ that is a weak retract of G and satisfies $\varphi(S) = S$.*

The subgraphs of a graph G that are images of weakly nonexpansive maps are exactly those subgraphs that have no "holes." This concept was first introduced under the name "gaps" by Nowakowski and Rival (1983).

Larose, Laviolette, and Tardif (1998) studied homomorphisms of Cartesian product graphs and proved numerous interesting results. Among other things, they showed the following: Let $G = \square_{i=1}^{k} G_i$ be a core (as defined in Exercise 3.9). Then there exists a homomorphism $G \,\square\, G \to G$ if and only if there exists a homomorphism $G_i \,\square\, G_i \to G_i$ for any i. Cores in Cartesian products were further studied in the paper of Che, Collins, and Tardif (2008), which we mentioned at the end of Section 6.4.

Nowakowski and Rival (1988) studied the question of for which Cartesian products all retracts are boxes. (See Exercise 6.10 for a related result.) Their work was later extended by Che, and Collins (2007). For instance, they proved that if G is a strongly $(2k+1)$-angulated graph (for the definition see p. 73) and H is a connected graph with odd girth at least $2k+1$, then any retract of $G \,\square\, H$ is a box. More precisely, any retract is of the form $S \,\square\, T$, where S is a retract of G and T a connected subgraph of H or $S = K_1$ and T is a retract of H. (Nowakowski and Rival (1988) proved the theorem for $k = 1$.)

16.4 Fixed Points of Several Nonexpansive Mappings

This section is concerned with results of Feder (2006) about finding a vertex in a Cartesian product that is the fixed point of several given nonexpansive mappings. Feder points out a dichotomy: either there exists a polynomial algorithm for solving the problem, or the problem is NP-complete. Because Part IV, which is as such reserved for algorithms, does not treat the complexity of finding fixed points of contractions, and because most prerequisites for the description of the results can be found in the present chapter, we outline the main results here.

The first result says that the problem is solvable in polynomial time for a Cartesian product $G_1 \square \cdots \square G_k$ if all factors satisfy the farthest point property. Otherwise there are Cartesian products of simple counterexamples to the farthest point property, for which one can construct two retractive mappings such that the problem of finding a common fixed point is NP-complete.

One says a graph satisfies the *farthest point property* if for any three vertices x, y, z there is a unique vertex t in $I(x,y) \cap I(x,z)$ that maximizes $d(x,t)$ over all t in $I(x,y) \cap I(x,z)$. Notice that cliques, cycles, and median graphs satisfy the farthest point property, and that Cartesian products and retracts of graphs satisfying the farthest point property also satisfy the farthest point property. The smallest graph not satisfying the farthest point property is $K_{2,3}$.

We now state the first result in detail. It is an extension of a theorem of Feder (1995) from one contraction to a finite number of contractions. Let $f_i : V(G) \to V(G)$, $1 \le i \le \ell$, be nonexpansive mappings on a Cartesian product $G = G_1 \square \cdots \square G_k$. Assume further that every f_i is given by a black box that can be queried in polynomial time. Then the following theorem holds:

Theorem 16.24 *Suppose each G_i satisfies the farthest point property. Then there is a polynomial algorithm that finds sets $S_{ij} \subseteq V(G_i \square G_j)$ for all $1 \le i < j \le k$ such that, given a partial assignment of values $a_i \in V(G_i)$ for $i \in S \subseteq \{1,2,\ldots,k\}$ with $|S| \ge 2$, there exists a common fixed point x of the nonexpansive mappings f_i such that $x_i = a_i$ for all $i \in S$ if and only if $a_i a_j \in S_{ij}$ for all $1 \le i < j \le k$ with $i,j \in S$. The partial assignment of values a_i can thus be extended to a common fixed point x with $f_i(x) = x$ for all $1 \le i \le \ell$ in polynomial time by considering the sets S_j.*

This theorem does not completely characterize the products for which a common fixed point can be computed in polynomial time. For such a characterization, another concept is needed. A family of graphs $\mathcal{G}$ has a *majority function* if each graph G_i in $\mathcal{G}$ has a function g_i such that $g_i(x_i,x_i,y_i) = g_i(x_i,y_i,x_i) = g_i(y_i,x_i,x_i)$ for all x_i, y_i in $V(G_i)$, and for every pair of graphs G_i, G_j in $\mathcal{G}$, if f is a nonexpansive mapping on $G_i \square G_j$, and x_ix_j, y_iy_j, z_iz_j are fixed points of f, then $g_i(x_i,y_i,z_i)g_j(x_j,y_j,z_j)$ is a fixed point of f as well. Examples are families of graphs that satisfy the farthest point property.

The following characterization holds:

Theorem 16.25 *The statement of Theorem 16.24 holds for a family $\mathcal{G}$ of graphs G_i if and only if $\mathcal{G}$ has a majority function.*

Part IV

Algorithms

Introduction to Part IV

THIS PART focuses on algorithms. We describe polynomial (sometimes linear) algorithms that find the prime factors of connected graphs over the commutative standard products, under suitable conditions such as connectedness. We also derive recognition and embedding algorithms for many classes of subgraphs of graph products, such as partial cubes, median graphs, partial Hamming graphs, and quasi-median graphs.

Two chapters treat the rudiments of algorithms. The first chapter of this part, Chapter 17, is a short introduction to graph algorithms and relevant data structures. It also compares the complexity of simple operations like vertex deletion and insertion for various data structures. The importance of this will become clear in the following chapters.

Chapter 20, on graph arboricity, is the other chapter on basic algorithms. Many of our graph classes are sparse, that is, they have few edges and small arboricity, and this helps bound their recognition complexity. The chapter also describes efficient algorithms that compute all triangles or all squares in a graph, an important task in recognizing median graphs.

Chapters 18, 19, 21, and 22 deal with recognizing hypercubes, Hamming graphs and their isometric subgraphs. These classes include partial cubes, median graphs, partial Hamming graphs, and quasi-median graphs. Chapter 22 also considers the role of quasi-median graphs in the solution of the dynamic location problem. Chapter 19 treats chemical graphs and the computation of the Wiener index, an important graph invariant, and concentrates almost entirely on applications.

The third group consists of Chapters 23 and 24, which provide polynomial factorization algorithms of graphs over the commutative standard products. We also explain why the lexicographic product does not fall into this category. Three algorithms pertain to Cartesian product decomposition. Two of them, one of complexity $O(mn)$ and one of complexity $O(m \log n)$, are treated in detail. The third is treated only cursorily; its complexity is linear but it requires a much more elaborate data structure than the other two algorithms. Interestingly, the simple algorithm of complexity $O(mn)$ uses a relation σ on the edge set of a graph, and our treatment of it also holds for infinite graphs. As a dividend, we get another proof for the uniqueness of the Cartesian prime factorization of connected graphs, one that points toward a theory of factoring infinite graphs.

A large part of Chapter 24 is concerned with computing Cartesian skeletons for the direct and strong products. In the first edition of this book, the Cartesian skeleton was defined algorithmically. Here the definition is nonalgorithmic, which makes it easier to handle. The definition, due to Hammack, will also be of theoretical importance in Chapter 31.

Another new feature in this book is the use of bitvectors for an $O(n^2)$ algorithm for the recognition of partial cubes. The method, due to Eppstein, makes strong use of the RAM model of computation.

Finally, special mention is appropriate for the close relationship between median graphs and triangle-free graphs established in Chapter 21. It leads to an interesting algorithm of complexity $O\big((m \log n)^{2\omega/(\omega+1)}\big) = O\big((m \log n)^{1.41}\big)$ for the recognition of median graphs. (Here ω denotes the coefficient of matrix multiplication.)

Concerning chapter dependencies, Part IV assumes a thorough knowledge of Part I. Also, the relation Θ is used extensively here, so Chapter 11 (in Part III) is an essential ingredient. Within Part IV, Chapters 17 through 20 form a core that is indispensable for the remaining chapters of Part IV. (Chapter 19, on chemical graph theory, could be considered optional except that the material is quite interesting!)

The remainder of Part IV makes use of relevant earlier parts of the book. For example, Chapter 21 (Recognizing Median Graphs) clearly assumes knowledge of Chapter 12 (Median Graphs). Likewise, the material on computing prime factors requires the corresponding prime factorization results from Part II. (The one exception is that Part IV uses Θ to develop a wholly different approach to prime factorization of the Cartesian product.)

Chapter 17

Graph Representation and Algorithms

A frequent question in graph theory is whether a given graph is a tree, bipartite, connected, a product, a median graph, a hypercube, or member of some other class of graphs. Our aim in Part IV is to provide efficient algorithms that answer these questions for the graphs considered in this book. On the way to answering these questions, we have to solve a number of lesser problems: We must represent graphs by appropriate data structures and exhibit efficient algorithms for basic operations, such as vertex insertion or deletion.

The chapter begins with a quick review of time and space complexity of algorithms and the model of computation. Then we discuss adjacency list representations of graphs. This is followed by an efficient method for checking whether a given map is an isomorphism, a short description of breadth-first search, and the complexity of union operations for disjoint sets. The chapter ends with matrix representations of graphs.

17.1 Time and Space Complexity

Recall that it is customary to express the running time of an algorithm in terms of the size of its input. We say that an algorithm runs in time $O(f(m))$ if for some constant $c > 0$ there exists an implementation of the algorithm that terminates after at most $cf(m)$ steps for all inputs of size m. The smallest function f such that the algorithm runs in time $O(f(m))$ is called the *(time) complexity* of the algorithm. The (time) complexity of a problem is the minimum (time) complexity of all algorithms solving the problem—if this minimum exists. One usually obtains reasonable upper bounds for the complexity of a problem by exhibiting a specific algorithm; lower bounds are harder to find. Clearly, the statements that an algorithm has complexity $O(\log_k n)$, $O(\log_2 n)$, or $O(\log n)$ are equivalent. We will therefore simply write them as $O(\log n)$, likewise $O(m \log n)$ instead of $O(m \log_2 n)$.

It makes sense to list the size of the input in O-notation too. In our case the input size will mostly be $O(m)$, $O(m+n)$, or $O(n^2)$, where m denotes the number of edges and n the number of vertices of the graph to be investigated.

An algorithm requires storage space, and we define its *space complexity* just as we defined its time complexity. Often there is a trade-off between time and space complexity.

We say that an algorithm is *linear* if its time and space complexity is $O(m+n)$. We will see later that hypercubes and Hamming graphs can be recognized by linear algorithms, and that the prime factorization of connected graphs over the Cartesian product can be found in linear time. This level of efficiency is a rather rare phenomenon.

Many of our algorithms have time complexity $O(mn)$ and are linear in space; some are polynomial, that is, their time (and space) complexity is $O(n^k)$ for some constant k. Efficient algorithms that find all squares and triangles in a graph are treated in Chapter 20. Depending on the type of graph, these complexities are between $O(m \log n)$ and $O(m\sqrt{n})$. Chapter 21 deals with recognizing median graphs. Planar median graphs can be recognized in linear time, but in general the complexity is only slightly better than $O(m\sqrt{n})$.

Our model of computation is the unit-cost random access machine (RAM). This allows random access, the use of arrays, unit-cost arithmetic, and bit-vector operations of arbitrarily large integers; see Aho, Hopcroft, and Ullman (1974).

We remind the reader that the obvious ways to perform operations on graphs are usually slower than more sophisticated ones. This makes little difference for small graphs, but for large graphs it may mean the difference between being able to run the algorithm in reasonable time or not at all.

17.2 Adjacency List

To describe a graph, it suffices to list the neighbors of every vertex. For example, consider the graph in Figure 17.1. Each line on the corresponding table is a list A_i of the vertices adjacent to v_i. We call this table the *adjacency list representation* of the graph.

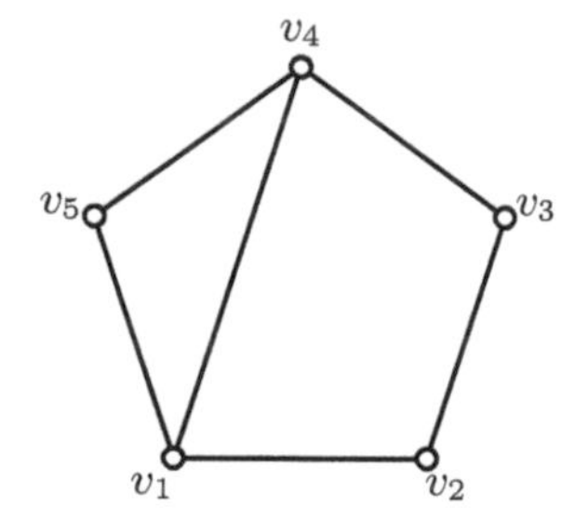

List	Contents
A_1	v_2, v_4, v_5
A_2	v_1, v_3
A_3	v_2, v_4
A_4	v_5, v_3, v_1
A_5	v_1, v_4

FIGURE 17.1 A graph and its adjacency list representation.

Each list A_i represents the vertex v_i. It is understood that A_i may be implemented so as to contain pertinent information (such as a label or color, etc.) about v_i. A similar remark applies to each item v_j in A_i, which can be interpreted as representing the edge v_iv_j.

If a graph has n vertices and m edges, then its adjacency list representation requires n lists, even if some are empty (as in the case of an isolated vertex), and every edge v_iv_j gives rise to two entries, one in A_i and the other in A_j. Thus the space needed for the representation is $O(m + n)$, or simply $O(m)$ if the graph is connected.

The simplest implementation is to store the above data sequentially as a single list and to provide an address for the head of every sublist A_i. We can store these addresses in an array of length n, so we have direct access to the entries in linear time. Note that changes in the list may require that it be rewritten completely; this may take $O(m + n)$ time.

Algorithms generally perform certain basic operations on graphs. Some of the most common are as follows:

- Check whether a pair v_iv_j is an edge.
- Mark (i.e., label) all neighbors of a vertex v_i.
- Mark each edge.
- Insert an edge v_iv_j.

- Delete an edge v_iv_j.
- Delete all edges incident with a vertex v_i.

With the above data structure of sequentially stored adjacency lists, the first two actions take $O(d(v_i))$ time, the others $O(m)$.

We wish to reduce the time needed for the basic actions as much as possible. For example, if the entries in the adjacency list are sorted, the complexity of checking whether v_iv_j is an edge reduces to $O(\log d(v_i))$.

Other time reductions require a refinement of our data structure, and with this goal in mind we now define a multiplylinked data structure called the *extended adjacency list representation* of a graph. This is illustrated in Figure 17.2. It is similar to the sequential adjacency list representation defined above, except that each entry v_j of the list A_i is replaced by an array of length 5 containing v_i plus some additional information. For example, the item v_4 of A_1 in Figure 17.1 is replaced with the array

$$p = (v_1, v_4,\ \&a,\ \&q,\ \&y)\,.$$

The entries of this array have the following meanings:

v_1	Vertex to which A_1 corresponds
v_4	Neighbor of v_1 that the array p represents
$\&a$	Pointer to the preceding array in list A_1 (or $\emptyset$ if p is the first array in A_1)
$\&q$	Pointer to the following array in list A_1 (or Λ if p is the last array in A_1)
$\&y$	Pointer to the array representing the neighbor v_1 of v_4

In addition to the $2m$ arrays $a, p, q, \ldots$, we must also store the addresses of the heads of the lists $A_1, \ldots, A_n$. This is best done with an array of length n. Thus the total space needed to represent a graph is still $O(m + n)$.

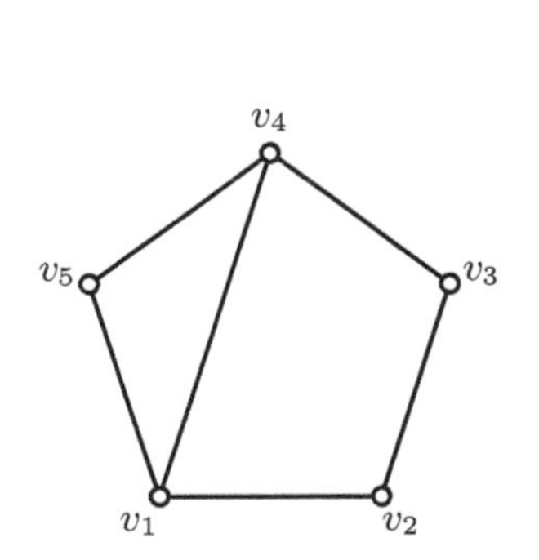

List	Name of Array	Array
A_1	a	$(v_1, v_2,\ \emptyset, \&p,\ \&b)$
	p	$(v_1, v_4, \&a, \&q,\ \&y)$
	q	$(v_1, v_5, \&p,\ \Lambda,\ \&e)$
A_2	b	$(v_2, v_1,\ \emptyset, \&r,\ \&a)$
	r	$(v_2, v_3, \&b,\ \Lambda,\ \&c)$
A_3	c	$(v_3, v_2,\ \emptyset, \&s,\ \&r)$
	s	$(v_3, v_4, \&c,\ \Lambda,\ \&x)$
A_4	d	$(v_4, v_5,\ \emptyset, \&x,\ \&z)$
	x	$(v_4, v_3, \&d, \&y,\ \&s)$
	y	$(v_4, v_1, \&x,\ \Lambda,\ \&r)$
A_5	e	$(v_5, v_1,\ \emptyset, \&z, \&x)$
	z	$(v_5, v_4, \&e,\ \Lambda,\ \&d)$

FIGURE 17.2 A graph and its extended adjacency list representation.

The complexities of basic operations for graphs described with this data structure are collected in Table 17.1. The first two columns of the table contain the complexities of these operations with respect to sequential adjacency list and the extended adjacency list. (The last two columns are for the adjacency matrix and the referenced adjacency matrix representations, which will be discussed in Section 17.4. For now we remark that the space complexity for these last two data structures is $O(n^2)$, which can be significantly larger than $O(m)$. For us the main advantage of the referenced adjacency matrix is that it can be initialized in $O(m)$ time.)

TABLE 17.1 Complexity of Basic Actions versus Data Structures.

Type of Action	Complexity			
	Sequential Adjacency List	Extended Adjacency List	Adjacency Matrix	Referenced Adjacency Matrix
– Check if v_iv_j is an edge	$O(d(v_i))$	$O(d(v_i))$	$O(1)$	$O(1)$
– Mark all neighbors v_i	$O(d(v_i))$	$O(d(v_i))$	$O(n)$	$O(d(v_i))$
– Mark all edges	$O(m)$	$O(m)$	$O(n^2)$	$O(m)$
– Insert an edge v_iv_j	$O(m)$	$O(1)$	$O(1)$	$O(1)$
– Delete an edge v_iv_j	$O(m)$	$O(d(v_i))$	$O(1)$	$O(1)$
– Delete all edges incident with v_i	$O(m)$	$O(d(v_i))$	$O(n)$	$O(d(v_i))$

Notice that the complexities of checking whether v_iv_j is an edge, marking all neighbors of a given vertex, and marking all edges remain unchanged when switching from sequential adjacency lists to extended lists. But with the extended adjacency lists an edge can be inserted in constant time, a single edge v_iv_j can be deleted in $O(d(v_i))$ time, and, perhaps surprisingly, all edges incident with a vertex v_i can be deleted in $O(d(v_i))$ time. Thus, in comparison with sequentially stored adjacency lists, the complexities of the first three basic actions remain unchanged, but the other three improve considerably.

To justify these assertions, first consider edge insertion. Suppose that we wish to add the edge v_2v_5 to the graph in Figure 17.2. We simply allocate and initialize two new arrays, say t and w, as

$$\begin{aligned} t &= (v_2, v_5, \emptyset, -, \&w), \\ w &= (v_5, v_2, \emptyset, -, \&t). \end{aligned}$$

Now append t and w to the beginning of lists A_2 and A_5: Access the first arrays b and e in A_2 and A_5, and update t and w as

$$\begin{aligned} t &= (v_2, v_5, \emptyset, \&b, \&w), \\ w &= (v_5, v_2, \emptyset, \&e, \&t). \end{aligned}$$

Then set the third entry of b equal to $\&t$, and the third entry of e to $\&w$. The edge v_2v_5 has now been inserted. Clearly, this can be done in constant time.

To delete an edge, say v_4v_3, we go to A_4 and scan its items sequentially until we find the item x whose second entry is v_3. This is the time-consuming part, its complexity being $O(d(v_i))$, because x could be at the end of the list. The rest is similar to the above and can be accomplished in constant time. Note the last entry $\&s$ of x allows for the quick deletion of s from A_3.

To delete all edges incident with v_i, we simply scan the $d(v_i)$ items of A_i, deleting each from A_i one by one. (In each iteration, we also use the last pointer to reference and delete the other occurrence of that edge.) This operation has complexity $O(d(v_i))$.

It is clear that this extended adjacency list can be constructed from a given sequentially stored adjacency list in $O(m+n)$ time. For most purposes it may suffice to do without the first entry in the edge array for every edge, and for many the last entry will not be needed. We then have doubly linked adjacency lists. If we also delete the third entry, we arrive at singly linked adjacency lists.

We will use extended adjacency lists for all complexity considerations in this book. (However, for actual implementation, a simpler data structure may be more appropriate.)

Sorting the lists A_i (by the second entry v_j) will improve the complexity of checking whether a pair v_iv_j is an edge. However, this makes the insertion of an edge more costly and will thus not be beneficial in general. Nonetheless, it allows a standard representation of a graph for a given ordering of the vertices. We will take advantage of this shortly.

If we sort these lists individually, it will cost us $O\big(d(v_i)\log d(v_i))\big)$ time for every list and thus $O(m\log n)$ time altogether. Luckily, there is a better way. Because the lists $A_1, A_2, \ldots, A_n$ are already arranged sequentially, their items are already sorted by the first entry. We can just transpose the first and second entries and move the corresponding arrays to the appropriate list.

Algorithm 17.1 Sorting adjacency lists of a graph

Input: The extended adjacency list of a graph G.
Output: A sorted extended adjacency list of G.

1: Allocate a new set $B_1, \ldots, B_n$ of empty adjacency lists.
2: **for all** adjacency lists A_i of G from A_n down to A_1 **do**
3:    **while** A_i is nonempty **do**
4:       Choose the first element of A_i, say, $(v_i, v_j, \&x, \&y, \&z)$.
5:       Transpose its first two entries: $(v_j, v_i, \&x, \&y, \&z)$.
6:       Insert it as the first element of B_j (updating $\&x$ and $\&y$ accordingly).
7:       Delete (unlink) it from A_i.
8:    **end while**
9: **end for**

Clearly, Steps 4 through 7 of Algorithm 17.1 take constant time. As the total number of arrays is $2m$, and because we create n lists B_i, the overall time complexity is $O(m+n)$. (See Golumbic (1980).) Note that this algorithm does not really need linked lists. In the case of sequentially stored lists, every A_i has length $d(v_i)$, and this is the space we have to allot to the B_i.

Algorithm 17.1 immediately yields a linear algorithm that checks whether a given bijection between the vertex sets $V(G)$ and $V(H)$ of two graphs G and H is an isomorphism.

Theorem 17.1 *Given graphs G and H and a bijection $\varphi : V(G) \to V(H)$, one can check in linear time and space whether φ is an isomorphism.*

Proof The idea is to sort the adjacency list of H according to the order imposed by the vertices of G under the mapping φ and to compare this list with the sorted adjacency lists of G.

Let $C_1, \ldots, C_n$ be the adjacency lists of H and C_i' be the list corresponding to the vertex $\varphi(v_i)$, where $v_i \in G$. We now use a variant of Algorithm 17.1. As before, we begin with n empty lists B_i, but in the next step we let the lists C_i' play the role of the A_i.

Now we rewrite the sorted list for G and the newly sorted list for H as sequentially stored lists L_G and L_H. If φ is an isomorphism, then these lists must be identical in the sense that L_H is obtained from L_G by replacing every v_i by $\varphi(v_i)$. If this is not the case, then φ cannot be an isomorphism.

Clearly, all operations can be performed in linear time and space. □

17.3 Breadth-First Search

This section is concerned with distances, connectedness, bipartiteness, and other standard graph properties. The basis of our considerations is the well-known *breadth-first search* (*BFS*) algorithm, which visits all vertices of a connected graph in order of their distances from a fixed vertex v_0.

The following algorithm is the prototype for breadth-first search. It scans the vertices of G in order of their distances from a fixed vertex v_0 and labels each by its distance from v_0.

Algorithm 17.2 Breadth-first search

Input: The adjacency list of a connected graph G and a vertex v_0.
Output: A labeling of the vertices v of G by integers $\ell(v) = d(v_0, v)$.

1: Start with a list L containing only v_0 and set $\ell(v_0) = 0$.
2: **while** L is nonempty **do**
3: Remove the first vertex w from L.
4: **for all** v in the adjacency list A_w of w for which $\ell(v)$ is undefined **do**
5: Set $\ell(v) = \ell(w) + 1$.
6: Append v to the end of L.
7: **end for**
8: **end while**

Notice that this algorithm labels every vertex of G. Indeed, the algorithm is structured so that any labeled vertex is appended to L; and when it is eventually removed from L, all its neighbors get labels. As L is eventually empty, this means all neighbors of a labeled vertex are labeled. Because G is connected, all its vertices get labeled.

We say that the order in which Algorithm 17.2 assigns labels to the vertices of G is a *BFS ordering* of $V(G)$ with respect to v_0. The following theorem asserts that a BFS ordering lists all vertices at distance 0 from v_0, followed by all vertices at distance 1 from v_0, then all vertices at distance 2, and so on.

Theorem 17.2 *Algorithm 17.2 labels each vertex v of G with the distance of v_0 from v. The labels in the resulting BFS ordering of $V(G)$ form a monotone nondecreasing sequence. The complexity of the algorithm is $O(m)$, where $m = |E(G)|$.*

Proof We first assert that as the algorithm runs, L is always a monotone nondecreasing sequence, and any two of its entries have labels that differ by at most 1. This is clearly true in the beginning, when L contains only v_0. Thereafter, elements are removed from the front of L, and others are appended to its end. As any appended element has a label that exceeds that of the most recently removed element by exactly 1, the assertion follows.

Whenever a vertex is labeled, it is immediately appended to the end of L. Because L is nonempty (until the algorithm terminates), the previous paragraph implies that whenever a vertex is labeled, its label is not less than that of the previously labeled vertex, and exceeds it by at most 1. Thus the labels in a BFS ordering are monotone nondecreasing.

We now claim that for each integer $k \geq 0$, any vertex v with $d(v_0, v) = k$ receives the label $\ell(v) = k$. This is certainly true for $k = 0$, for in this case $v = v_0$ is the only vertex labeled with 0. Assume that it is also true for all $k \leq K$ for some fixed K. Now consider how a vertex v gets the label $K + 1$: A vertex w with label $\ell(w) = K$ is removed from the head of L, an unlabeled neighbor $v \in N(w)$ is found, and the value $\ell(v) = K+1$ is assigned. By assumption, $\ell(w) = K = d(v_0, w)$. We must have $d(v_0, v) > K$, for otherwise v would have been labeled previously (by a value of at most K). But then because v is a neighbor of w, we have $d(v_0, v) = K + 1 = \ell(v)$. This proves the claim.

Notice that the algorithm examines every edge of G at most twice. This yields the assertion about complexity. □

It is important to note that Step 5 of the BFS algorithm can be replaced by some other processing of the vertex v. If this is done, we say the algorithm processes the vertices of G in BFS order. If Step 5 has complexity $O(\kappa)$, then the amended algorithm has complexity $O(m + n\kappa)$.

In particular, we can (in linear time) arrange for the algorithm to place the vertices of G into sets $L_0, L_1, \ldots, L_k$, where each L_i is the set of vertices at distance i from v_0. We call these sets the *distance levels* of G with respect to v_0.

Suppose that uv is an edge of G, where $u \in L_i$. We distinguish the cases in which $v \in L_{i-1}$, $v \in L_i$ or $v \in L_{i+1}$. In the first case, uv is called a *down-edge* (with respect to u); in the second, a *cross-edge*; and in the third, an *up-edge*. We can split every adjacency list into three sublists (possibly empty) containing the down-, cross-, and up-edges, respectively. This can be done in linear time and space. Note that the list of cross-edges may be empty for every vertex, that the list of down-edges is always nonempty with the sole exception of the list for v_0, and that all vertices in the highest level have an empty list of up-edges.

We define the *up-* and *down-degree* of vertices with respect to a given BFS-order in the obvious way. These degrees can be determined in linear time.

Slight adaptations of the BFS algorithm can determine all connected components of a graph in linear time and space and find the distances $d(v_i, v_j)$ between all pairs of vertices of G. Customarily, one stores these values in a matrix D with entries $d_{ij} = d(v_i, v_j)$, the so-called *distance matrix* of G.

Corollary 17.3 *Let G be a graph with n vertices and m edges. Then*

(i) *The connected components of G and their adjacency lists can be determined in linear time and space.*

(ii) *The distance matrix of G can be determined in $O(mn)$ time and $O(n^2)$ space.*

Proof To prove (i), we first observe that any component C of G can be determined in $O(|E(C)|)$ time by applying Algorithm 17.2 to one of its vertices. Then any remaining component can be found by applying the algorithm to an unlabeled vertex. As long as unlabeled vertices remain, we can find additional components. Thus to prove (i), it suffices to show that the effort of finding an unlabeled vertex is not larger than $O(|E(C)|)$, where C is the component that has been determined last.

This can be achieved with two auxiliary arrays P, Q and a running index k. The index k is initially 0, so that it references the first entry of Q. The array Q is initialized so that its entries are the vertices $v_0, v_1, \ldots, v_{n-1}$ of G, in order of their indexing. The array P is initialized so that each entry is a pointer to the corresponding entry of Q. From here on, we will maintain P so that that its ith pointer always points to the entry of Q that contains

v_i. We now run the BFS algorithm with respect to v_0. Each time the algorithm labels a vertex v_i, we interchange the kth entry v_{i_k} of Q with the entry of Q pointed to by the ith entry of P. Then we interchange the i_kth and ith entries of P and increment k. In this way, k always indexes an unlabeled vertex in Q, so once the algorithm labels the vertices of a component, it can immediately set to work on the next component.

If we also note the positions of the starting vertices for a new component, we can easily write the new adjacency lists within the claimed time and space complexity.

For (ii) we note that one run of Algorithm 17.2 determines the distances of all vertices from a given one in $O(m)$ time. Repeating this step for all n vertices results in an algorithm of time complexity $O(mn)$. Although every run requires only linear space, the overall space requirement is determined by the distance matrix, which needs $O(n^2)$ space. □

Proposition 17.4 *Bipartite graphs can be recognized in linear time and space.*

Proof By Corollary 17.3 (i) we can restrict attention to connected graphs. Use the BFS algorithm to determine the set R of all cross-edges in linear time. (Begin with $R = \emptyset$. Then, in Step 4, whenever w has a neighbor v with $\ell(w) = \ell(v)$, append wv to R.)

Observe that G is bipartite if and only if $R = \emptyset$. Indeed, if $R = \emptyset$, the distance levels $L_0 \cup L_2 \cup L_4 \cup \ldots$ and $L_1 \cup L_3 \cup L_5 \cup \ldots$ form a bipartition of G. On the other hand, if $wv \in R$, then $d(v_0, w) = d(v_0, v)$, and G has an odd closed walk containing wv. □

We have already seen that subgraphs of hypercubes have few edges. In fact, the Density Lemma 3.2 implies that the average degree of the vertices of such graphs is at most $\log_2 n$. Unfortunately, this does not bound the maximum degree. However, for partial cubes we have the following result.

Proposition 17.5 *Let G be a partial cube on n vertices, $L_0, L_1, \ldots, L_k$ the BFS-levels of G with respect to a vertex v_0, and $v \in L_i$. Then the down-degree of v is bounded by i.*

Proof Every down-neighbor of v has distance $i - 1$ from v_0 and is thus in the interval $I(v, v_0)$. If G is a hypercube, we infer from Proposition 3.1 (iii) that this interval is a hypercube of dimension i and thus regular of degree i. If G is a proper subgraph of a hypercube H, then, by isometry, $I_G(v, v_0)$ is a subgraph of $I_H(v, v_0)$ and the degrees in $I_G(v, v_0)$ are still bounded by i. □

Some of our algorithms will require a spanning tree of a graph, and we now present an algorithm that accomplishes this. In fact, given a vertex v_0 of a connected graph G, we are interested in a spanning tree T for which $d_T(v_0, v) = d_G(v_0, v)$ for every vertex $v \in V(G)$. Such a tree is called a *BFS-tree.*

Proposition 17.6 *A BFS-tree for a connected graph can be built in linear time and space.*

Proof Run the BFS algorithm and choose one down-edge at every vertex different from v_0. The resulting graph T consists of the n vertices of G, and has $n - 1$ edges. As every vertex $v \in L_i$ is connected by a path of length i with v_0, the new graph T is connected, and $d_T(v_0, v) = d_G(v_0, v)$ for every $v \in V(G)$. Also, T is a tree by Proposition 1.8, because it is connected and $|E(T)| = |V(T)| - 1$. □

In the sequel we will have to merge lists into larger ones in several steps. To do so, we also keep track of the size of the lists and always merge the smaller into the larger. This way we only have to change the back pointers of the smaller lists. Clearly, every vertex can have its back pointer changed at most $n \log_2 n$ times, where $n = |V(G)|$, because every time the size of its list at least doubles. For later reference we state this observation as a proposition.

Proposition 17.7 *Let G be a graph on n vertices. Then any sequence of merging operations of the components of G takes altogether at most $O(n \log n)$ time.*

17.4 Adjacency Matrix

The *adjacency matrix* $\mathrm{A}(G) = [a_{ij}]$ is another natural representation of a graph G with vertices $\{v_1, v_2, \ldots, v_n\}$. It is the $n \times n$ matrix for which $a_{ij} = 1$ if $v_i v_j \in E(G)$ and $a_{ij} = 0$ otherwise.

If $G \in \Gamma$, then $\mathrm{A}(G)$ is symmetric and all diagonal entries are 0. For graphs in Γ_0, every loop is represented by a 1 in the main diagonal.

Clearly, $A(G)$ depends on the indexing of the vertices. If there is a permutation π that relabels each v_i as $v_{\pi(i)}$, then adjacency matrix of G with this new indexing is $P^{-1}\mathrm{A}(G)P$, where P is the 0-1 permutation matrix with $p_{i\pi(i)} = 1$. Moreover, a map $\varphi : v_i \mapsto v_{\pi(i)}$ is an automorphism if and only if $\mathrm{A}(G) = P^{-1}\mathrm{A}(G)P$.

If the adjacency matrix is stored as an array, then we can check in constant time whether $v_i v_j$ is an edge, and we can insert and delete single edges in constant time as well. Thus we have two considerable improvements compared to extended adjacency lists at the expense of $O(n^2)$ space instead of $O(m)$. This is negligible for small or dense graphs, but enormous for large sparse ones.

Another drawback is that marking each vertex adjacent to a vertex v_i takes $O(n)$ time, as does marking or deleting all edges adjacent to a single vertex.

We can overcome this by combining the adjacency matrix with the extended adjacency list structure. Let $A(G)$ be an $n \times n$ array of pointers that are initially null. Scan the extended adjacency list, and for each $x = (v_i, v_j, \ldots)$, set a_{ij} to point to x. We call this structure the *referenced adjacency matrix* for G.

The complexities for basic operations are at least as good as for the extended adjacency list, but the addition of $A(G)$ provides some benefits. We can now check in constant time whether $v_i v_j$ is an edge by confirming that a_{ij} is not null. Also we can delete an edge $v_i v_j$ in constant time by accessing a_{ij} and a_{ji}, unlinking the arrays that they point to, and setting a_{ij} and a_{ji} to null. See the last column of Table 17.1.

One disadvantage of the above scheme is that the initialization of $A(G)$ to null pointers takes $O(n^2)$ time, even for sparse graphs where $m = O(n)$ or $O(n \log n)$. The initialization is necessary to eliminate any "stray" pointers from $A(G)$ that do not represent valid edges. But we can bypass the initialization of $A(G)$ by introducing a stack array $R(G)$ that allows us to determine whether entries of $A(G)$ correspond to valid edges. We scan the extended adjacency list of G. If entry $(v_i, v_j, \ldots)$ is the kth array encountered, we set the kth entry of $R(G)$ to (i, j) and set a (second) pointer from a_{ij} to the kth entry of $R(G)$. Then a_{ij} references a valid edge if and only if it points to an entry of $R(G)$ containing (i, j).

Of course, $R(G)$ must be updated when basic operations are performed on the overall data structure. When an edge $v_i v_j$ is deleted, we simply set to null the entries of $R(G)$ pointed to by a_{ij} and a_{ji}. If an edge $v_i v_j$ is added, we add corresponding entries (i, j) and (j, i) to the top of the stack $R(G)$, and arrange a_{ij} and a_{ji} to point to them. This extension of the referenced adjacency matrix does not change the complexities in Table 17.1, but it allows us to initialize it in $O(m)$ time. See Cormen, Leierson, and Rivest (1990, Exercise 12.1-4).

Notice that the (extended) adjacency list itself can play the role of the reference vector. If we only wish to initialize the ith line of $A(G)$, then we can use A_i as a reference vector.

Another feature of the adjacency matrix is that many graph properties can be read off it directly or can be determined by elementary matrix operations. For instance, the degree of v_i is the sum of the elements of the ith row (or column), although the complexity of determining it is $O(n)$ compared to $O(d(v_i))$ in the case of adjacency lists.

More importantly, powers of the adjacency matrix can count walks and find distances.

Theorem 17.8 *If $A(G)$ is the adjacency matrix of a graph G with vertices $v_1, v_2, \ldots, v_n$, then*

(i) *The i, j-entry of $A(G)^k$ is the number of walks of length k from v_i to v_j, and*

(ii) *The distance $d(v_i, v_j)$ is the least k for which the i, j-entry of $A(G)^k$ is nonzero.*

The proof is by induction and is omitted. We emphasize that (i) implies that the ith element in the main diagonal of $A(G)^2$ is the degree of v_i. For $A(G)^3$ this element is twice the number of triangles, because every triangle containing v_i gives rise to two walks—in opposite directions—of length 3 from v_i to v_i.

The assertion (ii) is true for $k = 0$ because $A(G)^0$ is the identity matrix. Also, if v_i and v_j are in different connected components, no finite power of $A(G)$ has nonzero i, j-entry.

We can therefore use matrix multiplication

(a) To determine whether a graph is triangle-free,

(b) To determine all distances $d(v_i, v_j)$, and

(c) To determine the connected components of G.

We have already seen how these actions can be executed with the aid of adjacency lists. Which method is faster depends very much on the graph. Let us consider the complexities of matrix methods.

We begin with the observation that Boolean matrix multiplication suffices to solve (a) through (c) because in these instances we are only interested in the existence of a v_i, v_j-walk and not in the number of such walks. Only in Chapter 20 will we also be interested in the number of triangles that contain a given edge.

Usual matrix multiplication of two $n \times n$ matrices requires n^3 multiplications, but it is well-known that this can be improved considerably. The smallest real number ω such that two $n \times n$ matrices can be multiplied in $O(n^\omega)$ steps is called the *exponent of matrix multiplication*. Currently, $2.376\ldots$ is the best bound for ω; see Coppersmith and Winograd (1990).

By Theorem 17.8 this implies that we can determine the number of triangles in a graph in time complexity $O(n^\omega)$. Surprisingly even (b) and (c) can be solved within this time complexity. See Romani (1980) for (b) and Munro (1971) as well as Fischer and Meyer (1971) for (c). As fast matrix multiplication will only be used in Section 21.3, we can rest content with these references.

Although matrix operations provide an easily understood tool to solve a multitude of important problems, they suffer from the drawback of the large amount of memory required for the adjacency matrix, the high complexity of matrix multiplication, and the requirement of random access. Because arithmetic is often unnecessary for graph algorithms and because random access is not really required for the adjacency list representation (if one replaces the vector containing the addresses of the first elements of the adjacency lists by a list), the purest graph algorithms use adjacency lists and only manipulate pointers. Only a rather weak model of computation is required for such algorithms. As the reader notices, we did not go to such extremes in our treatment of adjacency lists. In particular, we use random access when working with the distance matrix, although we may construct it by the methods of Corollary 17.3.

Chapter 18

Recognizing Hypercubes and Partial Cubes

In this chapter we show that hypercubes can be recognized in linear time. Deciding whether a given graph is an isometric subgraph of a hypercube is more difficult. We present three algorithms for the solution of this problem: a straightforward one of complexity $O(m^2)$, which relies on Theorem 11.8 (iii); a more refined one of complexity $O(mn)$; and one of complexity $O(n^2)$, which uses a different computational paradigm.

The second algorithm owes its efficiency to an algorithm of Feder (1995), that computes Θ^* in $O(mn)$ time.

The third algorithm is due to Eppstein (2008) and makes use of bitvectors of length $\log n$, where n is the order of the investigated graphs. It is based on the observation that our model of computation allows direct addressing of n addresses and thus supports arithmetic and bitwise Boolean operations on integers of at least $\log_2 n$ bits, as well as indexing operations in constant time.

The algorithm also makes use of the fact that the isometry of proper embeddings of a graph G into the hypercube can be checked in $O(n^2)$ time, where n is the number of vertices of G. The validity of this fact is shown, without the use of bitvectors, at the end of the chapter. Interestingly, it also leads to an algorithm that computes the distance matrix of a partial cube on n vertices in $O(n^2)$ time.

18.1 Hypercubes

As hypercubes are special partial cubes, we can use ideas from Chapter 11 to design a simple hypercube recognition algorithm, Algorithm 18.1. Its complexity is linear, hence optimal.

Later we will encounter two other algorithms that also recognize hypercubes in linear time. One is Algorithm 22.1 for the recognition of Hamming graphs, and the other is Algorithm 23.1, which computes the prime factors of connected graphs with respect to the Cartesian product. Both algorithms are linear, rely on different concepts, and are more elaborate than the one presented now.

Still another algorithm is the one of Bhat (1980). To our knowledge it was the first linear algorithm for the recognition of hypercubes.

We first recall a few basic facts about hypercubes. Let uv be an arbitrary edge of the r-cube Q_r. To fix ideas, let $u = 00 \ldots 0$ and $v = 10 \ldots 0$. Then the vertices of W_{uv}, namely the vertices closer to u than to v, are the vertices with first coordinate 0. Clearly, the subgraph $\langle W_{uv}\rangle$ they induce in Q_r is an $(r-1)$-cube. Analogously, the vertices with first coordinate 1 induce the $(r-1)$-cube $\langle W_{vu}\rangle$.

The edges of F_{uv} are the edges between W_{uv} and W_{vu}. They are of the form

$$(0x_2x_3\ \dots\ x_r)(1x_2x_3\ \dots\ x_r).$$

Clearly, they are a matching of Q_r, and this matching defines an isomorphism

$$\alpha : 0x_2x_3\ \dots\ x_r \mapsto 1x_2x_3\ \dots\ x_r$$

of $\langle W_{uv}\rangle$ onto $\langle W_{vu}\rangle$.

This information already suffices for an $O(n \log n)$ recognition algorithm. To see this, let G be a nontrivial connected graph on n vertices and m edges, given by its adjacency list. If it is a hypercube, its number m of edges must be $\frac{n}{2}\log_2 n = O(n \log n)$. We can check in that many steps whether G is bipartite. Then, choosing $uv \in E(G)$ arbitrarily, we can obtain the distances $d_G(u,x)$ and $d_G(v,x)$ for all $x \in V(G)$ in $2m$ steps. Thus $\langle W_{uv}\rangle$ and $\langle W_{vu}\rangle$ can be determined in $O(m)$ time and space. (By "determined" we mean that adjacency list representations of both of them are created.)

Within the same time and space complexity, we compute F_{uv} and thus a mapping from $\langle W_{uv}\rangle$ to $\langle W_{vu}\rangle$. Invoking Theorem 17.1, we can check in linear time and space whether it is an isomorphism. If not, we reject G.

If it is an isomorphism and if $n = 2$, then G is a hypercube. Otherwise, we repeat the procedure for $\langle W_{uv}\rangle$. Clearly, this process ends after at most $\log_2 n$ steps. To determine its complexity, let c be the constant of the procedure that determines $\langle W_{uv}\rangle$, $\langle W_{vu}\rangle$, and checks them for isomorphism. Then the total complexity is at most

$$cm + c\frac{m}{2} + c\frac{m}{4} + \cdots + c < 2cm\,.$$

This is best possible, because all edges must be checked. The following algorithm and theorem summarize it:

Algorithm 18.1 Hypercubes

Input: The adjacency list of a connected graph G.
Output: true if G is a hypercube, **false** otherwise.

1: **if** G is not bipartite or $m \neq \frac{n}{2}\log_2 n$, **then** return **false** and stop.
2: For an arbitrary edge uv compute W_{uv} and W_{vu}.
3: **if** the edges between $\langle W_{uv}\rangle$ and $\langle W_{vu}\rangle$ do not define a complete matching and an isomorphism between $\langle W_{uv}\rangle$ and $\langle W_{vu}\rangle$, **then** return **false** and stop.
4: **if** $\langle W_{uv}\rangle = K_2$, **then** return **true** and stop, **else** go to Step 2 with $\langle W_{uv}\rangle$ as input graph.

Theorem 18.1 *Hypercubes can be recognized in linear time and space.*

18.2 Partial Cubes

Theorem 11.8 (iii) asserts that a bipartite graph is a partial cube if and only if Θ is transitive. This elegant characterization of partial cubes is also practical from an algorithmic point of view. Suppose that we are given a graph G on n vertices with m edges and wish to check whether it is a partial cube. We first preprocess it by checking connectedness and bipartiteness. By Corollary 17.3 (i) and Proposition 17.4 this can be done in $O(m)$ time.

To determine Θ, it helps to have the distance matrix of G available, and this can be computed in $O(mn)$ time by Corollary 17.3 (ii). Then Θ can be determined by the brute-force method of checking all pairs of edges of G. There are $O(m^2)$ such pairs, and for every comparison we have to determine four distances, perform two additions, and one arithmetic comparison. With the distance matrix at hand, this can be done in $O(1)$ time. Hence Θ can be determined in $O(m^2)$ time.

To check Θ for transitivity, we invoke the following proposition:

Proposition 18.2 *If R is a symmetric and reflexive relation on a set X, then the equivalence classes of R^* can be determined in $O(|R|)$ time and space.*

Proof Let H be defined by $V(H) = X$ and $E(H) = \{xy \mid x, y \in X, \text{ where } xRy\}$, so H is an undirected graph with loops at every vertex, and the equivalence classes of R^* are its connected components. By Corollary 17.3 (i) they can be computed in $O(|E(H)| + |V(H)|)$ time and space. □

In other words, given the set of pairs of elements in R, we can compute the equivalence classes $E_1, E_2, \ldots, E_k$ of R^* in linear time. If we label every element of X with the index of the equivalence class in which it is contained, then we can determine in constant time whether a pair x, y of vertices is in R^*. From this point of view we have computed R^*, although we describe it as a list $E_1, E_2, \ldots, E_k$, whose total size is $|X|$, whereas the number of the pairs of elements in R^* is $\sum |E_i|^2$, which may be close to $|X|^2$.

Proposition 18.2 thus asserts that the complexity of the computation of Θ^* from Θ is bounded by the number of elements of Θ. Comparing the size of Θ with that of Θ^*, we check whether Θ is transitive. Altogether, we arrive at an algorithm of time and space complexity $O(m^2)$ for the recognition of partial cubes.

This approach depends on the knowledge of Θ, which we determined within $O(m^2)$ time. Interestingly, no faster algorithms are known for Θ. However, we will see in Section 13.1 that Θ^* can be computed within time and space complexity $O(mn)$.

We continue with an algorithm for the recognition of partial cubes that only requires the knowledge of Θ^*, and not of Θ. It makes use of the canonical isometric embedding α from Section 13.1, as well as Theorem 11.9, which asserts that a connected bipartite graph is a partial cube if and only if each $G - E_i$ has exactly two components, where E_i is an equivalence class of Θ^*.

In reading the algorithm it is helpful to recall the notation of Section 13.1, namely that G_i^* is the graph whose vertices are the components of $G - E_i$, and for which CC' is an edge of G_i^* precisely if some edge in E_i connects C to C'. Thus Theorem 11.9 can be interpreted as stating that G is a partial cube if and only if $G_i^* \cong K_2$ for each i. Compare Figure 18.1, which shows a partial cube G, together with the graphs G_i and G_i^* for each equivalence class E_i.

Algorithm 18.2 Partial cubes

Input: The adjacency list of a connected graph G.
Output: **true** and a Hamming labeling if G is a partial cube; **false** otherwise.

```
1: if G is not bipartite, then return false and stop.
2: Compute Θ*; denote the number of Θ-classes by k.
3: for i = 1 to k do
4:    Compute G_i and G_i^*.
5:    If G_i^* is not a K_2, then return false and stop.
6: end for
7: Select a BFS-tree T in G with root v_0.
8: for all v ∈ V(G) and i = 1 to k do
9:    if the path in T from v to v_0 contains an edge of E_i, then
10:      Set the ith coordinate of α(v) equal to 1.
11:   else
12:      Set the ith coordinate of α(v) equal to 0.
13:   end if
14: end for
15: Return true and the labeling of G obtained in Steps 10 and 12.
```

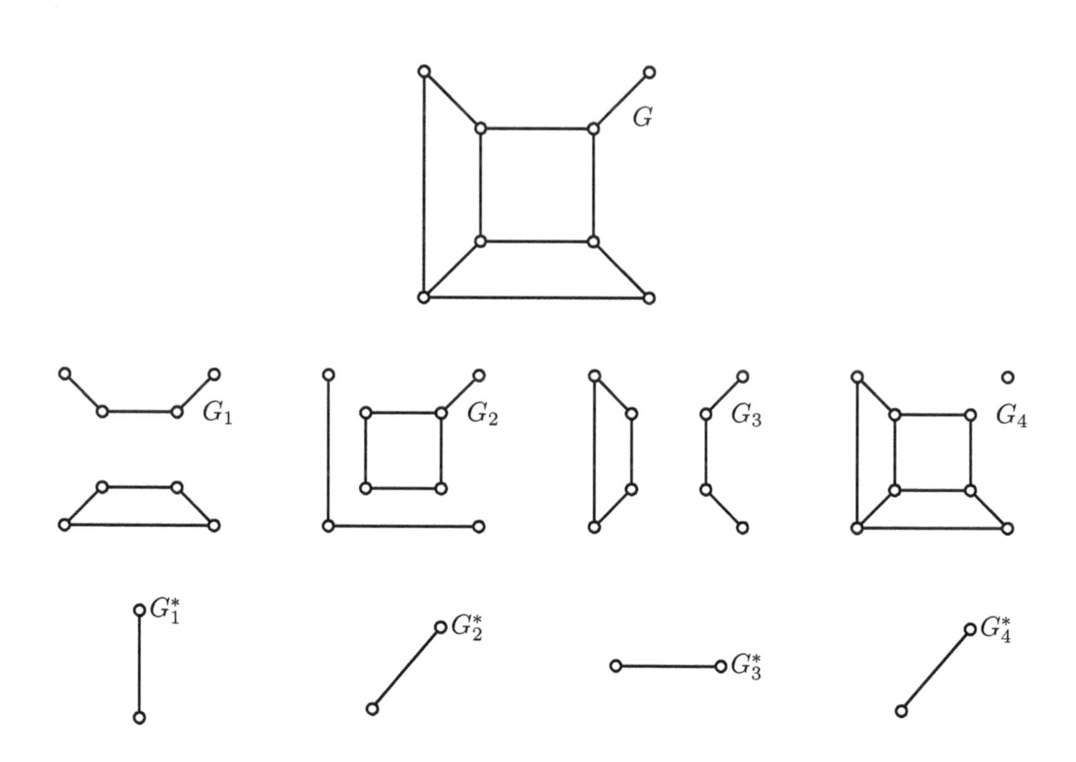

FIGURE 18.1 A graph G with G_i and G_i^* for each Θ^*-class.

Lemma 18.3 *Algorithm 18.2 correctly recognizes partial cubes and returns a Hamming labeling. If Θ^* can be computed in $O(mn)$ time and $O(m)$ space, then it runs in $O(mn)$ time, requiring $O(n^2)$ space.*

Proof Every graph G accepted by the algorithm is bipartite and every G_i has exactly two components. By Theorem 11.9 such graphs are partial cubes.

To see that α is a Hamming labeling, we first note that the algorithm sets all coordinates of $\alpha(v_0)$ equal to 0. Let P be the v, v_0-path in T. By Lemma 11.1, no two edges of P are in relation Θ. Because G is a partial cube, $\Theta^* = \Theta$ and all edges of P belong to different Θ^*-classes. By the definition of α, traversing an edge of E_i means a change in the ith coordinate, while all other coordinates remain unaltered. Thus the algorithm correctly determines α.

Now consider the complexity.

Step 1 can be implemented to run in $O(m)$ time and space.

The complexity $O(mn)$ for Step 2 holds by assumption. (Theorem 18.6 will show us that this is justified.)

For Step 3, note that G_i is constructed from G by removing the edges E_i. Assume that G is represented by an extended adjacency list. Then the removal of an edge $v_i v_j$ has complexity $O(d(v_i))$. Because $d(v_i)$ is bounded by n, the graph G_i can be determined in $O(n|E_i|)$ time and its components in $O(|E_i|)$ time; the space complexity is $O(m)$.

Thus the G_i and their components can be determined in altogether $O(n \sum |E_i|) = O(nm)$ time. If we do not keep the information about the G_i, then the overall space complexity is still $O(m)$.

For Step 4, observe that every $\alpha(v)$ can be computed in $O(n)$ time and space, because the length of the paths from v to v_0 in T is bounded by $n - 1$. Thus the overall time and space complexity of finding and storing the $\alpha(v), v \in V(G)$, is $O(n^2)$. □

18.3 Efficient Computation of Θ^*

We have just seen that the complexity of Algorithm 18.2 for the recognition of partial cubes is determined by the complexity of computing Θ^*. Moreover, Θ^* is also essential in computing the canonical embedding and, by Theorem 23.6, for a direct computational approach to the prime factorization of connected graphs with respect to the Cartesian product. Thus, the computational complexity of Θ^* plays an important role. The direct approach from Section 18.2 led to an $O(m^2)$ algorithm for Θ^*. We now present an idea of Feder (1992) that gives Θ^* in $O(mn)$ time.

Let T be a spanning tree of a graph G. Say two edges $e, e' \in E(G)$ are in *relation* Θ_1 if $e\Theta e'$ and at least one them belongs to T. Notice that $\Theta_1 \subseteq \Theta$, which implies $\Theta_1^* \subseteq \Theta^*$, and that the number of pairs of edges in Θ_1 is bounded by mn.

We show now that Lemma 13.1, an important structural result for Θ, also holds for Θ_1:

Lemma 18.4 *Let P be a shortest u, v-path in G and Q be a u, v-path. Then*

$$|P \cap E_i| \leq |Q \cap E_i|$$

for any Θ_1^*-class E_i.

Proof Let p be the pair (u, v) and P_T the path connecting u with v in T. Consistently direct P, Q and P_T from u to v, and assume that, in expressions such as $P \cap E_i$, the edges from E_i that are intersected by P are assigned the same orientation as P. (Conflicting orientations can only occur if an edge is in P and Q, but those we need not count.) Then

$$\begin{aligned}\mu(p, Q \cap E_i) &= \mu(P_T, Q \cap E_i) \\ &= \mu(P_T \cap E_i, Q \cap E_i) \\ &= \mu(P_T \cap E_i, Q) \\ &= \mu(P_T \cap E_i, p)\,.\end{aligned}$$

The first and the last equality hold by the same argument as in Lemma 13.1. The others follow because, if two edges e and f belong to different equivalence classes of Θ_1^* and at least one of them belongs to T, then $\mu(e, f) = 0$.

Furthermore, as in Lemma 13.1, we see that $\mu(p, P \cap E_i) = 2|P \cap E_i|$ and $\mu(Q \cap E_i, p) \leq 2|Q \cap E_i|$. Altogether we obtain

$$\begin{aligned}2|P \cap E_i| &= \mu(p, P \cap E_i) \\ &= \mu(P_T \cap E_i, p) \\ &= \mu(p, Q \cap E_i) \\ &\leq 2|Q \cap E_i|,\end{aligned}$$

which proves the lemma. □

As a consequence, we secure the key to fast computation of Θ^*:

Theorem 18.5 *Let G be a connected graph. Then $\Theta^* = \Theta_1^*$.*

Proof Notice that the proof of Theorem 13.2, which asserts that the canonical embedding is an isometry, depends only on Lemma 13.1. As we have just shown, this lemma also holds for Θ_1. Thus an isometric embedding can also be obtained from Θ_1.

Recall that $\Theta_1^* \subseteq \Theta^*$ and notice that the equivalence classes of Θ^* are unions of equivalence classes of Θ_1^*. By Theorem 13.3 (ii) and (iv), the canonical embedding α is the only embedding into the largest possible number of factors. So $\Theta^* = \Theta_1^*$. □

Theorem 18.5 yields the following algorithm:

Algorithm 18.3 The relation Θ^*

Input: The adjacency list of a connected graph G.
Output: The relation Θ^*.

1: Compute a spanning tree T of G.
2: **for all** edges $e = uv$ of T **do**
3: Compute the distances from u and v to all other vertices.
4: **end for**
5: Compute Θ_1^* with respect to T.

Theorem 18.6 *For a given graph G on n vertices and m edges, Algorithm 18.3 computes the equivalence classes of Θ^* of G. It can be implemented to run in $O(mn)$ time using $O(m)$ space.*

Proof The correctness of the algorithm follows by Theorem 18.5. Step 1 of the algorithm can be executed in linear time and Step 2 in $O(mn)$ time. Clearly, we can verify whether a given edge of T is in relation Θ_1 with any other edge in $O(m)$ time. For all edges of T, this takes $O(mn)$ time. Whenever we find two edges e and f that are in relation Θ_1, we merge the sets of edges that are already known to be in relation Θ_1 to e, respectively f. By Proposition 17.7, applied to Θ_1, this takes at most $O(m \log m)$ time. □

It is not hard to see that Algorithm 18.2 can easily be modified to compute the canonical isometric embedding and that the complexity is still determined by that of the computation of Θ_1. We thus have the following important consequence of Theorem 18.6:

Corollary 18.7 *The canonical isometric embedding of a graph on n vertices and m edges can be computed in $O(nm)$ time using $O(m)$ space.*

In Lemma 18.3 we assumed that Θ^* can be computed in $O(mn)$ time and $O(m)$ space for a graph on n vertices and m edges. Because this is indeed the case by the above, we have the following theorem:

Theorem 18.8 *Let G be a connected graph on n vertices and m edges. Algorithm* 18.2 *correctly recognizes whether G is a partial cube and can be implemented to run in $O(mn)$ time and $O(n^2)$ space. If G is a partial cube, then the algorithm returns a Hamming labeling.*

This reduces the time complexity of recognizing partial cubes from $O(m^2)$ to $O(mn)$, which does not look bad at a first glance. However, recall that partial cubes are sparse by the Density Lemma 3.2, that is, $m \leq \frac{1}{2} n \log_2 n$. Hence, for partial cubes, $O(m^2) = O(n^2(\log n)^2)$, and $O(mn)$ reduces to $O(n^2 \log n)$.

From a computational point of view, this is not much. But, as we will see, the same ideas are also applicable to the recognition of partial Hamming graphs, which are not sparse. There we have a substantial improvement.

Now to the space complexity. Consider P_n. The smallest hypercube into which P_n can be embedded isometrically is Q_{n-1}. In this case, every label has length $n-1$, and the total length of the labels for the vertices of P_n is $n(n-1)$. Hence, the space complexity really is $O(n^2)$ if we store all labels in full length.

One of the reasons for the introduction of Hamming labelings is that distances $d(u, v)$ can be determined by comparing $\alpha(u)$ with $\alpha(v)$. This takes k comparisons, where k is the number of Θ^*-classes.

Notice that the computation of the labels of u and v from the BFS-tree T and their comparison takes $O(k)$ time and space. This is of the same complexity as label comparison. Thus we might be content to prove that G is a partial cube and compute labels from a BFS-tree when needed. This can be done by deleting Step 4 of Algorithm 18.2. The modified algorithm still determines whether a graph is a partial cube in $O(mn) = O(n^2 \log n)$ time, but its space complexity reduces from $O(n^2)$ to $O(m)$, that is, to $O(n \log n)$. We formulate this as a corollary:

Corollary 18.9 *Partial cubes can be recognized in $O(n^2 \log n)$ time using $O(n \log n)$ space.*

We close the section by remarking that the first algorithm of complexity $O(mn)$ for the recognition of partial cubes is due to Aurenhammer and Hagauer (1995), although the one of Imrich and Klavžar (1993), which is presented here, appeared earlier in print.

18.4 Recognizing Partial Cubes in Quadratic Time

This section presents an algorithm that recognizes partial cubes in $O(n^2)$ time and space. It is due to Eppstein (2008) and based on the assumption that arithmetic and bitwise Boolean operations on integers of at least $\log_2 n$ bits, as well as indexing operations, are possible in $O(1)$ time. This is reasonable, because our algorithms depend on the RAM model of computation, and in this model any machine that is capable of storing addresses large enough to address the input to our problem has machine words with at least $\log_2 n$ bits.

It also means that we can determine in constant time whether a bitvector has exactly one nonzero bit and which one it is. This can be done by looking up that word in a table of size n that stores either the index of the nonzero bit if there is only one, or a flag value that indicates that there is more than one nonzero bit.

Because the length of our bitvectors may be close to n, we need some further considerations about the complexity of working with them. Following Eppstein (2008) we invoke the following lemma about operations with bitvectors of length k.

Lemma 18.10 *Let k be a natural number and $K = \lceil 1 + k/\log_2 n \rceil$. Then bitvectors of length k can be stored in $O(K)$ space per bitvector, and disjunction and symmetric difference*[1] *operations can be done in $O(K)$ time per operation. Moreover, one can determine in $O(K)$ time whether a bitvector contains nonzero bits. One can also determine in $O(K)$ time whether it has exactly one such bit, and find the index of that bit.*

Proof We store every bitvector in K words of $\log_2 n$ bits each. Disjunction and symmetric difference can be performed independently on each of the words.

To test for nonzero bits we test every one of the K words. To test whether a bitvector has exactly one nonzero bit, we again test every word. If there is just one that has nonzero bits, we use the precomputed table mentioned above to see whether there is only one nonzero bit and which one it is. □

Outline of the algorithm

Preprocessing The algorithm accepts connected, bipartite input graphs G on n vertices with m edges, where $m \leq \frac{n}{2} \cdot \log_2 n$.

Bitvector assignment The main part of the algorithm is Procedure 18.4. It begins with a vertex v_0 of maximum degree d and its neighbors $v_1, \ldots, v_d$. Then it computes bitvectors $w(v)$ of length d for every $v \in V(G)$, where the ith bit of $w(v)$ is 0 if $v \in W_{v_0,v_i}$, and 1 otherwise.

Classes and consistency The edges between W_{v_0,v_i} and W_{v_i,v_0} are the sets F_{v_0,v_i}. If G is a partial cube, then the F_{v_0,v_i} are the Θ-classes of the edges incident with v_0. Because Θ-classes are disjoint matchings in a partial cube, we discard all graphs that do not satisfy this consistency condition. Procedure 18.5 computes the F_{v_0,v_i}, checks for consistency and assigns labels to the edges of F_{v_0,v_i}.

Contraction Contraction of the edges in a Θ-class of a partial cube G (and replacement of double edges that may result as a consequence of the contraction) yields a partial cube again. By Exercise 11.7, the preimages (with respect to the contraction) of the Θ-classes of this partial cube are Θ-classes in G.

Hence, if G contains unlabeled edges, then we contract the labeled edges to single vertices, replace multiple edges by single ones, call the new graph G', and restart the algorithm with input G'.

This is repeated until we reach a graph with no unlabeled edges or until the total number of F-classes (we do not know whether they are Θ-classes) exceeds $n - 1$. In the latter case we discard G because it is not a partial cube.

---

[1] In this chapter and in Chapter 33, we use the notation Δ for symmetric difference, but in Chapter 29 the notation $+$ is more convenient; see pages 367 and 433.

Reexpansion In the first case, when all edges are labeled, we reexpand. That is, for every vertex $v' \in V(G')$, we concatenate the bitvectors $w'(v')$ with the bitvectors $w(v)$ of all vertices $v \in V(G)$ that are preimages of v' with respect to the contraction. This is repeated until the first input graph is reached.

Isometry Clearly, the assignment of bitvectors is a mapping of G into a hypercube whose dimension is the number of F-classes of G. If the mapping is isometric, then G is a partial cube. We will show that this can be done in $O(n^2)$ time.

Details, correctness, and complexity

Preprocessing needs no further explanations. **Bitvector assignment** is described in Procedure 18.4.

Procedure 18.4 Bitvector assignment

Input: A connected, bipartite graph G and a vertex v_0 of maximum degree d.
Output: Bitvectors $w(v)$ of length d for every $v \in V(G)$, where the ith bit of $w(v)$ is 0 if $v \in W_{v_0,v_i}$, and 1 otherwise.

1: Compute a BFS order of G with root v_0.
2: Reserve bitvectors $w(v)$ of length d for every $v \in V(G)$ and set $w(v) = (0, 0, \dots, 0)$ for all $v \in V(G)$.
3: For the elements in L_1, that is, the neighbors of v_0, set $w_i(v_i) = 1$.
4: Scan the vertices v of G in BFS order, beginning with the vertices in L_2. Set $w(v)$ equal to the disjunction of the bitvectors of all down-neighbors of v.

Correctness We have to show that the ith bit, say $w_i(v)$, of $w(v)$ is 0 if $v \in W_{v_0,v_i}$, and 1 otherwise. Clearly, this is the case for v_0 and the vertices in L_1.

Suppose that the bitvectors $w(v)$ have already been correctly assigned to the elements in L_k and that we wish to compute $w(u)$ for $u \in L_{k+1}$. If there is a shortest path from u to v_0 that passes through v_i, then $u \in W_{v_i,v_0}$ and $w_i(u) = 1$, otherwise $u \in W_{v_0,v_i}$ and $w_i(u) = 0$. Because all shortest paths from u to v_0 start with an edge from u to a down-neighbor of u, the bitvector $w(u)$ clearly is the disjunction of the bitvectors of the down-neighbors of u.

Complexity The identification of a vertex v_0 of maximum degree $d = d(v_0)$ and the execution of Step 1 take at most $O(m)$ time. Notice that $d \geq 2m/n$.

Step 2 takes at most $O(n(1 + d/\log_2 n))$ time and Step 3 takes $O(d)$.

In Step 4 we compute the disjunction of bitvectors of length d, one bitvector for every down-edge. This is at most m bitvectors altogether. Thus the time complexity of this step (and of the procedure as a whole) is $O(m(1+d/\log_2 n)) = O(d(m/d+m/\log_2 n)) = O(dn)$.

Remark If G is a partial cube, then all F_{v_0,v_i} for $v_i \in N_G(v_0)$ are correctly determined now. In particular, if G is a hypercube Q_r, then the bitvectors $w(v)$ computed in this first stage already define a coordinatization of G as a Cartesian product of r K_2's.

The graph of Figure 18.2 indicates how the procedure works. Clearly, the vertex v_0 is a vertex of maximal degree.

By Step 1 a BFS ordering with base v_0 is computed. The BFS-levels are $L_0 = \{v_0\}$, $L_1 = \{v_1, v_2, v_3\}$, $L_2 = \{v_4, v_5\}$, and $L_3 = \{v_6\}$. Every edge has been directed toward v_0.

By Steps 2 and 3, v_0 is assigned the bitvector $(0, 0, 0)$, and v_1, v_2, v_3 are assigned the vectors $(1, 0, 0), (0, 1, 0), (0, 0, 1)$. The other assignments are made in Step 4.

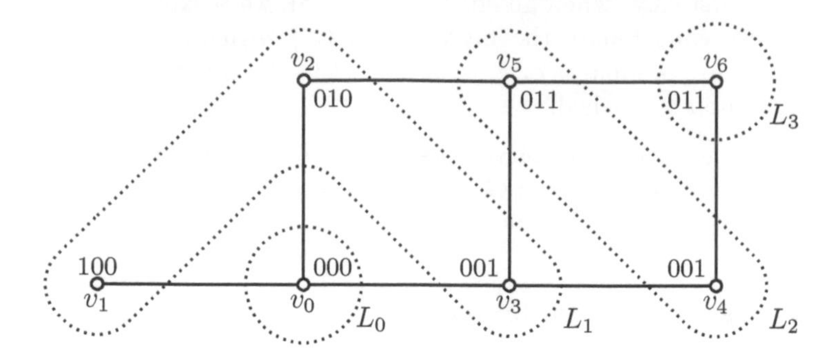

FIGURE 18.2 BFS-levels and bitvector assignment in a graph.

Classes and consistency are handled by Procedure 18.5.

Procedure 18.5 Classes and consistency

Input: A connected, bipartite graph G that passed the bitvector assignment Procedure 18.4.

Output: An edge-coloring that assigns color i to the edges of F_{v_0,v_i} if the F_{v_0,v_i} are disjoint matchings. If the F_{v_0,v_i} do not satisfy the condition, then G is discarded.

1: $\Delta(uv) = w(u) \,\Delta\, w(v)$ for all edges uv in G.
2: If there is $\Delta(uv)$ with more than one nonzero bit, discard G.
3: If the ith bit is the only nonzero bit of $\Delta(uv)$, assign color i to uv.
4: If $\Delta(uv)$ is zero, leave uv uncolored.
5: If there is an F_{v_0,v_i} that is not a matching, discard G.

Correctness If G is bipartite, then the set W_{v_0,v_i} consists exactly of the vertices u with $w_i(u) = 0$. Hence, F_{v_0,v_i} is the set of all edges uv, where $w(u)$ and $w(v)$ differ in the ith bit. By Proposition 11.7, F_{v_0,v_i} consists of all edges that are in the relation Θ to v_0v_i. Because Θ is transitive for partial cubes by Theorem 11.8, the set F_{v_0,v_i} is a Θ-class if G is a partial cube. Hence, if $w(u)$ and $w(v)$ differ in the ith bit, they must coincide in all the others, because Θ-classes are disjoint in partial cubes.

We thus consider the symmetric difference $\Delta(uv) = w(u) \,\Delta\, w(v)$ for all edges uv in G. If $\Delta(uv)$ has more than one nonzero bit, then G cannot be a partial cube. If just one bit, say the ith, is nonzero, then $uv \in F_{v_0,v_i}$ and we assign color i to uv. If $\Delta(uv)$ is zero, then uv is not in relation Θ to any of the edges v_0v_i and it is left uncolored.

Recall next that F_{v_0,v_i} is a matching if G is a partial cube. Thus, if a vertex of G is incident with more than one edge in one and the same F_{v_0,v_i}, then G is not a partial cube.

Complexity As in the case of Procedure 18.4, one shows that the complexity of Steps 1 through 4 is $O(dn)$. We leave it to the reader to spell out Step 5 in more detail and to show that the time complexity is $O(m)$. Hence the total time complexity is $O(dn)$.

Suppose the input to this procedure is the graph of Figure 18.2 with its bitvector assignment. Steps 1 and 2 go through without any problems, and Steps 3 and 4 color some edges and leave others uncolored; see Figure 18.3(a). Clearly, all F_{v_0,v_i} are matchings.

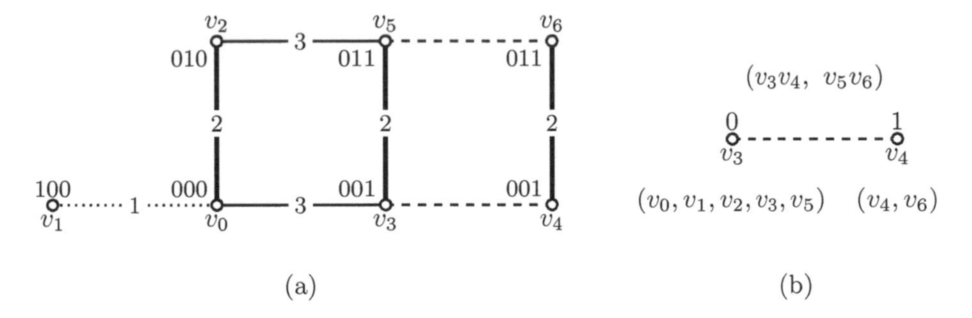

FIGURE 18.3 Assignment of edge-colors and contraction.

Contraction and reexpansion is treated more informally.

Correctness Given a partial cube G and a Θ-class F_{ab}, it is easily seen by Proposition 11.7 that contraction of every edge in F_{ab} to a single vertex and the replacement of double edges with single edges produces a partial cube again (Exercise 11.5). We do this for the d classes F_{ab} we have found so far, denoting the new graph by G'. If G' has a loop, then G cannot have been a partial cube and we terminate the algorithm.

We apply the algorithm again, obtaining some new Θ-classes if G' is a partial cube. If there are still uncolored edges, we iterate the procedure until there are no uncolored edges left or if more than $n-1$ classes are found. In the second case, G was not a partial cube, because no graph on n vertices can have more than $n-1$ Θ-classes.

At the end of the last contraction step, all edges of the resulting graph G^ν are colored and all vertices labeled; if G is a partial cube, then G^ν is a partial cube too and it is correctly colored and labeled. We reexpand. This is done iteratively as follows. Let G' be obtained from G as described above and suppose all edges of G' are colored and all vertices labeled. To form the color classes of G, we concatenate the color classes of G' (replacing edges by edges they correspond to on G) with the originally obtained color classes of G.

To form the vertex label of every v of G, consider the bitvector of the vertex v' corresponding to v in G' and concatenate it with the bitvector $w(v)$ already computed.

By Exercise 11.6, the coloring and labeling is correct.

Complexity By Exercise 18.7, the total complexity of the edge contractions necessary to construct G' from G is $O(m)$, and by Exercise 18.8 loops and multiple edges can be removed within the same time complexity. Altogether this is at most $O(dn)$ and can be subsumed with the complexity of coloring the edges of G by the bitvector assignment (Procedure 18.4) that prepares G for contraction.

In reexpansion, the color classes can clearly be computed in $O(m)$ time and the concatenation of the vertex labels going from G' to G can be effected in $O(dn)$ time.

For illustration we turn to Figure 18.3. Contraction of the colored edges in Figure 18.3(a) yields the graph of Figure 18.3(b). This figure also indicates the preimages of its vertices and of its edge. Clearly, the bitvector assigment for v_3 is (0) and (1) for v_4 if we choose v_3 as the root of the BFS ordering of this graph. The edge has color 4, because we add a fourth coordinate, in which the endpoints differ.

Reexpansion yields the graph of Figure 18.4 and assigns color 4 to edges v_3v_4 and v_5v_6.

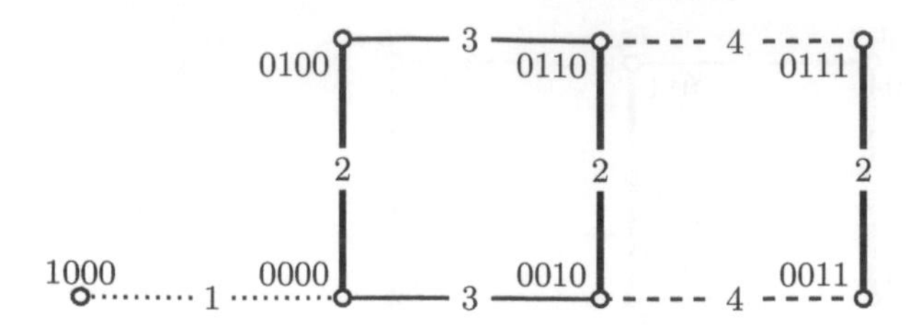

FIGURE 18.4 Reexpansion.

Isometry The graphs G that survive up to this stage include all partial cubes and are labeled by 0-1 vectors of length $r \leq n-1$. This labeling is an embedding into the hypercube Q_r, because the endpoints of every edge differ in exactly one coordinate.

The question is whether this embedding is an isometry. The graph G in Figure 18.5 shows that this need not be the case. With the depicted choice of base point v_0, all parts of the algorithm go through and produce the displayed embedding. However, G is not isometrically embedded in Q_3, because $d_G(v_4, v_7) = 3$, but $d_{Q_3}(v_4, v_7) = 1$.

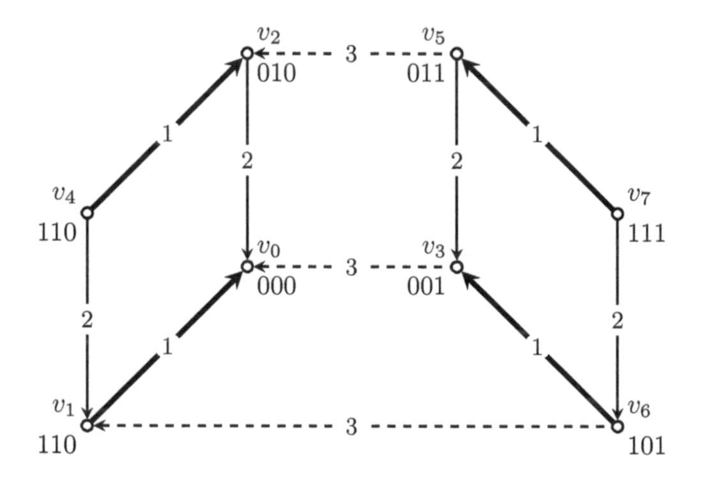

FIGURE 18.5 A nonisometrically embedded graph.

This is easily seen by inspection, but in general we will have to test for all $n(n-1)$ pairs of vertices u, v whether or not $d_G(u,v) > d_{Q_r}(u,v)$. In order to to this we will check for every u whether or not there exists a BFS-tree T_u of G with base u and the property that $d_G(u,v) = d_{Q_r}(u,v)$ for all $v \in V(G)$. If this is the case, then G is a partial cube, but not otherwise.

Such a tree T_u can be constructed in $O(n)$ time from a BFS-ordering with base u by arbitrarily choosing a single down-edge for every vertex $v \in V(G)$, $u \neq v$. Suppose the down-edge vz of T_u has color i and $w_i(v) = w_i(u)$. Then $w_i(z) \neq w_i(u)$ and z differs in one extra coordinate from u than v, has larger distance from u than v and thus cannot be on a shortest v,u-path in Q_r. Hence the v,u-path in T_u, which is a shortest G-path, is not a shortest Q_r-path, and G is not isometric in Q_r.

Choosing the notation $c(e)$ for the color of the edge e, we can thus say that G is not isometric if, for a given vertex v, $w_{c(vz)}(v) = w_{c(vz)}(u)$ for all edges $vz \in E(G)$. We will also say that the edge vz is oriented away from u if $w_i(v) = w_i(u)$ and $w_i(z) \neq w_i(u)$, otherwise we say that it is oriented toward u.

We now describe the algorithm that checks isometry. We begin with an arbitrary vertex u and compute the BFS-order of G. We also orient every edge vz of G. Because we already know the color of every edge vz, we only have to compare $w_{c(vz)}(v)$ with $w_{c(vz)}(u)$. For every vertex v we then create two lists of edges, the list $A_u(v)$ that contains the edges oriented toward u and the list $B_u(v)$ that contains the edges oriented away. This can be done in $O(m)$ time. We can also sort the lists by color within the same time complexity. Clearly, all edges incident with u are in $B_u(u)$, and $A_u(u) = \emptyset$.

If one of the $A_u(v)$ is empty for $v \neq u$, then G is not a partial cube and we are done. We thus assume that the $A_u(v)$ are not empty and choose the first edge of every $A_u(v)$ to obtain a BFS-tree that we call T_0 and rename u into u_0. The edges xy of T_0 are thus ordered pairs, where y is closer to v_0 than x.

Now we draw T_0 in the plane: It has one face and we follow the boundary of that face in counterclockwise direction. That way we reach every vertex twice, introducing the notation u_j for the vertex that we reach in j steps. We wish to construct a BFS-tree T_{u_j} for every j. Before that, we sort the edges in the A- and B-sets again, this time in the order just described. As our edges are ordered pairs, every ordered pair occurs exactly once in that order.

Let $j = 1$ and i be the color of the edge v_0v_1. We show how to construct T_{u_1}. The new root is v_1. We begin wih some preprocessing. All edges of color different from i keep their orientation; only the orientation of the edges in $F_{v_0v_1}$ is reversed. Consider v_0v_1. We set $A_{v_1}(v_1) = A_{v_0}(v_1) \setminus v_1v_0$ and $B_{v_1}(v_1) = B_{v_0}(v_1) \cup v_1v_0$, adding the edge v_1v_0 to the front of the list. Likewise we set $A_{v_1}(v_0) = A_{v_0}(v_0) \cup v_0v_1$, again adding v_0v_1 to the front of the list, and $B_{v_1}(v_0) = B_{v_0}(v_0) \setminus v_0v_1$. Similarly we proceed for all other edges $ab \in F_{v_0v_1}$.

To construct T_1, we now remove all edges of color $c(v_0v_1)$ from T_0 as they point away from v_1, but keep all the other edges, as they point toward v_1. If xy from T_0 was removed, then x has no outgoing edge in T_1 and we add the first edge of $A_{v_1}(x)$ to T_1. If $A_{v_1}(x) \neq \emptyset$ for all removed edges xy, then we obtain the desired BFS-tree T_{v_1}. Otherwise, that is if $A_{v_1}(x)$ is empty, G is not a partial cube.

Thus, to go from T_{v_0} to T_{v_1}, we have to consider the $n-1$ edges xy of T_{v_0}, remove them if they have color $C(v_0v_1)$, and, for x and y, delete a first element from a list, respectively add a first element to a list. Then first elements of such lists are used again to complete T_1. The complexity is $O(n)$. Because we do this at most $2(n-1)$ times, the overall complexity is $O(n^2)$ for the isometry check. This leads to the following result of Eppstein (2008):

Theorem 18.11 *Partial cubes on n vertices can be recognized in $O(n^2)$ time.*

Proof All that remains to show is the time complexity. Notice that we begin the algorithm with a vertex of maximal degree d and that the complexity of bitvector assignment, classes and consistency, and contraction and reexpansion is $O(dn)$. At the end of contraction, we continue with a graph G' that has exactly d color classes less than G. If G is a hypercube, then it has at most $n-1$ color classes. Hence, we stop whenever we reach more than $n-1$ color classes. Thus the total complexity for this part is $O(n^2)$.

This proves the theorem, because the isometry test also requires at most $O(n^2)$ time. □

Exercises

18.1. Given a graph G, show that every edge of G must be considered by any algorithm that checks whether G is a partial cube.

18.2. Write a pseudocode for the computation of the transitive closure of a relation R.

18.3. Given the distance matrix of a graph G and its adjacency list, write a pseudocode for the direct computation of Θ (complexity $O(m^2)$).

18.4. Using the results of Exercises 18.2 and 18.3, write a pseudocode for the recognition of partial cubes.

18.5. Given a BFS-tree of a partial cube G and its Θ-classes, write a pseudocode that computes the Hamming distance between any two vertices u, v of G in $O(k)$ time, where k is the number of Θ-classes of G.

18.6. Consider Algorithm 18.2. How can one check after Step 6 in $O(1)$ time whether the input graph G is a hypercube?

18.7. Let F be an arbitrary set of edges in a graph G given by its extended adjacency list. Show that one can contract every edge uv of F in $O(m)$ time.

18.8. Let G be a graph with multiple edges and loops. Show that all multiple edges can be replaced by single ones and all loops removed in $O(m)$ time.

18.9. Given the adjacency list of a connected graph G, write a pseudocode for the computation of the canonical isometric embedding α of G in $O(mn)$ time.

18.10. Consider the graph of Figure 18.5 and apply Procedures 18.4 and 18.5 with an arbitrary starting vertex different from v_0 and v_3. Verify that not all computed color classes are matchings.

18.11. (Foldes, 1977) Show that a connected bipartite graph G is a hypercube if and only if the number of shortest u, v-paths is $d(u, v)!$ for any two vertices u and v of G.

Chapter 19

Chemical Graphs and the Wiener Index

Graphs arising in chemistry are a primary source of examples for graph theory; chemical trees, benzenoid graphs, and fullerenes[1] are just a few of the prominent examples. After a molecule is represented as a graph, the primary goal of chemical graph theory is to investigate the graph and to predict the molecule's properties. This is frequently achieved by computing carefully selected graph invariants. The Wiener index, introduced by Wiener (1947), is the oldest such invariant.[2] It has been widely investigated; see for instance the extensive surveys on the Wiener index of trees (Dobrynin, Entringer, and Gutman, 2001) and of benzenoid graphs (Dobrynin, Gutman, Klavžar, and Žigert, 2002).

In this chapter we show how isometric embeddings into Cartesian products lead to a better understanding of the Wiener index. We first observe that benzenoid graphs are partial cubes and then use their isometric embeddings into hypercubes to obtain closed expressions for the Wiener index. We show that benzenoid graphs are also isometrically embeddable into the Cartesian product of three related trees, a result that leads to a linear algorithm for computing the Wiener index of benzenoid graphs. In the final section we show that the computation of the Wiener index of an arbitrary graph can be reduced to the computation of the weighted Wiener index of the quotients in the graph's canonical isometric embedding.

19.1 Benzenoid Graphs as Partial Cubes

Benzenoid graphs represent benzenoid hydrocarbons, a class of substances of great importance in chemistry. The number of known benzenoid hydrocarbons is about one thousand, and some of them play a major role in the chemical industry. For more information on these compounds as well as their graphs, we refer to the book by Gutman and Cyvin (1989).

Let Z be a circuit of the hexagonal lattice in the plane. A *benzenoid graph* or a *hexagonal system* is formed by the vertices and edges of the hexagonal lattice lying on and in the interior of Z. Figure 19.1 shows an example. Note that a benzenoid graph is bipartite, as it is a subgraph of a bipartite hexagonal lattice.

Contrary to the general practice in this book, the next definition is formulated in the language of Euclidean geometry. We thereby take advantage of the fact that benzenoid graphs are planar.

Let G be a benzenoid graph. A straight line segment $S = pq$ is called a *cut segment*

[1]Fullerenes are carbon molecules whose graphs are cubic, plane graphs with faces of size 5 and 6.

[2]In chemical literature, invariants used for predicting properties of molecules are commonly referred to as topological indices of graphs.

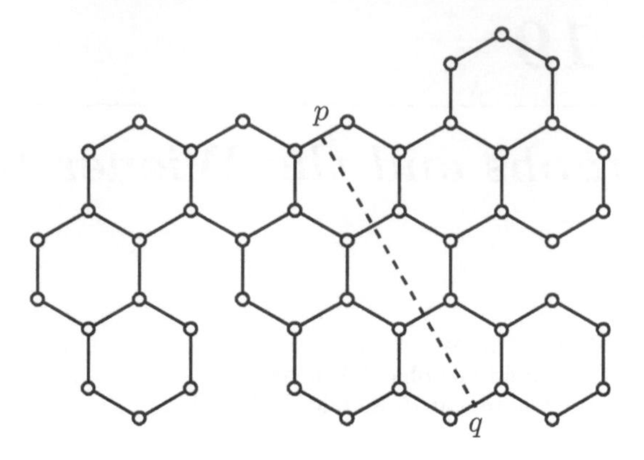

FIGURE 19.1 A benzenoid graph and one of its cut segments.

if it is a perpendicular bisector of two edges of the bounding cycle Z, and no point of S lies in the exterior region. The set of edges that S meets is called the *cut* corresponding to S. Note that the graph obtained from G by removing all such edges has exactly two connected components. Figure 19.1 shows a cut segment whose corresponding cut consists of four edges.

Let $e = uv$ be an arbitrary edge of a benzenoid graph G. If we remove the edges of the cut containing e, we obtain two connected components that are induced by the vertex sets W_{uv} and W_{vu}. (Recall that W_{uv} is the set of vertices closer to u than to v.) Moreover, it is clear that a shortest path between two vertices of W_{uv} cannot contain two edges of the cut containing e. It follows that W_{uv} induces a convex subgraph of G. By Theorem 11.8 (ii), we thus infer the following proposition due to Klavžar, Gutman, and Mohar (1995):

Proposition 19.1 *A benzenoid graph is a partial cube. Its Θ-classes coincide with its cuts.*

Proposition 19.1 can be applied to the computation of distance-related invariants of benzenoid graphs. We demonstrate it for the Wiener index; for a survey on corresponding results on related invariants, see Klavžar (2008).

The *Wiener index* $W(G)$ of a (molecular) graph G is defined as the sum of the distances between all pairs of vertices of G, that is,

$$W(G) = \frac{1}{2} \sum_{u \in V(G)} \sum_{v \in V(G)} d(u,v)\,.$$

Proposition 19.2 *Let G be a benzenoid graph on n vertices and $E_1, \ldots, E_k$ be its Θ-classes. For $i = 1, \ldots, k$, let $u_i v_i \in E_i$ and $n_i = |W_{u_i v_i}|$. Then*

$$W(G) = \sum_{i=1}^{k} n_i(n - n_i)\,.$$

Proof Let α be the isometric embedding of G into Q_k as in the proof of Theorem 11.8. Then

$$W(G) = \frac{1}{2} \sum_{u \in V(G)} \sum_{v \in V(G)} \sum_{i=1}^{k} \delta_i(u,v)\,,$$

where $\delta_i(u,v)$ is equal to 0 if the ith coordinates of $\alpha(u)$ and $\alpha(v)$ coincide; otherwise, $\delta_i(u,v)$ is equal to 1. Therefore

$$W(G) = \sum_{i=1}^{k} \left(\frac{1}{2} \sum_{u \in V(G)} \sum_{v \in V(G)} \delta_i(u,v) \right).$$

Because $\delta_i(u,v) = 1$ if and only if one of u and v is in $W_{u_i v_i}$ and the other is in $V(G) \setminus W_{u_i v_i}$, the inner sums equal $2n_i(n - n_i)$, and the proposition follows. □

Proposition 19.2 is particularly useful for deriving closed formulas for the Wiener index of benzenoid graph families. For instance, let H_k be the kth benzenoid graph from the so-called coronene/circumcoronene series. Figure 19.2 illustrates the first three such graphs. Historically the problem of finding a closed expression for $W(H_k)$ was considered challenging. Following Gutman and Klavžar (1996), we next show that Proposition 19.2 makes this routine.

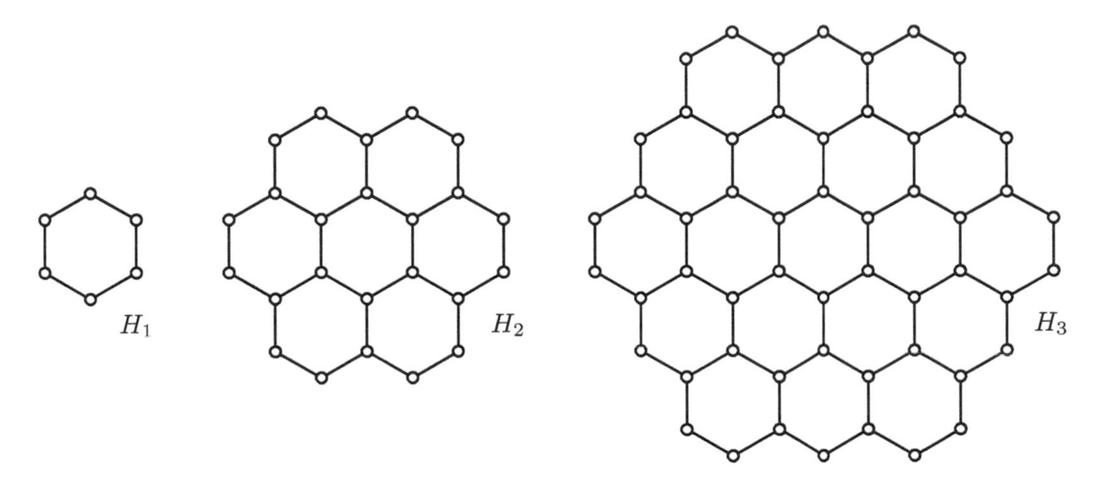

FIGURE 19.2 First elements of the coronene/circumcoronene series.

The relation $\Theta = \Theta^*$ partitions $E(H_k)$ into $3(2k-1)$ equivalence classes. Let $F_1, F_2, \ldots, F_{2k-1}$ be the classes corresponding to the horizontal cut segments. It is not difficult to verify that H_k has $6k^2$ vertices and that for $i = 1, 2, \ldots, k$, the two components of $G - F_i$ contain $i(2k+i)$ and $6k^2 - i(2k+i)$ vertices, respectively (Exercise 19.1). Therefore, by Proposition 19.2,

$$W(H_k) = 3\Big((3k^2)^2 + 2 \sum_{i=1}^{k-1} i(2k+i) \big[6k^2 - i(2k+i) \big] \Big),$$

which reduces to

$$W(H_k) = \frac{1}{5} \left(164k^5 - 30k^3 + k \right).$$

To close this section, we add that Proposition 19.2 extends naturally to all partial cubes. In addition, Chepoi, Deza, and Grishukhin (1997) extended it to all ℓ_1-graphs.

19.2 The Wiener Index of Benzenoid Graphs in Linear Time

We now turn our attention to isometric embeddings of benzenoid graphs into Cartesian products of trees, as introduced by Chepoi (1996), and show how to compute the Wiener index of benzenoid graphs in linear time.

Let G be a benzenoid graph and E_1, E_2, and E_3 the partition of $E(G)$ into sets of edges of the same direction. In other words, E_i is the union of Θ^* equivalence classes corresponding to cut segments of a given direction. For $i = 1, 2, 3$, set $G_i = G - E_i$. The connected components of the G_i are paths. Define a quotient graph T_i as usual, that is, as the graphs G_i^* in Section 11.1. The vertices of T_i are the connected components of G_i, and two such components P' and P'' are adjacent in T_i if some edge in E_i joins a vertex of P' to a vertex of P''. We then define a mapping

$$\gamma : V(G) \to V(T_1 \,\square\, T_2 \,\square\, T_3)$$

by $\gamma(v) = (v_1, v_2, v_3)$, where v_i is the vertex of T_i corresponding to the connected component of G_i containing v.

Note that every T_i is a tree, because a cycle in T_i would imply that G has a nonhexagonal interior face. (See Figure 19.3.) We call γ the *3-tree embedding* of G.

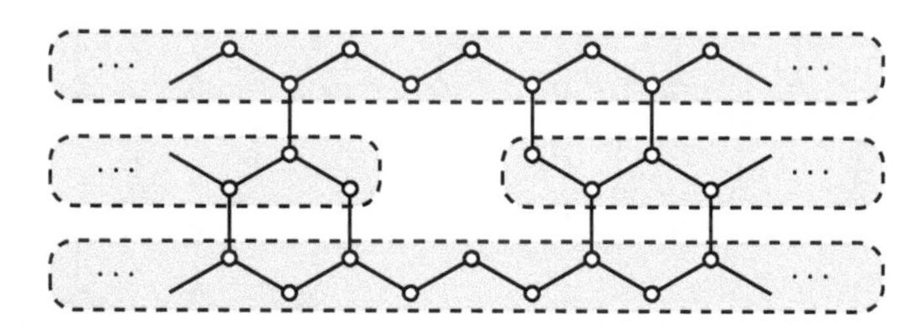

FIGURE 19.3 A cycle in T_i would yield a nonhexagonal interior face.

Theorem 19.3 *Let G be a benzenoid graph on n vertices. Then the 3-tree embedding γ is isometric and can be computed in $O(n)$ time.*

Proof Let $\gamma : V(G) \to V(T_1 \,\square\, T_2 \,\square\, T_3)$ be the 3-tree embedding and u, v be any two vertices of G. Furthermore, let $\gamma(u) = (u_1, u_2, u_3)$ and $\gamma(v) = (v_1, v_2, v_3)$. Select a shortest path P from u to v in G; and for $i = 1, 2, 3$, set $F_i = E(P) \cap E_i$. Observe that no two edges of F_i belong to the same cut, for otherwise P would not be a shortest path. It follows that $|F_i| = d_{T_i}(u_i, v_i)$. As $d_G(u, v) = |F_1| + |F_2| + |F_3|$, we infer by Lemma 5.2 that γ is indeed isometric.

As G is planar, Corollary 1.12 implies that it has $O(n)$ edges. It is thus straightforward to compute the trees T_1, T_2, and T_3 and the corresponding labels of the vertices of G in $O(n)$ time. □

Next we extend the definition of the Wiener index to weighted graphs. A *(vertex)-weighted graph* (G, w) is a graph G together with a function w from $V(G)$ into the set of positive integers. (In general, w can be a real valued function, but for our purposes integer-valued functions suffice.)

The Wiener index $W(G, w)$ of a weighted graph (G, w) is then defined as

$$W(G,w) = \frac{1}{2} \sum_{u \in V(G)} \sum_{v \in V(G)} w(u)\, w(v)\, d(u,v)\,.$$

Let G be a benzenoid graph and T_1, T_2, T_3 be the trees of its 3-tree embedding. We extend them to weighted trees (T_i, w_i) as follows: For $x \in V(T_i)$, we set $w_i(x)$ equal to the number of vertices in the connected component of G_i corresponding to x.

Proposition 19.4 *For a benzenoid graph G,*

$$W(G) = W(T_1, w_1) + W(T_2, w_2) + W(T_3, w_3)\,.$$

Proof The proof follows the proof of Proposition 19.2. Let $H = T_1 \Box T_2 \Box T_3$ and let γ be the 3-tree embedding of G. Furthermore, for $u \in V(G)$, let $\gamma(u) = (u_1, u_2, u_3)$. Then

$$\begin{aligned}
W(G) &= \frac{1}{2} \sum_{u \in V(G)} \sum_{v \in V(G)} d_H(\gamma(u), \gamma(v)) \\
&= \frac{1}{2} \sum_{u \in V(G)} \sum_{v \in V(G)} \sum_{i=1}^{3} d_{T_i}(u_i, v_i) \\
&= \sum_{i=1}^{3} \left(\frac{1}{2} \sum_{u \in V(G)} \sum_{v \in V(G)} d_{T_i}(u_i, v_i) \right) \\
&= \sum_{i=1}^{3} \left(\frac{1}{2} \sum_{u_i \in V(T_i)} \sum_{v_i \in V(T_i)} w_i(u) w_i(v) d_{T_i}(u_i, v_i) \right) \\
&= \sum_{i=1}^{3} W(T_i, w_i).
\end{aligned}$$

□

By Proposition 19.4, any linear algorithm that computes the Wiener index of a weighted tree provides a linear algorithm for $W(G)$. To show that such an algorithm exists, we invoke the following lemma. Its proof is similar to the proof of Proposition 19.2 and is left to the reader.

Lemma 19.5 *Let (T, w) be a weighted tree. For an edge e of T, let T_1 and T_2 be the connected components of $T - e$ and, for $i = 1, 2$,*

$$n_i(e) = \sum_{u \in V(T_i)} w(u)\,.$$

Then

$$W(T, w) = \sum_{e \in E(T)} n_1(e) n_2(e)\,.$$

Lemma 19.5 can be extended to arbitrary partial cubes; see Klavžar and Gutman (1997). Interestingly, Wiener (1947) attained Lemma 19.5 for unweighted trees.

We are now ready for the main result of this section. It is by Chepoi and Klavžar (1997).

Theorem 19.6 *The Wiener index of a benzenoid graph on n vertices can be computed in $O(n)$ time.*

Proof By Theorem 19.3, we can compute the trees T_1, T_2, and T_3 of the 3-tree embedding in $O(n)$ time. Therefore we can also obtain the weighted trees (T_i, w_i) within the same time complexity. By Proposition 19.4, it remains to show that we can compute the Wiener index of a weighted tree in linear time. Using Lemma 19.5 we proceed as follows:

Order the vertices of a given weighted tree (T, w) so that every vertex v is a pendant vertex in the subtree induced by the vertices with a larger index. Let u be the neighbor of v in this subtree. Then add the factor $w(v)(n - w(v))$ to the current sum and update $w(u)$ by setting $w(u) = w(u) + w(v)$. □

The above argument that the Wiener index of a weighted tree can be computed in linear time extends the linear algorithm of Mohar and Pisanski (1988) for the Wiener index of (unweighted) trees.

19.3 The Wiener Index via the Canonical Isometric Embedding

In this section we show that the Wiener index of an arbitrary connected graph can be expressed with the weighted Wiener index of (weighted) quotient graphs of the canonical isometric embedding of G. The reader should compare this approach to the one from Section 19.1 to realize that the present generalizes the former.

Recall from Chapter 13 that the canonical embedding α maps a connected graph G isometrically into $G/\Pi_1 \square G/\Pi_2 \square \cdots \square G/\Pi_k$, where G/Π_i is the quotient graph of $G - E_i$ (and E_i is a Θ-class of G). We now assign weights to the vertices of G/Π_i in the natural way: For $x \in V(G/\Pi_i)$, let $w_i(x)$ be the number of vertices in the connected component of $G - E_i$ corresponding to x. The following result is from Klavžar (2005):

Theorem 19.7 *Let G be a connected graph and let $\alpha : G \to G/\Pi_1 \square G/\Pi_2 \square \cdots \square G/\Pi_k$ be the canonical embedding. Then*

$$W(G) = \sum_{i=1}^{k} W(G/\Pi_i, w_i)\,.$$

Proof Set $H = G/\Pi_1 \square G/\Pi_2 \square \cdots \square G/\Pi_k$. By Theorem 13.2,

$$W(G) = \frac{1}{2} \sum_{u \in V(G)} \sum_{v \in V(G)} d_G(u, v) = \frac{1}{2} \sum_{u \in V(G)} \sum_{v \in V(G)} d_H(\alpha(u), \alpha(v))\,.$$

Then the Distance Formula (Corollary 5.2), in turn, implies that

$$\begin{aligned} W(G) &= \frac{1}{2} \sum_{u \in V(G)} \sum_{v \in V(G)} \sum_{i=1}^{k} d_{G/\Pi_i}(\alpha_i(u), \alpha_i(v)) \\ &= \sum_{i=1}^{k} \Big(\frac{1}{2} \sum_{u \in V(G)} \sum_{v \in V(G)} d_{G/\Pi_i}(\alpha_i(u), \alpha_i(v)) \Big)\,. \end{aligned} \tag{19.1}$$

Observe next that, by the definition of the weighted Wiener index,

$$W(G/\Pi_i, w_i) = \frac{1}{2} \sum_{u \in V(G)} \sum_{v \in V(G)} d_{G/\Pi_i}(\alpha_i(u), \alpha_i(v))\,. \tag{19.2}$$

Inserting (19.2) into (19.1) yields the result. □

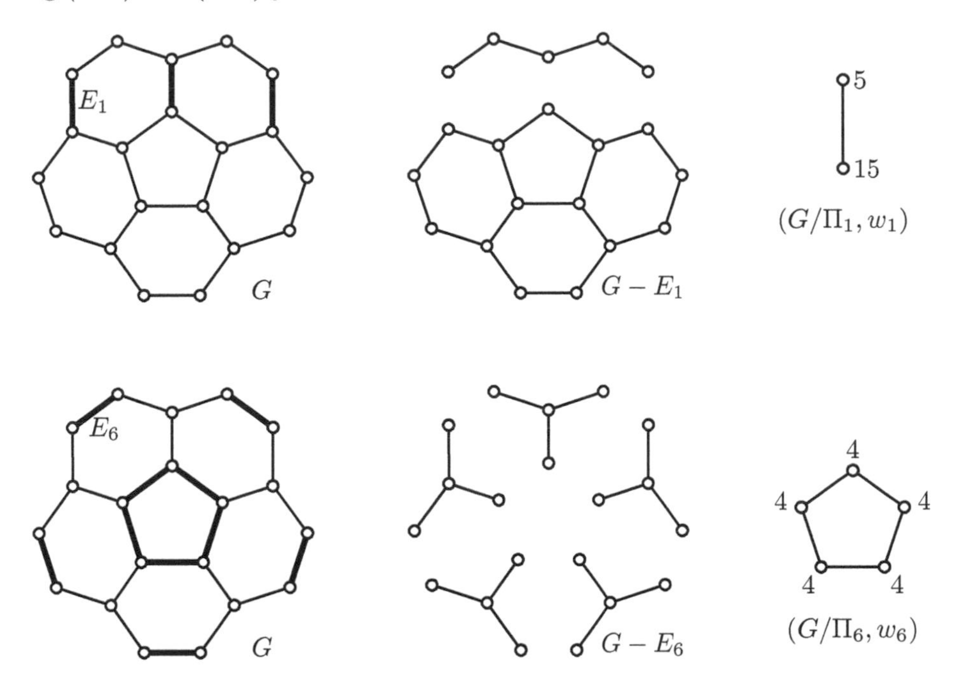

FIGURE 19.4 Computing the Wiener index of $C_{20}(2)$.

As an example, consider the chemical graph $G = C_{20}(2)$ of Figure 19.4. This graph has six Θ^*-classes $E_1, \ldots, E_6$. Five of them, say $E_1, \ldots, E_5$, consist of three edges, and the remaining class E_6 has five edges. The figure shows E_1 and E_6, as well as $G - E_1$, $G - F_6$, $(G/\Pi_1, w_1)$, and $(G/\Pi_6, w_6)$. Note that $W(G/\Pi_i, w_i) = 5 \cdot 15 = 75$ for $1 \leq i \leq 5$, and $W(G/\Pi_6, w_6) = 5 \cdot 4 \cdot 4 + 5 \cdot 2 \cdot 4 \cdot 4 = 240$. Therefore $W(C_{20}(2)) = 5 \cdot 75 + 240 = 615$.

Exercises

19.1. Verify that the graph H_k defined on p. 233 has $6k^2$ vertices and that for $i = 1, 2, \ldots, k$, the two components of $G - F_i$ contain $i(2k + i)$ and $6k^2 - i(2k + i)$ vertices, respectively. (F_i is also defined on p. 233.)

19.2. (Graovac and Pisanski, 1991; Yeh and Gutman, 1994) Show that for any connected graphs G and H, $W(G \,\square\, H) = |V(G)|^2 \cdot W(H) + |V(H)|^2 \cdot W(G)$.

19.3. Obtain closed formulas for $W(Q_d)$ and $W(P_n \,\square\, P_m)$, where $d, n, m \geq 1$.

19.4. Show that, among all trees with n edges, the Wiener index is minimized on the star $K_{1,n}$ and maximized on P_{n+1}.

19.5. The *linear chain* L_h is the benzenoid graph consisting of h hexagons arranged in a straight line. (For instance, the "middle horizontal level" of the graph H_k from p. 233 is isomorphic to L_{2k-1}.) Express $W(L_h)$ as a function of h.

19.6. Using Theorem 19.7, compute $W(G)$, where G is the graph from Figure 13.1.

Chapter 20

Arboricity, Squares, and Triangles

We now review the concept of the arboricity of a graph. This invariant plays a significant role in bounding the complexity of certain algorithms. For example, we will soon have need to find all squares or triangles of a graph G. These tasks can be accomplished with relatively simple algorithms whose running time is bounded by $O(|V(G)| \cdot a(G))$, where $a(G)$ is the arboricity of G. As $a(G)$ is quite small for many of the graphs we consider (such as subgraphs of hypercubes), these methods lead to significant improvements in running time.

The first section of the chapter presents a fundamental formula for the arboricity of a graph and proves several upper bounds on arboricity. This leads to the so-called Arboricity Lemma, a useful tool for bounding the complexity of certain edge-scanning algorithms.

The second section applies these results to algorithms that list the squares or triangles of graphs. We also show how these algorithms help to efficiently compute several relations defined on the edge set of a graph.

20.1 Arboricity

The arboricity $a(G)$ of a graph G is the minimum number of edge-disjoint spanning forests into which G can be decomposed. For example, $a(G) = 1$ if and only if G is acyclic.

We will shortly produce a formula for the arboricity of an arbitrary graph. In order to accomplish this, we first consider the problem of decomposing graphs into disjoint trees. For this purpose it will be helpful to admit graphs with *multiple edges*, that is, graphs for which distinct edges may have the same endpoints. Such graphs are called *multigraphs*.

Consider a multigraph G with a partition P of $V(G)$ into p disjoint nonempty sets $V_1, V_2, \ldots, V_p$. Let G/P denote the multigraph (without loops) whose vertices are the sets $V_1, V_2, \ldots, V_p$ and for which the number of edges joining V_i to V_j is the number of edges in G having one endpoint in V_i and the other in V_j. Note that G/P is defined like the quotient graphs in Chapter 11 (p. 159), with the exception that we admit multiple edges. If G has k edge-disjoint spanning trees, then $|E(G/P)| \geq k(p-1)$. Indeed, if T is a spanning tree of G, then $T/P \subseteq G/P$ is connected and thus has at least $p-1$ edges. Tutte (1961) and Nash-Williams (1961) independently proved that this necessary condition is also sufficient for the existence of k edge-disjoint spanning trees. The original proofs are rather long, and we only need the result for the special case in which $|E(G)| = k\,(|V(G)| - 1)$. So for this case, where every edge of G is an edge of one of the k spanning trees, we follow an approach of Bollobás (1978).

Lemma 20.1 *Let G be a multigraph with $k(|V(G)|-1)$ edges, such that*

$$|E(G/P)| \geq k\,(p-1)$$

for every partition $P = \{V_1, V_2, \ldots, V_p\}$ of $V(G)$. Then G has k edge-disjoint spanning trees.

Proof Assume that the hypotheses of the theorem hold. We consider systems $\mathcal{F} = \{F_1, F_2, \ldots, F_k\}$ of edge-disjoint spanning forests in G. Such systems always exist: take k copies of the totally disconnected graph on $V(G)$. We call such a system *maximal* if $\sum_{i=1}^{k} |E(F_i)|$ attains the maximum value over all $\mathcal{F}$. We have to show that such a maximal system consists of trees.

Suppose that this is not the case. Then, if $\mathcal{F}$ is maximal, there is an edge e of G that does not belong to any of the forests F_i. We call $R = \{e, \mathcal{F}\}$ a *maximal pair.* Because $F_i \cup e$ is not a forest, any F_i has a unique path $v_1 v_2 \ldots v_h$ joining the endpoints of e. Set $e' = v_j v_{j+1}$ for some $1 \leq j < h$. Consider the forest $F_i' = (F_i \cup e) - e'$ and the system

$$\mathcal{F}' = \{F_1, F_2, \ldots, F_{i-1}, F_i', F_{i+1}, \ldots, F_k\}.$$

Clearly, $R' = \{e', \mathcal{F}'\}$ is also a maximal pair. We say it is a *simple shift* of R and call a maximal pair $R^* = \{e^*, \mathcal{F}^*\}$ a *shift* of R if it is obtainable from R by sequence of simple shifts.

From now on we fix a maximal pair $R_0 = \{e_0, \mathcal{F}_0\}$, where $\mathcal{F}_0 = \{F_1, F_2, \ldots, F_k\}$, and call an edge a *shift edge* if it is the element e of a shift $\{e, \mathcal{F}\}$ of R_0. Let S be the set of all shift edges, and let V_0 be the set of their endpoints. Consider the subgraph $G_0 = (V_0, S)$. Now, G_0 is connected, as follows: The subgraph X formed by the shift edges resulting from simple shifts of R_0 is the union of the paths $P_i \subseteq F_i$ that join the endpoints of e_0. Clearly, X is connected. In turn, the shift edges resulting from simple shifts of any $e \in X$ are edges of paths joining the endpoints of e, and enlarging X to include them results in a connected subgraph. Continuing this process, we eventually arrive at the (connected) subgraph G_0.

We now show that the subgraph $\langle V_0 \rangle_{F_i}$ induced by V_0 on F_i is connected for each index i. To the contrary, suppose some $\langle V_0 \rangle_{F_i}$ is disconnected. Let U_0 be a component of $\langle V_0 \rangle_{F_i}$ and put $W_0 = V_0 \setminus V(U_0)$. Because R_0 is maximal, the edge e_0 does not join U_0 to W_0. Because G_0 is connected, there must be a shift $R' = \{e_0', \mathcal{F}_0'\}$ where e_0' joins U_0 with W_0. Then the unique path in F_i connecting the endpoints of e_0' has an edge joining U_0 to W_0. But this path consists of shift edges, so $\langle V_0 \rangle_{F_i}$ has an edge joining U_0 to W_0, a contradiction. Thus each $\langle V_0 \rangle_{F_i}$ is connected.

Because each $\langle V_0 \rangle_{F_i}$ is connected, we infer that $V_0 \neq V(G)$, otherwise each F_i is a tree, contrary to assumption. But each $\langle V_0 \rangle_{F_i}$ is a tree and has $|V_0|-1$ edges. Thus the subgraph $F_1 \cup F_2 \cup \cdots \cup F_k \cup \{e_0\}$ of $\langle V_0 \rangle_G$ has at least $k\,(|V_0|-1)+1$ edges. But this leaves at most $k\,(|V(G)|-1) - k\,(|V_0|-1) - 1 = k\,((|V(G)|-|V_0|+1)-1)-1$ edges between the $|V(G)|-|V_0|+1$ sets of the partition of $V(G)$ consisting of V_0 and the elements of $V(G) \setminus V_0$, contrary to the assumptions of the theorem. □

The more general problem of determining the arboricity of a graph was solved by Nash-Williams (1964). Here we follow the approach of Behzad and Chartrand (1971).

Theorem 20.2 *For any graph G with at least one edge,*

$$a(G) = \max_{H \subseteq G} \left\lceil \frac{|E(H)|}{|V(H)|-1} \right\rceil, \tag{20.1}$$

where the maximum is taken over all nontrivial subgraphs $H \subseteq G$.

Proof Let G be a graph with at least one edge. Because every forest of G intersects any $H \subseteq G$ at no more than $|V(H)| - 1$ edges, we have $a(G)\big(|V(H)| - 1\big) \geq |E(H)|$. Thus

$$a(G) \geq \max_{H \subseteq G} \left\lceil \frac{|E(H)|}{|V(H)| - 1} \right\rceil.$$

Let $k = \max_{H \subseteq G} \lceil |E(H)|/(|V(H)| - 1) \rceil$. It remains to show that $E(G)$ can indeed be decomposed into k subsets, each inducing a forest. The idea of the proof is to add edges to G until a new graph J is formed that is a union of edge-disjoint trees. Then deletion of the added edges yields the desired decomposition of G into spanning forests.

To this end we introduce some relevant notation and definitions. For any induced sub-multigraph M_1 of a multigraph M, we define the number

$$\psi(M_1) = k\big(|V(M_1)| - 1\big) - |E(M_1)|.$$

Also, we call a multigraph M *dense* if $\psi(M_1) \geq 0$ for every induced sub-multigraph M_1 of M. If $\psi(M_1) = 0$, then M_1 is called a *root of* ψ.

Note that G is dense because $\psi(H) \geq 0$ for all induced subgraphs H of G.

Next we show the existence of a dense multigraph J that is a root of ψ and for which G is a spanning subgraph of J. To begin, set $J = G$. If J already is a root of ψ, we are done. Otherwise, $\psi(J) > 0$, and we enlarge J as follows:

Let $u \in V(J)$. Note that the trivial subgraph $\langle \{u\} \rangle$ of J is a root of ψ. Suppose that the subgraphs G_1 and G_2 of J are both roots of ψ containing u. Consider $G_3 = \langle V(G_1) \cup V(G_2) \rangle$ and $G_4 = \langle V(G_1) \cap V(G_2) \rangle$. We show that G_3 and G_4 are also roots of ψ. From

$$|E(G_3)| + |E(G_4)| \geq |E(G_1)| + |E(G_2)|$$

and the definition of ψ, we infer that $\psi(G_3) + \psi(G_4) \leq \psi(G_1) + \psi(G_2) = 0$, and so $\psi(G_3) = 0 = \psi(G_4)$. But then the subgraph J' induced by the union of the vertex sets of the roots of ψ containing u also is a root of ψ containing u.

Because J is not a root of ψ, we have $V(J) \neq V(J')$. Hence there is a $v \in V(J) \setminus V(J')$ such that no root of ψ contains both u and v. Therefore the addition of an edge with the endpoints u and v to J yields a graph or multigraph J_1 with the property that $\psi(J_1) = \psi(J) - 1 \geq 0$. Note that J_1 is dense. To see this, let H_1 be an induced sub-multigraph of J_1 not containing both u and v; then H_1 is also an induced subgraph of J, and therefore $\psi(H_1) \geq 0$. On the other hand, if H_1 contains both u and v, then $H_1 - uv$ is an induced subgraph of J that is not a root of ψ, and thus $\psi(H_1 - e) \geq 1$, so $\psi(H_1) \geq 0$.

Note that $\psi(J') = \psi(J) - 1$. Set $J = J_1$. If J is not a root, then repeat the above process until J is a dense multigraph that is a root of ψ.

Let $P = V_1, V_2, \ldots, V_p$ be a partition of $V(J) = V(G)$. Then

$$|E(J/P)| = |E(J)| - \sum_{i=1}^{p} |E(\langle V_i \rangle)|. \tag{20.2}$$

Because J is a root of ψ, we have

$$|E(J)| = k(|V(J)| - 1). \tag{20.3}$$

Because J is dense,

$$k(|V_i| - 1) \geq |E(\langle V_i \rangle)| \tag{20.4}$$

for $1 \leq i \leq p$. Inserting the value for $|E(J)|$ from Equation (20.3) into Equation (20.2) and taking into account Inequality (20.4), we obtain

$$\begin{aligned}|E(J/P)| &\geq k\,(|V(J)|-1) - k\sum_{i=1}^{p}(|V_i|-1)\\ &= k(n-1) - k(n-p) = k(p-1).\end{aligned}$$

This means that J satisfies the assumptions of Lemma 20.1. It is thus the union of k pairwise edge-disjoint trees. Intersecting these trees with G, we obtain k mutually edge-disjoint spanning forests of G. □

Bounds on arboricity

It is not hard to derive general lower and upper bounds for arboricity. In particular,

$$\frac{\delta}{2} < a(G) \leq \Delta, \tag{20.5}$$

where δ and Δ are the minimum and maximum degrees of G; see Exercises 20.1 and 20.2.

Because many of the graphs treated here are sparse (in particular, hypercubes, their subgraphs and planar graphs), we are interested in bounds for them. Theorem 20.2 yields upper bounds on the arboricity of these classes of graphs. These bounds, together with a general estimate from Chiba and Nishizeki (1985), are the contents of the next theorem.

Theorem 20.3 *Let G be a graph on n vertices with m edges. Then*

(i) $a(G) \leq \left\lceil \frac{1}{2}\sqrt{2m+n}\,\right\rceil$ *in general,*

(ii) $a(G) \leq \left\lceil \frac{n}{2(n-1)}\log_2 n\right\rceil$ *if G is a subgraph of a hypercube, and*

(iii) $a(G) \leq 3$ *if G is planar.*

Proof We begin with the last assertion. Recall that a planar graph on n vertices has at most $3n-6$ edges, by Corollary 1.12. Thus $|E(H)| < 3\,(|V(H)|-1)$ for all such graphs. As each subgraph H of a planar graph G is planar, Theorem 20.2 gives $a(G) \leq 3$.

The second assertion is a consequence of the Density Lemma 3.2 that bounds the number of edges of any subgraph H of a hypercube by $\frac{1}{2}|H| \cdot \log_2 |H|$.

For (i), consider a subgraph $H \subseteq G$ for which the maximum in Equation (20.1) is attained. Put $p = |V(H)|$ and $q = |E(H)|$, so $p \leq n$ and $q \leq m$. Also, $q \leq p(p-1)/2$, because H cannot have more edges than K_p. Set $k = p(p-1)/2$ and consider two cases:

Suppose that $k \leq m$. Then

$$a(G) = \left\lceil \frac{q}{p-1}\right\rceil \leq \left\lceil \frac{k}{p-1}\right\rceil = \left\lceil \frac{p}{2}\right\rceil = \left\lceil \frac{\sqrt{2k+p}}{2}\right\rceil \leq \left\lceil \frac{\sqrt{2m+n}}{2}\right\rceil.$$

On the other hand, if $k \geq m$, then

$$\begin{aligned}a(G) &= \left\lceil \frac{q}{p-1}\right\rceil \leq \left\lceil \frac{m}{p-1}\right\rceil \leq \left\lceil \sqrt{\frac{mk}{(p-1)^2}}\right\rceil = \left\lceil \sqrt{\frac{m(p-1)+m}{2(p-1)}}\right\rceil \leq \\ &\left\lceil \sqrt{\frac{m}{2} + \frac{k}{2(p-1)}}\right\rceil = \left\lceil \frac{\sqrt{2m+p}}{2}\right\rceil \leq \left\lceil \frac{\sqrt{2m+n}}{2}\right\rceil.\end{aligned}$$

This completes the proof. □

Note that the proof gives $a(G) = \lceil \frac{1}{2}\sqrt{2m+n}\, \rceil$ if G is complete.

Theorem 20.3 yields good estimates for the complexity of algorithms that scan each edge uv of a graph G at a cost proportional to $\min(d(u), d(v))$. The next lemma, due to Chiba and Nishizeki (1985), places the running time of such algorithms at $O(a(G)\, m)$.

Lemma 20.4 (Arboricity lemma) *If G is a graph on n vertices and m edges, then*

$$\sum_{uv \in E(G)} \min(\, d(u), d(v)\,) \leq 2\, a(G)\, m.$$

Proof Let $F_1, F_2, \ldots, F_{a(G)}$ be edge-disjoint spanning forests of G whose union covers G. Each edge $e \in E(G)$ belongs to a component T of some F_i. (Clearly, T is a tree.) For each such T, choose a root $r_T \in V(T)$. Now for each edge e, let $h(e)$ be the endpoint of e that is further from r_T in T. Because each e is contained in exactly one F_i, we have

$$\begin{aligned} \sum_{uv \in E(G)} \min(\, d(u), d(v)\,) &\leq \sum_{i=1}^{a(G)} \sum_{e \in E(F_i)} d(h(e)) \\ &\leq \sum_{i=1}^{a(G)} \sum_{v \in V(G)} d(v) \\ &= 2\, a(G)\, m. \end{aligned}$$

□

20.2 Listing Squares and Triangles

From the definition of the Cartesian product, it is quite obvious that squares play a special role in all recognition and decomposition algorithms of graphs with respect to the Cartesian product. Consequently, the same holds for recognition and embedding algorithms of partial cubes, median graphs, and related classes of graphs.

It is not difficult to find all squares of a graph. One might consider all pairs v, w of vertices and all common neighbors $u_1, u_2, \ldots, u_k$. Then any choice u_i, u_j for $1 \leq i < j \leq k$ yields a square vu_iwu_j. We might even find it economical not to list all these squares individually but to store the triple $(v, w, \{u_1, u_2, \ldots, u_k\})$. As an example, consider K_5. The following triples describe its fifteen squares:

$$(v_1, v_2, \{v_3, v_4, v_5\}),\ (v_1, v_3, \{v_2, v_4, v_5\}),$$
$$(v_1, v_4, \{v_2, v_3, v_5\}),\ (v_1, v_5, \{v_2, v_3, v_4\}),$$
$$(v_2, v_3, \{v_4, v_5\}),\ (v_2, v_4, \{v_3, v_5\}),\ (v_2, v_5, \{v_3, v_4\}).$$

The first triple represents the squares $v_1v_3v_2v_4$, $v_1v_3v_2v_5$, and $v_1v_4v_2v_5$, while the last triple stands for the square $v_2v_3v_5v_4$.

Suppose that we are given a pair v, w of vertices in a graph G on n vertices. To find all common neighbors of v, w, it suffices to scan the vertices of G and to check whether they are adjacent to both v and w. Depending on the data structure, each one of these checks can be performed in constant time or in $O(\log n)$ time. Because G has $O(n^2)$ pairs of vertices, the overall complexity of this approach is $O(n^3)$ or $O(n^3 \log n)$. As the complete graph K_n has $O(n^4)$ squares, this may appear to be quite good already and supports the idea of storing the triples $(v, w, \{u_1, u_2, \ldots, u_k\})$.

A variant is to scan all neighbors of a vertex v, namely all vertices $u \in N(v)$, and to assign the label u to every neighbor w of u with $w \neq v$. Suppose, after completion of this procedure, that w has been assigned the labels $X_w = \{u_1, u_2, \ldots, u_k\}$. Then $(v, w, \{u_1, u_2, \ldots, u_k\})$ is the aforementioned triple. Next observe that we can remove v from G and continue with $G - v$, because all squares containing v have been found. Thus, in order to find an efficient algorithm, we should look for a good strategy to choose v. A natural approach is to order the vertices according to their degrees and to process them in this order. This leads to the following algorithm of Chiba and Nishizeki (1985):

Algorithm 20.1 Squares

Input: A connected graph G.
Output: A set of triples $(v, w, \{u_1, u_2, \ldots, u_k\})$ such that every square of G is of the form vu_iwu_j, where $1 \leq i < j \leq k$.

1: Sort the vertices of G such that $d(v_1) \geq d(v_2) \geq \cdots \geq d(v_n)$
2: **for all** $w \in V(G)$, initialize a set $X_w := \emptyset$.
3: **for** $i = 1$ to $n - 3$ **do**
4:    **for all** $u \in N_G(v_i)$ **do**
5:       **for all** $w \in N_G(u)$ with $w \neq v_i$ **do**
6:          $X_w := X_w \cup \{u\}$.
7:       **end for**
8:    **end for**   *(At this point $X_w = N_G(v_i) \cap N_G(w)$ for each w.)*
9:    **for all** $u \in N_G(v_i)$ **do**
10:       **for all** $w \in N_G(u)$ with $w \neq v_i$ **do**
11:          **if** $|X_w| \geq 2$, **then** store (v_i, w, X_w).
12:          $X_w := \emptyset$.
13:       **end for**
14:    **end for**
15:    Replace G by $G - v_i$.
16: **end for**

As an example of this algorithm's execution, consider the graph in Figure 20.1 with five squares. Its vertices are already indexed so that $d(v_1) \geq d(v_2) \geq \cdots \geq d(v_7)$. The algorithm produces the triples $(v_1, v_2, \{v_3, v_5\})$, $(v_1, v_3, \{v_2, v_7\})$, and $(v_1, v_4, \{v_2, v_5, v_6\})$.

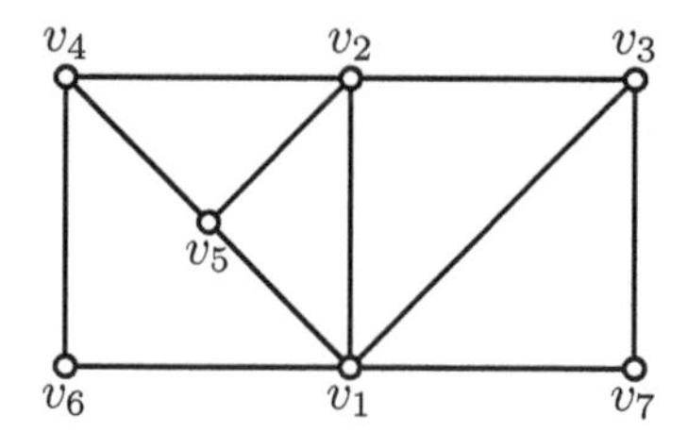

FIGURE 20.1 Graph with five squares.

Theorem 20.5 *If G is a connected graph with m edges, then Algorithm 20.1 correctly codes all squares of G, and can be implemented to run in $O(m\,a(G))$ time. The triples that code the squares contain at most $2m\,a(G)$ entries.*

Proof As noted above, the algorithm codes all squares of G. We need only compute its time complexity.

Clearly, the degrees can be computed in $O(m)$ time. Because the degree of every vertex is at most $n-1$, we can use bucket sort[1] to obtain the sequence $d(v_1) \geq d(v_2) \geq \cdots \geq d(v_n)$. Thus Line 1 of the algorithm needs $O(m+n) \leq O(m)$ time. Line 2 is $O(n) \leq O(m)$.

Next we move on to the outer for-loop in Lines 3 through16. In its ith iteration, our original graph G has been updated to $H = G - \{v_1, v_2, \ldots, v_{i-1}\}$. Consider the nested for-loop in Lines 4 through 7. These lines scan the neighbors w of neighbors u of v_i. Although v_i need not be of maximum degree in H, we still have

$$d_H(u) \leq d_G(u) \leq d_G(v_i)$$

for any $u \in N_G(v_i)$. Thus the number of times (for a given i) that Line 6 is executed is

$$\sum_{u \in N_H(v_i)} d_H(u) \leq \sum_{u \in N_H(v_i)} \min\{d_G(u), d_G(v_i)\}.$$

Consequently, over the entire runtime of the algorithm, Line 6 is executed no more than

$$\sum_{v_i \in V(G)} \sum_{u \in N_H(v_i)} \min\{d_G(u), d_G(v_i)\}$$

times. Because every edge of G is involved exactly once in the above double summation, we can rewrite it as

$$\sum_{uv \in E(G)} \min\{d(u), d(v)\},$$

whose value is at most $2m\, a(G)$ by Lemma 20.4. Thus the overall contribution of Lines 4–8 is $O\big(m\, a(G)\big)$.

By the same reasoning, the overall contribution of Lines 9 through 14 is $O\big(m\, a(G)\big)$. In addition, the total number of triples (v_i, w, X) is no more than $2m\, a(G)$.

Finally, consider Line 15. Here v_i is removed from G, and this can be done in $O(d(v_i))$ time. (To do so, one has to represent the graph as a doubly linked list as described in Chapter 17.) As Line 15 executes once for each vertex of G, its contribution to the total complexity is $O(m)$.

In summary, the overall complexity of the algorithm is $O\big(m\, a(G)\big)$. □

Relations associated with the squares of a graph

Here we are concerned with efficient computation of several relations on the edge set of a graph G. We begin with the definition of a binary relation δ. Two edges e, f are said to be in relation δ if they are identical or opposite edges of a square without diagonals. This relation and its transitive closure δ^* are important for the characterization and embedding of median and semi-median graphs (see p. 252). In view of Theorem 20.5, we should like to compute δ in $O\big(m\, a(G)\big)$ time, but the complete bipartite graph $K_{2,n}$ shows that this is not always possible. It has $\binom{n}{2}$ squares without diagonals and thus $n(n-1)$ pairs of edges that are in relation δ. However, its arboricity is 2, so $m \cdot a(K_{2,n}) \leq 4n$, which is an order of magnitude smaller than $|\delta| = n(n-1)$.

But a closer scrutiny of Algorithm 20.1 and Theorem 20.5 yields some improvements. The next proposition lists several.

---

[1] For the description of bucket sort and other sorting algorithms, see Aho et al. (1987).

Proposition 20.6 *Let G be a connected graph on n vertices with m edges.*

(i) *If G has no $K_{2,3}$ subgraph, then δ and δ^* can be computed in $O(m\,a(G))$ time.*

(ii) *If G is triangle-free, then δ^* can be computed in $O(m\,a(G))$ time.*

(iii) *If G is triangle-free and $K_{2,3}$-free, then it has at most $m\,a(G)/2$ squares.*

Proof (i) If G has no subgraph isomorphic to $K_{2,3}$, then Algorithm 20.1 produces only triples of the form $(v, w, \{u_1, u_2\})$. This means that δ can be computed for such graphs in $O(m\,a(G))$ time. Moreover, by Proposition 18.2, the transitive closure δ^* can also be determined in $O(m\,a(G))$ time.

(ii) Let G be triangle-free, and assume that it has a triple $(v, w, \{u_1, u_2, \ldots, u_k\})$, where $k \geq 3$. Then all squares vu_iwu_j with $1 \leq i < j \leq k$ are induced subgraphs without diagonals, so the vertices of the triple induce a $K_{2,k}$. Any two edges of this $K_{2,k}$ are in relation δ^*. Hence $\delta^*(G)$ can be computed in $O(m\,a(G))$ time.

(iii) Let G be triangle-free and $K_{2,3}$-free. By Theorem 20.5, Algorithm 20.1 produces at most $2m\,a(G)$ triples. Because each triple encodes exactly one square, G cannot have more than $m\,a(G)/2$ squares. □

As subgraphs of hypercubes are of special interest, we record the following special case:

Corollary 20.7 *If G is a connected subgraph of a hypercube and has n vertices and m edges, then δ and δ^* can be computed in $O(m \log_2 n)$ time. Moreover, G has at most*

$$\left\lceil \frac{mn}{4(n-1)} \log_2 n \right\rceil$$

squares.

Proof Because subgraphs of hypercubes contain no $K_{2,3}$, the first statement follows from Proposition 20.6 (i) and Theorem 20.3. For the second assertion, recall from Theorem 20.3 that the arboricity of subgraphs of hypercubes is bounded by $\lceil \frac{n}{2(n-1)} \log_2 n \rceil$. Now apply Proposition 20.6 (iii). □

Chapter 23 will introduce a relation τ that plays an important role in the prime factorization of graphs over the Cartesian product. Two edges vu, uw of a graph are in relation τ if u is the only common neighbor of v and w. Clearly, τ corresponds to all those triples $(v_i, w, N_G(v_i) \cap N_G(w))$ computed (and discarded) by Algorithm 20.1 for which $|N_G(v_i) \cap N_G(w)| = 1$ and $vw \notin E(G)$. Therefore, both τ and τ^* can be computed for any connected graph G by a slight modification of Algorithm 20.1 in $O(m\,a(G))$ time:

Proposition 20.8 *For any connected graph G, the relations τ and τ^* can be computed in $O(m\,a(G))$ time. The proposition holds for τ^*, but not for τ.*

Listing triangles in a graph

Finally we turn our attention to triangles in graphs. This topic is important because of a close connection between median graphs and triangle-free graphs that we will use in a recognition algorithm.

The following algorithm, Algorithm 20.2, which is also from Chiba and Nishizeki (1985), computes all triangles of a connected graph. Despite its similarity to Algorithm 20.1 for squares, it clearly finds all triangles of G. Its time complexity is $O(m\,a(G))$, as can be shown by the same arguments as were used for the proof of Theorem 20.5. We formulate this as a theorem.

Theorem 20.9 *If G is a connected graph with m edges, then Algorithm 20.2 finds and lists all triangles of G. It can be implemented to run in $O(m\,a(G))$ time.*

Algorithm 20.2 Triangles

Input: A connected graph G.
Output: The set of triangles of G.

1: Sort the vertices of G such that $d(v_1) \geq d(v_2) \geq \cdots \geq d(v_n)$.
2: **for** $i = 1$ to $n - 2$ **do**
3:     **for all** pairs of adjacent neighbors u and w of v_i **do**
4:         Store the triangle $v_i uw$.
5:     **end for**
6:     Replace G by $G - v_i$.
7: **end for**

Exercises

20.1. Show that $a(G) \leq \Delta(G)$ for any graph G, where $\Delta(G)$ denotes the maximum degree of G. Give examples where the bound is sharp.

20.2. Show that $\delta(G) < 2a(G)$ for any graph G, where $\delta(G)$ is the minimum degree of G.

20.3. Show that $a(K_n) = \lceil \frac{n}{2} \rceil$ and $a(K_{m,n}) = \lceil \frac{mn}{m+n-1} \rceil$.

20.4. Find a graph G for which $a(G) > \lceil \frac{|E(G)|}{|V(G)|-1} \rceil$.

20.5. Prove that any graph G with $\delta^* = \Theta$ is an isometric subgraph of a Cartesian product of triangles.

20.6. Show that any graph G with $\delta^* = \Theta$ is bipartite. (Such graphs are known as semi-median graphs; see p. 252.)

Chapter 21

Recognizing Median Graphs

Median graphs have been introduced as generalizations of trees and are used, for example, as median networks in genetics. (See Chapter 12.5.) As they are partial cubes and include all trees and hypercubes, which can be recognized in linear time, one might expect their recognition complexity to be somewhere between that of trees and partial cubes. This is indeed the case.

The first section of this chapter presents a simple recognition algorithm based on a corollary of the Convex Expansion Theorem 12.8 and the Convexity Lemma 11.6. It is simple to implement, but its complexity is not exciting; it recognizes whether a graph on n vertices and m edges is a median graph in $O(mn)$ time.

In the second section we observe that one can embed a median graph on n vertices isometrically into a hypercube if one knows that it is a median graph, apply the method to bipartite graphs, and then eliminate those graphs that violate important convexity properties. The complexity of the resulting recognition algorithm is $O(m\sqrt{n})$. This section is rather long and harder to read than the first section.

The last section is concerned with an intrinsic relationship between median graphs and triangle-free graphs. We show that the recognition complexities of these two classes are closely linked. This relationship is then used in the design of yet another, faster, recognition algorithm for median graphs. This algorithm also makes use of a method due to Alon, Yuster, and Zwick (1997) that is interesting in itself and can be read independently of the rest of this section.

21.1 A Simple Algorithm

Theorem 12.7, which is of a recursive nature, states that a bipartite graph G is a median graph if certain subgraphs are median graphs and convex in G. Interestingly, we can use the following nonrecursive corollary, due to Bandelt (1982), in designing a recognition algorithm for median graphs.

Corollary 21.1 *A connected bipartite graph G is a median graph if and only if $\langle U_{ab}\rangle$ and $\langle U_{ba}\rangle$ are convex for every edge ab of G.*

Proof Suppose G is a (connected) median graph. Theorem 12.4 says G is also a partial cube, so Theorem 11.8 implies that $\langle W_{ab}\rangle$ and $\langle W_{ba}\rangle$ are convex in G. Now, $\langle U_{ab}\rangle$ is convex in $\langle W_{ab}\rangle$ by Lemma 12.6, and as $\langle W_{ab}\rangle$ is convex in G, it readily follows that $\langle U_{ab}\rangle$ is convex in G also. By the same argument, $\langle U_{ba}\rangle$ is convex in G.

Conversely, suppose $\langle U_{ab}\rangle$ and $\langle U_{ba}\rangle$ are convex in G. We will use Theorem 12.7 to show that G is a median graph. Condition (ii) of the theorem is satisfied by our hypothesis, and (i) follows from (ii) and the fact that G is bipartite. Note that (i) also implies that $\langle W_{ab}\rangle$ and $\langle W_{ba}\rangle$ are convex.

For (iii), we show $\langle W_{ab}\rangle$ is a median graph by induction on $|V(G)|$. As $\langle W_{ab}\rangle$ is convex in G, the relation Θ_G restricted to $\langle W_{ab}\rangle$ is $\Theta_{\langle W_{ab}\rangle}$. Let $H = \langle W_{ab}^G\rangle$ and uv be an edge of H. Then $\langle W_{uv}^H\rangle = \langle W_{uv}^G\rangle \cap H$ and $\langle U_{uv}^H\rangle = \langle U_{uv}^G\rangle \cap H$. Because the intersection of convex subgraphs is convex, $\langle W_{ab}^G\rangle$ satisfies the assumptions of the corollary. By induction, $\langle W_{ab}\rangle$ is a median graph. □

For our algorithm we also need the following technical corollary to Theorem 11.8.

Corollary 21.2 *Let G be a partial cube, and ab and uv be a pair of edges in relation Θ. Then $U_{ab} = U_{uv}$.*

Proof Because Θ is transitive (by Theorem 11.8), we have $F_{ab} = F_{uv}$. Proposition 11.7 then implies that $W_{ab} = W_{uv}$ and $W_{ba} = W_{vu}$. Then $U_{ab} = U_{uv}$ by definition of U_{xy}. □

We are now ready for the recognition algorithm. The idea is to test whether a given graph is a partial cube and to check the conditions of Corollary 21.1 by invoking the Convexity Lemma 11.6.

Algorithm 21.1 Median graphs – simple

Input: The adjacency list of a connected graph G.
Output: **true** if G is a median graph, **false** otherwise.

1: If Algorithm 18.2 determines that G is not a partial cube, then return **false** and stop.
2: **for all** Θ^*-classes F_{ab} **do**
3: Determine the induced subgraphs $\langle U_{ab}\rangle$, $\langle U_{ba}\rangle$. If any one of them is disconnected return **false** and stop.
4: **for all** edges uv of $\partial\langle U_{ab}\rangle$ and $\partial\langle U_{ba}\rangle$ **do**
5: Check whether uv is in relation Θ with an edge of $\langle U_{ab}\rangle$ or $\langle U_{ba}\rangle$.
6: If this is the case, then return **false** and stop.
7: **end for**
8: **end for**
9: Return **true**.

Theorem 21.3 *Algorithm 21.1 correctly recognizes median graphs. It can be implemented to run in $O(mn) = O(n^2 \log n)$ time and $O(n^2)$ space.*

Proof Median graphs are partial cubes by Proposition 12.4. By Corollary 21.1, it suffices to check for every edge ab of the graph G being investigated that the subgraphs $\langle U_{ab}\rangle$ and $\langle U_{ba}\rangle$ are convex in G. By Corollary 21.2, it suffices to perform this check for only one edge of each Θ^*-class. Finally, Steps 4 through 7 are a reformulation of Lemma 11.6 for median graphs: The subgraph $\langle U_{ab} \rangle$ is convex if and only if no edge of $\langle U_{ab} \rangle$ is in relation Θ to an edge of $\partial\langle U_{ab} \rangle$. We conclude that Algorithm 21.1 correctly recognizes median graphs.

By Theorem 18.8, Step 1 runs in $O(mn)$ time and $O(n^2)$ space. Here Algorithm 18.2 determines all Θ^*-classes. We can implement it so that it also determines the distance matrix of G and store the Θ^*-classes as described in Chapter 17. We can then check in constant time in which class an edge is contained.

For Step 3 we first observe that the total number of vertices in all the sets U_{ab} (which we need to check) is $2m$, as in Exercise 21.1. We determine the complexity of finding the U_{ab} as follows. Consider the equivalence classes $E_1, E_2, \ldots, E_k$ of Θ^* one by one. At class E_i we begin by selecting an edge, say, ab. Then we initialize characteristic vectors of length n (with a place for every one of the n vertices of G) for the sets U_{ab} and U_{ba} by setting all entries equal to zero. For every edge $uv \in E_i$, we then check whether u or v is closer to a; that is, we determine to which set u and v belong and change the corresponding entry in the respective vector to 1. This takes $O(m)$ time and $O(n^2)$ space.

To find the edges in $\langle U_{ab} \rangle$ for a given U_{ab}, scan all edges of G and check whether both of their endpoints belong to U_{ab}. There are k pairs U_{ab}, U_{ab}, and hence the time complexity to determine all $\langle U_{ab} \rangle$s is $O(mk) = O(mn)$. Furthermore, since the total number of edges in the $\langle U_{ab} \rangle$s is $O(m \log n)$, the $\langle U_{ab} \rangle$ can be checked for connectedness within the same time complexity.

For Step 5 it is convenient to have a characteristic vector to every U_{ab} and U_{ba} that tells us which of the Θ^*-classes have nonempty intersection with $\langle U_{ab} \rangle$ and $\langle U_{ba} \rangle$. This can be achieved by scanning the edges of every E_i, determining whether both of their endpoints belong to U_{ab}, respectively, U_{ba}, and effecting appropriate entries in the characteristic vectors. The complexity is $O(mk) = O(mn)$.

For any vertex of a U_{ab}, we check for all incident edges e whether they are in U_{ab} or in the equivalence class of ab. If not, the edge e is in $\partial\langle U_{ba} \rangle$, and we have to check whether its equivalence class has nonempty intersection with $\langle U_{ba} \rangle$. With the above data structure we can perform every individual check in constant time. Thus the overall time complexity of the algorithm is $O(mn) = O(n^2 \log n)$. □

This median recognition algorithm is due to Imrich and Klavžar (1998). Jha and Slutzki (1989, 1992) presented two different approaches for recognizing median graphs, each yielding an $O(mn)$ algorithm. In the first paper they adapted Theorem 12.18, and in the second they used Theorem 12.7 in combination with Proposition 21.10. A third algorithm can be deduced from the work of Feder (1992), who presented an algorithm that finds the so-called canonical 2-isometric representation of a graph in $O(mn)$ time and $O(m)$ space.

21.2 A Fast Algorithm

We just showed how to recognize median graphs by a simple algorithm of complexity $O(mn) = O(n^2 \log n)$. In search of a faster algorithm we next show that median graphs can be embedded into hypercubes in $O(m \log n)$ steps. In other words, once it is known that a graph is a median graph, it can be embedded into a hypercube in $O(m \log n)$ time.

We mention in passing that the first subquadratic recognition algorithm for median graphs is due to Hagauer, Imrich, and Klavžar (1999). It has complexity $O(m\sqrt{n})$ but is quite different from the algorithm presented here. Nonetheless, the Embedding Lemma 21.4, which is one of the motivations for the algorithm outlined below, was also obtained there.

We now prepare for our $O(m\sqrt{n})$ recognition algorithm by exploring additional characteristic properties of median graphs.

Recognition versus embedding

Recall that two edges e, f of G are in relation δ if $e = f$ or e and f are opposite edges of a square in G. Clearly, δ is reflexive and symmetric. It is also contained in Θ, which implies that $\delta^* \subseteq \Theta^*$. For partial cubes we thus have $\delta^* \subseteq \Theta$ by Theorem 11.8.

We now show that any median graph G can be isometrically embedded into an r-cube in $O(m \log n)$ steps and that embedding is equivalent to finding the transitive closure of δ.

Suppose that every $\langle U_{ab} \rangle$ in a partial cube G is connected. Then any two edges e, f that are in relation Θ to the edge ab must be contained in F_{ab} and are therefore in relation δ^*. In this case we have $\delta^* = \Theta$. In particular, this is true for median graphs.

We must note, however, that the validity of

$$\delta^* = \Theta$$

in a bipartite graph G does not imply that G is median, as the example of Q_3^- shows. It is also easy to see that $\delta^* \neq \Theta$ in a hexagon, although a hexagon is a partial cube. Hence the class of bipartite graphs for which $\delta^* = \Theta$ constitutes a proper subclass of the class of partial cubes and properly contains the class of all median graphs. These graphs are interesting in their own right. We call them *semi-median graphs.* They were introduced by Imrich and Klavžar (1998).

Semi-median graphs and median graphs are partial cubes; therefore both δ and δ^* can be computed for them in $O(m \log n)$ time. Because knowing Θ^* for a partial cube is equivalent to embedding it into the hypercube, we have the following sequel to Corollary 20.7:

Lemma 21.4 (Embedding Lemma) *Let G be a semi-median graph or a median graph on n vertices with m edges. Then G can be embedded isometrically into a hypercube in $O(m \log n)$ time.*

Semi-median graphs can be embedded into hypercubes in $O(m \log n)$ time once they are recognized as semi-median graphs, but the present recognition complexity still is $O(mn)$, as they can be recognized by checking whether they are partial cubes for which $\delta^* = \Theta$.

For partial cubes we know of no shortcut for the embedding procedure, even if we already know that a graph is a partial cube. For median graphs the situation is similar to that of semi-median graphs: They can be recognized by checking whether they are partial cubes and subsequent verification whether the $\langle U_{ab} \rangle$ are convex; see Theorem 21.3.

But we reiterate that median graphs can be embedded in $O(m \log n)$ time by the Embedding Lemma 21.4. We use this embedding to derive an algorithm that recognizes median graphs in $O(m\sqrt{n})$ time, which is the main result of this section.

We often envisage the embedding process as a coloring procedure, and regard $F = F_{ab}$ as edges of a single and distinct *color.* If G is a partial cube, then we get an *edge-coloring* in the sense that every edge is assigned a color, and incident edges have different colors.[1]

By the same arguments, the equivalence classes of δ^* are a coloring if no two edges of the same color have a common endpoint. We will always ensure that this holds by eliminating graphs that violate this condition in the preprocessing parts of our algorithms.

By Corollary 21.1, a partial cube G is a median graph if and only if, for any edge ab,

[1]Similar to the chromatic number, one defines the *chromatic index.* It is the smallest number of colors needed to color the edges of a graph such that incident edges have different colors. Notice that the chromatic index in a partial cube can be much smaller than the number of Θ-classes.

the subgraphs $\langle U_{ab} \rangle$ and $\langle U_{ba} \rangle$ are convex. To check whether a given graph is a median graph G, we will therefore initially color it by the equivalence classes of δ^* and then test for selected properties of median graphs until we are sure that $\delta^* = \Theta$ and that all $\langle U_{ab} \rangle$ are convex.

Embedding versus coordinatization

We always claim that the embedding of a partial cube is complete when we have determined the Θ-classes, that is, colored all edges. Some authors speak of an embedding only after a coordinatization has been computed. This can make a big difference. Take a tree T on n vertices. It (and its Θ^*-classes) can be recognized in $O(n)$ steps, but its coordinatization takes $O(n^2)$ time and space. As such, that is, without the aid of bitvectors, it takes $n-1$ steps to look up the entries of a single coordinate vector of a given vertex once the coordinatization is completed. But we can find this vector with the same effort just from the Θ^*-classes.

To see this, suppose Θ of a partial cube G on n vertices has been determined. Compute a BFS-tree with respect to a root vertex v_0 and label the edges with the number of their Θ^*-class. Given a vertex v, one can then find its coordinate vector by following any shortest path from v to v_0, using the labels of the edges on that path for coordinatization. Notice that no two edges on a shortest path are in relation Θ, which is the same as Θ in partial cubes. Thus, the number of steps for the coordinatization of v is not larger than the number of Θ-classes, which is bounded by $n - 1$.

Therefore one does not improve the complexity of operations by the coordinatization, unless bitvectors are used as in Section 18.4. Moreover, the Θ-classes depend only on the graph, but the coordinatization depends on the basepoint and the order of the Θ-classes.

Hence, from our point of view the embedding complexity of a tree on n vertices and m edges is $O(m)$, whereas the complexity of coordinatizing its vertices is $O(n^2)$.

Compare this with a result of Cheng (2011), which we already mentioned on p. 191. Cheng computes the number of Θ-classes of median graphs in $O(n+m)$, without determining the classes themselves.

Preprocessing

As median graphs are connected, bipartite, and sparse, we have to screen our input graphs G for these properties. Moreover, we arrange the vertices of G in BFS-order with respect to a root v_0.

By Algorithm 17.2, a BFS-ordering of the vertices of a graph G given by its adjacency list can be computed in $O(m)$ time. Within the same time complexity we can then determine whether G is connected (Corollary 17.3) and bipartite (Proposition 17.4).

Sparseness requires a little more thought. Considering the edges of G as directed pairs of vertices, we recall that ab is an up-edge if $d(v_0, a) < d(v_0, b)$ and a down-edge if $d(v_0, a) > d(v_0, b)$. (Because G is bipartite, $d(v_0, a) \neq d(v_0, b)$.) Also the set

$$L_i = \{v \mid d(v_0, v) = i\}$$

is the ith *distance level* of G with respect to the root v_0. We set $l(v) = i$ if $v \in L_i$.

Recall that the down-degree of a vertex v in G is the number of neighbors of v in $L_{l(v)-1}$. By Lemma 22.6, the down-degree k of a vertex v in a median graph is at most $\log_2 |G|$, and v is contained in a k-cube that meets the levels $L_{l(v)}, L_{l(v)-1}, L_{l(v)-2}, \ldots, L_{l(v)-k}$.

We will mostly use this result for $k = 2$. Thus, to any two down-edges ab, ac of a median graph, there exists a vertex d in level $L_{l(a)-2}$ that is adjacent to both b and c. Moreover there exists exactly one such vertex d, for if a second one d' existed, then a, d, d' would have the two medians b and c. We call the property that exactly one such vertex exists *down-closure*.

Keeping in mind down-closure and the above degree limitations of a median graph, we eliminate all graphs that do not satisfy these properties, and compute δ and δ^*.

Algorithm 21.2 Preprocessing for median graph recognition

Input: The adjacency list of a graph G on n vertices and m edges.
Output: δ and δ^* for all admitted graphs G.

1: If $m > n \log_2 n$, **reject**.
2: Arrange the vertices in BFS-order with respect to an arbitrarily chosen root v_0.
3: If G is disconnected, nonbipartite, or has vertices of down-degree larger than $\log_2 n$, **reject**.
4: Find all squares.
5: If a $K_{2,3}$ is detected, **reject**.
6: Compute δ and δ^*.
7: If two distinct edges incident with the same vertex are in the same δ^*-class, **reject**.
8: If there is a square that meets only two BFS-levels, **reject**.
9: If there is a square that violates down-closure, **reject**.
10: Admit G and **return** δ and δ^*.

Lemma 21.5 *Algorithm 21.2 admits all median graphs and correctly determines δ and δ^*. All admitted graphs have at most $n \log_2 n$ edges, are connected and bipartite. Each square meets three levels with respect to a fixed, but arbitrarily chosen, BFS-order, and G satisfies down-closure with respect to this order.*

The time complexity of the algorithm is $O(m \log n)$.

Proof We first show that Algorithm 21.2 admits all median graphs. Then we consider its time complexity.

Correctness All steps except Steps 7 and 8 have already been motivated. For Step 7 notice that no two distinct edges of a hypercube that share one endpoint are in relation Θ and that we are looking for graphs with $\delta^* = \Theta$.

For Step 8 observe that every square of a median graph must meet three levels. To see this, suppose that the square $abcd$ just meets levels L_k and L_{k-1}, where $a, c \in L_k$ and $b, d \in L_{k-1}$. By down-closure we infer the existence of a vertex $v \in L_{k-2}$ that is adjacent to b and d. But then both b and d are medians of a, c, and v, which is not possible.

Hence, all median graphs are accepted and have the asserted properties.

Complexity With the exception of Steps 4 and 9, all steps of the algorithm are straightforward and have complexity $O(m \log n)$.

For Step 4 notice that no subgraph H of G has more than $|V(H)| \log_2 n$ edges. This is clearly so because the number of edges of H is the number of down-edges of G in H and thus bounded by $|V(H)| \log_2 n$. By Theorem 20.2, the arboricity of G is at most $2 \log_2 n$.

By Theorem 20.5 all squares of G can therefore be determined in $O(m \log n)$ time. The algorithm codes these squares as triples $(v, w, \{u_1, u_2, \ldots, u_i\})$, with no more than $2m\, a(G)$ entries. If $i > 2$, then G contains a $K_{2,3}$. Because $K_{2,3}$ is not a subgraph of a hypercube, we reject all graphs containing such triples. Clearly, this can be achieved in $O(m \log n)$ time.

Because all squares are then coded in the form $(v, w, \{u_1, u_2\})$, G has at most $2m\, a(G) \leq 2m \log_2 n$ squares. We reject G if it has more than $(mn/(4(n-1))) \log_2 n$ squares, because in this case it cannot be a subgraph of a hypercube by Corollary 20.7.

Notice that the squares contain no diagonal, because G is bipartite, and that the knowledge of all squares is equivalent to the knowledge of δ. Now an application of Proposition 18.2 shows that δ^* can also be computed in $O(m \log n)$ time.

For Step 9, note that Step 7 already eliminates all graphs in which three down-edges of a vertex have a common neighbor and those in which two down-edges have two common neighbors. Thus, to check down-closure, it suffices to count whether there are as many squares with top vertex v as there are distinct pairs of down-edges from v. □

Convexity

At a first glance it appears that we have almost completed our task after preprocessing, and that convexity of the U_{ab} (see Corollary 21.1) is the only problem left to deal with. However, we still do not know whether $\Theta = \delta^*$. It is not even clear whether the coloring we have obtained is an embedding into the hypercube and, if it is an embedding, whether it is an isometric embedding.

Although we conjecture that all graphs admitted by Algorithm 21.2 are properly embedded, it seems to be rather tedious to prove. Nonetheless, even if this conjecture is true, the admitted graphs need not be isometrically embedded. To see this, take Q_3, remove the edge (011)(111), and choose $v_0 = (000)$, as shown on left of Figure 21.1. Also note that this graph would not have been admitted if we had chosen (011) as a base, for then a square would be contained in two BFS-levels. One problem with this graph is that edges of the same color as an edge ab do not necessarily induce an isomorphism $\langle U_{ab} \rangle \to \langle U_{ba} \rangle$. (Take $ab = (000)(001)$.) To eliminate the problem, we might as well check directly whether the edges of each color induce such an isomorphism.

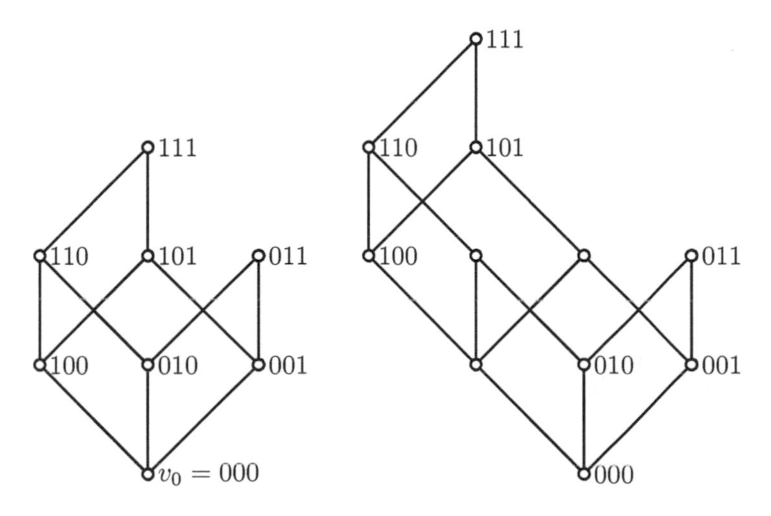

FIGURE 21.1 Two nonisometric embeddings.

To do this, we define F^*_{ab} as the set of edges in relation δ^* to ab. Suppose that ab is an up-edge. Then we set

$$U^*_{ab} = \{x \mid xy \in F^*_{ab} \text{ and } xy \text{ is an up-edge}\}$$

and define U^*_{ba} analogously (with down-edges). Because every vertex x is in $d(x)$ sets F^*_{ab}, the total number of vertices in the U^*_{ab}'s and U^*_{ba}'s is $2m$. Also, because every edge is a down-edge with respect to one endpoint and an up-edge with respect to the other, the total number of edges in the subgraphs $\langle U^*_{ab}\rangle$ and $\langle U^*_{ba}\rangle$ is bounded by the total number of vertices times the bound $\log_2 n$ for the down-degree; in other words, these graphs contain at most $m \log_2 n$ edges. By Theorem 17.1, we can thus check within complexity $O(m \log n)$ whether the sets F^*_{ab} induce isomorphisms between $\langle U^*_{ab}\rangle$ and $\langle U^*_{ba}\rangle$.

We have thus shown:

Lemma 21.6 *Without altering the time complexity, Algorithm 21.2 can be modified such that it also checks whether the F^*_{ab} induce isomorphisms between the $\langle U^*_{ab}\rangle$ and $\langle U^*_{ba}\rangle$.*

Note that any two edges uv and xy of F^*_{ab} with $u, x \in U^*_{ab}$ are either both up-edges or both down-edges. If this were not so, we would have two adjacent vertices u, x of this type and then the square $uvyx$ would either meet four or just two levels with respect to v_0, neither of which is possible.

Unfortunately, even then the graphs admitted need not be partial cubes. To illustrate this we alter our previous example as illustrated on the right of Figure 21.1.

To eliminate such graphs, we define down-convexity. We call a subgraph H of a graph G *down-convex* with respect to a root v_0 if it is connected and if no down-edge of ∂H has the same δ^*-color as an edge of H. (Note that $\langle U^*_{(000)(001)}\rangle$ is not down-convex in the second graph of Figure 21.1.)

Because all graphs properly embedded into a hypercube satisfy the property that removal of any F_{ab} disconnects them, we check whether removal of any single F^*_{ab} disconnects G.

Lemma 21.7 *For any graph G admitted by Algorithm 21.2, down-convexity can be checked in time and space complexity $O(m \log n)$. Moreover, removal of any δ^*-class disconnects G.*

Proof By Lemma 21.6, we can assume that we have already determined the $\langle U^*_{ab}\rangle$. For every vertex in $\langle U^*_{ab}\rangle$ we consider all of its down-edges. If a down-edge is not in $\langle U^*_{ab}\rangle$, we check whether the color of the down-edge is a color occurring in $\langle U^*_{ab}\rangle$. If this is so, we reject the graph G. We note that $2m = \sum |\langle U^*_{ab}\rangle|$ and that every vertex has at most $\log_2 n$ down-edges. Thus, at most $2m \log_2 n$ checks must be performed. Proceeding as in Algorithm 21.1 (Median Graphs – simple), we see that this can be done in constant time for every check and that we can stay within the limit of $O(m)$ space.

We show now that removal of any δ^*-class disconnects G. Suppose that this is not so and that there is a down-edge ab such that removal of F^*_{ab} does not disconnect G. Then there must be a shortest path P in $G - F^*_{ab}$ from a vertex u in U^*_{ab} to v_0. Let uv be the first edge of P. If it is a down-edge, then there also must be a down-edge $uw \in F^*_{ab}$ in G and, by down-closure, a vertex $x \in V(G)$ that is adjacent to both v and w. But then vx and uw have the same color, $vx \in F^*_{ab}$, and v is in U^*_{ab}.

If uv is an up-edge, we continue from v along P until we meet the first down-edge of P. Such an edge must exist because v_0 is in the lowest level. Let $P[u, w_k] = uvw_1w_2 \dots w_{k-1}w_k$ be the subpath of P consisting only of up-edges of P to the first down-edge $w_kw'_k$. By down-closure there are vertices $u', v', w'_1, \dots, w'_{k-1}$ such that $w_{k-1}w'_{k-1}$, $w_{k-2}w'_{k-2}$, $\dots, w_1w'_1$, vv', uu' all have the same color and that $v'w'_1 \dots w'_k$ is a path P' from v' to w'_k. Note that none of the w'_j can be equal to w_{j-2} or v because of the minimality of P. By the above, $v' \in U^*_{ab}$. Let i be the largest index such that w'_i is in U^*_{ab}. But then $P'[w'_i, w'_k] \cup P[w'_k, v_0]$ is a path from U^*_{ab} to v_0, which is shorter than P. Because all colors of P' also occur in P, this new path does not meet F^*_{ab} and contradicts the minimality of P. □

Lemma 21.8 *Let G be a graph admitted by Algorithm 21.2. Suppose that in G all pairs $\langle U^*_{ab}\rangle$ and $\langle U^*_{ba}\rangle$ are isomorphic and down-convex. Then G is a partial cube.*

Proof Let u, v be two arbitrary vertices in an admitted graph G. Then there is a shortest path P from u to v that consists of two parts $P[u,w]$ and $P[w,v]$, of which the first one contains only down-edges (in the direction to v) and the other one only up-edges. One of these parts can be empty.

To see this, suppose that Q is a shortest path from u to v that does not have the stated property. Then Q contains two down-edges yx and yz, and by down-closure there is a vertex y' adjacent to both x and z such that xy' and zy' are down-edges. Replacing the sequence xyz of Q by $xy'z$, we obtain a new shortest path Q' from u to v, and the sum of the levels of its vertices is lower than that of Q. Clearly, any path P for which the sum of the levels of its vertices is minimal satisfies the partition property claimed above, so the continuation of this process with Q eventually terminates with the required path P.

Next, consider two down-edges uv, xy in F^*_{ab}, where ab is a down-edge. We wish to show that $uv \,\Theta\, xy$.

If there is a shortest path P from u to x that contains no element of F^*_{ab}, we partition it into a down-path $P[u,w]$ and an up-path $P[w,x]$. (It is possible that w is equal to u or to x.) By down-closure, there are two neighbors v' and y' of w, such that uv and wv' are in relation Θ and δ^* and also the pair of edges xy and wy'. But then we must have $v' = y'$ and $uv \,\Theta\, xy$.

We claim that a shortest u,x-path contains neither v nor y.

Let uv, xy be a pair of edges for which a shortest path P from u to x contains both v and y. Let P be minimal with respect to this property. If $P[v,y]$ is completely contained in $\langle U^*_{ba}\rangle$, then $\langle U^*_{ba}\rangle$ and $\langle U^*_{ab}\rangle$ cannot be isomorphic. But then there is a shortest subpath Q of $P[v,y]$ that is not in $\langle U^*_{ba}\rangle$ but connects two vertices r, s of it. Let rr' be the first edge of Q and ss' the last. Because at most one vertex of P has no neighbors in lower levels, we can assume that rr' is a down-edge. By down-convexity, it cannot have a color that is the color of an edge in $\langle U^*_{ba}\rangle$. Hence there must be another edge tt' of the same color as rr' in Q. Let t' be closer to r' than t. Because there are no edges of the same color as that of rr' between r' and t', tt' is a down-edge. But this contradicts the minimality of the distance between uv and xy.

Suppose that P contains only v. Again we assume that uv and xy are of minimal distance with respect to this property. Partition P as before into a down-path $P[u,w]$ and an up-path $P[w,x]$. Neither $P[u,w]$ nor $P[w,x]$ contains an edge of the same color as uv. In the first case it would contradict the minimality of the distance of uv and xy, and in the second P would not be a shortest path. Hence there is a down-edge ww' in F^*_{ab}. But this is not possible because v and w must be either both in $\langle U^*_{ab}\rangle$ or both in $\langle U^*_{ab}\rangle$.

Thus $\delta^* \subseteq \Theta$.

Now suppose that uv and xy are in relation Θ but not δ^*. We can choose the notation such that there are shortest paths P from u to x and Q from v to y such that $uv \cup Q \cup yx \cup P$ is a cycle. No edge of P or Q can be in relation Θ to uv; otherwise, $uv \cup Q$ or $vu \cup P$ could not be a shortest path. But then they can also not be in relation δ^*, because δ^* is contained in Θ. It follows that F^*_{uv} does not separate G unless uv and xy are in relation Θ.

Hence, $F^*_{uv} = F_{uv}$, $\langle U^*_{ab}\rangle = \langle U_{ab}\rangle$, and G is a partial cube. □

We sum up our findings in the following lemma:

Lemma 21.9 *Algorithm 21.2 can be modified so that all accepted graphs are semi-median graphs in which the $\langle U_{ab}\rangle$ are down-convex with respect to the chosen BFS-ordering. The complexity of the algorithm remains $O(m \log n)$.*

It remains to check the $\langle U_{ab}\rangle$ for convexity. We already know from Algorithm 21.1 (Median graphs – simple) that a direct approach leads to time complexity $O(mn)$. We prefer to check for 2-convexity instead in some cases.

We call a subgraph H of a graph G *2-convex* if all shortest G-paths of length at most 2 between vertices of H are already in H. For bipartite graphs this is equivalent to the statement that the common neighbors of two vertices u, v in H are also in H.

This concept is equivalent to convexity for median graphs as the following proposition shows. It was observed by Bandelt, who communicated it to Jha and Slutzki; see Jha and Slutzki (1992).

Proposition 21.10 *Let G be a bipartite graph in which every triple of vertices has a median. Then a connected subgraph of G is convex if and only if it is 2-convex.*

Proof Because 2-convexity is weaker than convexity, we only have to show that every 2-convex connected subgraph H of a bipartite graph G is convex if every triple of vertices of G has a median. Suppose that this is not the case. Then there exists a 2-convex subgraph H of G with two vertices $u, v \in V(H)$ for which there exists a shortest path P from u to v that is not completely in H. Suppose that the graph H and the vertices u and v have been chosen such that $t = d_H(u, v)$ is smallest possible and that $d_G(u, v)$ is minimal with respect to this distance t.

Note that by 2-convexity, the length of P is at least 3 and can thus be written in the form $3k + e$ with $k \geq 1$ and $e \in \{0, 1, 2\}$.

Suppose first that there is a path Q from u to v in H that has the same length as P. Select vertices x and y on P and Q, respectively, with $d_G(u, x) = 2k + e$ and $d_G(u, y) = 2k$. Let w be a median of u, x, y. Clearly, there is a shortest path R of length $2k$ from u to y containing w and a shortest path of the same length from u to y that is completely in H. Because $2k < 3k + e$, the path R must also be completely contained in H. Thus $w \in V(H)$.

It is easy to see that the distance from w to x is at most $k + e$. Because x is on a shortest u, v-path and w on a shortest u, x-path, we note that $d_H(w, v) \leq 2k+e < 3k+e = d_H(u, v)$. Because both w and v are in H, the entire path must be in H by the minimality of $d_H(u, v)$. But then $x \in V(H)$, $d_H(u, w) + d_H(w, x) = 2k+e$, and $d_G(u, x) = 2k+e$. By the minimality of t, the part of P from u to x must be in H, just as the part from x to u. But then P is already in H.

We can therefore continue under the assumption that every path Q in H from u to v is longer than $d_G(u, v)$. Let z be a vertex on Q with $r = d_G(u, z) = d_G(v, z) + e$, where $e \in \{0, 1\}$. If $2r + e = d_G(u, v)$, then any shortest u, z-path and from v to z must be completely in H by the minimality of $d_H(u, v)$. Thus $2r + e > d_G(u, v)$.

Let w be a median of u, v, z. If w is not in H, we consider the path of length r from u via w to z. It is a shortest path not completely in H connecting two vertices of H of distance less than t in H, in contradiction to the minimality of $d_H(u, v)$. If it is in H, we note that then the v, z-path via w must also be in H. But this means that there is a path in H from u to v via w which has length $d_G(u, v)$, bringing us back to the first case; this is the case in which $d_G(u, v) = d_H(u, v)$. □

Unfortunately, we cannot apply this proposition in our situation because we do not know yet whether our graphs have medians; in fact, this is just what we wish to show. We use the following lemma instead:

Lemma 21.11 *For all partial cubes G with $\delta^* = \Theta$, an isometric subgraph is convex if and only if is 2-convex and down-convex.*

Proof It suffices to show that every 2-convex, down-convex, and down-closed subgraph H of G is convex. Let u, v be two vertices of minimum distance in G that are connected by a shortest path P that meets H only in its endpoints, and let Q be a shortest u, v-path in H. By isometry, every color of an edge of P is a color of an edge of Q and thus of H. If one of the edges of P incident with u or v is a down-edge, it must be in H by down-convexity, and P is completely in H by the minimality of the distance of u and v. If this is not the case, we transform it by the methods of the proof of Lemma 21.8 to a path P' satisfying this property. But then P' is in H. Now we recall that the transformation process used in the proof of the existence of such a path in Lemma 21.8 always changes only a pair of adjacent vertices. Reversing this process and going from P' to P, we see by 2-convexity that we proceed from one path in H to another one in H. Thus P is also in H. □

To check whether a graph G that has passed all previous tests is a median graph, we use Corollary 21.1. Thus we only have to check whether every $\langle U_{uv} \rangle$ is convex. Depending on the size of U_{uv}, we will either use the Convexity Lemma or check for 2-convexity.

To check for convexity by the Convexity Lemma, we scan the vertices x of U_{uv} one after the other and check for every color occurring in $\langle U_{ab} \rangle$ whether an edge xy in the boundary of U_{uv} has this color. As we have seen before, each of these tests can be performed in constant time with the proper data structure. Because $\langle U_{uv} \rangle$ cannot have more than $|U_{uv}|$ colors, the total time is bounded by $O(|U_{uv}|)^2$.

For a check using 2-convexity, we again scan all vertices x of U_{uv} and mark all neighbors of x not in U_{uv}. Then $\langle U_{uv} \rangle$ is 2-convex if and only if no vertex is marked twice. Because G has m edges, the total time is at most $O(m)$.

Consider the set S of all U_{uv}'s with $|U_{uv}| \leq \sqrt{n}$. Then the total time needed to check these graphs for convexity with the Convexity Lemma is

$$\sum_{U_{uv} \in \mathcal{S}} |U_{uv}|^2 \leq \sqrt{n} \sum_{U_{uv}} |U_{uv}| \leq n^{0.5} m,$$

as $\sum_{U_{uv}} |U_{uv}| = m$.

This inequality also shows that there are at most $m/\sqrt{n}$ sets U_{uv} containing more than $\sqrt{n}$ elements. Thus the total time for these checks is $O\big(n(m/\sqrt{n})\big) = O(m\sqrt{n})$, and we have derived the main result of this section:

Theorem 21.12 *For a graph G on n vertices and m edges, one can decide in $O(m\sqrt{n})$ time whether it is a median graph.*

21.3 Triangle-Free Graphs and Median Graphs

In this section we investigate the relationship between median graphs and triangle-free graphs. This relationship is the basis of an $O((m \log n)^{1.41})$ algorithm for the recognition of median graphs. (More precisely, the time complexity is $O((m \log n)^{2\omega/(\omega+1)})$, where ω is the exponent of matrix multiplication; see p. 216.)

Recognizing triangle-free graphs via median graphs

For a graph G, let $G^{\triangle}$ be the *2-simplex graph of* G. It is obtained from G by adding a new vertex z that joins all the vertices of G, and by replacing every edge $e = uv$ of G by a path of length 2. In other words,

$$V(G^{\triangle}) = V(G) \cup E(G) \cup \{z\},$$

and the edge set $E(G^{\triangle})$ of $G^{\triangle}$ is

$$\{zu \mid u \in V(G)\} \cup \{ue \mid u \in V(G) \text{ and is incident with } e \in E(G)\}.$$

This construction is illustrated by Figure 21.2, which shows that $C_3^{\triangle}$ is the vertex-deleted 3-cube Q_3^- and that $C_4^{\triangle} = P_3 \,\square\, P_3$.

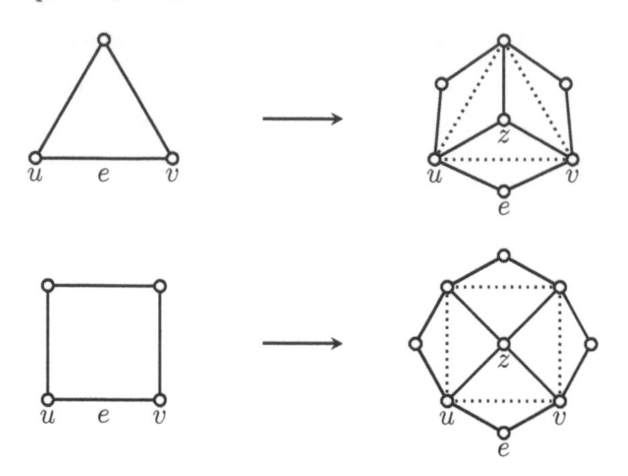

FIGURE 21.2 $C_3^{\triangle} = Q_3^-$ and $C_4^{\triangle} = P_3 \,\square\, P_3$.

Note that $C_3^{\triangle}$ is not a median graph, but $C_4^{\triangle}$ is. This is not a coincidence:

Theorem 21.13 *A graph G is triangle-free if and only if $G^{\triangle}$ is a median graph.*

Proof Suppose G is triangle-free. Now, $G^{\triangle}$ is connected and bipartite. By Corollary 21.1, it suffices to show that the subgraphs $\langle U_{ab}\rangle$ of $G^{\triangle}$ are convex.

To prove that each $\langle U_{ab}\rangle$ is convex, we need to consider two types of edges. First consider an edge of type uz, where u is a vertex of G (and thus also of $G^{\triangle}$). Our construction of $G^{\triangle}$ ensures that $\langle U_{uz}\rangle$ is the star consisting of all edges (other than uz) of $G^{\triangle}$ that are incident with u, and this is convex. Similarly, $\langle U_{zu}\rangle$ is a star with a central vertex z joining all $v \neq z$ for which $uv \in E(G)$; and the fact that G is triangle-free makes this star convex.

The other type of edge is ue, where u is a vertex of G incident with the edge e. Because the degree of e in $G^{\triangle}$ is 2, we infer that $\langle U_{ue}\rangle$ is again a star, with the center z. Similarly, $\langle U_{eu}\rangle$ is a star whose center is the neighbor of e different from u. As in the first case, $\langle U_{ue}\rangle$ and $\langle U_{eu}\rangle$ are convex. By Theorem 21.1, we conclude that $G^{\triangle}$ is a median graph.

For the converse, assume that G contains a triangle with edges e_1, e_2, and e_3. In $G^{\triangle}$, the vertices e_1, e_2, and e_3 are pairwise of distance 2, and because they have no common neighbor, this triple has no median. (See Figure 21.2.) Thus $G^{\triangle}$ is not a median graph. □

Corollary 21.14 *Let $M(m,n)$ be the complexity of recognizing median graphs with m edges and n vertices. Then the complexity of checking whether G is triangle-free is at most $O(M(m,m))$.*

Proof By Theorem 21.13, we can decide whether G is triangle-free by first constructing $G^{\triangle}$ and then checking whether $G^{\triangle}$ is median. Now, $G^{\triangle}$ has $2m+n$ edges and $n+m+1$ vertices, and we can construct it in linear time. To check whether it is a median graph, we need (recall that $G^{\triangle}$ is connected) $O(M(2m+n, n+m+1)) = O(M(m,m))$ time. □

By Corollary 21.14, the algorithm from Section 21.2 yields an $O(m^{3/2})$ algorithm for the recognition of triangle-free graphs.

We next take a closer look at the median graphs arising in Theorem 21.13. Clearly, not all median graphs appear. In order to describe the ones that do appear, the following definitions are helpful: Recall that the eccentricity $e(x)$ of a vertex x in a connected graph G is the maximum distance of x to any other vertex in G. The *radius* $r(G)$ of G is the minimum eccentricity in G, and a vertex x is a *central vertex* of G if $e(x) = r(G)$.

Let $\mathcal{M}^2_{n,m}$ be the class of median graphs with

1. Minimum degree 2
2. Radius 2
3. A single vertex, say w, of maximum degree n
4. Unique central vertex w
5. m vertices at distance 2 from w

In addition, let $\mathcal{T}^f_{n,m}$ stand for triangle-free graphs with n vertices, m edges, and without vertices of degree 1 and $n-1$. With these notions we can state the following result, which intuitively asserts that there are as many median graphs as there are triangle-free graphs:

Corollary 21.15 *For each n and m, the mapping $f : G \mapsto G^{\triangle}$ is a bijection between the graph classes $\mathcal{T}^f_{n,m}$ and $\mathcal{M}^2_{n,m}$.*

Proof We first note that f maps $\mathcal{T}^f_{n,m}$ into $\mathcal{M}^2_{n,m}$. Indeed, $G^{\triangle}$ is a median graph by Theorem 21.13, and it has minimum degree 2, because G has no vertex of degree 1. Also $G^{\triangle}$ has radius 2 because the vertex z is adjacent to all original vertices. In addition, z is the unique vertex of maximum degree n because G has no vertex of degree $n-1$. By the same reasoning, it is also the unique central vertex. Finally, there are m vertices at distance 2 from z, because G has m edges.

Let G_1 and G_2 be nonisomorphic graphs from $\mathcal{T}^f_{n,m}$. If $G_1^{\triangle}$ were isomorphic to $G_2^{\triangle}$, then any isomorphism would map the unique central vertex of maximum degree of $G_1^{\triangle}$ to the corresponding vertex of $G_2^{\triangle}$. But then G_1 and G_2 would be isomorphic as well.

To complete the proof, we need to show that f is surjective. So let H be a median graph from $\mathcal{M}^2_{n,m}$, and z be the unique central vertex that is also the unique vertex of maximum degree n. Let w be any vertex of H with $d(w, z) = 2$. Because H is bipartite, all neighbors of w must be adjacent to z. The degree of w is at least 2, because H is in $\mathcal{M}^2_{n,m}$. In fact, the degree of w is exactly 2 because H is a median graph, and therefore $K_{2,3}$-free. Because H has m vertices of distance 2 from z, and z has n neighbors, H has $1+n+m$ vertices and $n+2m$ edges. Now we construct the graph G on the set of all neighbors of z in H. We join two vertices of G by an edge if and only if they have a common neighbor in H that is different from z. Clearly, G has n vertices and m edges. Moreover, $H = G^{\triangle}$. Finally, because H is Q_3^--free, G is triangle-free. □

Recognizing median graphs via triangle-free graphs

We now ask whether algorithms for recognizing triangle-free graphs can help in the recognition of median graphs.

The starting point is the situation after Lemmas 21.9 and 21.11. We can assume that G is a semi-median graph whose vertices have been arranged in BFS-order such that all $\langle U_{ab} \rangle$ are down-convex, and that we only have to check for 2-convexity.

Suppose that there is a vertex x in $W_{ab} \setminus U_{ab}$ that has two neighbors in U_{ab}, say u and v. Because U_{ab} is isometric, there is a vertex $w \in U_{ab}$ that is adjacent to both u and v. Moreover, there are vertices u', v', and w' in U_{ba} that are adjacent to u, v, and w, respectively, such that these six vertices together with x induce a Q_3^-.

Let $L_0, L_1, \ldots$ be the distance levels of G with respect to the BFS-ordering, and assume that $x \in L_{i+1}$. The vertices u, v cannot be in the same level as x, as there are no cross-edges. Moreover, by down-convexity, the edges ux, vx cannot be down-edges. Thus u and v both belong to L_i.

The vertex w cannot be in L_{i+1}; otherwise, we would have a square that meets only two levels. Hence $w \in L_{i-1}$.

Because squares can meet at most three levels, all three edges uu', vv', ww' from u, v, w to U_{ba} must be up-edges, as in Figure 21.3, or all three must be down-edges. If all three are down-edges, we have a situation that contradicts the down-convexity of $\langle U_{vx} \rangle$.

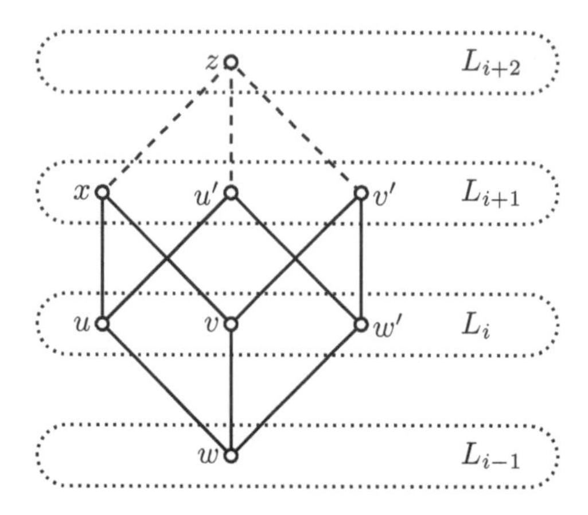

FIGURE 21.3 Testing 2-convexity.

We have thus shown that the violation of 2-convexity for a $\langle U_{ab} \rangle$ leads to the situation of Figure 21.3. If there exists a vertex $z \in L_{i+2}$ that is adjacent to x, u', and v', then $x \in U_{ab}$ and 2-convexity is not violated. If this is not the case, then x, u', and v' have no median.

We summarize our conclusions by the following lemma:

Lemma 21.16 *Algorithm 21.2 can be modified, without changing its complexity, such that only semi-median graphs are accepted and such that every accepted nonmedian graph has a Q_3^- that is embedded in the BFS-levels as in Figure 21.3.*

Recalling that $C_3^{\triangle} = Q_3^-$, we define graphs H_i on the vertex set L_i by letting two vertices of H_i be adjacent if they have a common neighbor in L_{i+1}. Clearly, H_i has at most $|L_{i+1}|(\log_2 n)^2$ edges. Thus all the graphs H_i have a total of at most $n(\log_2 n)^2$ edges. Moreover, if G is a median graph, there are not more than $n(\log_2 n)^3$ triangles in the H_i, because every triangle corresponds to a Q_3 embedded in G.

Suppose that we found all triangles in the H_i. For each such triangle of H_i, we only have to check whether the corresponding three vertices in L_{i+1} have a common neighbor z in L_{i+2}. This is easy, because G has already been embedded into a hypercube by the preprocessing. In other words, we know precisely the colors of the possible edges between z and the three vertices of L_{i+1}, and we can check this in constant time.

Suppose that we are given an algorithm of complexity $T(m, n)$ that finds all triangles in a given graph with n vertices and m edges. In our case $m = O(n \log n)$. Preprocessing requires $O(m \log n)$ time. Every triangle comes from a Q_3^-. To test whether these Q_3^- subgraphs are convex, we follow the approach above and take a total of $T(n(\log_2 n)^2, n)$ time. We have therefore proved the following theorem:

Theorem 21.17 *Let $T(m, n)$ be the complexity of finding all triangles of a given graph on* *n vertices and m edges. Then the complexity of checking whether a graph G on n vertices and m edges is a median graph is at most* $O(m \log n) + T(m \log n, n)$.

The best general algorithm known for listing all triangles of a graph has complexity $O(m^{3/2})$. By Theorem 21.17, we conclude that median graphs can be recognized in $O(m \log n + (m \log n)^{3/2})$ time. Because $m = O(n \log n)$, this reduces to $O(n^{3/2} \log^3 n)$, which only differs by factor $\log^2 n$ from the complexity of Theorem 21.12.

This is not what we had in mind, as we wish to reduce the complexity. We can use this theorem though to show that planar median graphs can be recognized in linear time.

The arguments leading to this result rely on the observation that the factor $\log n$ in the complexity of the preprocessing comes from the bound on the down-degree of the vertices in G. Because every vertex x of down-degree k in a median graph G is contained in a hypercube Q_k and because Q_k is nonplanar for $k > 3$, this implies that the down-degrees of planar median graphs are bounded by 3. Thus, preprocessing is linear in this case.

Corollary 21.18 *Planar median graphs can be recognized in linear time.*

Proof Let G be planar graph on n vertices with m edges. Because it is well known that planar graphs can be embedded into the plane in linear time, we assume that G is given together with such an embedding.

Let X_i be the subgraph spanned by L_{i+1} and L_i. In L_{i+1} there may be vertices of degree 1, 2, or 3 in X_i, but none of higher degree. Let w be a vertex in L_{i+1} of degree 3 and a, b, c be its neighbors in L_i. We split w into three vertices x, y, z and replace the edges aw, bw, cw by ax, ay, by, bz, cz, cx. We do this for every vertex of degree 3. Clearly, the new graph X_i' constructed this way is still planar. Moreover, every vertex of X_i not in L_i has degree 1 or 2. We now delete the vertices of degree 1 and replace every path $x_1x_2x_3$, where $x_1, x_3 \in L_i$ and $x_2 \in L_{i+1}$, by a single edge x_1x_3. Thus H_i is obtained by the same construction as the one used in the proof of Theorem 21.17.

Proceeding as in the proof of Theorem 21.17, we have to find all triangles in the H_i and perform certain checks. The complexity of these operations is determined by the complexity of finding all triangles. But the number of triangles in planar graphs can be found in linear time because planar graphs have arboricity 3.

The proof is completed by noting that H_i has at most $3n$ edges. □

The connections between triangle-free graphs and median graphs, in particular those stated in Theorem 21.13 and corollaries, are all from Imrich, Klavžar, and Mulder (1999), as is the linear recognition algorithm for planar median graphs.

Fast recognition of median graphs

One of the obstacles to improving the complexity of recognizing median graphs in the preceding considerations is the fact that we cannot determine all triangles of a given graph quickly enough. As we will see below, it is enough to be able to find for every edge the number of triangles that contain it.

As a corollary to a result of Alon, Yuster, and Zwick (1997), we show that we can do this slightly faster than listing all triangles. First the result of Alon, Yuster, and Zwick (1997).

Theorem 21.19 *It is possible to decide whether a directed or undirected graph G with n vertices and m edges contains a triangle, and to find one if it does, in* $O(m^{2\omega/(\omega+1)}) = O(m^{1.41})$ *time, where* ω *is the exponent of matrix multiplication.*

Proof Let $\Delta = m^{(\omega-1)/(\omega+1)}$. We say a vertex has *high degree* if its degree is larger than

Δ and *low degree* otherwise. Then the number of paths uvw, where v is of low degree, is at most $m\Delta$, and they can be found in $O(m\Delta)$ time. For each such path we check whether $uw \in E(G)$. This way we find all triangles containing a vertex of low degree.

If no triangles are found, all triangles are composed of vertices of high degree. Because there are at most $2m/\Delta$ vertices of high degree, we can use matrix multiplication to check for triangles in $O((m/\Delta)^\omega)$ time. Thus the total complexity is

$$O(m\Delta) + O(m/\Delta)^\omega) = O(m^{2\omega/(\omega+1)}).$$

□

Corollary 21.20 *Let G be an undirected graph with m edges. Then the complexity of determining in how many triangles every edge of G is contained is $O(m^{2\omega/(\omega+1)})$.*

Proof For vertices of low degree, all triangles containing them are actually constructed by the procedure above. We can thus determine within the given complexity in how many such triangles every edge is contained.

For vertices of high degree, we observe that matrix multiplication gives us to all pairs of vertices of high degree the number of common neighbors of high degree. If such a pair of vertices consists of the endpoints of an edge, this number is the number of triangles consisting only of vertices of high degree containing this edge.

Summing up the numbers of these different types of triangles that contain a given edge, we obtain the total number of triangles that contain this edge. □

With Lemma 21.16 we now prove a theorem of Imrich from the first edition of this book.

Theorem 21.21 *Let G be a graph with n vertices and m edges. Then one can decide in $O\big((m\log n)^{2\omega/(\omega+1)}\big) = O\big((m\log n)^{1.41}\big)$ time whether G is a median graph.*

Proof We begin with the situation described in Lemma 21.16 and define the graphs H_i as in the proof of Theorem 21.17. By the above we can determine the number of triangles in which every edge of H_i is contained.

On the other hand, if G is a median graph, every such triangle corresponds to a Q_3 in G. Let u be a given vertex. Consider all subcubes Q_3 of G that contain u and for which u is in the highest BFS-level of all vertices of Q_3. Because the down-degree of every vertex is bounded by $\log_2 n$, the number of such cubes is at most $(\log_2 n)^3$. Also one easily sees that the total number of cubes is at most $m(\log_2 n)^2$. Thus, under the assumption that G is a median graph, the complexity of finding for every edge the number of Q_3's that contain it is $O(m(\log n)^2)$.

If there is an edge e for which this number is different from the previously determined number of triangles which contain e, then G is not a median graph. □

Exercises

21.1. Given a partial cube on n vertices and m edges, show that the total number of vertices in all the sets U_{ab} is $2m$.

21.2. (Mulder, 1980a) Show that a connected graph G is a median graph if and only if it is triangle-free and any triple u, v, w of G with $d(u,v) = 2$ has a unique median.

21.3. (Imrich, Klavžar, and Mulder, 1999) Show that $\mathrm{Aut}(G^{\triangle}) = \mathrm{Aut}(G)$ if G is not a star.

21.4. (Imrich et al., 1999) Show that $G^{\triangle}$ is a Cartesian product if G is a star and determine the automorphism group of $G^{\triangle}$.

21.5. (Imrich, Klavžar, and Mulder, 1999) Let G be a triangle-free graph. Show that there is a Q_r and a mapping $j : V(G) \to V(Q_r)$ such that $d(j(u), j(v)) = 2$ for every edge uv of G.

Chapter 22

Recognizing Partial Hamming Graphs and Quasi-Median Graphs

Hamming graphs are Cartesian products of complete graphs and thus generalize the concept of hypercubes. In Chapter 14 we introduced partial Hamming graphs as isometric subgraphs of Hamming graphs, in analogy to partial cubes as isometric subgraphs of hypercubes, and quasi-median graphs as generalizations of median graphs.

This chapter is concerned with recognition algorithms for these classes of graphs. We begin with a simple, linear recognition algorithm for Hamming graphs and continue with an $O(mn)$ algorithm for partial Hamming graphs.

It turns out that quasi-median graphs have the same recognition complexity as median graphs, which is a highly nontrivial result of Hagauer (1995). We present the algorithm and the main part of the arguments for its validity.

Finally we provide an algorithm that determines the windex of a graph efficiently, which is important for the dynamic location problem of Chapter 14.

22.1 Hamming Graphs and Partial Hamming Graphs

The general question whether a graph is a Cartesian product of other graphs will be treated in Chapter 23. Here we provide a simple linear factorization algorithm for Hamming graphs.

Let G be a Hamming graph. Its vertices can be labeled with r-tuples $a_1a_2\ldots a_r$, where the alphabets from which the a_i are taken may depend on i. To simplify matters, we will always assume that $a_i \in \{0,1,\ldots,n_i-1\}$, where $r \geq 1$ and $n_i \geq 2$. Two vertices of G are adjacent if the corresponding tuples differ in precisely one position. We called such a labeling of vertices a *Hamming labeling*. For example, the Hamming graph on the left side of Figure 22.2 (on p. 270) is labeled with pairs a_1a_2, where $a_1 \in \{0,1,2\}$ and $a_2 \in \{0,1,2,3\}$.

Suppose that we are given a Hamming graph G without its labeling and wish to find it. We choose a vertex v_0 and consider the distance levels

$$L_k = \{u \in V(G) \mid d(v_0,u) = k\}$$

of a BFS-ordering of G. Label the components of $\langle L_1 \rangle$ as $C_1, C_2, \ldots C_r$. Notice that each $\langle V(C_i) \cup \{v_0\} \rangle$ is a layer $G_i^{v_0}$ of G (as a Cartesian product of r complete graphs), so the Hamming labels of G are r-tuples. We label v_0 with the r-tuple $000\ldots0$, and label the vertices of C_i with r-tuples $0\ldots0a_i0\ldots0$, where a_i assumes all values between 1 and n_i-1.

We label the remaining vertices in accordance with the following lemma. The labeling method is similar to that for partial cubes in Procedure 18.4, where the label $w(v)$ of v was computed as the disjunction of the labels of its down-neighbors.

Lemma 22.1 *Let G be a Hamming graph. Suppose that the vertices of the distance levels L_0 and L_1 of G with respect to a BFS-ordering of G have been labeled as described above. Then the vertices in the distance levels L_k, $k \geq 2$, can be labeled as follows: Suppose that we already know the labels of the vertices in L_{k-1}. Let $u \in L_k$. Then u has at least two neighbors v, w in L_{k-1}, and any two such neighbors differ in exactly two coordinates. If $v = b_1 b_2 \ldots b_r$ and $w = c_1 c_2 \ldots c_r$, then $u = a_1 a_2 \ldots a_r$, where $a_i = \max\{b_i, c_i\}$ for $i = 1, \ldots, r$.*

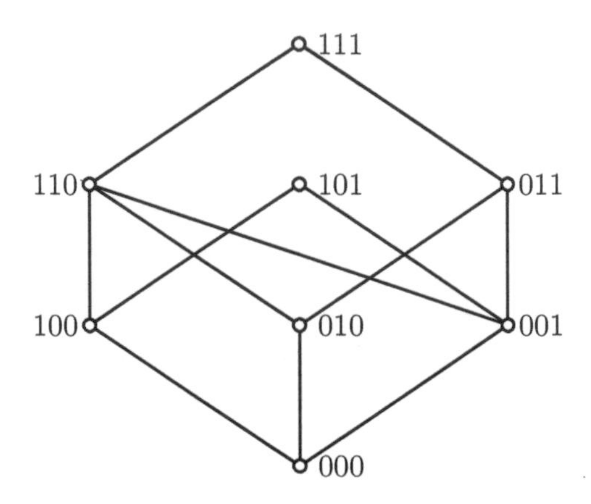

FIGURE 22.1 Graph with a labeling that is not a Hamming labeling.

If we do not know whether G is a Hamming graph but wish to check it, we also follow the above procedure. If it fails, then G is not a Hamming graph. On the other hand, G may not be a Hamming graph, even if the labeling procedure is successful; Figure 22.1 shows a non-Hamming graph with a labeling obtained by this method. (Note that it has the same number of vertices and edges as Q_3.) Thus we must check that the Hamming graph H associated with the labeling is indeed isomorphic to G. This yields the following algorithm.

Algorithm 22.1 Hamming graphs

Input: The adjacency list of a connected graph G.
Output: **true** if G is a Hamming graph, **false** otherwise.

1: Choose a vertex v_0 and find the connected components of $\langle N(v_0) \rangle$.
2: Let the components be $C_1, \ldots, C_r$, and suppose they have $n_1 - 1, \ldots, n_r - 1$ vertices, respectively.
3: Construct the Hamming graph H with vertices $a_1 \ldots a_r$ where $a_i \in \{0, \ldots, n_i - 1\}$.
4: Label v_0 with r zeros and the vertices of C_i with $0 \ldots 0 a_i 0 \ldots 0$, $a_i \in \{1, \ldots, n_i - 1\}$.
5: Label all other vertices of G according to Lemma 22.1. If this is not possible, then return **false** and stop.
6: Check whether the bijection between H and G given by the labelings is an isomorphism. If so, then return **true**; otherwise return **false**.

Algorithm 22.1 is an improved version of an algorithm of Imrich and Klavžar (1997). It labels a graph G according to Lemma 22.1 and checks whether it is a Hamming labeling.

Theorem 22.2 *Algorithm 22.1 correctly recognizes Hamming graphs. For a connected input graph with m edges, it can be implemented to run in $O(m)$ time and $O(m)$ space.*

Proof Concerning correctness, we observe that Steps 1, 2, 4, and 5 determine a Hamming labeling if G is a Hamming graph. This Hamming graph must be the Cartesian product of the $\langle C_i \cup v_0 \rangle$. This product is determined in Step 3 and denoted H. If G is a Hamming graph, then it must be isomorphic to H via the mapping identifying vertices with the same labels. This is checked in Step 6.

Now to the complexity of the algorithm. Because every vertex of G has at least r neighbors, we infer that $nr \leq 2m$. Hence the total length of the labels is $O(nr) = O(m)$. Clearly, Steps 1 to 4 of the algorithm can be executed within the claimed time. In Step 5 we select two neighbors $v, w \in L_{k-1}$ of an unlabeled vertex $u \in L_k$ and form the new label of u. This can be done in $O(r)$ time. Thus the complexity of this step is $O(nr) = O(m)$. For Step 6 we invoke Theorem 17.1.

For none of the steps we needed more than $O(m)$ space. □

Partial Hamming graphs

Partial Hamming graphs are isometric subgraphs of Hamming graphs. Let us assume that their embedding is irredundant, in other words, they have the minimal number of factors. By Corollary 13.7, the embedding is canonical, hence we can determine it efficiently by Corollary 18.7.

We thus arrive at the following algorithm for the recognition of partial Hamming graphs.

Algorithm 22.2 Partial Hamming graphs

Input: The adjacency list of a connected graph G.
Output: **true**, and a labeling, if G is a partial Hamming graph; **false** otherwise.

1: Compute Θ_1^*. {Recall that $\Theta^* = \Theta_1^*$.}
2: Denote the number of Θ^*-classes by k.
3: Compute G_i, $i = 1, 2, \ldots, k$, and $\alpha(v)$, $v \in V(G)$.
4: If for some i, $1 \leq i \leq k$, G/Π_i is not a complete graph, then return **false** and stop.
5: Return **true** and the labeling of G obtained in Step 3.

Theorem 22.3 *Algorithm* 22.2 *correctly recognizes partial Hamming graphs. It can be implemented to run in $O(mn)$ time with $O(m)$ space.*

Proof Steps 1 through 3 can be implemented by the method suggested in Exercise 18.9 for the design of a pseudocode for the computation of the canonical isometric embedding of a graph. By Corollary 18.7 this can be done in $O(mn)$ time and $O(m)$ space. It therefore remains to show that Step 4 can be implemented within this time and space complexity.

To see this, we note that every edge of a quotient graph G/Π_i corresponds to an edge of G and that this correspondence is injective. Hence $\sum_{i=1}^{k} |E(G/\Pi_i)| \leq m$. To implement Step 4 within the desired time and space, it therefore suffices to count the number of edges in the quotient graphs G/Π_i. □

Algorithm 22.2 is from Imrich and Klavžar (1993). Its complexity is the same as the complexity of the algorithm of Aurenhammer, Formann, Idury, Schäffer, and Wagner (1994). This algorithm was obtained before the one of Imrich and Klavžar (1993), although it appeared later; it is direct but difficult. Algorithm 22.2 was further simplified in Klavžar (2006), but the complexity remains the same.

22.2 Quasi-Median Graphs

The following algorithm reduces the recognition problem for quasi-median graphs to that of median graphs. For a graph G and a vertex $s \in V(G)$, the *skeleton* G_s of G (relative to s) is the graph obtained from G by removing all cross-edges uv of G, that is, all uv with $d(s,u) = d(s,v)$. Note that G_s is connected whenever G is connected. The next result is the key to our fast algorithm for the recognition of quasi-median graphs.

Theorem 22.4 *The skeletons of quasi-median graphs are median graphs.*

Proof Let G be a quasi-median graph that is isometrically embedded into a Hamming graph. We may assume that $s = 00\ldots0$. We now define a binary labeling of vertices of G_s. Let $u = u_1 \ldots u_t$ be a vertex of G. Then let the label $u' = u'_1 \ldots u'_t$ be defined as follows: Let the image of the ith positions of vertices in G be the set $\{0, \ldots, j_i\}$. Then u'_i is a string of length j_i. In such a string the jth bit is set to 1 if $u_i = j$. All the other bits are equal to 0. Figure 22.2 shows an example of such a labeling.

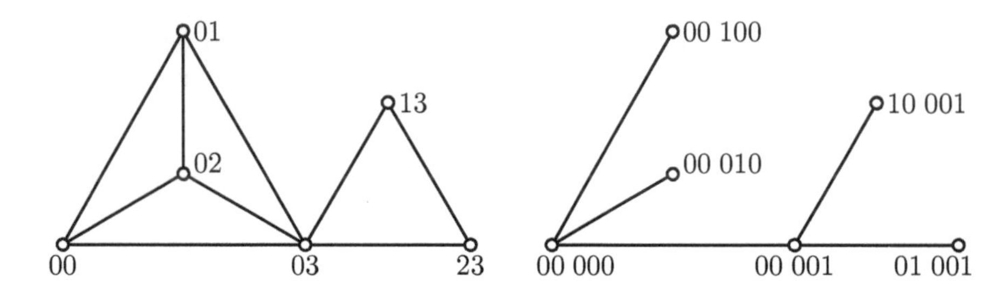

FIGURE 22.2 Quasi-median graph and its skeleton.

By Corollary 14.9, it suffices to prove that G_s is a median-closed induced subgraph of a hypercube.

We show first that the binary labeling we have defined embeds G_s as an induced subgraph into the hypercube. To distinguish between G and G_s, we represent vertices of G_s by their labels. Let $u'v'$ be an edge of G_s. Then uv is an edge of G, and thus u and v differ in exactly one position, say $u_i \neq v_i$. Because $d_G(s,u) \neq d_G(s,v)$, exactly one of u_i and v_i must be 0. Therefore u' and v' differ in one position. Because $H(u',v') \geq H(u,v)$, we have thus shown that G_s is an induced subgraph of a hypercube.

It remains to prove that G_s is median-closed. Let u', v', and w' be an arbitrary triple of vertices of G_s corresponding to the triple u, v, and w of G. Set

$$x = \operatorname{imp}\big(\operatorname{imp}(u,v;w), \operatorname{imp}(u,w;v); s\big) .$$

As G is a quasi-median graph, $x \in V(G)$. A straightforward check shows that if two values among u_i, v_i, and w_i are equal, then x_i also equals this value. Otherwise, $x_i = 0$. But this implies that $x' = \operatorname{imp}(u',v';w')$, in other words, that G_s is median-closed. □

The necessary condition of Theorem 22.4 is not sufficient to recognize quasi-median graphs. We must also verify whether the edges of G not in the skeleton fit into it. This is done via the following two relations.

Let s be a fixed vertex of a graph G. Then a subgraph H of G is called *s-gated* if there exists a vertex x of H such that $d(s,y) = d(s,x) + 1$ for any vertex $y \neq x$ of H. Then edges

e and f of G are in relation Σ if they belong to the same s-gated triangle. Let Σ^* denote the reflexive and transitive closure of Σ.

Assume that G_s is a median graph. Then we say that edges e and f of G_s are in relation T if there is an edge g of G_s such that $e\Sigma^* g$ and $g\Theta f$.

Theorem 22.5 *Let G be a connected graph and $s \in V(G)$. Then G is a quasi-median graph if and only if the following conditions hold:*

(i) *G_s is a median graph.*
(ii) *Each equivalence class of Σ^* on $E(G)$ induces an s-gated clique.*
(iii) *T is an equivalence relation on $E(G_s)$.*

We will only prove that the three conditions of Theorem 22.5 are necessary. The proof that they are also sufficient is rather lengthy and technical (Hagauer, 1995).[1]

Let G be a quasi-median graph. Because, by Lemma 14.19, every clique of G is gated and, by Theorem 14.13 (iv), G induces no $K_4 - e$ as a subgraph, condition (ii) holds. To show that T is an equivalence relation, recall that a quasi-median graph is a partial Hamming graph. Thus let G be equipped with a Hamming labeling in which the vertex s is labeled with a sequence of zeros. Then one can show that edges e and f of G_s are in relation T if and only if the Hamming labels of the endpoints of e and the endvertices of f differ in the same position. This immediately implies that T is an equivalence relation.

As an immediate consequence of Theorem 22.5, we infer that Algorithm 22.3 correctly recognizes quasi-median graphs.

Algorithm 22.3 Quasi-median graphs

Input: The adjacency list of a connected graph G.
Output: **true** if G is a quasi-median graph, **false** otherwise.

1: Compute G_s, where s is an arbitrary vertex of G.
2: If G_s is not a median graph, then return **false** and stop. Otherwise, compute a (binary) Hamming labeling.
3: If the equivalence classes of Σ^* on $E(G)$ do not induce s-gated cliques, then return **false** and stop.
4: If T is not an equivalence relation on $E(G_s)$, then return **false** and stop.
5: Return **true**.

Showing the time complexity requires some preparation. Recall that the down-degree of a vertex v in G is the number of neighbors of v in $L_{l(v)-1}$, and that the down-degree of a vertex v in a subgraph G of an r-cube is bounded by its level $l(v)$ by Proposition 17.5. For median graphs, this statement can be improved as follows:

Lemma 22.6 *Let G be a median graph and v a vertex of down-degree k with respect to a vertex v_0. Then $k \leq \log_2 |V(G)|$, and v is contained in a k-cube that meets the levels $L_{l(v)}, L_{l(v)-1}, L_{l(v)-2}, \ldots, L_{l(v)-k}$.*

Proof Because Q_k has 2^k elements, k cannot be larger than $\log_2 |V(G)|$ if $Q_k \subset G$. It therefore suffices to prove the second assertion.

We show that any k down-neighbors of v are contained in such a cube, even if the down-degree of v is higher. Let $v_1, v_2, \ldots, v_k$ be down-neighbors of v. We suppose that the edge

[1] This means that Algorithm 22.3, Theorem 22.7, and Theorem 22.9 are also only proved up to this sufficiency argument. These results are not invoked later.

vv_i is colored with color i.[2] By Exercise 22.11 it suffices to show that there is a vertex x in level $L_{l(v)-k}$ that is connected with v by $k!$ shortest paths containing $v_1, v_2, \ldots, v_k$.

For $k = 2$, consider the median x of v_1, v_2 and v_0. Because x is on a shortest v_0, v_1-path and $d(v_1, v_2) = 2$, we have $x \in L_{l(v)-2}$, and x has distance 1 from both v_1 and v_2. Thus it is a common down-neighbor of v_1 and v_2 and the assertion about the paths is true for $k = 2$.

Suppose now that $k > 2$ and that the assertion of the lemma is true for $k - 1$. We first note that any pair of vertices v_i, v_j, $1 \leq i, j \leq k$, has a common down-neighbor, say v_{ij}. Then the edge $v_i v_{ij}$ has the same color as the edge vv_j; that is, every v_i has $k - 1$ down-neighbors, all of which have colors different from that of vv_i. Let x_i be the vertex in level $L_{l(v)-k}$ that is connected with v_i by 2^{k-1} distinct shortest paths. Then v is connected with x_i by that many shortest paths whose first edge is vv_i. All these paths contain the colors $1, 2, \ldots, k$. Because this is true for all i, all vertices x_i must be identical, and we have found the $k!$ paths. □

Let $M(m, n)$ denote the complexity of recognizing median graphs on n vertices and m edges. We can now state:

Theorem 22.7 *For a graph G on n vertices and m edges, Algorithm 22.3 correctly recognizes quasi-median graphs. It can be implemented to run in $O(M(m, n) + m \log n)$ time.*

Proof The correctness of the algorithm has already been observed.

Clearly, Steps 1 and 2 can be implemented to run in the desired time and space. Moreover, in $O(m)$ time and space we can compute $d(s, u)$ for each $u \in V(G)$.

In Step 3 we have to compute the relation Σ^*. We begin by sorting the down-edges of each vertex. Altogether we sort at most m edges. Thus this can be done in $O(m \log m) = O(m \log n)$ time. Then, for each cross-edge uv of G, we do the following: We simultaneously scan the sorted lists of down-edges of u and v and determine the number of their common neighbors. If there is no such vertex or more than one, we reject the graph. (In both cases a clique containing uv is not s-gated.) If w is the only common neighbor of u and v, then we put the edges uv, uw, and vw into the same class using the simple merging operation of Proposition 17.7. By Lemma 22.6, we need to check at most $O(\log n)$ neighbors for each cross-edge; thus the overall complexity of computing Σ^* is $O(m \log n)$.

Now we must test whether the Σ^*-classes are s-gated cliques. Because we know all the distances $d(s, u)$, this can be done easily by checking these distances and counting the number of edges and vertices in each class.

Finally, we implement Step 4. Let uv be an edge of G_s. In Step 3 the edge uv is labeled by a unique clique to which it belongs (with respect to Σ^*). We may assume that in Step 2 the coordinate in which u and v differ with respect to the (binary) Hamming labeling is also computed. In other words, uv is also labeled by a color. We first sort the color list for each clique. This can be done in $O(m \log n)$ time.

By definition, two edges e and f are in relation T if there is an edge g from the same Σ^*-clique as e with the same color as f. Suppose that we have k colors. We form k boxes as follows: If a clique Q contains an edge (from G_s) of color i, then we put Q in the ith box. Now T is an equivalence relation if all the cliques of a box have the same color list. If Q and Q' are two cliques from the ith box, then we simultaneously scan the sorted color lists of the cliques to check whether they are the same. Any additional clique from the ith box must be compared only to the clique Q. Moreover, if j is a color of the clique Q, then the test of the ith box gives a test for the jth box as well. After the color lists of the cliques are sorted, the remaining test is linear in m. □

[2]In other words, we assume that vv_i is contained in the Θ^*-class E_i.

The presented recognition algorithm for quasi-median graphs is due to Hagauer (1995). Mulder (1980a); Chung, Graham, and Saks (1989); as well as Wilkeit (1992) observed earlier that quasi-median graphs can be recognized in polynomial time. For instance, one can first check whether a given graph is a partial Hamming graph, and then verify whether it is imprint-closed. Feder (1992) presented a method that yielded an $O(mn)$ algorithm.

Wilkeit (1986), see also Wilkeit (1992, p.218), proved that retracts of Hamming graphs are Cartesian products of median graphs and Hamming graphs. This yields a fast recognition algorithm for retracts of Hamming graphs: Given a connected graph, extract all complete factors and test whether the remainder is a median graph. By Section 23.4 and Theorem 21.12 this can be done in $O(m\sqrt{n})$ time.

Notice that quasi-median graphs are weak retracts and cannot be recognized by this algorithm.

22.3 Computing the Windex

The motivation to study quasi-median graphs was the dynamic location problem. It turned out that graphs of finite windex are quasi-median graphs. To compute the windex of a graph efficiently, we will apply the following result:

Proposition 22.8 *If $WX(G) < \infty$, then $WX(G) = \omega(G)$.*

Proof Let Q be a largest clique of G, meaning that $|V(Q)| = \omega(G)$. Then Q is a weak retract of G. (This is true for any clique of G.) Thus, by Lemmas 14.3 and 14.4, we have $\omega(G) = |V(Q)| = WX(Q) \leq WX(G)$.

For the converse, consider G as a connected, imprint-closed, irredundant subgraph of a Hamming graph $G_1 \Box G_2 \Box \cdots \Box G_n$. Let i be an arbitrary index, $1 \leq i \leq n$, and let u and v be adjacent vertices that differ in the ith coordinate. Such vertices exist because the embedding is irredundant and because G is connected. Because G is imprint-closed, the set of vertices $\{\mathrm{imp}(u, v; w);\ w \in V(G)\}$ forms a clique of size $|V(G_i)|$. It follows that $WX(G) \geq \omega(G)$. □

Algorithm 22.4 Windex

Input: The adjacency list of a connected graph G.
Output: $WX(G)$.

1: If G is not a quasi-median graph, then $WX(G) = \infty$.
2: Compute the quotient graphs G/Π_i of the canonical isometric embedding of G.
3: $WX(G) = \max_i \{|V(G/\Pi_i)|\}$.

Theorem 22.9 *If G is a connected graph on n vertices and m edges, then Algorithm 22.4 correctly computes $WX(G)$ and can be implemented to run in $O(mn)$ time.*

Proof Correctness follows from Theorem 14.13 and Proposition 22.8, and the complexity from Theorems 22.7 and 22.3. □

In addition to the canonical isometric representation described in Chapter 13, Feder (1992, 1995) considers canonical 2-isometric representations and canonical retract representations. In the case of Hamming graphs, the connected 2-isometric subgraphs coincide with the retracts. Because the canonical 2-isometric representation of a graph can be found in $O(mn)$ time, this yields alternative $O(mn)$ recognition algorithms for median graphs and quasi-median graphs as well as an algorithm for computing the windex of a graph within the same time complexity. On the other hand, the problem of finding the canonical retract representation of a graph is NP-hard. Along these lines, Feder also proved the following result: Let $V(K_2) = \{1, 2\}$, a be a vertex of a graph G, and H a subgraph of $G \,\Box\, K_2$ induced by the vertices $\{(u, 1) \mid u \in V(G)\} \cup (a, 2)$. Then the problem whether H is a retract of $G \,\Box\, K_2$ is NP-complete.

Chapter 23

Factoring the Cartesian Product

The Sabidussi-Vizing Theorem 6.6 states that connected graphs factor uniquely into primes with respect to the Cartesian product. The proof is surprisingly short. The first aim of this chapter is to provide an almost equally short argument that the prime factors of a connected graph on n vertices and m edges can be computed in $O(mn)$ time.

The preparations for the algorithm also lead to a new proof of the uniqueness of the prime factorization that also holds, with slight modifications, for infinite graphs.

The second goal is a fast algorithm for prime factorization. The main idea is to define a simple relation on the edge set of the graph to be factored that is finer than the product coloring (see p. 162). This relation is then made coarser in at most $\log_2 n$ steps until it equals the product coloring. The method—though conceptually simple—is technically tricky. We thus only present a relatively straightforward version of complexity $O(m \log n)$, but explain in the final section subtle changes that lead to a linear recognition algorithm.

23.1 Product Relation

Let $G = G_1 \,\square \cdots \square\, G_k$ be a connected Cartesian product. Recall that the product color $c(uv)$ of an edge $uv \in E(G)$ is i if u and v differ in coordinate i. We then say that two edges e and f of G are in relation $c(G_1 \,\square \cdots \square\, G_k)$ if $c(e) = c(f)$. We call the relation $c(G_1 \,\square \cdots \square\, G_k)$ a *product relation.*

Clearly, $c(G_1 \,\square \cdots \square\, G_k)$ is transitive, reflexive, and symmetric. By Lemma 13.5 (ii),

$$\Theta_G \subseteq c(G_1 \,\square \cdots \square\, G_k)\,.$$

Our goal is the computation of the product coloring of the prime factorization of G. In order to do this, we introduce a new relation τ on $E(G)$.

We say that edges $e = uv$ and $f = uw$ of G are in relation τ, in symbols $e\tau f$, if $e = f$, or $vw \notin E(G)$ and u is the only common neighbor of v and w. See Figure 23.1.

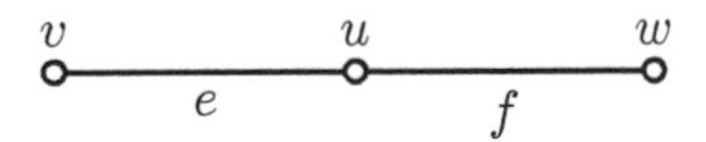

FIGURE 23.1 Edges e and f in relation τ.

Clearly, τ is symmetric. Furthermore, if $e = uv$ and $f = uw$ are in relation τ, e and f cannot both be edges of one and the same chordless square. In other words, $e\tau f$ implies that e and f have the same color with respect to any product coloring of G. Thus

$$\tau(G) \subseteq c(G_1 \square \cdots \square G_k) .$$

Because $c(G_1 \square \cdots \square G_k)$ is transitive, we infer that

$$(\Theta_G \cup \tau_G)^* \subseteq c(G_1 \square \cdots \square G_k). \tag{23.1}$$

It turns out that $(\Theta \cup \tau)^*$ is the product relation of the prime factorization of G. To show this, we show first that it is a product relation. To this end we extend the concept of the square property as defined on p. 66 to equivalence relations on $E(G)$ and prove a lemma.

We say an equivalence relation ρ on the edges of a graph G has the *square property* if for any two edges $e = uv$ and $f = uw$ that belong to different equivalence classes of ρ, there is a unique $x \in V(G)$ such that $uvxw$ is a diagonal-free square. (See Figure 23.2.)

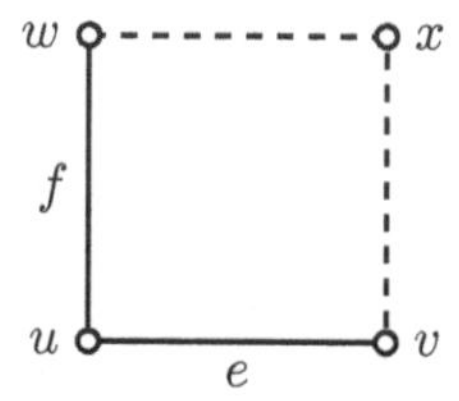

FIGURE 23.2 Square property.

Lemma 23.1 *The relation $\sigma = (\Theta \cup \tau)^*$ has the square property.*

Proof Suppose that $e = uv$ and $f = uw$ belong to different equivalence classes of σ. Then v and w are not adjacent; otherwise, e and f would be in relation Θ. Also e and f cannot be in relation τ. Thus there must exist another common neighbor, say x, of v and w. Clearly, x is not adjacent to u, for otherwise $e\Theta f$.

Suppose x is not unique and that there is another vertex, say x', that is adjacent to v and w. By the definition of Θ, we infer that

$$uv\Theta wx\Theta x'v\Theta uw.$$

But then $e\Theta^* f$, which is not possible. □

Theorem 23.2 *If G is connected, then $\sigma = (\Theta \cup \tau)^*$ is a product relation.*

As we shall see, this result immediately implies the unique prime factorization of connected graphs with respect to the Cartesian product, that is, Theorem 6.6 of Sabidussi-Vizing, and yields the $O(mn)$ Algorithm 23.1 for the computation of the prime factors. Moreover, it holds for infinite graphs. We present two proofs.

First proof Let $E_1, E_2, \ldots, E_k$ be the σ-equivalence classes of $E(G)$. For each index i, let G_i be the subgraph induced by $E(G) \setminus E_i$. Becaue every σ-class is a union of Θ^*-classes, Lemma 11.5 implies that the components of G_i are convex.

We claim that if C and C' are two adjacent components of G_i, then the set of edges in G joining C to C' is a matching that induces an isomorphism $C \to C'$. First note that each vertex of C is adjacent to at most one vertex of C'. Indeed, if $x \in V(C)$ were adjacent to $y, z \in V(C')$, then $yz \in E(C')$, by convexity of C'. But then $xy\, \Theta\, yz$, which is impossible.

Hence any vertex of C is adjacent to at most one vertex of C'. Now, if xx' joins C to C' and $xy \in E(C)$, then Lemma 23.1 guarantees a chordless square $xx'y'y$. By connectedness of C, we infer that the edges from C to C' induce an injective homomorphism $C \to C'$. Reversing the roles of C and C', we see that this homomorphism is an isomorphism.

Next observe that if C_i is a component of G_i for each index i, then $\bigcap_{i=1}^k C_i$ consists of at most one vertex. Indeed, this intersection has no edges because every edge is in some E_i but no edge of C_i belongs to E_i. As an intersection of convex subgraphs, it is connected, and thus a single vertex, provided that it is nonempty.

Actually, connectedness of G implies that it is nonempty: First take an $x \in V(G)$ and for each i let C_i be the component of G_i containing x, so $x \in \bigcap_{i=1}^k C_i$. Now, if for some i the component C_i is adjacent to a component C_i', then the previous paragraph guarantees that G has an edge xy with $y \in C_i'$. As the edge xy belongs to each C_j with $j \neq i$, we have $y \in C_1 \cap \cdots \cap C_i' \cap \cdots \cap C_k$. We infer that—by connectedness of G—any intersection $\bigcap_{i=1}^k C_i$ is nonempty. In summary, any $\bigcap_{i=1}^k C_i$ consists of exactly one vertex.

Now for each index i, let G_i^* be the graph whose vertices are the components of G_i and for which CC' is an edge precisely if G has an edge joining the components C and C'. For each vertex x of G, let $G_i(x)$ be the component of G_i that contains x. We define a map $\alpha : G \to G_1^* \,\Box\, G_2^* \cdots \,\Box\, G_k^*$ as

$$\alpha(x) = \big(G_1(x), G_2(x), \ldots, G_k(x)\big).$$

If α is an isomorphism, then the corresponding product relation on $E(G)$ equals σ, for then $xy \in E_i$ precisely if $\alpha(x)$ and $\alpha(y)$ differ in coordinate i.

Thus to finish the proof, we just need to show that α is an isomorphism. Clearly, it is a homomorphism: Given an edge xy of G, we have $xy \in E_i$ for some i. The definitions imply $G_i(x)G_i(y) \in E(G_i^*)$, and $G_j(x) = G_j(y)$ for $j \neq i$, so $\alpha(x)\alpha(y)$ is an edge.

The map is injective, for $\alpha(x) = \alpha(y)$ means that x and y are in the same component C_i of G_i for each i, so $x, y \in \bigcap_{i=1}^k C_i$. As the intersection is a single vertex, $x = y$. To see that it is surjective, note that any vertex $(C_1, \ldots, C_k)$ equals $\alpha(x)$, where $x = \bigcap_{i=1}^k C_i$.

Finally suppose $\alpha(x)\alpha(y) = (C_1, \ldots, C_i, \ldots, C_k)(C_1, \ldots, C_i', \ldots, C_k)$ is an edge. Note that $y = C_1 \cap \cdots \cap C_i' \cap \cdots \cap C_k$. Also $x \in C_i$ and $y \in C_i'$, where C_i is adjacent to C_i'. Our isomorphism $C_i \to C_i'$ guarantees an edge xz with $z \in C_i'$. Then $z \in C_1 \cap \cdots \cap C_i' \cap \cdots \cap C_k$, so $z = y$ and $xy \in E(G)$. □

Second proof Let $E_1, E_2, \ldots, E_k$ be the equivalence classes of $E(G)$ with respect to σ, and define the map α as in the first proof. Define the graphs G_i, G/Π_i, the mapping α of G into the Cartesian product of the G/Π_i as in Section 13.1. Notice that the sets $E_1, E_2, \ldots, E_k$ were equivalence classes with respect to Θ^* in Section 13.1, whereas they are equivalence classes with respect to σ now. To emphasize the difference, we set $G_i^* = G/\Pi_i$ and G^* for the Cartesian product of the G_i^*.

We wish to show that α is surjective. To this end, consider a vertex $u \in V(G)$ and an arbitrary edge $e \in E(\Box\, G_i^*)$ incident with $\alpha(u)$. If we can show that e is in $\alpha(G)$, then $G^* = \alpha(G)$ by induction, and we are done.

Let the other endpoint of e be y, and suppose that y differs from $\alpha(u)$ in the jth coordinate. Let C, respectively and C', be the components of $G - E_j$ corresponding to $\alpha_j(u)$, respectively to the jth coordinate y_j of y.

Because $\alpha_j(u)$ and y_j are adjacent in G_j^*, there is an edge bb' from C to C' in G. Consider a shortest path P from u to b in G, say $uu_1u_2 \ldots u_rb$. Because C is convex, P is in C. Let a be the vertex of P closest to u that has a neighbor a' in C'. If $a = u$, then $e \in E(\alpha(G))$. Otherwise, let u_s be the predecessor of a on P. Clearly, u_sa and aa' are not in relation σ. Thus there is a unique vertex x that is adjacent to u_s and a' but not to a. Then

$[u_s, a]\, \sigma\, [x, a']$. This means that x is in C' and u_s is adjacent to a vertex in C', contrary to the choice of a. □

With slight adjustments, the above proofs also hold for infinite graphs. We will see that G^* is not connected if σ has infinitely many equivalence classes. In this case the map α is not surjective. But, another look at the above proof shows that the words "If we can show that e is in $\alpha(G)$, then $G^* = \alpha(G)$ by induction, and we are done" actually mean that α is surjective on every component in its range. Notice that it can meet just one component because G is connected. The other parts of the proof remain valid, and show that α is an isomorphism of G onto a component of G^*. We formulate this as a corollary.

Corollary 23.3 *Let G be a connected, infinite graph. If α and G^* are defined as in the second proof of Theorem 23.2, then α is an isomorphism of G onto a connected component of G^*.*

Now we are ready for the main result of this section. The first part is a restatement of Theorem 6.6 of Sabidussi-Vizing, which we hereby reprove, and the second part, that $(\Theta \cup \tau)^*$ is the product relation of the unique prime factorization, is due to Feder (1992).

Theorem 23.4 (Sabidussi-Vizing and Feder) *Every connected graph has a unique prime factor decomposition over the Cartesian product and $(\Theta \cup \tau)^*$ is its product relation.*

Proof By Theorem 23.2, $\sigma = (\Theta \cup \tau)^*$ is a product relation. Moreover, Equation (23.1) implies that σ is contained in every other product relation; hence it is the finest product relation on G. It follows that it induces a decomposition into prime factors.

For uniqueness, let $G = A \,\Box\, B$, where A is prime. Let F_1, F_2 be the color classes of $c(A \,\Box\, B)$. Every class E_j, $1 \leq j \leq k$, of σ must be contained in F_1 or F_2, hence F_1 is the union of one or more E_j, say $F_1 = E_1 \cup E_2 \cup \cdots \cup E_s$. Clearly, the connected components of $H_i = (V(G), E_i)$ are layers of $G = G_1^* \,\Box\, G_2^* \,\Box\, \cdots \,\Box\, G_k^*$ with respect to G_i^*, so the connected components of the graph induced by F_1 are isomorphic to the Cartesian product $G_1^* \,\Box\, G_2^* \,\Box\, \cdots \,\Box\, G_s^*$, but then A_1 is not prime. Hence $F_1 = E_i$, for some i, $1 \leq i \leq s$. □

In his seminal paper on graph multiplication, Sabidussi (1960) used a so-called tower of equivalence relations on the edge set of a connected graph to decompose it into a Cartesian product of prime graphs. His aim was nonalgorithmic. Nonetheless, by following Sabidussi's approach, Feigenbaum, Hershberger, and Schäffer (1985) derived a polynomial algorithm that computes the prime factors of a connected graph; that is, they computed σ from the relation δ (as defined on p. 245).

Graham and Winkler's canonical isometric embedding opened the possibility of computing σ from Θ. Winkler (1987) and Feder (1992) followed this path, as have we here.

One can simplify Sabidussi's approach by invoking convexity properties of σ. That is to say, the relation σ can be obtained as the convex hull of $\delta \cup \tau$; see Imrich and Žerovnik (1994). (An equivalence relation γ with equivalence classes E_ι, $\iota \in I$ is called *convex* if for any $K \subseteq I$, every connected component of the graph induced on $\bigcup_{\iota \in K} E_\iota$ is convex.)

23.2 A Simple Algorithm

By the above, factoring a connected graph is equivalent to computing $\sigma = (\Theta \cup \tau)^*$. Notice that

$$\sigma = (\Theta \cup \tau)^* = (\Theta^* \cup \tau)^* = (\Theta_1^* \cup \tau)^* = (\Theta_1 \cup \tau)^*.$$

Theorem 18.6 tells us that Θ_1^* is computable in $O(mn)$ time and $O(m)$ space, so it remains to compute τ efficiently.

Proposition 23.5 *Given a graph with n vertices and m edges, τ can be computed in $O(mn)$ time, $O(m)$ space, and consists of at most mn pairs of edges.*

Proof Let v be a vertex of G. To find all edge pairs vu, uw that are in the relation τ, it suffices to scan all vertices w of G and to check how many neighbors of w are adjacent to v. If there is exactly one such neighbor, then the pair vu, uw is in τ. If there are no common neighbors, or more than one common neighbor, then there is no u such that vu, uw is in τ.

We thus scan all neighbors of w. When we find the first common neighbor, say u, we mark it. If no other neighbor is found, we add the pair vu,uw to τ.

For every w, the number of checks is $d(w)$, and every check can be executed in constant time with the aid of the adjacency matrix. Hence the cost for each vertex w is $O(d(w))$. Because n vertices must be scanned, the total cost for the vertex v is $O\big(\sum_{w\in V(G)} d(w)\big) = O(m)$. As this must be repeated for every vertex v of G, the total cost is $O(mn)$.

Clearly, the number of pairs vu, uw in τ is at most

$$\sum_{u\in V(G)} \frac{d(u)\big(d(u)-1\big)}{2} < \frac{1}{2} \sum_{u\in V(G)} d(u)\, n = mn\,.$$

If we use the full adjacency matrix, the space complexity is n^2, not m. However, when we check whether a neighbor is adjacent to v, we only need the line for v in the adjacency matrix. This line needs $O(n)$ space and can be generated with the aid of a reference vector in $O(d(v))$ time.

We do this once for every v and thus stay within the time and space limit. □

Theorem 23.6 *The prime factorization of a connected graph over the Cartesian product is computable in $O(mn)$ time and $O(m)$ space.*

Proof Because

$$\sigma = (\Theta \cup \tau)^* = (\Theta^* \cup \tau)^* = (\Theta_1^* \cup \tau)^*\,,$$

it suffices to compute $(\Theta_1^* \cup \tau)^*$. We first compute Θ_1^* in $O(mn)$ time with Algorithm 18.3. The outcome is a partition of $E(G)$.

Then we compute τ and scan the pairs e, f of edges in τ. If e and f are in different Θ_1^*-classes, then we merge the two classes (that is, we replace them with their union). Clearly, there at most $m-1$ possible merge operations.

If we consider the elements of Θ_1^* as edges of a graph $G_{\Theta_1^*}$ whose vertices are the edges of G, then every merge operation corresponds to a union of connected components of $G_{\Theta_1^*}$. By Proposition 17.7, the total time complexity is $O(m \log m) = O(mn)$.

Because the assertion about the space complexity is clear, we have proved the theorem. □

We close this section with a pseudocode for this procedure, but wish to remark that the algorithm does not provide the coordinates of the vertices of G with respect to the prime factors. We shall show later that they can easily be computed in $O(m)$ time and space.

Algorithm 23.1 Cartesian product decomposition

Input: The adjacency list of a connected graph G.
Output: The prime factors of G with respect to the Cartesian product.

1: Compute Θ_1^*.
2: Compute τ.
3: Compute $\sigma = (\Theta_1^* \cup \tau)^*$, that is, the equivalence classes $E_1, E_2, \dots, E_k$ of σ.
4: **for** $i = 1$ **to** k **do**
5: Compute an arbitrary connected component, say G_i^*, of $(V(G), E_i)$.
6: Return G_i^*. {The G_i^* are the prime factors of G.}
7: **end for**

23.3 Coordinatization

Once σ_G is known, we are confronted with the question of how to compute the coordinates of G's vertices relative to the prime factorization $G = G_1 \,\square \cdots \square\, G_k$. We now present a solution to this problem.

As usual, m and n denote the number of vertices and edges of G. Let $E_1, E_2, \dots, E_k$ be the equivalence classes of σ, and let v_0 be the root of a BFS-ordering of G. Choose the indexing so that $(V(G), E_i)$ corresponds to G_i. For each i, the component of $(V(G), E_i)$ that contains v_0 is called a *unit-layer*, and is denoted $G_i^{v_0}$. Thus $G_i^{v_0} \cong G_i$ and $G = G_1^{v_0} \,\square \cdots \square\, G_k^{v_0}$. In what follows, we coordinatize G relative to this product decomposition.

We first assign to each $v \in V(G)$ a BFS-number BFS(v), from 0 to $|V(G)|-1$, according to its order in a BFS traversal of G with root v_0. Thus BFS$(v) >$ BFS(u) whenever $d(v, v_0) > d(u, v_0)$. In particular, this assigns a label to each vertex of each unit-layer.

Every vertex v will have k coordinates $p_1(v), p_2(v), \dots, p_k(v)$, where each $p_i(v)$ is a label of a vertex in the unit-layer $G_i^{v_0}$. The coordinates of the unit-layers are as follows. If $v \in G_i^{v_0}$, then $p_i(v) = \text{BFS}(v)$, but all other coordinates of v are set to zero. Any v not in a unit-layer can now be given coordinates relative to the decomposition $G = G_1^{v_0} \,\square \cdots \square\, G_k^{v_0}$. Figure 23.3 shows an example. (Had the unit-layers been numbered consecutively from 1 to 4, and 1 to 2, respectively, we would have a different coordinatization. See Figure 6.1.)

We next develop a formula for the coordinates of the non-unit-layer vertices.

First, we claim that if all down-edges from v are in the same σ-class E_i, then v is in the unit-layer $G_i^{v_0}$. Indeed, the down-edges of the down-neighbors of v must all be in E_i, or otherwise the square property yields a down-edge from v that is not in E_i. By induction, there is a path from v to v_0 consisting of edges from E_i, and therefore v belongs to $G_i^{v_0}$.

In particular, this means that if v is not in a unit-layer, then it has two down-edges that are in different σ-classes.

Let v be a vertex in BFS-level L_k with two neighbors u, w of v in L_{k-1}, where vu and vw are in different σ-classes. It is easy to show (Exercise 23.2) that $p_i(v) = \max\big(p_i(u), p_i(w)\big)$ for $1 \le i \le k$. Denoting the down-neighbors of u by down(v), we thus have

$$p_i(v) = \max_{u \in \text{down}(v)} p_i(u), \quad 1 \le i \le k, \tag{23.2}$$

which we abbreviate as

$$p(v) = \max_{u \in \text{down}(v)} p(u)$$

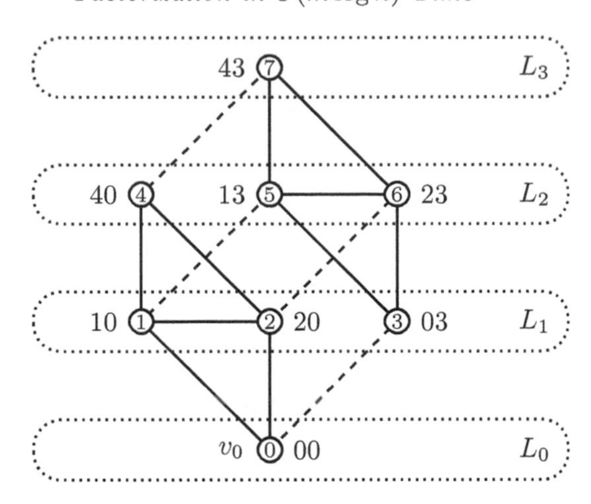

FIGURE 23.3 Prime factorization with coordinates.

for all vertices that are not unit-layer vertices. (See Exercise 23.3.)

Because the number of down-neighbors of v bounded by $d(v)$, this means that the coordinates of v can be found by $d(v) \cdot k$ comparisons from the coordinates of the down-neighbors of v. Hence all vertices of G that are not unit-layer vertices can be coordinatized in $O(mk)$ time. Now notice that k, that is, the number of factors of G, is bounded by $\log_2 n$, that the unit-layers can be computed in $O(m)$ time, and that the only nonzero coordinate of a unit-layer vertex is its BFS-number.

We have thus shown that one can use Equation (23.2) to coordinatize the vertices of G with respect to its prime factors in $O(m \log n)$ time from the equivalence classes of σ.

This simple method uses all down-neighbors. The complexity can be improved to $O(m)$ if we choose the down-neighbors carefully. See Exercise 23.4 for a proof of the next lemma.

Lemma 23.7 *Let G be a connected graph on n vertices and m edges. Given its σ-classes, one can coordinatize G with respect to its prime factors in $O(m)$ time.*

23.4 Factorization in $O(m \log n)$ Time

Algorithm 23.1 computes the product relation $\sigma = (\Theta \cup \tau)^*$ of a connected graph G on n vertices and m edges in $O(mn)$ time. Notice that Θ reflects both global and local properties of G, and τ only local ones. Recall that the best algorithm for the computation of Θ has complexity $O(m^2)$, and that even for Θ^* and τ, we know of no recognition algorithm of better complexity than $O(mn)$. Hence we have to look for other methods if we wish to improve the complexity of finding the prime factors and present the local approach of Aurenhammer, Hagauer, and Imrich (1992).

Given a connected graph G, we choose a vertex v_0 of minimal degree and arrange the vertices of G in BFS-order. We then assign coordinate vectors $f(v)$, or labels as we call them in this section, to the vertices of G in BFS-order in a way that we call product consistent, or simply consistent. (We also keep a list of the nonzero entries of $f(v)$, ordered by the place of the coordinate in the label vector.)

Before we define this concept, notice that $L_0 \cup L_1 \cup \cdots \cup L_j$ is the ball of radius j with center v_0, which abbreviates as $N_j(G, v_0)$. Suppose all vertices up to BFS-level j have been labeled. Then the labeling is called *consistent* if it satisfies the following properties:

(i) The endpoints u, v of each edge $uv \in N_j(G, v_0)$ differ in exactly one coordinate, whose number is called the *(temporary) color* of uv.

(ii) Define the ith unit-layer $H(i, j)$ of G up to level L_j as the maximal connected subgraph of $N_j(G, v_0)$ containing v_0 and with edges of (temporary) color i. Suppose that $N_j(G, v_0)$ has r colors. Set $H = H(1, j) \,\square\, H(2, j) \,\square\, \cdots \,\square\, H(r, j)$. Then

$$N_j(G, v_0) = N_j(H, v_0)\,.$$

(iii) The coloring of $N_j(G, v_0)$ is compatible with σ; in other words, edges of $N_j(G, v_0)$ from different σ-classes of G have different (temporary) colors.

This means we can use all properties of Cartesian products for $N_j(G, v_0)$ as soon as we have checked consistency.

Once all vertices in $N_j(G, v_0)$ have been consistently labeled, the idea of the algorithm is to label the vertices of L_{j+1} making use of Equation (23.2) and other properties of Cartesian products, check consistency, combine colors if it is violated, and relabel the vertices.

We terminate when the highest BFS-level has been consistently labeled. It consists of the vertices of maximal distance from v_0, and thus is $L_{e(v_0)}$, where $e(v_0)$ is the eccentricity of v_0.

For a pseudocode, see Algorithm 23.2. It uses the notation $f(v)$ for the label (coordinate vector) of v, $f(v, i)$ for the ith component of $f(v)$, and t for its length. Originally $t = d(v_0)$, but when the algorithm terminates, t equals the number k of prime factors of G and $f(v, i) = p_i(v)$, the projection of v into $G_i^{v_0}$.

Notice that the number of nonzero coordinates of any vertex $v \in G$ is bounded by $\log_2 n$. The reason is that the interval $I(v, v_0)$ is a box by Exercise 23.5. If v has j nonzero coordinates, then $I(v, v_0)$ has j factors and thus at least 2^j vertices. Because $I(v, v_0)$ cannot have more vertices than G, we infer that $j \leq \log_2 n$.

To combine colors and to check consistency it calls Procedure Combine 23.3 and Procedure Consistency Test 23.4.

Correctness of Algorithm Cartesian Product The initialization clearly starts with a coloring that is contained in σ (the product coloring).

Then the vertices of G are scanned in BFS-order, beginning with L_2. Since $f(u)$ and $f(v)$ of two adjacent vertices differ in exactly one coordinate Line 9 correctly labels v, with the exception of one coordinate, where it is still unclear which one.

If all down-neighbors of v are in one and the same unit-layer, then v is in the same unit-layer, and Line 11 clearly properly coordinatizes it. We should keep in mind though, that we have to combine colors when v has only one down neighbor and if this down-neighbor is not a unit-layer vertex, see Lines 29 to 32.

The standard case is treated in Lines 16 to 19: There is a second down-neighbor, say w, of v, and u and w differ in exactly two coordinates. Exercise 23.4 shows that Line 17 correctly coordinatizes the vertex v in this case.

If u and w differ in exactly one coordinate, say $f(u, i) \neq f(w, i)$, and if v is not a unit-layer vertex, then u it must have a nonzero coordinate, say $k \neq i$. By the unique square property v must have a down-edge, say vz, of color k. Clearly $f(v, i) = f(z, i)$ and $f(z, k) \neq f(u, k)$. This proves the correctness of Lines 20 to 27.

If u and w differ in more than two coordinates (or in no coordinate), then Combine is called.

Clearly we also have to check for consistency. Thus, the correctness depends on the correctness of the procedures Combine and Consistency test.

Algorithm 23.2 Cartesian product

Input: The adjacency list of a connected graph G on n vertices and m edges.
Output: A finest product labeling f of G with respect to the Cartesian product.

```
    Initialization
 1: Choose an arbitrary vertex v_0 of minimum degree.
 2: Modify the adjacency list of G such that v = BFS(v) with respect to v_0.
 3: For every v ∈ G, reserve a vector of length d(v_0) and a list of the same length for the
    label f(v) and for an ordered list of the nonzero entries of f(v).
 4: Set f(v_0, i) = 0 for 1 ≤ i ≤ t. {t = d(v_0).}
 5: Let v_1, ..., v_t be the vertices adjacent to v_0.
    Set f(v_i, i) = v_i, f(v_i, j) = 0 for 1 ≤ i, j ≤ t, i ≠ j.
 6: Call Consistency Test for L_1.
    Labeling
 7: for j = 2 to e(v_0) do
 8:   for all v ∈ L_j do
 9:     Choose the first vertex u in L_{j-1} adjacent to v and set f(v) = f(u).
10:     if all down-neighbors of v are unit-layer vertices of the same unit-layer i then
11:       Set f(v, i) = v.
12:       Continue with the next vertex in L_j.
13:     else
14:       if |L_{j-1} ∩ N(v)| > 1 then
15:         Choose w ∈ L_{j-1} ∩ N(v), w ≠ u.
16:         if u and w differ in exactly two components then
17:           f(v, i) = max(f(u, i), f(w, i)) for 1 ≤ i ≤ t.
18:           Continue with the next vertex in L_j.
19:         end if
20:         if u and w differ in exactly one component, say component i then
21:           Choose a k ≠ i, where f(u, k) ≠ 0.
22:           Scan all down-neighbors z ≠ u, w of v.
23:           if f(z, k) ≠ f(u, k) then
24:             Set f(v, i) = f(z, i).
25:             Continue with the next vertex in L_j.
26:           end if{Remaining case: f(z, k) = f(u, k) for all down neighbors z of v}
27:         end if{Remaining case: u, w differ in more than two components}
28:       end if{Remaining case: u = {L_{j-1} ∩ N(v)} and is not a unit-layer vertex}
29:       for all u' ∈ L_{j-1} ∩ N(v) do
30:         Call Combine(u, u'; i), where i is an arbitrary nonzero component of u.
31:       end for
32:       f(v, i) = v; f(v, k) = 0 for k ≠ i. {This labels v and all u' as unit layer vertices
          in the ith unit layer}
33:     end if{Now v is labeled, not necessarily consistently}
34:   end for{Now all vertices in L_j are labeled}
35:   Call Consistency Test for L_j.
36: end for
37: Call Consistency Test for L_{e(v_0)+1}.
```

Complexity of Algorithm Cartesian Product We consider it without procedure calls. Clearly, the initialization stays within $O(m)$ time and space.

We first show that the cost of one run through the loop from Line 7 to Line 34 is

$O(d(v_0))$. Since we have n vertices, the cost of coordinatization, without procedure calls, is $O(m)$.

For Line 10 we observe that there are at most $d(v) \leq d(v_0)$ down neighbors of v and that we can check in constant time whether a vertex is a unit-layer vertex (and the color of the unit-layer).

Since the labels have length $d(v_0)$ we can check in $O(d(v_0))$ time in how many coordinates two vertices differ and to write a coordinate vector. This takes care of lines Line 16 and Line 17.

For Lines 20 to 27 we note that we have to compare just one coordinate (coordinate k) of every down-edge of v with $f(z, k)$.

Thus the complexity of the entire algorithm will be determined by the Procedures Combine and Consistency test. For the number of calls, notice that $d(v_0)$ is the number of original colors and thus bounds the number of times that colors can be combined. In other words, it also bounds the number of calls of Procedure Combine 23.3. Procedure Consistency Test is called once for every BFS-level.

We now treat Procedure Combine.

Procedure 23.3 Combine

Require: Call *Combine*$(u, v; i)$ from *Algorithm Cartesian product* or *Procedure Consistency test.*

Ensure: A new labeling for all labeled vertices, u and v, and all vertices on a shortest path from u or v to v_0 are unit-layer vertices in the ith unit layer.

1: Calculate $Index := \{j \mid f(u,j) \neq 0 \text{ or } f(v,j) \neq 0 \text{ or } j = i\}$.
2: If $|Index| = 1$, then return (do nothing).
3: Calculate U, the set of all unit-layer vertices in the new unit layer.
$U = \{v \in V \mid v \text{ is labeled and } f(v,j) \neq 0 \Longrightarrow j \in Index\}$;
mark all vertices in U.
4: Assign a new label to each labeled vertex v by:
5: $f(v,j)$ remains unchanged for $j \notin Index$
6: $f(v,i) = v'$, where v' is the vertex in U with the property:
$f(v',j) = f(v,j)$ for all $j \in Index$. $\{f(v',k) = 0$ for all $k \notin Index.\}$
7: Adjust the label list and the label array.
{The new labeling of G has $t - |Index| + 1$ components.}

Figure 23.4 illustrates its action on a sample graph.

Correctness of Procedure Combine It is clear the procedure combines colors and correctly relabels the vertices. The only thing one has to ensure is that the procedure is only called if two vertices u and v have been recognized as unit-layer vertices. But this we have already done.

Complexity of Procedure Combine For the case that $|Index| = 1$ in Step 2, the complexity is bounded by the number of calls and therefore in $O(m)$. Otherwise $|Index| > 1$ and the number of components left reduces by $|Index| - 1$, so this can occur at most $d(v_0)$ times. Step 3 is performed by going through at most n label lists, each of length at most $\log n$; therefore the overall time complexity of Step 3 is $O\big(n \log n \, d(v_0)\big) = O(m \log n)$. In Step 4 the labeled vertices are sorted lexicographically by all components $j \in Index$ using a bucket sort algorithm. Then all vertices with identical components on $Index$ are in the same bucket. If each bucket contains exactly one unit-layer vertex v, which has been marked in Step 3, then the component $f(v',i) = v$ is assigned to each vertex v' in this bucket. Otherwise, if two or more unit-layer vertices are in the same bucket, then these vertices

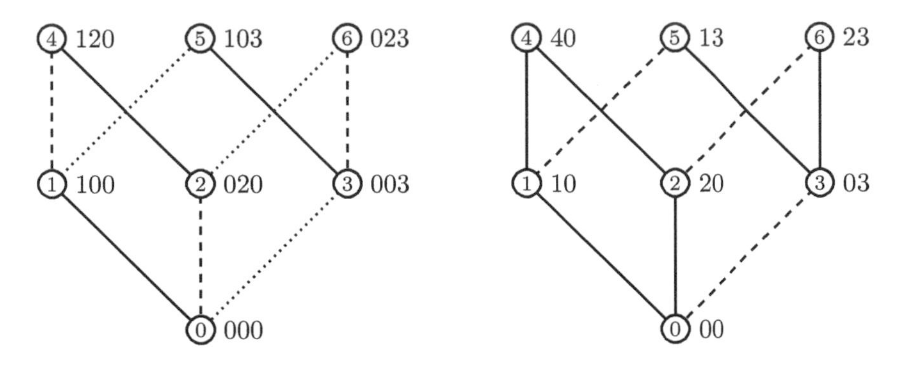

FIGURE 23.4 Graph before and after $Combine(1, 4; 1)$.

must be identically labeled vertices in the highest level that has been coordinatized and no other vertex can be in this bucket. The time complexity of Step 4 is determined by bucket sort and therefore in $O(n \cdot |Index|)$ for a single *Combine* and thus in $O(nd(v_0)) = O(m)$ for all *Combine*. Therefore the overall complexity of the *Combine* subroutine is $O(m \log n)$.

We continue with the consistency test. The main problems that may occur when one wishes to extend the product labeling are missing vertices, missing edges, or too many vertices or edges. Figure 23.5 depicts the case of a missing vertex and of a missing edge. They would be detected in Step 8 of Procedure Consistency test 23.4.

The case of too many vertices results in identically labeled vertices; see Figure 23.6. It shows identically labeled vertices, that is, vertices that are identically labeled by Algorithm Cartesian product, and the effect of Procedure Combine, that combines colors and assigns different labels to the previously identically labeled ones.

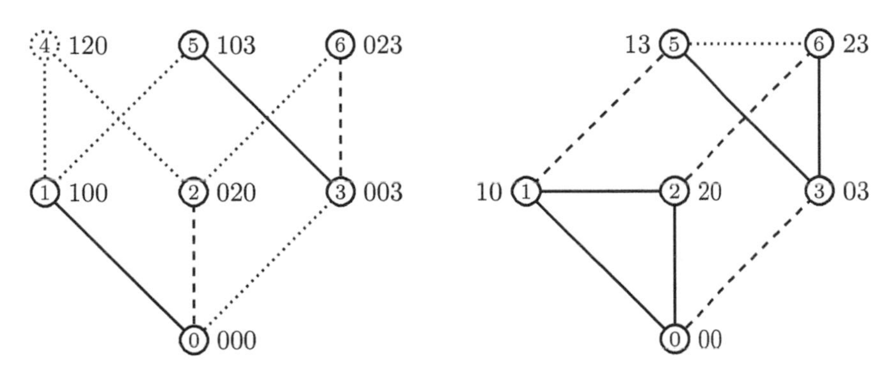

FIGURE 23.5 Missing vertex and missing edge.

Correctness of Consistency Test Suppose it is called from level L_j and, in Step 3 two labels $f(u)$ and $f(v)$ are found that differ in $\ell > 1$ components, so $d(u, v) \geq \ell$ by the distance formula. This can only be resolved if the ℓ colors in which $f(u)$ and $f(v)$ differ are merged.

If there is another nonzero coordinate of v, say the ith, then there must be a down-neighbor v' of v that differs from v only in the ith coordinate. If there is no down-neighbor u' of u with $f(u', i) = f(v', i)$, then the square property is violated and we have to combine coordinate i with the other ℓ coordinates that we just combined.

If there is such a down-neighbor, then v' and u' differ in ℓ coordinates (in the temporary product coloring of $N_{j-1}(G, v_0)$) and must have distance at least ℓ, but the mapping induced by vv',uu' (and their parallel edges) should be a layer isomorphism. So we have to combine again.

Notice that whenever we combine colors, these colors must be in the same σ-class.

Step 6 tests whether the mapping $u \mapsto f(u,i)$ is a homomorphisms from the ith layer through u into the ith unit-layer, and Step 8 tests whether the mapping is bijective and surjective. As before, one shows that the calls *Combine*$(u, v; i)$ and *Combine*$(u, v'; i)$ are justified.

The loop from Step 12 until Step 14 is treated similarly.

Procedure 23.4 Consistency test

Require: Call *Consistency* for L_j from *Algorithm Cartesian Product* 23.2.
Ensure: Consistency up to level L_j by combining colors and relabeling vertices.

1: **for all** $u \in L_{j-1}$ **do**
2: **for all** $v \in L_j \cap N(u)$ **do**
3: If $f(u)$ and $f(v)$ differ in more than one or no component, then *Combine*$(u', v; i)$ for all $u' \in L_{j-1} \cap N(v)$; i a nonzero component of v.
4: **end for**
5: **for all** actual components i **do**
6: Test for each edge uv in factor i, v in L_j whether there is an edge $u'v' = f(u,i)f(v,i)$.
7: If not, then *Combine*$(u, v; i)$.
8: Test for each edge $u'v'$ in factor i, $u' = f(u,i)$, v' in level $l+1$, u' in level l, whether there is exactly one edge uv, $v \in L_j$ and uv in factor i, such that $v' = f(v,i)$.
9: If not, then *Combine*$(u, v'; i)$. {v may not exist !}
10: **end for**
11: **end for**
12: **for all** $u \in L_j$ **do**
13: The same as in the first part, except that v' and u' are in the same level (in the counterpart of Line 8).
14: **end for**

Complexity of Consistency Test For the time bound in Step 3 take into account that we never have to compare more than $\log_2(n)$ components, and since we keep an ordered list of nonzero components, this can be done by $\log_2 n$ comparison per edge. This only has to be done for the down- and cross-edges of L_j, so altogether the time complexity of this part is $O(m \log n)$.

For the rest of the test, some preparations are needed. We split the adjacency list of u into at most $d(v_0)$ lists. The ith list contains all vertices w_j, where uw_j is in the ith factor. This list is sorted according to the ith component of the vertices w_j. For all vertices this takes $O(m \log n)$ time. Consider the vertex $f(u,i)$. The adjacency list is already split, but might not be sorted if a *Combine* step occurred. Observe that in a *Combine* step, the jth list of a labeled vertex must be linked to the ith list for all j in $Index = \{j \mid f(u,j) \neq 0)$, but must not be sorted to achieve the claimed time bound. If the lengths of the ith list of u and $f(u,i)$ differ, then a failure occurs and the lists need not be sorted. Otherwise sorting the ith list of $f(u,i)$ for all i takes as long as sorting all lists of u. Therefore the time complexity of sorting the lists for the vertices $f(u,i)$ is $O(m \log n)$. As soon as the ith lists of both u and $f(u,i)$ are sorted, one has to go through both lists simultaneously and to compare the corresponding components. The ith component of a vertex can be found in constant time using the label-array. At most $d(u)$ comparisons are needed; thus $O(m)$ time is spent for all comparisons. The time complexity of the Consistency Test is therefore determined by sorting and thus $O(m \log n)$.

It is not hard to show that the space complexities of algorithm Cartesian Product and procedures Combine and Consistency Test are linear. We thus have the following theorem:

Theorem 23.8 *For any connected graph G with n vertices and m edges, Algorithm 23.2 computes the prime factors of G with respect to the Cartesian product in $O(m \log n)$ time and $O(m)$ space.*

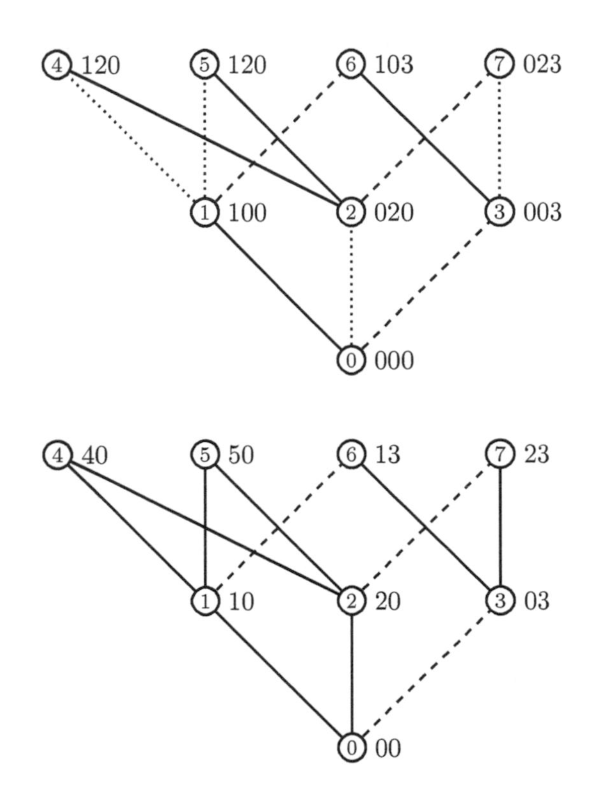

FIGURE 23.6 Identically labeled vertices before, and after, Combine.

23.5 Factorization in Linear Time and Space

The time complexity of Algorithm 23.2 depends on the complexities of the consistency test and the combine procedure. The complexity of the consistency test is determined by sorting. So the first question that arises is, whether one can replace the sorting algorithm by bucket sort.

This is possible, as we will explain, but then another problem comes up: that of checking whether two vertices are adjacent. With the adjacency list, one can check in $O(d(u))$ time whether two vertices u, v are adjacent. Taking into account the number of checks we have to make, this would raise the time complexity to $O(m \log n)$. If one checks with the aid of the (referenced) adjacency matrix, then the individual checks can be done in $O(1)$ time, but the space complexity becomes $O(n^2)$. This problem can be solved if one uses only one line of the adjacency matrix at any one time. However, because the initialization of the line of a vertex v in $\mathrm{A}(G)$ takes $O(d(v))$ time, one cannot do that too often without raising the time complexity. This requires very careful monitoring of the cases when the line of a vertex v in $\mathrm{A}(G)$ is generated.

Back to the sorting problem. This can be solved by ordering the edges of, say, color i that are incident with a vertex v of a unit-layer in a fixed way. That is, the down-edges of color i are sorted, the cross-edges of that color, and also the up-edges. When a vertex u is labeled, the order of the edges of color i incident with u is then determined by the order of the edges of color i in a down-neighbor v of u, where $c(uv) \neq i$. This is justified because adjacent layers are isomorphic.

Furthermore, it is not hard to see that the consistency test does not have to check isometry with unit-layers; it suffices to take the layer(s) through a down-neighbor.

Properly implemented, this method also does not need to relabel or recolor anything up to level L_j if the consistency test at level L_{j+1} requires the combination of colors. Finally, the combination of two colors becomes linear in n with this data structure. Nonetheless, one has to keep careful control of the lists of actual colors and the (original) colors that were combined with them.

For details, see Imrich and Peterin (2007).

Exercises

23.1. Let $G = G_1 \,\square\, G_2 \,\square\, \cdots \,\square\, G_k$ be a connected graph where $k \geq 2$, and let x be a vertex of G. Prove that the graph $G - x$ is prime.

23.2. Recalling the notation and definitions of Section 23.3, show that $p_i(v) = \max\left(p_i(u), p_i(w)\right)$ for $1 \leq i \leq k$, where vu and vw are down-edges of v from different σ-classes.

23.3. Prove the validity of Equation (23.2) for all vertices v that are not unit-layer vertices.

23.4. Show that one can coordinatize the vertices of G with respect to its prime factors in $O(m)$ time from the equivalence classes of σ.

23.5. Prove that every interval in a Cartesian product induces a box.

Chapter 24

Recognizing Direct, Strong, and Lexicographic Products

Any connected graph in Γ factors uniquely into primes over the strong product, and connected nonbipartite graphs in Γ_0 factor uniquely over the direct product. Unique prime factorization is not so common for the lexicographic product, but Chapter 10 described transformations that lead from one prime factorization to all others.

Here we are concerned with the computation and complexity of such factorizations. For the lexicographic product, Feigenbaum and Schäffer (1986) showed that prime factorization is at least as difficult as the graph isomorphism problem. But, as we shall see, prime factorizations over the direct and the strong product can be found in $O(mn^2)$ time.

Feigenbaum and Schäffer (1992) presented the first polynomial algorithm that determines the prime factorization of graphs over the strong product. Imrich (1998) adapted their approach to the direct product. Both methods involve the Cartesian Skeleton.[1] As we saw in Chapter 8, the skeleton $S(G)$ envelops a given graph G in such a way that $S(G \times H) = S(G) \,\square\, S(H)$. The present chapter will describe a variant $S[G]$ of $S(G)$, called the *closed Cartesian skeleton*, which obeys $S[G \boxtimes H] = S[G] \,\square\, S[H]$. These constructions link prime factorizations over $\times$ (respectively $\boxtimes$) to factorizations over $\square$, and algorithms for prime factorization over $\square$ are then applied to the Cartesian skeleton and transferred back to factorizations of G over $\times$ (respectively $\boxtimes$).

The main effort in this chapter will be devoted to computing the Cartesian skeleton $S(G)$, and the closed Cartesian skeleton $S[G]$, for a given graph G. Following Hammack and Imrich (2009), this gives rise to algorithms of varying complexities, the best one being $O(m\, a(G)\, \Delta)$ for the strong product. Notice that even this bound can be of the order $O(mn^2)$, which is just a little better than the complexity of the general algorithm of Feigenbaum and Schäffer (1992) for the strong product.

[1] Feigenbaum and Schäffer (1992), and Imrich (1998), define it algorithmically. We use the nonalgorithmic definition of Hammack and Imrich (2009), which is easier to handle.

24.1 Direct Product

To compute the prime factors of a connected, nonbipartite graph G with respect to the direct product, we proceed in several steps. We first factor R-thin graphs, continue with graphs that are not thin but contain no complete factor, and then treat the general case.

To factor an R-thin graph G, we first compute the Cartesian skeleton $S(G)$, decompose it into its prime factors with respect to $\Box$, and then apply Proposition 8.10 for the factorization of G with respect to the direct product. Recall that this proposition asserts that $S(H \times K) = S(H) \Box S(K)$ for any factorization $H \times K$ of G. By grouping the prime factors of $S(G)$ with respect to the Cartesian product appropriately, we will be able to compute the layers of the prime factors of G with respect to the direct product.

Algorithmic construction of $S(G)$

Recall from Definition 8.1 that the Cartesian skeleton $S(G)$ is formed from the Boolean square G^s by deletion of dispensable edges, where an edge xy is dispensable if it is a loop, or if there exists some $z \in V(G)$ for which both of the following statements hold:

$$\begin{array}{lllll} (1) & N_G(x) \cap N_G(y) \subset N_G(x) \cap N_G(z) & \text{or} & N_G(x) \subset N_G(z) \subset N_G(y) \\ (2) & N_G(y) \cap N_G(x) \subset N_G(y) \cap N_G(z) & \text{or} & N_G(y) \subset N_G(z) \subset N_G(x). \end{array}$$

We need a data structure that allows us to compute intersections of neighborhoods and to check for containment efficiently.

Because our graphs are most efficiently presented by the adjacency list data structure for G, let us briefly recall the main features. Fix an indexing $V(G) = \{g_1, g_2, \ldots, g_n\}$. Represent G as a table with n rows indexed by the vertices $g_1, g_2, \ldots, g_n$. Row i contains a list of the neighbors of g_i. This is illustrated in Figure 24.1.

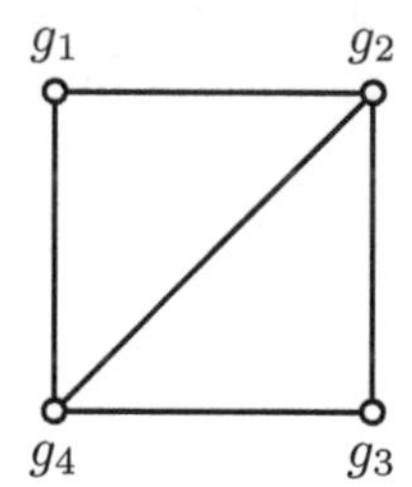

Vertex	Neighbors
g_1	g_2, g_4
g_2	g_1, g_3, g_4
g_3	g_2, g_4
g_4	g_1, g_2, g_3

FIGURE 24.1 Adjacency list representation of G.

Given a vertex g_i, we will often have to check for several vertices g_j whether they are in $N_G(g_i)$. With the adjacency matrix we could do it in constant time and $O(n^2)$ space. To reduce the space complexity, we recall a method of Cormen, Leierson, and Rivest (1990, Exercises 12.1 through 12.4) which allows us to check in constant time whether $g_j \in N_G(g_i)$. It involves some preprocessing that computes the line of g_i in A(G) in $O(d(g_i))$ time and $O(n)$ space.

It will be helpful to spell out the method in detail. First form a reference vector c_i of length n, where every $c_i(k)$ is an uninitialized pointer. Next form a vector ℓ_i of length $d(g_i)$, where every entry is a pointer. For each g_k in the adjacency list for g_i, do the following: If g_k is the pth vertex on the adjacency list for g_i, set $c_i(k)$ to point to $\ell_i(p)$ and set $\ell_i(p)$ to point back to $c_i(k)$. Then $g_j \in N(g_i)$ if and only if $c_i(j)$ points to a pointer in ℓ_i that points back to $c_i(j)$. This can be checked in constant time, while the effort to create c_i and ℓ_i takes $O(d(g_i))$ time and $O(n)$ space. For $x = g_i$ we will also write c_x and ℓ_x instead of c_i and ℓ_i.

This justifies the following two remarks concerning the computation of intersections of neighborhoods and the verification of containment properties. They express central ideas of our algorithms.

Remark 24.1 *Once c_i and ℓ_i have been created and linked, we can form a list representation of any intersection $N(g_i) \cap N(g_j)$ in $O(d(g_j))$ time as follows. Begin with an empty list I. Then for each $x \in N(g_j)$, check whether $x \in N(g_i)$, and if so, then append it to I.*

Remark 24.2 *If X and Y are finite sets, then $X \subset Y$ means $|X| = |X \cap Y|$ and $|X| < |Y|$. For instance, $N(x) \subset N(z)$ provided $|N(x)| = |N(x) \cap N(z)|$ and $|N(x)| < |N(z)|$. Similarly, $N(x) \cap N(y) \subset N(x) \cap N(z)$ if $|N(x) \cap N(y)| = |N(x) \cap N(y) \cap N(z)|$ and $|N(x) \cap N(y)| < |N(x) \cap N(z)|$.*

Thus we can decide if xy meets conditions (1) and (2) for dispensability by making such comparisons among the numbers $|N(x)|$, $|N(y)|$, $|N(z)|$, $|N(x) \cap N(y)|$, $|N(x) \cap N(z)|$, $|N(y) \cap N(z)|$, and $|N(x) \cap N(y) \cap N(z)|$.

Proposition 24.1 *Given an edge xy of G^s together with the reference vectors c_x, ℓ_x and c_y, ℓ_y, we can check the validity of dispensability conditions (1) and (2) for any vertex $z \in V(G) - \{x, y\}$ in $O(d(z))$ time.*

Proof We can assume that $z \neq y$ and compute the following sets and numbers, using c_x, ℓ_x and c_y, ℓ_y whenever appropriate:

(i) $|N_G(z)|$
(ii) $N_G(x) \cap N_G(z)$ and $|N_G(x) \cap N_G(z)|$
(iii) $N_G(y) \cap N_G(z)$ and $|N_G(y) \cap N_G(z)|$
(iv) $|N_G(x) \cap N_G(y) \cap N_G(z)|$

By comparing the cardinalities of the intersections computed above, we check if conditions (1) and (2) for the dispensability of xy hold. (See Remark 24.2.) All computations can be executed in $O(d(z))$ time. □

We continue with two algorithms for computing $S(G)$, both based on the above remarks. The first considers all triples of distinct vertices x, y, z.

Algorithm 24.1 Cartesian skeleton 1

Input: Adjacency list representation for graph G with n vertices.
Output: Adjacency list representation for $S(G)$.

1: **for all** pairs of distinct vertices x,y of G **do**
2: Compute c_x, ℓ_x, c_y, ℓ_y and check whether $xy \in G^s$. If not, then continue with the next pair x, y.
3: For all $z \in V(G) - \{x, y\}$, check the validity of dispensability conditions (1) and (2).
4: If these conditions fail for every z, add xy to the adjacency list of $S(G)$.
5: **end for**
6: Return the adjacency list representation for $S(G)$.

Proposition 24.2 *Given an input graph G with m edges and n vertices, the time complexity of Algorithm 24.1 to compute $S(G)$ is $O(mn^2)$. The space complexity is determined by the size of the output, that is, the number of edges in $S(G)$. It is between $O(n)$ and $O(n^2)$.*

Proof Step 2 takes $O(n)$ time. By Proposition 24.1, Step 3 is $O(d(z))$ for every z, contributing to a total of $\sum_{z \in G} O(d(z)) = O(m)$ time. Step 4 takes constant time. Because we have to perform Steps 2 through 4 at most n^2 times, we arrive at the asserted time complexity. □

In the above algorithm we took all pairs x, y and checked in Step 2 whether they were in $E(G^s)$. For the pair x, z, the check occurs in Step 3.

The next algorithm makes use of the fact that conditions (1) and (2) for dispensability can hold only if y and z are both at distance 2 from x. (We must have $d_G(x, y) = 2$ for $xy \in E(G^s)$, and $d_G(x, z) \neq 2$ implies $N_G(x) \cap N(z) = \emptyset$, whence none of the conditions hold. In fact, by condition (2), there must be a neighbor of x, say y', that is adjacent to both y and z.) The algorithm lets y and z run through the sets of neighbors of neighbors of x. To reach all vertices y of distance 2 from x we thus consider all neighbors y' of x and then all neighbors y of the y'. Because it may be possible to reach y from x on many distinct paths of length 2, and because we do not know all paths a priori, the complexity of this method may be high. However, if $\Delta^3 < n^2$, then it is better than $O(mn^2)$, as we shall see.

Note that we already observed that we can assume the existence of a $y' \in N(x)$ that is a neighbor of both y and z.

Algorithm 24.2 Cartesian skeleton 2

Input: Adjacency list representation for a graph G with n vertices.
Output: Adjacency list representation for $S(G)$.

1: **for all** $x \in V(G)$ **do**
2: Compute c_x, ℓ_x, and $|N_G(x)|$.
3: **for all** $y' \in N_G(x)$ **do**
4: **for all** $y \in N_G(y') - \{x\}$ **do**
5: Compute c_y, ℓ_y, and $|N_G(y)|$.
6: **for all** $z \in N_G(y') - \{x, y\}$ **do**
7: Check the dispensability conditions (1) and (2).
8: **end for**
9: If these conditions fail for all z, add xy to the adjacency list of $S(G)$.
10: **end for**
11: **end for**
12: **end for**
13: Return the adjacency list representation for $S(G)$.

Proposition 24.3 *Given an input graph G of size m, order n, and maximum degree Δ, the time complexity of Algorithm 24.2 to compute $S(G)$ is $O(m\Delta^3)$. The space complexity is between $O(n)$ and $O(n^2)$.*

Proof We have loops for x, y', y, z, and show first that the net complexity is given by the expression

$$\sum_{x \in V(G)} \left\{ O(|N(x)|) + \sum_{y' \in N(x)} \sum_{y \in N(y')} \left[O(|N(y)|) + \left(\sum_{z \in N(y')} O(|N(z)|) \right) + O(1) \right] \right\}.$$

The sum over $x \in V(G)$ stands for the loop from Steps 1 through 12 in the algorithm, and the term $O(|N(x)|)$ expresses the contribution of Step 2.

Step 3 comes next in the x-loop. It is a sum over all neighbors of x. For every such neighbor y', we must execute Step 4. It is a loop for all $y \in N(y')$. Every instance consists of three subinstances. The cost for Step 5 is $O(|N(y)|)$, that for Step 9 is $O(1)$, whereas Step 6 is a sum over $z \in N(y')$, every instance contributing a cost of $O(|N(z)|)$.

To evaluate this expression, note that the Step 6 is nested in all loops and has the largest contribution to the complexity. It therefore suffices to evaluate just the expression

$$\sum_{x\in V(G)} \sum_{y'\in N(x)} \sum_{y\in N(y')} \sum_{z\in N(y')} O(|N(z)|).$$

Clearly, $\sum_{y\in N(y')} \sum_{z\in N(y')} O(|N(z)|) = O(\Delta^3)$. Thus the total value is

$$\sum_{x\in V(G)} \sum_{y'\in N(x)} O(\Delta^3) = O(\Delta^3) \sum_{x\in V(G)} \sum_{y'\in N(x)} 1 = O(\Delta^3) \sum_{x\in V(G)} |N(x)| = O(\Delta^3)m.$$

□

Note that $O(m\,\Delta^3)$ is a better bound than $O(mn^2)$ when $\Delta^3 < n^2$, and that $m\Delta^3$ can be close to n for sparse graphs, say direct products of cycles or of cubic graphs.

24.2 Strong Product

We now describe a variation on $S(G)$, which we denote as $S[G]$. This modified skeleton is a subgraph of G (*not* of G^s) having the property $S[H \boxtimes K] = S[H]\Box S[K]$. We define it almost exactly as we defined $S(G)$, except that we use closed neighborhoods instead of open neighborhoods, and it is a subgraph of G rather than G^s.

We say an edge xy of a graph $G \in \Gamma$ is *dispensable* if there exists $z \in V(G)$ for which both of the following statements hold:

(1-strong) $N_G[x] \cap N_G[y] \subset N_G[x] \cap N_G[z]$ or $N_G[x] \subset N_G[z] \subset N_G[y]$,

(2-strong) $N_G[y] \cap N_G[x] \subset N_G[y] \cap N_G[z]$ or $N_G[y] \subset N_G[z] \subset N_G[x]$.

The *closed Cartesian skeleton* of G is the graph $S[G]$ obtained from G by removing all dispensable edges.

From here we easily obtain analogues of Lemmas 8.9 and 8.12, as well as Propositions 8.10 and 8.13. In the proofs, one simply replaces open neighborhoods with closed neighborhoods. We use $N_{H\boxtimes K}[(h,k)] = N[h] \times N[k]$ instead of $N_{H\times K}(h,k) = N(h) \times N(k)$. We also replace the condition $(N(x) = N(y)) \Longrightarrow (x = y)$ for R-thinness with the condition $(N[x] = N[y]) \Longrightarrow (x = y)$ for S-thinness. Reasoning exactly as we did for $S(G)$, we obtain the following results for $S[G]$. (The only substantial difference is that for connectivity, we no longer need G to be nonbipartite, as $S[G]$ is a subgraph of G, not of G^s. We can also remove the condition that G have no isolated vertices, as $N[x] \neq \emptyset$, even if x is isolated.)

Proposition 24.4 *If G is connected, then $S[G]$ is connected. Also, $S[H\boxtimes K] = S[H]\Box S[K]$ for S-thin graphs.*

Algorithmic construction of $S[G]$

We now adapt Algorithm 24.1 to $S[G]$. Note that the complexities of computing intersections of closed neighborhoods are the same as those for open neighborhoods. We use the the same notation c_i and ℓ_i for the reference vectors for closed neighborhoods.

Notice that conditions (1-strong) and (2-strong) for dispensability can hold only if y and z are both in $N[x]$. (We must have $y \in N(x)$ in order that $xy \in E(G)$, and $z \notin N[x]$ implies $N[x] \cap N[y] \not\subset N[x] \cap N[z]$, whence none of the conditions hold.) In other words, the dispensability conditions can hold only if x, y, and z induce a triangle in G. Thus in checking for dispensability of xy, the algorithm needs to consider only those z in $N[x]$.

We will also make use of the analog of Proposition 24.1 for strong products. Its validity is obvious by the above remarks.

Proposition 24.5 *Given distinct vertices x,y in $V(G)$ together with reference vectors c_x, ℓ_x, and c_y, ℓ_y, we can check the validity of dispensability conditions (1-strong) and (2-strong) for any vertex $z \in V(G) - \{x, y\}$ in $O(d(z))$ time.*

Algorithm 24.3 Closed Cartesian skeleton 1

Input: Adjacency list representation for graph a G with n vertices.
Output: Adjacency list representation for $S[G]$.

1: **for all** edges $xy \in E(G)$ **do**
2: Compute c_x, ℓ_x, c_y, and ℓ_y.
3: For each $z \in N(x)$, check the validity of the dispensability conditions.
4: If these conditions fail for all z, add xy to the adjacency list of $S[G]$.
5: **end for**
6: Return the adjacency list representation for $S[G]$.

Proposition 24.6 *If G has m edges, and maximum degree Δ, then the complexity of using Algorithm 24.3 to compute $S[G]$ is the minimum of $O(m^2)$ and $O(m\Delta^2)$.*

Proof Every instance of the loop from Step 1 to 5 has three subinstances. The first takes $O(\Delta)$ time, and the last constant time. We will bound the second in two ways.

On one hand, for every z, the cost of checking dispensability is $O(|N(z)|)$. Because $z \in N(x)$ and $N(x) \subseteq V(G)$, the time for the loop in Step 3 is bounded by $\sum_{z \in V(G)} O(|N(z)|) = O(m)$. Hence every instance of the first loop takes $O(\Delta) + O(m) + O(1)$ time. Because there are m edges, we arrive at a total complexity of $O(m^2)$.

On the other hand, the z are among the at most Δ neighbors of x, so the time needed for every z is bounded by $O(|N(z)|) = O(\Delta)$. This yields the bound of $O(\Delta^2)$ for Step 3, and a total of $O(m\Delta^2)$. □

A bound involving arboricity

Recall that the arboricity $a(G)$ of a graph G is the minimum number of forests into which its edges can be partitioned. It is a measure of density of G. For trees it is one, and in general Equation (20.5) gives the bounds

$$\frac{\delta}{2} < a(G) \leq \Delta .$$

From Theorem 20.9 we know that all triangles of a graph G can be listed in $O(m\,a(G))$

time and space. This is also important for us because the dispensability conditions (1-strong) and (2-strong) can hold only if x, y, z lie on a triangle. Thus, we will present a variant of Algorithm 24.3 for the computation of $S[G]$ that involves the arboricity. Its time complexity is $O(m\, a(G)\, \Delta)$.

Algorithm 24.4 Closed Cartesian skeleton 2

Input: Adjacency list representation for graph a G with n vertices.
Output: Adjacency list representation for $S[G]$.

1: Compute all triangles of G.
2: Initialize an empty list $t(e)$ for every edge $e \in E(G)$.
3: Scan all triangles $x_1x_2x_3$.
4: **for** every edge $e = x_ix_j$ of this triangle **do**
5: Add the third vertex x_k to $t(e)$.
6: **for all** edges $xy \in E(G)$ **do**
7: Compute c_x, ℓ_x, c_y, and ℓ_y.
8: For each $z \in t(xy)$, check the validity of the dispensability conditions.
9: If these conditions fail for every z, add xy to the adjacency list of $S[G]$.
10: **end for**
11: **end for**
12: Return the adjacency list representation for $S[G]$.

A complexity analysis along the lines of the previous ones yields the next proposition:

Proposition 24.7 *If G has m edges, arboricity $a(G)$, and maximum degree Δ, then $S[G]$ can be computed in $O(m\, a(G)\, \Delta)$ time and $O(m\, a(G))$ space.*

24.3 Factoring Thin Graphs

We now show that the complexity of computing $S(G)$, respectively $S[G]$, is closely related to the complexity of factoring thin graphs over the direct, respectively the strong, product. For the direct product the key to this result is Proposition 8.10, and for the strong one it is Proposition 24.4.

We consider the direct product first. Given an arbitrary factorization $G = H \times K$, Proposition 8.10 asserts that $S(H \times K) = S(H) \,\Box\, S(K)$. Because $S(H)$ and $S(K)$ are spanning subgraphs of H and K, this means that the $S(H)$-layers and $S(K)$-layers of $S(H) \,\Box\, S(K)$ have the same vertex sets as the H-layers and K-layers of $H \times K$.

Furthermore, if $G_1 \times G_2 \times \cdots \times G_k$ is the prime factorization of G, then $S(G) = S(G_1) \,\Box\, S(G_2) \,\Box\, \cdots \,\Box\, S(G_k)$.

Of course the $S(G_i)$ need not be prime with respect to the Cartesian product, but we can factor them further so that there is a prime factoring $S(G) = \Box_{i\in I} H_i$ and a partition $J_1 \cup J_2 \cup \cdots \cup J_k$ of the index set I, such that $S(G_i) = \Box_{j\in J_i} H_j$. Setting $H_J = \Box_{j\in J} H_j$ and $H_{I\setminus J} = \Box_{j\in I\setminus J} H_j$, we thus have $S(G) \cong H_J \,\Box\, H_{I\setminus J}$ and the vertex sets of the H_J-layers in $S(G)$ correspond to the vertex sets of G_i-layers in G if $J = J_i$. Clearly, every J_i is a minimal subset J of I with the property that the vertex sets of the H_J-layers in $S(G)$, together with the vertex sets of the $H_{I\setminus J}$-layers in $S(G)$, induce a factorization of G over the direct product.

This gives rise to Algorithm 24.5 below.

Algorithm 24.5 Direct product decomposition of R-thin graphs

Input: The adjacency list of a connected, nonbipartite thin graph G in Γ_0.
Output: The prime factorization $G_1 \times G_2 \times \cdots \times G_k$ of G.

1: Compute $S(G)$.
2: Compute the prime factorization $\Box_{j\in I} H_j$ of $S(G)$.
3: Find all minimal subsets J of the index set I such that the H_J-layers of $H_J \Box H_{I\setminus J}$, where $H_J = \Box_{j\in J} H_j$ and $H_{I\setminus J} = \Box_{j\in I\setminus J} H_j$, correspond to layers of a factor of G with respect to the direct product.
(This means the projection $p_J(G)$ onto $V(H_J) = V\big(\Box_{j\in I} H_j\big)$ is a prime factor of G.)

The algorithm is correct by the preceding considerations. For Step 1 we have the complexity $O(\min(mn^2, m\Delta^3))$ by Propositions 24.2 and 24.3.

The complexity of the next step is linear by Section 23.5.

Suppose that we take any subset J of I. We can then define graphs A and B by the projections $p_J(G)$ and $p_{I\setminus J}(G)$ onto the vertex sets $V\big(\Box_{j\in I} H_j\big)$ and $V\big(\Box_{j\in I\setminus J} H_j\big)$. If J is one of the sets J_i or a union of such sets, then $G = A \times B$.

Note that I has at most $2^{|I|}$ subsets. Construction of the graphs A and B requires the projection of $m = |E(G)|$ edges into the coordinate sets and has complexity $O(m\,|I|)$. It is clear that $E(G) \subseteq E(A \times B)$. Thus, $G = A \times B$ if $|E(G)| = |E(A)| \cdot |E(B)|$, and we can find all subsets of I such that $G = A \times B$ in $O\big(m\,2^{|I|}|I|\big)$ time.

If we scan the subsets of I by their size, first one-element subsets, then the two-element ones, then we can clearly determine all minimal subsets J with the desired properties in $O(m\,2^{|I|}|I|)$ time.

Notice that we do not need more than $O(m)$ space with this approach.

We still wish to estimate I. It is the number of prime factors of a subgraph of G^s with respect to the Cartesian product, which is bounded by $\log_2 n$. Then $2^{|I|} \le 2^{\log_2 n} = n$, and thus the total complexity of Step 3 is $O(m\,n\log n)$, which is smaller than $O(m\,n^2)$. Hence $O(m\,n^2)$ bounds the complexity of Algorithm 24.5.

However, if $2^{|I|}|I| \le \Delta^3 \le n^2$, then the complexity of Algorithm 24.5 is $O(m\Delta^3)$. We have thus shown:

Theorem 24.8 *Algorithm 24.5 correctly computes the prime factorization of connected, thin nonbipartite graphs G with respect to the direct product.*

If G has n vertices and m edges, then its time complexity is $O(mn^2)$, unless $S(G)$ has at most f prime factors with respect to Cartesian multiplication and $2^f f \le \Delta^3 \le n^2$, where Δ is the maximum degree of G. Then the complexity is $O(m\Delta^3)$.

For the strong product we similarly derive Algorithm 24.6. Clearly, the time complexity of Step 1 is $O(\min(m^2, ma(G)\Delta))$ by Propositions 24.6 and 24.7. For Step 3 we obtain, as in the case of the direct product, the time complexity $O(m\,2^{|I|}|I|)$.

For a bound on $|I|$, observe that $S[G]$ is a subgraph of G. Thus its minimum degree is bounded by the minimum degree δ of G. Of course this is a bound on $|I|$. Because $\delta < 2a(G)$ we have the bound $2a(G)$ on $|I|$. This is better than in the case of the direct product, but tedious to use for a general statement of the complexity of the algorithm. So we are content with the following theorem.

Theorem 24.9 *Algorithm 24.6 correctly computes the prime factorization of connected S-thin graphs G with respect to the strong product.*

If G has n vertices and m edges, then its time complexity is $O(mn \log n + m^2)$, unless $S(G)$ has at most q prime factors with respect to Cartesian multiplication and $2^q q \leq a(G)\Delta \leq m$, where Δ is the maximum degree of G. Then the complexity is $O(m\, a(G)\Delta)$.

Algorithm 24.6 Strong product decomposition of S-thin graphs

Input: The adjacency list of a connected S-thin graph G in Γ.
Output: The prime factorization $G_1 \boxtimes G_2 \boxtimes \cdots \boxtimes G_k$ of G.

1: Compute $S[G]$.
2: Compute the prime factorization $\square_{j \in I} H_j$ of $S[G]$.
3: Find all minimal subsets J of the index set I such that the H_J-layers of $H_J \square H_{I \setminus J}$, where $H_J = \square_{j \in J} H_j$ and $H_{I \setminus J} = \square_{j \in I \setminus J} H_j$, correspond to layers of a factor of G with respect to the strong product.
(This means the projection $p_J(G)$ onto $V(H_J) = V\big(\square_{j \in I} H_j\big)$ is a prime factor of G.)

We wish to point out that that Hellmuth (2011) presents an algorithm of complexity $O(n\Delta^6)$ for the prime factorization of connected S-thin graphs with respect to the strong product. His algorithm covers the given graph G with certain well-defined neighborhoods, which are decomposed with the help of the algorithms in this chapter. Notice that these neighborhoods need not be be thin, even if G is thin. Their factorizations are then used to factor G.

For fixed Δ and variable n, the algorithm is linear in n. One calls such algorithms *quasi-linear.* For a slightly more detailed description, compare Section 33.2.

24.4 Factoring Non-Thin Graphs

If G is not thin, then we first have to determine the relation R or S. For R this means that we have to partition $V(G)$ into maximal sets of vertices with the same open neighborhoods; for S we must do the same for closed neighborhoods.

We outline a method of finding the relation R. Given the adjacency list of a graph G, we can sort it (in linear time) with Algorithm 17.1. To be more precise, let the vertices of G be $v_1, v_2, \ldots, v_n$, ordered by their indices. If A_i is the list of vertices adjacent to v_i, then sorting means that every list A_i respects the order of the vertices of G.

If we order the lists lexicographically and subsequently group them by their lengths, then the lists corresponding to an equivalence class of R will form a contiguous block, which can be determined in $O(m)$ time. Interestingly, the lexicographic order of the lists can also be computed in linear time, as has been shown by Dahlhaus, Gustedt, and McConnell (2001). They use radix sort; the argument is subtle, but not difficult.

Because sorting by lengths is also linear, it is clear that R can be computed in $O(m)$ time. Similarly, using closed neighborhoods, one can compute S in linear time.

For the construction of the quotient graphs G/R or G/S, the adjacencies of the R-classes or S-classes must be checked, and this can also be be done in linear time. Clearly, the space complexity is linear too. We formulate this as a lemma.

Lemma 24.10 *If G is a connected graph with m edges, then the relations R, S, and the quotients G/R and G/S can be computed in $O(m)$ time and space.*

By Corollary 8.7, a graph G factors as $G = A \times K_p^s$ if and only if p divides the order of each R-class of G. Because greatest common divisors can be computed in linear time (with respect to the size of the numbers, not with respect to the lengths of their representations), we can extract maximal factors K_t^s from G in linear time, that is, in linear time we can find graphs K_t^s and G' such that $G = K_t^s \times G'$, where G' has no nontrivial K_k^s factor.

Because we also know how to factor G'/R, we are left with the problem of finding the prime factorization of G' from that of G'/R. (Notice that $G/R \cong G'/R$.)

Let us consider a decomposition $G = A \times B$, where G is connected, nonbipartite, and has no nontrivial K_k^s as a factor. Suppose we know G and have computed A/R and B/R. How do we find A and B? In other words, given an element $[x] \in A/R$, how do we find the the cardinality of $[x]$? Let $[v]$ be the R-class of $v \in V(G)$. It is an element of $A/R \times B/R$ and has projections into $V(A/R)$ and $V(A/R)$, say $[v_A]$ and $[v_B]$, where $[v_A] \in V(A/R)$ and $[v_B] \in V(B/R)$. Then $|[v]| = |[v_A]| \cdot |[v_B]|$.

Notice that the greatest common divisor of the numbers $|[x]|$, for $[x] \in V(A/R)$, is 1, otherwise G would have a nontrivial K_k^s factor. Now consider a vertex $[y]$ in B/R and the vertices in the A/R-layer through $([x],[y])$, where $[x] \in V(A/R)$. These are the vertices

$$([x_1],[y]),\ ([x_2],[y]),\ \ldots,\ ([x_{n_1}],[y]),$$

where $\{[x_1],[x_2],\ldots,[x_{n_1}]\} = V(A/R)$. Clearly, $|[y]|$ divides every $|[x_i]| \cdot |[y]|$, but because the gcd of the $|[x_i]|$ is 1, $|[y]|$ is the greatest common divisor of the cardinalities of the R-classes in the (A/R)-layer through $([x],[y])$.

This means that we can compute the size of the R-classes of the vertices in A/R and B/R in linear time.

It is of course possible (even likely) that G has fewer prime factors than G/R. Then we have to group them to get prime factors of G, just as we had to group prime factors of $S(G)$ to get prime factors of G when G was thin.

We thus have shown the correctness of the following algorithm:

Algorithm 24.7 Direct product decomposition

Input: The adjacency list of a connected, nonbipartite graph G in Γ_0.
Output: The prime factorization $G_1 \times G_2 \times \cdots \times G_k$ of G.

1: Represent G in the form $G' \times K_t^s$, where G' has no nontrivial factor isomorphic to K_k^s.
2: Determine the prime factorization of K_t^s, that is, of t.
3: Compute G/R.
4: Compute the prime factorization $Q_1 \times Q_2 \times \cdots \times Q_k$ of G/R.
5: Compute all minimal subsets J of $I = \{1,2,\ldots,k\}$ such that there are graphs A and B with $G = A \times B$, $A/R = \bigtimes_{i \in J} Q_i$ and $B = \bigtimes_{j \in I \setminus J} Q_j$.
(By the minimality of J, the A must be prime.)

Clearly, the complexity of Step 5 is the same as that of Step 3 in Algorithm 24.5. Notice also that G cannot have more factors than G/R and that G/R cannot have more factors than $S(G/R)$. We thus have the following theorem:

Theorem 24.11 *Algorithm 24.7 correctly computes the prime factors of connected nonbipartite graphs with respect to the direct product in $O(mn^2)$ time.*

For the strong product, similar reasoning leads to the following theorem:

Theorem 24.12 *One can compute the prime factors of connected graphs with respect to the strong product in $O(mn \log n + m^2)$ time.*

For the direct product, the situation becomes much more difficult if we are interested in the number of prime factorizations in Γ.

Beginning with the observation that $G \times H \notin \Gamma$ if and only if both G and H are in $\Gamma_0 \setminus \Gamma$, it is easy to verify the following corollaries to Theorem 8.17:

Corollary 24.13 *Let $G = Q_1 \times Q_2 \times \cdots \times Q_k$ be the prime factor decomposition in Γ_0 of a nonbipartite, connected graph $G \in \Gamma$. Furthermore, let the graphs G_i be defined by $G_i = \times_{j \in I_i} Q_j$, where the sets $I_1, I_2, \ldots, I_r$ form a partition of the index set $\{1, 2, \ldots, k\}$. Then $G_1 \times G_2 \times \cdots \times G_r$ is a prime factor decomposition of G in Γ if and only if every set $\{Q_j \mid j \in I_i\}$ contains exactly one element in Γ.*

With this corollary it is straightforward to find sequences of arbitrarily large nonbipartite, connected simple graphs G_k for which the number of prime factorizations in Γ is not bounded by a polynomial in $|V(G_k)|$. In other words, the number of representations of simple, connected nonbipartite graphs G as a direct product of prime simple graphs is not bounded by a polynomial in $|V(G)|$.

The next corollary shows that even the complexity of deciding whether a nonbipartite, connected simple graph has unique prime factorization in Γ is at least as hard as isomorphism testing of graphs and thus is most likely not polynomial.

Corollary 24.14 *Let $G = Q_1 \times Q_2 \times \cdots \times Q_k$ be the prime factor decomposition in Γ_0 of the nonbipartite, connected graph $G \in \Gamma$. Then G has unique prime factor decomposition with respect to the direct product if and only if one of the following conditions is satisfied:*

(i) *All factors Q_i are in Γ.*
(ii) *Only one of the Q_i is in $\Gamma \setminus \Gamma_0$, and all the other factors are pairwise isomorphic.*

24.5 Lexicographic Product

In this section we show that the recognition complexity of lexicographic products is polynomially equivalent to the graph isomorphism problem. We follow the approach of Feigenbaum and Schäffer (1986).

The *graph isomorphism* problem asks whether two graphs G and H are isomorphic. Let X be the disjoint union of graphs G and H. Clearly, $G \cong H$ if and only if $X = D_2 \,\square\, G = D_2 \boxtimes G = D_2 \circ G$. Thus, testing whether a disconnected graph is decomposable with respect to any of these three products is at least as hard as the graph isomorphism problem.

As we have seen in previous chapters, the recognition problem for *connected* composite graphs with respect to the Cartesian and the strong product is polynomial. Concerning connected lexicographic products, we note that $\overline{X} = \overline{D_2} \circ \overline{G} = K_2 \circ \overline{G}$, so we do not expect connected lexicographic products to behave better than disconnected ones.

Decomposition is not easier than isomorphism testing

Several rather strong restrictions of the graph isomorphism problems are known that are at least as hard as the general problem, for instance, regular graph isomorphism or complement isomorphism. For our purposes we need another restriction of the problem. To prove it, we invoke the following well-known number theoretic result.[2]

Lemma 24.15 *For every real number $x > 1$ there exists at least one prime in the open interval $(x, 2x)$.*

We now show that the graph isomorphism problem is equivalent to a restricted version that is more convenient for our purposes.

Lemma 24.16 *The graph isomorphism problem is polynomially reducible to the restricted version in which both graphs are connected graphs on p vertices, where p is an odd prime and where both graphs have maximum degree less than $p/2$.*

Proof Let G and H be graphs on $n \geq 2$ vertices. By Lemma 24.15, there is a prime in the open interval $(2n + 2, 4n + 4)$. Let p be the smallest prime in this interval. Clearly, p can be computed in time polynomial in n. (Note that we consider the input to be of size n and not the number of digits needed to represent n. This number is $\log_2 n$, and our algorithm would be exponential with respect to it.)

Select a vertex u of G and attach a path P of length $p - n$ to it. Join the neighbor u' of u on P by an edge to every vertex of G, and call the resulting graph G'. (See Figure 24.2.)

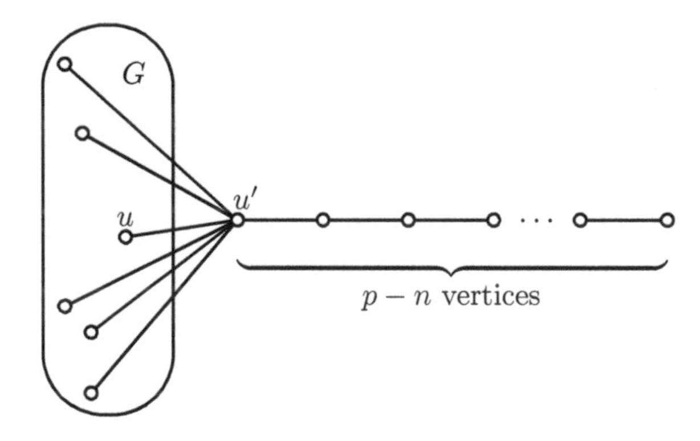

FIGURE 24.2 The graph G'.

Analogously, construct a graph H' from H, where v and v' are the corresponding selected vertices of H'. Clearly, G' and H' can be obtained from G and H in polynomial time.

If $G \cong H$, then $G' \cong H'$. Conversely, suppose that G' is isomorphic to H'. Then u' and v' are unique vertices of degree $n + 1$ in G' and H', respectively. Thus an isomorphism $G' \to H'$ maps u' to v' and, consequently, the subgraph G of G' onto the subgraph H of H'. It follows that G is isomorphic to H. Because both G' and H' are connected graphs on a prime number of vertices and because the maximum degree of both G' and H' is $n + 1 < p/2$, the lemma follows. □

Theorem 24.17 *Determining whether a connected graph is prime relative to the lexicographic product is at least as difficult as the graph isomorphism problem.*

[2]See, for instance, Theorem 8.7 in Niven, Zuckerman, and Montgomery (1991).

Proof We wish to reduce an instance $G \cong H$ of the graph isomorphism problem to an instance of decomposability testing. Lemma 24.16 tells us we may assume that G and H are connected, have a prime number p of vertices, and have maximum degrees less than $p/2$.

Choose a vertex of G, replace it by a copy of H, and replace each of the remaining $p-1$ vertices by a copy of G. For every edge uv of G, we add edges from each vertex of the graph corresponding to u to each vertex of the graph corresponding to v. Call the resulting graph X. Clearly, if $G \cong H$, then $X = G \circ G$. In this case X is decomposable.

Conversely, suppose that $X = G' \circ H'$. We wish to show that $G' \cong H' \cong G \cong H$. Because X is connected, G' is connected. Assume first that H' is connected as well. Because $|V(X)| = p^2$, we have $|V(G')| = |V(H')| = p$.

Interpret $G' \circ H'$ as being obtained from G' by replacing each vertex of G' with a copy of H', and adding all possible edges between copies of H' that correspond to adjacent vertices in G'. Let $u, v \in V(X)$ be arbitrary adjacent vertices. We claim that knowing $|N(u) \cap N(v)|$ allows us to decide whether u and v belong to the same copy of H'. We have constructed X so that $d_X(u) = d'p + d''$, where $d', d'' < p/2$ and u is adjacent to d'' vertices inside the copy of H' and to d' other copies. Now, if u and v belong to the same copy of H', then $|N(u) \cap N(v)| \geq d'p$. If they belong to different copies of H', then $|N(u) \cap N(v)| < (d'-1)p + p/2 + p/2 = d'p$. This proves the claim.

As H' is connected, we can detect the vertices that belong to the copy of H' containing u in polynomial time. This way we can find all p copies of H'. By the construction of X, one of these copies is isomorphic to H, and the others to G. Thus $G' \cong H' \cong G \cong H$.

In the second case, H' is not connected. By the above procedure we can find the connected component of H' containing u. If that component had fewer than p vertices, it would correspond to a component of G or H. However, G and H are connected, and so the second case cannot occur. □

Thus a polynomial algorithm for testing graph decomposability with respect to the lexicographic product would allow testing graph isomorphism in polynomial time. It is commonly believed that the existence of such an algorithm is very unlikely.

Decomposition is not harder than isomorphism testing

Now we show that the graph decomposition problem for the lexicographic product is not harder than the graph isomorphism problem.

Theorem 24.18 *Deciding whether a connected graph on n vertices is prime relative to the lexicographic product is not more difficult than solving a polynomial (in n) number of graph isomorphism problems, the size of each of which is also polynomial in n.*

Proof The proof is constructive. Assume there is an algorithm that solves the isomorphism problem. We show that we can decide if a given connected graph G on n vertices is prime over the lexicographic product by invoking the algorithm a polynomial number of times.

First note that there is a graph H for which $G = K_p \circ H$ if and only if $\overline{G}$ is a union of exactly p copies of a graph $\overline{H}$, if and only if the number of components of $\overline{G}$ in any given isomorphism class is a multiple of p. The latter condition can be confirmed with a polynomial number of isomorphism checks. Therefore we can determine if $G = K_p \circ H$ with a polynomial number of isomorphism checks.

Now consider the general problem of checking if G factors as $G = G_1 \circ G_2$. If there is such a factoring, then $n = n_1 n_2$, where $n_i = |V(G_i)|$. Given such n_1 and n_2, Theorem 10.8 implies that the corresponding factorization $G = G_1 \circ G_2$ is unique. Because there are only polynomially many different factorizations of n as a product of two integers, we may assume that n_1 and n_2 are fixed. We first assume that G_2 is connected; the other case is treated later. We try to find out which vertices of G belong to same G_2-layers. Let u be a vertex of G, and let v be an arbitrary neighbor of u. Let $G'(u,v)$ be the graph obtained from G by removing the neighbors of u different from v. Denote the connected component of G' containing these two vertices by $C(u,v)$. If u and v belong to the same G_2-layer, then $C(u,v)$ contains at most n_2 vertices. The same holds for $C(v,u)$. Otherwise, we label u and v differently, namely as vertices of different layers. We continue this process for every neighbor of u. If fewer that n_2 vertices remain unlabeled at the end, then we label them as vertices of G_2^u.

We repeat the above procedure for every vertex of G. As we have already noticed, if $G_2^u = G_2^v$, then $C(u,v)$ and $C(v,u)$ contain at most n_2 vertices. Thus u and v are not labeled as vertices of different copies. Let $G_2^u \neq G_2^v$. If uw is an edge of G and w is not adjacent to v, then $C(u,v)$ contains at least u, w and all the vertices of G_2^w. Thus $|V(C(u,v))| \geq n_2$, and u and v receive different labels. By symmetry, v and u receive different labels if v has a neighbor that is not a neighbor of u. Thus the only unmarked neighbors of u are the vertices from G_2^u and the vertices of layers whose projection to G_1 is a complete subgraph. In these layers we know all vertices but the vertices of G_2^u that are not adjacent to u. We add them as follows: Let U be the set of all unmarked neighbors of u and H be the subgraph of G induced by the vertices

$$\bigcup_{v \in U} (v \cup C(u,v))\,.$$

Let $m = |V(H)|/n_2$; then H must be isomorphic to $K_m \circ G_2$, and we can use the above algorithm for the case in which the first factor is complete.

If we arrive at a contradiction in the above procedure, for example, if a vertex receives two different labels or if $u \in V(G_2^v)$ but $v \notin V(G_2^u)$, then we stop. On the other hand, if the procedure consistently marks all vertices, then we have found the only possibility for the decomposition of G. In other words, we have found a candidate for G_2. Just as easily we can find the candidate for G_1. Finally, we check whether G is isomorphic to $G_1 \circ G_2$.

To complete the proof, we have to show how to proceed if G_2 is not connected. Let u be a vertex of G. Then the above factorization procedure does not find the entire G_2^u-layer but the connected component C_u of G_2^u that contains u. By a polynomial number of isomorphism tests, we can partition these components into isomorphism classes, where the number of components of each class must be divisible by n_1. To distribute these components to the G_2-layers, we first note that $G_2^u = G_2^v$ and $C_u \neq C_v$ imply that $C_u \cap C_v = \emptyset$ and that u and v have the same neighbors outside $C_u \cup C_v$. Moreover, if $C_u \cap C_v = \emptyset$ and if u and v have the same neighbors outside $C_u \cup C_v$, then we may, without loss of generality, put C_u and C_v into the same layer. □

Theorems 24.17 and 24.18, are the work of Feigenbaum and Schäffer (1986). In addition they showed that Theorem 24.18 also holds for disconnected G.

Part V

Invariants

Introduction to Part V

Any graph invariant can be studied on graph products. Here the standard question involves the relationship between the invariant of the product and the invariant of the factors. Not all invariants are equally interesting in this respect. Typically any invariant leads to interesting problems on specific products. For instance, while the chromatic number of the Cartesian product is almost trivial, it is quite exciting for the direct product, and also very interesting on the lexicographic one. On the other hand, the domination number leads to most interesting problems when applied to the Cartesian product; the independence number is most interesting on the strong product and the direct product.

We begin with a chapter on connectivity and edge-connectivity of products, which features many recent results. In particular, formulas for the connectivity of the Cartesian product, the strong product, and the lexicographic product are given.

This is followed by Chapter 26, on coloring. For the Cartesian product we easily establish that $\chi(G \,\square\, H) = \max\{\chi(G), \chi(H)\}$, while the corresponding question for the direct product is one of the deepest unsolved questions in graph theory: Hedetniemi's conjecture, which asserts that $\chi(G \times H) = \min\{\chi(G), \chi(H)\}$, has resisted proof for almost half a century. We present a proof by El-Zahar and Sauer (1985) that this conjecture holds for 4-chromatic graphs and the recent development by Zhu (2011) that the fractional version of the conjecture is true. Along the way we derive a number of bounds for the chromatic number and the circular chromatic number of various products. The lexicographic product plays a key role.

Chapter 27 investigates the independence number. Here the strong product offers many interesting questions. We investigate the Shannon capacity, for which Lovász's ϑ-function is of great importance. For the direct product we give central attention to products of vertex-transitive graphs.

Chapter 28 addresses the question of domination. We encounter yet another major unsolved problem in graph theory, namely Vizing's conjecture, which asserts that $\gamma(G \,\square\, H) \geq \gamma(G)\gamma(H)$. We offer a number of results that support this conjecture, and we prove its fractional version.

We round out Part V with a chapter on cycle bases of products, and a final chapter on selected results. The latter contains plenty of results on standard products related to one-factorizations and edge-colorings, problems involving Hamilton cycles, minors, reconstruction, topological embeddings, and nowhere-zero flows. We also address modeling of complex networks via the direct product.

For an extensive survey of invariants on graph products, we refer the reader to Nowakowski and Rall (1996).

Concerning chapter dependencies, Part V builds on Part I, but is otherwise entirely independent of the other portions of the book. We hasten to add, however, that whereas Part I can be digested by the novice graph theorist, Part V often assumes a deeper understanding of the subject.

Chapter 25

Connectivity

In Chapter 5 we characterized connectedness in the four standard products. The results were straightforward for all but the direct product. In this chapter we consider connectivity and edge-connectivity. Here the situation is more complex. In fact, despite some recent breakthroughs, the problem of determining the (edge-)connectivity of all products is not yet completely resolved.

In the next section we review connectivity notions and present formulas for the connectivity and edge-connectivity of arbitrary Cartesian products. Section 25.2 treats connectivity of lexicographic products, and the results are used for a construction that resolves a conjecture of Mader. The last section considers the strong and the direct product.

25.1 Cartesian Product

A Cartesian product of graphs is connected if and only if every one of its factors is connected (Corollary 5.3). In this section we present formulas for the connectivity and edge-connectivity of arbitrary Cartesian products. In the latter case, the structure of minimum disconnecting sets is described. Before doing this we recall the relevant concepts.

Let G be a graph. A subset $S \subseteq V(G)$ is a *separating set* if $G - S$ is disconnected. (Separating sets were defined for connected graphs in Chapter 1; the present definition allows S to be empty if G is disconnected.) The *connectivity* of G, denoted $\kappa(G)$, is the minimum size of $S \subseteq V(G)$ such that $G - S$ is disconnected or a single vertex. A separating set of size $\kappa(G)$ is called a *κ-set* of G. For any $k \leq \kappa(G)$, one says that G is k-connected.

Similarly, $S \subseteq E(G)$ is a *disconnecting set* if $G - S$ is disconnected. A graph is *k-edge-connected* if all its disconnecting sets have at least k edges. The *edge-connectivity* of G, denoted $\kappa'(G)$, is the maximum k for which G is k-edge-connected. A disconnecting set of size $\kappa'(G)$ is called a *κ'-set* of G.

Recall from Chapter 7 that for a graph H and a subgraph G' of G, the subproduct $G' \boxtimes H$ is called the H-tower over G'. We now define analogous concepts of towers over vertex or edge sets. Given a graph product $*$ and $S \subseteq V(G)$, the set $S \times V(H) \subseteq V(G * H)$ is called an *H-tower over S*. Note that if S is a separating set of G and $*$ is one of the four standard products, then the H-tower over S is a separating set of $G * H$. Similarly we may define G-towers over a separating set $S \subseteq V(H)$. (The only substantial difference is that such a tower may not be a separating set when $*$ is the lexicographic product.)

Similarly, for $S \subseteq E(G)$, the H-tower over S is the set $\{xy \in E(G{*}H) \mid p_G(x)p_G(y) \in S\}$.

This is a disconnecting set for $G * H$ if S is. In like fashion, we define a G-tower over $S \subseteq E(H)$, but this may not be a disconnecting set in the case of the lexicographic product.

The formula for the connectivity of Cartesian products has an interesting history. It was announced by Liouville (1978) but a proof never appeared. In subsequent decades, several partial results were obtained. Thirty years later, Špacapan (2008) ended the story by providing a proof for Liouville's formula:

Theorem 25.1 *Let G and H be graphs on at least two vertices. Then*

$$\kappa(G \square H) = \min\{\kappa(G)|V(H)|,\ \kappa(H)|V(G)|,\ \delta(G) + \delta(H)\}.$$

The proof that the stated minimum is an upper bound for $\kappa(G \square H)$ is reserved for Exercise 25.3. See Špacapan (2008) or Imrich, Klavžar, and Rall (2008) for the proof of the lower bound. A different proof of Theorem 25.1 is given by Xu and Yang (2010).

Theorem 25.1 has several interesting consequences. For instance, Exercise 25.4 states

$$\kappa(G \square H) \geq \kappa(G) + \kappa(H)$$

for any connected G and H. Interestingly, it was claimed several times that equality holds, a statement that is clearly wrong; consider for instance the example of Figure 25.1. Fitina, Lenard, and Mills (2010b) cited four instances of the incorrect claim in the literature, and resolved this issue by characterizing graphs for which equality holds:

Theorem 25.2 *If G and H are connected, nontrivial graphs, then $\kappa(G \square H) = \kappa(G)+\kappa(H)$ if and only if either $\kappa(G) = \delta(G)$ and $\kappa(H) = \delta(H)$ or one factor is complete and $\kappa = 1$ holds for the other factor.*

Another consequence of Theorem 25.1, proved in Klavžar and Špacapan (2008), asserts

$$\kappa(G^{\square,n}) = \delta(G^{\square,n}) = n\,\delta(G)$$

for all connected, nontrivial graphs G, and $n \geq 2$. In general, the minimum in Theorem 25.1 is not always attained on the minimum degree. For example, let G_n be the graph formed by joining two copies of K_n at a vertex. As $\kappa(G_k) = 1$, for any $m > n + 1 \geq 4$ we get

$$\min\{\kappa(G_n)|V(G_m)|,\ \kappa(G_m)|V(G_n)|\} = 2n - 1 < (n-1) + (m-1) = \delta(G_n) + \delta(G_m).$$

Theorem 25.1 gives the impression that every minimum separating set of $G \square H$ is either a tower over a minimum separating set of a factor or a neighborhood of a vertex of minimum degree. However, Figure 25.1 shows that this is not always so. Theorem 25.1 gives $\kappa(G_3 \square G_3) = 4$, but the figure shows a minimum separating set of $G_3 \square G_3$ that is neither a tower nor the neighborhood of a vertex. This phenomenon may explain why the proof of Theorem 25.1 was so elusive.

In order to describe a partial characterization of minimum separating sets in Cartesian products, we need the following concepts.

A graph is called *super-connected* if every minimum separating set is the neighborhood of some vertex. Note that the example of Figure 25.1 demonstrates that $G_3 \square G_3$ is not super-connected. This example can be generalized as follows. Call a graph G with at least one edge *locally complete* if it is not a complete graph and contains a complete block of order $\delta(G)+1$. (The graph G_n is locally complete as it contains two blocks of order $n = \delta(G_n) + 1$.) Then the Cartesian product of locally complete graphs is not super-connected (Exercise 25.6).

To characterize super-connected Cartesian product graphs, recall that $N[A]$ denotes the closed neighborhood of $A \subset V(G)$. We say that a connected graph G has *property* $\mathcal{P}$ if there

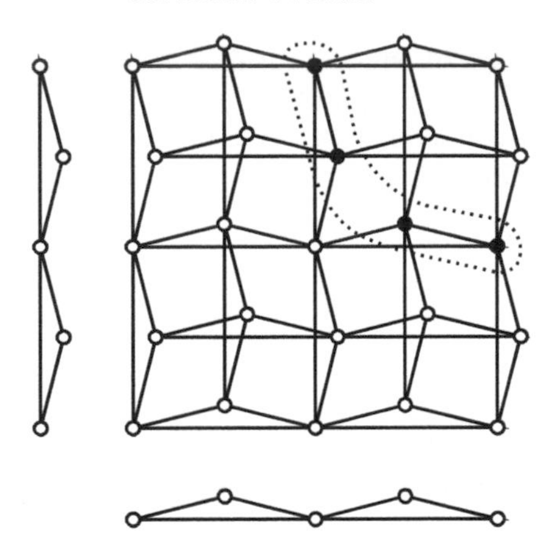

FIGURE 25.1 A noncanonical minimum separating set of $G_3 \,\square\, G_3$.

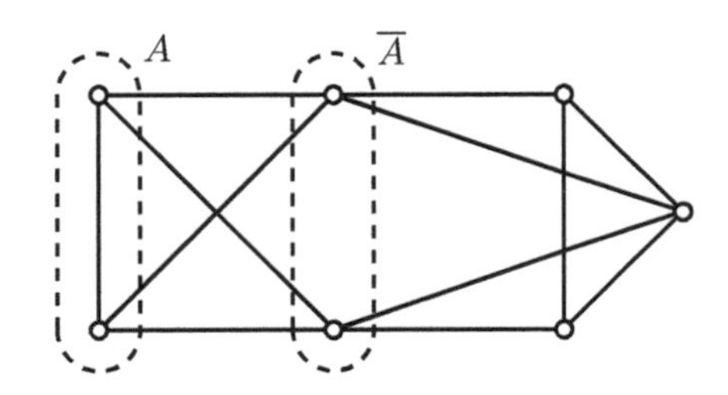

FIGURE 25.2 A graph with property $\mathcal{P}$.

is a set $A \subset V(G)$ such that $|N[A]| = \delta(G) + 1$. Note that in this case, $\overline{A} = N[A] \setminus A$ is a separating set, A induces a complete graph, and any vertex of A is adjacent to all vertices of $\overline{A}$. Figure 25.2 shows a graph with property $\mathcal{P}$. Now, if a graph G has property $\mathcal{P}$ and a graph H contains a pendant vertex, then $G \,\square\, H$ in not super-connected (Exercise 25.7).

We have thus observed that if G and H are locally complete graphs or one of them has property $\mathcal{P}$ and the other a pendant vertex, then $G \,\square\, H$ is not locally complete. Let us call such pairs G, H *excluded pairs*. Then Xu and Yang (2010) proved the following result.

Theorem 25.3 *Let G and H be connected graphs, each with at least one edge, and suppose that G, H is not an excluded pair. Then $G \,\square\, H$ is super-connected if and only if $\delta(G) + \delta(H) < \min\{\kappa(G)|V(H)|,\ \kappa(H)|V(G)|\}$ or $G \,\square\, H \in \{K_2 \,\square\, K_2, K_2 \,\square\, K_3\}$.*

By contrast, any minimum disconnecting set of a Cartesian product has a certain canonical form. This was proved by Klavžar and Špacapan (2008).

Theorem 25.4 *Suppose G and H are connected nontrivial graphs. If S is a minimum disconnecting set of $G \,\square\, H$, then one of the following holds:*

(i) *S consists of the edges incident to a vertex of $G \,\square\, H$ of minimum degree,*
(ii) *S is an H-tower over a κ'-set of G,*
(iii) *S is a G-tower over a κ'-set of H.*

Proof We note first that by Exercise 25.1, $|S| \leq \delta(G \,\square\, H) = \delta(G) + \delta(H)$. Moreover, the

H-tower over a disconnecting set of edges X of G is a disconnecting set in $G \,\Box\, H$. Thus $|S| \leq \kappa'(G)|V(H)|$ and analogously $|S| \leq \kappa'(H)|V(G)|$.

Because S is a minimum disconnecting set, $(G \,\Box\, H) - S$ consists of exactly two connected components C and $\overline{C}$. We distinguish three cases.

Case 1. Every G-layer intersects both C and $\overline{C}$.
In this case the removal of the edges of S from any G-layer disconnects the layer, hence $|S| \geq \kappa'(G)|V(H)|$. It follows that $|S| = \kappa'(G)|V(H)|$ and consequently no edges from H-layers are contained in S. Thus every H-layer is either contained in C or in $\overline{C}$. We conclude that S is an H-tower over a κ'-set of G.

Case 2. Every H-layer intersects both C and $\overline{C}$.
By the same argument as in Case 1 we infer that S is a G-tower over a κ'-set of H.

Case 3. There exist a G-layer and an H-layer that are connected in $(G \,\Box\, H) - S$.
These two layers are in the same connected component of $(G \,\Box\, H) - S$, say C. Let (a, x) be an arbitrary vertex of $G \,\Box\, H$ from $\overline{C}$.

We claim that $d(a, x) \leq |S|$. We prove this by assigning to each neighbor of (a, x) a unique edge from S. So let (a', x) be a neighbor of (a, x) in $G^{(a,x)}$. If $e = (a, x)(a', x) \in S$, we assign e to (a', x). If $(a, x)(a', x) \notin S$, then $(a', x) \in \overline{C}$, so the layer $H^{(a',x)}$ meets both $\overline{C}$ and C. (Recall that every layer meets C.) Then $H^{(a',x)}$ contains at least one edge from S and we assign such an edge to (a', x). We proceed analogously for a neighbor (a, x') of (a, x) that lies in $H^{(a,x)}$. Hence $|S| \geq d(a, x) \geq \delta(G \,\Box\, H)$, and the claim is proved.

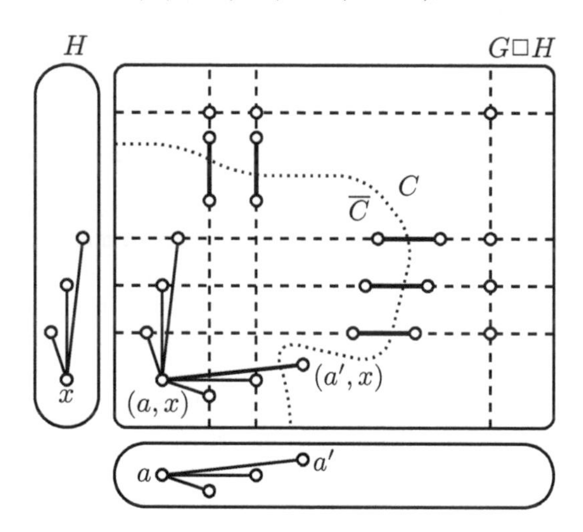

FIGURE 25.3 Illustration of Case 3. Bold edges are in S. Dashed lines represent layers.

As S is a minimum disconnecting set, the claim implies $|S| = d(a, x)$ for any vertex $(a, x) \in \overline{C}$. Suppose $|\overline{C}| > 1$. Then we may assume that $\overline{C}$ has a vertex (a', x) with $a' \neq a$ (for otherwise $\overline{C}$ would be disconnected provided that $G^{(a,x)}$ and $H^{(a,x)}$ would have only one vertex in $\overline{C}$). Then $(a, x)(a', x) \in E(G \,\Box\, H)$. Indeed, if this were not the case, then no edge of $H^{(a',x)}$ would be assigned to (a, x) in the argument that proved the claim. Because $H^{(a',x)}$ contains at least one edge of S, this would imply $|S| > d(a, x)$. By the same argument, $D = G^{(a,x)} \cap \overline{C}$ induces a complete graph.

Because D is complete and each vertex of D has the same degree restricted to the corresponding H-layer (and equal degree), it follows that each vertex in D has an equal number of neighbors in $G^{(a,x)} \cap C$. As G is connected, each vertex from D has at least one neighbor in $G^{(a,x)} \cap C$. Therefore, there is at least one edge in S that was not assigned to (a,x) in the above argument. We conclude that $|\overline{C}| = 1$ and thus S consists of the edges incident to a vertex of $G \,\Box\, H$ of minimum degree. □

Theorem 25.4 yields a formula for the edge-connectivity of a Cartesian product, due to Xu and Yang (2006). Note that it is completely parallel to the connectivity formula.

Corollary 25.5 *Let G and H be graphs on at least two vertices. Then*

$$\kappa'(G \,\Box\, H) = \min\{\kappa'(G)|V(H)|,\ \kappa'(H)|V(G)|,\ \delta(G) + \delta(H)\}\,.$$

Proof Note that the result is clearly true if G or H is not connected. If G and H are connected graphs on at least two vertices, the result follows from Theorem 25.4. □

Klavžar and Špacapan (2008) also prove that $\kappa'(G^{\Box,n}) = n\,\delta(G)$ for any connected graph on at least two vertices, and $n \geq 2$, a result entirely parallel to the situation for connectivity. Also, parallel to Theorem 25.2, we have the following theorem of Fitina, Lenard, and Mills (2010a) (see also Exercise 25.5):

Theorem 25.6 *If G and H are connected and nontrivial, then $\kappa'(G \,\Box\, H) = \kappa'(G) + \kappa'(H)$ if and only if either $\kappa'(G) = \delta(G)$ and $\kappa'(H) = \delta(H)$, or one factor is complete and $\kappa' = 1$ for the other factor.*

25.2 Critically Connected Graphs and the Lexicographic Product

Corollary 5.14 states that the connectedness of a lexicographic product depends only on the first factor: A lexicographic product $G_1 \circ G_2 \circ \cdots \circ G_k$ of nontrivial graphs is connected if and only if G_1 is connected. We now determine the connectivity of lexicographic products. We then use this result to construct ℓ-κ-critically connected graphs of diameter at least 3, the existence of which were conjectured by Mader.

Proposition 25.7 *If G and H are graphs and G is not complete, then $\kappa(G \circ H) = \kappa(G)|V(H)|$.*

Proof Let S be a κ-set of G. Then $S \times V(H)$ is a separating set of $G \circ H$, so $\kappa(G \circ H) \leq \kappa(G)|V(H)|$.

Let (a,x) and (b,y) be distinct vertices of $G \circ H$. If $a = b$, then there are $d_G(a)|V(H)| \geq \delta(G)|V(H)| \geq \kappa(G)|V(H)|$ internally disjoint $(a,x),(b,y)$-paths (each of length 2). If $a \neq b$, there are $\kappa(G)$ internally disjoint a,b-paths in G. Each such path can be easily extended to $|V(H)|$ internally disjoint $(a,x),(b,y)$-paths; hence we have at least $\kappa(G)|V(H)|$ such paths $G \circ H$. In any case, Menger's theorem implies $\kappa(G \circ H) \geq \kappa(G)|V(H)|$. □

The case when G is complete is left for Exercise 25.8.

Let $\ell \geq 0$. A graph $G \neq K_n$ is called *ℓ-κ-critical* if for any $X \subseteq V(G)$, $|X| \leq \ell$, we have $\kappa(G - X) = \kappa(G) - |X|$. Mader (1977) proved that a 3-κ-critical graph has diameter at most 4, and later Mader (1984) conjectured that there actually exist 3-κ-critical graphs of diameter 3 or 4. Using the lexicographic product, Kriesell (2006) was able to confirm the conjecture. To do this he first applied Proposition 25.7 to get:

Lemma 25.8 *Let G be an ℓ-κ-critical graph with $\kappa(G) = k$ and $|V(G)| \geq 2k+2$. Let $\widehat{G}$ be the graph obtained from the disjoint union of $G \circ K_2$ and two vertices, where one vertex is adjacent to all vertices of one G-layer and the other vertex to all vertices of the other G-layer. Then $\widehat{G}$ is an ℓ-κ-critical graph with $\kappa(\widehat{G}) = 2k+2$ and $\mathrm{diam}(\widehat{G}) \geq 3$.*

The graph $\widehat{G}$ from Lemma 25.8 is illustrated in Figure 25.4.

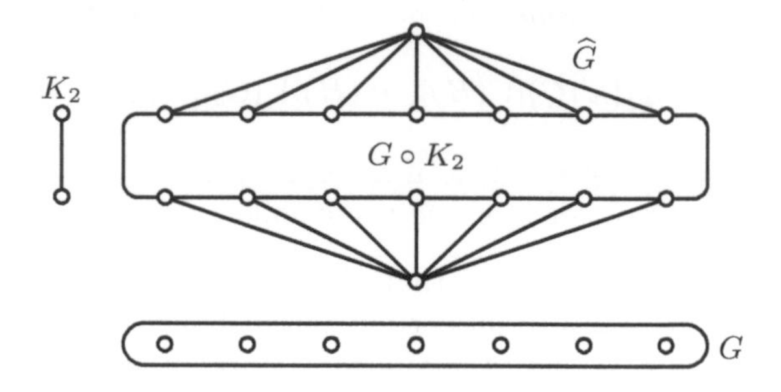

FIGURE 25.4 The construction of $\widehat{G}$.

With Lemma 25.8 in hand, we can now describe the construction of Kriesell (2006).

Theorem 25.9 *For every $\ell \geq 3$ there exists an ℓ-κ-critical graph of diameter at least 3.*

Proof Let $\ell \geq 3$ and let $G(n, \ell)$ be the complement of the ℓth direct power of K_n, that is,

$$G(n, \ell) = \overline{K_n^{\times, \ell}}.$$

Clearly, $|V(G(n, \ell))| = n^\ell$. Mader (1977) proved that $G(n, \ell)$ is an ℓ-critical graph with $\kappa(G(n, \ell)) = n^\ell - (n-1)^\ell - 1$. Now select an n large enough so that $|V(G(n, \ell))| = n^\ell \geq 2\kappa(G(n, \ell)) + 2$. Then Lemma 25.8 applied on $G(n, \ell)$ gives an ℓ-κ-critical graph with connectivity $2\kappa(G(n, \ell)) + 2$ and $\mathrm{diam}(\widehat{G}) \geq 3$. □

We close the section by noting that the lexicographic product plays an important role in some other problems related to connectivity. For instance, Meng (2003) proved that a connected vertex- and edge-transitive graph G is not super-connected if and only if G is $C_n \circ D_m$ ($n \geq 6$, $m \geq 1$) or $L(Q_3) \circ D_m$ ($m \geq 1$), where $L(Q_3)$ is the line graph of Q_3.

25.3 Strong and Direct Products

It is an easy consequence of Distance Formula for the strong product that a strong product of graphs is connected if and only if every factor is connected (Corollary 5.5). The next lemma is useful for determining the connectivity of strong products.

Lemma 25.10 *Let G and H be non-complete graphs and let S be a minimum separating set of $G \boxtimes H$. If S intersects every H-layer, then S is a G-tower over a κ-set of H.*

Proof We claim that S intersects any H-layer in at least $\kappa(H)$ vertices. Let $a \in V(G)$. Because we assumed that S intersects every H-layer, there exists a vertex $(a, x) \in S$.

Partition the vertices of $H^{(a,x)}$ into sets $S(a)$ and $\overline{S}(a)$ that contain (respectively do not contain) vertices of S. There is nothing to be proved if $|S(a)| = |V(H)|$. Hence let $\overline{S}(a) \neq \emptyset$, set $X = G \boxtimes H - S$, and consider the following two cases.

Case 1. $\overline{S}(a)$ is not contained in a single connected component of X.
In this case, $\overline{S}(a)$ induces a disconnected subgraph of $H^{(a,x)}$. It follows that $S(a)$ is a separating set of $H^{(a,x)}$ and hence has size at least $\kappa(H)$.

Case 2. $\overline{S}(a)$ is contained in a single connected component of X.
Let C be a component of X that does not contain $\overline{S}(a)$. (Such a component exists because S is a separating set.) Because S is a minimum separating set, (a, x) has at least one neighbor in C, say (a', x'). Clearly $a' \neq a$, but it is possible that $x' = x$. However, in any case $(a, x') \in S(a)$. Indeed, (a, x') is not in $\overline{S}(a)$, for otherwise (a', x'), being adjacent to (a, x'), would be in the same connected component of X as $\overline{S}(a)$. By the same argument, (a', x') has no neighbor in $\overline{S}(a)$. It follows that all the neighbors of (a, x') in $H^{(a,x)}$ lie in $S(a)$. Therefore, $|S(a)| \geq 1 + \delta(H) \geq 1 + \kappa(H)$.

We have thus shown that S intersects any H-layer in at least $\kappa(H)$ vertices. Therefore, $|S| \geq \kappa(H)|V(G)|$. On the other hand, it is straightforward to see that a G-tower over a separating set of H (such sets exist because H is not complete) is a separating set of $G \boxtimes H$. Therefore, $|S| \leq \kappa(H)|V(G)|$ and hence $|S| = \kappa(H)|V(G)|$. Moreover, this also implies that the conclusion $|S(a)| \geq 1 + \kappa(H)$ of Case 2 is not possible. It follows that for any vertex a of G, $\overline{S}(a)$ is contained in more than one connected component of X. To complete the argument, one now needs to show that for any vertices $a, b \in V(G)$, $p_H(S(a)) = p_H(S(b))$, which in turn implies that S is a G-tower over $p_H(S(a))$. Details are left to the reader. □

By symmetry, if S is a minimum separating set of $G \boxtimes H$ that has at least one vertex in every G-layer, then S is an H-tower over a κ-set of G. The last case to be considered is when S has empty intersection with at least one G-layer and with at least one H-layer. To describe S in such cases, we need the following definition.

Let S_G and S_H be separating sets of connected graphs G and H, and let G' and H' be arbitrary connected components of $G - S_G$ and $H - S_H$. Then the set of vertices

$$(S_G \times V(H')) \cup (S_G \times S_H) \cup (V(G') \times S_H)$$

is called a ⅂-set of $G \boxtimes H$; see Figure 25.5.

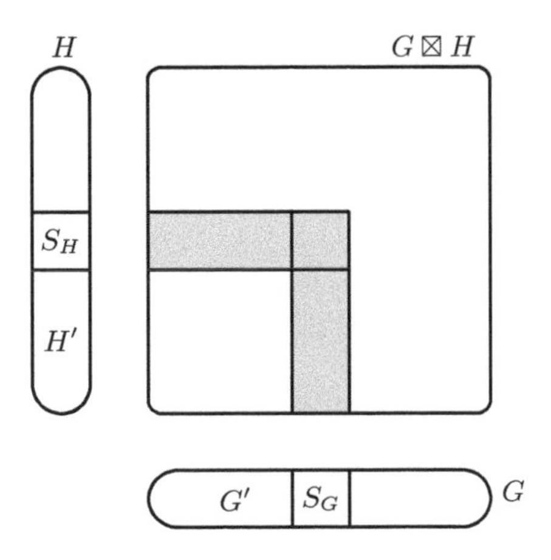

FIGURE 25.5 A ⅂-set of $G \boxtimes H$.

Špacapan (2010) gives a lengthy and technical proof that if a minimum separating set S has empty intersection with at least one G-layer and with at least one H-layer, then S is a $\urcorner$-set of $G \boxtimes H$. In conclusion, he proved the following result. (For the special case when one of G and H is complete, see Exercise 25.10.)

Theorem 25.11 *Let G and H be connected graphs, at least one not complete, and let S be a minimum separating set of $G \boxtimes H$. Then one of the following holds:*
(i) *S is a $\urcorner$-set in $G \boxtimes H$,*
(ii) *S is an H-tower over a κ-set of G,*
(iii) *S is a G-tower over a κ-set of H.*

To see that a $\urcorner$-set can be smaller than any of the towers over κ-sets of G and H, and that a $\urcorner$-set need not be induced by minimum separating sets of factors, consider the following example. Let $n \geq 3$ and let G_n be the graph obtained from the disjoint of K_n, K_{2n}, and K_{3n}, such that two vertices x and y of K_n are identified with two vertices of K_{2n}, and another vertex z of K_{2n} is identified with a vertex of K_{3n}. Now consider $G_n \boxtimes G_n$. Because $\{z\}$ is the unique κ-set of G_n, any tower over a κ-set of a factor of $G_n \boxtimes G_n$ contains $6n-3$ vertices. On the other hand, let X be the set of vertices of G_n induced by the vertices $V(K_n) \setminus \{x, y\}$; then

$$(\{x,y\} \times X) \cup (\{x,y\} \times \{x,y\}) \cup (X \times \{x,y\})$$

is a $\urcorner$-set of $G_n \boxtimes G_n$ of order $2(n-2) + 4 + (n-2)2 = 4n - 4 < 6n - 3$.

Theorem 25.11 immediately yields a formula for the connectivity of strong products.

Corollary 25.12 *Let G and H be connected graphs, at least one not complete. Set $\ell(G \boxtimes H)$ be the minimum size of a $\urcorner$-set of $G \boxtimes H$. Then*

$$\kappa(G \boxtimes H) = \min\{\kappa(G)|V(H)|, \kappa(H)|V(G)|, \ell(G \boxtimes H)\}.$$

Theorem 25.11 and Corollary 25.12 assume that at least one of the factors is not complete because otherwise their strong product is also complete.

For the edge-connectivity, Brešar and Špacapan (2007) proved the following result.

Theorem 25.13 *Let G and H be connected graphs. Then*

$$\kappa'(G \boxtimes H) = \min\{\kappa'(G)(|V(H)| + 2|E(H)|), \kappa'(H)(|V(G)| + 2|E(G)|), \delta(G \boxtimes H)\}.$$

It is again not difficult to see that $\kappa'(G \boxtimes H)$ is not bigger than the asserted minimum (Exercise 25.11). The other inequality requires several pages.

We conclude the section with a brief look at the direct product. Recall from Theorem 5.9 that if G and H are connected nontrivial graphs, then $G \times H$ is connected provided that at least one of G or H has an odd cycle; otherwise $G \times H$ has exactly two components. As this result is more involved than the corresponding results for other products, it is not surprising that the connectivity and the edge-connectivity of direct products is a difficult problem.

For the edge-connectivity of direct products, only some partial results are known. For instance, Brešar and Špacapan (2008) proved the following result.

Theorem 25.14 *Let G and H be nonbipartite graphs. Then*

$$\begin{aligned}
\kappa'(G \times H) &\geq \min\{\kappa'(G)|E(H)|, \kappa'(H)|E(G)|, \delta(G)\delta(H)\}, \\
\kappa'(G \times H) &\leq \min\{2\kappa'(G)|E(H)|, 2\kappa'(H)|E(G)|, \delta(G)\delta(H)\}.
\end{aligned}$$

The upper bound holds because $\delta(G \times H) = \delta(G)\delta(H)$ and because it is easy to verify that (i) the tower over a κ'-set of G is a disconnecting set of $G \times H$ and (ii) such a tower contains $2\kappa'(G)|E(H)|$ edges.

Even less is known about the connectivity of the direct product. Brešar and Špacapan (2008) give different upper bounds; see Exercise 25.12 for an example.

Exercises

25.1. Show that for any graph G, $\kappa(G) \leq \delta(G)$ and $\kappa'(G) \leq \delta(G)$.

25.2. Show that for any graph G, $\kappa(G) \leq \kappa'(G)$.

25.3. Show that for any graphs G and H on at least two vertices, $\kappa(G \,\Box\, H) \leq \min\{\kappa(G)|V(H)|, \kappa(H)|V(G)|, \delta(G) + \delta(H)\}$.

25.4. (Sabidussi, 1957) Show that for any connected graphs G and H, $\kappa(G \,\Box\, H) \geq \kappa(G) + \kappa(H)$.

25.5. Show that for any connected graphs G and H, $\kappa'(G \,\Box\, H) \geq \kappa'(G) + \kappa'(H)$.

25.6. Let G and H be connected, locally complete graphs. Show that then $G \,\Box\, H$ is not super-connected.

25.7. Let G be a connected graph with property $\mathcal{P}$ and let H be a connected graph with a pendant vertex. Show that $G \,\Box\, H$ is not super-connected.

25.8. Show that for any $n \geq 2$ and any graph H,

$$\kappa(K_n \circ H) = (n-1)|V(H)| + \kappa(H).$$

25.9. Determine the smallest order and connectivity of 3-κ-connected graphs constructed in Theorem 25.9.

25.10. Show that if G is not complete, then $\kappa(G \boxtimes K_n) = n\kappa(G)$.

25.11. Find disconnecting sets of edges of $G \boxtimes H$ that realize the minimum in the formula of Theorem 25.13.

25.12. (Brešar and Špacapan, 2008) For a graph G, let $\kappa_b(G)$ be the size of a smallest set X of vertices of G such that $G - X$ is bipartite. Show that

$$\kappa(G \times H) \leq \kappa_b(G)|V(H)| + \kappa_b(H)|V(G)| - \kappa_b(G)\kappa_b(H).$$

Chapter 26

Coloring and Hedetniemi's Conjecture

We begin with the chromatic number of Cartesian, strong, and lexicographic products. The lexicographic product plays a key role here, and Section 26.2 establishes upper and lower bounds for its chromatic number. These are then applied in three constructions involving lexicographic and strong products. The subsequent section considers the fractional and the circular chromatic number. The former can be expressed in terms of the lexicographic product, and is multiplicative on this product. The circular chromatic number has many appealing properties, especially in its relation to the lexicographic product.

The final sections treat Hedetniemi's conjecture, which asserts $\chi(G \times H) = \min\{\chi(G), \chi(H)\}$ and is the central open problem in product colorings. In Section 26.5 we prove it for the case where G is 4-colorable while in the final section the fractional version of the conjecture is proved.

We note that loops arise naturally in discussions of Hedetniemi's conjecture. We adopt the convention that $\chi(G) = \infty$ if G has a loop.

26.1 Product Coloring

We now develop formulas and bounds for the chromatic number of Cartesian products, strong products, and lexicographic products. The section concludes with a neat result about uniquely colorable direct products, although we delay a full treatment of the direct product until sections 26.4 and 26.5.

Sabidussi (1957) gave a complete and satisfactory answer to the question of the chromatic number of a Cartesian product. The result has been rediscovered several times.

Theorem 26.1 *For any graphs G and H, $\chi(G \,\Box\, H) = \max\{\chi(G), \chi(H)\}$.*

Proof Certainly $\chi(G \,\Box\, H) \geq \chi(G)$, because the G-layers of $G \,\Box\, H$ are isomorphic to G. Similarly $\chi(G \,\Box\, H) \geq \chi(H)$, so $\chi(G \,\Box\, H) \geq \max\{\chi(G), \chi(H)\}$.

Now consider the reverse inequality. By commutativity, we may assume $\chi(G) \geq \chi(H)$. Let $g : V(G) \to \{0, 1, \ldots, \chi(G)-1\}$ be a coloring of G, and $h : V(H) \to \{0, 1, \ldots, \chi(H)-1\}$ be a coloring of H. Define $f : V(G \,\Box\, H) \to \{0, 1, \ldots, \chi(G) - 1\}$ as

$$f(a, x) = g(a) + h(x) \pmod{\chi(G)}.$$

Given an edge in $G \square H$ of form $(a,x)(a,y)$ with $xy \in E(H)$, we have $f(a,x) \neq f(a,y)$. Any edge not of this type has form $(a,x)(b,x)$ with $ab \in E(G)$, and again $f(a,x) \neq f(b,x)$. Thus f is a $\chi(G)$-coloring of $G \square H$, so $\chi(G \square H) \leq \max\{\chi(G), \chi(H)\}$. □

Figure 26.1 illustrates the construction from the proof of Theorem 26.1.

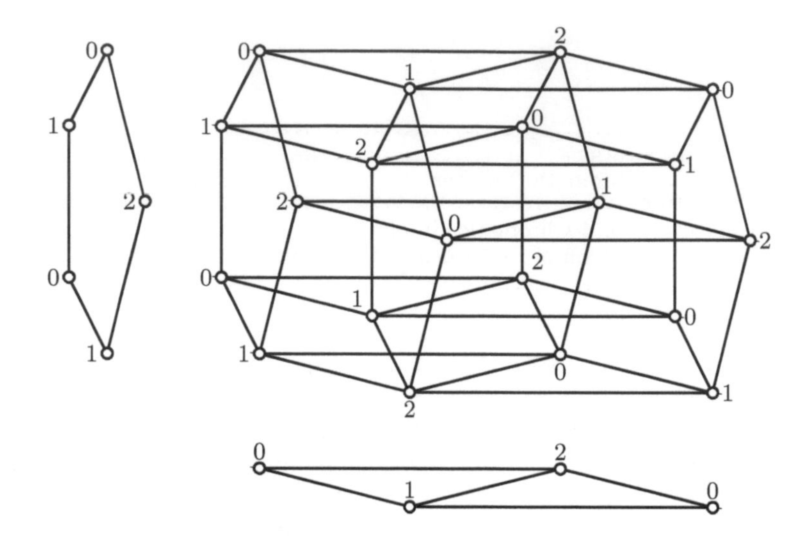

FIGURE 26.1 A 3-coloring of $(K_4 - e) \square C_5$.

In contrast to Theorem 26.1, there are no explicit formulas for the chromatic number of the strong or lexicographic products. We do, however, have the following bounds.

Proposition 26.2 *For any graphs G and H,*

$$\chi(G \boxtimes H) \leq \chi(G \circ H) \leq \chi(G)\,\chi(H)\,.$$

Proof Because $G \boxtimes H \subseteq G \circ H$, we infer that $\chi(G \boxtimes H) \leq \chi(G \circ H)$. For the other inequality, let g and h be minimal colorings of G and H, respectively. Given $(a,x) \in V(G \circ H)$, set

$$f(a,x) = (g(a), h(x))\,.$$

If $(a,x)(b,y)$ is an edge of $G \circ H$, then ab is an edge of G, or xy is an edge of H. Either way, $f(a,x) \neq f(b,y)$. Therefore f is a coloring of $G \circ H$ with $\chi(G)\chi(H)$ colors. □

To see that the inequalities of Proposition 26.2 are sharp, let G and H be graphs whose chromatic numbers equal their clique numbers. Because $K_{mn} = K_m \boxtimes K_n$, it follows that $G \boxtimes H$ has $K_{\chi(G)\chi(H)}$ as a subgraph, and the proposition yields $\chi(G \boxtimes H) = \chi(G)\,\chi(H)$. In particular, if G and H are bipartite with at least one edge, then $\chi(G \boxtimes H) = \chi(G \circ H) = 4$.

On the other hand, the inequalities can be strict, even simultaneously. Consider the lexicographic product $C_5 \circ K_2$ (which is isomorphic to $C_5 \boxtimes K_2$). It is easy to see that $\chi(C_5 \circ K_2) = 5$, so

$$5 = \chi(C_5 \circ K_2) < \chi(C_5)\chi(K_2) = 6\,.$$

Further, one can establish $\chi(C_5 \boxtimes C_5) = 5$ and $\chi(C_5 \circ C_5) = 8$, while $\chi(C_5)\chi(C_5) = 9$.

In general, it is difficult to compute chromatic numbers of strong and lexicographic products. However, the following theorem by Geller and Stahl (1975) implies that for the lexicographic product, it suffices to assume the second factor is complete.

Theorem 26.3 *If $\chi(H) = n$, then $\chi(G \circ H) = \chi(G \circ K_n)$ for any graph G.*

Proof Suppose $\chi(H) = n$. By Lemma 2.10 (i), there is a homomorphism $H \to K_n$ and hence also a homomorphism $G \circ H \to G \circ K_n$. Thus $\chi(G \circ H) \leq \chi(G \circ K_n)$.

Conversely, color $G \circ H$ with $\chi(G \circ H)$ colors. Because $\chi(H) = n$, any given H-layer has at least n different colors. Select n vertices of different colors from the H-layer. Connect every pair of selected vertices by an edge (if it is not already present). Repeating this procedure for all H-layers and discarding all unselected vertices, we arrive at $G \circ K_n$, and it inherits a coloring from $G \circ H$. Hence $\chi(G \circ H) \geq \chi(G \circ K_n)$. □

We now note two ways that coloring problems relate to the lexicographic product.

- Stahl (1976) introduced an *n-tuple coloring* of a graph as an assignment of a set of n colors to each vertex such that adjacent vertices receive disjoint sets of colors. (This is also known as a *set coloring*; see Bollobás and Thomason (1979).) The nth *chromatic number* $\chi_n(G)$ of G is the smallest number of colors needed to give G an n-tuple coloring. It is obvious that $\chi_n(G) = \chi(G \circ K_n)$. Thus the nth chromatic number is the chromatic number of a special lexicographic product.

- Here is another reformulation of the chromatic number of the lexicographic product, also due to Stahl (1976). Recall that, for $n \geq 2k$, the Kneser graph $K(n, k)$ has as vertices the $\binom{n}{k}$ k-subsets of $\{1, 2, \dots, n\}$, and two vertices are adjacent when they are disjoint. Consider an n-coloring c of $G \circ K_k$. Define a map $V(G) \to K(n, k)$ as $f(v) = \{c(v, i) \mid i \in V(K_k)\}$. This is clearly a homomorphism. Conversely, every such homomorphism induces an n-coloring of $G \circ K_k$. Therefore:

  Proposition 26.4 *The number $\chi(G \circ K_k)$ is the smallest integer n for which there is a homomorphism $G \to K(n, k)$.*

  This generalizes the fact that $\chi(G)$ is the smallest n for which there is a homomorphism $G \to K_n$.

In the wake of solving Kneser's conjecture, Lovász (1978) showed $\chi(K(n, k)) = n - 2k + 2$. His ingenious proof uses homotopy theory and Borsuk's theorem. We will have occasion to use Lovász's result later (as Proposition 26.4 may suggest) so we state it as a theorem. For a proof, see Hell and Nešetřil (2004).

Theorem 26.5 *If $n \geq 2k$, then $\chi(K(n, k)) = n - 2k + 2$.* □

We now shift our attention to the direct product. Recall that a graph G is *uniquely n-colorable* if any n-coloring determines the same partition of $V(G)$ into color classes. The following result is due to Greenwell and Lovász (1974). It will be applied in Section 26.2 to disprove Geller's conjecture.

Theorem 26.6 *If G is connected and $\chi(G) > n$, then $G \times K_n$ is uniquely n-colorable.*

Proof The map $(u, i) \mapsto i$ is a homomorphism $G \times K_n \to K_n$, so $G \times K_n$ is n-colorable.

Suppose $f : V(G \times K_n) \to \{1, \dots, n\}$ is an n-coloring of $G \times K_n$. Observe that there is a vertex $x \in V(G)$ for which the colors in the K_n-layer over x are all distinct: Indeed, suppose to the contrary that for every $x \in V(G)$ we have $f(x, i_x) = f(x, j_x)$ for some $i_x \neq j_x$. Denote this color as $g(x)$, so g is a map $V(G) \to \{1, \dots, n\}$. If $xy \in E(G)$, then the vertex (x, i_x) must be adjacent to either (y, i_y) or (y, j_y), so $g(x) \neq g(y)$. Therefore g is an n-coloring of G, contrary to assumption.

Thus G has a vertex x for which the K_n-layer above x contains n different colors. Let $xy \in E(G)$. Then (x, i) is adjacent to (y, j) whenever $i \neq j$. It follows that $f(x, i) = f(y, i)$. Because G is connected, we conclude that the color classes of $G \times K_n$ are all of form $V(G) \times \{i\}$, so $G \times K_n$ is uniquely n-colorable. □

Actually, Greenwell and Lovász (1974) proved Theorem 26.6 in a somewhat more general form, but the present version suffices for the next corollary.

Corollary 26.7 *For all $n, k \geq 3$ there is a uniquely n-colorable graph with no odd cycles shorter than k.*

Proof Apply Theorem 26.6 and the well-known fact that there are graphs with arbitrarily large chromatic number and without odd cycles shorter than a given constant. □

Theorem 26.6 was generalized by Zhu (1999) as follows: Given a core G (defined on p. 34), let $C(G)$ be the graph whose vertices are the maps $V(G) \to V(G)$ that are not automorphisms. Two such maps f and g are adjacent if $f(u)g(v) \in E(G)$ for all $uv \in E(G)$. Let H be a core, and suppose that G is connected and does not admit a homomorphism to $C(H)$. Then there is a unique homomorphism $G \times H \to H$.

26.2 Bounds and Three Applications

We continue with bounds on the chromatic number of lexicographic and strong products. These bounds are then used to construct counterexamples to conjectures by Geller and Hajós, and to prove a related result about retracts of strong products.

By Proposition 26.2, any upper bound for the lexicographic product is also an upper bound for the strong product, and any lower bound for the strong product is a lower bound for the lexicographic product. We will use these facts often.

Geller and Stahl (1975) proved a weaker version of our next result; the present formulation is by Klavžar (1993). Recall that a graph G is χ-critical if $\chi(G - v) < \chi(G)$ for every $v \in V(G)$. Any nontrivial graph G has a χ-critical subgraph G' with $\chi(G) = \chi(G')$.

Theorem 26.8 *If G is a χ-critical graph, then for any graph H,*

$$\chi(G \boxtimes H) \leq \chi(G \circ H) \leq \chi(H)\,(\chi(G) - 1) + \left\lceil \frac{\chi(H)}{\alpha(G)} \right\rceil .$$

Proof Let $\{C_0, C_1, \ldots, C_{\chi(H)-1}\}$ be the color classes of a minimal coloring of H, and let $S = \{a_0, a_1, \ldots, a_{\alpha(G)-1}\}$ be an independent set in G. Because G is χ-critical, each $G - a_i$ can be colored with $\chi(G)-1$ colors. For each $a_i \in S$, let $f_i : V(G) - a_i \to \{1, 2, \ldots, \chi(G)-1\}$ be such a coloring.

For each $0 \leq j \leq \chi(H) - 1$, write $j = p_j \cdot \alpha(G) + q_j$, where $0 \leq q_j < \alpha(G)$. Then set

$$f(a, x) \;=\; \begin{cases} j\chi(G) + f_{q_j}(a) & \text{if } x \in C_j \text{ and } a \neq a_{q_j}\,, \\ p_j\chi(G) & \text{if } x \in C_j \text{ and } a = a_{q_j}\,. \end{cases}$$

It is straightforward to see that f is a coloring of $G \circ H$ that attains the upper bound of the theorem. □

Finding a good upper bound for the chromatic number of a product involves finding a good construction. Lower bounds can be more problematic. Nonetheless, we present three such results. The first is by Stahl (1976); see also Bollobás and Thomason (1979). Our proof is from Klavžar and Milutinović (1994).

Theorem 26.9 *If G is a graph with at least one edge, and H is a graph with $\chi(H) = n$, then*

$$\chi(G \circ H) = \chi(G \boxtimes K_n) \geq \chi(G) + 2n - 2\,.$$

Proof The equality follows from Theorem 26.3 and the fact that $G \circ K_n \cong G \boxtimes K_n$.

To prove the inequality, first set $s = \chi(G \boxtimes K_n) - 2n + 2$. Because G has at least one edge, $G \boxtimes K_n$ has a subgraph isomorphic to $K_2 \boxtimes K_n = K_{2n}$, and hence $\chi(G \boxtimes K_n) \geq 2n$, so $s > 0$. If we can produce a coloring of G with at most s colors, then $\chi(G) \leq s = \chi(G \boxtimes K_n) - 2n + 2$, and the theorem will follow.

We now construct this coloring. Let $f : V(G \boxtimes K_n) \to \{1, 2, \ldots, \chi(G \boxtimes K_n)\}$ be a coloring of $G \boxtimes K_n$. Put $V(K_n) = \{x_1, \ldots, x_n\}$; and for each $a \in V(G)$, set $m_a = \min\{f(a, x_1), \ldots, f(a, x_n)\}$. Define a map $g : V(G) \to \{1, 2, \ldots, s\}$ as follows:

$$g(a) = \begin{cases} m_a & \text{if } m_a \leq s - 1\,, \\ s & \text{if } m_a > s - 1\,. \end{cases}$$

We prove g is a coloring of G by showing $g(a) = g(b)$ implies $ab \notin E(G)$. If $g(a) = g(b) \leq s - 1$, then $m_a = m_b$, and $G \boxtimes K_n$ has vertices (a, x_i) and (b, x_j) with the same color m_a. Hence $(a, x_i)(b, x_j) \notin E(G \boxtimes K_n)$, so $ab \notin E(G)$. On the other hand, suppose $g(a) = g(b) = s$. Then the $2n$ vertices in $\{a, b\} \times V(K_n) \subseteq V(G \boxtimes K_n)$ are colored with colors from $\{s, s+1, \ldots, \chi(G \times K_n)\}$. There are only $\chi(G \times H) - s + 1 = 2n - 1$ such colors. Thus $ab \notin G$, for otherwise $\{a, b\} \times V(K_n)$ induces a K_{2n} colored with $2n - 1$ colors. □

Theorems 26.5 and the following corollary imply that the lower bound in Theorem 26.9 is sharp. However, Bollobás and Thomason (1979) improved the bound to $\chi(G) + 2n - 1$ in the case where G is uniquely colorable. (Geller (1976) proved this in the case $n = 2$.)

Corollary 26.10 *For any n and k, $\chi(K(n, k) \circ K_k) = n$.*

Proof By Theorems 26.5 and 26.9, we have

$$\chi(K(n, k) \circ K_k) \geq (n - 2k + 2) + 2k - 2 = n\,.$$

On the other hand, we can color $K(n, k) \circ K_k$ as follows: For any $u \in V(K(n, k))$, color the vertices $\{u\} \times K_k$ with the k elements of the set corresponding to u. This clearly is a n-coloring of $K(n, k) \circ K_k$. □

We continue with two more lower bounds. The first was proved in Linial and Vazirani (1989).

Theorem 26.11 *For any graphs G and H,*

$$\chi(G \circ H) \;\geq\; \frac{(\chi(G) - 1)\, \chi(H)}{\ln |V(G)|}\,.$$

Proof Set $\chi(G) = m$ and $\chi(H) = n$. Let $\chi(G \circ H) = k$, and let $C_1, C_2, \ldots, C_k$ be the color classes of some k-coloring.

For each i, the set $X_i = p_G(C_i)$ is independent in G, so $\bigcup_{i \in S} X_i \neq V(G)$ for every index set S with $|S| = m - 1$. Hence $\bigcap_{i \in S} \overline{X_i} \neq \emptyset$ when $|S| = m - 1$. For each index set S with $|S| = m - 1$, choose a representative from $\bigcap_{i \in S} \overline{X_i}$.

Now, for any representative a, the H-layer in $G \circ H$ over a is isomorphic to H, so there are at least n color classes C_i that intersect it. Hence there are *at most* $k - n$ color classes that *do not* intersect this H-layer. Consequently there are $\binom{k-n}{m-1}$ index sets S that could have a as a representative. Now reason as follows.

$$\begin{aligned} |V(G)| &\geq \text{Number of distinct representatives} \\ &\geq \frac{\text{Number of choices of } S}{\text{Maximal number of sets } S \text{ with same representative}} \\ &\geq \binom{k}{m-1}\binom{k-n}{m-1}^{-1} \\ &\geq \left(\frac{k}{k-n}\right)^{m-1} = \left(1-\frac{n}{k}\right)^{-(m-1)} \\ &\geq \exp\left(\frac{(m-1)n}{k}\right). \end{aligned}$$

Taking logarithms, we see that the theorem follows. □

The next lower bound was obtained by Stahl (1976). Our proof is due to Zhu (personal communication). See Exercise 26.11 for yet another approach.

Theorem 26.12 *Let G be a nonbipartite graph. Then for any graph H,*

$$\chi(G \circ H) \geq 2\chi(H) + \left\lceil \frac{\chi(H)}{k} \right\rceil,$$

where $2k+1$ is the length of a shortest odd cycle in G.

Proof Because the chromatic number of a graph is at least as big as the chromatic number of any of its subgraphs, we may assume $G = C_{2k+1}$. Let C_1, C_2, ..., $C_{\chi(G \circ H)}$ be the color classes of a coloring of $G \circ H$. For each $a \in V(G)$, let n_a denote the number of these color classes that have a nonempty intersection with the H-layer above a. Because each H-layer is isomorphic to H, we have

$$n_a \geq \chi(H).$$

Now, the sum $\sum_{a \in V(G)} n_a$ gives a crude counting of the number of color classes C_i. Because each $p_G(C_i)$ is an independent set in $G = C_{2k+1}$, we have $|p_G(C_i)| \leq k$, and it follows that no C_i gets counted more than k times. Therefore

$$k\,\chi(G \circ H) \geq \sum_{a \in V(G)} n_a \geq (2k+1)\chi(H),$$

and the result follows. □

Let us now examine three applications of the above results. We will use them to disprove two conjectures and to answer a question about retracts of strong products. The Kneser graphs $K(n,k)$ play an essential role in two cases.

Geller's conjecture. Geller conjectured that $\chi(G \circ K_k) = nk$ for every uniquely n-colorable graph G. This was disproved by Bollobás and Thomason (1979) with the following subtle construction. Let

$$H = (K(n+2k-1,k) \times K_n) \circ K_k.$$

By Theorem 26.5 we have $\chi(K(n+2k-1,k)) = n+1$, and hence Theorem 26.6 implies that $K(n+2k-1,k) \times K_n$ is uniquely n-colorable. Thus Geller's conjecture would imply $\chi(H) = nk$.

The projection from $K(n+2k-1,k) \times K_n$ onto $K(n+2k-1,k)$ is a homomorphism, so Proposition 26.4 implies that $\chi(H) \le n+2k-1$. But $n+2k-1 < nk$ except for very small n and k, and this disproves Geller's conjecture.

Hajós conjecture. This well-known conjecture had been open for almost thirty years before it was disproved by Catlin in 1979. The counterexample requires the following straightforward corollary to Theorems 26.8 and 26.12.

Corollary 26.13 *For any $k \ge 1$ and any $n \ge 1$, we have*

$$\chi(C_{2k+1} \circ K_n) = 2n + \left\lceil \frac{n}{k} \right\rceil .$$

A *subdivided* K_n is a graph obtained by replacing edges uv of the complete graph K_n by disjoint u,v-paths. In the early 1950s, Hajós posed the following conjecture:

If $\chi(G) = n$, then G contains a subdivided K_n.

This conjecture is stronger than the well-known Hadwiger's conjecture from 1943 that asserts that every n-chromatic graph has a subgraph that is contractible to K_n. Hadwiger's conjecture is trivial for $n=2$ and $n=3$, and Dirac (1952) proved it for $n=4$. The case $n=5$ is equivalent to the Four Color Theorem. Robertson, Seymour, and Thomas (1993) proved it for $n=6$, but the general case is still open.

Hajós's conjecture is also trivially true for $n=2$ and $n=3$. (For the latter case, just observe that an odd cycle is a subdivided K_3.) Dirac (1952) proved it for $n=4$. The general case was disproved by Catlin (1979), whose construction follows.

Consider the lexicographic product $C_{2n+1} \circ K_{n+1}$ for $n \ge 2$. Corollary 26.13 implies $\chi(C_{2n+1} \circ K_{n+1}) = 2n+4$. We claim $C_{2n+1} \circ K_{n+1}$ contains a subdivided K_{2n+3} but no subdivided K_{2n+4}, that is, $C_{2n+1} \circ K_{n+1}$ is a counterexample to Hajós's conjecture.

Let X be a subset of $V(C_{2n+1} \circ K_{n+1})$ with $|X| = 2n+4$. Because $n \ge 2$, at least two vertices of X, say u and v, lie in nonadjacent K_{n+1}-layers. Therefore there exist two K_{n+1}-layers whose union separates u and v. Because u and v can be separated by a vertex set of cardinality $2n+2$, we deduce that X cannot be the original vertices of a subdivided K_{2n+4}. Hence $C_{2n+1} \circ K_{n+1}$ contains no subdivided K_{2n+4}. On the other hand, it's easy to check that two adjacent K_{n+1}-layers together with a vertex adjacent to all vertices of one of the layers give rise to a subdivided K_{2n+3}.

If G is any counterexample to Hajós's conjecture, then the join $G \oplus K_1$ is also a counterexample with $\chi(G \oplus K_1) = \chi(G)+1$. As the smallest counterexample described above is $C_5 \circ K_3$ with chromatic number 8, we now have counterexamples for every $n \ge 8$. Moreover, Catlin observed that a counterexample for $n=7$ is obtained by removing two nonadjacent vertices from $C_5 \circ K_3$.

Hajós's conjecture is still open for $n=5$ and 6 (see Jensen and Toft (1995)). Progress was made by X. Yu and Zickfeld (2006) who proved that a minimal counterexample (with respect to the number of vertices) for $n=5$ must be 4-connected.

Retracts of strong products. Our third application uses Corollary 26.10. By Theorem 15.11, every retract of $G \boxtimes H$ has form $G' \boxtimes H'$, where G' is a weak retract of G, and H' is a weak retract of H. The question arises as to whether G' and H' also must be retracts of G and H. Klavžar and Milutinović (1994) disproved this.

Theorem 26.14 *For every $n \ge 2$ there is an infinite sequence of pairs of graphs G and G' such that G' is not a retract of G, but $G' \boxtimes K_n$ is a retract of $G \boxtimes K_n$.*

Proof For $k \geq 2$, let $H(n,k)$ be the graph obtained from a copy of the Kneser graph $K(2n+k,n)$ and a copy of K_{k+1} by joining a vertex of one graph to a vertex of the other. We can color $H(n,k) \boxtimes K_n$ with $n(k+1)$ colors as follows: By Corollary 26.10, there exists a $(2n+k)$-coloring of $K(2n+k,n) \boxtimes K_n$. Such a coloring can easily be extended to the subgraph $K_{k+1} \boxtimes K_n$ of $H(n,k) \boxtimes K_n$ in such a way that not more than $n(k+1)$ colors are used altogether. It follows that $\chi(H(n,k) \boxtimes K_n) = n(k+1)$. Thus we have a retraction from $H(n,k) \boxtimes K_n$ onto the subgraph $K_{k+1} \boxtimes K_n$. On the other hand, because $\chi(K(2n+k,n)) = k+2$, we infer that $\chi(H(n,k)) = k+2$. Now Proposition 3.10 implies that there is no retraction from $H(n,k)$ to K_{k+1}. □

Theorem 26.14 was first proved by Klavžar (1992) for $n = 2$. The question of characterizing graphs G, H, G', and H', for which $G' \boxtimes H'$ is a retract of $G \boxtimes H$, and where G' is not a retract of G, remains open.

The lexicographic product has often been used in situations similar to the above examples. See, for instance, Thomason (1989) and Shearer and Watkins (1987).

26.3 Fractional and Circular Chromatic Number

The fractional chromatic number was introduced by Hilton, Rado, and Scott (1973; 1975). It is also known as the *multichromatic number*, *set-chromatic number*, and *ultimate chromatic number*. To adequately define it, we must first describe the chromatic number in terms of linear programming.

Given a graph G, let $V(G) = \{v_1, v_2, \ldots, v_n\}$, and let $\mathcal{I}(G) = \{I_1, I_2, \ldots, I_m\}$ be the set of all independent sets of G. The *independence matrix* $I(G)$ of G (relative to this indexing) is the $n \times m$ matrix whose i,j-entry is 1 if $v_i \in I_j$ and 0 otherwise. Let $\mathbf{1}_n$ and $\mathbf{0}_n$ denote the n-dimensional vectors containing only ones and zeros, respectively. Furthermore, let $\mathbf{x} \geq \mathbf{y}$ denote the situation that each entry of the vector $\mathbf{x}$ is greater than or equal to the corresponding entry of the vector $\mathbf{y}$. Then the chromatic number of G is the solution of the following integer linear program:

$$\chi(G) = \min_{\mathbf{x}} \mathbf{1}_m^T \mathbf{x}$$
$$\text{subject to } \mathbf{x} \geq \mathbf{0}_m,\ \mathbf{x} \in \mathbb{Z}^m, \text{ and } I(G)\mathbf{x} \geq \mathbf{1}_n. \quad (1)$$

The condition $I(G)\mathbf{x} \geq \mathbf{1}_n$ requires that every vertex v of G belong to at least one independent set I_i, where the ith component x_i of $\mathbf{x}$ is 1. Hence an optimal solution of (1) is the minimum number of independent sets that cover $V(G)$. This is indeed the chromatic number of G.

The *fractional chromatic number* $\chi_f(G)$ of G is the optimal solution of the linear program (1) in which we relax the condition $\mathbf{x} \in \mathbb{Z}^m$ to $\mathbf{x} \in \mathbb{R}^m$:

$$\chi_f(G) = \min_{\mathbf{x}} \mathbf{1}_m^T \mathbf{x}$$
$$\text{subject to } \mathbf{x} \geq \mathbf{0}_m,\ \mathbf{x} \in \mathbb{R}^m, \text{ and } I(G)\mathbf{x} \geq \mathbf{1}_n. \quad (2)$$

Note that the fractional chromatic number as well as an optimal feasible vector $\mathbf{x}$ are rational. (This follows from the theory of linear programming on regions of feasible solutions using the fact that $I(G)$ is an integer matrix.)

Reinterpreting the above linear program yields an equivalent description of $\chi_f(G)$: A *fractional coloring* of G is a map $f : \mathcal{I}(G) \to [0,1]$ such that for each $v \in V(G)$ we have

$$\sum_{I \in \mathcal{I}(G), v \in I} f(I) \geq 1\,.$$

The *weight* of f is the number $w(f) = \sum_{I \in \mathcal{I}} f(I)$. The fractional chromatic number $\chi_f(G)$ is then the minimum of the weights of all fractional colorings of G. Note that this minimum remains unchanged if we replace the above condition by $\sum_{I \in \mathcal{I}, v \in I} f(I) = 1$.

Theorem 26.15 *For any graph G there exists an m such that*

$$\chi_f(G) = \frac{\chi(G \circ K_m)}{m}.$$

Proof Let c be a p-coloring of $G \circ K_n$, and $C_1, C_2, \ldots, C_p$ be the corresponding color classes. Each projection $p_G(C_i)$ is an independent set of G, though different color classes may project onto the same independent set of G. Let $D_1, D_2, \ldots, D_q$ be the distinct projections of the color classes. For each $i = 1, 2, \ldots, q$, let d_i be the number of different color classes C_j for which $p_G(C_j) = D_i$. Define a map $f : \mathcal{I}(G) \to [0, 1]$ as $f(D_i) = d_i/n$ for $i = 1, 2, \ldots, q$ and $f(D) = 0$ for any other $D \in \mathcal{I}(G)$. It is easy to check that f is a fractional coloring of G, and its weight is p/n. It follows that

$$\chi_f(G) \leq \frac{\chi(G \circ K_n)}{n}.$$

Next, let f be a fractional coloring of G. We may assume that for all $C \in \mathcal{I}(G)$, the denominator of $f(C)$ is fixed, say m. Then, similar to what was done above, we can find an $m\chi_f(G)$-coloring of $G \circ K_m$. Hence

$$\chi_f(G) \geq \frac{\chi(G \circ K_m)}{m},$$

which completes the argument. □

Combining Theorem 26.15 with Proposition 26.4, we get the following corollary:

Corollary 26.16 *For any graph G,*

$$\chi_f(G) = \min\left\{\frac{n}{k} \mid \textit{there is a homomorphism } G \to K(n,k)\right\}.$$

It is interesting to add that $\chi_f(G)$ also equals

$$\lim_{n \to \infty} \sqrt[n]{\chi(G^{\circ, n})};$$

see Hell and F. S. Roberts (1982).

Recall that $\chi(G \circ H) \leq \chi(G)\chi(H)$, and the inequality can be strict. The situation is different for the fractional chromatic number, for equality always holds in that case. Before proving this, we need a new concept. The *fractional clique number* $\omega_f(G)$ of G is the optimal solution of the linear program

$$\begin{aligned} &\omega_f(G) = \max_{\mathbf{y}} \mathbf{1}_n^T \mathbf{y} \\ &\text{subject to } \mathbf{y} \geq \mathbf{0}_n,\ \mathbf{y} \in \mathbb{R}^n, \text{ and } I(G)^T \mathbf{y} \leq \mathbf{1}_m. \end{aligned} \tag{3}$$

(Notice this would give the usual clique number $\omega(G)$ if we imposed the restriction $\mathbf{y} \in \mathbb{Z}^n$.) Because (3) is the dual of the linear program (2), we have $\chi_f(G) = \omega_f(G)$ by the duality theorem of linear programming.

Now we can proceed with the above-mentioned result. We give two proofs; the first is the original from Gao and Zhu (1996), and the second is by Tardif (personal communication).

Theorem 26.17 *For any graphs G and H we have $\chi_f(G \circ H) = \chi_f(G)\,\chi_f(H)$.*

First proof Let f_G and f_H be fractional colorings of G and H, respectively. Define a map $f : \mathcal{I}(G \circ H) \to [0,1]$ as follows: Notice that if I is an independent set of G, and J an independent set of H, then $I \times J$ is an independent set of $G \circ H$. Set $f(I \times J) = f_G(I) f_H(J)$. For any other independent set L of $G \circ H$, set $f(L) = 0$. If $(a,x) \in G \circ H$, then $\sum_{I \in \mathcal{I}(G), a \in I} f_G(I) \geq 1$ and $\sum_{I \in \mathcal{I}(H), x \in I} f_H(J) \geq 1$. Therefore

$$\sum_{I \in \mathcal{I}(G \circ H), (a,x) \in I} f(I) \geq 1\,,$$

so f is a fractional coloring of $G \circ H$. Moreover, notice that $w(f) = w(f_G)\,w(f_H)$, and hence $\chi_f(G \circ H) \leq \chi_f(G)\,\chi_f(H)$.

For the reverse inequality, let g be a fractional clique of G of weight $\omega_f(G)$ and h be a fractional clique of H of weight $\omega_f(H)$. Then it can be easily verified that $\phi(a,x) = g(a)h(x)$ defines a fractional clique of $G \circ H$ of weight $\omega_f(G)\omega_f(H)$. □

Second proof Let $\chi_f(G) = n/k$ and $\chi_f(H) = n'/k'$. By Corollary 26.16, there are homomorphisms $G \to K(n,k)$ and $H \to K(n',k')$. Combining them coordinatewise, we obtain a homomorphism

$$G \circ H \to K(n,k) \circ K(n',k')\,.$$

Together with a natural homomorphism

$$K(n,k) \circ K(n',k') \to K(nn',kk')\,,$$

we infer the existence of a homomorphism from $G \circ H$ into $K(nn',kk')$. This means that $\chi_f(G \circ H) \leq (nn')/(kk')$, and we conclude that $\chi_f(G \circ H) \leq \chi_f(G)\chi_f(H)$.

Conversely, let $\chi_f(G \circ H) = n/k$ so there is a homomorphism $G \circ H \to K(n,k)$. Then

$$n = \chi((G \circ H) \circ K_k) = \chi(G \circ (H \circ K_k)).$$

Set $t = \chi(H \circ K_k)$. Then there are homomorphisms $G \to K(n,t)$ and $H \to K(t,k)$. It follows that

$$\chi_f(G)\chi_f(H) \leq \frac{n}{t} \cdot \frac{t}{k} = \chi_f(G \circ H)\,,$$

and we are done. □

It is interesting to note that Theorem 26.17 implies that $\chi_f(G \circ H) = \chi_f(H \circ G)$, although, as we know, the lexicographic product is not commutative.

We now turn our attention to the circular chromatic number. After defining it and stating some of its properties, we then prove several results that relate it to the lexicographic product, and which can be seen as generalizations of results presented earlier in this chapter. The graphs whose fractional and circular chromatic numbers are equal are then characterized via the lexicographic product.

The circular chromatic number $\chi_c(G)$ of a graph G may be one of the most natural generalizations of the chromatic number. It was introduced by Vince (1988) under the name *star chromatic number* as follows: Let k and d be two integers with $k \geq 2d$. A map $c : V(G) \to \{0, 1, \ldots, k-1\}$ is a *(k,d)-coloring* of G if $d \leq |c(u) - c(v)| \leq k - d$ whenever $uv \in E(G)$. The *circular chromatic number* of G is then

$$\chi_c(G) = \inf \left\{ \frac{k}{d} \mid G \text{ has a } (k,d)\text{-coloring} \right\}.$$

The name stems from a connection with r-circular colorings: Let G be a graph, and C a circle of circumference r. A map f that assigns to every vertex of G an open arc of unit length on C, such that adjacent vertices are mapped to disjoint arcs, is called an *r-circular coloring* of G. A graph G is *r-circular colorable* if it admits an r-circular coloring. The circular chromatic number can also be expressed as

$$\chi_c(G) = \inf\{r \mid G \text{ is } r\text{-circular colorable}\}.$$

See Zhu (1992b, 2001) for a proof that the two definitions are indeed equivalent.

For $k \geq 2d$, let G_k^d be the graph on the vertex set $\{0, 1, \ldots, k-1\}$, where i and j are adjacent if $d \leq |i - j| \leq k - d$. Figure 26.2 shows some examples.

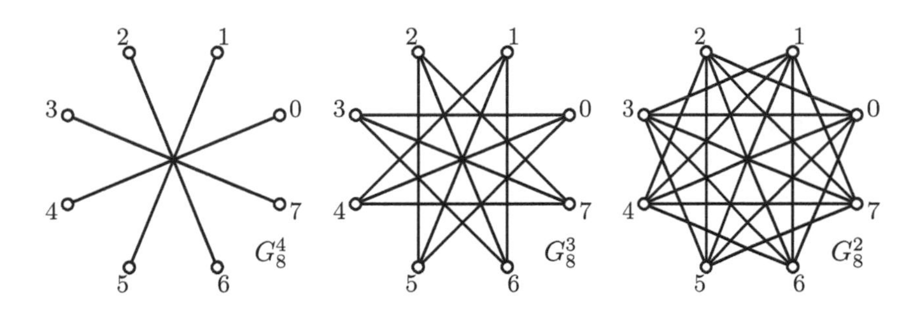

FIGURE 26.2 Graphs G_8^4, G_8^3, and G_8^2.

We now list several basic properties of circular chromatic numbers. For proofs, see the original paper by Vince (1988) or the excellent survey of Zhu (2001). (Note, however, that the upper bound of (ii) follows immediately from the definition, and (iii) is a trivial consequence of (ii). Also, any $\chi(G)$-coloring of G is a $(\chi(G), 1)$-coloring, so $\chi_c(G) \leq \chi(G)$.)

Theorem 26.18 *For any graph G, we have*

(i) $\chi_c(G) = k/d$ *for some k and d with* $2d \leq k \leq n = |V(G)|$.

(ii) $\chi(G) - 1 < \chi_c(G) \leq \chi(G)$.

(iii) $\chi(G) = \lceil \chi_c(G) \rceil$.

(iv) $\chi_c(G_k^d) = k/d$.

Theorem 26.19 *For any graph G, we have* $\chi_f(G) \leq \chi_c(G) \leq \chi(G)$.

Proof By Theorem 26.18 (ii), we only need to show $\chi_f(G) \leq \chi_c(G)$. Let $\chi_c(G) = k/d$, and let c be a (k, d)-coloring of G. For $i = 0, 1, \ldots, k-1$, let I_i be the set of vertices of G colored with colors $\{i, i+1, \ldots, i+d-1\}$ (addition is modulo k). Then each I_i is an independent set of G. Define a map $f : \mathcal{I}(G) \to [0, 1]$ by $f(I_i) = 1/d$ for $i = 0, 1, \ldots, k-1$ and $f(D) = 0$ for any other independent set $D \in \mathcal{I}(G)$. Then f is a fractional coloring of G, and its weight is k/d. Therefore $\chi_f(G) \leq k/d = \chi_c(G)$. □

In view of Theorem 26.19, it is natural to ask which graphs satifsy $\chi_f(G) = \chi_c(G)$. The answer requires some preparation.

Lemma 26.20 *For any graphs G and H we have* $\chi_c(G \circ H) \leq \chi_c(G)\,\chi(H)$.

Proof Let $\chi_c(G) = k/d$, and f_G be a (k, d)-coloring of G. Moreover, let $\chi(H) = n$ and $f_H : V(G) \to \{0, 1, \ldots, n-1\}$ be an n-coloring of H. For any $(a, x) \in V(G \circ H)$, set

$$f(a, x) = f_G(a) + k f_H(x).$$

Then f is a map $V(G \circ H) \to \{0, 1, \ldots, kn-1\}$, and it is easy to see that f is a (kn, d)-coloring of $G \circ H$. Therefore $\chi_c(G \circ H) \le kn/d = \chi_c(G)\chi(H)$. □

Theorem 26.21 *If G is a graph with at least one edge, and $\chi(H) = n$, then*

$$\chi_c(G \circ H) = \chi_c(G \circ K_n).$$

Proof Let c be a (k, d)-coloring of $G \circ K_n$. Then c can be extended to a (k, d)-coloring of $G \circ H$ as follows: Let $V(K_n) = \{1, 2, \ldots, n\}$, and $C_1, C_2, \ldots, C_n$ be the color classes of an n-coloring of H. For a vertex $(a, x) \in V(G \circ H)$ with $x \in C_i$, set $c'(a, x) = c(a, i)$. Clearly, c' is a (k, d)-coloring of $G \circ H$, hence $\chi_c(G \circ H) \le \chi_c(G \circ K_n)$.

Conversely, let c be a (k, d)-coloring of $G \circ H$, and ab be an edge of G. We may assume, without loss of generality, that for some vertex x of H we have $c(a, x) = 0$. To prove that $G \circ K_n$ is (k, d)-colorable, it is enough to show that there are vertices $y_1, y_2, \ldots, y_n$ of H such that $d \le |c(b, y_i) - c(b, y_j)| \le k - d$ for any distinct vertices y_i and y_j.

We define vertices $y_1, y_2, \ldots, y_n$ inductively as follows: Let y_1 be a vertex of H that minimizes $\{c(b, x) \mid x \in V(H)\}$. After y_i has been defined, we select y_{i+1} as a vertex for which the minimum of

$$\{c(b, x) \mid x \in V(H) \text{ and } c(b, x) \ge c(b, y_i) + d\}$$

is attained. Clearly, if $y_1, y_2, \ldots, y_n$ are well defined, then $d \le |c(b, y_i) - c(b, y_j)| \le k - d$ holds. In particular, because (a, x) is adjacent to (b, y_n), we infer $c(b, y_n) \le k - d$ and therefore the condition also holds for y_1 and y_n. Hence we are done if we can show that the y_i's are well defined.

Assume not; that is, for some $i \le n - 1$, we cannot define y_{i+1}. In other words, there is no vertex $y \in V(H)$ with $c(b, y) \ge c(b, y_i) + d$. For any $x \in V(H)$, set $f(x) = j$ if $c(b, y_j) \le c(b, x) < c(b, y_j) + d$. Now f is a map $V(H) \to \{1, 2, \ldots, i\}$. Moreover, if $f(x) = f(x')$, then x and x' cannot be adjacent. It follows that f is an i-coloring of H with $i < n$, a contradiction. □

Note that Theorem 26.21 together with Theorem 26.18 (iii) implies Theorem 26.3. (The case when G is totally disconnected is trivial.)

Combining the above results with Theorem 26.17, we obtain the following theorem:

Theorem 26.22 *If G has at least one edge and $\chi_c(G) = \chi_f(G)$, then $\chi_c(G \circ H) = \chi_c(G)\chi(H)$ for every H.*

Proof Say $\chi(H) = n$, and argue as follows:

$$\begin{array}{rcll} \chi_c(G \circ H) & = & \chi_c(G \circ K_n) & \text{(by Theorem 26.21)} \\ & \le & \chi_c(G)n & \text{(by Lemma 26.20)} \\ & = & \chi_f(G)n & \text{(by assumption)} \\ & = & \chi_f(G)\chi_f(K_n) & \text{(obvious)} \\ & = & \chi_f(G \circ K_n) & \text{(by Theorem 26.17)} \end{array}$$

Theorem 26.19 gives $\chi_f(G \circ K_n) \le \chi_c(G \circ K_n)$, so the above yields $\chi_c(G \circ H) = \chi_c(G)n$. □

Note that because $\chi_c(C_{2k+1}) = \chi_f(C_{2k+1}) = 2 + 1/k$ (Exercise 26.8), Theorem 26.22 implies Corollary 26.13.

It would be interesting to find a general characterization of graphs G for which $\chi_c(G) = \chi_f(G)$. The next theorem describes such graphs in terms of the lexicographic product.

Theorem 26.23 *If a graph G has at least one edge, then the following are equivalent:*

(i) $\chi_f(G) = \chi_c(G)$.
(ii) *For any graph H we have $\chi_c(G \circ H) = \chi_c(G)\chi(H)$.*

Proof Theorem 26.22 asserts that (i) implies (ii). Conversely, suppose $\chi_c(G \circ H) = \chi_c(G)\chi(H)$ for every H. Recall (Theorem 26.15) that $\chi_f(G) = \chi(G \circ K_m)/m$ for some m. Combining this with Theorem 26.19 gives

$$\chi_c(G \circ K_m) \leq \chi(G \circ K_m) = \chi_f(G)m \leq \chi_c(G)m = \chi_c(G)\chi(K_m).$$

By assumption, the left- and right-hand terms are equal, hence $\chi_f(G) = \chi_c(G)$. □

Theorem 26.21 is by Zhu (1992b), while Theorems 26.17, 26.22, and 26.23 are from Gao and Zhu (1996).

Klavžar and H.-G. Yeh (2002) observed that $\chi(G) = \chi_f(G)$ if and only if $\chi(G \circ H) = \chi(G)\chi(H)$ for all H. (cf. Exercise 26.10.) See Zhu (2001) for a discussion of the case $\chi(G) = \chi_c(G)$.

As Kneser graphs have played such a prominent role in this chapter, it is fitting to add several remarks about their circular chromatic number. Johnson, Holroyd, and Stahl (1997) conjectured that the circular chromatic number of a Kneser graph equals its chromatic number. (See Theorem 26.5.) They proved this conjecture for $K(2n+1, n)$ and $K(2n+2, n)$ for $n \geq 1$, and $K(n, 2)$ for $n \geq 4$, while Hajiabolhassan and Zhu (2003) proved it for every $n \geq 2k^2(k-1)$. Meunier (2005), and independently Simonyi and Tardos (2006), used topological tools to prove it for every even n. Finally, P.-A. Chen (2011) closed the story by establishing the following result.

Theorem 26.24 *If $n \geq 2k \geq 2$, then $\chi_c(K(n,k)) = \chi(K(n,k)) = n - 2k + 2$.* □

26.4 Hedetniemi's Conjecture

We now examine the chromatic number in the context of direct products of graphs. Here there is a clear and immediate upper bound. Because projections $G \times H \to G$ and $G \times H \to H$ are homomorphisms, Proposition 2.10 implies $\chi(G \times H) \leq \min\{\chi(G), \chi(H)\}$. Indeed, given an n-coloring g of G, the map $f(a, x) = g(a)$ is an n-coloring of $G \times H$, as illustrated schematically in Figure 26.3. Notice that the bound $\chi(G \times H) \leq \min\{\chi(G), \chi(H)\}$ holds even in the class Γ_0, for $G \times H$ has a loop if and only if both G and H do.

This simple bound led Hedetniemi (1966) to make the following conjecture:

Conjecture 26.25 *For any graphs G and H,*

$$\chi(G \times H) = \min\{\chi(G), \chi(H)\}.$$

The conjecture is very natural, and is a perfect companion to Theorem 26.1. However, it has withstood all attempts at proof. It is the most challenging problem in product coloring; only special cases have been solved affirmatively.

We begin with several equivalent formulations of the conjecture. One of them uses the concept of a coloring graph, an important tool in many approaches to Hedetniemi's conjecture. We will use coloring graphs in this section, and also in the next section, to prove Hedetniemi's conjecture for 4-chromatic graphs. Coloring graphs have also been used by Lovász (1967) and Vesztergombi (1978).

FIGURE 26.3 Canonical coloring of a direct product.

If G is a graph and n is a positive integer, then the *n-coloring graph* $\mathcal{C}_n(G)$ *of* G has vertex set

$$V(\mathcal{C}_n(G)) = \{ f \mid f : V(G) \to \{1, 2, \ldots, n\} \},$$

where functions f and g are adjacent if $f(u) \neq g(v)$ for every $uv \in E(G)$. (Equivalently, we may view $\mathcal{C}_n(G)$ as the set of maps $G \to K_n$, where $fg \in E(\mathcal{C}_n(G))$ provided $f(u)g(v) \in E(K_n)$ whenever $uv \in V(G)$.) Notice that ff is a loop in $\mathcal{C}_n(G)$ if and only if f is a (proper) n-coloring of G. For example, $\mathcal{C}_2(P_3)$ consists of an edge, two vertices with a loop, and four isolated vertices. The vertices with a loop are the two 2-colorings of P_3, and the endpoints of the edge are the two constant maps. Figure 26.4 shows two more examples.

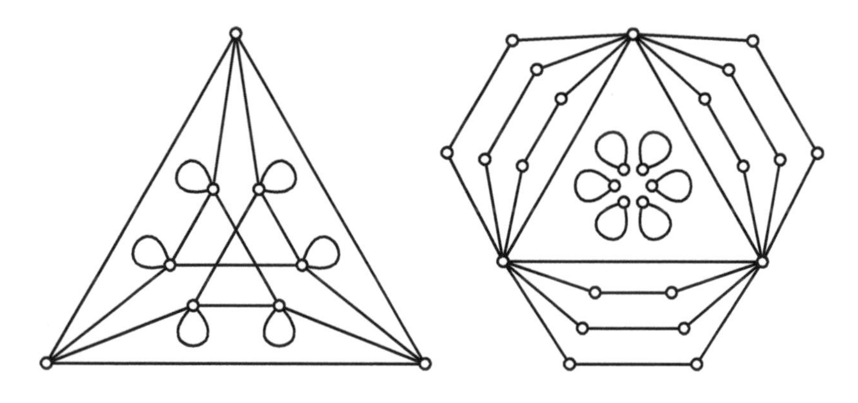

FIGURE 26.4 Coloring graphs $\mathcal{C}_3(K_2)$ and $\mathcal{C}_3(K_3)$.

From Figure 26.4 we see that $\mathcal{C}_3(K_2)$ and $\mathcal{C}_3(K_3)$ each contains exactly one triangle. The vertices of these triangles are the constant maps. More generally, the constant maps form a K_n in $\mathcal{C}_n(G)$, and hence any graph G obeys $\chi(\mathcal{C}_n(G)) \geq n$. We summarize these remarks as a lemma.

Lemma 26.26 *For any graph G,*

(i) $\chi(\mathcal{C}_n(G)) \geq n$ *and*
(ii) $\mathcal{C}_n(G)$ *has loops if and only if* $\chi(G) \leq n$.

The remainder of this section uses the following useful notation: If f is an n-coloring of $G \times H$, then for $x \in V(H)$ there is a function

$$f_x : V(G) \to \{1, \ldots, n\}$$

defined as

$$f_x(a) = f(a, x).$$

Notice that $f_x \in V(\mathcal{C}_n(G))$, and the map $x \mapsto f_x$ is a homomorphism $H \to \mathcal{C}_n(G)$. For $a \in V(G)$ we define $f_a : V(H) \to \{1, \ldots, n\}$ analogously, and $a \mapsto f_a$ is a homomorphism $G \to \mathcal{C}_n(H)$.

Theorem 26.27 *The following statements are equivalent:*

(i) $\chi(G \times H) = \min\{\chi(G), \chi(H)\}$ *for all graphs G and H.*
(ii) $\chi(G \times H) = \min\{\chi(G), \chi(H)\}$ *for all graphs G and H with $\chi(G) = \chi(H)$.*
(iii) *For all graphs G, H, and $n \geq 0$, if $\chi(G) > n$ and $\chi(H) > n$, then $\chi(G \times H) > n$.*
(iv) *For any $n \geq 0$, and any graph G, $\chi(G) > n$ implies $\chi(\mathcal{C}_n(G)) = n$.*

Proof The implication (i) $\Rightarrow$ (ii) is trivial. Because every graph G with $\chi(G) > n$ contains an $(n+1)$-chromatic subgraph, (ii) implies (iii).

(iii) $\Rightarrow$ (iv). Suppose (iv) is false, so there is a graph G with $\chi(G) > n$ and $\chi(\mathcal{C}_n(G)) \neq n$. Then $\chi(\mathcal{C}_n(G)) > n$, by Lemma 26.26 (i), and $\mathcal{C}_n(G)$ has no loops, by Lemma 26.26 (ii). Define a coloring of $G \times \mathcal{C}_n(G)$ as follows. Give each vertex (a, f) of $G \times \mathcal{C}_n(G)$ the color $f(a)$. This is an n-coloring of $G \times \mathcal{C}_n(G)$, so (iii) is false.

(iv) $\Rightarrow$ (i). Suppose (i) is false. Then there exist graphs G and H for which $\chi(G) \leq \chi(H)$ but $\chi(G \times H) = n < \chi(G)$. Let f be a n-coloring of $G \times H$. It follows that the map $x \mapsto f_x$ is a homomorphism $H \to \mathcal{C}_n(G)$. Then $n < \chi(G) \leq \chi(H) \leq \chi(\mathcal{C}_n(G))$, so (iv) is false. □

Condition (iv) above indicates the important connection between Hedetniemi's conjecture and coloring graphs. The following equivalent formulation of the conjecture, due to Larose and Tardif (2000), is of a different nature.

Proposition 26.28 *Hedetniemi's conjecture is equivalent to the following statement: For any graphs G and H and any $n \geq 1$, if K_n is a retract of $G \times H$, then K_n is a retract of G or K_n is a retract of H.*

Proof Suppose that Hedetniemi's conjecture holds. Let K_n be a retract of $G \times H$. Then $\chi(G \times H) = \chi(K_n) = n$ by Proposition 3.10. Therefore $\chi(G) = n$ or $\chi(H) = n$ by our assumption. On the other hand, because $G \times H$ contains K_n as a subgraph, this subgraph projects to a K_n in G as well as to a K_n in H (Exercise 26.1). We conclude (using Exercise 3.7) that K_n is a retract of G or K_n is a retract of H.

Conversely, suppose that the condition of the statement holds. Let G and H be arbitrary graphs and let $\chi(G \times H) = n$. Now consider the graph

$$X = (G + K_n) \times (H + K_n) = (G \times H) + (G \times K_n) + (K_n \times H) + (K_n \times K_n).$$

Because $\chi(G \times H) = n$ and any of the graphs $G \times K_n$, $K_n \times H$, and $K_n \times K_n$ is n-colorable, we have $\chi(X) = n$. Moreover, $K_n \times K_n$ contains K_n as a subgraphs. Hence (using Exercise 3.7 again) K_n is a retract of $X = (G + K_n) \times (H + K_n)$. Because we have assumed the condition of the statement, K_n is a retract of $G + K_n$ or K_n is a retract of $H + K_n$. It follows that $\chi(G) \leq n$ or $\chi(H) \leq n$ and thus

$$\min\{\chi(G), \chi(H)\} \leq n = \chi(G \times H).$$

As $\chi(G \times H) \leq \min\{\chi(G), \chi(H)\}$ always holds, the argument is complete. □

We next examine two conditions under which Hedetniemi's conjecture holds, but a preliminary lemma is needed. In Lemma 26.26 we noted the constant maps form a complete subgraph of $\mathcal{C}_n(G)$. The next lemma asserts that this is the only K_n in $\mathcal{C}_n(G)$ when G is connected and $(n+1)$-chromatic.

Lemma 26.29 *If G is connected and $(n+1)$-chromatic, then $\mathcal{C}_n(G)$ contains a unique complete subgraph on n vertices (namely the one induced by the constant maps).*

Proof Suppose G is connected and $(n+1)$-chromatic, and f_1, f_2, ..., f_n are vertices of a complete subgraph of $\mathcal{C}_n(G)$.

Our strategy is to show now G has a vertex u for which the values $f_1(u), f_2(u), \ldots, f_n(u)$ are pairwise different. Suppose for a moment that this is true. Then take an edge $uw \in E(G)$ and observe that any $f_i(w)$ must equal some $f_j(u)$. But then $i = j$, for otherwise f_i and f_j are not adjacent. Thus $f_i(w) = f_i(u)$ for each i, and the values $f_1(w), f_2(w), \ldots, f_n(w)$ are pairwise different. As G is connected, we get $f_i(x) = f_i(u)$ for each i and any $x \in V(G)$, so each f_i is constant.

It remains to produce some $u \in V(G)$ for which $f_1(u), f_2(u), \ldots, f_n(u)$ are pairwise different. Let H be a χ-critical subgraph of G with $\chi(H) = n+1$, and take any $u \in V(H)$. Suppose the $f_1(u), f_2(u), \ldots, f_n(u)$ are *not* pairwise different, and, without loss of generality, say $f_1(u) = f_2(u)$. Because $\chi(H-u) = n$, the vertices of $H-u$ can be partitioned into n independent sets X_1, X_2, ..., X_n. Define a map $c : V(H) \to \{1, 2, \ldots, n\}$ as

$$c(x) = \begin{cases} f_1(x) & \text{if } x = u\,, \\ f_i(x) & \text{if } x \in X_i\,. \end{cases}$$

Given $uw \in E(H)$, if $w \in X_1$, we have $f_1(u) = f_2(u) \neq f_1(w)$, which means $c(u) \neq c(v)$. If $w \in X_j$ for $j \neq 1$, then $c(u) \neq c(w)$ because f_1 is adjacent to f_j. Similarly, $c(v) \neq c(w)$ for any other $vw \in E(H)$. Thus c is an n-coloring of H, a contradiction. We conclude that the values $f_i(u)$ are pairwise different. □

The following two theorems demonstrate the utility of coloring graphs. The first result was obtained by Welzl (1984) and independently by Duffus, Sands, and Woodrow (1985).

Theorem 26.30 *If G and H are connected, $(n+1)$-chromatic graphs that both contain a complete subgraph on n vertices, then $\chi(G \times H) = n+1$.*

Proof If G and H are as stated, then $G \times H$ has a K_n as a subgraph, so $n \leq \chi(G \times H) \leq n+1$. Suppose f is an n-coloring of $G \times H$. Let a_1, a_2, ..., a_n be the vertices of a K_n in G. By Lemma 26.26 (ii), the graph $\mathcal{C}_n(H)$ has no loops, and hence maps f_{a_1}, f_{a_2}, ..., f_{a_n} are pairwise distinct and form a complete subgraph of $\mathcal{C}_n(H)$. By Lemma 26.29, these are constant maps. Likewise, if x_1, x_2, ..., x_n are the vertices of a K_n in H, then f_{x_1}, f_{x_2}, ..., f_{x_n} are constant. Then

$$f(a_1, x_1) = f_{a_1}(x_1) = f_{a_1}(x_2) = f_{x_2}(a_1) = f_{x_2}(a_2) = f(a_2, x_2),$$

contrary to f being a coloring. Thus $G \times H$ has no n-coloring, so $\chi(G \times H) = n+1$. □

Hedetniemi's conjecture is thus true for $(n+1)$-chromatic factors that contain complete subgraphs on n vertices. The next theorem, due to Burr, Erdős, and Lovász (1976), places a condition on only one factor, but it is stronger than the one of Theorem 26.30.

Theorem 26.31 *If G and H are $(n+1)$-chromatic graphs and each vertex of G is contained in a complete subgraph on n vertices, then $\chi(G \times H) = n+1$.*

Proof Suppose $\chi(G \times H) < n+1$. Let f be an n-coloring of $G \times H$. Now, the map $a \mapsto f_a$ is a homomorphism $G \to \mathcal{C}_n(H)$. As $\mathcal{C}_n(H)$ has no loops (Lemma 26.26 (ii)), complete subgraphs of G map onto complete subgraphs of $\mathcal{C}_n(G)$. By Lemma 26.29, G is mapped onto the unique complete subgraph on n vertices in $\mathcal{C}_n(G)$, contrary to $\chi(G) = n+1$. □

We next briefly discuss two different approaches to Hedetniemi's conjecture. Duffus,

Sands, and Woodrow (1985) proposed a method that is based on a result of Hajós (1961). The approach uses a new concept, the so-called multiplicativity property. In both cases, (generalized) coloring graphs play an important role.

The *Hajós sum* of graphs G and H, with respect to edges $ab \in E(G)$ and $xy \in E(H)$, is the graph obtained from the disjoint union of G and H by identifying the vertices a and x, removing the edges ab and xy, and adding the edge by. Note that although G and H are undirected graphs, the order of vertices in the edges ab and xy is important in the definition. Hajós (1961) proved that every graph G with $\chi(G) > n$ can be constructed from copies of K_{n+1} by the following three operations: Hajós sum, addition of vertices and edges, and contraction of nonadjacent vertices. For brevity, let us refer to these as "the three operations." It is easy to see that none of the three operations decreases the chromatic number.

Fix an integer n. Call a graph G with $\chi(G) > n$ *persistent* if $\chi(G \times H) = n + 1$ for any graph H with $\chi(H) = n + 1$. Clearly, Hedetniemi's conjecture (for fixed n) is equivalent to the assertion that every graph G with $\chi(G) > n$ is persistent. Furthermore, because every graph G with $\chi(G) > n$ can be constructed from copies of K_{n+1} by the three operations, it is enough to prove that each graph constructed in this manner is persistent. Duffus, Sands, and Woodrow (1985) obtained the following result.

Theorem 26.32 *Let G be constructed from copies of K_{n+1} by executing the three operations in such a way that contractions of nonadjacent vertices are executed after all other operations. Then G is persistent.* □

Call a graph *strongly persistent* if it is persistent and its Hajós sum with any other persistent graph is again persistent. The following (highly nontrivial) result is due to Sauer and Zhu (1992).

Theorem 26.33 *Let G be constructed from copies of K_{n+1} by executing the three operations with at most one contraction. Then G is strongly persistent. Furthermore, the Hajós sum of two strongly persistent graphs is strongly persistent.* □

It is easy to see that a persistent graph remains persistent after contraction of two nonadjacent edges. Thus Theorem 26.33 extends Theorem 26.32 such that in addition to the given operations, we may perform one contraction at any time.

The second approach to the conjecture leads to a less optimistic perception. Häggkvist, Hell, Miller, and Neumann-Lara (1988) claim that it seems conceivable that the conjecture may be false for large n. Let us write $G \not\to H$ if there exists no homomorphism $G \to H$. A graph G is called *multiplicative* if $G_1 \not\to G$ and $G_2 \not\to G$ imply $G_1 \times G_2 \not\to G$ for all graphs G_1 and G_2. Hedetniemi's conjecture is equivalent to the assertion that complete graphs are multiplicative. This approach to the conjecture stems from Häggkvist et al. (1988), although the concept of multiplicativity is not new; see, for example, Nešetřil and Pultr (1978). Häggkvist, Hell, Miller, and Neumann-Lara proved the following theorem:

Theorem 26.34 *Each cycle C_n is multiplicative.* □

Finally, we mention another result that suggests the conjecture may be false for graphs with large chromatic number. Poljak and Rödl (1981) introduced the function

$$f(n) = \min\{\chi(G \times H) \mid \chi(G) = \chi(H) = n\}.$$

It is surprising that it is not even known whether $f(n)$ tends to infinity for $n \to \infty$. Poljak and Rödl (1981) showed that if f is bounded, then $f(n) \leq 16$ for all n. This result can be improved to the following theorem.

Theorem 26.35 *Let n be a natural number. Then the minimum chromatic number of the direct product of two n-chromatic graphs is either bounded by 9 or tends to infinity.* □

Theorem 26.35, an improvement from 16 to 9, was discovered independently by Poljak, Schmerl, and Zhu. For a proof, see Poljak (1991).

Multiplicativity is also a natural concept for digraphs. It was introduced and studied under the name *productivity* by Nešetřil and Pultr (1978). They prove that all directed paths are multiplicative, and that a directed cycle $\vec{C}_n$ is multiplicative for prime n. It was also conjectured that $\vec{C}_n$ is multiplicative if n is a power of a prime. This conjecture has been confirmed by Häggkvist, Hell, Miller, and Neumann-Lara (1988). Their proof uses homotopy theory, while a simple combinatorial proof is due to Zhu (1992a). Zhou (1991a,b) obtained new classes of multiplicative and nonmultiplicative digraphs. A complete classification of multiplicative directed cycles is due to Hell, Zhou, and Zhu (1994). New classes of multiplicative digraphs have also been obtained by Zhou and Zhu (1997).

Two extensive surveys on Hedetniemi's conjecture were written by Zhu (1998) and Sauer (2001). The latter survey in particular gives an emphasis on the ordered sets that arise in the context. Another, more recent survey due to Tardif (2008) is shorter but nevertheless very informative and gives a clear picture of the state of the art.

26.5 Hedetniemi's Conjecture for 4-Chromatic Graphs

In this section we prove that Hedetniemi's conjecture holds for 4-chromatic graphs, that is, we will show that if G is 4-chromatic and H is arbitrary, then $\chi(G\times H) = \min\{\chi(G), \chi(H)\}$. The proof is due to El-Zahar and Sauer (1985), and in a sense it presents the strongest support yet for the conjecture.

Note first that the conjecture is true if G or H is 2-chromatic (i.e., bipartite), for then $G \times H$ is also bipartite. It also holds if one factor is 3-chromatic: Indeed, if G and H both have odd cycles, then so does $G \times H$, so its chromatic number is at least 3.

Therefore, to prove the conjecture is true when G is 4-chromatic, it suffices to show that the condition $4 = \chi(G) \leq \chi(H)$ implies $\chi(G \times H) = 4$. For this, the following lemma is key.

Lemma 26.36 *If $n = \chi(G) \leq \chi(H)$, and $\chi(\mathcal{C}_{n-1}(G)) = n - 1$, then $\chi(G \times H) = \chi(G)$.*

Proof Suppose $n = \chi(G) \leq \chi(H)$, and $\chi(\mathcal{C}_{n-1}(G)) = n - 1$, but $\chi(G \times H) < \chi(G)$. Let f be an $(n-1)$-coloring of $G \times H$. Then $x \mapsto f_x$ is a homomorphism $H \to \chi(\mathcal{C}_{n-1}(G))$, so $\chi(G) \leq \chi(H) \leq \chi(\mathcal{C}_{n-1}(G)) < \chi(G)$, a contradiction. □

Our proof of Hedetniemi's conjecture for 4-chromatic graphs uses Lemma 26.36. We will prove that $\chi(G) = 4$ implies $\chi(\mathcal{C}_3(G)) = 3$. The coloring graph $\mathcal{C}_3(G)$ will enable us to reduce certain arguments to odd cycles. Although this is a splendid idea, it will be clear that it can hardly be generalized to graphs that are more than 4-chromatic.

Given a cycle C_n with vertices $v_0, v_1, \ldots, v_{n-1}$, we will always understand that its edges are v_iv_{i+1}, where the addition is computed modulo n.

Consider the 3-coloring graph $\mathcal{C}_3(C)$ of a cycle C, and let $f \in V(\mathcal{C}_3(C))$. A vertex v_i of C is a *fixed vertex* for f, if $f(v_{i-1}) \neq f(v_{i+1})$. We say f is *odd/even* if it has an odd/even number of fixed vertices. We need two lemmas concerning parity of odd cycles.

Lemma 26.37 *Let $fg \in E(\mathcal{C}_3(C_{2k+1}))$. Then f and g have the same parity.*

Proof The direct product $K_2 \times C_{2k+1}$ is isomorphic to the even cycle C_{4k+2}. Because $fg \in E(\mathcal{C}_3(C_{2k+1}))$, we can (properly) 3-color $K_2 \times C_{2k+1}$ by applying f and g to the two C_{2k+1}-layers, respectively. Denote this coloring by h, and note that the number of fixed vertices for h is the sum of the numbers of fixed vertices for f and g. To complete the proof, it suffices to show that h is even. We prove this for 3-colorings of even cycles by induction on length.

Suppose h is a 3-coloring of the even cycle C_{2n}. If $n = 2$, then h is clearly even. Suppose now that $n \geq 3$. If all vertices are fixed, we are done. We may thus assume, without loss of generality, that vertex v_1 is not fixed. We may also assume that $h(v_1) = 1$ and $h(v_2) = h(v_0) = 2$. Let C_{2n-2} be the cycle that we get from C_{2n} by deleting v_1 and identifying v_0 with v_2. Clearly, C_{2n-2} is properly colored by the restriction of h. Note also that if $h(v_{2n}) = h(v_3) = 3$, then the number of fixed vertices decreases by two. In all the other cases, the number is unchanged. Induction completes the argument. □

By Lemma 26.37, all vertices of a connected component of $\mathcal{C}_3(C_{2k+1})$ have the same parity, which is called the *parity of a component* of $\mathcal{C}_3(C_{2k+1})$. Recall that for any n-coloring f of $G \times H$ and any $x \in V(H)$, there is a vertex $f_x \in \mathcal{C}_n(G)$ defined as $f_x(a) = f(a, x)$, and $x \mapsto f_x$ is a homomorphism $H \to \mathcal{C}_n(G)$. (Similar remarks hold for any $a \in V(G)$.)

Lemma 26.38 *If f is a proper 3-coloring of $C_{2k+1} \times C_{2\ell+1}$ and $(a, x) \in V(C_{2k+1} \times C_{2\ell+1})$, then f_a and f_x have opposite parity.*

Proof Put $V(C_{2k+1}) = \{v_0, v_1, \ldots, v_{2k}\}$, and $V(C_{2\ell+1}) = \{u_0, u_1, \ldots, u_{2\ell}\}$. Set $n = 2k+1$ and $m = 2\ell + 1$.

For any i and j, let M_i and N_j be the number of fixed vertices in the induced colorings f_{v_i} and f_{u_i}, respectively. Because the map $v_i \mapsto f_{v_i}$ is a homomorphism $C_n \to \mathcal{C}_3(C_m)$, all colorings f_{v_i} lie in the same component of $\mathcal{C}_3(C_m)$, and thus, by Lemma 26.37, they all have the same parity. Similarly, all the N_j's have the same parity. In order to show that the M_i's and M_j's have different parity, it is enough to show that the following number is even:

$$a = nm - \sum_{i=1}^{n} M_i - \sum_{j=1}^{m} N_j\,.$$

To this end we consider 4-cycles of $C_n \times C_m$. Let Q_{ij} denote the 4-cycle induced by the vertices (v_{i-1}, u_j), (v_i, u_{j+1}), (v_{i+1}, u_j), (v_i, u_{j-1}), namely, the 4-cycle "centered" at (v_i, u_j). Observe that there are nm such 4-cycles.

Because f is a 3-coloring, each Q_{ij} is colored with two or three colors. If Q_{ij} is colored with three colors, then exactly one of v_i or u_j is a fixed vertex of f_{u_j} or f_{v_i}, respectively. If Q_{ij} is colored with two colors, then neither v_i nor u_j is a fixed vertex of f_{u_j} or f_{v_i}, respectively. It follows that $\sum_{i=1}^{n} M_i + \sum_{j=1}^{m} N_j$ equals the total number of Q_{ij} that are colored with three colors. Hence a is the number of Q_{ij} that are colored with two colors. To complete the proof, we show that the number of 2-colored Q_{ij} is even.

Direct the edges of $C_n \times C_m$ so that arrows go from color 1 to 2, from 2 to 3, and from 3 to 1. Simple checking shows that the arrows of a Q_{ij} are directed "in parallel" if 3 colors are used, and in opposite directions otherwise. Consider the following sequence of 4-cycles:

$$Q_{00}, Q_{11}, \ldots, Q_{i-1,j-1}, Q_{i,j}, Q_{i+1,j+1} \cdots, Q_{00}\,.$$

For instance, for $n = m = 5$, the sequence is Q_{00}, Q_{11}, Q_{22}, Q_{33}, Q_{44}, Q_{00}. For $n = 3$ and $m = 5$, it is Q_{00}, Q_{11}, Q_{22}, Q_{03}, $\ldots$, Q_{24}, Q_{00} and lists all 4-cycles. In general we can partition the set of all Q_{ij} into $\gcd(n, m)$ disjoint sequences of this type. Moreover, any two consecutive 4-cycles in such a sequence share an edge. The number of 2-colored 4-cycles in a sequence equals the number of orientation reversals of common edges as we traverse the sequence. This number must be even, because first and last common edges are the same. □

We continue the proof with two lemmas for 4- and 3-chromatic graphs.

Lemma 26.39 *If $\chi(G) = 4$, and the restriction of some $f \in V(\mathcal{C}_3(G))$ to each odd cycle of G has odd parity, then f is an isolated vertex of $\mathcal{C}_3(G)$.*

Proof To the contrary, suppose G and f are as stated, but assume fg is an edge of $\mathcal{C}_3(G)$.

Let H be the subgraph of G induced on the vertices u that have a neighbor v with $f(v) = f(u)$. We claim that H has an odd cycle. Indeed, if H were bipartite, with partite sets V_1 and V_2, we could define a map $c : G \to \{1, 2, 3\}$ as

$$c(u) = \begin{cases} f(u) & \text{if } u \in V(G) \setminus V_1\,, \\ g(u) & \text{if } u \in V_1\,. \end{cases}$$

Using $fg \in E(\mathcal{C}_3(G))$, it is easy to verify that c is a 3-coloring of G, a contradiction.

We now know H has an odd cycle C. Notice that no three consecutive vertices $v_{i-1}v_iv_{i+1}$ are assigned pairwise different colors by f. If this were the case, then from $g(v_i) \neq f(v_{i-1})$ and $g(v_i) \neq f(v_{i+1})$ we would deduce $g(v_i) = f(v_i)$. By definition, H would have an edge v_iw, not on C, with $f(v_i) = f(w)$. But then $g(v_i) = f(w)$, violating $fg \in E(\mathcal{C}_3(G))$.

Now partition $V(C)$ into maximal monochromatic intervals $v_i, v_{i+1}, \ldots, v_j$. Notice that the fixed vertices of f on C are precisely the endpoints of the monochromatic intervals of length greater than 1. There are an even number of such endpoints, so f has an even number of fixed vertices on C, so its restriction to C has even parity. This contradiction proves the lemma. □

It is interesting to add that it is not known whether there is a 4-chromatic G and a $f \in V(\mathcal{C}_3(G))$ whose restriction to every odd cycle of G is odd; see El-Zahar and Sauer (1985).

Lemma 26.40 *Every component of $\mathcal{C}_3(C_{2k+1})$ with even parity is at most 3-chromatic.*

Proof For the sake of contradiction, suppose $\mathcal{C}_3(C_{2k+1})$ has a component T with even parity, and with $\chi(T) > 3$. Then T has a connected 4-chromatic subgraph H. Certainly, the graph $C_{2k+1} \times H$ is 3-chromatic, for the map $c(v, h) = h(v)$ is a 3-coloring of it. Moreover, for any $h \in V(H)$, we have $c_h(v) = c(v, h) = h(v)$. Hence $c_h = h \in V(T)$ is even.

By Lemma 26.38, any c_v has odd parity on each odd cycle of H. By Lemma 26.39, c_v is an isolated vertex of $\mathcal{C}_3(H)$. Consider an edge $vw \in E(C_{2k+1})$. For any $fg \in E(\mathcal{C}_3(H))$, we have $f(v) \neq g(w)$, and so $c_v(f) \neq c_w(g)$, which means $c_vc_w \in E(\mathcal{C}_3(H))$, a contradiction. □

We have reached the main result of this section. Combined with Lemma 26.36, it proves Hedetniemi's conjecture for 4-chromatic graphs.

Theorem 26.41 *If G is 4-chromatic, then $\mathcal{C}_3(G)$ is 3-chromatic.*

Proof Suppose, to the contrary, that $\chi(G) = 4$, but $\chi(\mathcal{C}_3(G)) > 3$. Then $\mathcal{C}_3(G)$ has a 4-chromatic connected subgraph H. Let $h \in V(H)$. By Lemma 26.39, G has an odd cycle C for which the restriction of h to C is even. Let $\psi : H \to \mathcal{C}_3(C)$ be the map sending each $h \in V(H)$ to its restriction to C. Clearly, ψ is a homomorphism. Moreover, ψ maps H into a component T of $\mathcal{C}_3(C)$ with even parity. Thus, $\chi(T) \geq \chi(H)$, contrary to Lemma 26.40. □

26.6 Circular and Fractional Version of Hedetniemi's Conjecture

We conclude the chapter with a discussion of the circular and the fractional version of Hedetniemi's conjecture. Because $\chi(G) = \lceil \chi_c(G) \rceil$, the following conjecture of Zhu (1992b) generalizes Hedetniemi's conjecture.

Conjecture 26.42 *For any graphs G and H,*

$$\chi_c(G \times H) = \min\{\chi_c(G), \chi_c(H)\}.$$

The next result by Zhu lends credibility to this conjecture.

Theorem 26.43 *Let G and H be graphs, and $k \geq 1$. If $\min\{\chi_c(G), \chi_c(H)\} > 2+1/k$, then $\chi_c(G \times H) > 2 + 1/k$.*

Proof A graph G has a (k, d)-coloring if and only if there is a homomorphism $G \to G_k^d$ (Exercise 26.9). Thus, by Theorem 26.18 (iv), we infer that $\chi_c(G) \leq k/d$ if and only if there is a homomorphism $G \to G_k^d$. For our situation this means that neither G nor H admits a homomorphism into G_{2k+1}^k. Note now that $G_{2k+1}^k = C_{2k+1}$. Theorem 26.34 completes the argument. □

In support of Conjecture 26.42, Larose and Tardif (2002) and Tardif (2005b) prove results analogous to Theorems 26.30 and 26.41. The latter result asserts that Conjecture 26.42 holds for all graphs G and H with $\min\{\chi_c(G), \chi_c(H)\} \leq 4$ and is the strongest result so far on the circular version of Hedetniemi's conjecture.

Consider now the factional version of Hedetniemi's conjecture. As with Hedetniemi's conjecture, the inequality $\chi_f(G \times H) \leq \min\{\chi_f(G), \chi_f(H)\}$ is easy to establish (Exercise 26.17). Zhu (2002) asked if equality holds for all G and H. The same paper answered the question affirmatively for the case when one factor is a circulant graph or a Kneser graph. Tardif (2005a) proved that

$$\chi_f(G \times H) \geq \frac{1}{4} \min\{\chi_f(G), \chi_f(H)\}$$

for all graphs G and H.[1] Moreover, because $\chi_f(G) = \frac{|V(G)|}{\alpha(G)}$ for any vertex-transitive graph, Theorem 27.15 answers Zhu's question affirmatively for vertex-transitive graphs. These efforts culminated in the following theorem from Zhu (2011):

Theorem 26.44 *For any graphs G and H,*

$$\chi_f(G \times H) = \min\{\chi_f(G), \chi_f(H)\}.$$

Before Zhu proved this, Paul and Tardif (2011) observed that it implies the Burr-Erdős-Lovász conjecture concerning chromatic Ramsey numbers posed in Burr, Erdős, and Lovász (1976). We do not go into details here, but the interested reader can consult Paul and Tardif (2011) or Zhu (2011).

The remainder of the section is devoted to proving Theorem 26.44.

As noted on p. 325, $\chi_f(G)$ is the minimum of the weights of all fractional colorings of G, where a fractional coloring is a map $f : \mathcal{I}(G) \to [0, 1]$ such that for each $v \in V(G)$,

[1] In fact, he proved a stronger result: $\chi_f(G \times H) \geq \min\{\chi_f(G), \chi_f(H)\}/4$ for any **digraphs** G and H.

$\sum_{I\in\mathcal{I}(G),v\in I} f(I)\geq 1$. (Recall that $\mathcal{I}(G)$ is the set of all independent sets of G.) Recall also (see p. 325) that $\chi_f(G)=\omega_f(G)$, where $\omega_f(G)$ is the fractional clique number of G. Because these two concepts are dual, the fractional clique number can also be defined as follows: $\omega_f(G)$ is the maximum of the weights of all fractional cliques of G, where a *fractional clique* is a map $f: V(G)\to[0,1]$ such that $\sum_{x\in I} f(x)\leq 1$ for each $I\in\mathcal{I}(G)$.

Because we already know that $\chi_f(G\times H)\leq\min\{\chi_f(G),\chi_f(H)\}$ and because $\chi_f(G\times H)=\omega_f(G\times H)$, it suffices to prove that

$$\omega_f(G\times H)\geq\min\{\omega_f(G),\omega_f(H)\}\,.$$

For $A\subset V(G)$, let $N[A]=\cup_{u\in A}N[u]$ be the *closed neighborhood* of A.

Lemma 26.45 *Let G be a graph, $I\in\mathcal{I}(G)$, and g a maximum fractional clique of G. Then*

$$g(I)\leq\frac{g(N[I])}{\omega_f(G)}\,.$$

Proof Let $G'=G-N[I]$ and $I'\in\mathcal{I}(G')$. Then $I\cup I'\in\mathcal{I}(G)$, so $g(I\cup I')=g(I)+g(I')\leq 1$. If $g(I)=1$, then $g(u)=0$ for any $u\in V(G')$, and it follows that $\omega_f(G)=g(N[I])$, and we are done.

Assume $g(I)<1$. Define $g': V(G')\to[0,1]$ as $g'(u)=g(u)/(1-g(I))$. Then for $I'\in\mathcal{I}(G')$ we have

$$g'(I')=\sum_{u\in I'}\frac{g(u)}{1-g(I)}=\frac{g(I')}{1-g(I)}\leq 1\,.$$

Hence g' is a fractional clique of G'. It follows that

$$g(V(G'))=g'(V(G'))(1-g(I))\leq\omega_f(G')(1-g(I))\leq\omega_f(G)(1-g(I))\,.$$

Using this inequality and the fact that $\omega_f(G)=g(V(G'))+g(N[I])$, we get

$$g(N[I])=\omega_f(G)-g(V(G'))\geq\omega_f(G)-\omega_f(G)(1-g(I))=\omega_f(G)g(I)\,,$$

and the argument is complete. □

Here is the key idea of the proof of Theorem 26.44. Let g and h be maximum fractional cliques of G and H, respectively. Define $\varphi_{g,h}: V(G\times H)\to[0,1]$ as $\varphi_{g,h}(a,x)=g(a)h(x)$. We will show that

$$\frac{\varphi_{g,h}}{\max\{\omega_f(G),\omega_f(H)\}} \tag{26.1}$$

is a fractional clique of $G\times H$. Then its weight is $\frac{\omega_f(G)\omega_f(H)}{\max\{\omega_f(G),\omega_f(H)\}}$, and this implies that

$$\omega_f(G\times H)\geq\frac{\omega_f(G)\omega_f(H)}{\max\{\omega_f(G),\omega_f(H)\}}=\min\{\omega_f(G),\omega_f(H)\}\,,$$

and the theorem follows.

To show that (26.1) is indeed a fractional clique of $G\times H$, we need to prove that

$$\varphi_{g,h}(I)\leq\max\{\omega_f(G),\omega_f(H)\}$$

for any $I\in\mathcal{I}(G\times H)$. Assume without loss of generality that $\omega_f(G)\geq\omega_f(H)$ and partition I into subsets A and B where

$$A=\{(a,x)\in I\mid aa'\in E(G)\Rightarrow(a',x)\notin I\}$$

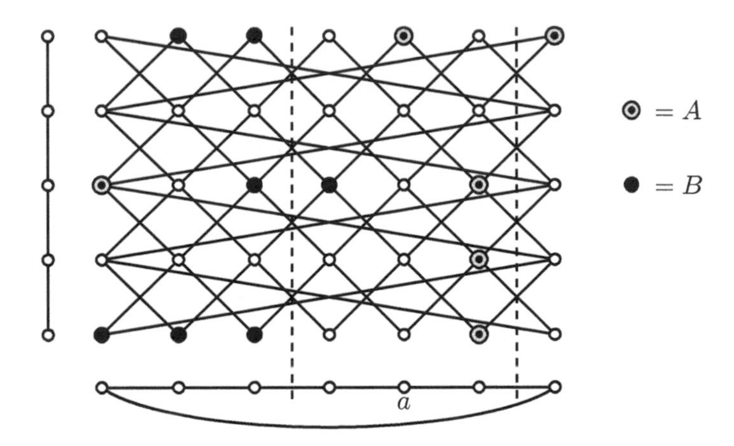

FIGURE 26.5 An independent set of $C_7 \times P_5$ partitioned into subsets A and B.

and $B = I \setminus A$. Note that B is the set of all pairs $(a, x) \in I$ for which there exists a pair $(a', x) \in I$ such that $aa' \in E(G)$. This is illustrated in Figure 26.5.

For a vertex $x \in V(H)$ let

$$A(x) = \{a \in V(G) \mid (a, x) \in A\},$$

that is, $A(x)$ is the projection on G of the intersection of A with the G-layer above x. Similarly, for a vertex $a \in V(G)$ let

$$B(a) = \{x \in V(H) \mid (a, x) \in B\}.$$

Now, $A(x)$ is an independent set of G by the definition of A. Moreover, $B(a)$ is an independent set of H by the definition of B and the definition of the direct product.

For $a \in V(G)$ we also define

$$C(a) = \{x \in V(H) \mid N[a] \cap A(x) \neq \emptyset\}.$$

Notice that

$$C(a) = \{x \in V(H) \mid a \in N[A(x)]\}.$$

For example, in Figure 26.5 the set $C(a)$ consists of all but one vertex of $H = P_5$.

By Lemma 26.45,

$$\varphi_{g,h}(B) = \sum_{a \in V(G)} g(a)h(B(a)) \leq \frac{1}{\omega_f(H)} \sum_{a \in V(G)} g(a)h(N[B(a)]).$$

Similarly, keeping in mind that $\omega_f(G) \geq \omega_f(H)$, we see that

$$\begin{aligned}
\varphi_{g,h}(A) &= \sum_{x \in V(H)} g(A(x))h(x) \leq \frac{1}{\omega_f(G)} \sum_{x \in V(H)} g(N[A(x)])h(x) \\
&= \frac{1}{\omega_f(G)} \sum_{x \in V(H), a \in N[A(x)]} g(a)h(x) \\
&= \frac{1}{\omega_f(G)} \sum_{a \in V(G)} g(a)h(C(a)) \\
&\leq \frac{1}{\omega_f(H)} \sum_{a \in V(G)} g(a)h(C(a)).
\end{aligned}$$

Because I partitions into A and B, we thus obtain

$$\begin{aligned} \varphi_{g,h}(I) &= \varphi_{g,h}(A) + \varphi_{g,h}(B) \\ &\leq \frac{1}{\omega_f(H)} \sum_{a \in V(G)} g(a)\,(h(C(a)) + h(N[B(a)]))\,. \end{aligned}$$

Using the definitions of $B(a)$ and $C(a)$, it is not hard to see that $C(a) \cap N[B(a)] = \emptyset$. It follows that

$$h(C(a)) + h(N[B(a)]) \leq h(V(H)) = \omega_f(H)$$

and therefore $\varphi_{g,h}(I) \leq \omega_f(G)$.

Exercises

26.1. Show $\omega(G \times H) = \min\{\omega(G), \omega(H)\}$ for any graphs G and H in Γ.

26.2. Suppose $G \in \Gamma$. Prove that K_n is a retract of $G \times K_n$ if and only if $n \leq \omega(G)$.

26.3. (Vesztergombi, 1978) Show that $\chi(G \boxtimes K_2) \geq \chi(G) + 2$ if G has at least one edge.

26.4. (Vesztergombi, 1978) Show that $\chi(C_{2k+1} \boxtimes C_{2n+1}) = 5$ for $k \geq 2$ and $n \geq 2$.

26.5. Show that $\chi(G \circ K_{n+m}) \leq \chi(G \circ K_n) + \chi(G \circ K_m)$ for all natural numbers $n, m \geq 1$.

26.6. Determine $\chi(C_{2k+1} \circ C_{2n+1})$ for all $k \geq n \geq 2$.

26.7. Show that $\chi_f(G) \geq |V(G)|/\alpha(G)$ for any graph G.

26.8. Show that $\chi_f(C_{2k+1}) = \chi_c(C_{2k+1}) = 2 + 1/k$ for any $k \geq 1$.

26.9. Show that G has a (k, d)-coloring if and only if there is a homomorphism $G \to G_k^d$.

26.10. (Klavžar and H.-G. Yeh, 2002) Show $\chi(G \circ H) \geq \chi_f(G)\chi(H)$ for any graphs G and H.

26.11. Using Exercise 26.10, give another proof of Theorem 26.12.

26.12. (Bondy and Hell, 1990) Show that if a graph G has a (k, d)-coloring, then it also has a (k', d')-coloring, where k' and d' are positive integers such that $k/d \leq k'/d'$.

26.13. (Zhu, 1992b) Show that $\chi_c(G \,\square\, H) = \max\{\chi_c(G), \chi_c(H)\}$.

26.14. (Zhu, 1992b) Show that the graphs G_k^d are circular critical, that is, $\chi_c(G_k^d - u) < k/d$ for any vertex u of G_k^d.

26.15. Show that $\chi(G \times G) = \chi(G)$.

26.16. With the aid of Lemma 26.29, verify Hedetniemi's conjecture for 3-chromatic graphs.

26.17. Show that for any graphs G and H, $\chi_f(G \times H) \leq \min\{\chi_f(G), \chi_f(H)\}$.

Chapter 27

Independence Number and Shannon Capacity

This chapter is concerned with the independence number for the four standard products. The question is simple for the lexicographic product, and Exercise 27.1 gives a complete answer. The central concept of the first section, the Shannon capacity, involves the strong product, and in this case the determination of the independence number is extremely difficult. Section 27.2 studies the independence number of direct products, with a special emphasis on vertex-transitive graphs. We also treat the ultimate direct independence ratio, a concept parallel to the Shannon capacity. The final section investigates the independence number for the Cartesian product and introduces the ultimate Cartesian independence ratio.

27.1 Shannon Capacity

If I_G and I_H are independent sets of G and H, respectively, then $I_G \times I_H$ is (by the definition of the strong product) an independent set of $G \boxtimes H$. Hence

$$\alpha(G \boxtimes H) \geq \alpha(G)\,\alpha(H)\,.$$

However, Exercise 27.3 shows that a largest independent set of $G \boxtimes H$ need not be a subproduct. In fact, in the course of this section it will become clear that the determination of $\alpha(G \boxtimes H)$ is a notoriously difficult problem.

The primal motivation for the independence number of the strong product is an information-theoretical concept introduced by Shannon (1956). Suppose that we wish to transmit messages through a channel and have a finite alphabet for this purpose. Due to noise in the channel, some pairs of letters may be "confoundable." We say that two words of equal length are *confoundable* if for any i, the ith letters of the words are either the same or confoundable. We wish to find sets of words for which no pair is confoundable. Let G be a graph whose vertices are letters of the alphabet and in which two vertices are adjacent whenever the corresponding letters are confoundable. Clearly, the maximum number of one-letter messages that can be sent such that no confusion occurs equals $\alpha(G)$. Moreover, by the definition of the strong product, two words of length n are confoundable if and only if the corresponding vertices in $G^{\boxtimes,n}$ are adjacent. It follows that the largest set of pairwise nonconfoundable words of length n has $\alpha\left(G^{\boxtimes,n}\right)$ elements. The *Shannon capacity* of G is then defined as

$$\Theta(G) = \lim_{n\to\infty} \sqrt[n]{\alpha\left(G^{\boxtimes,n}\right)}.$$

Because $\alpha(G^{\boxtimes,m+n}) \geq \alpha(G^{\boxtimes,m})\,\alpha(G^{\boxtimes,n})$, the limit exists by Fekete's lemma. (Fekete's lemma states that if $f : \mathbb{N} \to \mathbb{N}$ satisfies $f(m+n) \geq f(m)f(n)$ for all m and n, then $\lim_{n\to\infty} \sqrt[n]{f(n)}$ exists. See Lemma 11.6 of van Lint and Wilson (1992).)

As $\alpha\left(G^{\boxtimes,n}\right) \geq \alpha(G)^n$, we have $\Theta(G) \geq \alpha(G)$. In fact, if n is fixed, then for any $k \in \mathbb{N}$,

$$\sqrt[nk]{\alpha\left(G^{\boxtimes,nk}\right)} \geq \sqrt[nk]{\alpha(G^{\boxtimes,n})^k} = \sqrt[n]{\alpha(G^{\boxtimes,n})}. \tag{27.1}$$

Combined with the definition of Θ, this implies $\Theta(G) \geq \sqrt[n]{\alpha(G^{\boxtimes,n})}$ for every $n \in \mathbb{N}$. Thus

$$\Theta(G) = \sup_{n\in\mathbb{N}} \sqrt[n]{\alpha\left(G^{\boxtimes,n}\right)}.$$

We consider the Shannon capacity only in this chapter, so there is no danger of confusing it with the relation Θ from previous chapters.

Attaining the capacity

Let $K(G)$ be the smallest integer n for which $\Theta(G) = \sqrt[n]{\alpha\left(G^{\boxtimes,n}\right)}$, if such an integer exists; otherwise, set $K(G) = \infty$. The above considerations imply $K(G) = 1$ if and only if $\alpha(G^{\boxtimes,n}) = \alpha(G)^n$ for all n, that is, if and only if $\Theta(G) = \alpha(G)$. Also, $K(G) = \infty$ if and only if $\Theta(G) > \sqrt[n]{\alpha(G^{\boxtimes,n})}$ for all n. As we will see below, $K(G) = \infty$ is indeed possible.

It is known that $\Theta(G) = \alpha(G)$ if G is perfect, so $K(G) = 1$ in this case. In fact, for every graph G where $\Theta(G)$ is known, $K(G)$ is either $1, 2$ or ∞; see Alon and Lubetzky (2006). We now present a construction by Guo and Watanabe (1990) of a graph with $K(G) = \infty$.

Recall that $G + H$ denotes the disjoint union of graphs, and observe that $\alpha(G+H) = \alpha(G) + \alpha(H)$. Recall also that nG denotes the disjoint union of n copies of G. By the distributive and commutative properties of the strong product, we have

$$\begin{aligned}(G+H)\boxtimes(G+H) &= (G\boxtimes G) + (G\boxtimes H) + (H\boxtimes G) + (H\boxtimes H)\\ &= G^{\boxtimes,2} + 2(G\boxtimes H) + H^{\boxtimes,2},\end{aligned}$$

and, more generally, the binomial theorem

$$(G+H)^{\boxtimes,n} = \sum_{i=0}^{n} \binom{n}{i} G^{\boxtimes,i} \boxtimes H^{\boxtimes,n-i}.$$

Call a graph H *universal* if $\alpha(G\boxtimes H) = \alpha(G)\alpha(H)$ holds for any graph G. We can now state the theorem of Guo and Watanabe.

Theorem 27.1 *If G satisfies $K(G) > 1$ and H is universal, then $K(G+H) = \infty$.*

Proof Let $X = G + H$. Because H is universal,

$$\begin{aligned}\alpha\left(X^{\boxtimes,2n}\right) &= \alpha\left(\left(\sum_{i=0}^{n}\binom{n}{i}G^{\boxtimes,i}\boxtimes H^{\boxtimes,n-i}\right)\boxtimes\left(\sum_{j=0}^{n}\binom{n}{j}G^{\boxtimes,j}\boxtimes H^{\boxtimes,n-j}\right)\right)\\ &= \alpha\left(\sum_{i=0}^{n}\sum_{j=0}^{n}\binom{n}{i}\binom{n}{j}G^{\boxtimes,i+j}\boxtimes H^{\boxtimes,2n-i-j}\right)\\ &= \sum_{i=0}^{n}\sum_{j=0}^{n}\binom{n}{i}\binom{n}{j}\alpha\left(G^{\boxtimes,i+j}\right)\alpha\left(H^{\boxtimes,2n-i-j}\right).\end{aligned}$$

Similarly, we obtain

$$\alpha\left(X^{\boxtimes,n}\right)^2 = \sum_{i=0}^{n}\sum_{j=0}^{n}\binom{n}{i}\binom{n}{j}\alpha\left(G^{\boxtimes,i}\right)\alpha\left(G^{\boxtimes,j}\right)\alpha\left(H^{\boxtimes,2n-i-j}\right).$$

Combining the above equations, $\alpha\left(X^{\boxtimes,2n}\right) - \alpha\left(X^{\boxtimes,n}\right)^2$ equals

$$\sum_{i=0}^{n}\sum_{j=0}^{n}\binom{n}{i}\binom{n}{j}\alpha\left(H^{\boxtimes,2n-i-j}\right)\left(\alpha\left(G^{\boxtimes,i+j}\right) - \alpha\left(G^{\boxtimes,i}\right)\alpha\left(G^{\boxtimes,j}\right)\right).$$

Thus for any n we have $\alpha\left(X^{\boxtimes,2n}\right) \geq \alpha\left(X^{\boxtimes,n}\right)^2$, because $\alpha\left(G^{\boxtimes,i+j}\right) \geq \alpha\left(G^{\boxtimes,i}\right)\alpha\left(G^{\boxtimes,j}\right)$. Moreover, as $K(G) > 1$, there are integers I and J with $\alpha\left(G^{\boxtimes,I+J}\right) > \alpha\left(G^{\boxtimes,I}\right)\alpha\left(G^{\boxtimes,J}\right)$. Therefore $\alpha\left(X^{\boxtimes,2n}\right) > \alpha\left(X^{\boxtimes,n}\right)^2$ for all $n \geq k_0 = \max\{I, J\}$. In other words,

$$\sqrt[2n]{\alpha\left(X^{\boxtimes,2n}\right)} > \sqrt[n]{\alpha\left(X^{\boxtimes,n}\right)}$$

for all $n \geq k_0$. Combining this with Inequality (27.1), we see that, for any integer n,

$$\Theta(X) \geq \sqrt[2nk_0]{\alpha\left(X^{\boxtimes,2nk_0}\right)} > \sqrt[nk_0]{\alpha\left(X^{\boxtimes,nk_0}\right)} \geq \sqrt[n]{\alpha\left(X^{\boxtimes,n}\right)}.$$

This means $K(X) = \infty$. □

By Exercise 27.4, the even cycles C_{2m} are universal. Moreover, Hales (1973) proved that $\alpha\left(C_{2k+1}^{\boxtimes,2}\right) > \alpha(C_{2k+1})^2$ for $k \geq 2$, so $K(C_{2k+1}) > 1$ in these cases. Thus the graphs $X = C_{2k+1} + C_{2m}$ with $k, m \geq 2$ satisfy $K(X) = \infty$.

With a probabilistic approach, Alon and Lubetzky (2006) proved two theorems that show the situation is much worse than we might expect from the above discussion. Their first theorem asserts, roughly, that for every large integer k, there exists a $\delta = \delta(k) > 0$ and a graph G such that all the values $\alpha(G)$, $\sqrt[2]{\alpha\left(G^{\boxtimes,2}\right)}, \ldots, \sqrt[k]{\alpha\left(G^{\boxtimes,k}\right)}$ are at least $|V(G)|^\delta$ away from $\Theta(G)$. Their second result shows that the series $\sqrt[n]{\alpha\left(G^{\boxtimes,n}\right)}$ can increase significantly in an arbitrary number of terms. More precisely, they proved:

Theorem 27.2 *For any $j_1 < j_2 < \cdots < j_s \in \mathbb{N}$ and any $\epsilon > 0$, there exists a graph G such that for all $k < j_i$,*

$$\sqrt[k]{\alpha\left(G^{\boxtimes,k}\right)} < \sqrt[j_i]{\alpha\left(G^{\boxtimes,j_i}\right)}^{\epsilon}$$

holds for $i = 1, 2, \ldots, s$.

We thus cannot necessarily expect that we have a good approximation to $\Theta(G)$ if the series $\sqrt[n]{\alpha\left(G^{\boxtimes,n}\right)}$ remains roughly the same in several consecutive terms. Alon and Lubetzky (2006) go further and pose:

Question 27.3 *Is the problem of deciding whether $\Theta(G)$ of a given graph G exceeds a given value decidable?*

Lovász's ϑ-function

To investigate the Shannon capacity, Lovász (1979) introduced a concept that is now known as Lovász's ϑ-function. The approach, in which eigenvalues play a crucial role, leads to a computation of $\Theta(C_5)$.

Let G be a graph on n vertices. An *orthonormal representation* of G is a system $(\mathbf{v}_1, \mathbf{v}_2, \dots, \mathbf{v}_n)$ of unit vectors in some $\mathbb{R}^m$, such that each vertex of G corresponds to a unique vector $\mathbf{v}_i$, and two vectors $\mathbf{v}_i$ and $\mathbf{v}_j$ are orthogonal whenever the corresponding vertices of G are nonadjacent. Clearly, G has an orthonormal representation, namely any orthonormal basis of $\mathbb{R}^n$. (But unless $G = D_n$, it has orthonormal representations in some $\mathbb{R}^m$ with $m < n$.) The *value* of an orthonormal representation $(\mathbf{u}_1, \mathbf{u}_2, \dots, \mathbf{u}_n)$ of G is defined as the minimum over all unit vectors $\mathbf{c} \in \mathbb{R}^m$ of the quantity

$$\max\left\{\frac{1}{(\mathbf{c}^T\mathbf{u}_1)^2}, \frac{1}{(\mathbf{c}^T\mathbf{u}_2)^2}, \dots, \frac{1}{(\mathbf{c}^T\mathbf{u}_n)^2}\right\}.$$

A vector $\mathbf{c}$ for which the minimum is attained is called a *handle* of the representation. *Lovász's ϑ-function*, $\vartheta(G)$, is defined as the minimum value over all orthonormal representations of G. Several equivalent definitions of Lovász's ϑ-function can be found in the literature. Often one finds a definition based on the following theorem of Lovász (1979).

Theorem 27.4 *Let G be a graph with vertex set $\{1, 2, \dots, n\}$. Let $\mathcal{A}(G)$ be the set of all symmetric $n \times n$ matrices $A = (a_{ij})$ such that $a_{ij} = 1$ whenever i is not adjacent to j. Let $\lambda_{\max}(A)$ denote the largest eigenvalue of the matrix A. Then*

$$\vartheta(G) = \min\{\lambda_{\max}(A) \mid A \in \mathcal{A}(G)\}.$$

We will present two computations of $\Theta(C_5)$, one with and one without reference to Theorem 27.4. To this end we next describe some basic properties of the ϑ-function. For vectors $\mathbf{u} = (u_1, u_2, \dots, u_n)^T \in \mathbb{R}^n$ and $\mathbf{v} = (v_1, v_2, \dots, v_m)^T \in \mathbb{R}^m$, let

$$\mathbf{u} \otimes \mathbf{v} = (u_1v_1, \dots, u_1v_m,\ u_2v_1, \dots, u_2v_m,\ u_nv_1 \dots, u_nv_m)^T$$

be their tensor product in $\mathbb{R}^{nm}$. The next lemma uses the tensor product to construct an orthonormal representation of the strong product of graphs.

Lemma 27.5 *For any $\mathbf{u}, \mathbf{v} \in \mathbb{R}^n$ and $\mathbf{x}, \mathbf{y} \in \mathbb{R}^m$, we have $(\mathbf{u} \otimes \mathbf{x})^T(\mathbf{v} \otimes \mathbf{y}) = (\mathbf{u}^T\mathbf{v})(\mathbf{x}^T\mathbf{y})$. Moreover, suppose $(\mathbf{u}_1, \mathbf{u}_2, \dots, \mathbf{u}_n)$ and $(\mathbf{v}_1, \mathbf{v}_2, \dots, \mathbf{v}_m)$ are orthonormal representations of the graphs G and H, respectively. Then*

$$(\mathbf{u}_1 \otimes \mathbf{v}_1, \dots, \mathbf{u}_1 \otimes \mathbf{v}_m,\ \mathbf{u}_2 \otimes \mathbf{v}_1, \dots, \mathbf{u}_2 \otimes \mathbf{v}_m,\ \dots\ , \mathbf{u}_n \otimes \mathbf{v}_1 \dots \mathbf{u}_n \otimes \mathbf{v}_m)$$

is an orthonormal representation of $G \boxtimes H$.

Proof For the first statement, let $\mathbf{u} = (u_1, \dots, u_n)^T$, $\mathbf{v} = (v_1, \dots, v_n)^T$, $\mathbf{x} = (x_1, \dots, x_m)^T$, and $\mathbf{y} = (y_1, \dots, y_m)^T$. Direct computations show that both sides of the equation equal $\sum_{i=1}^{n}\sum_{j=1}^{m} u_iv_ix_jy_j$.

For the second statement, suppose $a_i \in V(G)$ corresponds to $\mathbf{u}_i$, and $b_j \in V(H)$ corresponds to $\mathbf{v}_j$. Notice that $(\mathbf{u}_i \otimes \mathbf{v}_j)^T(\mathbf{u}_k \otimes \mathbf{v}_\ell) = (\mathbf{u}_i^T\mathbf{u}_k)(\mathbf{v}_j^T\mathbf{v}_\ell)$. This equals 1 if $(a_i, b_j) = (a_k, b_\ell)$, and it equals 0 if (a_i, b_j) and (a_k, b_ℓ) are not adjacent in $G \boxtimes H$. □

Lovász proved that $\vartheta(G \boxtimes H) = \vartheta(G)\vartheta(H)$, but for our purposes the first statement of the following lemma will suffice.

Lemma 27.6 *Let G and H be any graphs. Then*

(i) *$\vartheta(G \boxtimes H) \le \vartheta(G)\vartheta(H)$ and*
(ii) *$\alpha(G) \le \vartheta(G)$.*

Proof (i) Let $(\mathbf{u}_1, \mathbf{u}_2, \ldots, \mathbf{u}_n)$ and $(\mathbf{v}_1, \mathbf{v}_2, \ldots, \mathbf{v}_m)$ be optimal orthonormal representations of G and H, and let $\mathbf{c}$ and $\mathbf{d}$ be corresponding handles, respectively. By Lemma 27.5 the vectors $\mathbf{u}_i \otimes \mathbf{v}_j$ form an orthonormal representation of $G \boxtimes H$. Moreover, $\mathbf{c} \otimes \mathbf{d}$ is a unit vector, as Lemma 27.5 gives $(\mathbf{c} \otimes \mathbf{d})^T(\mathbf{c} \otimes \mathbf{d}) = (\mathbf{c}^T\mathbf{c})(\mathbf{d}^T\mathbf{d}) = 1$. We now obtain

$$\begin{aligned} \vartheta(G \boxtimes H) &\leq \max_{i,j} \frac{1}{((\mathbf{c} \otimes \mathbf{d})^T(\mathbf{u}_i \otimes \mathbf{v}_j))^2} \\ &= \max_{i,j} \frac{1}{(\mathbf{c}^T \otimes \mathbf{u}_i)^2} \cdot \frac{1}{(\mathbf{d}^T \otimes \mathbf{v}_j)^2} \\ &= \vartheta(G)\vartheta(H). \end{aligned}$$

(ii) Let $\alpha(G) = k$, and $(\mathbf{u}_1, \mathbf{u}_2, \ldots, \mathbf{u}_n)$ be an optimal orthonormal representation of G with handle $\mathbf{c} \in \mathbb{R}^m$. We may assume, without loss of generality, that the vectors $\mathbf{u}_1, \mathbf{u}_2, \ldots, \mathbf{u}_k$ correspond to a largest independent set of G and are thus pairwise orthogonal. Thus these vectors can be extended to an orthonormal basis B of $\mathbb{R}^m$, say $B = \{\mathbf{v}_1, \mathbf{v}_2, \ldots, \mathbf{v}_m\}$, where $\mathbf{u}_i = \mathbf{v}_i$ for $i = 1, 2, \ldots, k$. Let $\mathbf{c} = \sum_{i=1}^m \lambda_i \mathbf{v}_i$. Then $\mathbf{c}^T\mathbf{c} = \sum_{i=1}^m \lambda_i^2$ and $\sum_{i=1}^k (\mathbf{c}^T\mathbf{u}_i)^2 = \sum_{i=1}^k (\mathbf{c}^T\mathbf{v}_i)^2 = \sum_{i=1}^k \lambda_i^2$. Hence

$$1 = \mathbf{c}^T\mathbf{c} \geq \sum_{i=1}^k (\mathbf{c}^T\mathbf{u}_i)^2 \geq \frac{k}{\vartheta(G)} = \frac{\alpha(G)}{\vartheta(G)}.$$

□

Here is the fundamental relation between Lovász's ϑ-function and the Shannon capacity:

Theorem 27.7 *For any graph G, we have $\Theta(G) \leq \vartheta(G)$.*

Proof By Lemma 27.6 (ii), we have $\alpha(G^{\boxtimes,n}) \leq \vartheta(G^{\boxtimes,n})$, and by Lemma 27.6 (i), we get $\vartheta(G^{\boxtimes,n}) \leq \vartheta(G)^n$. So, for any n, we have $\sqrt[n]{\alpha(G^{\boxtimes,n})} \leq \vartheta(G)$, and the theorem follows. □

We now use Theorem 27.7 to compute the Shannon capacity of the 5-cycle.

Theorem 27.8 *The Shannon capacity of the 5-cycle is $\Theta(C_5) = \sqrt{5}$.*

Proof By Exercise 27.3, $\alpha(C_5 \boxtimes C_5) = 5$, so $\Theta(C_5) \geq \sqrt{5}$. To complete the proof, it suffices to show that $\vartheta(C_5) \leq \sqrt{5}$.

Let $(\mathbf{u}_1, \mathbf{u}_2, \mathbf{u}_3, \mathbf{u}_4, \mathbf{u}_5)$ be an orthonormal representation of C_5 with handle $\mathbf{c} \in \mathbb{R}^3$, as follows: the vectors $\mathbf{u}_i$ are the five ribs of an umbrella opened such that the maximum angle between them is $\pi/2$, and $\mathbf{c}$ is the umbrella's handle. (Figure 27.1.) Let C be the circle of radius r defined by the endpoints of the $\mathbf{u}_i$. As each $\mathbf{u}_i$ has length 1, we have $\mathbf{c}^T\mathbf{u}_i = \sqrt{1 - r^2}$. Let T be the point where $\mathbf{c}$ intersects the plane containing C, and consider the triangle $\mathbf{u}_1\mathbf{u}_3T$. By the cosine theorem, we have $2 = r^2 + r^2 - 2r^2\cos(4\pi/5)$. Hence $r^2 = (5 - \sqrt{5})/5$, so $\mathbf{c}^T\mathbf{u}_i = \sqrt{1 - r^2} = 1/\sqrt[4]{5}$. Then $\vartheta(C_5) \leq \max_i 1/(\mathbf{c}^T\mathbf{u}_i)^2 = \sqrt{5}$. □

Lovász (1979) asked whether $\vartheta(G) = \Theta(G)$ holds for all graphs G. Haemers (1979) settled this question negatively by showing that the Shannon capacity of the Schläfli graph is strictly less than its ϑ-function.

Theorem 27.8 is one of the main achievements in the area. It can also be proved with the following Corollary of Theorem 27.4.

Corollary 27.9 *Let G be a d-regular graph on n vertices, and let $\lambda_{\min}$ be the smallest eigenvalue of the adjacency matrix A of G. Then*

$$\vartheta(G) \leq \frac{-n\lambda_{\min}}{d - \lambda_{\min}}.$$

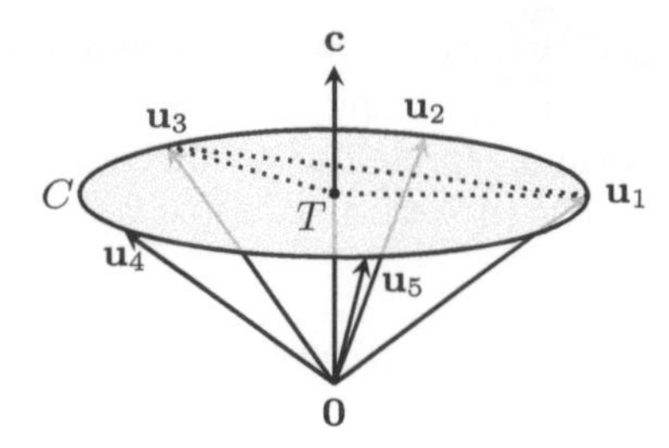

FIGURE 27.1 The construction in Theorem 27.8.

Proof Let J be the $n \times n$ matrix, all entries of which are 1, and

$$M = J - \frac{n}{d - \lambda_{\min}} A\,.$$

Note that $m_{ij} = 1$ whenever the corresponding vertices of G are not adjacent. By Theorem 27.4, it suffices to prove that the largest eigenvalue of M equals $(-n\lambda_{\min})/(d - \lambda_{\min})$.

Because G is d-regular, $A\mathbf{1}_n = d\mathbf{1}_n$. We thus infer

$$M\mathbf{1}_n = \left(n - \frac{n}{d - \lambda_{\min}} d\right) \mathbf{1}_n\,.$$

Let $\mu \neq d$ be an eigenvalue of A, with corresponding eigenvector $\mathbf{x}$. Because A is symmetric, we may assume (by the Spectral Theorem) that $\mathbf{x}$ is orthogonal to $\mathbf{1}_n$. Thus $J\mathbf{x} = 0$, so

$$M\mathbf{x} = -\frac{n\mu}{d - \lambda_{\min}} \mathbf{x}\,.$$

Hence the eigenvalues of M are $n - [n/(d - \lambda_{\min})]d$ and $-n\mu/(d - \lambda_{\min})$, where μ is an eigenvalue of A. The latter quotient is largest for $\mu = \lambda_{\min}$. Now an easy computation shows that $-n\lambda_{\min}/(d - \lambda_{\min}) \geq n - [n/(d - \lambda_{\min})]d$, and so $-n\lambda_{\min}/(d - \lambda_{\min})$ is indeed the largest eigenvalue of M. □

For C_5 we have $\lambda_{\min} = (-1 - \sqrt{5})/2$, so Corollary 27.9 yields $\vartheta(C_5) \leq \sqrt{5}$. Invoking Theorem 27.7 and the fact that $\Theta(C_5) \geq \sqrt{5}$, we have a second proof of Theorem 27.8.

Schrijver (1979) and McEliece, Rodemich, and Rumsey (1978) found alternative forms of Theorem 27.7 for graphs that appear in association schemes; see also Miklós (1996). In particular, their approach yields yet another proof that $\Theta(C_5) = \sqrt{5}$.

The simplest graphs for which the Shannon capacity is not known are odd cycles of length at least 7 and their complements. Let us have a closer look at C_7. Combining Theorem 27.7 with the fact that for any odd k, $\vartheta(C_k) = (k\cos(\pi/k))/(1 + \cos(\pi/k))$ (see Lovász (1979)), we infer that

$$\Theta(C_7) \leq \vartheta(C_7) = (7\cos(\pi/7))/(1 + \cos(\pi/7)) < 3.3177\,.$$

Baumert, McEliece, Rodemich, Rumsey, Stanley, and H. Taylor (1971), and, independently, Vesel (1998), showed that $\alpha(C_7^{\boxtimes,3}) = 33$. Thus $\Theta(C_7) \geq \sqrt[3]{33} > 3.2075$. Vesel and Žerovnik (2002) found an independent set of $C_7^{\boxtimes,4}$ with 108 vertices, improving the lower bound to

$$\Theta(C_7) > 3.2237\,.$$

However, Theorem 27.2 says that these facts give no clue as to how close we are to the actual value of $\Theta(C_7)$. In this respect we add that Hales (1973) and independently Sonnemann and Krafft (1974) proved the following result. (See Exercise 27.3.)

Theorem 27.10 *For $1 \le k \le n$,*

$$\alpha(C_{2k+1} \boxtimes C_{2n+1}) = kn + \left\lfloor \frac{k}{2} \right\rfloor .$$

Hell and F. S. Roberts (1982) (see also Farber (1986)) studied several analogues of the Shannon capacity involving the strong product, the lexicographic product, and the so-called disjunction, which is the complementary product of the strong product.

27.2 Independence in Direct Products

We know from Section 26.4 that direct product graphs have small chromatic numbers. Consequently they must have large independent sets. This is confirmed by our next result.

Proposition 27.11 *For any graphs G and H,*

$$\alpha(G \times H) \ge \max\{\alpha(G)\,|V(H)|, \alpha(H)\,|V(G)|\} .$$

Proof If I is an independent set of G, then $I \times V(H)$ is an independent set of $G \times H$. □

On the other hand, Špacapan (2011) proved the following nontrivial upper bound:

Theorem 27.12 *For any graphs G and H,*

$$\alpha(G \times H) \le \alpha(G)\,|V(H)| + \alpha(H)\,|V(G)| - \alpha(G)\alpha(H) .$$

Proof Let I be an independent set of $G \times H$. Partition I into

$$J = \{(a,x) \in I \mid \text{there exists } (a,x') \in I \text{ such that } xx' \in E(H)\}$$

and $K = I \setminus J$. For any $a \in V(G)$, let K^a be the intersection of K with the H-layer over a; and for any $x \in V(H)$, let J^x the intersection of J with the G-layer over x. In addition, let I_H be a largest independent set of H, let $Y = (V(G) \times I_H) \cap J$, and for $a \in V(G)$ let Y^a be the intersection of Y with the H-layer over a. Clearly, $p_H(Y^a)$ is an independent set of H because $p_H(Y^a) \subseteq I_H$. By definition of J and K, we also infer that $p_H(K^a)$ is independent, $p_H(Y^a) \cap p_H(K^a) = \emptyset$, and that $p_H(K^a) \cup p_H(Y^a)$ is independent. Therefore, $\alpha(H) \ge |K^a| + |Y^a|$. Note also that $p_G(J^x)$ is an independent set of G so that $\alpha(G) - |J^x| \ge 0$ for any $x \in V(G)$. Putting all this together, we have

$$\begin{aligned}
&\sum_{a\in V(G)} (\alpha(H) - |K^a|) + \sum_{x\in V(H)} (\alpha(G) - |J^x|) \ge \\
&\sum_{a\in V(G)} |Y^a| + \sum_{x\in I_H} (\alpha(G) - |J^x|) = \\
&\sum_{x\in I_H} |J^x| + \sum_{x\in I_H} (\alpha(G) - |J^x|) = \\
&\alpha(G)\alpha(H) .
\end{aligned}$$

The result now follows from that fact that I is partitioned into J and K and hence $|I| = \sum_{a\in V(G)} |K^a| + \sum_{x\in V(H)} |J^x|$. □

Graphs for which equality holds in Proposition 27.11 were studied in a series of papers by Jha and Klavžar (1998), Jha (2000), and Paulraja and Varadarajan (2004). For instance, equality holds if G is a bipartite graph with a perfect matching and H has a Hamilton path.

Sharpness of the bound of Proposition 27.11 is also demonstrated by a result due to Larose and Tardif (2002) involving powers of vertex-transitive graphs. (See Exercise 27.7 for a specific case.) Their proof uses the No-Homomorphism Lemma 2.13; the reader is invited to rediscover their arguments in Exercise 27.6. Here we follow the elegant, self-contained probabilistic proof by Alon, Dinur, Friedgut, and Sudakov (2004).

Theorem 27.13 *Let G be a vertex-transitive graph. Then for any $n \geq 1$,*

$$\alpha\left(G^{\times,n}\right) = \alpha(G)\,|V(G)|^{n-1}\,.$$

Proof Let $V(G) = \{v_1, v_2, \ldots, v_{|V(G)|}\}$ and $n \geq 1$. Select a maximum independent set I of $G^{\times,n}$.

Because G is vertex-transitive, $|\mathrm{Aut}(G)| \geq |V(G)|$. Hence we may randomly, independently, and uniformly select automorphisms $\alpha_1, \alpha_2, \ldots, \alpha_{|V(G)|}$ of G. Then to each vertex v_i of G, $1 \leq i \leq |V(G)|$, assign a random vertex x_i of $G^{\times,n}$ with

$$x_i = (\alpha_1(v_i), \alpha_2(v_i), \ldots, \alpha_{|V(G)|}(v_i))\,.$$

By Lemma 2.14, the vertices x_i are uniformly distributed among the vertices of $G^{\times,n}$ and hence the expected size of $I \cap \{x_1, x_2, \ldots, x_{|V(G)|}\}$ is $|I| \cdot |V(G)|/|V(G^{\times,n})|$. Now, because the α_i's are automorphisms, $x_i = (\alpha_1(v_i), \alpha_2(v_i), \ldots, \alpha_{|V(G)|}(v_i))$ is adjacent to $x_j = (\alpha_1(v_j), \alpha_2(v_j), \ldots, \alpha_{|V(G)|}(v_j))$ if and only if v_i is adjacent to v_j. Therefore, $x_1, x_2, \ldots, x_{|V(G)|}$ induce a subgraph isomorphic to G. This implies $|I \cap \{x_1, x_2, \ldots, x_{|V(G)|}\}| \leq \alpha(G)$, and consequently

$$|I|\,\frac{|V(G)|}{|V(G^{\times,n})|} = \alpha\left(G^{\times,n}\right)\frac{|V(G)|}{|V(G)|^n} \leq \alpha(G)\,.$$

On the other hand, writing $G^{\times,n} = G \times G^{\times,n-1}$, Proposition 27.11 implies

$$\alpha\left(G^{\times,n}\right) \geq \alpha(G)\,|V(G)|^{n-1}$$

and the argument is complete. □

With Theorem 27.13 in hand, it is natural to ask what can be said about the structure of largest independent sets in powers of vertex-transitive graphs and, more generally, of regular graphs. To this end, we say that a direct product $G \times H$ is *MIS-normal* (maximum-independent-set-normal) if all of its largest independent sets are *canonical*, that is, of the form $I \times V(H)$ or $V(G) \times I$, where I is a maximum independent set of G or H, respectively. These definitions extend naturally to direct products of more than two factors. In particular, $G^{\times,n}$ is MIS-normal if for any maximum independent set I of $G^{\times,n}$ there exists an i, $1 \leq i \leq n$, and a maximum independent set I_G of G such that

$$I = V(G^{\times,i-1}) \times I_G \times V(G^{\times,n-i})\,.$$

Powers of many classes of graphs are MIS-normal. This is the case for complete graphs, as was proved by Greenwell and Lovász (1974) and independently by Müller (1979). It is also true for powers of circular graphs and of Kneser graphs, as proved in Larose and Tardif (2002); see also Valencia-Pabon and Vera (2006). The latter result includes complete graphs as a special case and follows from a more general result of Ahlswede, Aydinian, and

Khachatrian (1998). On the other hand, Larose and Tardif (2002) showed that the powers of truncated simplices (which are vertex-transitive) are not MIS-normal.

Alon, Dinur, Friedgut, and Sudakov (2004) used spectral techniques and Fourier analysis on Abelian groups to obtain several interesting results on the structure of maximal independent sets in direct powers of regular graphs. They first reproved the above-mentioned result of Greenwell and Lovász, and then proved that an independent set that is close to being a maximum is close to being canonical. (This was further developed in Dinur, Friedgut, and Regev (2008).) From the spectral point of view, they proved the following theorem. (Compare it with Corollary 27.9.)

Theorem 27.14 *Let G be a connected, d-regular graph on r vertices and let $\lambda_{\min}$ be the smallest eigenvalue of the adjacency matrix of G. If*

$$\alpha(G) = |V(G)|\frac{-\lambda_{\min}}{d-\lambda_{\min}},$$

then for any $n \geq 1$,

$$\alpha(G^{\times,n}) = |V(G^{\times,n})|\frac{-\lambda_{\min}}{d-\lambda_{\min}}.$$

Moreover, if G is not bipartite, then G^n is MIS-normal.

There exist connected, bipartite, vertex-transitive graphs whose second powers are MIS-normal, but whose higher powers are not. (See Exercise 27.10.) On the other hand, Ku and McMillan (2009) proved that if G is a connected, vertex-transitive, nonbipartite graph, and $G^{\times,2}$ is MIS-normal, then so is any power $G^{\times,n}$. H. Zhang (2011) proved the same result independently, and in a more general setting. Moreover, Ku and McMillan (2009) showed that the Cayley graph of the symmetric group S_4 generated by the permutations with at most one fixed point is a connected, nonbipartite, vertex-transitive graph that does not fulfill the eigenvalue assumption of Theorem 27.14, yet all its powers are MIS-normal.

Theorem 27.13 (and Proposition 27.11) also led to the following problem posed in Tardif (1998): Is it true that for any vertex-transitive graphs G and H,

$$\alpha(G \times H) = \max\{\alpha(G)\,|V(H)|, \alpha(H)\,|V(G)|\}\,?$$

The problem was solved in H. Zhang (2010), where the structure of largest independent sets is also clarified. To state H. Zhang's theorem, we need one more concept. An independent I of a graph G is called *IS-imprimitive* if $|I| < \alpha(G)$ and $\frac{|I|}{|N[I]|} = \frac{\alpha(G)}{|V(G)|}$. Moreover, G is *IS-imprimitive* if it has an IS-imprimitive independent set. H. Zhang (2010) proved:

Theorem 27.15 *If G and H are vertex-transitive graphs for which $\frac{\alpha(G)}{|V(G)|} \geq \frac{\alpha(H)}{|V(H)|}$, then $\alpha(G \times H) = \alpha(G)|V(H)|$. Moreover, one of the following holds:*

(i) *$G \times H$ is MIS-normal,*

(ii) *$\frac{\alpha(G)}{|V(G)|} = \frac{\alpha(H)}{|V(H)|}$ and one of G and H is IS-imprimitive,*

(iii) *$\frac{\alpha(G)}{|V(G)|} > \frac{\alpha(H)}{|V(H)|}$ and H is disconnected.*

Ultimate direct independence ratio

The *independence ratio* $i(G)$ of a graph G is defined as

$$i(G) = \frac{\alpha(G)}{|V(G)|}.$$

In the spirit of the Shannon capacity, Brown, Nowakowski, and Rall (1996) introduced the *ultimate direct independence ratio* $I_\times(G)$ of a graph G as

$$I_\times(G) = \lim_{n\to\infty} i\left(G^{\times,n}\right) .$$

This parameter is well-defined. Indeed, from Proposition 27.11 we easily infer that $i\left(G^{\times,n}\right) \geq i\left(G^{\times,n-1}\right)$ for any n. Hence the sequence $i\left(G^{\times,n}\right)$ is nondecreasing, and it is clearly bounded above by 1, so the limit exists. Brown, Nowakowski, and Rall (1996) proved the following fundamental and rather surprising result.

Theorem 27.16 *Let G be a graph. Then $I_\times(G) \in \left(0, \frac{1}{2}\right] \cup \{1\}$.*

Proof Because $i(G) > 0$ and the sequence $i\left(G^{\times,n}\right)$ is nondecreasing, $I_\times(G) > 0$. If $i\left(G^{\times,n}\right) \leq \frac{1}{2}$ for any n, then $I_\times(G) \leq \frac{1}{2}$. Suppose next that there is a k such that $i\left(G^{\times,k}\right) > \frac{1}{2}$. Selecting k to be the smallest such index, we distinguish two cases.

Case 1. $k = 1$. In this case, G has an independent set I with $|I| = \alpha(G) > |V(G)|/2$. Consider the subset

$$J = \bigcup_X \{(x_1, \ldots, x_n) \mid x_i \in I \text{ if } i \in X, \quad x_i \in V(G) \setminus I \text{ if } i \notin X\}$$

of $G^{\times,n}$, where the union is taken over all subsets X of $\{1, 2, \ldots, n\}$ with $|X| > n/2$. Then for any vertices x and y of J, there is a coordinate i for which $x_i, y_i \in I$. Thus x and y are not adjacent, so J is an independent set. Notice that

$$|J| = \sum_{\frac{n}{2}<k\leq n} \binom{n}{k} |I|^k \left(|V(G)| - |I|\right)^{n-k},$$

from which we obtain

$$\frac{|J|}{|V(G)|^n} = \sum_{\frac{n}{2}<k\leq n} \binom{n}{k} i(G)^k \left(1 - i(G)\right)^{n-k}.$$

This equals the probability of getting more than $n/2$ successes out of n trials of an event that has probability $i(G) > 1/2$. Therefore the value approaches 1 as n tends to infinity. We conclude $I_\times(G) = 1$.

Case 2. $k > 1$. Set $H = G^{\times,k}$. Then by Case 1, $I_\times(H) = 1$. Because for any $s \geq 1$, $H^{\times,s} = G^{\times,ks}$, and having in mind that the sequence $i\left(G^{\times,n}\right)$ is nondecreasing,

$$I_\times(G) = \lim_{n\to\infty} i\left(G^{\times,n}\right) = \lim_{s\to\infty} i\left(G^{\times,ks}\right) = \lim_{s\to\infty} i\left(H^{\times,s}\right) = I_\times(H) = 1 .$$

□

Brown, Nowakowski, and Rall (1996) also proved that for any independent set I of G,

$$I_\times(G) \geq \frac{|I|}{|I| + |N(I)|},$$

where $N(I) = \bigcup_{v\in I} N(v)$ is the neighborhood of I. This lower bound and Theorem 27.16 led Alon and Lubetzky (2006) to define

$$i_\times(G) = \max\left\{\frac{|I|}{|I| + |N(I)|} \mid I \text{ is an independent set of } G\right\},$$

and to set

$$i^*_\times(G) = \begin{cases} i_\times(G) & \text{if } i_\times(G) \le 1\,, \\ 1 & \text{if } i_\times(G) > 1\,. \end{cases}$$

They posed the question of whether every graph G satisfies $I_\times(G) = i^*_\times(G)$. A positive answer would reduce the computation of the seemingly much more complicated invariant $I_\times(G)$ to $i^*_\times(G)$. Nevertheless, no negative example to the question is known. We conclude the section by noting that Tóth (2009) determined the ultimate direct independence ratio of complete multipartite graphs.

27.3 Independence in Cartesian Products

For the independence number of Cartesian products, Vizing (1963) observed:

Theorem 27.17 *For any graphs G and H,*
(i) $\alpha(G \,\Box\, H) \le \min\{\alpha(G)\,|V(H)|,\, \alpha(H)\,|V(G)|\}$,
(ii) $\alpha(G \,\Box\, H) \ge \alpha(G)\,\alpha(H) + \min\{|V(G)| - \alpha(G),\, |V(H)| - \alpha(H)\}$.

Proof (i) Layers of the Cartesian product are isomorphic to the factors; hence there are at most $\alpha(G)$ and $\alpha(H)$ independent vertices in any G-layer and H-layer, respectively.

(ii) If I and J are independent sets in G and H, respectively, then $I \times J$ is an independent set of $G \,\Box\, H$. Let I and J be largest independent sets of G and H. By commutativity, we may assume that $|V(H)| - \alpha(H) \le |V(G)| - \alpha(G)$. Say $V(H) \setminus J = \{x_1, \ldots, x_k\}$, and take a subset $\{a_1, \ldots, a_k\} \subseteq V(G) \setminus I$. Then $(I \times J) \cup \{(a_i, x_i) \mid 1 \le i \le k\}$ is an independent set of $G \,\Box\, H$ with $\alpha(G)\,\alpha(H) + \big(|V(H)| - \alpha(H)\big)$ vertices. (This set is shown schematically in Figure 27.2.) □

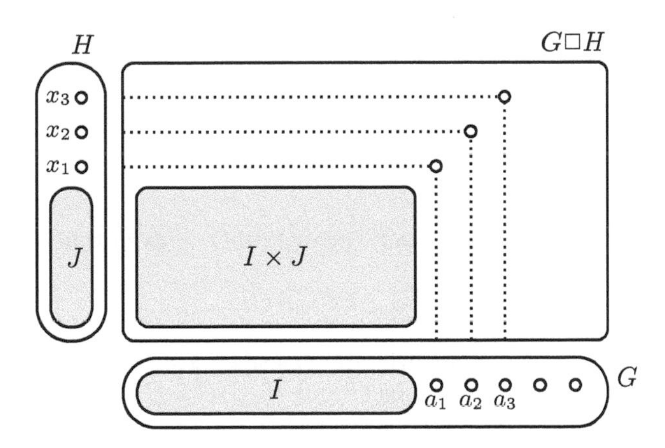

FIGURE 27.2 Bound (ii) of Theorem 27.17.

It is easy to verify that $\alpha(P_3 \,\Box\, P_3) = 5$, so equality can hold in Theorem 27.17 (ii). The bound (i) also cannot be improved in general, as the following result shows.

Proposition 27.18 *If $1 \le k \le n$, then*

$$\alpha(C_{2k+1} \,\Box\, C_{2n+1}) = k(2n+1).$$

Proof Theorem 27.17 (i) yields $\alpha(C_{2k+1} \,\square\, C_{2n+1}) \leq k(2n+1)$. In fact, there exists an independent set I with exactly $k(2n+1)$ vertices. This is indicated in Figure 27.3, where the vertices of I are darkened. Let $V(C_m) = \{0, 1, \ldots, m-1\}$. For each $i \in V(C_{2n+1})$, with $0 \leq i \leq 2k$, the set I has k vertices

$$(i, i), (i, i+2), (i, i+4), \ldots, (i, i+2k-2)$$

in the C_{2k+1}-layer over i. (The arithmetic is done modulo $2k+1$.) For the remaining $2n-2k$ vertices of C_{2n+1}, the first two columns in the diagram are repeated $n-k$ times. It is easy to check that I is independent. □

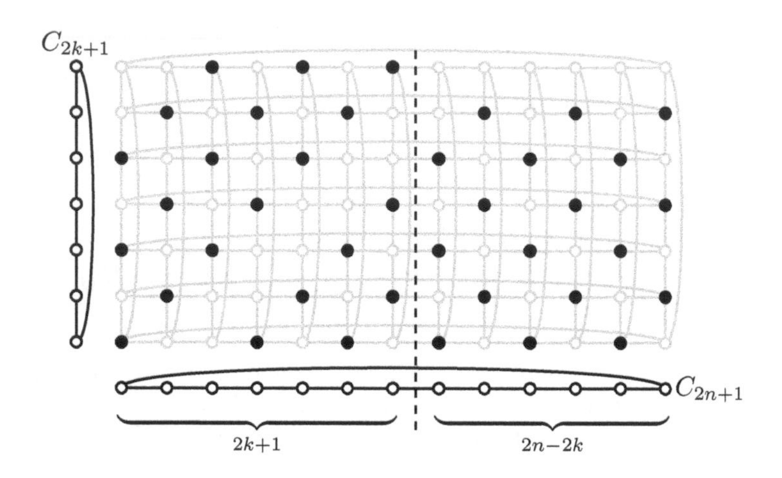

FIGURE 27.3 A maximum independent set in $C_{2n+1} \,\square\, C_{2k+1}$.

As Klavžar (2005) observed, Theorem 27.17 (i) generalizes (Exercise 27.14) as follows:

Theorem 27.19 *Let H be a graph and let $\{V_1, V_2, \ldots, V_k\}$ be a partition of $V(H)$. Then for any graph G,*

$$\alpha(G \,\square\, H) \leq \sum_{i=1}^{k} \alpha(G \,\square\, \langle V_i \rangle)\,.$$

Abay-Asmerom, Hammack, Larson, and Taylor (2011) proved that

$$\alpha(G \,\square\, H) \geq 2r(G)r(H)\,,$$

where $r(G)$ is the radius of G (cf. page 261) and classified graphs for which equality holds. They also proved that for nontrivial graphs G and H, $\alpha(G \,\square\, H) = r(G \,\square\, H)$ if and only if one factor is a K_2 and the other a nontrivial complete graph.

The *independence graph* of G has the maximum independent sets of G as vertices, two vertices being adjacent if the corresponding independent sets are disjoint. Hell, X. Yu, and H. Zhou (1994) observed that the equality in Theorem 27.17 (i) holds if and only if there exists a homomorphism from one of the factors to the independence graph of the other. The independence graph was further studied in Brešar and Zmazek (2003). For instance, they proved that every graph is the independence graph of some graph.

Ultimate Cartesian independence ratio

Hell, X. Yu, and H. Zhou (1994) introduced the *ultimate Cartesian independence ratio* $I_\Box(G)$ of a graph G as

$$I_\Box(G) = \lim_{n\to\infty} i(G^{\Box,n}).$$

By Theorem 27.17 (i), $i(G^{\Box,2}) \leq i(G)$, and, more generally, $i(G^{\Box,n+1}) \leq i(G^{\Box,n})$. Thus the sequence $i(G), i(G^{\Box,2}), \ldots$ is decreasing and bounded by 0, so $I_\Box(G)$ always exists.

Hahn, Hell, and Poljak (1995) proved the following result, which is similar to the No-Homomorphism Lemma:

Theorem 27.20 *If there exists a homomorphism from G to H, then $I_\Box(G) \geq I_\Box(H)$.*

Hahn, Hell, and Poljak (1995) in addition proved that if H is a subgraph of a graph G, then $I_\Box(H) \geq I_\Box(G)$.

Theorem 26.1 immediately implies $\chi(G^{\Box,n}) = \chi(G)$. The fact that $\chi(G) \geq |V(G)|/\alpha(G)$ then yields $i(G^{\Box,n}) = \alpha(G^{\Box,n})/|V(G)|^n \geq 1/\chi(G^{\Box,n}) = 1/\chi(G)$. Hence the inequalities $1/\chi(G) \leq I_\Box(G) \leq i(G)$ always hold. These bounds can be improved as follows. The lower bound is by Zhu (1996); the upper bound is by Hahn, Hell, and Poljak (1995).

Theorem 27.21 *For any graph G,*

$$\frac{1}{\chi_c(G)} \leq I_\Box(G) \leq \frac{1}{\chi_f(G)}.$$

We refer to the extensive survey of Hahn and Tardif (1997) for more information on the ultimate independence ratio. Simonyi (2006) studied a related concept called the Hall-ratio, with respect to various graph products. It turns out that the ultimate Cartesian Hall-ratio is essentially the same concept as the ultimate Cartesian independence ratio.

Exercises

27.1. (Geller and Stahl, 1975) Show that $\alpha(G \circ H) = \alpha(G)\,\alpha(H)$ for all graphs G and H.

27.2. (Jha and Slutzki, 1994) Prove that $\alpha(G \circ H) \leq \alpha(G \boxtimes H) \leq \alpha(G \,\Box\, H) \leq \alpha(G \times H)$ for any G and H.

27.3. Show that $\alpha(C_5 \boxtimes C_5) = 5$.

27.4. Show that even cycles are universal.

27.5. (Vesztergombi, 1978) Show that

$$2\alpha(G \boxtimes C_{2k+1}) \leq (2k+1)\alpha(G)$$

for arbitrary G and any $k \geq 1$.

27.6. Use No-Homomorphism Lemma 2.13 to give another proof of Theorem 27.13.

27.7. (Greenwell and Lovász, 1974) Show directly (that is, not using Theorem 27.13) that $\alpha(K_n^{\times,k}) = n^{k-1}$ for any $n, k \geq 1$.

27.8. (Alon, Dinur, Friedgut, and Sudakov, 2004) Suppose that a red-amber-green traffic light is controlled with n ($n \geq 1$) three-position switches. The color of the traffic light changes whenever all the switches are changed. Show that the traffic light is controlled by only one of the switches.

27.9. (Jha and Slutzki, 1994) Show that

$$\alpha(G \times H) \;\leq\; |V(G)|\,|V(H)| - \omega(G)\,\omega(H) + \max\{\omega(G),\, \omega(H)\}\,.$$

27.10. (Ku and McMillan, 2009) Show that for any $m \geq 2$, $K_{m,m}^{\times,2}$ contains only canonical maximum independent sets, and that when $n \geq 3$, $K_{m,m}^{\times,n}$, also contains noncanonical maximum independent sets.

27.11. Show that for any graphs G and H,

$$\alpha(G \times H) \leq \frac{2\,|E(G)|\,|E(H)|}{\delta(G)\,\delta(H)}\,.$$

27.12. Show that $\alpha(G \times K_n) \leq n|E(G)|/\delta(G)$ for any graph G.

27.13. Show that $I_\times(G) = i^*_\times(G)$ holds for any vertex-transitive graph G.

27.14. Deduce Theorem 27.17 (i) as a consequence of Theorem 27.19.

27.15. Let $\alpha_2(G)$ be the size of a largest bipartite subgraph of G and let $\tau(G)$ be the size of its largest matching (that is, independent set of edges). Show that for any graphs G and H,

$$\alpha(G \,\square\, H) \leq \tau(H)\,\alpha_2(G) + (|V(H)| - 2\tau(H))\,\alpha(G)\,.$$

27.16. (Hahn, Hell, and Poljak, 1995) Show that $I_\square(G) = 1/2$ if G is a bipartite graph with at least one edge.

Chapter 28

Domination and Vizing's Conjecture

Vizing's conjecture, the central open problem of domination theory, concerns the domination number of Cartesian products. We investigate this conjecture in Sections 28.1 and 28.2. Then, in Section 28.3, we turn to the the fractional domination number and prove that the fractional version of Vizing's conjecture is true. Along the way we obtain several results about the domination number for strong products. The final section treats domination of direct products.

28.1 Vizing's Conjecture

Vizing's conjecture regards the domination number of the Cartesian product. Before stating it, we note the following upper bound.

Proposition 28.1 *For any graphs G and H,*

$$\gamma(G \,\square\, H) \leq \min\{\gamma(G)\,|V(H)|,\ \gamma(H)\,|V(G)|\}.$$

Proof Let D be a minimum dominating set of G. Then $D \times V(H)$ dominates $G \,\square\, H$, so $\gamma(G \,\square\, H) \leq \gamma(G)\,|V(H)|$. Similarly, $\gamma(G \,\square\, H) \leq \gamma(H)\,|V(G)|$. □

The following conjecture of Vizing (1968) is the central problem of domination theory. It asserts a lower bound that appears natural, but has eluded all attempts at proof.

Conjecture 28.2 *If G and H are graphs, then $\gamma(G \,\square\, H) \geq \gamma(G)\gamma(H)$.*

Given a graph G, we say that *Vizing's conjecture holds for G* if $\gamma(G \,\square\, H) \geq \gamma(G)\gamma(H)$ for every graph H. We present in this section large classes of graphs for which Vizing's conjecture holds. We will need the following simple but useful result due to Rall; see Theorem 7.3 in Hartnell and Rall (1998).

Proposition 28.3 *Let G be a graph for which Vizing's conjecture holds, and v be a vertex of G for which $\gamma(G - v) = \gamma(G) - 1$. Then Vizing's conjecture holds for $G - v$.*

Proof Let $G' = G - v$, let H be arbitrary, and let D be a minimum dominating set of $G' \,\square\, H$. We can dominate $G \,\square\, H$ by combining D with a minimum dominating set for the H-layer above v. Hence $\gamma(G' \,\square\, H) + \gamma(H) \geq \gamma(G \,\square\, H) \geq \gamma(G)\gamma(H)$, from which follows $\gamma(G' \,\square\, H) \geq \gamma(G)\gamma(H) - \gamma(H) = \gamma(G')\gamma(H)$. □

The next observation is also useful in determining when Vizing's conjecture holds for various graphs.

Proposition 28.4 *Suppose $\gamma(G \,\square\, H) \geq \gamma(G)\gamma(H)$, and let G' be a spanning subgraph of G with $\gamma(G') = \gamma(G)$. Then $\gamma(G' \,\square\, H) \geq \gamma(G')\gamma(H)$.*

Proof We have $\gamma(G' \,\square\, H) \geq \gamma(G \,\square\, H) \geq \gamma(G)\gamma(H) = \gamma(G')\gamma(H)$, where the first inequality holds because G' is a spanning subgraph of G, and therefore $G' \,\square\, H$ is a spanning subgraph of $G \,\square\, H$. □

We now turn to a fundamental result. A graph G is called *decomposable* if $V(G)$ can be partitioned into $\gamma(G)$ sets, each of which induces a complete subgraph of G.

Theorem 28.5 *If G' is a spanning subgraph of a decomposable graph G, and $\gamma(G') = \gamma(G)$, then Vizing's conjecture holds for G'.*

Proof By Proposition 28.4, it suffices to prove the theorem only for the case $G' = G$. Partition $V(G)$ into sets $Q_1, Q_2, \ldots, Q_{\gamma(G)}$, where each Q_i induces a complete subgraph of G. Take an arbitrary H, and let D be a minimum dominating set for $G \,\square\, H$. For each $i \in \{1, 2, \ldots, \gamma(G)\}$, form the set $D_i = D \cap (Q_i \times V(H))$. For each index i and vertex $v \in V(H)$, let K_{iv} denote the complete subgraph of $G \,\square\, H$ induced on $Q_i \times \{v\}$. Let $\mathcal{K} = \{K_{iv} \mid 1 \leq i \leq \gamma(G),\ v \in V(H)\}$, and form the following disjoint subsets of $\mathcal{K}$:

$$\begin{aligned} \mathcal{R} &= \{K_{iv} \in \mathcal{K} \mid \text{ no vertex of } K_{iv} \text{ is dominated by a vertex of } D_i\}, \\ \mathcal{S} &= \{K_{iv} \in \mathcal{K} \mid D \cap V(K_{iv}) \neq \emptyset\}. \end{aligned}$$

Note that if $K_{iv} \in \mathcal{R}$, then (by definition of the Cartesian product) any vertex of K_{iv} is dominated by an element of D that is in some K_{jv}, with $j \neq i$.

To prove $\gamma(G \,\square\, H) \geq \gamma(G)\gamma(H)$, we first show $|\mathcal{R} \cup \mathcal{S}| \geq \gamma(G)\gamma(H)$, then $|D| \geq |\mathcal{R} \cup \mathcal{S}|$.

For each $1 \leq i \leq \gamma(G)$, let $\mathcal{R}_i = \{K_{iv} \in \mathcal{R} \mid v \in V(H)\}$ and $\mathcal{S}_i = \{K_{iv} \in \mathcal{S} \mid v \in V(H)\}$. Observe that for fixed i, the set $\{v \in V(H) \mid K_{iv} \in \mathcal{R}_i \cup \mathcal{S}_i\}$ dominates H and has cardinality $|\mathcal{R}_i \cup \mathcal{S}_i|$, so $|\mathcal{R}_i \cup \mathcal{S}_i| \geq \gamma(H)$. Summing over all i, we get $|\mathcal{R} \cup \mathcal{S}| \geq \gamma(G)\gamma(H)$.

Fix $v \in V(H)$, and put $D_v = D \cap (V(G) \times \{v\})$ and $\mathcal{R}_v = \{K_{iv} \in \mathcal{R} \mid 1 \leq i \leq \gamma(G)\}$ and $\mathcal{S}_v = \{K_{iv} \in \mathcal{S} \mid 1 \leq i \leq \gamma(G)\}$. Every K_{iv} in $\mathcal{R}_v$ or $\mathcal{S}_v$ is a subgraph of $\langle p_H^{-1}(v) \rangle \cong G$, and every vertex of such a K_{iv} is dominated by a vertex in D_v. Now notice that there are $\gamma(G) - |\mathcal{R}_v \cup \mathcal{S}_v|$ subgraphs K_{iv} that are not in $\mathcal{R}_v \cup \mathcal{S}_v$. We can extend D_v to a set that dominates $\langle p_H^{-1}(v) \rangle$ by appending to it one vertex from each of these extra subgraphs. Thus

$$|D_v| + (\gamma(G) - |\mathcal{R}_v \cup \mathcal{S}_v|) \geq \gamma(G),$$

so $|D_v| \geq |\mathcal{R}_v \cup \mathcal{S}_v|$. Summing over all $v \in V(H)$, we get $|D| \geq |\mathcal{R} \cup \mathcal{S}| \geq \gamma(G)\gamma(H)$. □

Theorem 28.5 was proved by Barcalkin and German (1979). Consequently, any graph satisfying its hypothesis is called a *BG-graph*. The next corollary illustrates their utility.

Corollary 28.6 *Vizing's conjecture holds for graphs with domination number 2.*

Proof Suppose G' is a graph with $\gamma(G') = 2$. Form another graph G by adding to G' the greatest possible number of edges for which we still have $\gamma(G) = 2$. By Theorem 28.5, we just need to show that G is decomposable.

Let $Q_1 \subseteq V(G)$ be a maximal vertex set inducing a complete subgraph of G. Then Q_1 must be a proper subset of $V(G)$, for otherwise $\gamma(G) = 1$. Set $Q_2 = V(G) \setminus Q_1$. We claim that Q_2 induces a complete subgraph. To see this, take an arbitrary pair $vw \notin E(G)$ and note that v and w are not both in Q_2: Surely, because $\gamma(G) = 2$ and $\gamma(G \cup vw) = 1$, either

v or w dominates all of $G \cup vw$. Thus, if v and w were both in Q_2, then one of them would dominate every vertex of Q_1, contrary to our choice of Q_1. It follows that any two vertices of Q_2 are adjacent. □

Sun (2004) proved Vizing's conjecture for graphs with domination number 3. His proof is very technical and a more intuitive proof would be welcome.

Theorem 28.5 appeared in an obscure journal, in Russian, and remained largely unnoticed until 1991. Hartnell and Rall (1991) finally pointed out its importance and gave an alternative proof, which we followed above. In the meantime, several corollaries were obtained by authors who were unaware of Barcalkin and German's work.

Before listing notable examples of BG-graphs, we define an important concept. The 2-*packing number* $P_2(G)$ of G is the largest cardinality of a set $S \subseteq V(G)$ such that $d(u, v) \geq 3$ for any distinct vertices u, v of S. (In other words, S is a largest subset of $V(G)$ for which $N[u] \cap N[v] = \emptyset$ for any pair of distinct vertices $u, v \in S$.)

- Jacobson and Kinch (1986) independently proved that Vizing's conjecture holds for any graph whose 2-packing number equals its domination number (Exercise 28.1). This is in particular true for trees, as proved first by Meir and Moon (1975).
- El-Zahar and Pareek (1991) independently proved Vizing's conjecture for cycles (Exercise 28.3).
- Vizing's conjecture was proved for graphs with a certain vertex-partition property by Faudree, Schelp, and Shreve (1990) and by G. Chen, Piotrowski, and Shreve (1996). Hartnell and Rall (1998) showed that the first class is a proper subclass of the BG-graphs while the second is the same as the class of BG-graphs.

Two classes of graphs that fulfill Vizing's conjecture and properly contain BG-graphs are known. They are the largest such classes of graphs currently known:

1. Hartnell and Rall (1995) introduced so-called graphs of Type $\mathcal{X}$, and proved that this class fulfills Vizing's conjecture and properly contains BG-graphs. Further, they proved that this implies the validity of Vizing's conjecture for graphs G with $\gamma(G) \leq P_2(G) + 1$. For a rather technical definition of Type $\mathcal{X}$ graphs, see Hartnell and Rall (1995), as well as Hartnell and Rall (1998).
2. Brešar and Rall (2009) introduced the concept of fair domination, and proved that a graph G with a fair reception of size $\gamma(G)$ fulfills Vizing's conjecture, and that BG-graphs are particular examples of such graphs. Moreover, they used this approach to prove Vizing's conjecture for chordal graphs. (For an alternative approach to chordal graphs, see Section 28.2.)

Hartnell and Rall (1991) proposed another approach to Vizing's conjecture. The idea is to start with some class of graphs for which the conjecture is known to be true. Then certain operations that preserve the conjecture's validity are applied to these graphs. One would have to show that all graphs can be obtained in this fashion.

Several exact domination numbers of particular Cartesian products have been computed by Jacobson and Kinch (1984); Cockayne, Hare, Hedetniemi, and Wimer (1985); T. Y. Chang and Clark (1993); as well as Klavžar and Seifter (1995).

A subset $S \subseteq V(G)$ is a *perfect code* of G if the closed neighborhoods of the vertices in S form a partition of $V(G)$. Not every graph contains a perfect code. However, if S is a perfect code of G, then $|S| = \gamma(G)$ (Exercise 28.4). Perfect codes (and hence the domination number) have been studied in various graph products in Abay-Asmerom, Hammack, and Taylor (2009); Gravier, and Mollard (1997); Jerebic, Klavžar, and Špacapan (2005); Jha (2002, 2003); Klavžar, Špacapan, and Žerovnik (2006); and Taylor (2009). J. Žerovnik, Perfect codes in direct products of cycles—a complete characterization, Adv. in Appl. Math., 41 (2008), 197–205.

28.2 Clark and Suen's Approach

In this section we prove $\gamma(G \square H) \geq \frac{1}{2}\gamma(G)\gamma(H)$ for all graphs G and H, thus proving Vizing's conjecture up to a factor of 2. The approach, due to Clark and Suen, is called the *double-projection argument*, and nicely incorporates the product structure of $G \square H$. The idea yields additional results. In particular we will show that Vizing's conjecture holds for chordal graphs.

Let $\{h_1, \ldots, h_{\gamma(H)}\}$ be a dominating set of a graph H. Choose a partition $\{\pi_1, \ldots, \pi_{\gamma(H)}\}$ of $V(H)$ with $h_i \in \pi_i$ and $\pi_i \subseteq N[h_i]$ for each i. Let $G_i = V(G) \times \pi_i$. For $g \in G$, the set $\{g\} \times \pi_i$ is called a *cell*. (See the left-hand side of Figure 28.1.)

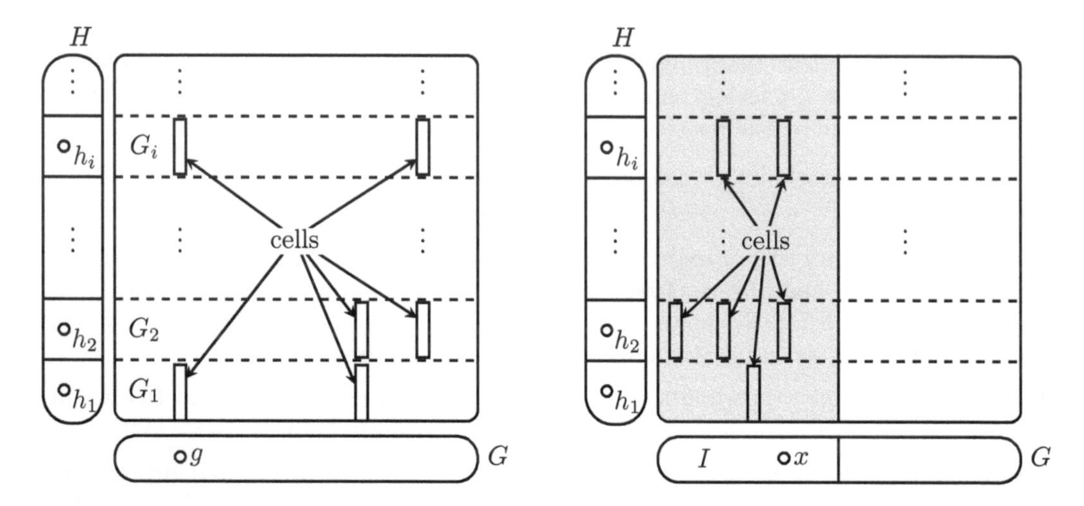

FIGURE 28.1 Clark-Suen partition (left) and Aharoni-Szabó partition (right).

Let D be a minimum dominating set of $G \square H$. For $i = 1, \ldots, \gamma(H)$, let n_i be the number of cells in G_i for which all vertices are dominated from the same H-layer as the cell. Then, considering the projection $p_G(D \cap G_i)$, it follows that $|D \cap G_i| + n_i \geq \gamma(G)$, and thus

$$|D| + \sum_{i=1}^{k} n_i \geq \gamma(G)\gamma(H)\,. \tag{28.1}$$

On the other hand, for each $g \in V(G)$, set $D_g = D \cap (\{g\} \times V(H))$, and let m_g denote the number of cells $\{g\} \times \pi_i$ for which every vertex is dominated by an element of D_g. We can form a dominating set of the H-layer above g by appending to D_g the $\gamma(H) - m_g$ vertices (g, h_i) that do not belong to the above-mentioned cells. Thus $|D_g| + (\gamma(H) - m_g) \geq \gamma(H)$, so $|D_g| \geq m_g$. Hence

$$|D| \geq \sum_{g \in G} m_g\,. \tag{28.2}$$

The sums $\sum_{i=1}^{k} n_i$ and $\sum_{g \in G} m_g$ are equal because they represent two ways of counting the same set of cells. Inequalities (28.1) and (28.2) now yield the result of Clark and Suen (2000):

Theorem 28.7 *For all graphs G and H,*

$$\gamma(G \,\square\, H) \geq \frac{1}{2}\gamma(G)\gamma(H).$$

The factor 1/2 of Theorem 28.7 comes from a double counting of the vertices of the minimum dominating set D. To avoid this problem, Aharoni, and Szabó (2009) modified Clark and Suen's approach as follows.

Let M be a set of vertices in a graph G. The smallest cardinality of a set D that dominates M (i.e., such that $M \subseteq N[D]$) is denoted $\gamma_G(M)$. Let $\gamma^i(G)$ denote the maximum, over all independent sets M in G, of the value $\gamma_G(M)$.

Let I be an independent set of G that is dominated by at least $\gamma^i(G)$ vertices. Let $\{\pi_1, \ldots, \pi_{\gamma(H)}\}$ be a partition of $V(H)$ as above. Let D be a (minimum) set that dominates $I \times V(H)$. In this modified approach we consider only cells of form $\{x\} \times \pi_i$, where $x \in I$. (See the right-hand side of Figure 28.1.) Otherwise we follow Clark and Suen's approach. Let m be the number of cells that are dominated from their corresponding H-layers. Considering the projection onto G, we infer that

$$|D \cap (V(G) \setminus I \times V(H))| \geq \gamma^i(G)\gamma(H) - m. \tag{28.3}$$

Projection onto H gives

$$|D \cap (I \times V(H))| \geq m, \tag{28.4}$$

so Inequalities (28.3) and (28.4) imply $|D| \geq \gamma^i(G)\gamma(H)$. Because any dominating set of $G \,\square\, H$ also dominates $I \times V(H)$, we have $\gamma(G \,\square\, H) \geq |D|$, hence the following theorem.

Theorem 28.8 *If G and H are graphs, then $\gamma(G \,\square\, H) \geq \gamma^i(G)\gamma(H)$.*

Clearly, $\gamma^i(G) \leq \gamma(G)$. Unfortunately, $\gamma^i(G)$ can be arbitrarily smaller than $\gamma(G)$ (Exercise 28.7). On the positive side, Aharoni, E. Berger, and Ziv (2002) proved that if M is any set of vertices in a chordal graph G and if $x, y \in M$ are adjacent, then at least one of $\gamma_G(M - \{x\})$ or $\gamma_G(M - \{y\})$ equals $\gamma_G(M)$. By starting with $M = V(G)$ and applying this repeatedly, it follows that $\gamma(G) = \gamma^i(G)$ when G is a chordal graph. This combined with Theorem 28.8 yields the main result of Aharoni, and Szabó (2009):

Theorem 28.9 *If G is a chordal graph, then G satisfies Vizing's conjecture.*

If G is a graph—such as a tree—that has a 2-packing of cardinality $\gamma(G)$, it is easy to see that $\gamma(G) = \gamma^i(G)$.

We conclude our discussion of Vizing's conjecture by noting that in general $\gamma(G \,\square\, H)$ is most likely much larger than $\gamma(G)\gamma(H)$. Nevertheless, there are infinite families of graphs for which equality holds. Such families occur implicitly in Payan and Xuong (1982) and explicitly in Fink, Jacobson, Kinch, and J. Roberts (1985), as well as in Jacobson and Kinch (1986) and Hartnell and Rall (1991). El-Zahar, Khamis, and Nazzal (2007) proved that $\gamma(C_n \,\square\, G) = \gamma(C_n)\gamma(G)$ holds only if $n \equiv 1 \pmod 3$. They also characterized graphs that satisfy the equality when $n = 4$ and provided infinite classes of such graphs for general $n \equiv 1 \pmod 3$.

For additional information on Vizing's conjecture, see the extensive survey by Brešar et al. (2011) that also contains some new results and proofs.

28.3 Fractional Version of Vizing's Conjecture

We now prove the so-called fractional version of Vizing's conjecture. It follows immediately from the multiplicativity of the fractional domination number on the strong product. We adopt the approach of D. C. Fisher (1994) and of D. C. Fisher, Ryan, Domke, and Majumdar (1994).

The domination number and the fractional domination number can be expressed as solutions to certain linear programs. To begin, let us rephrase the domination number of a graph G into the language of weight functions. There is a bijective correspondence between subsets $X \subset V(G)$ and *weight functions* $w : V(G) \to \{0,1\}$, where $w(x) = 1$ exactly when $x \in X$. Note that X dominates G if and only if $\sum_{x \in N[v]} w(x) \geq 1$ for each $v \in V(G)$. Moreover, $\gamma(G)$ is the minimum, over all such w, of the sum $\sum_{x \in V(G)} w(x)$.

By analogy, a weight function $w : V(G) \to \mathbb{R}$ is called a *fractional domination* of G provided that for any vertex v we have $w(v) \geq 0$ and

$$\sum_{x \in N[v]} w(x) \geq 1\,.$$

The *fractional domination number* of G, denoted $\gamma_f(G)$, is the minimum, over all fractional dominations, of the sum $\sum_{x \in V(G)} w(x)$.

If G has adjacency matrix $A(G)$ relative to some ordering of its vertices, then its *neighborhood matrix* is $N(G) = A(G) + I$, where I is the identity matrix. Any $X \subset V(G)$ can be encoded as 0-1 vector $\mathbf{x} \in \mathbb{Z}^n$, whose ith entry is 1 precisely when the ith vertex of G is in X. Then

$$\begin{aligned} &\gamma(G) = \min_{\mathbf{x}} \mathbf{1}_n^T \mathbf{x} \\ &\text{subject to } N(G)\mathbf{x} \geq \mathbf{1}_n,\ \mathbf{x} \geq \mathbf{0}_n,\ \text{and } \mathbf{x} \in \mathbb{Z}^n\,. \end{aligned} \tag{28.5}$$

Because we search for a minimum, conditions $\mathbf{x} \geq \mathbf{0}_n$ and $\mathbf{x} \in \mathbb{Z}^n$ imply that the solution $\mathbf{x}$ is a 0-1 vector. The condition $N(G)\mathbf{x} \geq \mathbf{1}_n$ ensures that for any vertex v, we find in $N[v]$ at least one vertex of $\mathbf{x}$. It follows that a solution to the above linear program is indeed $\gamma(G)$. Similarly, the linear program below determines the fractional domination number of G:

$$\begin{aligned} &\gamma_f(G) = \min_{\mathbf{x}} \mathbf{1}_n^T \mathbf{x} \\ &\text{subject to } N(G)\mathbf{x} \geq \mathbf{1}_n,\ \mathbf{x} \geq \mathbf{0}_n,\ \text{and } \mathbf{x} \in \mathbb{R}^n\,. \end{aligned} \tag{28.6}$$

Likewise, the following integer linear program returns the 2-packing number of G:

$$\begin{aligned} &P_2(G) = \max_{\mathbf{x}} \mathbf{1}_n^T \mathbf{x} \\ &\text{subject to } N(G)\mathbf{x} \leq \mathbf{1}_n,\ \mathbf{x} \geq \mathbf{0}_n,\ \text{and } \mathbf{x} \in \mathbb{Z}^n\,. \end{aligned} \tag{28.7}$$

It will be convenient to use the dual version of the program (28.6):

$$\begin{aligned} &\gamma_f(G) = \max_{\mathbf{x}} \mathbf{1}_n^T \mathbf{x} \\ &\text{subject to } N(G)\mathbf{x} \leq \mathbf{1}_n,\ \mathbf{x} \geq \mathbf{0}_n,\ \text{and } \mathbf{x} \in \mathbb{R}^n\,. \end{aligned} \tag{28.8}$$

A feasible solution of the linear program (28.8) is called a *fractional 2-packing*.

Figure 28.2 illustrates the concepts of fractional domination and fractional 2-packing.

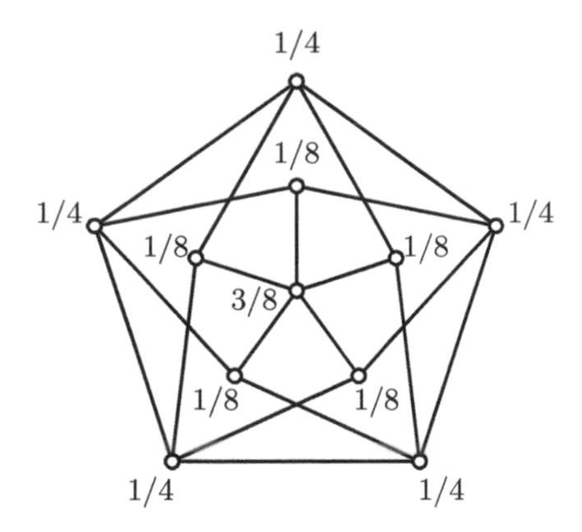

FIGURE 28.2 A fractional domination and a fractional 2-packing of Mycielski's graph.

(This graph M is known as the Mycielski or Grötsch graph.) The weights are a fractional domination. Hence the linear program (28.6) gives $\gamma_f(M) \leq 9/4$. On the other hand, $\gamma_f(M) \geq 9/4$ as the weights are also a fractional 2-packing. Thus $\gamma_f(M) = 9/4$.

The linear programming approach yields the following result of Domke, Hedetniemi, and Laskar (1988).

Proposition 28.10 *For any graph G, we have $P_2(G) \leq \gamma_f(G) \leq \gamma(G)$.*

Proof The maximum of (28.7) is searched on a subset of the set used in (28.8); thus $P_2(G) \leq \gamma_f(G)$. Comparing (28.5) with (28.6), we get $\gamma_f(G) \leq \gamma(G)$. □

Let's apply the above approach to the strong product. Suppose G and H are graphs with vertex sets $V(G) = \{g_1, g_2, \ldots, g_n\}$ and $V(H) = \{h_1, h_2, \ldots, h_m\}$, respectively. Label the i,j-entry of the neighborhood matrix $N(G)$ as g_{ij}, so that $g_{ij} = 1$ precisely when $g_i g_j \in E(G)$ or $i = j$. By the same rule, label each i,j-entry of $N(H)$ as h_{ij}.

Let $\mathbb{Z}(n,m)$ denote the set of $n \times m$ integer matrices, and $\mathbb{R}(n,m)$ denote the set of $n \times m$ real matrices. We regard such a matrix $Z = [z_{ij}]$ as an assignment of the number z_{ij} to each vertex (g_i, h_j) of $G \boxtimes H$. If Z is a 0-1 matrix, we view it as a characteristic vector for the subset $X_Z = \{(g_i, h_j) \;:\; z_{ij} = 1\} \subseteq V(G \boxtimes H)$.

Now consider the product $N(G)ZN(H)$. Notice that its i,j-entry is

$$(N(G)ZN(H))_{ij} = \sum_{k=1}^{n} g_{ik} \sum_{s=1}^{m} z_{ks} h_{sj} = \sum_{k=1}^{n} \sum_{s=1}^{m} g_{ik} h_{js} z_{ks}\,. \tag{28.9}$$

Observe that each term in this sum has value

$$g_{ik} h_{js} z_{ks} = \begin{cases} z_{ks} & \text{if } (g_i, h_j)(g_k, h_s) \in E(G \boxtimes H) \text{ or } (g_i, h_j) = (g_k, h_s)\,, \\ 0 & \text{otherwise}\,. \end{cases} \tag{28.10}$$

In the case where Z is a 0-1 matrix, it follows that $g_{ik} h_{js} z_{ks} = 1$ precisely if the vertex (g_i, h_j) of $G \boxtimes H$ is dominated by the vertex $(g_k, h_s) \in X_Z$. Comparing with Equation (28.9), we see that X_Z dominates $G \boxtimes H$ if and only if $(N(G)ZN(H))_{ij} \geq 1$ for each i and j. Moreover, the cardinality of the dominating set is the sum of the entries of Z.

We can thus express $\gamma(G \boxtimes H)$ as the solution of the integer linear programming problem of minimizing the sum of the z_{ij} subject to the constraints

$$\begin{aligned} \textstyle\sum_{k=1}^{n}\sum_{s=1}^{m} g_{ik}h_{js}z_{ks} &\geq 1\,, \\ z_{ks} &\geq 0\,. \end{aligned} \tag{28.11}$$

We can write these constraints as $N(G)ZN(H) \geq \mathbf{1}_n\mathbf{1}_m^T$, and $Z \geq O$, where O is the $n \times n$ zero matrix. In summary, for graphs G and H of order n and m, respectively, we have

$$\begin{aligned} &\gamma(G \boxtimes H) = \min_Z \mathbf{1}_n^T Z \mathbf{1}_m \\ &\text{subject to } N(G)\,Z\,N(H) \geq \mathbf{1}_n\mathbf{1}_m^T,\ \ Z \geq O,\ \ Z \in \mathbb{Z}(n,m)\,. \end{aligned} \tag{28.12}$$

By the same kind of reasoning, we also have

$$\begin{aligned} &\gamma_f(G \boxtimes H) = \min_Z \mathbf{1}_n^T Z \mathbf{1}_m \\ &\text{subject to } N(G)\,Z\,N(H) \geq \mathbf{1}_n\mathbf{1}_m^T,\ \ Z \geq O,\ \ Z \in \mathbb{R}(n,m)\,, \end{aligned} \tag{28.13}$$

$$\begin{aligned} &P_2(G \boxtimes H) = \max_Z \mathbf{1}_n^T Z \mathbf{1}_m \\ &\text{subject to } N(G)\,Z\,N(H) \leq \mathbf{1}_n\mathbf{1}_m^T,\ \ Z \geq O,\ \ Z \in \mathbb{Z}(n,m)\,. \end{aligned} \tag{28.14}$$

Finally, by examining the constraints (28.11) and using the symmetry of $N(G)$ and $N(H)$, we can formulate the dual of the program (28.13) as

$$\begin{aligned} &\gamma_f(G \boxtimes H) = \max_Z \mathbf{1}_n^T Z \mathbf{1}_m \\ &\text{subject to } N(G)\,Z\,N(H) \leq \mathbf{1}_n\mathbf{1}_m^T,\ \ Z \geq O,\ \ Z \in \mathbb{R}(n,m)\,. \end{aligned} \tag{28.15}$$

Theorem 28.11 *If G and H are graphs, then $\gamma_f(G \boxtimes H) = \gamma_f(G)\gamma_f(H)$.*

Proof Suppose G and H have orders n and m, respectively. Let $\mathbf{x}$ and $\mathbf{y}$ be solutions of the linear program (28.8) for G and H, respectively. Then

$$N(G)\mathbf{x}\mathbf{y}^T N(H) = (N(G)\mathbf{x})(N(H)\mathbf{y})^T \leq \mathbf{1}_n\mathbf{1}_m^T\,.$$

We infer that $\mathbf{x}\mathbf{y}^T$ is a feasible solution of (28.15). Therefore

$$\gamma_f(G \boxtimes H) \geq \mathbf{1}_n^T\mathbf{x}\mathbf{y}^T\mathbf{1}_m = (\mathbf{1}_n^T\mathbf{x})(\mathbf{1}_m^T\mathbf{y})^T = \gamma_f(G)\gamma_f(H)\,.$$

Conversely, let $\mathbf{z}$ and $\mathbf{w}$ be solutions of (28.6) for G and H, respectively. As above we conclude that $\mathbf{z}\mathbf{w}^T$ is a feasible solution of (28.13). Therefore

$$\gamma_f(G \boxtimes H) \leq \mathbf{1}_n^T\mathbf{z}\mathbf{w}^T\mathbf{1}_m = (\mathbf{1}_n^T\mathbf{z})(\mathbf{1}_m^T\mathbf{w})^T = \gamma_f(G)\gamma_f(H)\,.$$

□

Note the similarity between Theorems 28.11 and 26.17.

Because $G \,\Box\, H$ is a spanning subgraph of $G \boxtimes H$, we have $\gamma_f(G \,\Box\, H) \geq \gamma_f(G \boxtimes H)$. From this, Theorem 28.11 immediately implies the next corollary.

Corollary 28.12 (Fractional version of Vizing's conjecture) *If G and H are graphs, then $\gamma_f(G \,\Box\, H) \geq \gamma_f(G)\gamma_f(H)$.*

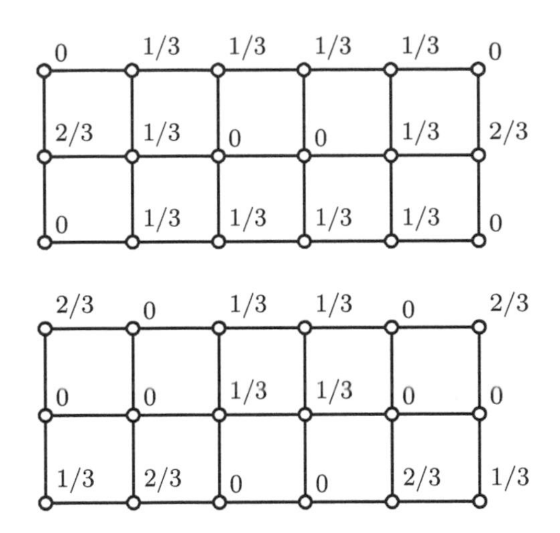

FIGURE 28.3 Fractional domination and fractional 2-packing of $P_3 \,\square\, P_6$.

Consider, for instance, the graph $P_3 \,\square\, P_6$, in Figure 28.3. The top shows a fractional domination, and the bottom shows a fractional 2-packing. Because both sum to 14/3, we infer $\gamma_f(P_3 \,\square\, P_6) = 14/3$. As $\gamma_f(P_3) = 1$ and $\gamma_f(P_6) = 2$, we see that the inequality in Corollary 28.12 can be strict. The Hamming graph $K_3 \,\square\, K_3$ is a similar example, as $\gamma_f(K_3 \,\square\, K_3) = 9/5$. (Assign the weight 1/5 to every vertex of $K_3 \,\square\, K_3$.) The inequality in Corollary 28.12 is also the best possible, as $\gamma_f(P_4) = 2$ and $\gamma_f(P_4 \,\square\, P_4) = 4$.

The above technique yields additional results. Here is an example.

Theorem 28.13 *For any graphs G and H,*

$$\gamma(G)\gamma_f(H) \leq \gamma(G \boxtimes H) \leq \gamma(G)\gamma(H)\,.$$

Proof If X and Y are dominating sets of G and H, respectively, then $X \times Y$ is a dominating set of $G \boxtimes H$, and the upper bound follows. (Of course this bound could also be derived by linear programming methods, as above.) The lower bound requires new arguments.

Let $Z \in \mathbb{Z}(n,m)$ be a solution of the program (28.12). Then

$$N(G)\,Z\,N(H) \geq \mathbf{1}_n \mathbf{1}_m^T\,.$$

For the jth column, we have

$$(N(G)\,Z\,N(H))_j \geq \mathbf{1}_n\,,$$

and hence

$$(N(G)\,Z\,N(H))_j = N(G)(Z\,N(H))_j \geq \mathbf{1}_n\,.$$

Thus $(Z\,N(H))_j$ is a feasible solution of the program (28.5); that is, it represents a dominating set of G. Hence $\mathbf{1}_n^T(Z\,N(H))_j \geq \gamma(G)$ holds for any column j, and thus

$$\mathbf{1}_n^T Z\,N(H) \geq \gamma(G)\mathbf{1}_m^T\,.$$

Transposing this inequality and using $N(H) = N(H)^T$, we find

$$N(H)\left(\frac{1}{\gamma(G)} Z^T \mathbf{1}_n\right) \geq \mathbf{1}_m\,.$$

By program (28.6), this implies that $Z^T\mathbf{1}_n/\gamma(G)$ is a fractional domination of H. Hence

$$\gamma_f(H) \le \frac{1}{\gamma(G)}(\mathbf{1}_m^T Z^T \mathbf{1}_n) = \frac{1}{\gamma(G)}(\mathbf{1}_n^T Z \mathbf{1}_m)^T = \frac{1}{\gamma(G)}\gamma(G \boxtimes H),$$

so $\gamma(G)\gamma_f(H) \le \gamma(G \times H)$. □

Theorem 28.13 has several consequences. This one is immediate:

Corollary 28.14 *If $\gamma_f(G) = \gamma(G)$, then $\gamma(G \boxtimes H) = \gamma(G)\gamma(H)$.*

The next corollary follows from Theorem 28.13 and Proposition 28.10. Nowakowski and Rall (1996) give a direct proof.

Corollary 28.15 *If $P_2(G) = \gamma(G)$, then $\gamma(G \boxtimes H) = \gamma(G)\gamma(H)$.*

We conclude the section with one more related result, which the reader can verify by imitating the proof of Theorem 28.13.

Theorem 28.16 *If G and H are graphs, then $P_2(G)P_2(H) \le P_2(G \boxtimes H) \le P_2(G)\gamma_f(H)$.*

28.4 Domination in Direct Products

Proposition 28.1 gives a natural and obvious upper bound for the domination number of a Cartesian product. The corresponding bound for the direct product is less intuitive. Nevertheless, our next theorem presents both upper and lower bounds. The lower bound involves the following natural concept. A *total dominating set* of G is a subset $S \subseteq V(G)$ for which every vertex of G is adjacent to a vertex in S. (If G has no isolated vertices, then $V(G)$ is a total dominating set.) The *total domination number* $\gamma_t(G)$ of G is the minimum cardinality of a total dominating set. The lower bound in our next theorem is from Nowakowski and Rall (1996); the upper bound is from Brešar, Klavžar, and Rall (2007).

Theorem 28.17 *For any graphs G and H with no isolated vertices,*

$$\max\{P_2(G)\gamma_t(H), P_2(H)\gamma_t(G)\} \le \gamma(G \times H) \le 3\gamma(G)\gamma(H).$$

Proof Suppose D is a dominating set for $G \times H$. To prove the lower bound, it suffices (by commutativity) to show $P_2(G)\gamma_t(H) \le |D|$. Let S be a 2-packing of G, so the sets $N[a] \times V(H)$ for $a \in S$ are pairwise disjoint. The desired inequality will follow as soon as we establish $\gamma_t(H) \le |D \cap (N[a] \times V(H))|$ for each $a \in S$. This is easy to do: Set $M = D \cap (N[a] \times V(H))$. Each vertex $(a,h) \in \{a\} \times V(H)$ is dominated by some vertex from $D \cap (N[a] \times V(H))$. If a vertex $(a,h) \in D$ is dominated only by itself, then remove it from M and append to M a vertex (a',h'), where $(a,h)(a',h') \in E(G \times H)$. After doing this for each such (a,h), we arrive at a set M with $|M| \le |D \cap (N[a] \times V(H))|$. It is straightforward to verify that $p_H(M)$ is a total dominating set for H, so $\gamma_t(H) \le |p_H(M)| \le |M| \le |D \cap (N[a] \times V(H))|$.

For the upper bound, select minimum dominating sets S and T of G and H, respectively. Enlarge S to a set $\overline{S}$ by including a neighbor of any vertex $x \in S$ that has no neighbor in S. Then $\overline{S}$ is a total dominating set of G and $|\overline{S}| \le 2|S|$. In the same way, enlarge T to a total dominating set $\overline{T}$ of H. Then $(S\times\overline{T})\cup(\overline{S}\times T)$ dominates $G\times H$, by Exercise 28.11. As $|S| = \gamma(G)$, and $|T| = \gamma(H)$, and $|\overline{S}| \leq 2|S|$, and $|\overline{T}| \leq 2|T|$, it follows that $|(S\times\overline{T})\cup(\overline{S}\times T)| = |S\times\overline{T}| + |\overline{S}\times T| - |S\times T| \leq 2|S||T| + 2|S||T| - |S||T| = 3\gamma(G)\gamma(H)$. □

Because $\gamma(G) \leq \gamma_t(G)$ holds for any graph G, the lower bound of Theorem 28.17 implies that $\max\{P_2(G)\gamma(H), P_2(H)\gamma(G)\} \leq \gamma(G\times H)$.

Given arbitrary connected graphs G and H, subdivide each of their edges by two vertices to get G' and H'. Let G'' and H'' be further obtained from G' and H' by attaching two pendant vertices to each vertex of their minimum dominating sets. Then $\gamma(G''\times H'') = 3\gamma(G'')\gamma(H'')$, so the upper bound of Theorem 28.17 is the best possible.

Gravier and Khelladi (1995) made the following Vizing-like conjecture:

$$\gamma(G\times H) \geq \gamma(G)\gamma(H).$$

It was disproved by a counterexample (Exercise 28.9) in Nowakowski and Rall (1996). Also, Klavžar and Zmazek (1996) showed that for any $k \geq 0$, there is a graph G with $\gamma(G\times G) \leq \gamma(G)^2 - k$.

The next result is due to Mekiš (2010) (cf. Exercise 28.10):

Proposition 28.18 *If* $G = \times_{i=1}^{r} K_{n_i}$, *with* $r \geq 3$ *and* $n_i \geq r+1$ *for all* i, *then* $\gamma(G) = r+1$.

This was used for the following nice construction. Let $n \geq 3$ and set $G = \times_{i=1}^{n} K_{2n+1}$. Then Proposition 28.18 gives $\gamma(G)\gamma(G) - \gamma(G\times G) = (n+1)(n+1) - (2n+1) = n^2$.

Exercises

28.1. Show that $\gamma(G\,\square\,H) \geq P_2(G)\gamma(H)$ for any graphs G and H.

28.2. (Jacobson and Kinch, 1984) Show that for any graphs G and H,

$$\gamma(G\,\square\,H) \geq \frac{|V(H)|}{\Delta(H)+1}\gamma(G).$$

28.3. Show that the n-cycle C_n, $n \geq 3$, is a spanning subgraph of a decomposable graph G with $\gamma(G) = \lceil n/3 \rceil$.

28.4. Let S be a perfect code of a graph G. Show that $|S| = \gamma(G)$.

28.5. (Jacobson and Kinch, 1984) Show that $\gamma(P_2\,\square\,P_n) = \lceil (n+1)/2 \rceil$.

28.6. (Klavžar and Seifter, 1995) Show that $\gamma(C_3\,\square\,C_n) = n - \lfloor n/4 \rfloor$ for $n \geq 4$.

28.7. (Roy Meshulam, see Aharoni and Szabó (2009)) Let $n \geq 1$ and let G_n be the complement of the Kneser graph $K(n, n^2)$. Show that $\gamma^i(G_n) = 1$ and that $\gamma(G_n) = n$.

28.8. Show that $\gamma(K_n \times K_m) = 3$ for any $n, m \geq 3$.

28.9. (Nowakowski and Rall, 1996) Let G be the graph on six vertices obtained from K_6 by removing three independent edges (that is, a perfect matching). Show that $\gamma(G\times G) = 3$.

28.10. Let G be as in Proposition 28.18. Find a dominating set of G of size $r+1$.

28.11. Complete the proof of Theorem 28.17 by verifying that the set $S \times \overline{T} \cup \overline{S} \times T$ from the proof is a dominating set of $G \times H$.

28.12. Let H be a graph with $\gamma(H) = 1$ (that is, a graph with a vertex of degree $|V(H)|-1$). Show $\gamma(G \circ H) = \gamma(G)$ for any G.

Chapter 29

Cycle Spaces and Bases

There are numerous interesting and useful graph invariants associated with the cycle space of a graph. Of particular interest are certain "efficient" bases called *minimum cycle bases.* These ideas have a long history in applied discrete mathematics, going back at least as far as G. Kirchhoff's (1847) treatise on electrical networks. More recently, Berger, Flamm, Gleiss, Leydold, and Stadler (2004) describe an application of minimum cycle bases to the problem of characterizing molecular graphs. See Kaveh (1995) for applications to structural flexibility analysis. Kaveh and Mirzaie (2008) apply minimum cycle bases of Cartesian and strong products to the force method of frame analysis.

This chapter is primarily concerned with constructing minimum cycle bases of product graphs in terms of the cycle structures of the factors. We begin with a review of cycle spaces.

29.1 The Cycle Space of a Graph

Given a graph G, let $\mathcal{E}(G)$ be the power set of $E(G)$, that is, the set of all subsets of $E(G)$, including the empty set. This is a vector space over the two-element field $\mathrm{GL}(2) = \{0, 1\}$, where the zero vector is $0 = \emptyset$ and the sum $X + Y$ of elements $X, Y \in \mathcal{E}(G)$ is symmetric difference; that is, $X + Y = (X \cup Y) \setminus (X \cap Y)$. We call $\mathcal{E}(G)$ the *edge space* of G. We identify any element $X \in \mathcal{E}(G)$ with the subgraph of G whose edge set is X and whose vertices are endpoints of edges in X. Thus $\mathcal{E}(G)$ is the set of all subgraphs of G that have no isolated vertices (plus the empty subgraph 0). Therefore $E(G)$ is a basis for $\mathcal{E}(G)$, and $\dim(\mathcal{E}(G)) = |E(G)|$.

Similarly, the *vertex space* $\mathcal{V}(G)$ of G is the power set of $V(G)$ viewed as a vector space over $\mathrm{GL}(2)$. (Again the sum is symmetric difference.) It is the set of all edgeless subgraphs of G. Its dimension is $|V(G)|$.

There is a linear *boundary map* $\delta : \mathcal{E}(G) \to \mathcal{V}(G)$ defined on the basis $E(G)$ as $\delta(xy) = x+y$. The subspace $\mathcal{C}(G) = \ker(\delta)$ is called the *cycle space* of G. Notice that the cycle space consists precisely of the subgraphs X in $\mathcal{E}(G)$ whose vertices all have positive even degree. Each component of such a subgraph is Eulerian, so it can be decomposed into edge-disjoint cycles (each of which belongs to $\mathcal{C}(G)$), so it follows that $\mathcal{C}(G) \subseteq \mathcal{E}(G)$ is spanned by the cycles in G. Figure 29.1 shows the cycle space of a graph. In this instance the cycle space is two-dimensional; the sum of any two nonzero elements equals the remaining nonzero element.

The dimension of $\mathcal{C}(G)$, denoted $\beta(G)$, is called the *first Betti number*, or the *cyclomatic number*, of G. It has a simple formula. By Exercise 29.1, the image of δ has dimension $|V(G)| - c(G)$, where $c(G)$ is the number of components of G. The rank theorem yields

$$\beta(G) = |E(G)| - |V(G)| + c(G). \tag{29.1}$$

A basis for the cycle space is called a *cycle basis* for G. For the graph in Figure 29.1, any two nonzero elements form a cycle basis. (However, in general, the elements of a cycle basis need not be cycles.)

FIGURE 29.1 The cycle space of a graph.

There is a simple construction for a cycle basis of G: Let F be a maximal spanning forest of G, so the set $S = E(G) - E(F)$ has $|E(G)| - (|V(G)| - c(G)) = \beta(G)$ edges. For each $e \in S$, let C_e be the unique cycle in $F + e$. Then $\mathcal{B} = \{C_e \mid e \in S\}$ is linearly independent, as each element possesses an edge that belongs to no other element. (Such an edge cannot be canceled in a nontrivial sum of the cycles C_e.) As $\mathcal{B}$ has cardinality $\beta(G)$, it is a basis.

The elements of a cycle basis are naturally weighted by their number of edges. The *total length* of a cycle basis $\mathcal{B}$ is the number $\ell(\mathcal{B}) = \sum_{C \in \mathcal{B}} |C|$. Given a graph G, a cycle basis with the smallest possible total length is called a *minimum cycle basis*, or *MCB* for G. As an example, the graph in Figure 29.1 has exactly one MCB, namely the basis consisting of the square and the triangle.

The cycle space is a weighted matroid where each element C has weight $|C|$. Hence the Greedy Algorithm (see Oxley (1992)) always terminates with an MCB. (That is, begin with $\mathcal{M} = \emptyset$; then append shortest cycles to it, maintaining independence of $\mathcal{M}$, until no further shortest cycles can be appended; then append next-shortest cycles, maintaining independence, until no further such cycles can be appended; and so on, until $\mathcal{M}$ is a maximal independent set. Then $\mathcal{M}$ is an MCB.)

Here is our primary criterion for determining if a cycle basis is an MCB. For a proof, see Exercise 29.4.

Proposition 29.1 *A cycle basis $\mathcal{B} = \{B_1, B_2, \ldots, B_{\beta(G)}\}$ for a graph G is an MCB if and only if every $C \in \mathcal{C}(G)$ is a sum of basis elements whose lengths do not exceed $|C|$.*

We close this section with some remarks that will be essential for our later arguments. Any weak homomorphism $f : G \to H$ induces a linear map $f^* : \mathcal{E}(G) \to \mathcal{E}(H)$ defined on the basis $E(G)$ as $f^*(xy) = f(x)f(y)$ if $f(x) \neq f(y)$ and $f^*(xy) = 0$ if $f(x) = f(y)$. One checks that this restricts to a linear map $f^* : \mathcal{C}(G) \to \mathcal{C}(H)$ between cycle spaces. Clearly, if f is a graph isomorphism, then f^* is a vector space isomorphism.

We also remark that if H is a subgraph of G, then $\mathcal{C}(H)$ is a subspace of $\mathcal{C}(G)$.

29.2 Minimum Cycle Bases for Cartesian and Strong Products

We now describe a means of forming MCBs of Cartesian and strong products in terms of MCBs of their factors. However, in both cases it is necessary to include some cycles that do not correspond to cycles in the factors. For instance, a Cartesian product $G \square H$ contains many squares of form $e \square f$, where e and f are edges in G and H. We would expect such short cycles to appear in an MCB, and this is indeed the case.

Let $\mathcal{S}(G \square H) \subseteq \mathcal{C}(G \square H)$ be the subspace spanned by $\{e \square f \mid e \in E(G), f \in E(H)\}$. We call this the *square space* of $G \square H$. The next lemma produces a large independent set of squares in the square space. It, and the theorems that follow, are adapted from Imrich and Stadler (2002).

Lemma 29.2 *Suppose G and H are graphs with maximal spanning forests $T \subseteq G$ and $U \subseteq H$. Then the following set of squares is linearly independent in $\mathcal{S}(G \square H) \subseteq \mathcal{C}(G \square H)$:*

$$\mathcal{B} = \big\{e \square f \mid e \in E(G), f \in E(U)\big\} \cup \big\{e \square f \mid e \in E(T), f \in E(H) \setminus E(U)\big\}.$$

(Note: $\mathcal{B}$ consists of all $e \square f$ except *those with $e \in E(G) \setminus E(T)$ and $f \in E(H) \setminus E(U)$.)*

Proof Let B_L denote the set on the left of the union and B_R the set on the right. It suffices to show that B_L and B_R are linearly independent, and $\text{Span}(B_L) \cap \text{Span}(B_R) = \{0\}$.

To check the independence of B_L, consider an arbitrary nonempty collection $\{e_i \square f_i \mid i \in I\} \subseteq B_L$, indexed over some set I. We will show that the linear combination $S = \sum_{i \in I} e_i \square f_i$ is nonzero. Take a pendant edge ab of the subforest $\bigcup_{i \in I} f_i \subseteq U$, such that b is an end vertex. Only the terms of the sum that have the form $e_i \square ab$ have an edge in the G-layer $p_H^{-1}(b)$. Each $e_i \square ab$ contributes a unique such edge, so S contains edges in a G-layer and hence is not zero.

By the same argument (with the roles of the factors reversed), we see that B_R is independent, and any nontrivial linear combination of its elements has edges in an H-layer.

Finally, suppose $S \in \text{Span}(B_L) \cap \text{Span}(B_R)$. If S were nonzero, the previous paragraph would guarantee that S had some edge $\{x\} \square f$ in an H-layer. But then the definitions of B_L and B_R would imply $f \in E(U)$ and $f \in E(H) \setminus E(U)$. Hence $S = 0$. $\square$

With the lemma proved, we can now give a construction for a MCB of $G \square H$ in terms of MCBs of layers $G^x \cong G$ and $H^x \cong H$. For simplicity, we impose the added constraint that the factors are triangle-free. In reading the proof, the reader may opt to assume that both G and H are connected; this makes the dimension count simpler. (And it does restrict the generality of the result: The proof can be completed with the observation that an MCB of a disconnected graph is the union of MCBs of its components.)

Theorem 29.3 (An MCB for $\mathbf{G \square H}$) *For any $x \in V(G \square H)$, there is a direct sum decomposition*

$$\mathcal{C}(G \square H) = \mathcal{C}(G^x) \oplus \mathcal{S}(G \square H) \oplus \mathcal{C}(H^x).$$

Let $\mathcal{G}$ be an MCB for G^x, let $\mathcal{H}$ be an MCB for H^x, and let $\mathcal{B}$ be as defined in Lemma 29.2. If G and H are triangle free, then $\mathcal{G} \cup \mathcal{B} \cup \mathcal{H}$ is an MCB for $G \square H$.

Proof We first check that the three spaces in the sum are independent. Suppose $A+B+C = A' + B' + C'$, where $A, A' \in \mathcal{C}(G^x)$, $B, B' \in \mathcal{S}(G \square H)$ and $C, C' \in \mathcal{C}(H^x)$. We must show $A = A'$, $B = B'$ and $C = C'$. Notice that $p_G^*(e \square f) = 0 = p_H^*(e \square f)$ for any square $e \square f$, so $0 = p_G^*(B) = p_G^*(B') = p_H^*(B) = p_H^*(B')$. Also $p_G^*(C) = 0 = p_G^*(C')$ and $p_H^*(A) = 0 = p_H^*(A')$. Applying p_G^* to both sides of $A + B + C = A' + B' + C'$ gives $p_G^*(A) = p_G^*(A')$. Now, $p_G : G^x \to G$ is an isomorphism, so $p_G^* : \mathcal{C}(G^x) \to \mathcal{C}(G)$ is too. We conclude $A = A'$. Similarly $C = C'$, hence $B = B'$.

A routine computation involving Equation (29.1) reveals $\dim(\mathcal{C}(G \square H)) = |\mathcal{G}|+|\mathcal{B}|+|\mathcal{H}|$, so $\mathcal{G} \cup \mathcal{B} \cup \mathcal{H}$ is a basis. To see that it is a minimum cycle basis, we use Proposition 29.1. Take any $C \in \mathcal{C}(G \,\Box\, H)$, and write it as

$$C = \sum_{i \in I} G_i + \sum_{j \in J} B_j + \sum_{k \in K} H_k \,,$$

where the G_i are from $\mathcal{G}$, the B_j are from $\mathcal{B}$, and the H_k are from $\mathcal{H}$. According to Proposition 29.1, it suffices to show that C has at least as many edges as any term in this sum. Certainly C is not shorter than any square B_j (by the triangle-free assumption). To see that it is not shorter than any G_i in the sum, apply p_G^* to the above equation to get

$$p_G^*(C) = \sum_{i \in I} p_G^*(G_i) \,.$$

Now, because $p_G^* : \mathcal{C}(G^x) \to \mathcal{C}(G)$ is an isomorphism, the terms $p_G^*(G_i)$ are part of an MCB for G, and thus $|p_G^*(C)| \geq |p_G^*(G_i)| = |G_i|$ for each i, by Proposition 29.1. But also $|C| \geq |p_G^*(C)|$ (as some edges may cancel in the projection) so $|C| \geq |G_i|$. Similarly, $|C| \geq |H_i|$. □

Finding an MCB for $G \,\Box\, H$ is not quite so simple if the factors have triangles, for an MCB should capitalize on the triangles in the layers at the expense of squares in the square space. Imrich and Stadler (2002) prove that in general an MCB for $G \,\Box\, H$ can be obtained by starting with $\mathcal{G} \cup \mathcal{H}$, then appending to this the triangles in the MCBs of the other layers, and finally appending a "suitable" subset of $\mathcal{B}$. However, the exact details of this construction are not spelled out.

We now consider the problem of constructing an MCB for the strong product. The construction is quite similar to that of the Cartesian product, and does not require the factors to be triangle-free. For each pair of edges e and f of G and H, there is a subgraph $e \boxtimes f \subseteq G \boxtimes H$. Each such subgraph is isomorphic to K_4, and thus contains four triangles. Notice that for each such triangle K, we have $p_G^*(K) = 0 = p_H^*(K)$. Letting $\mathcal{T}(G \boxtimes H) \subseteq \mathcal{C}(G \boxtimes H)$ be the subspace spanned by all such triangles, we immediately get a direct sum

$$\mathcal{C}(G^x) \oplus \mathcal{T}(G \boxtimes H) \oplus \mathcal{C}(H^x)$$

of subspaces of $\mathcal{C}(G \boxtimes H)$. Parallel to the case of the Cartesian product, an MCB for the strong product hinges on finding a basis of triangles for $\mathcal{T}(G \boxtimes H)$.

Such a basis can be found with the aid of the set $\mathcal{B}$ from Lemma 29.2. Each $e \,\Box\, f$ in $\mathcal{B}$ corresponds to a complete subgraph $e \boxtimes f$ in $G \boxtimes H$. For each such $e \,\Box\, f$, let $\mathcal{T}_{ef}$ be a basis of (three) triangles for $\mathcal{C}(e \boxtimes f)$. Now, the set $\mathcal{T} = \bigcup_{e \,\Box\, f \in \mathcal{B}} \mathcal{T}_{ef}$ is independent, for any sum of its elements either contains a diagonal of some square $e \,\Box\, f$ (and is hence nonzero), or it is a nontrivial sum of squares in the linearly independent set $\mathcal{B}$ (thus nonzero).

But additional triangles can append to $\mathcal{T}$ without violating independence. For each of the $\beta(G)\beta(H)$ squares $e \,\Box\, f$ that is *not* in $\mathcal{B}$, we can append to $\mathcal{T}$ two triangles of $e \boxtimes f$ whose sum is not $e \,\Box\, f$. (That is, one triangle contains one diagonal of $e \,\Box\, f$, and the other contains the other diagonal.) The extended set $\mathcal{T}$ is still independent, for each of the new triangles contains an edge that does not belong to any other element of $\mathcal{T}$. (Hence it cannot be canceled in a linear combination of elements of $\mathcal{T}$.)

Therefore we have a set $\mathcal{T}$ of $3|\mathcal{B}| + 2\beta(G)\beta(H)$ independent triangles in $\mathcal{T}(G \boxtimes H)$. The proof of the following proposition mirrors that of Proposition 29.3.

Theorem 29.4 (An MCB for $\mathbf{G \boxtimes H}$) *Let G and H be graphs, and $x \in V(G \boxtimes H)$. Let $\mathcal{G}$ be an MCB for G^x, and $\mathcal{H}$ an MCB for H^x. Then $\mathcal{G} \cup \mathcal{T} \cup \mathcal{H}$ is an MCB for $G \boxtimes H$, where $\mathcal{T}$ is as follows.*

Take maximal spanning forests $T \subseteq G$ and $U \subseteq H$. For each subgraph $e \boxtimes f \subseteq G \boxtimes H$ with $e \in E(G) \setminus E(T)$ and $f \in E(H) \setminus E(U)$, let the set $\mathcal{T}_{ef}$ consist of two linearly independent triangles in $\mathcal{C}(e \boxtimes f)$ whose sum is not $e \,\square\, f$. For all other $e \boxtimes f \subseteq G \boxtimes H$, let $\mathcal{T}_{ef}$ be a basis of $\mathcal{C}(e \boxtimes f)$ consisting of three triangles. Define $\mathcal{T}$ to be the union of all the $\mathcal{T}_{ef}$.

29.3 Minimum Cycle Bases for the Lexicographic Product

Because lexicographic products typically have an abundance of triangles, their minimum cycle bases generally consist mostly of triangles. Given graphs G and H, let $\mathcal{T}(G \circ H)$ be the subspace of $\mathcal{C}(G \circ H)$ spanned by those triangles of $G \circ H$ that have exactly one edge in some H-layer. (In other words, it is spanned by the triangles with vertices (a, y), (b, u), and (b, v), where $ab \in E(G)$ and $uv \in E(H)$.) This subspace plays a major role in our construction of an MCB for $G \circ H$, and in the next several paragraphs we develop a large (in fact, maximal) independent set of triangles in $\mathcal{T}(G \circ H)$.

Let us break the problem into smaller parts. Assume H is connected and take a spanning tree U of H. Give G an arbitrary but fixed orientation, so that any edge ab under discussion will be oriented from a to b. For each $ab \in E(G)$, form the following set of triangles in $\mathcal{T}(G \circ H)$:

$$\mathcal{T}_{ab} = \{\langle (a,y),(b,u),(b,v)\rangle \mid y \in V(U), uv \in E(U)\}\,.$$

This set is easily seen to be independent. (Use induction on the order of H.) Now fix $y_0 \in V(H)$ and append to $\mathcal{T}_{ab}$ all triangles of the form $\langle (a,u),(a,v),(b,y_0)\rangle$ with $uv \in E(U)$. The appended set $\mathcal{T}_{ab}$ is still independent, for each new triangle contains an edge $(a,u)(a,v)$ that belongs to no other element of $\mathcal{T}_{ab}$. Observe also $|\mathcal{T}_{ab}| = |E(U)| \cdot |V(U)| + |E(U)| = |V(H)|^2 - 1$.

Now, put $\mathcal{T}' = \bigcup_{ab \in E(G)} \mathcal{T}_{ab}$. Notice that this union is independent, as follows: Each $\mathcal{T}_{ab}$ is a set of triangles in $ab \circ U$. For a fixed edge ab, any nontrivial linear combination of triangles in $\mathcal{T}_{ab}$ cannot lie in the acyclic graph $U^a + U^b$, and therefore it contains an edge of form $(a,y)(b,z)$. Such an edge cannot be canceled by any triangles in $\mathcal{T}_{a'b'}$ for $a'b' \neq ab$.

Next, for each $a \in V(G)$, fix an edge $ab \in E(G)$, and let

$$\mathcal{T}_a = \{\langle (a,u),(a,v),(b,y_0)\rangle \mid uv \in E(H) - E(U)\}\,.$$

Put $\mathcal{T}'' = \bigcup_{a \in V(G)} \mathcal{T}_a$. This is an independent set of triangles, because any triangle in it contains an edge $(a,u)(a,v)$ that belongs to no other triangle of $\mathcal{T}''$. But, then, for the same reason, $\mathcal{T}' \cup \mathcal{T}''$ is linearly independent.

We have now constructed a set $\mathcal{T} = \mathcal{T}' \cup \mathcal{T}''$ of linearly independent triangles for which

$$\begin{aligned} |\mathcal{T}| &= |E(G)| \cdot \big(|V(H)|^2 - 1\big) + |V(G)| \cdot \big(|E(H)| - |V(H)| + 1\big) \\ &= \big(|E(G)| \cdot |V(H)|^2 + |V(G)| \cdot |E(H)|\big) - |V(G)| \cdot |V(H)| + c(G \circ H) - \beta(G) \\ &= \beta(G \circ H) - \beta(G)\,. \end{aligned}$$

(Here we used $c(G \circ H) = c(G)$.) This can be extended to a cycle basis of $G \circ H$ by appending just $\beta(G)$ more elements. In fact, we have the following proposition, modeled after F. Berger (2004) and Jaradat (2008).

Theorem 29.5 (An MCB for $\mathbf{G \circ H}$) *Suppose G and H are graphs, and H is connected. Let $x \in V(G \circ H)$, and $\mathcal{G}$ be an MCB for the layer G^x. Then $\mathcal{G} \cup \mathcal{T}$ is an MCB for $G \circ H$, where $\mathcal{T}$ is the set of triangles defined above.*

Proof As was noted above, $|\mathcal{G} \cup \mathcal{T}| = \beta(G \circ H)$. Given that $\mathcal{G}$ and $\mathcal{T}$ are independent, it will follow that $\mathcal{G} \cup \mathcal{T}$ is a basis as soon as we show $\text{Span}(\mathcal{G}) \cap \text{Span}(\mathcal{T}) = \{0\}$. To this end, suppose A is in the intersection of the spans. Now, because $A \in \text{Span}(\mathcal{T})$, we have $p_G^*(A) = 0$, as p_G^* sends every triangle in $\mathcal{T}$ to zero. But $p_G^*(A) = 0$ implies $A = 0$, because $A \in \text{Span}(\mathcal{G})$, and p_G^* restricts to an isomorphism on $\mathcal{C}(G^x) \subseteq \mathcal{C}(G \circ H)$. Thus $\mathcal{G} \cup \mathcal{T}$ is a basis.

We use Proposition 29.1 to show that it is an MCB. For any $C \in \mathcal{C}(G \circ H)$, write

$$C = \sum_{i \in I} G_i + \sum_{j \in J} T_j ,$$

where the G_i are from $\mathcal{G}$ and the T_j are from $\mathcal{T}$. Certainly C is not shorter than any triangle T_j. To see that it is not shorter than any G_i in the above sum, apply p_G^* to the sum to get

$$p_G^*(C) = \sum_{i \in I} p_G^*(G_i) .$$

Now, because $p_G^* : \mathcal{C}(G^x) \to \mathcal{C}(G)$ is an isomorphism, the terms $p_G^*(G_i)$ are part of an MCB for G, and thus $|p_G^*(C)| \geq |p_G^*(G_i)| = |G_i|$ for each i, by Proposition 29.1. But also $|C| \geq |p_G^*(C)|$ (as some edges may cancel in the projection) so $|C| \geq |G_i|$. □

This proof breaks down if H is disconnected, for in that case the set $\mathcal{T}$ is too small. (Indeed, at one extreme, if H is totally disconnected, then $\mathcal{T} = \emptyset$.) Indeed, this situation is somewhat more complex; for details see Hellmuth, Ostermeier, and Stadler (2010).

29.4 Minimum Cycle Bases for the Direct Product

The problem of constructing an MCB for a direct product in terms of the cycle structures of its factors is difficult. This is primarily so because the layers in a direct product need not be isomorphic to the factors, so the cycle spaces of the factors are not subspaces of the cycle space of the product, as was the case for the other products. Even the problem of finding an MCB for the direct product of two complete graphs is surprisingly nontrivial and specialized—see Hammack (2007) and Bradshaw and Jaradat (2009).

This section presents a brief survey of several results.

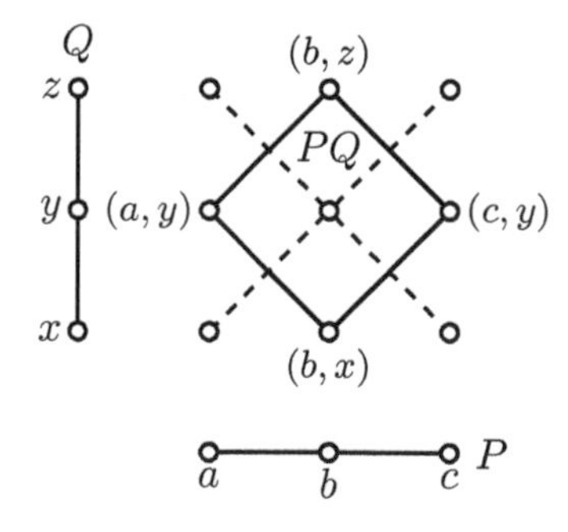

FIGURE 29.2 A diamond in $G \times H$.

Direct products tend to contain an abundance of a particular kind of square. Given paths P and Q of length 2 in G and H, respectively, one of the two components of $P \times Q$ is a square. (See Figure 29.2.) We call such a square a *diamond* in $G \times H$; the subspace of $\mathcal{C}(G \times H)$ spanned by all diamonds is called the *diamond space* of $G \times H$.

If $G \times H$ is bipartite, then diamonds are shortest cycles, and we expect them to play a role in any minimum cycle basis. In fact, we have the following result. See Hammack (2006) for a proof, as well as a construction for the basis $\mathcal{D}$ of the diamond space.

Theorem 29.6 *Suppose G and H are connected bipartite graphs, and let e and f be edges of G and H, respectively. Thus $G \times f$ consists of two disjoint copies of G, and $e \times H$ is two disjoint copies of H. Let $\mathcal{G}$ and $\mathcal{G}'$ be MCBs for the two copies of G; let $\mathcal{H}$ and $\mathcal{H}'$ be MCBs for the two copies of H. Let $\mathcal{D}$ be a basis for the diamond space of $G \times H$. Then $\mathcal{G} \cup \mathcal{G}' \cup \mathcal{D} \cup \mathcal{H} \cup \mathcal{H}'$ is a minimum cycle basis for $G \times H$.*

Despite the simplicity of Theorem 29.6, removing the assumption of bipartiteness makes the problem hard. In this case $G \times f$, rather than being two copies of G, is a connected graph whose cycle structure may bear little resemblance to that of G. Bradshaw and Hammack (2009) address this problem and present a construction for a minimum cycle basis for $G \times C_p$, where G has an odd cycle and $p > 3$ is odd. To underscore the subtlety of the general situation, we present their homological construction for a minimum cycle basis for the product $C_p \times C_q$ of two odd cycles, where $3 < p \leq q$.

A minimum cycle basis for the product of odd cycles

Let $3 < p \leq q$. Observe that $C_p \times C_q$ can be embedded on the torus with pq square regions whose boundaries are the diamonds of $C_p \times C_q$. This is illustrated for $C_5 \times C_9$ in Figure 29.3(a) and for $C_5 \times C_{11}$ in Figure 29.3(b). In each case the torus is an identification space obtained by identifying paths A (of length p), paths B (of length p), and the zig-zag path C (of length $q - p$). The general case is illustrated in Figures 29.4(a) and 29.4(b).

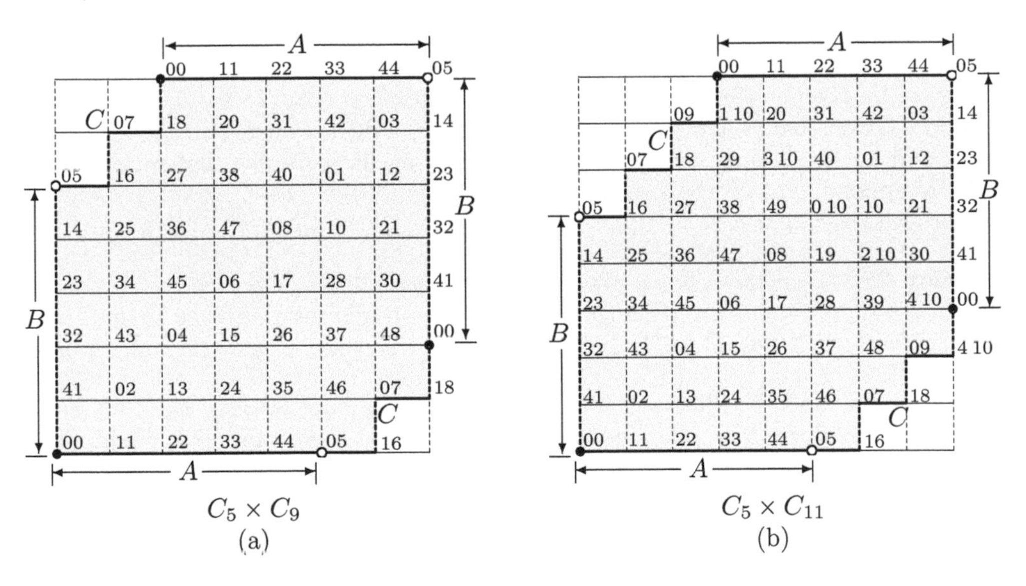

$C_5 \times C_9$
(a)

$C_5 \times C_{11}$
(b)

FIGURE 29.3 Identification space for $C_p \times C_q$ on the torus.

The set of all pq diamonds is linearly dependent, for if they are all added together their edges will cancel pair-by-pair. But it is not hard to see that any $pq - 1$ of them form an independent set. Because $C_p \times C_q$ has no triangles, a set $\mathcal{B}$ of $pq-1$ of these diamonds must be a part of an MCB. And because $\beta(C_p \times C_q) = pq+1$, there are just two more cycles to append to $\mathcal{B}$ in order to get an MCB.

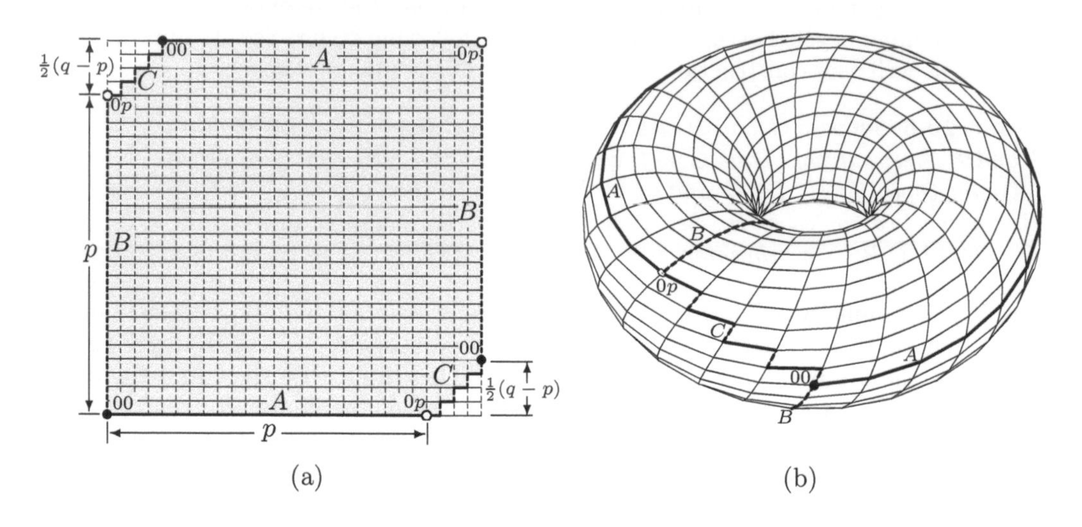

FIGURE 29.4 The product $C_p \times C_q$ of odd cycles embedded on the torus.

In Figure 29.3 the edges of the products are colored solid and dashed according to whether they run horizontally or vertically in the grids. With this coloring, every diamond has two edges of each color, so any element of Span($\mathcal{B}$) has an even number of edges of each color. Now, the even cycle $A+B$ has p (odd) edges of each color, so $A+B \notin$ Span($\mathcal{B}$). Further, $A+C$ and $B+C$ are cycles of odd length q, so they are certainly not in Span($\mathcal{B}$). Because $A+B = (A+C)+(B+C)$, it follows that appending to $\mathcal{B}$ any two elements of $\{A+B, A+C, B+C\}$ will produce a basis.

Moreover, by Exercise 29.7, any even cycle of length less than $2p = |A+B|$ is a sum of diamonds. Given that $A+C$ and $B+C$ are shortest odd cycles in $C_p \times C_q$ (they have length q), and an MCB must contain at least one odd cycle, it follows that we can obtain an MCB by appending to $\mathcal{B}$ the two shortest cycles in $\{A+B, A+C, B+C\}$. Therefore we have proved the following.

Proposition 29.7 *Suppose p and q are odd integers, with $3 < p \leq q$.*
If $q < 2p$, then $C_p \times C_q$ has an MCB consisting of $pq-1$ squares and two q-cycles.
If $2p < q$, then $C_p \times C_q$ has an MCB consisting of $pq-1$ squares, a $2p$-cycle and a q-cycle.

Figures 29.3(a) and (b) illustrate this proposition. In Figure 29.3(a), $C_5 \times C_9$ has an MCB consisting of forty-four diamonds, and two 9-cycles $A+C$ and $B+C$. In Figure 29.3(b), $C_5 \times C_{11}$ has an MCB consisting of fifty-four diamonds, one 10-cycle $A+B$, and one 11-cycle $A+C$.

Given such sensitivity to the relative sizes of p and q, one expects that the structure of a minimum cycle basis of $G \times H$ (relative to the factors) to be quite complex in general.

Exercises

29.1. For a graph G, show that the image of the boundary map $\delta : \mathcal{E}(G) \to \mathcal{V}(G)$ is the set $V = \{X \in \mathcal{V}(G) \mid X$ has an even number of vertices in each component of $G\}$. Moreover, show that $\dim(V) = |V(G)| - c(G)$. Deduce Equation (29.1).

29.2. Describe the cycle spaces $\mathcal{C}(K_4)$ and $\mathcal{C}(K_5)$.

29.3. Find formulas for $\beta(G \,\square\, H)$, $\beta(G \boxtimes H)$, $\beta(G \times H)$, and $\beta(G \circ H)$.

29.4. Prove Proposition 29.1.

29.5. Describe an MCB of $K_m \,\square\, K_n$.

29.6. Show that $K_3 \times K_3$ has an MCB consisting of six triangles and four squares. Show that $K_3 \times K_4$ has an MCB consisting of twenty-three triangles and two squares.

29.7. Consider the direct product $G = C_p \times C_q$, where $3 < p \leq q$, and p, q are both odd. If $Z \in \mathcal{C}(G)$, $|Z|$ even and $|Z| < 2p$, then Z is a sum of diamonds.

29.8. Consider the direct product $G = C_p \times C_q$, where p is even and q is odd. Show that G has an MCB consisting of $pq - 1$ squares, one p-cycle, and one $2q$-cycle.

29.9. Consider the direct product $G = C_p \times C_q$, where p and q are both even. Show directly that G has an MCB consisting of $pq - 2$ squares, two p-cycles, and two q-cycles.

Chapter 30

Selected Results

Chapters 25 through 29 considered in detail the relationships between invariants of graph products and invariants of their factors. The current chapter collects further results of a similar nature. It would be impossible to include all known results that in one way or another relate products to their factors, so our selection is necessarily subjective. However, we have selected mostly among topics of recent interest.

We begin with classical concepts of one-factorizations and edge-colorings; one of the highlights of this section is the solution of Kotzig's problem. We continue with problems involving Hamilton cycles. Much work has been done in the area, though three central conjectures are still open. Section 30.3 treats clique minors of Cartesian products, mostly in connection with the famous Hadwiger conjecture. The subsequent section touches on the Reconstruction conjecture, topological embedding, and nowhere-zero flows. The chapter concludes with a recent intriguing application of the direct product to modeling complex networks.

30.1 One-Factorization and Edge-Coloring

We now present several results about the existence of one-factors (also known as perfect matchings) in graph products, their extensions to edge-colorings, and k-extendable graphs.

One-factorizations

One-factorizations arise in many branches of discrete mathematics, but also in measure theory and fields such as the theory of ferromagnetism. See Seshu and M. B. Reed (1961), Lovász and Plummer (1986), and Wallis (1997). It is a lively area of research, for one-factorizations of complete graphs see Pasotti and Pellegrini (2010) and references therein, including the classical survey paper of Mendelsohn and Rosa (1985).

A *one-factor* or a *perfect matching* in a graph is a set of independent edges whose endpoints cover the graph's vertices. Clearly, a graph with a one-factor has an even number of vertices. A *one-factorization* of a graph is a partition of its edges into one-factors. Thus for a given one-factorization of a graph, each edge belongs to a unique one-factor. Clearly, a graph with a one-factorization is regular and has an even number of vertices. The Petersen graph shows that this condition is not sufficient; see Wallis (1997).

Concerning the Cartesian product, Kotzig (1979) proved the following result:

Theorem 30.1 *If $G_1, \ldots, G_k$ are regular graphs with at least one edge, then the product $G_1 \,\square \cdots \square\, G_k$ has a one-factorization provided that one of the following conditions holds:*

(i) *At least one of the graphs G_i has a one-factorization.*
(ii) *At least two graphs G_i and G_j have a one-factor.*

Borowiecki and Szelecka (1993) extended this theorem to so-called generalized Cartesian products. (A *generalized Cartesian product* is a product in which layers with respect to a factor have the same number of vertices but are not necessarily isomorphic.)

Some of the remaining cases are taken care of by another result of Kotzig (1979).

Theorem 30.2 *If G is a cubic graph, then $G \,\square\, C_n$ has a one-factorization whenever $n \geq 4$.*

This implies that neither condition (i) nor (ii) of Theorem 30.1 is necessary. Indeed, consider $Y \,\square\, C_5$, where Y is the cubic graph in Figure 30.1. Neither Y (Exercise 30.1) nor C_5 has a one-factor, but $Y \,\square\, C_5$ has a one-factorization by Theorem 30.2.

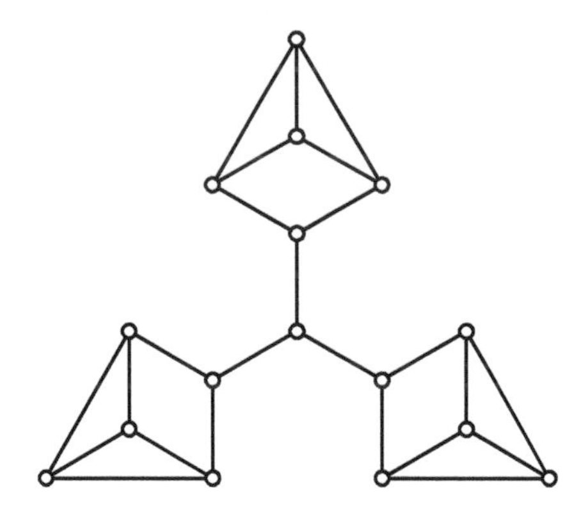

FIGURE 30.1 The graph Y.

The simple case $G \,\square\, C_3$ for a cubic graph G is not covered by Theorem 30.2. If G contains a *bridge* (i.e., an edge whose removal yields a graph with more components than G), then $G \,\square\, C_3$ cannot have a one-factorization (Exercise 30.2). Kotzig asked whether $G \,\square\, C_3$ necessarily has a one-factorization if G is a bridgeless cubic graph, and this question became known as *Kotzig's problem.* It generated many partial results (Wallis and Wang, 1985, 1987) before Horton and Wallis (2002) gave a definitive answer:

Theorem 30.3 *If G is a bridgeless cubic graph, then $G \,\square\, C_3$ has a one-factorization.*

Horton and Wallis' proof invokes Theorem 20.1 concerning the existence of edge-disjoint spanning trees in multigraphs.

Partitions of Cartesian products into graphs other than K_2 have also been studied. Notably, Bryant, El-Zanati, and Vanden Eynden (2001, 2007) proved the following result:

Theorem 30.4 *If $k, n, m \geq 3$, then $C_n^{\square,m}$ admits a partition into stars $K_{1,k}$ if and only if $(k+1)|n^m$ and $k|m$.*

Mohar, Pisanski, and Shawe-Taylor (1981) and Pisanski, Shawe-Taylor, and Mohar (1983) explored the existence of one-factors for the other standard products. The lexicographic product yielded a result analogous to Theorem 30.1. It was later generalized to the X-join by Borowiecki and Szelecka (1993).

The following theorem, which is similar to Theorem 30.2, is also due to Mohar, Pisanski, and Shawe-Taylor (1981).

Theorem 30.5 *If G is a cubic graph and C_n a cycle of length greater than 3, then $C_n \circ G$ has a one-factorization.*

Mohar and Pisanski (1983) conjectured that the lexicographic product $G \circ D_m$ has a one-factorization if and only if G is regular and $|V(G \circ D_m)|$ even. The conjecture was based on results of Mohar and Pisanski (1983) and earlier results of Laskar and Hare (1972) as well as Parker (1973). (See Exercise 30.3 for Parker's result.) Nevertheless, Truszczynski (1983) disproved the conjecture.

Partitions of lexicographic products into other graphs (notably cycles of uniform length) have also been explored. In a series of papers, Manikandan and Paulraja (2006, 2007, 2010a) proved that if p is a prime number that divides the number of edges of $K_m \circ D_n$, and $5 \leq p \leq mn$, and $(m-1)n$ is even, then $K_m \circ D_n$ can be decomposed into cycles of length p. (Interestingly, the direct product turned out to be an indispensable ingredient of the proof.) Smith (2008) followed by considering decompositions into cycles of length $2p$ (p prime) and proved that if $mn \geq 2p$, then, under the same conditions as above, $K_m \circ D_n$ can be decomposed into cycles of length $2p$. For products $C_n \circ D_m$, Smith and Cavenagh (2010) proved the following theorem, which they then used to deduce additional results on decompositions of $K_n \circ D_m$.

Theorem 30.6 *The lexicographic product $C_n \circ D_m$ decomposes into cycles of odd length k if and only if $n \leq k \leq mn$, n is odd, and $k|nm^2$.*

For further results on decompositions of lexicographic and direct products into cycles (and paths) see Muthusamy and Paulraja (1995), Cavenagh (1998), and Billington, Cavenagh, and Smith (2010).

Concerning the direct and strong products Mohar, Pisanski, and Shawe-Taylor (1981) and Pisanski, Shawe-Taylor, and Mohar (1983) proved the following theorem, which was rediscovered by Zhou (1989).

Theorem 30.7 *Suppose $G_1, \ldots, G_k$ are regular graphs that are not totally disconnected. If at least one G_i has a one-factorization, then both $G_1 \times \cdots \times G_k$ and $G_1 \boxtimes \cdots \boxtimes G_k$ have one-factorizations.*

Alspach and George (1990) also studied one-factorizations of direct products. Among other results, they proved the following.

Theorem 30.8 *If G is p-regular, where p is a prime power, then $G \times K_p$ has a one-factorization.*

One-factorizations and their enumerations in (Cartesian) product graphs have significant applications, particularly in the Ising model; see Lu, Zhang, and Lin (2010). The Ising model in ferromagnetism (Kasteleyn, 1963) is related to the statistics of nonoverlapping dimers, each occupying two neighboring sites of a lattice graph, cylinder, or torus. Yan and F. Zhang (2006) present additional results on the enumeration of one-factors in Cartesian products.

Edge-colorings

Many of the above results concerning one-factorizations of regular graphs can be extended to results about edge-colorings of arbitrary graphs. By Vizing's theorem,[1] every graph is either of *Class* 1 (if the chromatic index equals the largest degree) or of *Class* 2 (if the chromatic index equals the largest degree plus one). A regular graph is of Class 1 if and only if it has a one-factorization. Thus the following result of Mahmoodian (1981) generalizes part (i) of Theorem 30.1:

Theorem 30.9 *A product $G_1 \,\square \cdots \square\, G_k$ is of Class 1 if at least one of its factors is of class one and not totally disconnected.*

Mohar (1984) generalized several of the above theorems about one-factorizations to graphs of Class 1. For example, Theorem 30.1 extends as follows:

Theorem 30.10 *A product $G_1 \,\square \cdots \square\, G_k$ is of Class 1 if either of the following conditions holds:*

(i) *At least one of the graphs G_i is of Class 1 (and not totally disconnected).*
(ii) *At least two graphs G_i, G_j have a one-factor.*

See Jaradat (2005) for results on edge-colorings of other graph products. He proves among other things that if G or H is of Class 1, then so is $G \boxtimes H$.

k-Extendable graphs

A connected graph is called *k-extendable* if it has a one-factor and every independent set of k edges can be extended to a one-factor. Győri and Plummer (1992) proved:

Theorem 30.11 *If G is k-extendable and H is ℓ-extendable, then their Cartesian product is $(k+\ell+1)$-extendable.*

Győri and Imrich (2001) deduced a similar theorem for the strong product:

Theorem 30.12 *If G is k-extendable and H is ℓ-extendable, then their strong product is $\lfloor (k+1)(\ell+1)/2 \rfloor$-extendable.*

Limaye and Sarvate (1997) studied k-extendability on hypercubes. They showed that Q_n is k-extendable for every k with $1 \le k \le n-1$.

---

[1]Vizing's theorem states that the chromatic index of a graph equals either the maximal degree or the maximal degree plus one.

30.2 Hamilton Cycles and Hamiltonian Decompositions

A graph is called *Hamiltonian* if it has a Hamilton cycle, that is, a cycle that contains every vertex of the graph. A graph is said to be *Hamilton decomposable* if either (i) it is $2d$-regular and its edge set can be partitioned into d Hamilton cycles, or (ii) it is $(2d+1)$-regular and its edge set can be partitioned into d Hamilton cycles and a one-factor. Hamilton cycles and Hamilton decompositions of graph products have been studied extensively. This section presents a brief survey.

Cartesian products

Three conjectures, all still open, will lead us through this subsection. We begin with a conjecture of Bermond (1978).

Conjecture 30.13 *If G and H decompose into Hamilton cycles, then so does $G \,\square\, H$.*

Stong (1991) made a major step toward a proof by establishing the validity of the conjecture under mild restrictions on the factors:

Theorem 30.14 *Suppose G and H decompose into m and n Hamilton cycles, respectively, where $m \le n$. Then $G \,\square\, H$ decomposes into $m+n$ Hamilton cycles if one of the following conditions holds:*

(i) $n \le 3m$,
(ii) $m \ge 3$,
(iii) *G has an even number of vertices, or*
(iv) $|V(H)| \ge 6\lceil n/m \rceil - 3$.

Stong's approach subsumes many prior results by Kotzig (1973), Foregger (1978), as well as Aubert and Schneider (1981, 1982). In particular, Theorem 30.14 implies Kotzig's theorem that the Cartesian product of two cycles decomposes into two Hamilton cycles, as well as Foregger's extension of this to three cycles. Kotzig's result was generalized to the Cartesian product of two multicycles (i.e. cycles in which each edge has a fixed multiplicity) by Mellendorf (1997).

For hypercubes, Alspach, Bermond, and Sotteau (1990) proved:

Theorem 30.15 *For $n \ge 2$, the hypercube Q_n is Hamilton decomposable.*

Aubert and Schneider (1981) proved a similar result for products of complete graphs:

Theorem 30.16 *If $m, n \ge 2$, then $K_m \,\square\, K_n$ is Hamilton decomposable.*

We next consider prisms over 3-connected graphs. Tutte proved that every 4-connected planar graph G is Hamiltonian. It is then easy to see (Exercise 30.4) that $G \,\square\, K_2$ is Hamiltonian as well. On the other hand, the prism over the complete bipartite graph $K_{2,5}$ (which is 2-connected and planar) is not Hamiltonian. For the remaining 3-connected case, Rosenfeld and Barnette (1973) made the following conjecture:

Conjecture 30.17 *If G is a 3-connected planar graph, then $G \,\square\, K_2$ is Hamiltonian.*

Alspach and Rosenfeld (1986) followed with the following related conjecture:

Conjecture 30.18 *If G is 3-connected and cubic, then $G \,\square\, K_2$ is Hamilton decomposable.*

Note that, as $G \,\square\, K_2$ is 4-regular when G is cubic, Alspach and Rosenfelds' conjecture asserts that $G \,\square\, K_2$ can be decomposed into two Hamilton cycles. Conjectures 30.17 and 30.18 led to the following investigations.

- If G is planar, cubic, and 2-connected, then $G \,\square\, K_2$ is Hamiltonian, as proved by Fleischner (1989). He first proved that if G is cubic, then the prism $G \,\square\, K_2$ is Hamiltonian if and only if G contains a so-called BEPS-graph. Then he proved that if G is planar and 2-connected, then G indeed contains such a graph.
- Paulraja (1993) characterized the existence of Hamilton cycles in $G \,\square\, K_2$ for any G. We refer the reader to the paper for the exact statement. In fact, no simple characterization can be expected because Fleischner (1989)[p.169] observed that the decision problem of whether the prism over G is Hamiltonian is NP-complete. Paulraja also proved that $G \,\square\, K_2$ is Hamiltonian if G is 3-connected and cubic. Balakrishnan and Paulraja (2005) extended this to a subclass of cubic 2-connected graphs that are not necessarily planar and conjectured that the prism over every 2-connected cubic graph is Hamiltonian.
- Goddard and Henning (2001) proved that the prism over a 3-connected cubic graph G has cycles of every even length. Moreover, if G contains a triangle, then the prism has cycles of every length.
- Čada, Kaiser, Rosenfeld, and Ryjáček (2004) proved Conjecture 30.18 for the following cases:

  (i) G is a planar bipartite graph (and, of course, 3-connected and cubic); and

  (ii) G is the dual of a planar triangulation that can be obtained from K_4 by repeatedly adding a vertex and joining it to the three vertices of some face.

  They also obtained a simple proof of Paulraja's theorem asserting that $G \,\square\, K_2$ is Hamiltonian if G is 3-connected and cubic.
- Biebighauser and Ellingham (2008) proved Conjecture 30.17 for bipartite (3-connected planar) graphs and for triangulations of the plane, the projective plane, the torus, and the Klein bottle.

We continue with prisms over specific important classes of graphs.

- The famous middle-levels problem (see Shields, Shields, and Savage (2009) and the references therein) asks for a Hamilton cycle in the middle levels of the $(2d+1)$-cube, that is, in the subgraph of Q_{2d+1} induced by vertices containing d and $d+1$ zeros. In this respect, Horák, Kaiser, Rosenfeld, and Ryjáček (2005) proved that prisms over the middle-levels graphs are Hamiltonian.
- Kaiser, Ryjáček, Kráľ, Rosenfeld, and Voss (2007) deduced that prisms over generalized Halin graphs and over 2-connected line graphs are Hamiltonian. They also proved that the existence of a Hamiltonian prism is a stronger property than having a 2-tree (that is, a Hamilton path) and a weaker property than having a so-called 2-walk. Intuitively, this means that graphs with Hamiltonian prisms are close to being Hamiltonian, a fact that adds value to the above theorem asserting that the prism over a middle-levels graph is Hamiltonian.

- A celebrated result of Fleischner (1974) asserts that the square of any 2-connected graph is Hamiltonian, where the square of a graph G is obtained from G by adding edges between vertices at distance 2. Kaiser et al. (2007) also proved that the 2-connectedness assumption is not needed for prisms, that is, the prism over the square of any connected graph is Hamiltonian. In fact, they proved a stronger result: For any nontrivial tree, the prism over its square is Hamiltonian.
- Ozeki (2009) provides a sufficient condition (in terms of the minimum degree sum of an independent set of three vertices) for a graph to have a Hamiltonian prism.
- Bueno and Horák (2011) proved that Kneser graphs $K(2k+1,k)$ (also called *odd graphs*) have Hamiltonian prisms for even k.

Čada, Flandrin, and Li (2009) studied Cartesian products that generalize prisms. Among other things, they established the following:

- If G is a connected graph with $\Delta(G) \leq n$, then $G \,\square\, C_n$ is Hamiltonian.
- If G is a bridgeless graph and $n \geq 3$, then $G \,\square\, P_n$ is Hamiltonian.

The following particularly nice result in this spirit goes back to Batagelj and Pisanski (1982).

Theorem 30.19 *Let T be a tree on at least three vertices and $n \geq 3$. Then $T \,\square\, C_n$ is Hamiltonian if and only if $\Delta(T) \leq n$.*

Strong and lexicographic products

Conjecture 30.13 becomes a theorem when the strong product replaces the Cartesian product.

Theorem 30.20 *If G and H have decompositions into Hamilton cycles, then so does $G \boxtimes H$.*

Zhou (1989) proved this for the case where at least one factor has odd order, and the remaining case was done by Fan and Liu (1998).

Ramachandran and Parvathy (1996) characterized the existence of Hamilton cycles in strong products $G \boxtimes K_2$. They also proved a theorem about pancyclic strong products. (A graph is *pancyclic* if its vertices can be labeled $v_1, v_2, \ldots, v_n$ such that the subgraph induced by $v_1, v_2, \ldots, v_k$ contains a cycle of length k for each $3 \leq k \leq n$.)

Theorem 30.21 *To every connected graph G on at least two vertices, there exists an integer m such that the strong product of m copies of G is pancyclic.*

Theorem 30.21 extends a result of Bermond, Germa, and Heydemann (1979) asserting the existence of an integer m to any nontrivial connected graph G such that the strong product of m copies of G is Hamiltonian. In the same paper it is conjectured that m can be selected as the maximum degree of G; in other words that $G^{\boxtimes,\Delta(G)}$ is Hamiltonian for any nontrivial connected graph G. Kráľ, Maxová, Šámal, and Podbrdský (2005) confirmed this with the following stronger result.

Theorem 30.22 *If $G_1, G_2, \ldots, G_\Delta$ are nontrivial connected graphs with maximum degree at most $\Delta \geq 2$, then $G_1 \boxtimes G_2 \boxtimes \cdots \boxtimes G_\Delta$ is Hamiltonian.*

Kráľ, Maxová, Podbrdský, and Šámal (2004) further proved that for any $c > \ln \frac{25}{12} + \frac{1}{64}$ there exists c' such that the strong product of at least $\lfloor c\Delta \rfloor + \lceil \log_2 \Delta \rceil + c'$ connected graphs of degree at most Δ is pancyclic for $\Delta > 32$. This result asymptotically improves Theorem 30.22 for large values of Δ and at the same time extends Theorem 30.21. The result was further improved by Kráľ and Stacho (2008) as follows:

Theorem 30.23 *If* $\Delta \geq 1$, *the strong product of* $(\ln 2)\Delta + (10 + \ln 4)\sqrt{\Delta} + 0.5 \log_2 \Delta + O(1)$ *nontrivial connected graphs of maximum degree at most* Δ *is pancyclic.*

The constant $\ln 2$ in the linear term is best possible because the graph $K_{1,\Delta}^{\boxtimes, \lfloor (\ln 2)\Delta \rfloor}$ is not Hamiltonian, a result that goes back to Zaks (1974). The reason is that these graphs contain independent sets of size greater than half their order.

For the lexicographic product, Baranyai and Szasz (1981) proved the following analogue of Theorem 30.20:

Theorem 30.24 *If* G *and* H *can be decomposed into* g *and* h *Hamilton cycles, respectively, and* $|V(H)| = n$, *then* $G \circ H$ *can be decomposed into* $h + gn$ *Hamilton cycles.*

Kriesell (1997) derived sufficient conditions for the existence of Hamilton cycles in lexicographic products. For instance, if a cubic graph G is 2-edge connected and H has at least two vertices, then $G \circ H$ has a Hamilton cycle.

It is an open question whether $G \circ H$ is Hamilton decomposable when G is connected and Hamilton decomposable; see Alspach, Bermond, and Sotteau (1990). On the other hand, Kaiser and Kriesell (2006) proved:

Theorem 30.25 *If* G *and* H *are graphs, each with at least one edge, then* $G \circ H$ *has a cycle of every length between the lengths of its shortest and longest cycle.*

Direct products

In the first paper about Hamilton cycles in direct products that we know of, Borowiecki (1972) gives necessary and sufficient conditions on G and H such that $G \times H$ is Hamiltonian. (For this result as well as for a survey of Hamilton properties of products of graphs and digraphs, see the book of Schaar, Sonntag, and Teichert (1988).) This is applied to prove that for every Hamiltonian graph G and $n \geq 3$, there is a Hamiltonian graph H on n vertices such that $G \times H$ is Hamiltonian.

Along these lines, Gravier (1997) characterized Hamiltonian products $G \times H$ under the assumption that both G and H have a Hamilton cycle. In addition, Balakrishnan and Paulraja (1998) give a (rather complicated) characterization of graphs G for which $G \times K_2$ has a Hamilton cycle.

As the direct product of two bipartite graphs is disconnected, we do not expect the direct product of two Hamilton decomposable graphs to be Hamilton decomposable. Moreover, the direct product of two Hamilton decomposable graphs need not be Hamilton decomposable even if the product itself is connected, as Balakrishnan and Paulraja (1998) have shown.

Sufficient conditions are due to Zhou (1989), Jha (1992), and Jha, Agnihotri, and Kumar, (1996). Zhou (1989) proved that if G and H can be decomposed into Hamilton cycles and at least one of them has an odd number of vertices, then $G \times H$ can be decomposed into Hamilton cycles. Hence direct products of complete graphs of odd order have such decompositions.

Balakrishnan, Bermond, Paulraja, and Yu (2003) extended this to all complete graphs, thus proving a result parallel to Theorem 30.16:

Theorem 30.26 *For any $m, n \geq 3$, the product $K_m \times K_n$ is Hamilton decomposable.*

Manikandan and Paulraja (2008) proved that the direct product of two multipartite graphs, each having at least three partite sets, all of the same order, is Hamilton decomposable. Manikandan and Paulraja (2010b) proved:

Theorem 30.27 *For any $r \geq 2$, $m \geq 3$ and $n \geq 1$, $K_{r,r} \times (K_m \circ D_n)$ is Hamilton decomposable.*

Paulraja and Sivasankar (2009) obtained also related results for digraphs. Denoting with G^* the digraph obtained from G by replacing each edge with a symmetric pair of arcs, they proved, among others, the following:

Theorem 30.28 (i) *Let $m, n \geq 3$, $n \neq 4$, and $m \notin \{4, 6\}$. Then $K_m^* \times K_n^*$ is directed Hamilton decomposable.*

(ii) *Let $r, n \geq 3$, $n \neq 4$, and $r \notin \{4, 6\}$. Then $(K_r \circ D_s)^* \times K_n^*$ is directed Hamilton decomposable.*

Analogous to Batagelj and Pisanski's work on Hamilton cycles in Cartesian products of cycles and trees, Jha, Agnihotri, and Kumar (1997) investigated the same question for the direct product. A typical result asserts that if m and n are both odd, then the length of a longest cycle in $C_m \times P_n$ is $m(n-1)$, and $C_m \times P_n$ can be decomposed into two longest cycles. See Exercises 30.7 and 30.8.

30.3 Clique Minors in Cartesian Products

A graph H is a *minor* of a graph G if it can be obtained by contracting edges of a subgraph of G. In particular, K_k is a minor of G if G contains k connected, pairwise disjoint subgraphs, any two of which are joined by an edge.

Recall from Section 26.2 that Hadwiger's conjecture asserts that every n-chromatic graph has a subgraph that is contractible to K_n. In other words, Hadwiger conjectured that K_n is a minor of every n-chromatic graph. The *Hadwiger number* $\eta(C)$ is the maximum k such that K_k is a minor of G. Thus Hadwiger's conjecture can be reformulated as the assertion

$$\eta(G) \geq \chi(G)$$

for every graph G.

Although Hadwiger's conjecture is open for Cartesian product, several special cases are settled. In what follows, "Hadwiger's conjecture holds for G" is abbreviated as HC(G).

- Chandran and Sivadasan (2007) proved that HC($G^{\square,k}$) for any G and $k \geq 3$. The result was exteneded to $k = 2$ by Chandran, Kostochka, and Raju (2008). A further generalization was obtained by Wood (2007) who proved that $|\chi(G) - \chi(H)| \leq 2$ implies HC($G \,\square\, H$). His theorem also implies HC($G \,\square\, H$) when $\chi(G) = \chi(H)$, a result first proved in Chandran, Kostochka, and Raju (2008).
- Wood (2007) proved that $|\chi(G) - \chi(H)| \leq 2$ implies HC($G \,\square\, H$). In particular, this yields HC($G^{\square,k}$) for any G and $k \geq 2$. The latter result was first proved in Chandran, Kostochka, and Raju (2008), and in turn earlier proved for $k \geq 3$ in Chandran and Sivadasan (2007). Wood's theorem also implies HC($G \,\square\, H$) when $\chi(G) = \chi(H)$, a result first proved in Chandran, Kostochka, and Raju (2008).

- Chandran, Kostochka, and Raju (2008) proved that there is a constant c' for which HC(G) for any connected G with prime factorization $G = G_1 \,\square\, G_2 \,\square \cdots \square\, G_k$, where $k \geq 2\log_2 \log_2 \chi(G) + c'$. This theorem extends an earlier result for $k \geq 2\log_2 \chi(G) + 3$ in Chandran and Sivadasan (2007).

- Let $\chi(G) \geq \chi(H)$. Wood (2007) proved that HC($G \,\square\, H$) if (i) G has treewidth $\mathrm{tw}(G) \geq 2^{4\chi(G)^4}$ or (ii) H has at least $\chi(G) + 1$ vertices. That (i) suffices for HC($G \,\square\, H$) follows from the more general result asserting that $\mathrm{tw}(G) \geq 2^{4\ell^4}$ implies $\eta(G \,\square\, K_2) \geq \ell$. This exponential bound was greatly improved by B. A. Reed and Wood (2008) to the following polynomial bound: For some constant c, if $\mathrm{tw}(G) \geq c\ell^4\sqrt{\log_2 \ell}$, then $\eta(G \,\square\, K_2) \geq \ell$.

There are numerous results giving either bounds or exact values for the Hadwiger number of Cartesian products. A sample follows.

- One of the earliest results goes back to Miller (1978), though it was stated there without proof. If T is a tree with at least one edge, and $n \geq 1$, then $\eta(T \,\square\, K_n) = n+1$. Wood (2007) gives a proof.

- Another early result by Ivančo (1988) states $\eta(K_{1,m} \,\square\, K_{1,n}) = m + 2$ for $2 \leq m \leq n$.

- Exercise 30.9 asks for the Hadwiger number of Cartesian products of two paths. Archdeacon, Bonnington, Pearson, and Širáň (1997) compute it for the Cartesian products of two cycles.

- Kotlov (2001) proved that for every bipartite G, the strong product $G \boxtimes K_2$ is a minor of $G \,\square\, C_4$. As a consequence, he deduced that

$$\eta(Q_d) \geq \begin{cases} 2^{(d+1)/2} & \text{if } d \text{ is odd}, \\ 3 \cdot 2^{(d-2)/2} & \text{if } d \text{ is even}. \end{cases}$$

- Wood (2007) proved that for every bipartite graph G and every connected graph H, the lexicographic product $G \circ H$ is a minor of $G \,\square\, H \,\square\, H$. This can be viewed as a generalization of the above result of Kotlov because $G \boxtimes K_2$ is isomorphic to $G \circ K_2$ (and as $G \,\square\, C_4 = G \,\square\, K_2 \,\square\, K_2$). Wood also proved that for every graph H,

$$\eta(K_n \circ H) = \left\lfloor \frac{n}{2} \left(|V(H)| + \omega(H)\right) \right\rfloor .$$

- Chandran, Kostochka, and Raju (2008) proved that for $m \leq n$,

$$n\sqrt{m}(1 - o(1)) \;\leq\; \eta(K_m \,\square\, K_n) \;\leq\; n\sqrt{m} + m .$$

(Exercise 30.11 asks for a related upper bound.) The lower bound implies that $\eta(G \,\square\, H) \geq \eta(H)\sqrt{\eta(G)}(1 - o(1))$ whenever $\eta(G) \leq \eta(H)$. The upper bound cannot be generalized in this way, that is, $\eta(G \,\square\, H) \leq \eta(H)\sqrt{\eta(G)} + \eta(G)$ is generally false. To see this, recall that $\eta(K_{1,m} \,\square\, K_{1,n}) = m + 2$ for $2 \leq m \leq n$. Because no tree has a K_3-minor, $\eta(K_{1,m}) = 2$, hence $\eta(K_{1,n})\sqrt{\eta(K_{1,m})} + \eta(K_{1,m}) = 6 < m + 2$ for any $m \geq 5$. Wood (2007) obtained the exact values $\eta(K_2 \,\square\, K_n) = n + 1$ ($n \geq 1$) and $\eta(K_3 \,\square\, K_n) = n + 2$ ($n \geq 1$), as well as bounds $3\lfloor \frac{n}{2} \rfloor \leq \eta(K_4 \,\square\, K_n) \leq \frac{3}{2}n + 2$ ($n \geq 2$).

- Wood (2007) obtained best-known upper and lower bounds for grid graphs and Cartesian products of complete graphs (i.e., Hamming graphs). The latter case is as follows.

Theorem 30.29 *If $n_1 \geq \cdots \geq n_k \geq 2$, then*

$$\left\lfloor \frac{n_1}{2} \right\rfloor \prod_{i=2}^{k} \lfloor \sqrt{n_i} \rfloor \leq \eta(K_{n_1} \,\square\, \cdots \,\square\, K_{n_k}) < \sqrt{k}\, n_1 \prod_{i=2}^{k} \sqrt{n_i} + 3\,.$$

In addition to the results listed above, the seminal paper by Wood (2007) also contains a proof of Exercise 30.12, as well as additional results not mentioned here. More importantly, it gives a structural characterization of Cartesian products with bounded Hadwiger number. The characterization is similar to that of Robertson and Seymour (2003) for general graphs but is more focused due to the special structure of Cartesian products. Informally, Wood's characterization says that for connected G and H, $\eta(G \,\square\, H)$ is bounded if and only if

(i) One factor has bounded treewidth and the other factor has bounded order, or

(ii) Both factors have bounded hangover.

Here the *hangover* is defined as follows. A path P in a connected graph G is called *clean* if every internal vertex of P has degree two (in G), and every edge of P is disconnecting. (Thus each vertex of G is a clean path P_1.) The hangover of a path or a cycle is defined to be zero. In general, the hangover of G is the minimum, taken over all clean paths P in G, of the maximum number of vertices in a component of $G - E(P)$.

Goldberg (2009) studied the so-called Colin de Verdière number μ of Cartesian products, and conjectured that $\mu(G \,\square\, H) \geq \mu(G) + \mu(H)$ for connected G and H. The conjecture is closely related to the Hadwiger number of products, because $\mu(G) \geq \eta(G) - 1$. Thus Goldberg's theorem, asserting that $\mu(G \,\square\, H) \geq \mu(G) + \eta(H) - 1$, can be considered a weak version of his conjecture.

To close the section we add that minors of direct products have been considered to some extent. Bottreau and Métivier (1998) proved that every connected graph G is a minor of $G \times H$, where H is an arbitrary connected nonbipartite graph. On the other hand, they demonstrated that G need not be a minor of $G \times K_2$ (Exercise 30.13), thus disproving a conjecture from Jha and Slutzki (1993).

30.4 Reconstruction, Topological Embeddings, and Flows

In this section we consider three additional topics that are of interest in graph products and begin with reconstruction.

The set (or multi-set) of vertex-deleted subgraphs of a graph is called its *deck*. A graph is said to be *reconstructible* if it is uniquely determined by its deck. Note that K_2 and D_2 have the same deck (consisting of two copies of K_1) so K_2 and D_2 are not reconstructible. On the other hand, the famous Reconstruction conjecture[2] asserts that any simple graph with at least three vertices is reconstructible. The conjecture is still open, though it is known to be true for certain classes of graphs. In particular, it is true for Cartesian products, as proved by Dörfler (1973):

Theorem 30.30 *Any nontrivial Cartesian product graph is reconstructible.*

[2]Many authors call it Ulam's conjecture though it was stated by Kelly in 1942 and by Ulam in 1960.

Theorem 30.30 also holds for infinite graphs (Imrich and Žerovnik, 1996).

No graph can be reconstructed from a single vertex-deleted subgraph. But, if one knows that a graph belongs to a certain class of graphs, then one may be able to reconstruct it from a single vertex-deleted subgraph. For example, this is the case for nontrivial Cartesian products, which was first proved by Sims and Holton (1978). Independently, Imrich and Žerovnik (1996) extended Dörfler's work as follows.

Theorem 30.31 *Let G_1 and G_2 be connected Cartesian product graphs with $x \in V(G_1)$ and $y \in V(G_2)$. If $G_1 - x$ and $G_2 - y$ are isomorphic, then so are G_1 and G_2.*

Thus a one-vertex-deleted subgraph suffices to reconstruct Cartesian products of at least two nontrivial factors, this is called *weak reconstruction.* Hagauer and Žerovnik (1999) designed an algorithm that reconstructs a Cartesian product graph G from any vertex-deleted subgraph in $O(mn(\Delta(G)^2 + m \log n))$ time. The term $m \log n$ comes from the complexity of factoring a graph over the Cartesian product; hence in view of Section 23.5, it can be replaced by m.

McAvaney (1980) conjectured that every connected nontrivial Cartesian product graph is also uniquely determined by each of its two-vertex-deleted subgraphs. This claim turned out to be false; see Exercise 30.14 for a counterexample. Imrich, Zmazek, and Žerovnik (2003) extended this to an infinite family of counterexamples where one of the factors is P_3. On the other hand, they proved that a Cartesian product is uniquely determined by any one of its k-vertex-deleted subgraphs if it has at least $k+1$ prime factors, each on at least $k+1$ vertices. They also posed the following problem: Is it true that any connected Cartesian product graph with $k \geq 2$ prime factors, each with more than $\max\{3, k\}$ vertices, is uniquely determined by each of its k-vertex deleted subgraphs?

For the strong product, Dörfler (1973) proved:

Theorem 30.32 *Let $X = G \boxtimes H$ be a nontrivial strong product. If at least one factor has two vertices in relation S, then X is reconstructible.*

Zmazek and Žerovnik (2007) proved that all nontrivial strong products can be reconstructed provided that their product structure is assumed. They also proved:

Theorem 30.33 *A connected nontrival S-thin strong product graph is weakly reconstructible. In other words, it is uniquely determined by each of its one-vertex-deleted subgraphs.*

For the lexicographic product we have the following result, also from Dörfler (1973):

Theorem 30.34 *If $X = G \circ H$ is a nontrivial lexicographic product with $|V(H)| \geq 3$, then X is reconstructible.*

We note in passing that Feigenbaum and Haddad (1989) considered related problems for the Cartesian product, namely minimal factorable extensions and maximal factorable subgraphs. They showed that neither minimal factorable extensions nor maximal factorable subgraphs are unique and that finding them is NP-hard.

We now turn to the second topic of this section: topological embeddings of product graphs.

A graph is *embeddable* into a surface if it can be drawn on the surface such that no two of its edges intersect, except possibly at endpoints. (See the definition of planar graphs in Chapter 1, p. 11.) The *genus of a graph* is the minimum number of handles to be attached to a sphere so that the graph can be embedded into the resulting surface. For nonorientable surfaces, one likewise defines the *nonorientable genus* of a graph, by adding Möbius bands (rather than handles).

Embeddings of graph products into surfaces have been studied extensively. Here we merely give suggestions for further reading.

Ringel (1955) determined the genus of the simplest Cartesian product, the hypercube. This is now a consequence of the following theorem of Pisanski (1980), which generalizes results of White (1970).

Theorem 30.35 *If G and H are connected, bipartite r-regular graphs, then the genus of $G \,\Box\, H$ is* $1 + |V(G)||V(H)|(r-2)/4$.

Embeddings of Cartesian products were also investigated by Mohar, Pisanski, Škoviera, and White (1985); Mohar, Pisanski, and White (1990); Craft and Schwenk (1993); and Bonnington and Pisanski (2004).

For the work on embeddings of direct and lexicographic products up to 1993, we refer the reader to Dakić, Nedela, and Pisanski (1995). More recently, Abay-Asmerom published a series of papers on embeddings of direct products. See, for example, Abay-Asmerom (1998).

We close the section with a brief overview of the research on nowhere-zero flows in product graphs.

An *orientation* D of a graph G is obtained from G by turning each edge of G into an arc. Let $f : E(G) \to \mathbb{Z}$ be a function with $-k < f(e) < k$ for each edge e. The pair (D, f) is called a *k-flow of G* if Kirchhoff's condition holds at every vertex v of G: The flow into v is equal to the flow out of v. A k-flow (D, f) is called *nowhere-zero* if $f(e) \neq 0$ for every $e \in E(G)$.

The theory of flows in graphs is rich with beatiful theorems and open problems. As an entry into this area we suggest the book by West (1996). Here we point out the *3-flow conjecture* due to Tutte, asserting that every 4-edge-connected graph has a nowhere-zero 3-flow.

Flows in product graphs seem to have been first considered by Imrich and Škrekovski (2003) who proved:

Theorem 30.36 *If G and H are nontrivial connected graphs, then $G \,\Box\, H$ admits a nowhere-zero 4-flow. Moreover, if G and H are bipartite, then $G \,\Box\, H$ admits a nowhere-zero 3-flow.*

Their work has been extended in two ways. Shu and C.-Q. Zhang (2005) extended the second part of Theorem 30.36 by proving that the 3-flow conjecture is true for products $G \,\Box\, H$ unless each block of G is an odd cycle and H has a bridge. On the other hand, Rollová and Škoviera (2009) extended the first part of Theorem 30.36 by proving that any nontrivial Cartesian graph bundle admits a nowhere-zero 4-flow. (See Section 33.1 for the definition of graph bundles.)

For the direct product, Z. Zhang, Y. Zheng, and Mamut (2007) introduced a class of graphs $\mathcal{G}$ containing $K_2 \in \mathcal{G}$ and closed with respect to joining any two of its members by an edge, and proved:

Theorem 30.37 *If G is a graph with $\delta(G) \geq 2$ and H a graph not in $\mathcal{G}$, then $G \times H$ admits a nowhere-zero 3-flow.*

Y. Zheng, Z. Zhang, and Mamut (2009) also considered the lexicographic product and proved that $G \circ H$ admits a nowhere-zero 3-flow as soon as G is a nontrivial graph and H a connected graph of order at least 3.

For the strong product, Imrich, Peterin, Špacapan, and C.-Q. Zhang (2010) proved:

Theorem 30.38 *There exists an algorithm that computes a nowhere-zero k-flow in* $G \boxtimes H$ *for any two connected nontrivial graphs* G *and* H*, where*

(i) $k = 2$ *if* G *and* H *are Eulerian graphs,*

(ii) $k = 3$ *unless* G *is a tree and* $H = K_2$*,*

(iii) $k = 4$ *if* G *is a tree and* $H = K_2$*.*

The complexity of the algorithm of Theorem 30.38 is linear, except in Case (ii), when G is not a tree. Then the complexity is $O\big(|E(G)| + |E(H)| \cdot |V(H)| \cdot \Delta_H\big)$.

30.5 Modeling Complex Networks

Networks arise in many different areas, such as biology, ecology, mathematical chemistry, software technology, and operations research. Nonetheless, the investigation of very complex graphs and networks became an important research topic only in the last decade, coinciding with increased interest in the Internet, social networks, citation networks, and neural networks, to name just a few topics.

Some networks are very large and most of them grow extremely fast. For example, in 2009 the United States had over 383 million Internet hosts,[3] up from just four connected mainframes in 1969. (These four formed ARPANET, the precursor of the present Internet.)

Networks are studied from numerous points of view, ranging from performance, reliability, stability, robustness, growth, and self-organization to virus propagation and its prevention. But all these require that the network first be mapped. This is a formidable task in itself, and has led to discoveries of several fundamental properties that are shared by many real-life networks: the power law degree distribution, the small-world property, self-similarity, and the fact that the eigenvalues of the adjacency matrix also follow a power law. These are known as *static graph properties*, and we discuss them in more detail in the next subsection.

As networks grow, their parameters relative to these properties change. For instance, as a network becomes denser, its diameter decreases. Such traits, known as *temporal graph properties*, were discovered 2005 by Leskovec, Kleinberg, and Faloutsos (2005). They showed that at any time t, the numbers $V(t)$ of vertices and $E(t)$ of edges follow the densification power law

$$E(t) \propto V(t)^a, \text{ where } 1 \leq a \leq 2\,. \tag{30.1}$$

Formerly it was assumed that a was always 1. They observed that the diameter shrinks and stabilizes over time; previously logarithmic or double logarithmic growth was assumed.

In order to study the properties addressed above, it is necessary to generate realistic sequences of growing graphs that obey the static and temporal graph patterns.

It has been common to use the random graph model of Erdős and Rényi (1959, 1960) or that of Barabási, and Albert (1999). But these and other random graph models do not obey all properties, or it cannot be proved that they obey them. Moreover, it usually takes much effort and time to generate such random graphs. Hence new methods are needed.

[3] https://www.cia.gov/library/publications/the-world-factbook/fields/2184.html

Kronecker graphs

Leskovec, Chakrabarti, Kleinberg, Faloutsos, and Ghahramani (2010) suggested an appealing step in this direction with a model using the direct product of graphs (or, equivalently, the Kronecker product of matrices).[4]

Given a graph G, the adjacency matrix of the direct product $G^{\times,k}$ is the kth power of the adjacency matrix of G with respect to the Kronecker product. We will call such graphs *Kronecker graphs* and show first that they obey the above-mentioned static graph properties.

Static properties. We begin with the small-world property. It states that most real-world graphs have small diameter. This idea goes back to Milgram (1967), who first addressed the phenomenon of "six degrees of separation" for residents of the United States. Clearly, direct powers of nonbipartite graphs satisfy this property, as the diameter of $G^{\times,k}$ is bounded by $2\,\mathrm{diam}(G) + c$, where c is the length of a shortest odd cycle of G. (See Exercise 5.14.)

The power law degree distribution (Pareto distribution) says that the probability that a vertex has degree k is $P(k) \simeq ck^{-\gamma}$, where $2 < \gamma < 3$ in our cases. This causes the variance of the degrees to be large and makes the graph in a sense scale-free, which is related to the fractal structure.

We do not show the power law directly, but observe that the degree of a vertex $v \in V(G^{\times,k})$ is $p_1(v)p_2(v)\cdots p_k(v)$. Hence, if the degrees of the vertices in G are $d_1, d_2, \ldots, d_n$, then the degrees in $G^{\times,k}$ are $d_1^{i_1} d_2^{i_2} \cdots d_n^{i_n}$, where $\sum_{j=1}^{n} i_j = k$. The degree probabilities will therefore be proportional to $\binom{k}{i_1 i_2 \ldots i_n}$, that is, we have a multinomial degree distribution. Now, a careful choice of the degrees of the vertices of G causes the multinomial distribution to behave like a power law degree distribution; see Bi, Faloutsos, and Korn (2001).

We already mentioned the close connection between the power law for the degree distribution with so-called scale-freeness and fractal structure. But, as the fractal structure is obvious in our case, we simply illustrate it with Figure 30.2, which shows a graph G and the positions of the nonzero elements in the adjacency matrix of $G^{\times,4}$.

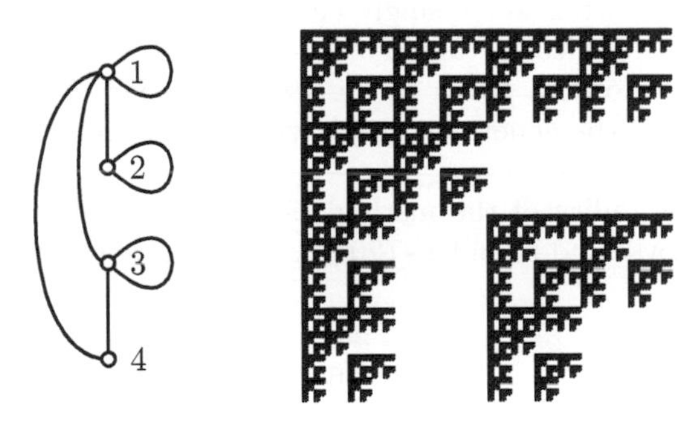

FIGURE 30.2 A graph G and nonzero elements in $A(G^{\times,4})$.

By the properties of the Kronecker product of matrices the eigenvalues of $A(G^{\times,k})$ are $\lambda_1^{i_1}\lambda_2^{i_2}\cdots\lambda_n^{i_n}$, where $\lambda_1, \lambda_2, \ldots, \lambda_n$ are the eigenvalues of $A(G)$ and $\sum_{j=1}^{n} i_j = k$. Hence, the power law for the distribution of the eigenvalues of $A(G^{\times,k})$ follows by the same arguments as for the degree distribution.

Temporal properties. Consider a graph G on n vertices and m edges. Then $G^{\times,k}$ has n^k

[4]Leskovec et al. (2010) assume that every vertex of the network has a loop; thus they are really working with the strong product, but it suffices to work with the direct product of connected, nonbipartite graphs.

vertices and m^k edges. Thus, setting $V(k) = |V(G^{\times,k})|$ and $E(k) = |E(G^{\times,k})|$, we have

$$E(k) \propto V(k)^a, \quad \text{where } a = \log(E(1))/\log(V(1)).$$

Thus the densification power law 30.1 is satisfied.

Clearly all graphs $G^{\times,k}$ have the same diameter for $k > 1$.

Stochastic Kronecker graphs

Start with a square probability matrix $\mathcal{P}_1$ whose i,j-entry represents the probability that an edge joins vertex i to vertex j, and compute the Kronecker kth power $\mathcal{P}_k$. Then (an instance of) a *stochastic Kronecker graph* is obtained from $\mathcal{P}_k$ by including an edge between two vertices with probability as given in $\mathcal{P}_k$. Considering the size of the networks modeled, it is extremely important that stochastic Kronecker graphs K on N nodes can be generated in $O(E)$ time, where E is the (expected) number of edges in K (Leskovec et al., 2010). This is not obvious; naively we would expect $O(N^2)$ time, where N is the number of vertices in $\mathcal{P}_k$.

Stochastic Kronecker graphs, if generated appropriately, satisfy the static and temporal properties described above. We wish to point out, though, that prior construction of the kth power of a graph with respect to the direct product and subsequent uniform and random selection of edges would result in a binomial degree distribution.

Because the properties of the direct product of graphs (respectively the Kronecker product of matrices) are well understood and generally not difficult to handle, it is not surprising that they can be studied using analytical tools. This allows us to fit them, by appropriate choice of $\mathcal{P}_1$, to real-world networks. To be more precise, $\mathcal{P}_1$ can be chosen to fit such parameters as diameter, the densification power law exponent a in Equation (30.1), the constants c and γ in the degree distribution power law, among others.

Leskovec et al. (2010) show that one can effectively find a stochastic Kronecker graph that is "similar" to any real-world network. They give examples where, starting from a random initial matrix, 100 steps of the so-called "gradient descent" produced stochastic Kronecker graphs with properties surprisingly close to the given network. Of course, this approach involves several technical details. For instance, the number of vertices of a real-world network is not likely to be a power of the size of the base graph. One solution is to pad the adjacency matrix of the generated graph with zeros, but not all such problems are equally easy to deal with.

To illustrate the applicability of the methods and models outlined in this section, we close with a result of Leskovec and Faloutsos (2007). They showed that the simple generating matrix

$$\begin{pmatrix} .98 & .58 \\ .58 & .06 \end{pmatrix}$$

yields a Kronecker graph that fits the Internet (at the autonomous system level) fairly well.

Random dot product

Another approach to the representation of random graphs with prescribed properties is the *random dot product*. It uses vectors to represent the vertices of a graph and a random function on the inner product (the dot product) to determine adjacency.

Let $G \in \Gamma$ and $k \in \mathbb{N}$. Then G is a *dot product graph* if there is a function $x : V(G) \to \mathbb{R}^k$ that assigns to each vertex v a vector x_v such that two vertices v, w are adjacent if $x_v \cdot x_w \geq 1$. Notice that "$\cdot$" denotes the standard inner product in $\mathbb{R}^k$.

We have already met the assignment of vectors to vertices in Section 27.1, the section on the Shannon capacity. On p. 344, following Lovász (1979), we considered an orthonormal representation of the vertices of a graph G by vectors in $\mathbb{R}^k$, and then used the tensor product, which is the same as the Kronecker product in this case, to construct an orthonormal representation of the vertices of the strong product of graphs.

The dot product also generalizes the intersection representation of graphs, and the *dot product dimension* of a graph G, that is, the smallest possible length of the vectors that represent the vertices of G is a pendant to the intersection number. To establish the connection, let us mention that an *intersection representation* of a graph G is an assignment of a set S_v to each vertex v such that $vw \in E(G)$ if $S_v \cap S_w \neq \emptyset$. The *intersection number* of a graph is the smallest size of the union of the sets assigned to the vertices in an intersection representation; see Erdős, Goodman, and Pósa (1966). If one replaces the sets S_v by characteristic $0,1$-vectors $\mathbf{v}$, then the condition $S_v \cap S_w \neq \emptyset$ can be replaced by $\mathbf{v} \cdot \mathbf{w} \geq 1$.

For further motivations for the dot product and numerous results, compare Fiduccia, Scheinerman, Trenk, and Zito (1998).

The model is then generalized as follows. In addition to the mapping x, let $f : \mathbb{R} \to [0, 1]$ be a function that maps the dot product of the vectors into probabilities. Then the *random dot product graph* G is defined as a graph of a given order, where $u, v \in V(G)$ are adjacent with probability $f(x_u \cdot x_v)$.

In this model we are given the vectors x_v to every $v \in V(G)$ a priori; and based on these vectors, a probability function f is chosen. However, an important variant is to draw the vectors x_v independently from a random distribution and then generate G with an appropriate choice of f.

If this is done properly, the temporal and static properties of large networks as described above can be modeled. In other words, if the vectors are taken properly, then the random graphs exhibit properties akin to those of social and communication networks including clustering, low diameter, and power-law distribution of degrees. The inverse problem of how to model a given set of graphs as random dot product graphs also becomes accessible. For details, see Young, and Scheinerman (2008), and Nickel (2007).

Exercises

30.1. Show that the graph Y of Figure 30.1 contains no one-factor.

30.2. Show that $G \,\square\, C_3$ has no one-factorization if G is a cubic graph with a bridge.

30.3. (Parker, 1973) Show that $C_n \circ D_m$ has a one-factorization if and only if at least one of n and m is even.

30.4. Show that $G \,\square\, K_2$ is Hamiltonian if G has a Hamilton path.

30.5. Show that $P_m \,\square\, P_n$, $m, n \geq 2$, has a Hamilton path, and that it has a Hamilton cycle if and only if m or n is even.

30.6. (Behzad and Mahmoodian, 1969) Show that $G \,\square\, H$ is Hamiltonian if G and H are both Hamiltonian.

30.7. (Jha, Agnihotri, and Kumar, 1997) Let T be a tree and $m \geq 3$. Show that $C_m \times T$ is Hamiltonian if and only if m is odd and $T = K_2$.

30.8. (Jha, Agnihotri, and Kumar, 1997) Let T be a tree of order n with p pendant vertices. Show that for $m \geq 3$, the length of a longest cycle of $C_m \times T$ is at most $mn - p$.

30.9. Show that for any $m, n \geq 3$, $\eta(P_m \square P_n) = 4$ and that $\eta(P_2 \square P_n) = 3$ for any $n \geq 2$.

30.10. Show that if M_G is a minor of G and M_H is a minor of H, then $M_G \square M_H$ is a minor of $G \square H$.

30.11. Let $1 \leq m \leq n$. Show that $\eta(K_m \square K_n) \leq n\sqrt{2m}$.

30.12. (Wood, 2007) Show that Hadwiger's conjecture holds for any nontrivial product $G \square H$, where $\chi(G) \geq \chi(G)$ if and only if Hadwiger's conjecture holds for $G \square K_2$.

30.13. Find a connected graph G such that G is not a minor of $G \times K_2$.

30.14. Find vertices x and y of $K_{1,3} \square P_3$ and vertices z and w of $P_4 \square P_3$ such that $K_{1,3} \square P_3 - \{x, y\}$ and $P_4 \square P_3 - \{z, w\}$ are isomorphic.

30.15. (Imrich and Žerovnik, 1996) A product square of a Cartesian product is a square in which the two pairs of opposite edges are in different σ-classes. Suppose that G is a connected nontrivial Cartesian product and that $G - v$ contains no product square (of G). Show that $G = K_{1,n} \square K_{1,m}$ for some $n, m \in \mathbb{N}$.

Part VI

Related Concepts

Introduction to Part VI

UNTIL now we have dealt almost exclusively with finite undirected graphs, their products, and structures involving retracts or subgraphs of their products. This ground has been so abundantly fertile that we have scarcely felt the need to extend it to the domain of infinite or directed graphs. In this final part of the book we entertain such notions.

Chapter 31 introduces the theory of infinite graphs. Almost immediately we meet a new concept—the free product—that has no finite analogue. This construction is then employed in the characterization of certain classes on infinite median graphs. The chapter concludes with a selection of prime factorization and cancellation properties of infinite graphs over the four standard products.

The free product of graphs, in particular of rooted graphs, is not new. It has been extensively studied from the point of view of growth, random walks, spectral properties, and even applications to quantum probability; see, for example, Pisanski and Tucker (2002), and Accardi, Lenczewski, and Sałapata (2007). Definition and notation is not uniform, but for vertex transitive graphs all definitions of the free product seemingly coincide. In particular, the free product of Cayley graphs is compatible with the free product of groups; see p. 403. We chose a definition that suited our investigation of vertex transitive median graphs with finite blocks.

We also define a generalized free product, which gives rise to the structure Theorem 31.5 of infinite vertex transitive graphs with cut vertices. This simple theorem seems to be new, as are Theorems 31.20 and 31.21 about the weak direct and the weak strong product. The latter results depend on Hammack's nonalgorithmic definition of the Cartesian skeleton.

In Chapter 32 we consider products of digraphs, with an eye toward developing results parallel to some of our earlier theorems for undirected graphs. We consider connectedness, prime factoring, and cancellation for the four standard products. For the lexicographic product, transitive tournaments play a surprising role.

Finally, in Chapter 33, we take up the issue of product-like structures. The first part treats the notion of graph bundles. These graphs—which need not be products but have local product structures—are the graph-theoretical analogues of topological fiber bundles.

The second part of the chapter is an overview of the currently developing theory of approximate graph products, that is, graphs that are very close to being products. We give brief mention to some of the promising potential applications of this idea and a a brief account of sophisticated heuristics for the investigation of approximate strong products. Interestingly, one of the side-products of these investigations is a quasi-linear algorithm for the factorization of connected S-thin graphs with respect to the strong product.

The chapter ends with remarks about graph spectra, the significance of the second largest eigenvalue of a graph in many parts of mathematics and computer science, and the definition of the zig-zag and the replacement product, which were introduced for the construction of expander graphs.

The reader will find various earlier chapters useful here. For example, the discussion of median graphs in Chapter 31 presupposes the material from Chapter 12 (on median graphs). Likewise, full appreciation of the prime factorization of infinite or directed graphs probably requires a good knowledge of the corresponding material (for finite graphs) in Part II.

Chapter 31

Infinite Graphs

The first part of this chapter pertains to infinite median graphs, the second to infinite products. In the case of products, both the factors and the number of factors may be infinite. On the way we introduce the free product of graphs, ends of graphs, and the growth rate of graphs. There are many similarities with finite graphs, which we will exploit, and many striking differences, which we will describe.

Infinite graphs are a common concept. Everyone is familiar with the integer lattice in one, two, or three dimensions. In dimension one, it is the *two-sided infinite path* P_Z defined on $V(P_Z) = \mathbb{Z}$, where each edge is a pair of successive integers. In dimension two, it is $P_Z \,\square\, P_Z$, and in dimension three $P_Z^{\square,3}$.

Infinite graphs are defined exactly as are finite graphs, except that the vertex and edge sets are allowed to be infinite. Concepts such as paths, cycles, connected components, trees, and bipartiteness carry over immediately to infinite graphs. The same is true for homomorphisms, retractions, and automorphisms.

But the situation is different for maximal spanning forests (or spanning trees of connected graphs), as we need the axiom of choice or transfinite induction to prove their existence. Interestingly, the *axiom of choice* is equivalent to the existence of maximal spanning forests in infinite graphs; see Exercise 31.1. Thus we may as well *assume the existence of maximal spanning forests in infinite graphs as an axiom.* It is not worse than accepting the axiom of choice and makes life easier in graph theory.

Still we face the question of which other properties of finite graphs hold for infinite graphs. The answer depends very much on how "finite" an infinite graphs is. There are several restrictions—*finiteness conditions*—that may be placed on an infinite graph.

Connectedness is one such condition. A connected infinite graph has a spanning tree, and every vertex can be reached from a given one by a path (of finite length). To appreciate the restrictions that connectivity implies, it is instructive to consider the infinite hypercube, whose vertices are infinite sequences of 0's and 1's. (As usual, edges join sequences that differ in exactly one place.) This graph is disconnected, for no path joins the vertex $000\ldots$ to any vertex that has infinitely many 1's.

The vertices of a connected graph can be arranged in BFS-order. If all degrees are finite, then we do not even need the well-ordering theorem (which is also equivalent to the axiom of choice) to accomplish this. This leads to another finiteness condition: A graph is *locally finite* if every vertex has finite degree. Clearly, a locally finite graph has countably many vertices if it is connected.

31.1 Growth Rate and Ends

Even if a connected graph is locally finite, the number of vertices in the distance levels L_i of a BFS-tree may grow very fast. This motivates the concept of a graph's *growth rate.* A locally finite, connected graph G has *polynomial growth* if there is a vertex v in G and a polynomial f such that

$$|N_n(v)| < f(n)$$

for all natural numbers n. This property is independent of the choice of v; see Exercise 31.2.

It is clear what *linear* and *quadratic growth* mean, and that $P_{\mathbb{Z}}$ and $P_{\mathbb{Z}} \square P_{\mathbb{Z}}$ have linear and quadratic growth, respectively.

If $|N_n(v)| \geq c^n$ for some constant $c > 1$, then we say G has *exponential growth.* If $|N_n(v)|$ grows faster than any polynomial, but not exponentially, the growth is *intermediate.* We consider graphs of polynomial growth to be barely infinite.

It is more than a good pastime to discover properties of finite graphs that carry over to infinite graphs of polynomial growth. Many deep and new results have been found; we will mention a few later.

Growth rate was originally defined for groups by Adelson-Velsky and Šreĭder (1957). The definition immediately extends from groups to Cayley graphs. For graphs that are not Cayley graphs, it seems to have been used first by Trofimov (1984).

Restrictions other than the growth rate include properties involving embeddability into surfaces (such as the plane, or orientable or nonorientable surfaces of finite or infinite genus). In such cases the number of vertices is bounded by the cardinality of the number of points of the surface. In all these cases, further natural requirements, such as transitive group action or fixed-point free action, lead to interesting, intriguing problems and results.

If no limits are placed on the cardinalities of the vertex sets or the degrees, however, one rapidly moves into the theory of infinite cardinals.

We now define ends of a graph. Our definition follows Halin (1964), but the concept goes back to Freudenthal (1931). See Diestel and Kühn (2003) for other definitions and the relationship to topological ends.

In geometry, topology, and function theory, we learn at an early stage how to close a space by adding objects at infinity. Such objects are often defined as limits of sequences. In infinite graphs these objects are called *ends* and are defined in terms of rays.

A *ray* is a one-sided infinite path $P_{\mathbb{N}}$, defined on $\mathbb{N}$ analogously to $P_{\mathbb{Z}}$. Two rays R_1 and R_2 of a graph G are *equivalent,* $R_1 \sim R_2$, if G has a ray R_3 that meets both R_1 and R_2 infinitely often. It is easy to see that $\sim$ is an equivalence relation. Its equivalence classes are the *ends* of G.

As an example, consider $P_{\mathbb{Z}}$. Clearly, it has two ends, as does every $Q_d \square P_{\mathbb{Z}}$. Contrariwise, the regular tree T_d (in which every vertex has the same degree d) has infinitely many ends if $d > 2$. But the infinite star consisting of three rays originating from a common vertex has just three ends. For a connection to the growth rate, we mention that infinite vertex-transitive graphs of linear growth have just two ends; see Imrich and Seifter (1989).

We continue with theorem about infinite trees that indicates that ends can be regarded as points (or vertices) at infinity. If v is a vertex of a tree, then every end has a representative ray with endpoint v. We can therefore think of the ends of a tree as the set of rays emanating from a given vertex v. The fact that every automorphism of a tree stabilizes an edge or fixes a vertex has the following analogue for automorphisms of infinite trees, as shown by Tits (1970).

Proposition 31.1 *If α is an automorphism of an infinite tree T, then exactly one of the following statements holds:*

(i) *α fixes a vertex.*

(ii) *α interchanges two adjacent vertices.*

(iii) *There exists a unique two-sided infinite path P_Z in T on which α acts as a translation, that is, $\alpha(P_Z) = P_Z$ but α fixes no vertex of G.*

Note that only (i) or (ii) is possible if T has no rays. In this case the result is the same as for finite graphs. If α acts as a proper translation on a P_Z, then it fixes the two ends defined by P_Z. One can justly say that α fixes two vertices at infinity.

We conclude the section with some remarks about rayless median graphs. Infinite median graphs are defined as in the finite case by the condition that every triple of vertices has a unique median. They are tree-like by their very construction. This should nourish the suspicion that rayless infinite median graphs share fixed-cube properties with their finite relatives. This is indeed the case. We state just three examples. The first is by Polat and Sabidussi (1994).

Proposition 31.2 *Every rayless tree has a vertex or an edge that is invariant under every automorphism.*

Note that the simplest infinite rayless tree is a star with $\aleph_0$ edges incident with a central vertex. Note also that any ray in a tree is isometric, though this need not be true for graphs in general. For median graphs, Tardif (1996) proved the following theorem:

Theorem 31.3 *Every nonexpansive map of a median graph without isometric rays contains a fixed cube.*

For quasi-median graphs we only have the following, slightly weaker result of Chastand and Polat (1996), as rays, and not just isometric rays, are excluded.

Theorem 31.4 *Every nonexpansive map of a quasi-median graph without rays contains a fixed Hamming graph.*

For numerous other results we refer to the insightful survey of Polat (1995).

31.2 Free Product

This section is motivated by the desire to construct vertex-transitive graphs from intransitive building blocks. Thus, for us the main property of the free product $G * H$ of two connected graphs will be the fact that it is vertex-transitive, even if its factors are not.

Lot G and H be connected graphs with k and ℓ orbits, respectively. Pick a vertex r_i from each of the k orbits of G, and let $G(r_i)$ denote the graph G with root r_i. Define rooted graphs $H(s_j)$ analogously.

The *free product* $G * H$ is a connected graph made from copies of G and H, where all cycles of $G * H$ are contained in a copy of G or H, and where any two copies of G or H are either disjoint or meet at a vertex, as follows: All vertices of $G * H$ are cut vertices. Given such a vertex v, consider every copy of G or H that is incident with v as a rooted graph

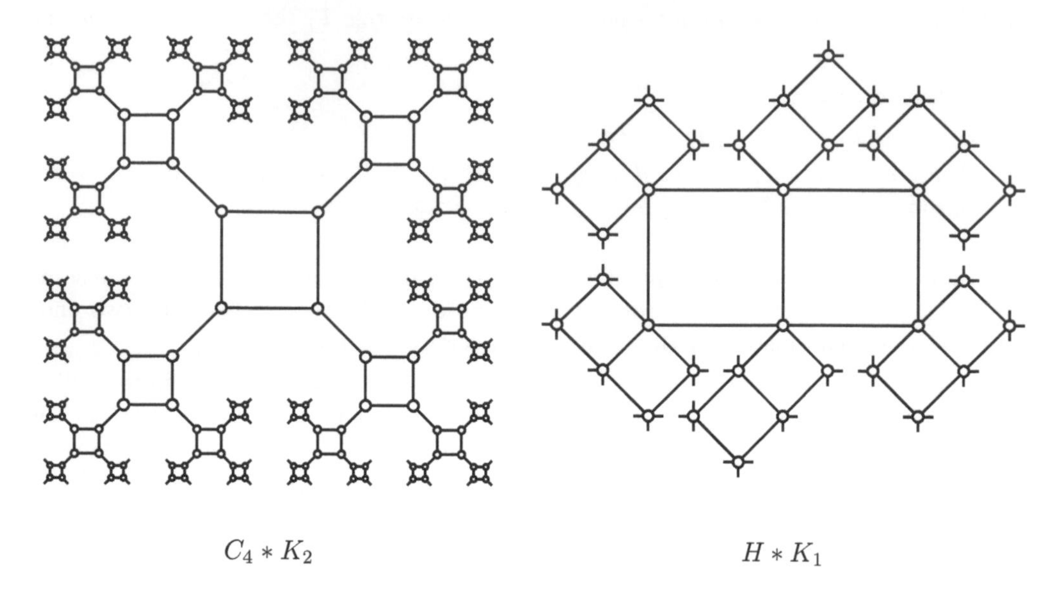

FIGURE 31.1 Free products of graphs.

with root v. Then $G * H$ is defined such that v is incident with exactly one copy of $G(r_i)$ and one of $H(s_j)$ for all $i \in \{1, 2, \ldots, k\}$ and $j \in \{1, 2, \ldots, \ell\}$.

For a simple example of a free product, note that $K_2 * K_2 = P_Z$. Figure 31.1 (left) shows the free product $C_4 * K_2$ of a square with an edge. For yet another example, the right side of Figure 31.1 shows $H * K_1$, where H is a hexagon with one diagonal.

For a more elaborate example, consider the infinite graph G of Figure 31.2 (top). It has two orbits, so every vertex of $G * K_1$ is incident with two blocks. A part of $G * K_1$ is depicted in the lower part of the figure.

From the definition and examples, it should be evident that the free product is commutative and associative. It does not have a unit, unless we restrict to vertex-transitive graphs, in which case K_1 is a unit.

Moreover, $G * H$ has transitive automorphism group, even if G or H is not transitive. Notice also that the definition easily extends to the case when G and H have infinitely many orbits. In such a case $G * H$ is not locally finite.

We now vary our definition of the free product slightly. As before, suppose G and H have k and ℓ orbits, respectively. Define multiplicities m_i and n_j, where $i \in \{1, 2, \ldots, k\}$ and $j \in \{1, 2, \ldots, \ell\}$ and to each vertex v attach m_i copies of $G(r_i)$ and n_j copies of $H(s_j)$. Again care is taken that all cycles of the new graph are in a copy of G or in a copy of H. The resulting graph is called a *generalized free product.*

The above remarks about associativity, etc., apply also to generalized free products.

Theorem 31.5 *Any connected, infinite vertex-transitive graph with a cut vertex is a nontrivial generalized free product.*

Proof If such a graph G has a cut vertex, then every vertex is a cut vertex by vertex-transitivity. Consider the collection of blocks incident with an arbitrary vertex v. We can partition this collection into sets of pairwise isomorphic blocks. Let G_ι, $\iota \in I$, be the set of representatives for these sets of blocks. If a G_ι is not vertex-transitive, then we form the rooted graphs $G_\iota(r)$, where r is from a set of representatives of the orbits of G_ι.

Let $m(G_\iota(r))$ be the number of times $G_\iota(r)$ occurs as block incident with v. (Notice that every $m(G_\iota(r))$ must be at least 1.) Then G is the generalized free product of the G_ι with the multiplicities $m(G_\iota(r))$. □

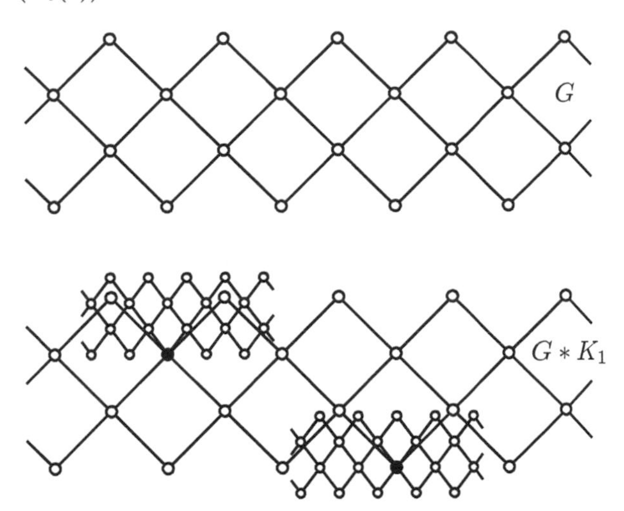

FIGURE 31.2 Infinite median graph G with two orbits (top) and part of $G * K_1$ (bottom).

If all factors of a generalized free product are vertex-transitive, and if the multiplicity m is greater than 1 for one of the factors G, then one can consider the product as a free product in which G occurs m times as a factor.

In graph theory the free product of graphs is a very underestimated concept, despite its great potential. The earliest publications (on the free product of rooted graphs) are probably due to Teh and Gan (1970) and Znoĭko (1975). For recent work, see Accardi, Lenczewski, and Sałapata (2007). This paper also treats the *comb product*, the *star product*, and the *orthogonal product*. The comb product product has the same vertex set as the Cartesian product, but cannot be described as a binary operation on the set $\{\Delta, 1, 0\}$ as the products treated in Chapter 4. The vertex sets of the other two products are smaller than that of the comb product.

In combinatorial group theory the free product of groups, and indirectly of graphs, plays an important role. Usually the factors are Cayley graphs and thus vertex-transitive, and there is no need for a generalized free product.

Observe that the free product of Cayley graphs is a Cayley graph again. In particular, if $G = \Gamma(A, S_A)$ and $H = \Gamma(B, S_B)$, then $G * H$ is the Cayley graph $\Gamma(A * B, S_A \cup S_B)$ of the free product $A * B$ of the groups A and B.[1] In this setting many important subgroup theorems can be derived very easily. They include those of Nielsen-Schreier, Kurosh, and Grushko.

Notice that the homomorphism $a_1b_1a_2b_2 \cdots a_kb_k \mapsto (a_1a_2 \cdots a_k, b_1b_2 \cdots b_k)$ of $A * B$ onto the product $A \times B$ of groups corresponds to a homomorphism of $\Gamma(A, S_A) * \Gamma(B, S_B)$ onto $\Gamma(A, S_A) \square \Gamma(B, S_B)$.

For an introduction to Combinatorial Group Theory we recommend the reprint of the classic work of Lyndon and Schupp (2001), originally published in 1977, and the second edition of the textbook of Stillwell (1993), first edition 1980.

It may seem surprising, but the free product $G * H$ of graphs is an isometric subgraph of the weak Cartesian product of infinitely many copies of G and H (Aleksandra Jędrzejaszek, Free Products of Graphs, Master Thesis, AGH Cracow, 2011).

[1] This allows an alternative definition of the free product of groups.

31.3 Transitive Median Graphs with Finite Blocks

Observe that every connected infinite graph without rays must have at least one vertex of infinite degree. This is a consequence of a result of König (1936), known as König's lemma, which asserts that every vertex in a locally finite connected infinite graph is contained in a ray Z_N (actually in an isometric ray). Therefore Proposition 31.2 and Theorem 31.4, which pertain to rayless graphs, are void for locally finite, connected infinite median, or quasi-median graphs.

Hence, if we wish to consider connected, locally finite infinite graphs, we have to admit rays. We will show that locally finite, connected infinite median graphs have numerous intriguing properties. We begin with their characterization under the assumption that they have only finite blocks. (Recall that a block of a graph is a maximal subgraph without cut vertices.) If every block of G is a median graph, then G is a median graph too. This is easily verified; see Exercise 31.3.

In the finite case, the class of regular median graphs is not very rich; it consists of all hypercubes, by Exercise 12.8. Notice that vertex-transitivity is not even needed as an assumption, as it follows from regularity. For infinite graphs the situation changes significantly.

For instance, any finite or infinite median graph G with largest degree d gives rise to a d-regular median graph as follows: Let u be an arbitrary vertex of G of degree smaller than d. Then, at u, attach to G an infinite rooted tree in which the root is of degree $d - d(u)$ and any other vertex is of degree d. In fact, Bandelt and Mulder (1983) observed that there are $2^{\aleph_0}$ cubic median graphs. This is the reason that we impose vertex-transitivity.

We thus have to construct transitive graphs form finite blocks, which need not be transitive. We will take advantage of the free product, but begin with an observation about the number of orbits in finite median graphs.

First we observe that median graphs of bounded maximum degree have arbitrarily many orbits if they are sufficiently large. This result is a consequence of the tree-like structure of median graphs, and in particular of Fixed Cube Theorem 12.21.

Proposition 31.6 *Every finite median graph of largest degree k on at least $(2k)^k$ vertices has at least k orbits.*

Proof Let G be a finite median graph with $\Delta(G) \leq k$ and Q_c a subcube that is invariant under all automorphisms of G. Then $0 \leq c \leq k$. Furthermore, no automorphism sends a vertex of Q_c to a vertex in $V(G) \setminus V(Q_c)$. Thus, $V(Q_c)$ consists of one or more orbits under the action of the automorphism group of G. This implies that the set L_1 of vertices at distance 1 from Q_c consists of one or more orbits too. By induction, the same holds for the set L_r of vertices of any distance r from Q_c.

If G has fewer than k orbits, then we infer from the above that G has at most

$$|V(Q_c)| + |L_1| + \cdots + |L_{k-2}|$$

vertices. Because $|V(Q_c)| \leq 2^k$ and $|L_r| \leq |V(Q_c)| \cdot (k-1)^r$, we obtain the estimate

$$|V(G)| \leq 2^k \left[1 + (k-1) + (k-1)^2 + \cdots + (k-1)^{k-2}\right] < 2^k \cdot k^{k-1} < (2k)^k .$$

□

Note that this result is not true for general isometric subgraphs of hypercubes, as the cycle C_{2r} shows. It has degree 2, can be arbitrarily large, but has only one orbit.

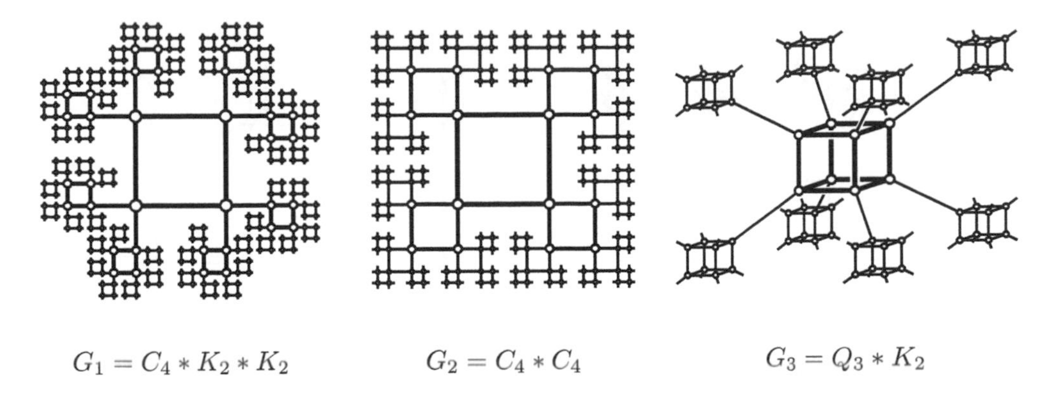

$G_1 = C_4 * K_2 * K_2$ $G_2 = C_4 * C_4$ $G_3 = Q_3 * K_2$

FIGURE 31.3 Some 4-regular, vertex-transitive median graphs with finite blocks.

Now we are ready for a characterization of locally finite, connected vertex-transitive median graphs without infinite blocks.

Theorem 31.7 *Every connected vertex-transitive median graph without infinite blocks is a generalized free product of finite median graphs without cut vertices. This implies that, for fixed k, there are only finitely many k-regular, vertex-transitive median graphs without infinite blocks.*

Proof Suppose G is a connected vertex-transitive median graph without infinite blocks. Then all blocks are finite and, as G is a median graph, they are median graphs. Now the first assertion follows from Theorem 31.5.

If G is k-regular, then no free factor of G can have a vertex of degree larger than k or more than k orbits. By Proposition 31.6 there are only finitely many median graphs of maximum degree k with at most k orbits. Thus, the number of ways to represent G as a generalized free product is finite. □

As an application we construct the 4-regular, vertex-transitive median graphs with finite blocks.

Proposition 31.8 *The 4-regular, vertex-transitive median graphs with finite blocks are Q_4, the infinite regular tree of degree 4, and the graphs $K_2 * K_2 * C_4$, $C_4 * C_4$, and $K_2 * Q_3$.*

Proof We will apply Theorem 31.7. Thus we need all median graphs without cut vertices, and degree at most 4. The regular ones are the hypercubes Q_2, Q_3, and Q_4.

Any other has a vertex of degree at least 3 and at least two orbits. As its minimum degree is at least 2, a free product with such a graph as a factor is at least 5-regular (see, for instance, Figure 31.1, right) and cannot be used in our construction.

Thus the only possibilities are $K_2^{*,4} = K_2 * K_2 * K_2 * K_2$, $K_2 * K_2 * C_4$, $C_4 * C_4$, $K_2 * Q_3$, and Q_4. Notice that $K_2^{*,4}$ is the infinite regular tree of degree 4. The graphs $G_1 = K_2 * K_2 * C_4$, $G_2 = C_4 * C_4$, and $G_3 = K_2 * Q_3$ are shown in Figure 31.3. □

The next theorem lists all 4-regular, vertex-transitive median graphs. For the proof we refer to Imrich and Klavžar (2009). It follows the lines of the arguments in this section.

Theorem 31.9 *Any vertex-transitive, 4-regular median graph must be one of the following: Q_4, the 4-regular tree, $P_Z \Box P_Z$, or one of the graphs $G_1, \ldots, G_{10}$ from Figures 31.3 and 31.4.*

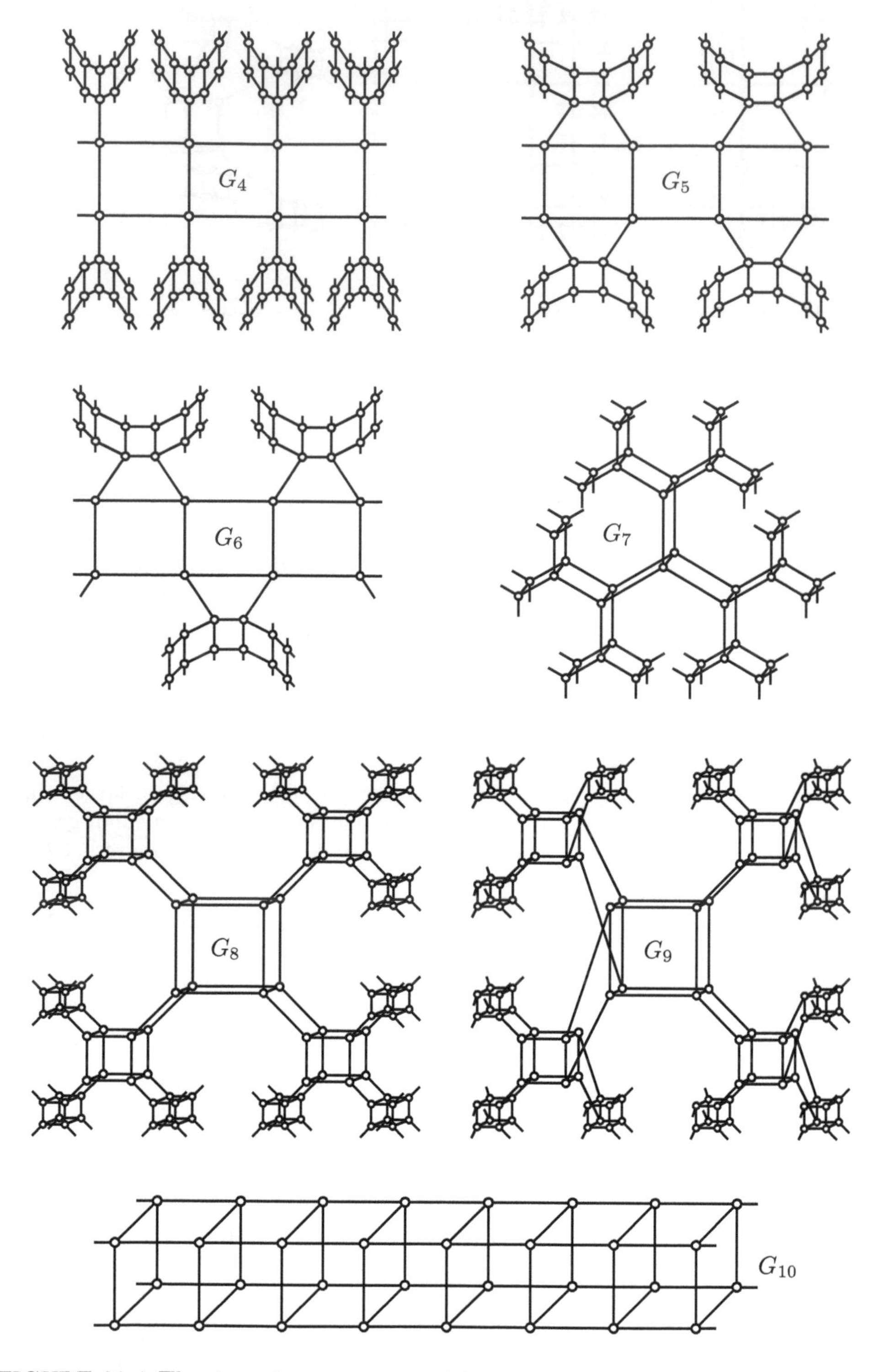

FIGURE 31.4 The 4-regular, vertex-transitive median graphs without finite blocks.

31.4 Two-Ended Median Graphs

We just considered infinite, vertex-transitive median graphs and proved that there are only finitely many such graphs of a given finite degree with finite blocks. We also constructed an infinite family of vertex-transitive median graphs with finite intransitive blocks, and determined all vertex-transitive median graphs of degree 4.

The latter result extends the work of Bandelt and Mulder (1983), who showed that there are exactly three cubic vertex-transitive median graphs: the 3-cube, the 3-regular tree, and the *infinite ladder* $K_2 \,\Box\, P_{\mathbb{Z}}$. By their work, infinite ladders are characterized as two-ended vertex-transitive cubic median graphs. In this section we prove that the result also holds without the condition of vertex-transitivity; see Theorem 31.13.

As the infinite ladder is $Q_1 \,\Box\, P_{\mathbb{Z}}$, one might ask whether any regular median graph with two ends must be $Q_n \,\Box\, P_{\mathbb{Z}}$. We will show that this is not the case by constructing a 4-regular median graph that has a cut vertex. Also, this graph will have cubic growth, whereas Cartesian products of hypercubes by $P_{\mathbb{Z}}$ have only linear growth.

Thus the question arises how the growth rate is reflected in the structure of regular median graphs. This leads to Theorem 31.14, the main result of this section. It characterizes two-ended, regular median graphs of linear growth as the class of Cartesian products of finite hypercubes by $P_{\mathbb{Z}}$. Because vertex-transitive graphs of linear growth are regular and have only two ends, this also characterizes vertex-transitive graphs of linear growth as Cartesian products of hypercubes by $P_{\mathbb{Z}}$.

Theorem 31.14 and the other main results of the present section are from Imrich and Klavžar (2011). We continue with several lemmas needed in the proofs. For brevity, we often blur the distinction between vertex sets such as W_{ab} and the subgraphs they induce. Thus, depending on context, W_{ab}, etc. can be either a vertex set or a subgraph.

Lemma 31.10 *Let F_{ab} be a Θ-class of a k-regular median graph. If U_{ab} is not peripheral, then neither is U_{ba}, and both $W_{ab} \setminus U_{ab}$ and $W_{ab} \setminus U_{ab}$ induce subgraphs with isometric rays.*

Proof If U_{ab} is not peripheral, there must be a vertex in $W_{ab} \setminus U_{ab}$, say c_1. By the connectedness of W_{ab} we can assume that c_1 is adjacent to a vertex in U_{ab}, say a.

Because $d_G(a) = k$ we infer that $d_{U_{ab}}(a) \leq k-2$ and hence, as $U_{ab} \cong U_{ba}$, we conclude that $d_{U_{ba}}(b) \leq k-2$ as well. But then U_{ba} cannot be peripheral either.

Now we consider F_{ac_1}, U_{ac_1}, and U_{c_1a}. Because U_{ac_1} does not contain ab and ac_1, it is clear that U_{ac_1} is not peripheral. By the above, U_{c_1a} is also not peripheral.

Proceeding in this manner we arrive at a sequence of vertices $c_1, c_2, c_3, \ldots$ with the property that neither $U_{c_ic_{i+1}}$ nor $U_{c_{i-1}c_i}$ is peripheral. It is easily confirmed (by definition of W_{uv}, etc.) that any $c_1c_2c_3\ldots c_\ell$ is a shortest path in W_{ab}. Hence W_{ab} contains an isometric ray. By the same arguments this holds for W_{ba} too. □

Lemma 31.11 *Let G be a finite, connected graph in which all vertices are of degree d or $d-1$. If the vertices of degree $d-1$ induce a convex subgraph, then G is not median.*

Proof Let $v_1, v_2, \ldots, v_k$ be the vertices of degree $d-1$ in G. Take an isomorphic copy G' of G that is disjoint from G and form a new graph H that consists of $G \cup G'$ together with the edges v_iv_i', for $i = 1, \ldots, k$.

If G is a median graph, then H is a median graph too and the k edges v_iv_i' form a Θ-class. (This follows from Mulder's Convex Expansion Theorem 12.8.) By Exercise 12.8, G is a hypercube. But then edges in every Θ-class meet all vertices, which clearly is not the case here. □

We continue with another simple but important fact about median graphs.

Lemma 31.12 *If an edge of a median graph lies on a cycle, then it lies on a square.*

Proof Suppose an edge $e = uv$ lies on a cycle C of a median graph. By Lemma 11.3, there is another edge $f = xy$ on C with $e\Theta f$. Say $x \in U_{uv}$. As U_{uv} is connected, it has a u, x-path. Consequently, there is a ladder from e to f. Then e together with the first "rung" of the ladder after e induce the required square. □

We begin with the cubic case and prove the following result that extends the classification of cubic vertex-transitive median graphs.

Theorem 31.13 *If G is a 2- or 3-regular median graph with two ends, then either $G \cong P_Z$ or $G \cong Q_1 \Box P_Z$.*

Proof If G is 2-regular, then clearly $G \cong P_Z$. Thus assume that G is 3-regular.

Let v be a vertex in some U_{ab}. It is adjacent to a vertex in U_{ba}, and thus its degree in U_{ab} can only be 1 or 2. As U_{ab} is connected, it must be a finite path, a ray, a P_Z, or a cycle.

Now, U_{ab} is a P_Z or a cycle if and only if it it 2-regular, if and only if it is peripheral. If $U_{ab} \cong P_Z$, then $G \cong Q_1 \Box P_Z$, and we are done. If U_{ab} is a cycle, then G is a finite regular median graph and has zero ends.

Thus it remains to investigate the cases in which each U_{ab} is a finite path or a ray and is not peripheral. In what follows we show that each of these scenarios is impossible.

Suppose some U_{ab} is a ray R_a with origin a. Then U_{ba} is a ray R_b with origin b and $U_{ab} \cup F_{ab} \cup U_{ba}$ is an infinite ladder L with first rung ab. With the exception of a and b, all vertices of L have degree 3 in L. It follows that any path joining $W_{ab} \setminus U_{ab}$ to L must contain a, and any path from $W_{ba} \setminus U_{ba}$ to L must contain b. By Lemma 31.10, $W_{ab} \setminus U_{ab}$ and $W_{ab} \setminus U_{ab}$ have rays R_1 and R_2, respectively. Consider the three rays R_1, R_2, and R_a. Notice that any path joining two of these rays pass through a or b. It follows that no ray can meet two of them infinitely often. Thus rays R_1, R_2, and R_a belong to three distinct ends. As G has only two ends, we conclude that no U_{ab} is a ray.

Only one case remains: Every U_{ab} is a finite path that is not peripheral. Then $U_{ab} \cup F_{ab} \cup U_{ba}$ is a finite ladder L. Let ab be the first rung and xy the last, where $x \in U_{ab}$. By Lemma 31.10, a, b, x, and y are origins of infinite rays, say P_a, P_b, P_x, and P_y. Now, P_a and P_x are separated from P_b and P_y by the finite sets U_{ab} and U_{ba}. Because we have only two ends, P_a and P_x must be in the same end. Thus there must be a third ray R that meets both of them infinitely often. Let P be a subpath of R from P_a to P_x that does not meet U_{ab} and let P'_a and P'_x be the corresponding starting sections of P_a and P_x, respectively. Then

$$U_{ab} \cup P'_a \cup R \cup P'_x$$

is a closed walk in W_{ab}. Clearly, U_{ab} is contained in a subcycle of this walk. But then every edge of U_{ab} is in a square in W_{ab} by Lemma 31.12. This is only possible if U_{ab} consists of a single edge. It follows that F_{ab} (and by the same argument, every F_{uv}) has exactly two edges. Thus G is an even cycle, in contradiction to its degree. □

For degree 4 the situation is quite different; there exist 4-regular, two-ended median graphs not of the form $Q_3 \Box P_Z$. To construct such an example (Figure 31.5), we begin with the Cartesian product H of a ray by itself. Clearly this is a median graph. We can consider its vertices as nonnegative integer lattice points (i, j) on the plane. For brevity put $u_i = (i, 0)$ and $v_j = (0, j)$.

For every integer $r \geq 0$ we make a copy H_r of H, and introduce the notation $u_{i,r}$ and $v_{j,r}$ for its vertices u_i and v_j.

Furthermore, for every $r \geq 0$ we identify the path $u_{1,r}u_{2,r}u_{3,r}\ldots$ of H_r with the path

$v_{0,r+1}v_{1,r+1}v_{2,r+1}\dots$ of H_{r+1}. Let the new graph be A. We take a copy of A, say A', and identify the path $v_{0,0}v_{1,0}v_{2,0}\dots$ of A with the path $v'_{0,0}v'_{1,0}v'_{2,0}\dots$ of A'.

All vertices of the new graph, say B, have degree 4, with the sole exception of the vertex $(0,0)$, which has degree 3. Finally, let B' be a copy of B and M the graph obtained by joining the vertices $(0,0)$ of B and $(0,0)'$ of B' by an edge. (See Figure 31.5.) Clearly, M is 4-regular and has two ends.

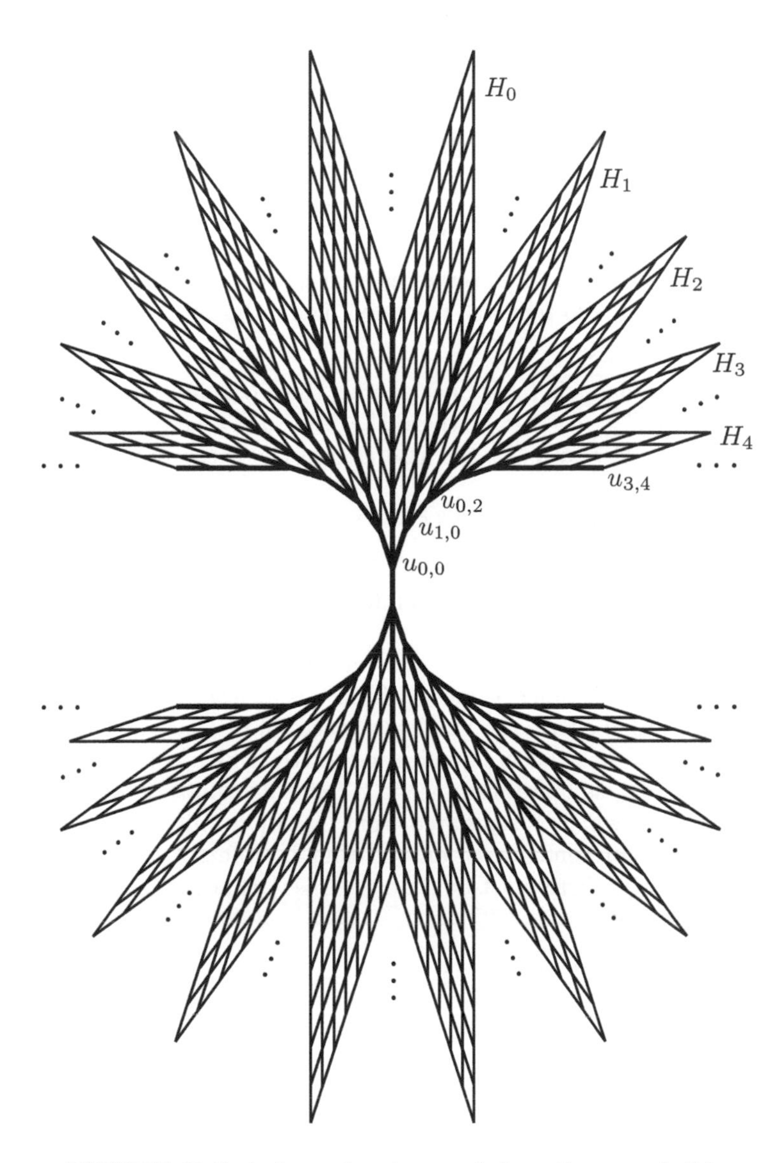

FIGURE 31.5 A 4-regular, two-ended median graph M.

One way to see that it is a median graph is as follows. The only isometric cycles of M are 4-cycles. Moreover, M contains no $K_{2,3}$, so the convex closure of any C_4, that is, the smallest convex subgraph of M containing the C_4, is the C_4 itself. Now apply a theorem of Bandelt (1982) (see also Klavžar and Mulder (1999)), asserting that a connected graph is median if and only if the convex closure of any isometric cycle is a hypercube. It is not difficult to check that M does not have linear growth, its growth rate is cubic, and it is planar.

The construction of M can be modified so that the resulting 4-regular median graph has arbitrarily many ends. If we match the ray $u_{k,0}u_{k+1,0}u_{k+2,0}\dots$ of H_r with $v_{0,1}v_{1,1}v_{2,1}\dots$ of H_1, and do not change the rest of the construction, then we obtain a graph, say B^k, with $2k-1$ vertices of degree 3. They can be joined by edges to the vertices of degree 3 in $2k-1$ copies of B.

Another easy modification shows that the sets that separate the ends of these examples can be of arbitrary (finite) cardinality.

We are now ready for the main result of the section:

Theorem 31.14 *If G is a two-ended, d-regular, median graph of linear growth, then*

$$G \cong Q_{d-2} \,\square\, P_Z\,.$$

Proof We induct on d. Theorem 31.13 serves as a basis, so we assume that $d \geq 4$ and that the theorem holds for all $(d-1)$-regular two-ended median graphs with linear growth.

Case 1: G has an infinite Θ-class.
Let F_{ab} be an infinite Θ-class. If U_{ab} is peripheral, then U_{ba} must also be peripheral by Lemma 31.10. But then $G = K_2 \,\square\, U_{ab}$, where U_{ab} is a $(d-1)$-regular two-ended median graph with linear growth. For such U_{ab} the assertion of the theorem is true, and hence

$$G = K_2 \,\square\, U_{ab} = K_2 \,\square\, (Q_{d-3} \,\square\, P_Z) = (K_2 \,\square\, Q_{d-3}) \,\square\, P_Z = Q_{d-2} \,\square\, P_Z\,.$$

Therefore assume that U_{ab} is not peripheral. By Lemma 31.10 there is an isometric ray $P_a = ac_1c_2\dots c_ic_{i+1}\dots$ in $W_{ab}\setminus U_{ab}$. As U_{ab} is infinite and locally finite, it too must contain at least one end.

Because P_a is isometric the sets $F_{c_{i-1}c_i}$ are mutually disjoint. If they are all infinite, then the number of edges at distance at most n from a is at least $n(n-1)/2$. Because every vertex is incident with just d edges, the number vertices of distance at most n from a is at least at $n(n-1)/2d$ and G has at least quadratic growth, which is not possible.

Hence, one of the $U_{c_ic_{i+1}}$ is finite. Because it is finite it cannot be peripheral. Thus the set $W_{c_{i+1}c_i}$ contains a ray, which is separated from U_{ab} by a finite cutset. In other words, $ac_1c_2\dots c_ic_{i+1}\dots$ is the representative of an end that is different from the representative of any end in U_{ab}. Similarly, there must be a finite $U_{c'_ic'_{i+1}}$ that separates an end from U_{ba}. But then G is not two-ended.

Case 2: All Θ-classes of G are finite.
Let U_{ab} be arbitrarily chosen. Neither U_{ab} nor U_{ba} can be peripheral, otherwise G would be finite. Using Lemma 31.10 again, there exist two rays P_a and P_b, originating from a, resp. b, that are separated by the finite set U_{ab} and thus represent the two ends of G.

Suppose a has a neighbor a' in U_{ab} that is in turn adjacent to a vertex that is neither in U_{ab} nor in U_{ba}. Then a' is the origin of a ray $P_{a'}$, which must represent the same end as P_a. By Lemma 31.12, there is a square $aa'c'_1c_1$, where c'_1, c_1 are not in $U_{ab} \cup U_{ba}$.

Now we replace a, a' by c_1, c'_1. Clearly, both c_1 and c'_1 have neighbors that are not in $U_{c_1a} \cup U_{ac_1}$. Thus we can find a square $c_1c'_1c'_2c_2$, where c'_2, c_2 are not in $U_{c_2c_1} \cup U_{c_1c_2}$.

Continuing in this manner we get an infinite ladder whose first rung is aa'. We thus have an infinite Θ-class. Thus any neighbor of a is adjacent only to vertices in U_{ab} or U_{ba}.

Let a'' be such a neighbor and consider $F_{aa''}$.

If a vertex $v \neq a$ of P_a meets an edge e of $F_{aa''}$, then there is a ladder L in W_{ab} from aa'' to e. It is possible that aa'' is not the only rung of L that is in U_{ab}. Let xy be the last rung of L in U_{ab}. Thus both vertices x, $y \in U_{ab}$ have neighbors outside of U_{ab} and U_{ba}. But then we can use the arguments we used for aa' for the edge xy to see that L is infinite. Hence a is the only vertex of P_a that meets $F_{aa''}$.

Analogously we show that b is the only vertex of P_b that meets $F_{aa''}$. This means that $U_{aa''}$ separates the two ends P_a and P_b from $W_{a''a}$.

Because $U_{a''a}$ is finite, it is not peripheral. Hence $W_{a''a}$ contains an infinite ray. This ray is separated from the P_a and P_b by a finite set, which implies that G has at least three ends, which is not possible. □

31.5 Cartesian Product

To define products of infinitely many factors, it is convenient to replace the vector of coordinates by a function from an index set for the factors into the vertex sets of the factors. To be precise, let I be an index set and G_ι, $\iota \in I$, be a family of graphs. The Cartesian product

$$G = \mathop{\square}_{\iota \in I} G_\iota$$

is defined on the set of all functions $x : \iota \mapsto x_\iota$, $x_\iota \in V(G_\iota)$, where two vertices x, y are adjacent if there exist a $\kappa \in I$ such that $x_\kappa y_\kappa \in E(G_\kappa)$ and $x_\iota = y_\iota$ for $\iota \in I \setminus \kappa$.

Not to have to change the terminology, we will still speak of the x_ι as coordinates of x.

For finite I this coincides with the usual definition. For products of infinitely many nontrivial graphs G_ι, we note a fundamental difference to the finite case. If we have only finitely many factors, then the product is connected if and only if the factors are, whether they are finite or not. However, if we have infinitely many factors, there are vertices that differ in infinitely many coordinates. One cannot connect them by paths of finite length, because the endpoints of every edge differ in just one coordinate.

We call the components of G *weak Cartesian products.* To identify a component, it suffices to know a single vertex of it. Thus the weak Cartesian product

$$G = \mathop{\square}_{\iota \in I}^{a} G_\iota$$

is the connected component of $\square_{\iota \in I} G_\iota$ containing the vertex a. Instead of $\square^{a}_{\iota \in I} G_\iota$ we sometimes also write $\square_{\iota \in I}(G_\iota, a_\iota)$. Clearly, $\square^{a}_{\iota \in I} G_\iota = \square^{b}_{\iota \in I} G_\iota$ if and only if a and b differ in only finitely many coordinates.

Theorem 31.15 *Every connected graph is uniquely representable as a weak Cartesian product of connected prime graphs.*

We provide two proofs. For the first proof, the results on the Cartesian product form Part I and II suffice; for the second proof, the product relation σ from Chapter 23 in Part IV is needed.

First proof We begin with the observation that in a product $G_1 \square G_2$, every edge is either contained in a G_1-layer or a G_2-layer.

Let $v_0 \in V(G)$. For any edge e that is incident with v_0, let G_e be the intersection of the layers containing e in all decompositions $G = G_1 \square G_2$. This intersection is not empty, because it contains e.

We claim that two such subgraphs G_e, G_f are either identical or have just v_0 in common. To see this, assume that there is a decomposition $G = G_1 \square G_2$ such that $e \in G_1^{v_0}$ and

$f \in G_2^{v_0}$. Then $G_1^{v_0} \cap G_2^{v_0} = v_0$. Because $G_e \subseteq G_1^{v_0}$ and $G_f \subseteq G_2^{v_0}$ we also have $G_e \cap G_f = v_0$. On the other hand, if no such decomposition exists, then e and f are always in the same layer for any representation $G_1 \Box G_2$ of G. In other words, e and f are either both in $G_1^{v_0}$ or both in $G_2^{v_0}$, and thus $G_e = G_f$.

Let $G_\iota, \iota \in I$, be the set of all distinct G_e's. We wish to show that

$$G = \mathop{\Box}_{\iota \in I}^{v_0} G_\iota \,.$$

We show first that every G_e is a factor of G. If $G_e \neq G$, then there exists a vertex $v \notin V(G_e)$. To such a vertex there is a factorization $G_1 \Box G_2$ of G where $G_e \subseteq G_1^{v_0} \neq G_1^v$. As G is connected, we can even assume that v is adjacent to a vertex of G_e. Notice that the edges between $G_1^{v_0}$ and G_1^v induce an isomorphism between them. Hence, these edges also induce an isomorphism between G_e and its neighbors in G_1^v. We denote this copy by G_e^v and call it a layer. Let e' be the image of e in G_e^v under this isomorphism. We leave it to the reader to show that G_e^v is the intersection of all layers with respect to any decomposition of G with respect to $\Box$ that contain e'.

Notice that G_e and $G_{e'}$ are intersections of convex subgraphs and thus convex too. It is easily seen that the subgraph of G induced by these two layers is also convex.

Continuing in this way, we cover all vertices of G with G_e-layers.

Now we define the projections $p_\iota(v)$ for the vertices of G. Recall that to any vertex v in G and any layer G_1^u in a representation $G = G_1 \Box G_2$, there is a unique vertex z in G_1^u of shortest distance from v. Moreover, if x is any vertex in G_1^u, then z is on a shortest v, x-path. (Compare Exercise 4.6.)

For any $\iota \in I$, we will define $p_\iota(v)$ to be the vertex $x \in G_\iota$ of shortest distance from v. Such a vertex exists, because G is connected, so we only have to show that it is unique. Suppose there is another vertex y with this property. But then there exists a vertex w of maximal distance from v that is on shortest v, x- and v, y-paths. If we can show that $w \in G_\iota$, then $x = w = y$. Suppose $w \notin G_\iota$. Then there is a decomposition $G = G_1 \Box G_2$ such that $G_\iota \subseteq G_1^{v_0}$, but $w \notin G_1^{v_0}$. By convexity of the layers, v cannot be in $G_1^{v_0}$. Hence the unique vertex $z \in G_1^{v_0}$ of shortest distance from w is on shortest paths from x and y to w, and hence also to v, contrary to the definition of w. Thus $x = y$.

For any vertex u of G_e and every layer G_e^v, there thus exists a unique vertex y_v of shortest distance from u in G_e^v. Let H_e^u be the subgraph induced by these vertices. It is easily seen that any two H_e^x, H_e^y are disjoint or identical.

Notice that we can use the square property for any two incident edges a,b, where one edge is in a G_e^v and the other in H_e^v, because there must be a decomposition $G_1 \Box G_2$ of G, where a and b have different product colors. With this property it is easily seen that all H_e^w are isomorphic and that $G \cong G_e \Box H_e^{v_0}$.

Given an edge f incident with v_0, but not in G_e, we infer by the same arguments that G_f factors $H_e^{v_0}$, and therefore $G_e \Box G_f$ factors G by associativity. By induction, this holds for all products $G_e \Box G_{e_1} \Box G_{e_2} \Box \cdots \Box G_{e_k}$, where the e_i are incident with v_0.

Proceeding by induction again, it is now easy to show that for every vertex v of distance k there are edges $e_1, \ldots, e_{k'}$, $k' \leq k$, that are incident with v_0, such that v is in the v_0-layer of $G_e \Box G_{e_1} \Box G_{e_2} \Box \cdots \Box G_{e_{k'}}$.

Hence the union of these products is G and $G = \Box_{\iota \in I}^{v_0} G_\iota$ because all edges incident with v_0 are needed. □

This result is due to Miller (1970b) and Imrich (1971). The above proof has similarities with the one of Imrich, whereas Miller extends the methods of Sabidussi (1960).

Second proof By the second proof of Theorem 23.2 and Corollary 23.3, σ is a product relation. Furthermore, there is an isomorphism α from G onto a connected component of

G^*. In our language, α is an isomorphism from G onto a weak Cartesian product whose factors are components of the graphs induced by the σ-classes of G, one component from each class.

Notice that $\sigma = \Theta \cup \tau^*$ as in the finite case, and that Equation (23.1) also holds. That is, if $G = G_1 \Box \cdots \Box G_k$, then $\sigma \subseteq c(G_1 \Box \cdots \Box G_k)$. This means that σ is finer than any product relation of G with respect to any factorization of G. Hence the components of the graphs induced by the σ-classes of G are prime.

Uniqueness is shown as in the proof of Theorem 23.4, with the sole exception that the cardinals s and k can be infinite, and that the connected components of the graph induced by F_1 are isomorphic to a connected component of the Cartesian product $G_1^* \Box G_2^* \Box \cdots \Box G_s^*$, and not to the product, if s is infinite. □

We continue with the connection between prime factorization and the refinement property for products. We say that two factorizations $A \Box B = C \Box D$ of a graph G have a common refinement if there are graphs A_1, A_2, B_1, B_2 such that $A = A_1 \Box A_2$, $B = B_1 \Box B_2$, $C = A_1 \Box B_1$, and $D = A_2 \Box B_2$, and hence

$$A \Box B = (A_1 \Box A_2) \Box (B_1 \Box B_2) = (A_1 \Box B_1) \Box (A_2 \Box B_2) = C \Box D\,.$$

Such refinements need not always exist. For example, $N \times \Delta = K_2 \times A$ on p. 90 have no common refinement in Γ with respect to the direct product, but have one in Γ_0. For arbitrarily many factors we must be more precise. We say the *refinement property* holds if, for any two factorizations

$$\mathop{\Box}_{\iota \in I} A_\iota = \mathop{\Box}_{\kappa \in K} B_\kappa$$

of G, there exist graphs $C_{\iota,\kappa}$ for which

$$A_\iota = \mathop{\Box}_{\kappa \in K} C_{\iota,\kappa} \text{ and } B_\kappa = \mathop{\Box}_{\iota \in I} C_{\iota,\kappa}\,.$$

If I or K are infinite, then these formulas have entirely different meanings for the Cartesian and the weak Cartesian product. But, if a graph is connected, then it has the refinement property with respect to both the Cartesian product and the weak Cartesian product. This is a consequence of the fact that there exists a finest product relation, namely σ.

Clearly, the refinement property implies unique factorization for finite graphs. For infinite graphs this is not the case. For example, the weak hypercube $Q_{\mathfrak{n}}$ of any infinite dimension $\mathfrak{n}$, although uniquely factorable into prime graphs with respect to the weak Cartesian product, is not representable as a Cartesian product of prime graphs. Nonetheless, it satisfies the refinement property with respect to the Cartesian product.

It is an interesting exercise though to show that the Cartesian product $G = \Box_{\iota \in I} K_2$, where $|I| = \aleph_0$, has no other representations as a product of prime graphs. Notice that G has uncountably many connected components. See Exercises 31.12 and 31.13.

We conclude the section with the connection between the automorphism group of G and the groups of the prime factors. As such, it is same as in the finite case; see Theorem 6.10. This was observed by Miller (1970a) and Imrich (1971). Some precautions are appropriate though. For infinite graphs it is not true that they have transitive automorphism group if and only if each prime factor has transitive group. To see this, let the graphs $P_i, i \in \mathbb{Z}$, be isomorphic to the path xyz and set

$$G = \mathop{\Box}_{i} (P_i, a_i)\,, \tag{31.1}$$

where $a_i = x$ when i is even and $a_i = y$ otherwise. Evidently the path xyz is not vertex-transitive. Nevertheless, G is. Note first that every vertex b of G differs from a in only finitely many coordinates; thus all three coordinates x, y and z occur infinitely often in b. With this in mind, it is not hard to find an automorphism of G that maps a into b; see Exercise 31.14.

This construction is due to Imrich (1989). Notice the similarity with the construction of infinite, vertex-transitive graphs from finite, intransitive blocks in Section 31.2.

The factorization results in this section imply cancellation properties for the weak Cartesian product and describe the structure of the automorphism groups of infinite Hamming graphs (if we define them as weak Cartesian products of complete graphs).

The canonical isometric embedding of a connected graph G into a Cartesian product also extends to the infinite case. Thus infinite connected graphs can be canonically embedded into weak Cartesian products. Theorem 13.8 holds without restrictions, but we have a slightly different version of Theorem 13.9.

Theorem 31.16 *Let G be a connected graph with transitive automorphism group and α the canonical isometric embedding of G into the weak Cartesian product $G^* = \square^{a}_{\iota \in I} G^*_\iota$. Suppose that to every G^*_ι there are at most finitely many G^*_κ that are isomorphic to G^*_ι. Then G^* has transitive automorphism group.*

For a proof see Imrich (1989). Even Theorem 2.8 holds for graphs with arbitrary cardinality if infinite cubes are understood to be weak Cartesian products; see Imrich (1969b).

Distinguishing number

We now turn to the distinguishing number of infinite products and of infinite graphs of general interest. The definition of the distinguishing number in Section 6.2 naturally extends to the infinite case. Many results are similar, but there are some exceptions when it comes to large cardinalities.

Watkins and Zhou (2007) showed that all infinite, locally finite trees without vertices of degree 1 have distinguishing number 2, unless they are asymmetric (then their distinguishing number is 1). Imrich, Klavžar, and Trofimov (2007) extended the result to tree-like graphs of cardinality not greater than $\mathfrak{c}$, but the result is no longer true for larger cardinalities.

Imrich et al. (2007) also show that the countable random graph has distinguishing number 2, and that $D(K_{\mathfrak{n}} \square K_{\mathfrak{n}}) = D(K_{\mathfrak{n}} \square K_{2^{\mathfrak{n}}}) = 2$ for any infinite cardinal $\mathfrak{n}$. But if $2^{\mathfrak{n}} < \mathfrak{m}$, then

$$D(K_{\mathfrak{n}} \square K_{\mathfrak{m}}) > \mathfrak{n}\,.$$

For the weak infinite cube $Q_{\mathfrak{n}}$ of any dimension $\mathfrak{n}$, it is shown that $D(Q_{\mathfrak{n}}) = 2$, as in the finite case. However, the proof is rather tricky. It proceeds by transfinite induction, and limit cardinals turn out to be a big hindrance.

31.6 Strong and Direct Product

In this section we begin with general results about the strong and direct product. Then we introduce the weak direct product and illustrate some of its surprising features.

For the definition of the direct product we assume that our graphs are in Γ_0. The vertex set of the direct product

$$G = \bigtimes_{\iota \in I} G_\iota$$

is the same as that of the Cartesian one. Two vertices x, y are adjacent if $x_\iota y_\iota \in E(G_\iota)$ for every $\iota \in I$. If all factors G_ι are in Γ, then xy can be an edge of G only if x and y differ in all coordinates.

The strong product is then defined as

$$\boxtimes_{\iota \in I} G_\iota = \mathcal{N}(\bigtimes_{\iota \in I} \mathcal{L}(G_\iota)),$$

where the G_ι are in Γ, and where $\mathcal{N}$ is an operator that removes all loops. (See Exercise 8.7 for its definition.) In this case it is not necessary that x and y differ in all coordinates if they are adjacent.

As in the finite case, the projections p_ι of G into the factors are homomorphisms for the direct product and weak homomorphisms for the strong product.

We note in passing that neither the direct nor the strong product of finite graphs is connected if there is no finite bound on the diameter of the factors. To see this, let $G = \bigtimes_{i=1}^{\infty} P_i$, and consider the vertices x, y, where $x_i = 1$ and $y_i = i$ for all $i \in \mathbb{N}$. If there exists an x, y-path P of finite length k, then the projection $p_i(P)$ of P into the ith factor must have length at most k. However, if $k + 1 < i$, then $k < d\big(p_i(x), p_i(y)\big) \leq |P| = k$, which is not possible.

We continue with several deep results from McKenzie (1971), who is still the most authoritative source for the infinite case. The main topics of the paper are refinement, absorption, and cancellation results, where refinement is defined as for the Cartesian product; see Section 31.5. From our point of view, the main decomposition result is the following theorem:

Theorem 31.17 (Refinement Theorem) *Every connected, nonbipartite thin graph G has the refinement property. The requirement that G is thin can be dropped if G is finite or if it has a vertex that is not in relation R to any other vertex of G.*

Note that the R-equivalence class of a vertex u is trivial if and only if uRv implies $u = v$. For finite graphs this theorem implies Theorem 8.17, the uniqueness of the prime factorizations of connected, nonbipartite graphs with respect to the direct product.

The main problems with infinite direct products are the extraction of complete factors and the factorization of complete graphs (with loops at every vertex). For example, if $\mathfrak{n}$ is an infinite cardinal, then $\big(\mathcal{L}(K_2)\big)^{\mathfrak{n}} \cong \big(\mathcal{L}(K_3)\big)^{\mathfrak{n}}$, but there is no common refinement. However,

$$\mathcal{L}(K_{\aleph_0})$$

has the refinement property.

If the common refinement property holds, one does not know a priori whether it can be continued indefinitely. If it stops, that is, if there is a prime factorization of the graph in question, then, as McKenzie (1971) shows, it is unique for thin, connected graphs with a loop at every vertex. In other words, connected, simple thin graphs that allow a prime factor decomposition with respect to the strong product have unique prime factorizations.

Another important result is the following absorption theorem:

Theorem 31.18 *If $A \cong B \times C \times A$, where A is countable and B finite, then $A \cong B \times A \cong C \times A$.*

The implication $A \times B \cong A \times C \Longrightarrow B \cong C$ is a cancellation law. It is a consequence of unique prime factorization but it may also hold in a more general situation.

Theorem 31.19 *Let $A, B, C \in \Gamma$. If A is finite and connected, then $A \boxtimes B \cong A \boxtimes C$ implies that $B \cong C$.*

These compare with the renowned Proposition 9.7 and Theorem 9.10 due to Lovász (1967).

One of the problems left open by McKenzie (1971) was the existence of countable graphs G with the property that $G \cong G^3$ but $G \not\cong G^2$. This was solved affirmatively by Trnková (1976) for, roughly speaking, the strong product, and (1984) for connected simple graphs with respect to the direct product. With Koubek, Trnková also considered representations of countable commutative semigroups by Cartesian, strong, or direct products. See Trnková and Koubek (1978). Later she generalized these results to products of metric, uniform, and topological spaces; see, for example, Trnková (1990).

Weak direct and weak strong product

We first introduce the weak direct product and present conditions that ensure unique prime factorization of graphs with respect to it.

The fact that the Cartesian product $G = \Box_{\iota \in I} G_\iota$ of infinitely many nontrivial connected factors is disconnected was the motivation to call the connected components of G weak Cartesian products. The vertex set of such a component is then easily characterized as the set of all vertices that differ in only finitely many coordinates from any given vertex in that component.

If we use this without safeguards to define the vertex set of the weak direct product, we run into connectedness problems again. In addition to bipartiteness, we have already encountered two other obstacles to the connectedness of the weak direct product: The first is a lack of loops in the factors, and the other the lack of a finite upper bound on the diameter of the factors.

Therefore we define the weak direct product

$$G = \mathop{\times}_{\iota \in I}^{a} G_\iota$$

only if at most finitely many of the factors do not have loops at every vertex, and if there exists a bound N such that $\operatorname{diam}(G_\iota) < N$ for all $\iota \in I$. It is the subgraph of $\times_{\iota \in I} G_\iota$ induced by all vertices that differ from $a \in V(G)$ in only finitely many coordinates. Clearly, G is disconnected if more than one factor is bipartite.

The key to the prime factorization is the Cartesian skeleton. We define it as in the finite case. In analogy to Proposition 8.10 in the finite case, we have

$$S(\times_{\iota \in I} G_\iota) = \Box_{\iota \in I} S(G_\iota),$$

if $\times_{\iota \in I} G_\iota$ has no isolated vertices. (And under the usual assumption of R-thinness.) Connectedness is crucial again; unfortunately, we have no good general criteria to decide whether $S(G)$ is connected. However, suppose the Cartesian skeleton $S(G)$ of an infinite, thin graph G is connected. Then $S(G)$ has unique prime factorization with respect to the weak Cartesian product. We show below how it can be used to obtain the (unique) prime factorization of G with respect to the weak direct product:

Theorem 31.20 *Let $G \in \Gamma_0$ be a connected, R-thin, infinite graph with connected Cartesian skeleton $S(G)$. Then G is uniquely representable as a weak direct product of prime graphs.*

Proof Slightly modifying the arguments from the second proof of Theorem 31.15, we choose a vertex v_0 and a Cartesian pair v_0v. Because v_0v is an edge in $S(G)$, we set $e = v_0v$.

Given any factorization $G = G_1 \times G_2$, e is an edge in an $S(G_1)$- or an $S(G_2)$-layer of $S(G) = S(G_1) \Box S(G_2)$. Let $e \in S(G_1)^{v_0}$, and given another decomposition $G = H_1 \Box H_2$, let $e \in S(H_1)^{v_0}$. By the Refinement Theorem 31.17, the vertex set of $S(G_1)^{v_0} \cap S(H_1)^{v_0}$ is (the vertex set of) a layer of a direct factor of G.

Let G_e denote the intersection of all such layers. It has more than one vertex because it contains e, and one can show, just as in the case of the weak Cartesian product, that its vertex set is the vertex set of a layer of a prime factor of G with respect to the direct product. □

The relationship between the automorphism groups of the factors and that of the product is the same as for weak Cartesian products.

As an example, consider $G = \times^{a}_{\iota \in I} G_\iota$, where $G_\iota = \mathcal{L}(Q_3)$ for each index and where a is an arbitrary vertex of $\times_{\iota \in I} G_\iota$. Clearly $Q_3 = S(\mathcal{L}(Q_3))$ and $S(G) = \Box_{\iota \in I} K_2$ for infinite I. Notice that $S(G)$ depends only on G and not on the representation used to define it. Because $\mathcal{L}(Q_3)$ is prime with respect to the direct product, it is not hard to infer with the refinement property that the factorization $G = \times_{\iota \in I} G_\iota$ is the only prime factor decomposition of G with respect to the weak direct product; see Exercises 31.17 and 31.18.

Notice that $G = G^{\times, k}$ for any k, and that any presentation of G as a product of finitely many factors has a refinement. The example shows again that indefinite continuation of the refinement process and uniquely defined prime factors are no contradiction.

We should like to remark that the Cartesian skeleton of $\times_{\iota \in I} Q_3$ consists of uncountably many components, each of which is isomorphic to the infinite weak hypercube of dimension $|I|$, and that each component induces an edgeless subgraph of $\times_{\iota \in I} Q_3$. See Exercise 31.19.

From the above it should be clear how to define the weak strong product $\boxtimes^{a}_{\iota \in I} G_\iota$. It is connected if all factors are connected and if $\max_{\iota \in I}(\mathrm{diam}(G_\iota))$ is finite.

As we have already mentioned in Section 15.1, the weak strong product is used by Bonato, Hahn, and Tardif (2010) for the construction of infinite k-cop-win graphs, that is, of graphs G with $c(G) = k$. For every infinite cardinal $\mathfrak{n}$, they construct $2^{\mathfrak{n}}$ nonisomorphic k-cop-win graphs satisfying additional properties, such as vertex-transitivity, or having universal endomorphism monoid and automorphism group. Vertex-transitivity is achieved by the method proposed by Imrich (1989) for the weak Cartesian product; see also Equation (31.1). The paper also makes use of the X-join, a generalization of the lexicographic product due to Sabidussi. (See p. 128 for the definition.) For some of its properties, see the remarks after Theorem 31.22 in Section 31.7.

We conclude with the analogue to Theorem 31.20.

Theorem 31.21 *Let G be a connected, infinite graph with connected strong Cartesian skeleton $S[G]$. Then G is uniquely representable as a weak strong product of prime graphs.*

Again the relationship between the automorphism groups of the factors and that of the product is the same as for weak Cartesian products.

31.7 Lexicographic Product

The investigation of the lexicographic product of infinite relational structures goes back to Chang and Morel (1960). The following theorem states one of their results in the language of graphs.

Theorem 31.22 *If A, B, C, and D are graphs, each with at least one edge, and $A \circ B \cong C \circ D$, then the following assertions hold:*

(i) *If $|V(B)| = |V(D)| < \aleph_0$, then $A \cong C$.*

(ii) *If A and C are finite, then one of the graphs B or D is isomorphic to a subgraph of the other.*

Of course, Theorem 31.22 (i) implies Theorem 10.8. Part (ii) is important, because it already pinpoints a possibility that allows the automorphism group of a lexicographic product to properly contain the wreath product of its factors, even if the conditions of Theorem 10.13 are satisfied. This was treated by Hemminger (1968), who gave a complete characterization of the situation in which the automorphism group of a lexicographic product is the wreath product of the groups of the factors, respectively, when an X-join of graphs has only "natural" automorphisms.

Note that in the result above, the number of factors is finite, although the factors, or at least some of them, are allowed to be infinite. To allow infinitely many factors, it is natural to order them and to define a weak lexicographic product. For steps in that direction, see Sabidussi (1959). The concept dates back to Birkhoff (1940).

Corollary 10.17, which asserts that vertex-transitive graphs G are retracts of Cayley graphs, still holds in the infinite case. This is so because Theorem 10.16 remains valid for infinite graphs; one only has to replace the natural number n of the statement by the order of the stabilizer of a vertex of G. This approach is not very satisfactory if the order is infinite, because one may view the smallest n for which $G \circ D_n$ admits a regular group of automorphisms as a measure of how close G is to a Cayley graph. In the attempt to find out how close vertex-transitive graphs are to Cayley graphs, does there exist a general procedure for finding a smaller n, say a finite one, such that the assertion is true? Or some other method to show that they are close?

One answer is due to Trofimov (1984), whose result we provide below.

Let us recall first that a system of imprimitivity with respect to the action of a permutation group on a set V is a partition of V into disjoint sets called blocks such that every element of the group induces a permutation of these blocks. Moreover, a group is finitely generated if there exists a finite set S such that every element of A can be written as a product of elements in S. The set S is called a set of generators. If A is infinite, then arbitrarily long—but finite—words in the elements of S must be admitted. Also, A is countable in this case.

Now the result of Trofimov (1984):

Theorem 31.23 *Let G be an infinite, vertex-transitive, connected locally finite graph. Then G has polynomial growth if and only if there exists an imprimitivity system γ of $\mathrm{Aut}(G)$ on $V(G)$ with finite blocks such that $\mathrm{Aut}(G/\gamma)$ is finitely generated and the stabilizer in $\mathrm{Aut}(G/\gamma)$ of every vertex in the graph G/γ is finite.*

Together with Theorem 10.16 of Sabidussi, this theorem shows that vertex-transitive graphs of polynomial growth are closely related to Cayley graphs.

Theorem 31.23 actually states that $\mathrm{Aut}(G/\gamma)$ also has polynomial growth, where the growth rate of a finitely generated group is defined as the growth rate of any of its Cayley graphs with respect to a finite generating set. Moreover, Gromov (1981) characterized groups of polynomial growth as finitely generated groups with a nilpotent[2] subgroup of finite index. One can view Gromov's theorem as the fundamental result about Cayley graphs of polynomial growth and Trofimov's theorem as the extension to vertex-transitive graphs.

[2]Nilpotent groups include all Abelian groups and are solvable. See Hall (1976).

Trofimov's proof is deep in itself; it is not an easy consequence of Gromov's characterization of groups of polynomial growth.

Exercises

31.1. Prove that the existence of a maximal spanning forest in an infinite graph is equivalent to the axiom of choice.

31.2. Show that the growth rate of a locally finite, connected graph is independent of the choice of the base point.

31.3. Suppose all blocks of a connected graph G are median graphs. Show that G is a median graph.

31.4. Show that every tree in which all vertices have the same degree $d > 2$ has exponential growth.

31.5. What is the growth rate of the integer lattice of dimension k?

31.6. For a given natural number k, find a tree with growth rate n^k. Is this still possible for real k?

31.7. Let v be a vertex of an infinite tree T. For two rays P, Q emanating from v, set $d(P,Q) = 1/(|E(P) \cap E(Q)| + 1)$. Show that d is a metric on the set of ends of T.

31.8. Let T be a tree in which every vertex has the same finite degree. Show that the metric space consisting of the set of ends $\mathcal{E}(T)$ of T and the metric from Exercise 31.7 is compact.

31.9. Prove Proposition 31.1.

31.10. Give an example of a median graph that contains rays but no isometric ones.

31.11. Let G be a connected graph. If $\sigma_G = (\Theta_G \cup \tau_G)^*$ has infinitely many equivalence classes, then G has infinitely many prime factors G_ι, $\iota \in I$. How does one identify the component of $\Box_{\iota \in I} G_\iota$ to which G is isomorphic?

31.12. Show that the graph $G = \Box_{\iota \in I} K_2$, where $|I| = \aleph_0$, has uncountably many connected components.

31.13. Show that the graph G of Exercise 31.12 has unique prime factorization with respect to the Cartesian product if one admits only connected factors.

31.14. Show that the graph G in Equation (31.1) has transitive automorphism group.

31.15. Show that a connected infinite graph can have transitive automorphism group even if all prime factors with respect to the Cartesian product have trivial groups.

31.16. Prove or disprove: The Cartesian product of infinitely many connected prime graphs has transitive group if and only if every factor has transitive automorphism group.

31.17. Show that $Q_3 = S(\mathcal{L}(Q_3))$.

31.18. Prove that $G = \bigtimes_{\iota \in I} G_\iota$ is the only prime factor decomposition of G with respect to the weak direct product.

31.19. Let $G = \bigtimes_{\iota \in I} Q_3$. Show that $S(G)$ consists of uncountably many components, each of which is isomorphic to the infinite weak hypercube of dimension $|I|$, and that each component of $S(G)$ induces an edgeless subgraph of $\bigtimes_{\iota \in I} Q_3$

31.20. Show that every bipartite graph G is a zero-divisor.

Chapter 32

Products of Digraphs

Our choice of topics in this chapter is governed by an interest in extending to digraphs some of the themes of earlier chapters. We first note that the four standard products can be interpreted as products on digraphs. Then we develop properties of connectedness, prime factorization, and cancellation. For the lexicographic product, these issues (as well as that of commutativity) are intimately tied to transitive tournaments, and one section is devoted to this.

32.1 Definitions

Recall from Chapter 1 that a digraph G is a pair $G = (V(G), A(G))$, where $A(G) \subseteq V(G) \times V(G)$. An arc $(x, y) \in A(G)$ is abbreviated as xy and is visualized as an arrow pointing from x to y. As usual, a reflexive arc xx is called a *loop*.

For example, the *directed cycle* $\overrightarrow{C}_n$ is the digraph with vertices $\{0, 1, 2, \ldots, n-1\}$ and arcs $\{01, 12, 23, \ldots, (n-1)0\}$. In particular, $\overrightarrow{C}_1$ consists of a single vertex with a loop. The *directed path* $\overrightarrow{P}_n$ is $\overrightarrow{C}_n$ with one arc removed. (Note $\overrightarrow{P}_1 = K_1$.)

Any graph G can be identified as a symmetric digraph, that is, one for which $xy \in A(G)$ if and only if $yx \in A(G)$. (For example, $\overrightarrow{C}_2 = K_2$.) In this sense, the entire theory of graphs falls under the umbrella of the theory of digraphs. Consequently, many notions involving graphs apply equally well to digraphs. For instance, a *homomorphism* $\varphi : G \to H$ between digraphs is a map $V(G) \to V(H)$ for which $\varphi(x)\varphi(y) \in A(H)$ whenever $xy \in A(G)$. Similarly, the notions of weak homomorphism and isomorphism—which we defined earlier for graphs—are formulated for digraphs exactly as they were for graphs.

In this vein, the four standard products can be understood as products on digraphs; we simply interpret pairs such as xy and $(x, u)(y, v)$ as *arcs* rather than edges. For example, the direct product $G \times H$ of digraphs has vertex set $V(G) \times V(H)$, and $(x, u)(y, v)$ is an arc in the product precisely if xy and uv are arcs in G and H, respectively. Figure 32.1 shows examples of the Cartesian, direct, and strong products of digraphs.

In Figure 32.1, some edges of the direct product are dotted to emphasize the fact that $\overrightarrow{C}_4 \times \overrightarrow{C}_6$ is two copies of $\overrightarrow{C}_{12}$. This illustrates the general formula

$$\overrightarrow{C}_p \times \overrightarrow{C}_q \cong \gcd(p, q)\, \overrightarrow{C}_{\mathrm{lcm}(p,q)}, \tag{32.1}$$

whose proof is Exercise 32.1.

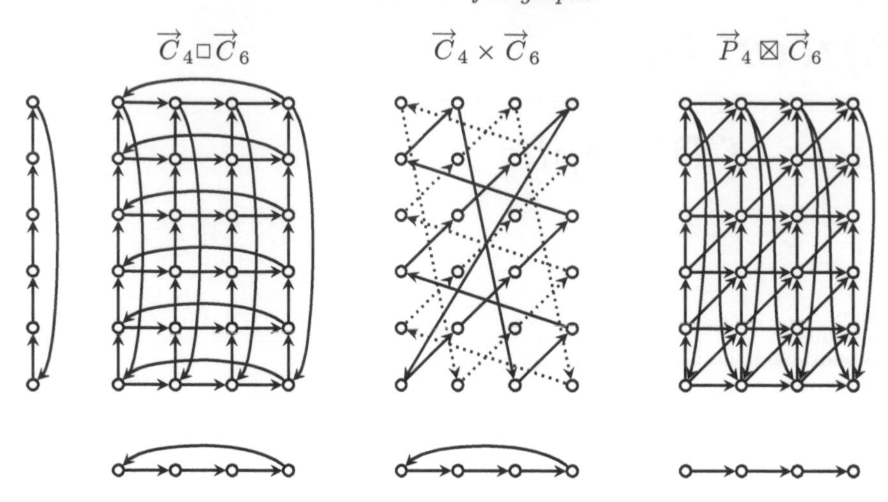

FIGURE 32.1 Products of digraphs.

As in the case for graphs, the four products of digraphs are associative and (except for the lexicographic product) commutative, and have the usual units.

A remark is in order. When dealing with the Cartesian, strong, or lexicographic products, we often assume tacitly that the digraphs involved have no loops. The reason is that if we admit loops, then the projections may not be weak homomorphisms, and layers may not be isomorphic to factors. (Consider $K_1 \square C_1$.) This can lead to theoretical difficulties. By contrast, the direct product poses no such obstacles, so we may admit loops in that case.

32.2 Connectedness

Digraphs have several notions of connectedness. To review them, we first recall the notion of walks. A sequence of vertices $x_1, x_2, \ldots, x_n$ in a digraph is called a *directed walk* if $x_i x_{i+1}$ is an arc for each $1 \leq i < n$. On the other hand, if $x_i x_{i+1}$ or $x_{i+1} x_i$ is an arc for $1 \leq i < n$, we say there is a *walk* from x to y. If $x_1 = x_n$, the walk (or directed walk) is *closed.* (Notice that a walk may not be uniquely determined by its vertex sequence, as the sequence gives no information on orientation of arcs. This causes no difficulty in what follows.)

A digraph is said to be *strongly connected* provided that for each pair x, y of its vertices, there is a directed walk from x to y and a directed walk from y to x. The maximal strongly connected sub-digraphs of a digraph are called its *strong components.*

A digraph is *unilaterally connected* if for each pair x, y of its vertices, it has a directed walk from x to y *or* a directed walk from y to x. (Because this relation on vertices is not generally symmetric, there is no notion of unilateral components.) Finally, a digraph is *connected* if for any two vertices x, y there is a walk from x to y. A maximal connected sub-digraph is called a *component* of the digraph.

The straightforward proofs of the following propositions are left as exercises.

Proposition 32.1 *If $G = \prod_{i=i}^{k} G_i$ denotes the Cartesian or strong product of digraphs G_i, then G is strongly connected (respectively connected) if and only if each factor G_i is strongly connected (respectively connected).*

Proposition 32.2 *A lexicographic product $G_1 \circ G_2 \circ \cdots \circ G_k$ of nontrivial digraphs is strongly connected (respectively connected) if and only if the first factor G_1 is strongly connected (respectively connected).*

See Exercise 32.4 for a characterization of unilaterally connected Cartesian and strong products, and Exercise 32.5 for the lexicographic product.

Connectedness for the direct product is a much more delicate issue. For a digraph G, let $d(G)$ be the greatest common divisor of the lengths of all the closed directed walks in G. (Any such walk has a decomposition into directed cycles, so $d(G)$ is also the gcd of the lengths of all directed cycles in G.) See McAndrew (1963) for a proof of the following theorem.

Theorem 32.3 *If $G_1, G_2, \ldots, G_k$ are strongly connected digraphs, then the direct product $G_1 \times G_2 \times \cdots \times G_k$ has*

$$\frac{d(G_1) \cdot d(G_2) \cdots d(G_k)}{\operatorname{lcm}\big(d(G_1), d(G_2), \ldots, d(G_k)\big)}$$

strong components.

Notice how this theorem agrees with the number of components in Equation (32.1).

Any directed walk from $(x_1, x_2, \ldots, x_k)$ to $(y_1, y_2, \ldots, y_k)$ in $G = G_1 \times G_2 \times \cdots \times G_k$ projects to each G_i as a directed walk from x_i to y_i. It follows that if G is strongly connected, then so is each factor. Combining this with Theorem 32.3, we get:

Corollary 32.4 *A direct product $G_1 \times G_2 \times \cdots \times G_k$ is strongly connected if and only if each G_i is strongly connected and the numbers $d(G_1), d(G_2), \ldots, d(G_k)$ are relatively prime.*

We conclude this section with a characterization of unilaterally connected direct products, due to Harary and Trauth (1966).

Theorem 32.5 *A product $G_1 \times G_2 \times \cdots \times G_k$ is unilaterally connected if and only if*

(i) *At most one factor, say G_1, is unilaterally connected but not strongly connected,*
(ii) *$G' = G_2 \times G_3 \times \cdots \times G_k$ is strongly connected, and*
(iii) *For each strong component C of G_1, $C \times G'$ is strongly connected.*

32.3 Tournaments and the Lexicographic Product

Recall that a *tournament* is an orientation of K_n—in other words, a directed graph with no loops, and such that any two distinct vertices are connected by exactly one arc. A tournament is *transitive* if whenever xy and yz are arcs, then xz is an arc too. For each positive integer n, there is (up to isomorphism) exactly one transitive tournament on n vertices, and we denote it as T_n.

The Cartesian, strong, or direct product of two nontrivial tournaments is not a tournament. (Consider the product of T_2 with itself.) But the lexicographic product of two tournaments is again a tournament; moreover, the product is transitive if and only if each factor is transitive (Exercise 32.7). Thus

$$T_m \circ T_n \cong T_{mn} .$$

In particular, $T_m \circ T_n \cong T_n \circ T_m$. This suggests that the commutativity properties for the lexicographic product of digraphs are richer than those laid out for graphs in Theorem 10.9. In fact, Dörfler and Imrich (1972) prove the following:

Theorem 32.6 *Two digraphs commute with respect to the lexicographic product if and only if both are powers of one and the same digraph, or both are complete graphs, or both are completely disconnected graphs, or both are transitive tournaments.*

We now summarize other results from Dörfler and Imrich (1972) that lead to theorems concerning prime factorization and cancellation. We use the notation and definitions of Chapter 10, suitably adapted to digraphs (without loops). For example, the *join* of two digraphs is defined as $X \oplus Y = \overline{\overline{X} + \overline{Y}}$. The following equations hold not just for graphs, but also for digraphs:

$$D_n \circ (X + Y) = D_n \circ X + D_n \circ Y\,,$$
$$K_n \circ (X \oplus Y) = K_n \circ X \oplus K_n \circ Y\,.$$

Theorem 32.7 *Suppose there is an isomorphism $\varphi : X \circ Y \to A \circ B$, where $|V(Y)| \nmid |V(B)|$ and $|V(B)| \nmid |V(Y)|$. If Y is indecomposable with respect to $+$ and $\oplus$, then there exist a digraph G and transitive tournaments T_n and T_m such that $X \cong G \circ T_n$, and $A \cong G \circ T_m$, and φ maps the $(T_n \circ Y)$-layers of $G \circ (T_n \circ Y)$ onto the $(T_m \circ B)$-layers of $G \circ (T_m \circ B)$. Furthermore,*

$$n = \frac{|V(B)|}{\gcd(|V(Y)|, |V(B)|} \quad \textit{and} \quad m = \frac{|V(Y)| \cdot |V(B)|}{\gcd\left(|V(Y)|, |V(B)|\right) \cdot |V(A)|}\,.$$

This leads to a prime factorization theorem, and we now introduce several ideas that are necessary for its statement. If q is prime and if $D_q \circ X + D_m$ is prime, then

$$(D_q \circ X + D_m) \circ D_q = D_q \circ (X \circ D_q + D_m)$$

are two different prime factorizations of the same graph. We say they are related by a *transposition of a totally disconnected graph.* Analogously,

$$(K_q \circ X \oplus K_m) \circ K_q = K_q \circ (X \circ K_q \oplus K_m)$$

are two different prime factorizations of the same graph, and we say they are related by a *transposition of a complete graph.* Also, we call the transition from $T_m \circ T_n$ to $T_n \circ T_m$ a *transposition of transitive tournaments.*

Here is the main theorem concerning prime factorings over the lexicographic product.

Theorem 32.8 *Any prime factorization of a digraph over the lexicographic product can be transformed into any other prime factorization by transpositions of totally disconnected graphs, transpositions of complete graphs, and transpositions of transitive tournaments.*

Parallel with Chapter 10, development along these lines gives the following particularly strong cancellation property for the lexicographic product of digraphs, which is completely analogous to the corresponding results for graphs (Proposition 10.7 and Theorem 10.8).

Theorem 32.9 *If $X \circ Y \cong A \circ B$ and $|V(Y)|$ divides $|V(B)|$, then Y is right divisor of B. If $|V(Y)| = |V(B)|$, then $Y \cong B$ and $X \cong A$.*

32.4 Prime Factorings

Theorem 32.8 describes prime factorings of digraphs over the lexicographic product: Any two prime factorings differ by transpositions of K_n's, D_n's and T_n's. We now examine the corresponding problem for the other products.

For the Cartesian product, we have the following analogue of our Theorem 6.6, due to Feigenbaum (1986), who also describes a polynomial algorithm for finding the prime factors.

Theorem 32.10 *Every finite connected digraph has a unique representation as a Cartesian product of prime digraphs, up to isomorphism and order of factors.*

The proof uses the observation that any Cartesian-prime factoring of a digraph induces a factoring (not necessarily prime) of the underlying graph. It is then possible to use the unique prime factorization of the underlying graph to complete the proof.

For the direct product, a special kind of connectedness is required. An *anti-walk* is a walk for which any two successive arcs have opposite orientations. Anti-walks of even length are divided into two types: those whose first and last arcs are directed toward the endpoints, and those whose first and last arcs are directed away from the endpoints. See Figure 32.2.

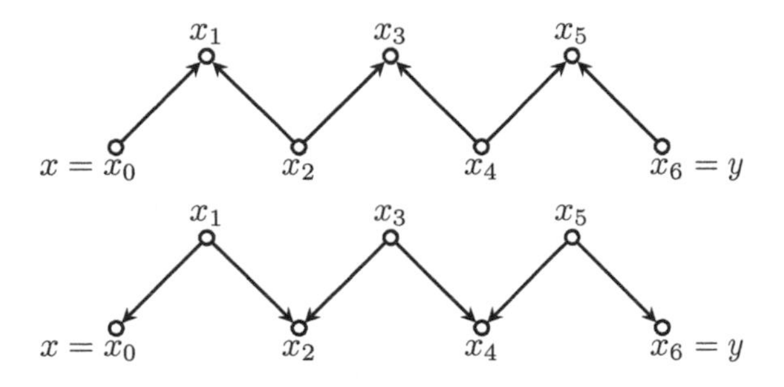

FIGURE 32.2 The two types of anti-walks.

For the purposes of this section, let us say that a digraph is *anti-connected* if any pair of its vertices is joined by even anti-walk *of both types.*[1]

The fundamental work of McKenzie (1971) on relational structures yields the following as a corollary.

Theorem 32.11 *Every finite anti-connected digraph has a unique representation as a direct product of prime digraphs (with loops allowed), up to isomorphism and order of factors.*

To see that prime factorization may fail if the hypotheses of this theorem are not met, let G be the closed even anti-walk on six vertices, which is not anti-connected. Indeed, we have the nonunique prime factorization

$$G \cong \overrightarrow{P}_2 \times K_3 \cong \overrightarrow{P}_2 \times H\,,$$

where H is the path of length two with loops at each end.

Theorem 32.11 leads almost immediately to a parallel theorem for the strong product. For a digraph G, let $\mathcal{L}(G)$ be the digraph obtained from G by adding a loop to each vertex.

[1]This differs from standard usage, which requires only that every two vertices be joined by some anti-walk.

Notice that if G is connected, then $\mathcal{L}(G)$ is automatically anti-connected, and it has neither sources nor sinks. Mimicking the alternative proof of Theorem 7.14 from Section 8.7 (which uses the operator $\mathcal{L}$), we get the following:

Theorem 32.12 *Every finite connected digraph (without loops) has a unique representation as a strong product of prime digraphs, up to isomorphism and order of the factors.*

There is no general polynomial algorithm that decomposes a finite, anti-connected thin graph into its prime factors with respect to the direct product. A step in this direction was taken by Imrich and Klöckl (2007).

They consider graphs where any two vertices are connected by a path of the type depicted at the top of Figure 32.2 and call such graphs N^+-connected. If no two vertices have the same out-neighborhood, then such a graph is R^+-*thin* in their terminology. They show that the prime factor decomposition of every finite N^+-connected, R^+-thin graph is unique, and that it can be computed in polynomial time.

Notice that this class encompasses some graphs that are not anti-connected and R-thin. In other words, graphs that do not satisfy the conditions of Theorem 32.11. Imrich and Klöckl relax connectivity, but strengthen the thinness condition.

32.5 Cancellation

We now extend the investigations of Chapter 9 to digraphs. Given that $A * C \cong B * C$ for some product $*$, we seek conditions on the digraphs A, B, and C that guarantee $A \cong B$.

For the lexicographic product, Theorem 32.9 gives a complete and satisfactory answer. Cancellation for the Cartesian product of connected digraphs is an immediate consequence of the uniqueness of prime factorization; for the disconnected case, we need only apply the argument from Section 6.4 to see that cancellation holds in general.

By contrast, cancellation fails for the direct product. A digraph C is said to be a *zero divisor* if there exist nonisomorphic digraphs A and B for which $A \times C \cong B \times C$. For example, Figure 32.3 shows that $\overrightarrow{C_3}$ is a zero divisor: If $A = \overrightarrow{C_3}$ and $B = 3\overrightarrow{C_1}$, then $A \ncong B$, yet $A \times \overrightarrow{C_3} \cong B \times \overrightarrow{C_3}$. (Both products are isomorphic to $3\overrightarrow{C_3}$.)

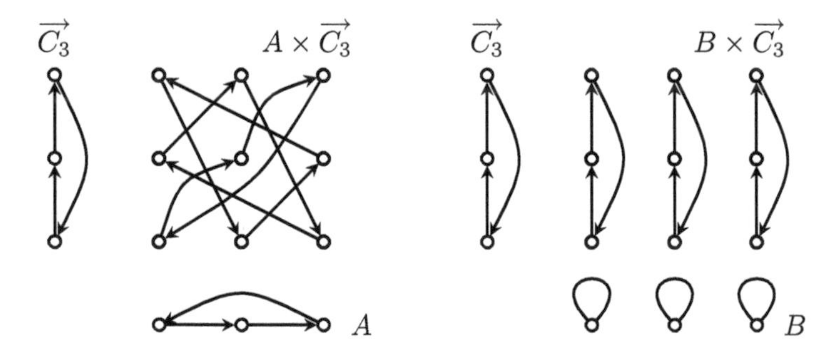

FIGURE 32.3 Example of a zero divisor.

In the category of graphs, C is a zero divisor if and only if it is bipartite. This follows from Theorem 9.10 and the discussion preceding it. As the above example indicates, the situation for digraphs is more complex. The following characterization is due to Lovász (1971).

Theorem 32.13 *A digraph C is a zero divisor if and only if there is a homomorphism $C \to \overrightarrow{C}_{p_1} + \overrightarrow{C}_{p_2} + \overrightarrow{C}_{p_3} + \cdots + \overrightarrow{C}_{p_k}$ for prime numbers $p_1, p_2, \ldots, p_k$.*

Thus $\overrightarrow{C}_n$ is a zero divisor if $n > 1$, as there is a homomorphism $\overrightarrow{C}_n \to \overrightarrow{C}_p$ for any prime divisor p of n. Also, each $\overrightarrow{P}_n$ is a zero divisor, as there are homomorphisms $\overrightarrow{P}_n \to \overrightarrow{C}_p$.

Given a digraph A and a zero divisor C, a natural problem is to determine all digraphs B for which $A \times C \cong B \times C$. Hammack and Toman (2010) give some partial results in this direction. Let $\mathrm{Perm}(V(A))$ be the set of permutations of $V(A)$. For $\alpha \in \mathrm{Perm}(V(A))$, let A^α be the digraph with $V(A^\alpha) = V(A)$, and with an arc $x\alpha(y)$, for each arc xy of A. If $A \times C \cong B \times C$, then $B = A^\alpha$ for some $\alpha \in \mathrm{Perm}(V(A))$. This condition is also sufficient if there is a homomorphism $C \to \overrightarrow{P}_2$. For more general zero divisors, more stringent conditions must be placed on the permutation α. (For an indication of this, see Exercises 32.8, 32.9, and 32.10.) A complete solution to this problem has yet to be realized.

We now turn to the strong product. Here cancellation fails in general; indeed, $\overrightarrow{C}_1 \boxtimes \overrightarrow{C}_1 \cong K_1 \boxtimes \overrightarrow{C}_1$. However, if we disallow loops, then Theorem 32.13 yields a positive result.

Theorem 32.14 *If $A \boxtimes C \cong B \boxtimes C$ for digraphs A, B, and C without loops, and C is nonempty, then $A \cong B$.*

Proof Exactly as for graphs in Section 8.7, it is immediate that $\mathcal{L}(G \boxtimes H) \cong \mathcal{L}(G) \times \mathcal{L}(H)$ for any digraphs G and H. Then from $A \boxtimes C \cong B \boxtimes C$, we get $\mathcal{L}(A) \times \mathcal{L}(C) \cong \mathcal{L}(B) \times \mathcal{L}(C)$. By Theorem 32.13, $\mathcal{L}(C)$ is not a zero divisor, so $\mathcal{L}(A) \cong \mathcal{L}(B)$, whence $A \cong B$. □

For another proof, see Culp and Hammack (2010).

Exercises

32.1. Prove (32.1) directly, without the aid of Theorem 32.3.

32.2. Prove Proposition 32.1.

32.3. Prove Proposition 32.2.

32.4. Show that a Cartesian product of digraphs is unilaterally connected if and only if one factor is unilaterally connected and the others are strongly connected. Show that this is also true for the strong product.

32.5. Prove that a lexicographic product of digraphs is unilaterally connected but not strongly connected if and only if each factor is unilaterally connected, and the first factor is not strongly connected.

32.6. Prove that each strong component in a Cartesian product of digraphs is a Cartesian product of strong components of the factors. Do the same for the strong product.

32.7. Prove that a lexicographic product of two digraphs is a tournament if and only if each factor is a tournament. Moreover, the product is transitive if and only if both factors are.

32.8. The notion of a graph factorial (Section 9.3) can also be defined for digraphs. Given any digraph G, there is a corresponding digraph $G!$ whose vertices are the permutations of $V(G)$. There is an arc from a permutation α to a permutation β provided that $\alpha(x)\beta(y) \in A(G) \iff xy \in A(G)$, for all pairs $x, y \in V(G)$.

For a permutation α, define G^α as $V(G^\alpha) = V(G)$ and $A(G^\alpha) = \{x\alpha(y) : xy \in A(G)\}$.

Show that if α, β are in the same connected component of $G!$, then $G^\alpha \cong G^\beta$.

32.9. Prove that $A \times \overrightarrow{P}_n \cong B \times \overrightarrow{P}_n$ if and only if $B = A^\alpha$, where α is on a directed walk of length $n - 2$ in the factorial of A.

32.10. Suppose a digraph C is homomorphically equivalent to $\overrightarrow{C}_2$. Prove that $A \times C \cong B \times C$ if and only if $B = A^\alpha$, where α has the property that xy is an arc of A if and only if $\alpha(x)\alpha^{-1}(y)$ is an arc of A. (This is the anti-automorphism property of Section 9.3.)

Chapter 33

Near Products

Several constructions that extend the utility of product graphs to graphs with a product-like structure have been proposed. This chapter introduces three meaningful examples: graph bundles, approximate graph products, and the zig-zag product. It also contains a brief account of spectra and their relationship to graph products, approximate graph products and the zig-zag product.

33.1 Graph Bundles

Graph bundles were first introduced as generalizations of the Cartesian product. We begin this section with two equivalent formulations of the original definition, and we remark how this idea can be extended to generalize products other than the Cartesian product. Then we briefly overview results concerning invariants, recognition, and characteristic polynomials of graph bundles.

Graph bundles were introduced in an unpublished manuscript by Pisanski and Vrabec (1982) and first appeared in print in Pisanski, Shawe-Taylor, and Vrabec (1983).

Let B and F be arbitrary graphs. A graph G is a *graph bundle* with *base* B and *fiber* F if there exists a weak homomorphism $p : G \to B$ for which

(i) For any $u \in V(B)$, the subgraph $p^{-1}(u)$ is isomorphic to F, and
(ii) For any $e \in E(B)$, the subgraph $p^{-1}(e)$ is isomorphic to $K_2 \,\square\, F$.

(Here we interpret $p^{-1}(u)$ topologically, as the preimage of the point u, so $p^{-1}(u)$ is identified with the subgraph $\langle p^{-1}(u)\rangle$. A similar remark applies to $p^{-1}(e)$, the preimage of an edge.) The triple (G, p, B) is called a *representation of G as a graph bundle*, and the map p is called the *(natural) projection* of the bundle G onto its base B. The edges of G that are mapped by p to vertices are called *degenerate*; the other edges are *nondegenerate*. In other words, the degenerate edges of G are those belonging to the copies of F.

Note that if G is a graph bundle with base B and fiber F, then $V(G) = V(B) \times V(F)$. Observe also that the Cartesian product $G = B \,\square\, F$ is a graph bundle. If F has no edges, then G is an instance of the so-called *covering graph*. Hence, graph bundles at the same time generalize Cartesian products and covering graphs. The graph $K_{3,3}$ as presented in

Figure 33.1 (with dotted lines indicating p) is the standard example of a bundle that is neither a product nor a covering graph.

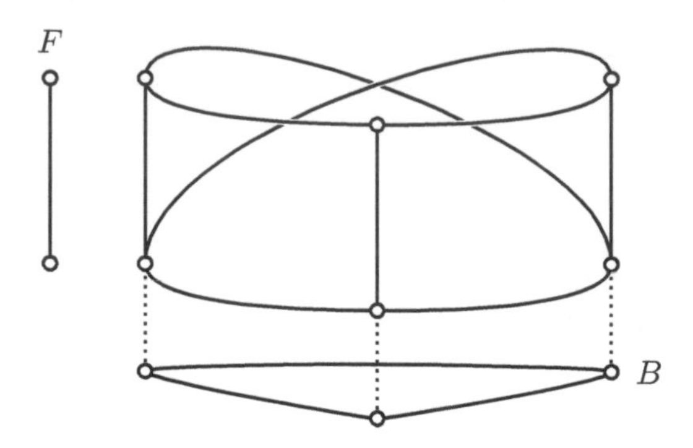

FIGURE 33.1 $K_{3,3}$ as a graph bundle with base C_3 and fiber K_2.

It is clear that if C is a connected component of B, then $p^{-1}(C)$ is a graph bundle with base C and fiber F. Thus we will assume henceforth that the base graphs are connected.

Graph bundles can be equivalently described in the following, more algebraic way. A graph G is a graph bundle with (connected) base B and fiber F if the following hold:

(i) $V(G) = V(B) \times V(F)$.

(ii) (u, x) and (u, y) are adjacent in G if and only if x and y are adjacent in F.

(iii) For each edge $e = uv \in E(G)$, there is an orientation, say from u to v, and an automorphism $\varphi_e \in \mathrm{Aut}(F)$ for which $(u, x)(v, \varphi_e(x)) \in E(G)$ for all $x \in V(F)$.

The automorphisms φ_e are called the *voltages* of the bundle. This algebraic representation of G as a graph bundle is denoted by $G = B \,\square^{\varphi} F$. The projection $G \to B$ is denoted as p_φ.

Note that the edge orientations are used only implicitly: If the orientation from v to u were selected, the voltage φ_e^{-1} would yield the same graph bundle. If all voltages are the identity, then $G = B \,\square\, F$. Moreover, it is easy to show that any bundle is isomorphic to one constructed by selecting a spanning tree T of B, setting $\varphi_e = \mathrm{id}$ for any $e \in E(T)$, and assigning the remaining voltages as appropriate. It follows that if B is a tree or $\mathrm{Aut}(F) = \{\mathrm{id}\}$, then every bundle with base B and fiber F is just the Cartesian product $B \,\square\, F$.

As we know, any connected graph has a unique prime factorization with respect to the Cartesian product. On the other hand, there can be different representations of a given graph G as a bundle with base B and fiber F. To make this more precise, we say that graph bundles $B \,\square^{\varphi} F$ and $B \,\square^{\varphi'} F$ are *isomorphic* if there exists an isomorphism $F : B \,\square^{\varphi} F \to B \,\square^{\varphi'} F$ and $f \in \mathrm{Aut}(B)$ such that the diagram

$$\begin{array}{ccc} B \,\square^{\varphi} F & \xrightarrow{F} & B \,\square^{\varphi'} F \\ \downarrow p_\varphi & & \downarrow p_{\varphi'} \\ B & \xrightarrow{f} & B \end{array}$$

commutes. Kwak and Lee (1990) gave a characterization of when two bundles (with the same base and fiber) are isomorphic and derived counting formulas for the number of nonisomorphic bundles. This line of research has been extended by Hong, Kwak, and Lee (1999), where bipartite graph bundles with connected fibers are studied, in particular the number of nonisomorphic bundles is computed for the case where the fiber is a path or a cycle.

Graph bundles as defined above generalize the Cartesian product operation, but we will shortly modify the definition to obtain similar structures that generalize other products. Hence as we have defined them, graph bundles would be more precisely named *Cartesian graph bundles*; but because the majority of research was done from the Cartesian point of view, it seems appropriate to call then simply graph bundles.

The definition of graph bundles with respect to other products is analogous to that of graph bundles. For instance, *strong graph bundles* are defined just as graph bundles except that (in the first of the two definitions given) we require that $f^{-1}(e)$ is isomorphic to $K_2 \boxtimes F$. Figure 33.2 shows a strong graph bundle with base C_5 and fiber P_3 that is not isomorphic to $C_5 \boxtimes P_3$.

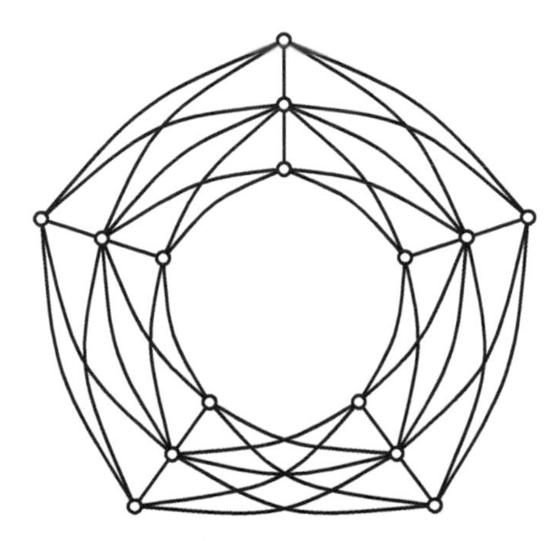

FIGURE 33.2 A strong graph bundle with base C_5 and fiber P_3.

Invariants of graph bundles

The first paper on graph bundles (Pisanski, Shawe-Taylor, and Vrabec (1983)) examined the problem of edge-colorings. The main result is the following extension of Theorem 30.10.

Theorem 33.1 *Let G be a graph bundle with base B and fiber F. Then G is of class* 1 *if at least one of the following conditions is satisfied:*

(i) *B is of class* 1 *and not totally disconnected.*
(ii) *F is of class* 1 *and not totally disconnected.*
(iii) *B and F both contain a one-factor.*

The chromatic number of graph bundles, strong graph bundles, and direct graph bundles was considered by Klavžar and Mohar (1995b). The chromatic number of bundles is much different from the chromatic number of products. For instance, the following result stands in stark contrast to Theorem 26.1.

Theorem 33.2 *For any $m, k \geq 1$, there is a graph B_k with $\chi(B_k) = k$ and voltages φ such that*

$$\chi(B_k \,\square^{\varphi} K_m) = km\,.$$

Klavžar and Mohar (1995a) determined the chromatic numbers of (strong, direct) graph bundles whose base and fiber are cycles. Concerning direct graph bundles, we also remark that Exercise 33.2 shows that Hedetniemi's conjecture 26.25 cannot be extended to bundles.

The domination number of graph bundles was considered by Zmazek and Žerovnik (2006). If G_k it the *corona* of K_k (i.e., the graph obtained from K_k by attaching a pendant edge to each of its vertices), then the following holds:

Theorem 33.3 *For any $k \geq 1$ there exist voltages φ such that*

$$\chi(C_4 \,\square^{\varphi}\, G_{3k+4}) = \gamma(C_4)\gamma(G_{3k+4}) - 2k\,.$$

Hence also, Vizing's conjecture 28.2 cannot be extended to graph bundles.

Several other graph invariants of bundles have been considered: Mohar, Pisanski, and Škoviera (1988) studied maximum genus of graph bundles, Kwak, Lee, and Sohn (1996) their isoperimetric numbers; Banič and Žerovnik (2006) their fault-diameter; and Banič, Erveš, and Žerovnik (2009) their edge fault-diameter. We also point out that the result of Theorem 30.19 has been extended to graph bundles by Pisanski and Žerovnik (2009).

Recognizing graph bundles

The recognition of covering graphs, that is, of bundles with totally disconnected fibers, is difficult. (See Abello, Fellows, and Stillwell (1991).) Hence in order to be able to efficiently recognize graph bundles, it makes sense to restrict to connected fibers. There are two main recognition algorithms for such graphs.

- Imrich, Pisanski, and Žerovnik (1997) proved that it is possible to recognize in polynomial time whether a given graph is a graph bundle with a triangle-free base and a connected fiber. Pisanski, Zmazek, and Žerovnik (2001) showed that the complexity of this algorithm is $O(mn)$, and, moreover, that all such representations can be determined in $O(mn^2)$ time. The main idea of the algorithm is that the transitive closure δ^* of the relation δ separetes degenerate and nondegenerate edges, provided that the base of a bundle is triangle-free. That this approach cannot be applied in general is addressed in Exercise 33.5.

- Zmazek and Žerovnik (2002) extended the above algorithm by proving that it is possible to recognize in polynomial time whether a given graph is a graph bundle with a $(K_4 - e)$-free base and a connected fiber.

Žerovnik (2000) also designed an algorithm for recognizing strong graph bundles with connected fiber over a triangle-free base. In general, it is not polynomial but becomes such for graphs with small clique number. It applies the above algorithm for graph bundles with connected fiber over a triangle-free base.

Characteristic polynomials of graph bundles

Chae, Kwak, and Lee (1993) and Kwak and Lee (1992) initiated a series of papers on characteristic polynomials (that is, characteristic polynomials of adjacency matrices) of graph bundles. In the first paper characteristic polynomials are computed for graph bundles with fibers K_2 and $\overline{K}_2$. The second paper considers the case where the voltages belong to an Abelian subgroup of $\mathrm{Aut}(F)$. The latter line of research was continued in Sohn and Lee (1995) and in H. K. Kim and J. Y. Kim (1996). Kwak and Kwon (2001) computes characteristic polynomials of graph bundles whose fibers are circulant graphs. We note that the Laplacian spectra of graph bundles has also been considered: See J. Y. Kim (1996) and D. Kim, H. K. Kim, and J. Lee (2008).

33.2 Approximate Graph Products

Certain applications require that observable phenomena be organized and modeled by a graph G. Typically the collected data is incomplete or prone to error, and we may thus regard G as being an approximation to some ideal graph H.

Such is the case, for example, in theoretical biology, where the set of phenotypes of an organism may be represented by a vector $\mathbb{X}$ of characters, such as color, presence of certain bones, etc. Typically, certain pairs of instantiations of $\mathbb{X}$ are known or expected to be interconvertible, and this endows $\mathbb{X}$ with a graph structure. Figure 33.3, which is adapted from Stadler, and Stadler (2004), is a schematic (if somewhat simplified) representation of this idea. A key biological question is whether the various characters can vary independently of each other; if this is the case, then we would expect the system of observable phenotypes $\mathbb{X}$ to have the structure of a product graph, or, more realistically, an approximation of a product graph. It is therefore becomes imperative to determine the extent to which $\mathbb{X}$ resembles a product graph.

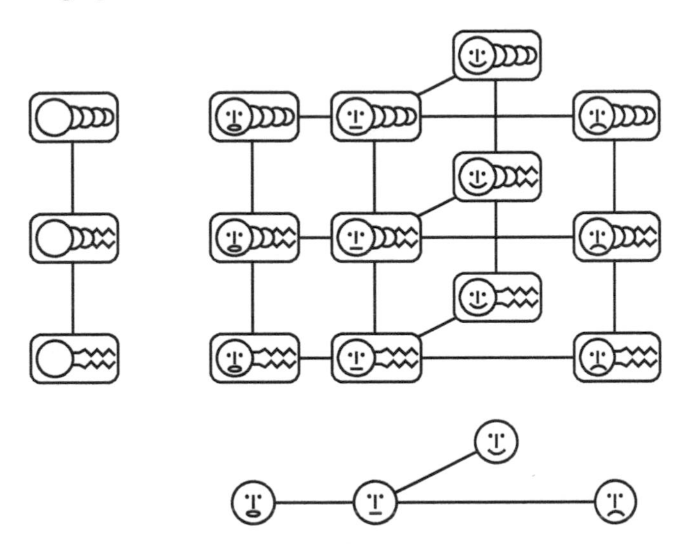

FIGURE 33.3 Phenotype graph as a product of character graphs for faces and bodies.

This type of investigation has instigated a theory of approximate graph products, and we now briefly introduce the rudiments of this subject.

Of fundamental importance is the idea of the "distance" between two graphs. Let Δ denote the symmetric difference of sets, that is $X \Delta Y = (X \cup Y) \setminus (X \cap Y)$. Given two graphs G and H, the *distance* $d(G, H)$ between them is the smallest number k such that G and H have representations G' and H' for which

$$|V(G')\Delta V(H')| + |E(G')\Delta E(H')| \leq k\,.$$

Thus, for example, $d(G, H) = 0$ if and only if $G \cong H$. If G and H both have n vertices, their distance is bounded by $\binom{n}{2}$. By Exercise 33.6, the number of graphs at distance at most k from G is $O(|V(G)|^{2k})$.

A graph G is a *k-approximate product* if there is a product graph H for which $d(G, H) \leq k$. For computational purposes, we assume tacitly that H is a nontrivial product of connected factors.

Figure 33.4 shows a 3-approximate Cartesian product; it differs from $P_4 \Box P_3$ by one missing edge, one extraneous edge, and one extraneous vertex. (It is also a 7-approximate Cartesian product relative to $P_5 \Box P_3$.) Clearly, we can also consider approximate strong, direct, or lexicographic products. Currently only the strong and Cartesian cases have been considered in the literature.

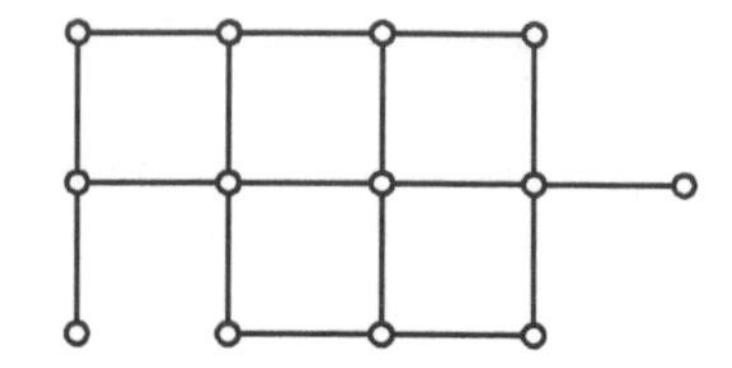

FIGURE 33.4 An approximate Cartesian product.

It is perhaps surprising that it is decidable in polynomial time whether a given graph is a k-approximate strong or Cartesian product. To see how this can be done, let $G = (V, E)$ be a graph on n vertices. As noted above, there are $O(n^{2k})$ graphs H within distance k from G. These can be easily computed (possibly with some redundancy) as $H = (V \Delta V', E \Delta E')$, where $|V'| + |E'|$ is bounded by k. (Some care must be taken that V' and E' are chosen such that H is actually a graph, but these details are easily disposed of. See Hellmuth, Imrich, Klöckl, and Stadler (2009a) for details.) Those H that are connected can now be factored in polynomial time.

If the value of k is not fixed, then the problem for the Cartesian product is NP-complete, as shown by Feigenbaum, and Haddad (1989). However, polynomial algorithms can be attained under mild restrictions on the graphs. We now outline a simple approach for the strong product by Hellmuth et al. (2009a).

The key idea is that subgraphs induced on closed neighborhoods of strong products are strong products of subgraphs induced on neighborhoods of the factors, that is,

$$\langle N_{G \boxtimes H}[(x,y)] \rangle = \langle N_G[x] \rangle \boxtimes \langle N_H[y] \rangle.$$

Thus, because the product structure of a graph is reflected locally, one might reasonably expect to approximate a graph with a product if enough of its neighborhoods have nontrivial factorizations that "fit together" in a manner that suggests a global product.

To glean an idea of how this might work, consider the graph in Figure 33.5. The labels $v_1, v_2, \ldots, v_6$ mark the vertices whose neighborhoods are S-thin and have nontrivial factorings. Each such neighborhood can be factored quickly (and uniquely), and its Cartesian edges are independent of the factoring. We then give the Cartesian edges of such a neighborhood a product coloring induced by its factoring.

Now, we can obtain a partial product-like coloring of the entire graph as follows. Begin with the coloring of one neighborhood, say $X = \langle N[v_1] \rangle$. If some neighborhood (say $\langle N[v_2] \rangle$) shares a Cartesian edge with X, then we recolor the edges of $\langle N[v_2] \rangle$ to match the coloring of X, and set $X := X \cup \langle N[v_2] \rangle$. This process is continued until no additional neighborhoods $N[v_i]$ can be absorbed into X. In Figure 33.5, we obtain the 2-coloring indicated by bold and dashed lines.

At this point there is enough information to reconstruct the factors P_7 and P_5 of an approximate product: We can (at least in this example), extract the largest components form each color class (e.g., the layers through v_3), and the given graph is approximated by the strong product of these components.

However, a glance at Figure 33.6 reveals that this method runs into difficulties when

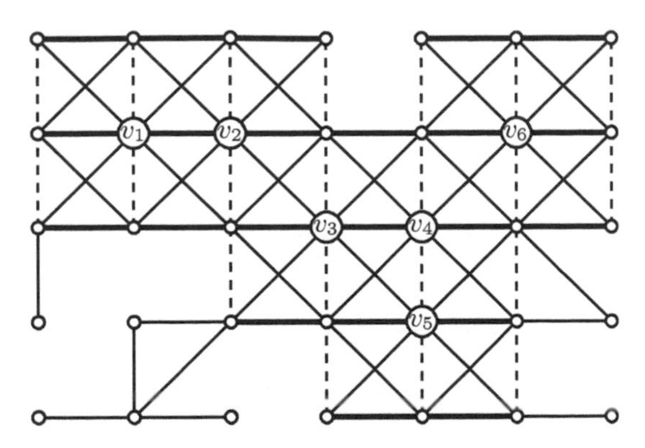

FIGURE 33.5 Simple approximate strong product.

the vertex neighborhoods are not thin. This is possible, even when the entire graph is a thin strong product. To deal with such situations Hellmuth, Imrich, Klöckl, and Stadler (2009b) introduced the backbone of a graph, and a so-called $S1$-condition under which Cartesian edges in a neighborhood can be detected, even when the neighborhood is not thin. Using these concepts together with edge-neighborhoods and enlarged edge-neighborhoods, Hellmuth (2011) presented a local, quasi-linear algorithm for the prime factorization of connected graphs with respect to the strong product.

It is not our aim to present the technical details of this approach, but we will say a few explanatory words about what we mean by a local approach, quasi-linearity, the backbone, and the significance of this result for the recognition of approximate strong products.

First of all, it must be said that a method that is capable of recognizing approximate graph products must also recognize products. But then we are also interested in its complexity. The complexity of Hellmuth's algorithm for the factorization of a thin graph of order n and maximal degree Δ is $O(n\Delta^6)$. If one considers only graphs whose maximal degree is bounded by a predetermined constant k, then the complexity of the algorithm is linear. (Such algorithms are called *quasi-linear.*) Let us compare its complexity with that of Algorithm 24.6 for the factorization of thin graphs over the strong product. By Theorem 24.9 it has complexity $O(mn \log n + m^2)$ unless several special conditions are met. Hence, for given Δ and large n, the new algorithm outperforms Algorithm 24.6.

The approach is local in the sense that at a given time, only neighborhoods, edge-neighborhoods, or enlarged edge neighborhoods are factored. It is reasonable to use it for the recognition of approximate graph products, as shown by the above example. However, the search for k-approximate graph products can produce many different, nonisomorphic solutions, even with different numbers of factors. Thus, in general, one has to understand the approach as a heuristic that needs additional information for the selection of a solution.

Computer experiments show that the proposed heuristics delivers excellent results, even for approximate products that were obtained from given products by random deletion of a proportionally large number of vertices and edges; see Hellmuth (2010).

The *backbone* of a graph is the set of vertices v with strictly maximal $N[v]$. In a thin graph the backbone vertices are a dominating set and they play special role in the recognition of products and strong products. For example, consider Figure 33.6. Its backbone vertices are marked $0, 1, \ldots, 5$ and x. The subgraphs induced by the neighborhoods of the vertices marked by an x are prime, but the other neighborhoods suffice to construct Cartesian edges (indicated by bold and broken lines in the figure) by the methods of Hellmuth (2011). The bold and broken lines suggest an edge-coloring of the Cartesian edges of the suspected original product.

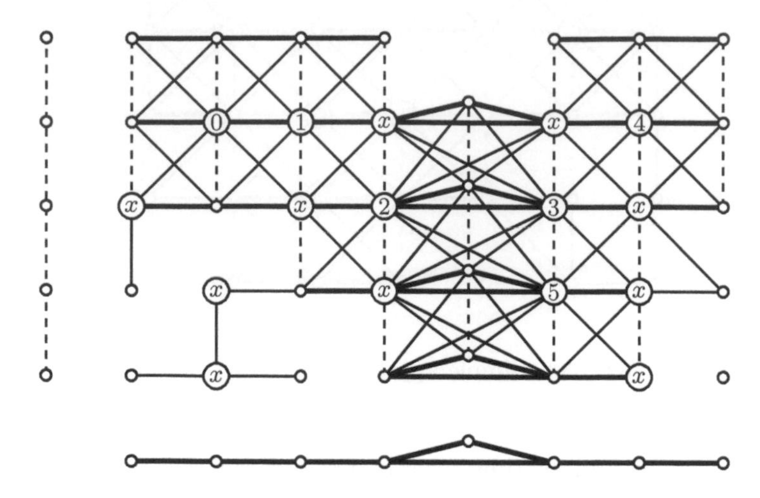

FIGURE 33.6 Approximate strong product.

One of the potential applications of approximate graph products is visualization. If a graph is a product, its regular structure lends itself for visualization, be it of the graph as it is, or of its factors. To describe an approximate graph product, it may then suffice and be instructive to mark just those places where the product structure is disturbed. For an intriguing step in this direction, see Jänicke, Heine, Hellmuth, Stadler, and Scheuermann (2010).

In addition to applications in visualization and the problems in theoretical biology that motivated our investigations, approximate graph products are helpful in engineering and structural mechanics. For results in this direction, we mention Kaveh (2006).

33.3 Graph Spectra

The *spectrum* of a graph G is the spectrum of its adjacency matrix $\mathrm{A}(G)$. It is a link between the discrete and the continuous. Although it does not completely define a graph, it reveals many surprising properties.

In this short section we mention basic results, establish a connection to graph products, and say just enough about the relevance of the gap between the first and the second eigenvalue to motivate the zig-zag product.

We have met the adjacency matrix several times. Its formal definition and its main properties were the topic of Section 17.4. It was a useful tool in the recognition of triangle-free graphs and median graphs. In Section 5.3 we saw that $\mathrm{A}(G \times H)$ is the Kronecker product $\mathrm{A}(G) \otimes \mathrm{A}(H)$ and that adjacency matrices of bipartite graphs have block structure.

We begin with a simple observation about spectra and show first that the maximum degree of a graph is an upper bound for its eigenvalues. To fix ideas, let G be a graph on n vertices $\{v_1, v_2, \ldots, v_n\}$, and $\mathbf{x} = (x_1, x_2, \ldots, x_n)^T$ an eigenvector with eigenvalue λ. That is,

$$\mathrm{A}(G)\,\mathbf{x} = \lambda\mathbf{x}\,,$$

or equivalently,

$$\lambda x_j = \sum_{v_i v_j \in E(G)} x_i, \quad \text{where} \quad j \in \{1, 2, \ldots, n\}.$$

Lemma 33.4 *If a graph has maximum degree Δ, then $|\lambda| \leq \Delta$ for each eigenvalue λ.*

Proof Suppose $\mathrm{A}(G)\,\mathbf{x} = \lambda\mathbf{x}$. Select j so that $|x_j| = \max(|x_1|, |x_2|, \ldots, |x_n|)$. Then

$$|\lambda|\,|x_j| \leq \sum_{v_i v_j \in E(G)} |x_i| \leq \Delta\,|x_j|,$$

and thus $|\lambda| \leq \Delta$, because $x_j \neq 0$. □

Another useful observation is that the vector $\mathbf{1} = (1, 1, \ldots, 1)^T$ is an eigenvector (with eigenvalue d) of G if and only if G is regular (of degree d).

To describe the relationship between the adjacency matrix of a product and that of its factors we introduce the notation I_n for the identity matrix of order n.

Theorem 33.5 *The adjacency matrices of the Cartesian, the strong, and the direct product of two graphs G, H are as follows:*

$$\begin{aligned}
A(G \,\Box\, H) &= A(G) \otimes I_{|V(H)|} + I_{|V(G)|} \otimes A(H), \\
A(G \boxtimes H) &= A(G) \otimes I_{|V(H)|} + I_{|V(G)|} \otimes A(H) + A(G) \otimes A(H), \\
A(G \times H) &= A(G) \otimes A(H).
\end{aligned}$$

The straightforward proof is omitted. Of course these relations allow the computation of the eigenvalues of the product from those of the factors. However, the relationship is better described by means of the eigenvalues of the Laplacian of G. The *Laplacian* $\mathrm{L}(G)$ of a graph G is defined as

$$\mathrm{D}(G) - \mathrm{A}(G),$$

where $\mathrm{D}(G)$ is a diagonal matrix of the same order as G, with the entries $d_{ii} = d(v_i)$. Notice that loops have no influence on $\mathrm{L}(G)$. The Laplacian has the advantage that the unit vector $\mathbf{1}$ always is an eigenvector, even when G is not regular. Clearly the corresponding eigenvalue is zero, it is the smallest eigenvalue, and its multiplicity is the number of components of G.

If G is regular of degree d with eigenvalues $\lambda_1 \geq \lambda_2 \geq \cdots \geq \lambda_n$, then the Laplacian has eigenvalues $d - \lambda_1 \leq d - \lambda_2 \leq \cdots \leq d - \lambda_n$. We call them the *Laplacian eigenvalues* and set $\nu_i = d - \lambda_i$.

The relationship between the Laplacian spectrum of a product and that of the factors is very difficult except for the Cartesian product, unless the graphs are regular. We shall not explore this further, but wish to point out a result of Fiedler (1973) about the Cartesian product and some of its consequences.

Proposition 33.6 *The Laplacian eigenvalues of $G \,\Box\, H$ are of the form $\nu(G) + \nu(H)$, where $\nu(G)$ and $\nu(H)$ are Laplacian eigenvalues of G and H, respectively. Furthermore, if $\mathbf{x}(G)$ and $\mathbf{x}(H)$ are Laplacian eigenvectors to $\nu(G)$ and $\nu(H)$, respectively, then*

$$\mathbf{x}(G) \otimes \mathbf{x}(H)$$

is a Laplacian eigenvector of $G \,\Box\, H$ to the eigenvalue $\nu(G) + \nu(H)$.

For illustration, consider the case where the smallest nonzero eigenvalue of H, say ν, is smaller than that of G. Suppose it is simple and $\mathbf{x}$ is an eigenvector to ν. Then ν is the smallest eigenvalue of $G \,\Box\, H$; it is simple if H is connected, and $\mathbf{y} = \mathbf{1}_{|V(G)|} \otimes \mathbf{x}$ is an eigenvector.

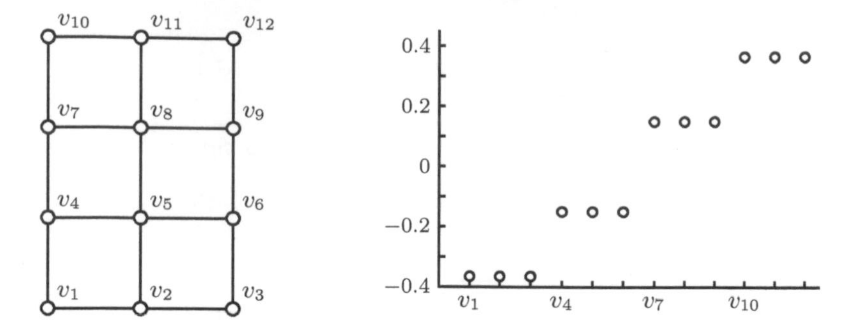

FIGURE 33.7 Product and Fiedler vector.

Figure 33.7 provides an example for $P_3 \,\Box\, P_4$. In this case, $\nu = .5858$, and

$$\mathbf{y}/|\mathbf{y}| = (-.577, -.577, -.577)^T \otimes (+.653, +.270, -.270, -.653)^T.$$

The right side of the figure plots the values of y_i over v_i. Clearly one can easily recover the layers of the second factor from the eigenvector $\mathbf{y}$. Even if an edge is removed from the graph we can recover the layer structure of the original product; see Figure 33.8.

This indicates that a wealth of information is contained not only in the spectrum, but also in the eigenvectors. In particular, this holds for the eigenvector to the second-smallest eigenvalue of the Laplacian. It is also known as a Fiedler vector, in recognition of the fact that Fiedler was the first to realize its significance in graph partitioning.

The influence of the second-smallest eigenvalue of the Laplacian on the shape of a graph seems to have been first documented by Bussemaker, Čobeljić, Cvetković, and Seidel (1976).

Since then, its significance and relation to numerous graph invariants has been extensively investigated. These invariants include connectivity, expanding properties, isoperimetric number, maximum cut, independence number, genus, diameter, mean distance, and bandwidth-type parameters. For surveys, see Mohar (1991) and the book by Chung (1997).

There are also numerous publications on load balancing and diffusion schemes in products of graphs that make use of spectral methods. They are beyond the scope of this book. Let us mention Elsässer, Monien, Preis, and Frommer (2004) for at least one contribution to this topic.

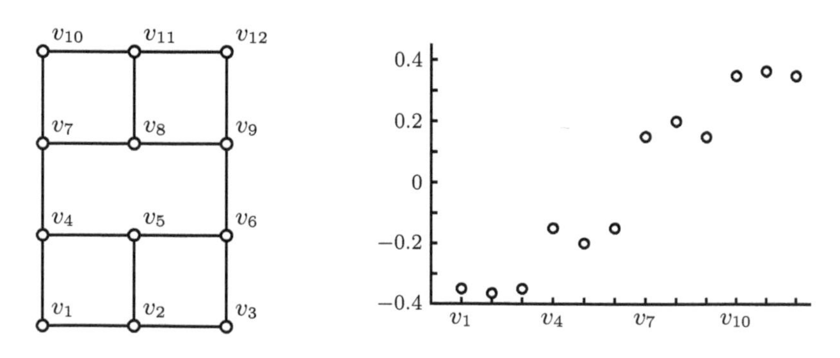

FIGURE 33.8 Approximate product and Fiedler vector.

33.4 Zig-Zag Product

In this section we consider regular graphs again and normalize the adjacency matrix. That is, given a d-regular graph G, then we consider $\mathrm{A}(G)/d$. This matrix has largest eigenvalue 1, and the second-largest eigenvalue, which we will simply denote $\lambda(G)$, plays the role of the second-smallest eigenvalue of the Laplacian. Clearly

$$\lambda(G) = \max_{|\mathbf{v}|=1, \mathbf{v} \perp \mathbf{1}} \left\{ \frac{1}{d} |\mathrm{A}(G)\mathbf{v}| \right\},$$

and $\lambda(G) \in [0,1]$.

A family $\mathcal{G}$ of graphs is then called an *expander family* if there is an $\alpha < 1$ such that $\lambda(G) \leq \alpha$ for all $G \in \mathcal{G}$.

There is enormous interest in the construction of large expanders of small degree because of their applicability in computer science and pure mathematics. In computer science, expanders are applied in the areas of circuit complexity, error correcting codes, and communication networks. In pure mathematics, they have been used in topology, group theory, measure theory, and number theory. To this quite arbitrarily assembled list one has to add, of course, graph theory and combinatorics.

Until the seminal paper of Reingold, Vadhan, and Wigderson (2002), most construction relied on Cayley graphs, but this paper introduced a new construction. Upon a suggestion of Peter Winkler, it is now called *zig-zag product.* We will briefly describe it and show how a variant can be used to define semidirect products.

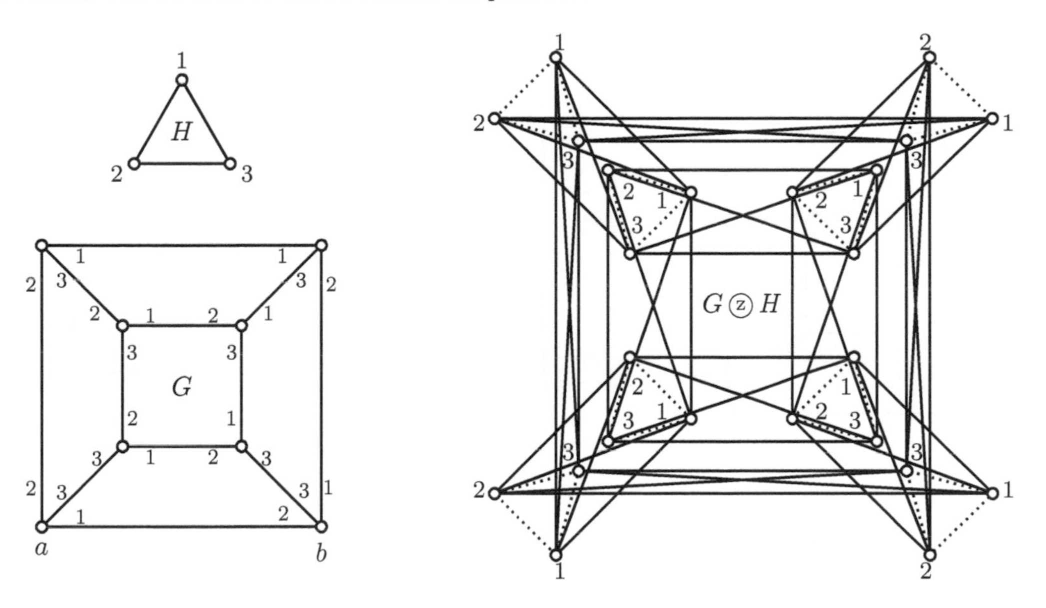

FIGURE 33.9 Zig-zag product Q_3 ⓩ K_3

First a definition. A graph G is an $[n, d]$-*graph* if it has order n and degree d. It is called an $[n, d, \alpha]$-*graph* if $\lambda(G) \leq \alpha$. Given an $[n, m, \alpha]$-graph G and an $[m, d, \beta]$-graph H, we will construct an $[nm, d^2, \gamma]$-graph G ⓩ H, the *zig-zag product* of G and H.

Let G be an $[n, m, \alpha]$-graph and H an $[m, d, \beta]$-graph. Then G ⓩ H is a graph whose vertex set is $V(G) \times V(H)$ and whose edge set is defined as follows:

Set $Q = D_n \square H$ and let M be a complete matching of $G \circ D_m$ with the property that contraction of each D_m-layer to a single vertex produces no double edges, that is, the contraction produces an isomorphic copy of G. Then

$$E(G \textcircled{z} H) = \bigcup_{ab \in M} (N_Q(a) \oplus N_Q(b)),$$

where $\oplus$ denotes the join of graphs. Notice that the product depends on the choice of M. Figure 33.9 shows an example of $Q_3 \textcircled{z} K_3$. Notice that the numbers on the edges identify the matching edges. For example, the edge from a to b corresponds to the matching edge $(a,1)(b,2)$.

That $G \textcircled{z} H$ has mn vertices and degree d^2 is easily seen. Concerning its expanding property, Reingold, Vadhan, and Wigderson (2002) showed that $G \textcircled{z} H$ is an $[nm, d^2, \gamma]$-graph, where $\gamma \leq \alpha + \beta + \beta^2$ and that $\gamma < 1$ if $\alpha, \beta < 1$.

We shall exploit a variant, the *replacement product* $G \textcircled{R} H$. It has the same vertex set as the zig-zag product, but its edge set is defined as

$$E(G \textcircled{R} H) = M \bigcup E(Q).$$

Then $G \textcircled{R} H$ is an $[nm, d+1, f(\alpha, \beta)]$-graph, where $f(\alpha, \beta) < 1$ if $\alpha, \beta < 1$; see Kelley, Sridhara, and Rosenthal (2008). Figure 33.10 shows $Q_3 \textcircled{R} K_3$.

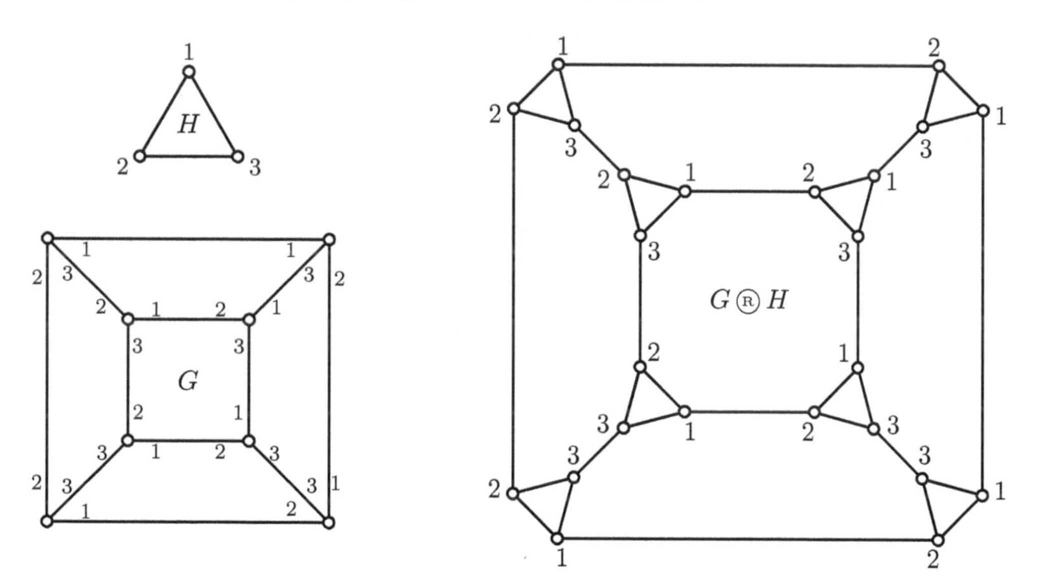

FIGURE 33.10 Replacement product $Q_3 \textcircled{R} K_3$.

We cannot pursue the enormous range of applications of the zig-zag and the replacement product, but conclude with an observation pertaining to the semidirect product of groups. Let A, B be groups and let B act on A as a group of automorphisms. Let a^b denote the action of b on a. Assume that S, T are generating sets of A, B and that $S = s^B$; in other words, S is a single B-orbit. Then

$$\Gamma(A, S) \textcircled{R} \Gamma(B, T)$$

is a Cayley graph of a group C with generating set $\{s\} \cup T$. As shown by Alon, Lubotzky, and Wigderson (2001), this group C is the semidirect product $A \rtimes B$. (The construction can be extended to more than one orbit.)

For a graph theorist, this means that the semidirect product can be defined via groups acting on replacement product graphs, just as one can define the free product of groups by groups acting on the free product of graphs.

Exercises

33.1. Determine all nonisomorphic graph bundles with base K_3 and fiber K_3.

33.2. Show that for direct graph bundles, $\chi(B \times^{\varphi} F)$ need not equal $\min\{\chi(B), \chi(F)\}$.

33.3. Show that $\chi(B \boxtimes^{\varphi} F) \leq \chi(B)\chi(F)$.

33.4. Show that $\gamma(B \boxtimes^{\varphi} F) \leq \gamma(B)\gamma(F)$.

33.5. Find an example of a graph bundle with a nontrivial base and a nontrivial fiber such that δ^* has a single equivalence class.

33.6. If G is a graph on n vertices, then there are $O(n^{2k})$ graphs whose distance from G is at most k.

Appendix

Hints and Solutions to Exercises

Part I

Chapter 1

Exercise 1.1 Apply Theorem 1.1.

Exercise 1.2 Routine.

Exercise 1.3 Routine.

Exercise 1.4 Suppose G is disconnected. Then it has a component X with at most $\frac{n}{2}$ vertices. Any vertex in this component has a degree of at most $\frac{n}{2} - 1 < \frac{1}{2}(n-1)$.

Exercise 1.5 Hint: Show that the homomorphic image of an odd cycle is a closed walk that contains an odd cycle.

Exercise 1.6 Every cubic graph has an even number of vertices. To construct such a graph on $2k$ vertices, take a disjoint union of two copies of C_k and connect the corresponding vertices by an edge. (As we will see later, this graph is the Cartesian product $C_k \,\Box\, K_2$.)

Exercise 1.7 $\binom{m}{2} \cdot \binom{n}{2}$.

Exercise 1.8 Such a graph is clearly a complete bipartite graph, and hence it must be of the form $K_{x,n-x}$ for some $1 \le x \le n-1$. This graph has $x(n-x)$ edges by Theorem 1.1. As a real-valued function, $x(n-x)$ has a maximum at $x = n/2$. Thus the answer to the question is $\lfloor n/2 \rfloor \cdot \lceil n/2 \rceil$.

Exercise 1.9 The number of vertices of degree 1 exceeds that of degree 3 by 2.

Exercise 1.10 Hint: Treat the case $|L| = 1$ first. For $|L| > 1$, consider two vertices x, y with $d(v,x) = d(v,y)$ for all $v \in L$. Next, choose two elements $u, v \in L$ such that $d(x,u) + d(x,v)$ is minimum. Show then that $x = y$ by considering the paths between u, v, x, and y in T.

Actually Kel′mans showed that the vertices of a tree with s vertices of degree 1 are uniquely characterized by the distances from any $s-1$ of them.

Exercise 1.11 If a graph G is not a forest, then it has a cycle C. Let e be an edge of C. Notice that $C - e$ is a connected subgraph of G that is not induced.

Conversely, suppose a graph G has a connected subgraph H that is not induced. This means H has two vertices x and y for which $xy \in E(G)$ but $xy \notin E(H)$. Let P be an x,y-path in H. By appending the edge xy to P, we get a cycle in G, so G is not a forest.

Exercise 1.12 Hint: Find a subdivision of $K_{3,3}$ in the Petersen graph.

Chapter 2

Exercise 2.1 Hint: Use the idea of the proof of Theorem 2.5.

Exercise 2.2 Routine.

Exercise 2.3 Hint: Automorphisms preserve distances.

Exercise 2.4 Routine.

Exercise 2.5 Because Q_r is vertex-transitive, we may assume that φ fixes $v_0 = (0,0,\ldots,0)$ and all of its neighbors $u_i, 1 \le i \le r$. Let the notation be chosen such that the all coordinates of the u_i are zero, except the ith. Consider the neighbors of u_1. They are v_0 and the vertices $w_{1,j}$ that differ from u_1 in the jth coordinate, $2 \le j \le r$. It is routine to show that $w_{1,j}$ is the only vertex of Q_r that is adjacent to u_1 and u_j. Hence it must also be fixed by φ. Thus φ fixes all vertices in $N(u_1)$. By the connectedness of Q_r, the proof can now be completed by induction.

Exercise 2.6 Let $\varphi \in \mathrm{Aut}(Q_r)$ and set $v = \varphi(v_0)$, where $v_0 = (0,0,\ldots,0)$. Suppose v differs from v_0 in the coordinates $i_1, i_2, \ldots, i_j$. Then $\psi = \psi_{i_1}\psi_{i_2}\ldots\psi_{i_j}$ maps v into v_0 and $\psi\varphi$ is an automorphism that fixes v_0 and permutes the neighbors of v_0. Let π be this permutation. Clearly there is a product, say β, of the $\psi_{i,j}$ that produces the same permutation π of the neighbors of v_0 as $\psi\varphi$. Thus $\beta^{-1}\psi\varphi$ fixes v_0 and all its neighbors. By Exercise 2.5 it is the identity mapping, and thus $\varphi = \psi\beta$.

Exercise 2.7 Hint: Use Exercise 2.6.

Exercise 2.8 Hint: By Theorem 2.8 every such graph G consists of a Q_k, where $k = 2$ or 3, and additional edges. For $k = 2$ it is easily seen by inspection that no matter how additional edges are inserted, $\mathrm{Aut}(G)$ has nontrivial vertex stabilizers.

For $k = 3$ we invoke Exercise 2.6 and note that the group of G must be generated by the ψ_i, $1 \leq i \leq 3$, and that with every edge $e \in E(G)$, the edges in $\mathrm{Aut}(G)(e)$, that is, the edges in the orbit of $\mathrm{Aut}(G)$, also are in $E(G)$. For example, if $(0,0,0)(1,1,0) \in E(G)$, then the edges $(1,0,0)(0,1,0)$, $(0,0,1)(1,1,1)$, and $(1,0,1)(0,1,1)$ will also be in G. Use this to show that the orbit of any edge of G contains one or two edges whose endpoints are vertices of the hexagon $(1,0,0)(1,1,0)(0,1,0)(0,1,1)(0,0,1)(1,0,1)$.

Now show that the subgraph of G induced by these six vertices has at least one nontrivial automorphism and that this extends to an automorphism of G.

Exercise 2.9 Hint: Keep deleting vertices that do not decrease the chromatic number until the desired subgraph is obtained.

Exercise 2.10 Hint: Show that any such group must have an element of order 2, and that every automorphism of the Petersen graph that has order 2 fixes a vertex.

Exercise 2.11 Hint: Use the No-Homomorphism Lemma and the fact that $\alpha(G) \geq |V(G)|/2$.

Exercise 2.12 Hint: For $K = K_1$ this is the No-Homomorphism Lemma; mimic the proof of the No-Homomorphism Lemma.

Chapter 3

Exercise 3.1 Routine.

Exercise 3.2 Hint: It suffices to verify if for $r = 4$.

Exercise 3.3 We may without loss of generality assume that $v = 11\ldots1$. If a shortest x,y-path in Q_r passes v, then v can be replaced by an appropriate vertex to obtain an x,y-path in $Q_r - v$ of the same length. Thus $Q_r - v$ is a partial cube. To see that $Q_r - v$ is not median, consider the vertices $0111\ldots1$, $1011\ldots1$, and $1101\ldots1$.

Exercise 3.4 Hint: Label two antipodal vertices of C_{2r} with $00\ldots0$ and $11\ldots1$. Proceed accordingly.

Exercise 3.5 Apply Proposition 3.11 and Theorem 3.7.

Exercise 3.6 Let $\varphi : G \to H$ be a retraction. Because H is a subgraph of G, we have $\omega(H) \leq \omega(G)$. On the other hand, consider a complete subgraph K in G. Because any two of its vertices are adjacent, their images under the homomorphism φ must be unequal and adjacent. Thus the image $\varphi(K)$ is a complete subgraph of H with the same number of vertices as K. Hence $\omega(H) \geq \omega(G)$.

Next, let C be the shortest odd cycle in G. The image $\varphi(C)$ must be a nonbipartite subgraph of H. The shortest odd cycle in this image cannot be shorter than C, so we infer that C and $\varphi(C)$ are cycles of the same length. Thus H has a cycle whose length equals that of the shortest odd cycle of G. Conversely, as $H \subseteq G$, the shortest odd cycle in H is also in G. The assertion follows.

Exercise 3.7 Suppose $\chi(G) = \omega(G) = n$. Take a coloring $\varphi : V(G) \to \{1,2,\ldots,n\}$. Let K_n be a clique in G. Its vertices must be colored with n distinct colors, so label the vertices by their colors, that is $V(K_n) = \{1,2,\ldots,n\}$. Now the map $\varphi : G \to K_n$ satisfies $\varphi^2 = \varphi$. Also it is a homomorphism, for if $xy \in E(G)$, we have $\varphi(x) \neq \varphi(y)$, so $\varphi(x)\varphi(y) \in E(K_n)$. Thus K_n is a retract of G.

Conversely, suppose K_n is a retract of G. In particular, this means K_n is a subgraph of G, so $\chi(G) = n \leq \omega(G)$. But obviously $\chi(G) \geq \omega(G)$, so we are done.

Exercise 3.8 Consider a retraction $\varphi : G \to H$ of a χ-critical graph G. If φ is not the identity, then H is a subgraph of any vertex deleted subgraph $G - v$, where v is not in the image of φ. Thus $\chi(H) \leq \chi(G - v) < \chi(G)$. But Proposition 3.10 says $\chi(H) = \chi(G)$, so φ is the identity. Hence G itself is the only retract of G.

Exercise 3.9 Suppose G is not a core, so there is a retraction $\varphi : G \to H$, where H is a proper subgraph of G. Then φ is a homomorphism of G that is not an automorphism.

Conversely, suppose it is not the case that every homomorphism of G is an automorphism, and let φ be such a homomorphism. Then $\varphi(G)$ is a proper subgraph of G. Now for each $x \in V(G)$, there is an integer k_x for which either $\varphi^{k_x}(x) = x$, or $\varphi^m(x) = \varphi^{k_x}(x)$ for all $m \geq k_x$. (That is, either the sequence $x, \varphi(x), \varphi^2(x), \ldots$ is periodic, with period k_x, or it stabilizes after k_x terms.) Let ℓ be the least common multiple of the k_x, over all $x \in V(G)$. Then φ^ℓ is a nonidentity homomorphism of G that satisfies $(\varphi^\ell)^2 = \varphi^\ell$. Hence G has a proper retract and is not a core.

Exercise 3.10 Consider a nonexpansive map $f : Q_r \to Q_r$ that fixes each of u, v, and w. Now, f is a weak homomorphism by Proposition 3.8. Let the median x lie on shortest u,v-, u,w-, and

v, w-paths P, Q, and R. Because f is a weak homomorphism the subgraphs $f(P)$, $f(Q)$, and $f(R)$ are connected. Also, $f(P)$ contains u and v and has at most $|P| = d(u,v)$ edges, so it is a shortest u, v-path. Similarly, $f(Q)$ is a shortest u, w-path, and $f(R)$ is a shortest v, w-path. Because $f(x)$ lies on each of these paths, we infer that it is a median of u, v, w. But Q_r is a median graph, so $f(x) = x$.

Exercise 3.11 Suppose $z = z_1 z_2 \ldots z_r = (u \vee v) \wedge (u \vee w) \wedge (v \vee w)$. Then for each i we have $z_i = \min\{\max\{u_i, v_i\}, \max\{u_i, w_i\}, \max\{v_i, w_i\}\}$. Now, if any two of u_i, v_i, w_i are 0, then one of the maximums is 0, and $z_i = 0$. If any two of u_i, v_i, w_i are 1, then all of the maximums are 1, and $z_i = 1$. Thus, in the language of the proof of Proposition 3.7, z_i is determined by "majority rule." Following the proof of that proposition, we see that z is a median of u, v, w. Argue similarly for $(u \wedge v) \vee (u \wedge w) \vee (v \wedge w)$.

Chapter 4

Exercise 4.1 Routine.

Exercise 4.2 $|E(G \times H)| = 2 \cdot |E(G)| \cdot |E(H)|$, $|E(G \Box H)| = |V(G)| \cdot |E(H)| + |V(H)| \cdot |E(G)|$, $|E(G \boxtimes H)| = |E(G \Box H)| + |E(G \times H)|$, $|E(G \circ H)| = |V(G)| \cdot |E(H)| + |E(G)| \cdot |V(G)|^2$.

Exercise 4.3 Hint: Consider two cases: a cubic graph with no triangle and a cubic graph with a triangle.

Exercise 4.4 Hint: First count the number of triangles in $K_3 \times K_3$.

Exercise 4.5 This is an easy consequence of definitions of the direct product and open neighborhoods.

Exercise 4.6 Proof is an easy consequence of definitions of the strong product and closed neighborhoods.

Exercise 4.7 Routinc. One approach uses Exercise 4.6.

Exercise 4.8 Routine.

Exercise 4.9 Hint: Let $V(K_3) = \mathbb{Z}_3$. Show that the bijection $V(K_3 \Box K_3) \to V(\overline{K_3 \Box K_3})$ defined as $(i,j) \mapsto (i+j, i-j)$ is an isomorphism.

Exercise 4.10 Because the factors are complete, we can argue as follows:

$$\begin{aligned}
(x,y)(x',y') \in E(\overline{K_m \Box K_n}) &\iff \\
(x,y)(x',y') \notin E(K_m \Box K_n) &\iff \\
x \neq x' \text{ and } y \neq y' &\iff \\
(x,y)(x',y') \in E(K_m \times K_n)\,. &
\end{aligned}$$

Exercise 4.11 Use Exercises 4.9 and 4.10.

Exercise 4.12 Use Exercises 4.9 and 4.11.

Exercise 4.13 Routine.

Exercise 4.14 Routine.

Exercise 4.15 Consider products $(G * H) * K$ and $G * (H * K)$. We make the following observations:

$$\begin{aligned}
\delta\big(((g,h),k),((g',h'),k')\big) &= \\
\delta\big((g,h),(g',h')\big) * \delta(k,k') &= \\
\big[\delta(g,g') * \delta(h,h')\big] * \delta(k,k'), & \\
\delta\big((g,(h,k)),(g',(h',k'))\big) &= \\
\delta(g,g') * \delta\big((h,k),(h',k')\big) &= \\
\delta(g,g') * \big[\delta(h,h') * \delta(k,k')\big]\,. &
\end{aligned}$$

Suppose the table for $*$ is associative. The above implies $\delta\big(((g,h),k),((g',h'),k')\big) = 1$ if and only if $\delta\big((g,(h,k)),(g',(h',k'))\big) = 1$. This means $((g,h),k)((g',h'),k')$ is an edge of $(G * H) * K$ if and only if $(g,(h,k))(g',(h',k'))$ is an edge of $G*(H*K)$. Thus the map $((g,h),k) \mapsto (g,(h,k))$ is an isomorphism, so the graph product $*$ is associative.

Conversely, if the graph product $*$ is associative, then the above chain of reasoning can be reversed.

Exercise 4.16 Hint: Show that both $(G \overline{*} H) \overline{*} K$ and $G \overline{*} (H \overline{*} K)$ are equal to $\overline{\overline{G} * \overline{H} * \overline{K}}$.

For the second statement, begin with $G \overline{\overline{*}} H$. Define $\overline{\overline{*}}$ from $\overline{*}$ and $\overline{*}$ from $*$.

Exercise 4.17 One possible approach is as follows. Consider the product $G *' H = \overline{G} * \overline{H}$. Argue that the table for $*'$ is the table for $*$ with the second and third rows (and columns) interchanged. Now note $G \overline{*} H = \overline{G *' H}$.

Exercise 4.18 Hint: One approach is to use Exercise 4.17.

Exercise 4.19 Outline: Use Exercise 4.17 and the hypothesis about the projections to argue that the table for $*$ must have at least one 0 in the second row (and column), and at least one 1 in the third row (and column).

In turn, reason that the following table entries are forced:

$*$	Δ	1	0
Δ	Δ	1	0
1	1		
0	0		

Now show the assumption $1^2 = 1$ forces the table for the modular product; and $1^2 = 0$ forces the table for its complementary product.

Exercise 4.20 It is easy to see that the conditions are sufficient and that the assertion is true if one factor is trivial.

Suppose the modular product $G \diamondsuit H$ of the nontrivial graphs G, H is disconnected. Notice that $E(G \diamondsuit H) = E(G \square H) \cup E(G \times H) \cup E(\overline{G} \times \overline{H})$.

Because the edge set of $G \diamondsuit H$ contains $E(G \square H)$, at least one of the factors G, H must be disconnected.

Case 1. Only one factor is disconnected. Without loss of generality, let H be connected and $G = G_1 + G_2 + \cdots + G_k$, where the G_i are the connected components of G. Then the connected components of $G \square H$ are $G_1 \square H, G_2 \square H, \ldots, G_k \square H$.

Clearly there are no edges between any two connected components of $G \square H$ in $G \times H$.

Also, because $E(G \diamondsuit H)$ contains $E(\overline{G} \times \overline{H})$, any two connected components of $G \square H$ will be adjacent in $G \diamondsuit H$ unless H is complete.

Case 2. Both G and H are disconnected. Let the disconnected components of H be $H_1, H_2, \ldots, H_\ell$. Notice that the components of $G \square H$ are $G_i \square H_j$. If one of the components of G, say G_1, is not complete, then there are edges between $G_1 \square H_1$ and every $G_1 \square H_j$ in $\overline{G} \times \overline{H}$ for $j > 1$. Because $\overline{G} \times \overline{H}$ contains edges between $G_i \square H_j$ and $G_{i'} \square H_{j'}$ for $i \neq i'$ and $j \neq j'$, all components of G and H must be complete.

Finally, again considering $\overline{G} \times \overline{H}$, it is clear that $G \diamondsuit H$ is connected if G or H has more than two (complete) connected components.

Exercise 4.21 From the definition of the modular product, we have $E(G \diamondsuit H) = E(G \boxtimes H) \cup E(\overline{G} \times \overline{H})$. If $H = K_n$, then $E(\overline{G} \times \overline{H}) = \emptyset$, so $G \diamondsuit H = G \boxtimes H$. As $G \circ K_n = G \boxtimes K_n$, we have $G \circ H \cong G \diamondsuit H$.

Conversely, suppose $G \circ H \cong G \diamondsuit H$. Suggestion: Count the edges of $G \circ H$ and $G \diamondsuit H$. The resulting equation yields $\binom{|V(H)|}{2} = E(H)$.

Exercise 4.22 $\overline{K}_2 \diamondsuit \overline{K}_2 = K_2 \diamondsuit \overline{K}_2$.

Exercise 4.23 No. Observe that $(1 \nabla \Delta) \nabla 0 \neq 1 \nabla (\Delta \nabla 0)$.

Exercise 4.24 If $G \cong H$, it is immediate that the set $\{(x, x) | x \in V(G)\}$ induces a clique on n vertices. Conversely, let $\{(x_1, y_1), (x_2, y_2), \ldots, (x_n, y_n)\}$ be the vertices of an n-vertex clique in $G \nabla H$. One easily establishes that the first (respectively, second) coordinates are distinct. Thus there is a bijection $f : x_i \mapsto y_i$ from $V(G)$ to $V(H)$. Now, because $(x_i, y_i)(x_j, y_j) \in E(G \nabla H)$ for any indices i and j, it follows that either $x_i x_j \in E(G)$ and $y_i y_j \in E(H)$, or $x_i x_j \notin E(G)$ and $y_i y_j \notin E(H)$. Therefore f is an isomorphism.

Exercise 4.25 Hint: It is not difficult to exclude all other possibilities by considering the example where both factors are an orientation of K_2.

Exercise 4.26 Hint: In this case, a 2×2 multiplication table suffices.

Exercise 4.27 Hint: Expand the multiplication table.

Exercise 4.28 Hint: Distinguish the direction of an edge with the use of $+1$ and -1.

Chapter 5

Exercise 5.1 Hint: Let $V(C_{2k+1}) = \mathbb{Z}_{2k+1}$ and $E(C_{2k+1}) = \{i(i+1) \mid i \in \mathbb{Z}_{2k+1}\}$. Next, define $\varphi : V(C_{2k+1} \square C_{2k+1}) \to V(C_{2k+1} \times C_{2k+1})$, where $\varphi((i, j)) = (i + j, i - j)$. Now verify that φ is an isomorphism.

The result does not hold for even cycles. For instance, $C_{2k} \square C_{2k}$ is connected but the corresponding direct product is not. (In this case φ is not injective.)

Exercise 5.2 The proof is identical to the proof of Proposition 5.3, except that we construct two internally disjoint $(u, v), (x, y)$-paths rather than one.

Exercise 5.3 Hint: Adapt argument in the proof of Theorem 5.9.

Exercise 5.4 Hint: First show that $I_{G \square H}((u, v), (x, y)) = I_G((u, x)) \times I_H((v, y))$.

Exercise 5.5 Hint: For $u \in V(G_i)$, let $\overline{u}$ be a vertex of G with $p_i(\overline{u}) = u$. Let $\overline{v}$ be an antipode of $\overline{u}$ in G. Then show that $p_i(\overline{v})$ is an antipode of u in G_i.

Exercise 5.6 Set x to be the unique vertex of G^u with $p_G(v) = p_G(x)$ and apply the Distance Formula (Corollary 5.2).

Exercise 5.7 It is easy to see that $P_m \square P_n$ and $P_m \square C_n$ are planar for any $m, n \geq 3$. To see that $C_m \square C_n$ is not planar for $m, n \geq 3$, find a subdivision of $K_{3,3}$ in $C_m \square C_n$. This takes care of the case involving factors G with $\Delta(G) = 2$.

Suppose $\Delta(G) \geq 3$. Then $G \square H$ contains $K_{1,3} \square P_3$ as a subgraph. Again find a subdivision of $K_{3,3}$ in $K_{1,3} \square P_3$ to conclude that $G \square H$ is not planar.

Exercise 5.8 Hint: Consider an outerplanar embedding of G with exterior cycle C. Observe that $C \square K_2$ can be embedded in $\mathbb{R}^3$ as a cylinder. Now put copies of the outerplanar embedding of G on the top and bottom of the cylinder.

Exercise 5.9 Hint: Using Kuratowski's theorem, one can show that a graph is outerplanar if and only if it has no subgraph that is a subdivision of K_4 or $K_{2,3}$. Now proceed as in Exercise 4.7.

Exercise 5.10 Hint: First deal with the case when G is disconnected.

Exercise 5.11 Consider $\boxtimes_{i=1}^{k} G_i$. For each i, choose $x_i, y_i \in V(G_i)$ with $d_{G_i}(x_i, y_i) = \operatorname{diam}(G_i)$. Apply Corollary 5.5 to get $d((x_1, \ldots, x_k), (y_1, \ldots, y_k)) = \max_{1 \le i \le k} \operatorname{diam}(G_i)$. Hence the diameter of the product is at least the maximum of the diameters of the factors. Conversely, if x_i and y_i are arbitrary, then $d((x_1, \ldots, x_k), (y_1, \ldots, y_k)) = \max_{1 \le i \le k} d_{G_i}(x_i, y_i) \le \max_{1 \le i \le k} \operatorname{diam}(G_i)$.

Exercise 5.12 Hint: It suffices to consider strong products of two factors.

Exercise 5.13 Hint: Notice that $P_n \boxtimes K_2$ can be embedded on a tube (or annulus), as shown in Figure A.1. This tube can be capped at one or both ends to get an embedding on the disk or sphere. Thus $P_n \boxtimes K_2$ is planar. Use induction to show that $T \boxtimes K_2$ is planar for any tree T.

Finding a planar embedding of $P_3 \boxtimes P_3$ is trivial.

Conversely, consider a product $G \boxtimes H$ of connected graphs. If this is not one of the graphs mentioned above, then $G \boxtimes H$ has a subgraph with one of the following forms: $P_3 \boxtimes P_4$, or $P_3 \boxtimes K_{1,3}$, or $K_2 \boxtimes C_k$. Use Kuratowski's theorem to show that $G \boxtimes H$ is not planar.

Exercise 5.14 Hint: Show that between any two vertices of G there is a path of arbitrary parity of length at most $2\operatorname{diam}(G) + c$.

Exercise 5.15 Show first that the direct product of two cycles is not planar.

Exercise 5.16 The solution is not difficult but involves many cases.

Exercise 5.17 Let $V(K_2) = \{0, 1\}$. Plane drawings of $G \times K_2$ and $H \times K_2$ are shown in Figure A.2.

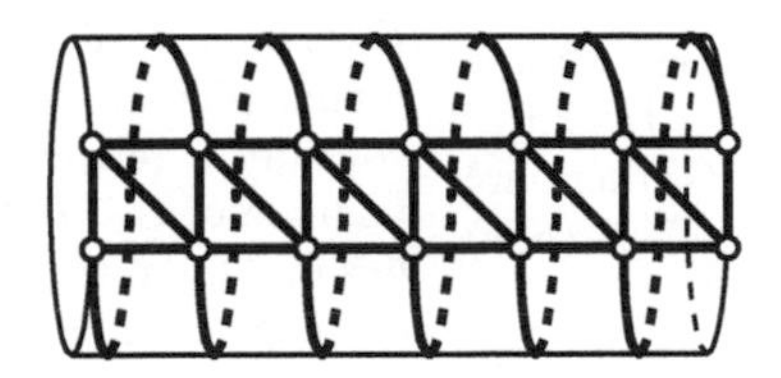

FIGURE A.1 $K_2 \boxtimes P_n$ is planar.

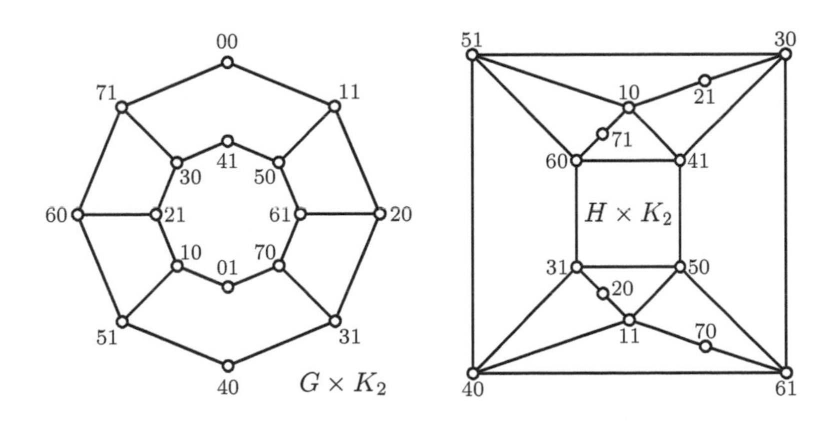

FIGURE A.2 Solution to Exercise 5.17.

Part II

Chapter 6

Exercise 6.1 For the converse, let $W = U_1 \Box \cdots \Box U_k$, where each U_i is convex in G_i. Take vertices $a = (a_1, \ldots, a_k)$ and $b = (b_1, \ldots, b_k)$ in W, and let $x = (x_1, \ldots, x_k)$ be on a shortest a,b-path in G. We must show x belongs to W. By Corollary 5.1 and the triangle inequality, we have

$$\begin{aligned} d_G(a,b) &= d_G(a,x) + d_G(x,b) \\ &= \sum_{i=1}^{k} d_{G_i}(a_i, x_i) + \sum_{i=1}^{k} d_{G_i}(x_i, b_i) \\ &\geq \sum_{i=1}^{k} d_{G_i}(a_i, b_i) = d_G(a,b)\,. \end{aligned}$$

Thus all terms in the above expression are equal, so $d_{G_i}(a_i, x_i) + d_{G_i}(x_i, b_i) = d_{G_i}(a_i, b_i)$ for each i. Therefore each x_i is on a shortest path between $a_i, b_i \in V(U_i)$, so $x_i \in U_i$. Thus $x = (x_1, \ldots, x_k)$ belongs to $U_1 \Box \ldots \Box U_k = W$, so W is convex.

Exercise 6.2 If G is not prime, then it has a nontrivial factoring $G = H \Box K$, with H and K connected. For any vertex (h,k) of G, there are edges $hh' \in E(H)$ and $kk' \in E(K)$. Then (h,k) belongs to the square $hh' \Box kk'$ in G.

To show that the converse is not true, consider the 7-vertex graph G obtained by joining two squares at a common vertex. Then every vertex of G is on a square, but G is prime.

Exercise 6.3 Hint: Consider the minimum degree of a vertex in complement of a Cartesian product of two graphs and observe that the degree of a vertex in a Cartesian product $G \Box H$ is at most $|V(G)| + |V(H)| - 2$.

Exercise 6.4 Hint: Apply Theorem 6.13.

Exercise 6.5 Hint: Use Theorem 6.13.

Exercise 6.6 Hint: Consider $G \Box \overline{G}$ and observe that at least one of the factors is prime by Exercise 6.3.

Exercise 6.7 Hint: Use the No-Homomorphism Lemma 2.13.

Exercise 6.8 Let $\mathcal{T}(G)$ denote the tree graph on G. List the blocks of G as $B_1, B_2, \ldots, B_k$. We first make several easily confirmed observations. (1) If T is a tree in G, then each subgraph $T \cap B_i$ is a tree in B_i. (2) If T_i is a tree in B_i for each i, then $T_1 \cup \cdots \cup T_k$ is a tree in G. (3) If TS is an edge of $\mathcal{T}(G)$, and $T = (S - e) \cup f$, then e and f are on a cycle in G, and therefore belong to a common block of G.

Define $\varphi : V(\mathcal{T}(G)) \to V(\mathcal{T}(B_1)) \times \cdots \times V(\mathcal{T}(B_k))$ as $\varphi(T) = (T \cap B_1, \ldots, T \cap B_k)$. This is well-defined by (1) and surjective by (2). It is obviously injective.

Now, if ST is an edge of $\mathcal{T}(G)$, and $T = (S - e) \cup f$, then (3) implies that e and f are in the same block, say B_i. Moreover, $T \cap B_j = S \cap B_j$ for any $j \neq i$. From this it follows that $\varphi(T)\varphi(S)$ is an edge of $\mathcal{T}(B_1) \Box \cdots \Box \mathcal{T}(B_k)$. Therefore φ is a homomorphism from G to $\mathcal{T}(B_1) \Box \cdots \Box \mathcal{T}(B_k)$.

Finally, suppose $(T_1, \ldots, T_k)(S_1, \ldots, S_k)$ is an edge of $\mathcal{T}(B_1) \Box \cdots \Box \mathcal{T}(B_k)$. Then $T_j = S_j$ for all but a single index i, and $T_i = (S_i - e) \cup f$ for some edges e and f of B_i. Let $T = T_1 \cup \cdots \cup T_k$ and $S = S_1 \cup \cdots \cup S_k$. Then $T = (S - e) \cup f$, so TS is an edge of $\mathcal{T}(G)$ that maps to $(T_1, \ldots, T_k)(S_1, \ldots, S_k)$. We have now verified that φ is an isomorphism.

Exercise 6.9 Hint: Show that $\mathcal{T}(G)$ is complete if G is a cycle.

Exercise 6.10 Hint: First, note that the homomorphic image of a triangle is a triangle. Now consider a retraction $\varphi : G \Box H \to G \Box H$. Any triangle of $G \Box H$ lies in a G-layer, so its image lies in a G-layer. Because G is connected and every edge is on a triangle, it follows that φ maps G-layers to G-layers.

Exercise 6.11 Suppose G has p connected components and is not prime, say $G = A \Box B$; then one factor, say A, must be connected and the other one must have p connected components, say $B = B_1 + B_2 + \cdots + B_p$. But then the $A \Box B_i$ are the connected components of G, and A is a common divisor.

Clearly G is not prime when its connected components have a common nontrivial factor.

Exercise 6.12 Let $G_1, G_2, \ldots, G_p$ be the connected components of G. If G is not prime, then the G_i must have one or more common factors. Because the prime factorization of the G_i is unique, the common prime factors of the G_i are uniquely determined. Let Z be their product, and set $G_i = Z \Box H_i$. Then $G \cong Z \Box (H_1 + H_2 + \cdots + H_p)$. By Theorem 6.21, $H = H_1 + H_2 + \cdots + H_p$ is also uniquely determined, and because the H_i are relatively prime, their sum H is prime by Exercise 6.11.

Exercise 6.13 By Exercise 6.12 it suffices to treat the case in which G has four components.

If all prime factorizations of G have a prime factor with four connected components, then the prime factorization is unique by the solution to Exercise 6.12. (In that exercise we used the fact that the number of components of G was prime to ensure that one factor had as many components as G, and all other factors just one. But here we already assume that we always have a factor with the same number of components as G.)

Hence, there must be a prime factor with two connected components, which is only possible if there is a second factor with two components. Suppose

$$\begin{aligned} G &= (A+B) \,\square\, (C+D) \\ &= A \,\square\, C + A \,\square\, D + B \,\square\, C + B \,\square\, D \\ &= G_1 + G_2 + G_3 + G_4 \,. \end{aligned}$$

If the G_i have a common, nontrivial prime factor, say H, then H must either be a factor of A and B or of C and D. This is easy to see, because every connected graph has unique prime factorization with respect to the Cartesian product. On the other hand, if either A and B or C and D have a common factor, then the G_i are not relatively prime.

We can thus assume that the G_i are relatively prime. If G, that is, their disjoint union, is prime, there is nothing to show. Thus let

$$(A+B) \,\square\, (C+D) \text{ and } (A'+B') \,\square\, (C'+D')$$

be two prime factorizations of G. Then a component of the first product must be isomorphic to a component of the second, say $A \,\square\, C \cong A' \,\square\, C'$. Let $A = A_1 \,\square\, A_2$, where A_1 is the product of the prime factors of A that are prime factors of A', and $C = C_1 \,\square\, C_2$, where C_1 is thc product of the prime factors of C that are prime factors of C'. Then $A \,\square\, C = A' \,\square\, C'$ implies that

$$A_1 \,\square\, A_2 \,\square\, C_1 \,\square\, C_2 = A_1 \,\square\, A_2' \,\square\, C_1 \,\square\, C_2' \,.$$

Hence, $A_2 \,\square\, C_2 = A_2' \,\square\, C_2'$. Because A_2 and A_2' are relatively prime, we infer $A_2' = C_2$. But then we also have $C_2' = A_2$. Therefore

$$\begin{aligned} & (A+B) \,\square\, (C+D) \\ = \; & (A_1 \,\square\, A_2 + B) \,\square\, (C_1 \,\square\, C_2 + D) \\ = \; & A \,\square\, C + A_1 \,\square\, A_2 \,\square\, D + \\ & C_1 \,\square\, C_2 \,\square\, B + B \,\square\, D, \end{aligned}$$

and

$$\begin{aligned} & (A'+B') \,\square\, (C'+D') \\ = \; & (A_1 \,\square\, C_2 + B') \,\square\, (A_2 \,\square\, C_1 + D') \\ = \; & A \,\square\, C + A_1 \,\square\, C_2 \,\square\, D' + \\ & A_2 \,\square\, C_1 \,\square\, B' + B' \,\square\, D'. \end{aligned}$$

Because A and B are relatively prime, $A_1 \,\square\, C_2 \,\square\, D' \cong A_1 \,\square\, A_2 \,\square\, D$. Then C_2 must be a divisor of A_2; and because $A_2 \,\square\, C_1 \,\square\, B' \cong C_1 \,\square\, C_2 \,\square\, B'$, we infer that A_2 is a divisor of C_2. But then $A_2 \cong C_2$, and $D' \cong D$, $B' \cong B$. So the two decompositions are the same.

Exercise 6.14 These graphs are clearly $\boxtimes$-S-prime and $\circ$-S-prime. On the other hand, any graph on $n \geq 3$ vertices is a nontrivial subgraph of $K_2 \boxtimes K_{n-1} = K_2 \circ K_{n-1} = K_{2n-2}$.

Exercise 6.15 Suppose G is a nontrivial subgraph of $G \,\square\, H$. Then G is also a nontrivial subgraph of $K_{|V(G)|} \,\square\, K_{|V(H)|}$. The other implication is clear.

Chapter 7

Exercise 7.1 It is routine to check that S is an equivalence relation. If x and y are in the same equivalence class, then xSy, so $N[x] = N[y]$. Hence x and y are adjacent (or equal), so equivalence classes induce complete subgraphs.

Consider distinct classes U and V. Suppose there is an edge uv with $u \in U$ and $v \in V$. We now show any $x \in U$ is adjacent to any $y \in V$. Because xSu we have $N[x] = N[u]$, but also $v \in N[u] = N[x]$, so $xv \in E(G)$. Thus $x \in N[v] = N[y]$, so $xy \in E(G)$.

Exercise 7.2 Obviously, P_1 is S-thin, as it has only one vertex. Clearly P_2 is not S-thin, as each of its vertices has the same closed neighborhood. Consider P_k for $k > 2$, and list its vertices as $1, 2, 3, \ldots, k$. Then $N[1] = \{1, 2\}$, $N[k] = \{k-1, k\}$ and $N[i] = \{i-1, i, i+1\}$ whenever $1 < i < k$. These sets are pairwise distinct, so P_k is S-thin.

Exercise 7.3 Let G be a triangle with an extra edge appended to one of its vertices. The only nontrivial automorphism of G interchanges the two vertices of degree 2, which form an S-class. This induces the trivial automorphism of G/S. Consequently, $\mathrm{Aut}(G)/S$ is trivial. But $G/S \cong P_3$, which has a nontrivial automorphism. Therefore $\mathrm{Aut}(G)/S$ is a proper subgroup of $\mathrm{Aut}(G/S)$.

Exercise 7.4 If $m = n = 2$, then the automorphism group of $P_m \boxtimes P_n = K_4$ is the symmetric group S_4 and has twelve elements. The group of $P_m \,\square\, P_n = C_4$ is the dihedral group D_4, with eight automorphisms.

If m, n are both larger than 2, then the factors of $P_m \boxtimes P_n$ are S-thin. Hence any automorphism is induced by automorphisms of the factors and the transposition of isomorphic factors (possible if $m = n$), by Theorem 7.17. Hence there are four automorphisms if $m \neq n$ and eight if $m = n$, the same number as for $P_m \square P_n$.

It remains to consider the case $K_2 \boxtimes P_n$, with $n > 2$. This has n S-classes, each of form $\{(0, x), (1, x)\}$ with $x \in P_n$. Any automorphism must preserve S-classes and also induce an automorphism of $(K_2 \boxtimes P_n)/S \cong P_n$. It follows that any automorphism of $K_2 \boxtimes P_n$ is a composition of an automorphism induced by P_n, followed by an arbitrary permutation of the vertices in each S-class. Thus $|\mathrm{Aut}(K_2 \boxtimes P_n)| = 2^{n+1}$. By contrast, $K_2 \square P_n$ has just four automorphisms.

Exercise 7.5 Hint: Compare the proof of Corollary 6.18.

Exercise 7.6 Hint: Observe that a square without diagonals must always be contained in a layer with respect to the strong product, and that a K_4 must always be contained in a layer with respect to the Cartesian product.

Exercise 7.7 Take $P_3 \boxtimes P_3$, which is of course composite. Let x be one of its vertices of degree three. Form a new graph G as follows. Remove x and replace it with a copy of K_3. Add edges running from each vertex of K_3 to the former neighbors of x. Then G is prime because it has eleven vertices. But also $G/S = P_3 \boxtimes P_3$, which is composite.

Exercise 7.8 Straightforward.

Chapter 8

Exercise 8.1 A vertex of a direct product has a loop if and only if all its projections carry loops.

Exercise 8.2 Hint: Show that any bijection $\{1, 2, \ldots, p\} \times \{1, 2, \ldots, p\} \to \{1, 2, \ldots, pq\}$ is an isomorphism $K_p^s \times K_q^s \to K_{pq}^s$.

Exercise 8.3 The factor $K_1^s + K_2^s + (K_2^s)^{\times,2}$ is prime because it has a prime number of vertices (seven). Now consider the factor $G = K_1^s + (K_2^s)^{\times,3}$. Because it has nine vertices, it could only factor nontrivially as $G = H \times K$, where H and K each have three vertices. Moreover, because G has loops at each vertex, H and K would have loops at each vertex. Because G is nonbipartite and disconnected, Weichsel's theorem implies one of H or K is disconnected (say H) and the other (say K) is connected. We infer that each K-layer of $H \times K$ is isomorphic to K, and thus each component of G has at least three vertices, a contradiction. The other factors can be treated similarly.

Exercise 8.4 Routine.

Exercise 8.5 Suppose two R-classes $X, Y \in V(G/R)$ have the same neighborhoods $\{U_1, U_2, \ldots, U_k\}$ in G/R. Let $x \in X$ and $y \in Y$. By Lemma 8.2, $N_G(x) = N_G(y) = U_1 \cup U_2 \cup \cdots \cup U_k$. Thus $x R_G y$, so $X = Y$. Hence G/R is R-thin.

Exercise 8.6 Suppose G is connected and nonbipartite. Then argue that each pair of vertices u and v in G is connected by a u, v-walk of even length. Therefore G^s has a u, v-walk as well and is thus connected. Because the Boolean square of an odd cycle is an odd cycle (of the same length), G^s is nonbipartite.

For the converse, use contrapositive.

Exercise 8.7 Every K_n, $n \geq 3$, does the job.

Exercise 8.8 Hint: Every vertex of a 4-regular graph of order 6 is nonadjacent to exactly one vertex.

Exercise 8.9 Label the vertices of the C_4-layers of $C_3 \times C_4$ with $1, 2, 3, 4$, with $1', 2', 3', 4'$, and with $1'', 2'', 3'', 4''$, respectively. Assume that the vertices of $K_{3\times 2}$ are ordered such that there is no edge between the first two vertices, the third and the fourth vertices, and between the last two vertices, and label the corresponding $K_{3\times 2}$-layers with a, b, c, d, e, f and a', b', c', d', e', f', respectively. Then $1 \mapsto a, 2 \mapsto a', 3 \mapsto b, 4 \mapsto b', 1' \mapsto c, 2' \mapsto c', 3' \mapsto d, 4' \mapsto d', 1'' \mapsto e, 2'' \mapsto e', 3'' \mapsto f, 4'' \mapsto f'$ is an isomorphism $C_3 \times C_4 \to K_2 \times K_{3\times 2}$.

Exercise 8.10 H is 4-regular and has twelve vertices. Hence as a direct product it must be a product of two regular graphs, say G_1 and G_2. By the commutativity, we only need to consider two possibilities: $|V(G_1)| = 2$ and $|V(G_2)| = 6$, or $|V(G_1)| = 3$ and $|V(G_2)| = 4$. In the first case, G_2 must be a 4-regular graph (on six vertices). By Exercise 8.8, $G_2 = K_{3\times 2}$. In the second case, G_1 and G_2 must be 2-regular graphs, that is, $G_1 = C_3$ and $G_2 = C_4$. Hence, using Exercise 8.9, if H is a direct product of simple graphs, we must have $H = C_3 \times C_4$.

To show that $H = C_3 \times C_4$ indeed holds, select an arbitrary vertex a of H. It has exactly two vertices, say b and c, at distance 3 from a. Then a, b, c must be in the same C_4-layer. The fourth vertex of this layer is then a vertex $d \neq a$ at distance 3 from c (or, equivalently, at distance 3 from b). See Figure A.3.

Exercise 8.11 Suppose, say, that G is bipartite with bipartition $V(G) = G_0 \cup G_1$. Then the definition of the direct product implies that $G \times H$ has no edge $(g, h)(g', h')$ with both endpoints in $G_0 \times V(H)$ or both endpoints in $G_1 \times V(H)$.

It follows that $G_0 \times V(H)$ and $G_1 \times V(H)$ form a bipartition of $G \times H$, so $G \times H$ is bipartite.

Exercise 8.12 Hint: Confirm that one component of G_1 of $G \times G$ is induced on the vertices (x, y) for which x and y are in different partite sets of G. Then show that the map $(x, y) \mapsto (y, x)$ is the desired involution.

Exercise 8.13 Take odd cycles $C_G = g_0g_1g_2 \dots g_mg_0$ and $C_H = h_0h_1h_2 \dots h_nh_0$ in G and H, respectively. Let $P = g_0g_1 \dots g_0$ be a closed walk that travels $\text{lcm}(m, n)/m$ times around C_G. Likewise let $Q = h_0h_1 \dots h_0$ be a closed walk that travels $\text{lcm}(m, n)/n$ times around C_H. Then we have a cycle $(g_0, h_0)(g_1, h_1) \dots (g_0, h_0)$ in $G \times H$ of odd length $\text{lcm}(m, n)$.

Exercise 8.14 Hint: Write an explicit isomorphism from $G \times K_2$ to $G + G$.

Exercise 8.15 Write $K_2 \times \cdots \times K_2 \times K_4$ as $(K_2 \times \cdots \times K_2) \times K_4$. By a simple induction it follows that $K_2 \times \cdots \times K_2$ is the disjoint union of 2^{n-1} copies of K_2. Because $K_2 \times K_4 = Q_3$ (cf. Exercise 20), it follows that $K_2 \times \cdots \times K_2 \times K_4$ consists of 2^{n-1} connected components each isomorphic to Q_3.

Exercise 8.16 Suppose $G = H_1 \times H_2$, where H_1 and H_2 are nontrivial graphs. As $|V(G)|$ is odd, so are $|V(H_1)|$ and $|V(H_2)|$ and thus each of H_1 and H_2 has at least three vertices. Being connected, each of them contains a vertex of degree at least 2. But then $\Delta(G) \geq 4$, a contradiction.

Exercise 8.17 Apply Exercise 8.16.

Exercise 8.18 Hint: The graph H must be 2-regular.

Exercise 8.19 Note P_{2n+1} is prime by Exercise 8.16. Also $P_{2n} = K_2 \times P_n^*$, where P_n^* denotes the path on n vertices with a loop added to one endpoint.

Exercise 8.20 Hints: Note first that H must be 3-regular. Observe next that Q_3 can be represented as $K_{4,4}$ with a perfect matching removed.

Exercise 8.21 $K_2 \times K_{3 \times 2} = K_2 \times (K \times K_3) = (K_2 \times K) \times K_3 = C_4 \times K_3 = C_3 \times C_4$.

Exercise 8.22 Suppose $G \times H \cong G \,\Box\, H$. Because $\Delta(G \times H) = \Delta(G)\Delta(H)$ and $\Delta(G \,\Box\, H) = \Delta(G) + \Delta(H)$ we infer that $\Delta(G) = \Delta(H) = 2$. Similarly we get that $\delta(G) = \delta(H) = 2$. It follows that G and H must be cycles. If at least one of them is of even length, then $G \times H$ is bipartite but $G \,\Box\, H$ is not. So G and H are both odd cycles. Finally note that the products would have different lengths of shortest odd cycles if G and H were of different lengths.

The converse is proved in Exercise 4.1.

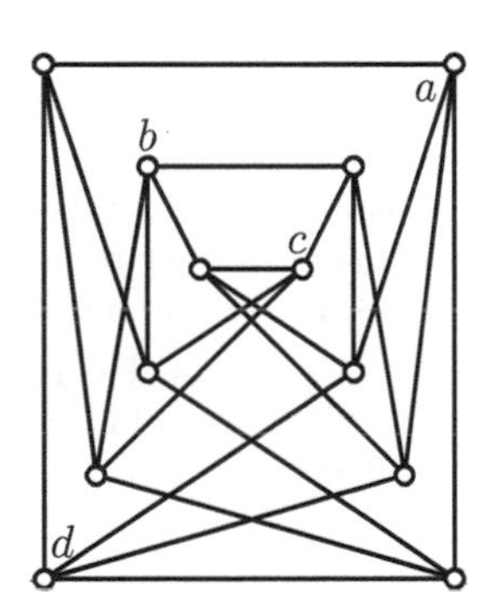

FIGURE A.3 Graph H.

Chapter 9

Exercise 9.1 Hint: This is probably most easily done without the aid of Theorem 9.15. Use Proposition 9.7 to reduce to the case $C_{10} \times K_2 \cong B \times K_2$.

Exercise 9.2 Any homomorphism $f : X \to A \times C$ has component form $f = (p_A f, p_C f)$. As the projections p_A and p_C are homomorphisms on $A \times C$, the compositions $P_A f : X \to A$ and $P_C f : X \to C$ are homomorphisms. Thus each $f : X \to A \times C$ corresponds to a pair $(p_A f, p_C f)$

Conversely, to any pair (f_A, f_C) of homomorphism $f_A : X \to A$ and $f_C : X \to C$ there is a homomorphism $f : X \to A \times C$, where $x \mapsto (f_A(x), f_C(x))$, and $(f_A, f_C) = (p_A f, p_C f)$. Hence, $\text{hom}(X, A \times C) = \text{hom}(X, A) \cdot \text{hom}(X, C)$. Because projections p_A and p_C are weak homomorphisms on $A \boxtimes C$, the same argument gives the companion formula $\hom_w(X, A \times C) = \hom_w(X, A) \cdot \hom_w(X, C)$.

Exercise 9.3 The hypothesis implies $\hom(X,A) \cdot \hom(X,C) = \hom(X,B) \cdot \hom(X,C)$. If $\hom(X,C) \neq 0$, we get $\hom(X,A) \cdot \hom(X,D) = \hom(X,B) \cdot \hom(X,D)$, and hence $\hom(X, A \times D) = \hom(X, B \times D)$. On the other hand, if $\hom(X,C) = 0$, then also $\hom(X,D) = 0$, for otherwise there is a homomorphism $X \to D \to C$. Again, $\hom(X,A) \cdot \hom(X,D) = \hom(X,B) \cdot \hom(X,D)$, and hence $\hom(X, A \times D) = \hom(X, B \times D)$. Theorem 9.3 now gives $A \times D \cong B \times D$.

Exercise 9.4 We first prove by induction that given any odd walk $(\mu_1)(\mu_2)(\mu_3)\ldots(\mu_{2p})$ in $A!$, the pair $(\mu_1)(\mu_{2p})$ is an edge of $A!$. This is trivial if $p = 1$. If $p > 1$, the induction hypothesis guarantees $(\mu_3)(\mu_{2p}) \in E(A!)$, so $(\mu_1)(\mu_2)(\mu_3)(\mu_{2p})$ is a walk in $E(A!)$. Using the fact that the edges of this walk are edges in $A!$, we get

$$\begin{gathered} aa' \in E(A) \iff \\ \mu_1(a)\mu_2(a') \in E(A) \iff \\ \mu_3^{-1}\mu_1(a)\mu_2^{-1}\mu_2(a') \in E(A) \iff \\ \mu_3\mu_3^{-1}\mu_1(a)\mu_{2p}\mu_2^{-1}\mu_2(a') \in E(A) \iff \\ \mu_1(a)\mu_{2p}(a') \in E(A)\,. \end{gathered}$$

Therefore $(\mu_1)(\mu_{2p}) \in E(A!)$.

Now, if C is a component of $A!$ that happens to be bipartite, then there is an odd path between any vertices α and β that are in different partite sets of C. Thus $(\alpha)(\beta) \in E(A!)$, so C is a complete bipartite graph. On the other hand, if C has an odd cycle (possibly just a loop), then there is an odd walk joining any pair of its vertices, so all pairs of vertices in C are adjacent, and thus $C \cong K_p^s$.

Exercise 9.5 This is reflexive because $\alpha = \mathrm{id}\,\alpha\,\mathrm{id}^{-1}$. It is symmetric, for given that $\alpha \simeq \beta$, we have $\alpha = \lambda\beta\mu^{-1}$ for $(\lambda)(\mu) \in E(A!)$. But then $\beta = \lambda^{-1}\alpha\mu$, and $(\lambda^{-1})(\mu^{-1}) \in E(A!)$, so $\beta \simeq \alpha$. To check transitivity, suppose $\alpha \simeq \beta$ and $\beta \simeq \gamma$. Then $\alpha = \lambda\beta\mu^{-1}$ and $\beta = \nu\gamma\xi^{-1}$ for edges $(\lambda)(\mu)$ and $(\nu)(\xi)$ in $E(A!)$, so $\alpha = \lambda\nu\gamma\xi^{-1}\mu^{-1} = (\lambda\nu)\gamma(\mu\xi)^{-1}$. But $(\lambda\nu)(\mu\xi) \in E(A!)$ because $xy \in E(A) \Leftrightarrow \nu(x)\xi(y) \in E(A) \Leftrightarrow \lambda\nu(x)\mu\xi(y) \in E(A)$. Therefore $\alpha \simeq \gamma$.

Exercise 9.6 Hint: Apply Theorem 9.15.

Exercise 9.6 was reformulated in the errors part because it was based on a wrong interpretation of Proposition 1 in Hammack (2008), which appears in the abstract of that paper. However, Proposition 1, the main result of the paper, is entirely correct and the basis for Theorem 9.15.

For correct interpretations compare the new formulation of Exercise 9.6 and the new Exercises 9.8 and 9.9.

Exercise 9.7 Hint: For the first law, consider the map $f \mapsto (f|_A, f|_B)$. For the second, consider $f \mapsto (p_A f, p_B f)$. For the third, use $f \mapsto [c \mapsto f(\cdot, c)]$.

Exercise 9.8 Hint: Use Corollary 9.8.

Exercise 9.9 Hint: Use Theorem 9.15 and Corollary 9.8.

Chapter 10

Exercise 10.1 Let $d = \gcd(m,n) > 1$. Then for $m = rd$ and $n = sd$,

$$\begin{aligned} D_n \circ G + D_m &= D_{sd} \circ G + D_{rd} \\ &= (sd)G + (rd)K_1 \\ &= d(sG + rK_1) \\ &= K_d \circ (sG + rK_1)\,. \end{aligned}$$

Proceed similarly for $G \circ D_n + D_m$.

Exercise 10.2 If G has a proper retraction $f : G \to G$, then any lexicographic product $H \circ G$ has a proper retraction $(a,x) \mapsto (a, f(x))$.

Conversely, consider a proper retraction $f : K_n \circ G \to K_n \circ G$. We claim that f preserves G-layers, that is, $f(a,x) = (a, \varphi(a,x))$. Indeed, suppose $f(a,x) = (b,y)$, where $a \neq b$. Then, $(a,x)(b,y)$ is an edge of $K_n \circ G$, but $f(a,x) = f(b,y)$, violating the homomorphism property of f.

Now fix a and check that $x \mapsto \varphi(a,x)$ is a proper retraction of G.

Adapt this argument for $C_{2n+1} \circ G$.

Exercise 10.3 Let $H \subseteq G$ be the smallest subgraph for which there is a retraction $g : G \to H$. (Possibly $H = G$.) Now, given a retraction $f : H \to H'$, the composition fg is a retraction of G to H', so $H' = H$ by choice of H. Therefore H is a core of G.

Next, take two cores H and H' of G, and retractions $g : G \to H$ and $g' : G \to H'$. Consider the restrictions $g : H' \to H$ and $g' : H \to H'$. By Exercise 3.9, their compositions gg' and $g'g$ are automorphisms of H and H', respectively. It follows that the homomorphisms $g : H' \to H$ and $g' : H \to H'$ are bijective, hence $H \cong H'$.

Exercise 10.4 Let G' be the core of G and let $r : G \to G'$ be a retraction (where G' is now a subgraph of G). Let $f : G \to H$ and $g : H \to G$ be homomorphisms. Then the composition rgf is a homomorphism and maps G' onto G'. As G' is a core, $rgf : G' \to G'$ is an automorphism. Then $rg : H \to G'$ is a retraction, so G' is also the core of H.

Exercise 10.5 Identify the core of H with a subgraph H' of H. There exist homomorphisms $G \circ H \to G \circ H'$ and $G \circ H' \to G \circ H$. Hence by Exercise 10.4 it suffices to show that the core of $G \circ H'$ has the asserted representation. Let $(G \circ H')'$ be the core of $G \circ H'$ and identify it with its subgraph. Let $r : G \circ H' \to (G \circ H')'$ be a retraction. Then show that r maps each H'-layer into itself. Because H' is a core we conclude that $(G \circ H')' = G' \circ H'$ for some subgraph G' of G.

Exercise 10.6 Note that $C_n = \Gamma(\mathbb{Z}_n, \{1\})$ and $K_n = \Gamma(\mathbb{Z}_n, \mathbb{Z}_n \setminus \{0\})$.

Exercise 10.7 Hint: According to the definitions, $G \times G'$ and $\Gamma(A \times A', S \times S')$ both have vertex set $A \times A'$. Show that the identity map on $A \times A'$ is the required isomorphism.

Exercise 10.8 Hint: Let $S = \{(1,0,\ldots,0),(0,1,\ldots,0),\ldots,(0,0,\ldots,1)\} \subseteq \mathbb{Z}_2^n$. Show $Q_n = \Gamma(\mathbb{Z}_2^n, S)$.

Exercise 10.9 Hint: The graphs are in fact equal. Both graphs have vertex set $A \times A'$. Following the definitions, any edge of $G \circ G'$ has one of the following forms:

(i) $[(a,b),(as,c)] = [(a,b),(a,b)(s,b^{-1}c)]$, with $(s,b^{-1}c) \in S \times A'$,

(ii) $[(a,b),(a,bs)] = [(a,b),(a,b)(1,s)]$, with $(1,s) \in \{1\} \times A'$.

In either case we have an edge of $\Gamma(A \times A', (\{1\} \times S') \cup (S \times A'))$. Conversely, any edge of $\Gamma(A \times A', (\{1\} \times S') \cup (S \times A'))$ also has one of these forms.

Exercise 10.10 Fix a vertex $v \in V(G)$ and let Γ_v be its stabilizer. For each $x \in V(G)$, choose an $\alpha_x \in A = \mathrm{Aut}(G)$ for which $\alpha_x(v) = x$. Define a map $\varphi : V(G) \times \Gamma_v \to A$ as $\varphi(x,\gamma) = \alpha_x\gamma$. It is not hard to check that this is a bijection.

Let $S = \{\alpha \in A \mid \alpha(v) \in N_G[v],\ \alpha \neq \mathrm{id}\}$. Identify the vertex set of K_n with Γ_v. Now φ is a bijection $G \circ K_n \to \Gamma(A,S)$. We claim that it is an isomorphism.

Suppose $[(x,\gamma),(y,\delta)] \in E(G \circ K_n)$, so $[x,y] \in E(G)$, or $x = y$ and $\gamma \neq \delta$. Applying φ to this edge, we get

$$[\alpha_x\gamma, \alpha_y\delta] = [\alpha_x\gamma,\ (\alpha_x\gamma)\big((\alpha_x\gamma)^{-1}\alpha_y\delta\big)]\,,$$

and to show this is an edge of $\Gamma(A,S)$, we must check that $(\alpha_x\gamma)^{-1}\alpha_y\delta \in S$. If $x = y$ and $\gamma \neq \delta$, then $(\alpha_x\gamma)^{-1}\alpha_y\delta = \gamma^{-1}\delta$ is a nonidentity element of Γ_v, so it is in S. If $[x,y] \in E(G)$, then $[x,y] = [\alpha_x\gamma(v), \alpha_y\delta(v)]$. Applying the automorphism $(\alpha_x\gamma)^{-1}$ to this edge produces an edge $[v, (\alpha_x\gamma)^{-1}\alpha_y\delta(v)]$, so $(\alpha_x\gamma)^{-1}\alpha_y\delta(v) \in N_G(v)$, hence $(\alpha_x\gamma)^{-1}\alpha_y\delta \in S$. Therefore φ is a bijective homomorphism.

Conversely, suppose $[\alpha, \alpha s] \in E(\Gamma(A,S))$. Treat the cases $s \in \Gamma_v$ and $s \in A \setminus \Gamma_v$ separately. In the first case, $[\alpha, \alpha s]$ is the image of an edge of form $[(x,\gamma),(x,\delta)]$. In the second case, it is an image of an edge $[(x,\gamma),(y,\delta)]$ with $[x,y] \in E(G)$. The details are left to the reader.

Part III

Chapter 11

Exercise 11.1 Hint: Write down the distances from the endpoints of f and f' to the endpoints of e.

Exercise 11.2 Hint: Consider the subgraph $H = P_4 \subseteq C_4$.

Exercise 11.3 Hint: If vertices u and v of G belong to two different components, then show that every shortest u,v-path passes through a cut vertex of a block in which u lies (and through a cut vertex of a block in which v lies).

Exercise 11.4 Let G be a connected, bipartite graph in which every edge is contained in at most one cycle. Then the blocks of G are even cycles and edges. Hence G is a partial cube by Exercise 11.3.

Exercise 11.5 Hint: Consider a coordinatization of the vertices of H by 0-1 vectors and remove the coordinate corresponding to the edge ab, that is, the coordinate in which a and b differ.

Exercise 11.6 Hint: Add a new coordinate to the coordinates of the isometric embedding of G into a hypercube Q_r. Set the new coordinate equal to zero for the vertices in $V_1 \setminus V_2$ and for every vertex v_1 created from $v \in V_1 \cap V_2$ by (i). For all other vertices let the new coordinate be equal to 1. Show that this coordinatization gives rise to an isometric embedding of H into Q_{r+1}.

Exercise 11.7 Hint: By Exercise 11.5 and Exercise 11.6, an expansion is the inverse of a contraction. It thus suffices to show that every partial cube can be reduced to K_1 by a sequence of contractions.

Exercise 11.8 Hint: Apply Theorem 11.8 one way or another.

Exercise 11.9 Suppose G and H isometrically embedd into Q_d and $Q_{d'}$, respectively. Then find a natural isometric embedding of $G \,\square\, H$ into $Q_{d+d'}$.

Exercise 11.10 Hint: Consider the degrees in factors of a possible factorization of G.

Exercise 11.11 First observe that intervals in hypercubes are subcubes and hence convex. Then use the definition of partial cubes.

Exercise 11.12 Let $\mathcal{A} = \{H_1, \ldots, H_r\}$. The main idea is to (naturally) embed $G_{\mathcal{A}}$ into Q_r as follows. Let H_i' and H_i'' be the open half spaces separated by H_i. For a vertex u of $G_{\mathcal{A}}$, set $\ell(u) = b_1 b_2 \ldots b_r$, where $b_i = 0$ if the region of u lies in X_i', and $b_i = 1$ if it lies in X_i''.

Observe first that if u and v are vertices of the regions that meet along a $(d-1)$-dimensional face, then $\ell(u)$ and $\ell(v)$ differ in exactly one coordinate. Therefore, $d_{Q_r}(\ell(u), \ell(v)) \leq d_{G_{\mathcal{A}}}(u, v)$. To prove the reverse inequality, select vertices p_u and p_v in the regions of u and v such that every point on the straight line connecting p_u with p_v lies on at most one hyperplane of $\mathcal{A}$. (Such points exist because for any point p of the region of u, the set of points p' that do not satisfy the condition forms a subset of measure zero of the region of v.) Because this line crosses only those hyperplanes H_i for which $\ell(u) \neq \ell(v)$, we get $d_{G_{\mathcal{A}}}(u, v) \leq d_{Q_r}(\ell(u), \ell(v))$.

Exercise 11.13 For each $1 \leq i \leq n$, let $\mathbf{e}_i = (0, 0, \ldots, 1, \ldots, 0)$ be the vertex of Q_n with a 1 only in the ith position. Any two $\mathbf{e}_i, \mathbf{e}_j$ are the endpoints of a length-2 path in Q_n whose intermediate vertex $\mathbf{e}_{i,j}$ has 1's only in the ith and jth positions. It is straightforward to check that the vertices $\mathbf{e}_i$ together with the $\mathbf{e}_{i,j}$ induce an isometric subdivided K_n. In other words, the graph obtained from K_n by subdividing each of its edges exactly once is isomorphic to the (isometric) subgraph of Q_n induced by the vertices containing exactly one or two 1's.

Exercise 11.14 Let G be 5-gonal and assume that $\langle W_{uv} \rangle$ is not convex for some edge uv of G. Select vertices x, y from W_{uv} such that there exists a shortest x, y-path containing a vertex $z \notin W_{uv}$. Consider $\{u, z\}$ and $\{v, x, y\}$ and compute: $d(u, x) + d(x, z) + d(u, v) + d(v, z) + d(u, y) + d(y, z) \leq (d(v, x) - 1) + (d(x, y) - d(y, z)) + 1 + (d(u, z)) + (d(v, y) - 1) + d(y, z) < d(v, x) + d(x, y) + d(u, z) + d(v, y)$. Hence the sets $\{u, z\}$ and $\{v, x, y\}$ violate the condition.

Exercise 11.15 Combining Exercise 11.14 with Theorem 11.8 (ii), we infer that bipartite 5-gonal graphs are partial cubes. For the converse, verify that hypercubes (as Cartesian products of K_2's) are 5-gonal and hence are their isometric subgraphs.

Exercise 11.16 Hint: All vertices of G (embedded in Q_d) lie in the same bipartition set of Q_d.

Chapter 12

Exercise 12.1 Routine.

Exercise 12.2 Suppose G is median and let $e = uv$ lie on 4-cycles $uvwz$ and $uvz'w'$. Then $z \neq z'$ and $w \neq w'$ because G is bipartite. Because G contains no $K_{2,3}$, we have $z \neq w'$ and $w \neq z'$. Similarly, z is not adjacent to z' and w is not adjacent to w'. Assume that $G - e$ is also median. Then take the median of v, w', z and find a $K_{2,3}$ in G, a contradiction.

Argue similarly that if $G - e$ is median, then G cannot be such.

Exercise 12.3 For one direction assume that G is a connected subgraph of the Cartesian product of two paths with n vertices, m edges, and $m - n + 1$ squares. If $m - n + 1 = 0$, then G is a tree and hence median. Let $m - n + 1 > 0$ and e an edge of G in a square S of G. If e lies in two squares, select the opposite edge on S and continue until an edge f of G is found that lies in exactly one square. By induction, $G - f$ is median. Deduce now that G is also median.

The other direction follows by Proposition 12.14.

Exercise 12.4 Combine Corollary 12.13 with Proposition 12.14 to deduce that G has $2n - m - 2$ Θ-classes. Then prove by induction on the number of 4-cycles of G that the length of the outer face is twice this number.

Exercise 12.5 Hint: Use Lemma 12.20.

Exercise 12.6 Let G be a median graph. By Proposition 12.4 we may consider G as an isometric subgraph of some hypercube, say Q_d. Let $u, v \in V(G)$ and assume without loss of generality that they differ in coordinates $1, 2, \ldots, k$, where $k = d_G(u, v)$. Let $x, y, z \in I(u, v)$, then $x_i = y_i = z_i = u_i = v_i$ for any $i > k$. Then the median w of x, y, z in G also has the same coordinates w_i for $i > k$ because it is obtained by the majority rule in each coordinate. Now deduce that $w \in I(u, v)$.

Alternatively, by Exercise 11.11 intervals of median graphs are convex, so they must be median.

Exercise 12.7 Hint: Use arguments similar to those from the proof of Theorem 12.7.

Exercise 12.8 Clearly the n-cube is an n-regular median graph.

In the other direction we proceed by induction on n. Clearly the assertion is true for $n = 1$. Let it be true for $n - 1 \geq 1$ and let G be an n-regular median graph. By Lemma 12.16 it has a peripheral W_{ab}, and by Theorem 12.7 F_{ab} induces an isomorphism between $\langle W_{ab} \rangle = \langle U_{ab} \rangle$

and $\langle U_{ba}\rangle$. Clearly $\langle U_{ab}\rangle$ and $\langle U_{ab}\rangle$ are $(n-1)$-regular, so $\langle W_{ba}\rangle$ must be peripheral not to violate regularity. Hence $G \cong K_2\Box\langle W_{ab}\rangle$. The assertion follows, because $\langle W_{ab}\rangle \cong Q_{n-1}$ by the induction hypothesis.

Exercise 12.9 For instance, even cycles of length at least 6 will do.

Exercise 12.10 Hint: Apply Mulder's Convex Expansion Theorem 12.8 and induction.

Exercise 12.11 Hint: Show that all convex cycles of median graphs have length 4.

Exercise 12.12 Let H be a median subgraph of a median graph G. By Theorem 12.18, G is a retract of some Q_d and H a retract of some $Q_{d'}$, where $d \geq d'$. Because $Q_{d'}$ is a retract of Q_d, H is also a retract of Q_d. Let $f : Q_d \to H$ be a retraction; then the restriction of f to G is a retraction $G \to H$.

Exercise 12.13 Let H be a convex subgraph of a median graph G. Let $V(K_2) = \{0, 1\}$ and consider the Cartesian product $X = G \Box K_2$. Let G' be the subgraph of X induced by $\{(u, 0) \mid u \in V(G)\} \cup \{(u, 1) \mid u \in V(H)\}$. Note that G' is isomorphic to the peripheral expansion of G over H, hence G' is a median subgraph of X. Because X is median (being the Cartesian product of median graphs), there exists a retraction $r : X \to G'$ by Exercise 12.12. Define $f : G \to G$ with $f(u) = p_G(r(u, 1))$. Now verify that f has the required properties.

Chapter 13

Exercise 13.1 Routine.

Exercise 13.2 If G has more than one Θ^*-class, then the canonical embedding of G has more than one factor. Hence G is not irreducible by Theorem 13.3 (i).

Suppose that there exists an irredundant isometric embedding of G into a Cartesian product of more than one factor. Then by Theorem 13.3 (ii), the canonical embedding has at least two factors and hence Θ^* has that many classes.

Exercise 13.3 Combine Exercise 13.2 with the fact that edges from different blocks of a graph are not in relation Θ.

Exercise 13.4 For instance, odd cycles.

Exercise 13.5 By Exercise 13.2 it suffices to show that almost all graphs have a single Θ^*-class.

Let G be a random graph on n vertices and let u, v and u', v' be disjoint pairs of vertices. Let x, y be an arbitrary additional pair of vertices. We say that $X = \{u, v, u', v', x, y\}$ form an H-graph provided that they are connected precisely with the edges ux, vy, xy, xu', yv', and the edges uv and $u'v'$ may be present or not. Note that if uv and $u'v' \in E(G)$, then $uv\Theta u'v'$ when X forms an H-graph. Because there are fifteen edges (and non-edges) to be fixed in an H-graph, the probability that X does not form an H-graph is at most $1 - 2^{-15}$. For $n > 9$ there are at least n disjoint possibilities to select the pair x, y once u, v and u', v' have been chosen, hence the probability that none of them gives an H-graph is at most $(1 - 2^{-15})^n$. There are less that n^4 possibilities for the selection of the pairs u, v and u', v'. Therefore, the probability that for some pair we find no H-graph is at most $n^4(1 - 2^{-15})^n$, which tends to 0 when $n \to \infty$. We conclude that the probability that we find an H-graph for all disjoint pairs of vertices tends to 1. Hence almost all graphs are irreducible.

Exercise 13.6 Routine.

Chapter 14

Exercise 14.1 Routine.

Exercise 14.2 Let $x = (x_1, \ldots, x_r)$ and $y = (y_1, \ldots, y_r)$ be vertices of $G = \Box_{i=1}^{r} K_{k_i}$. Let $x_j \neq y_j$ for $j = i_1, \ldots, i_s$. Then $d_G(x, y) = s$. Moreover, $z \in I(x, y)$ if and only if z_j, $j = i_1, \ldots, i_s$, either equals x_j or y_j, and $z_j = x_j = y_j$ for the other indices j. It is now straightforward that $I(x, y)$ induces Q_s.

Exercise 14.3 The assertion is clear for complete graphs. It transfers to their Cartesian products and to isometric subgraphs of such products.

Exercise 14.4 Let u, v, w be vertices of a median graph G with $d(u, w) = d(v, w)$ and $d(u, v) = 2$. Then the median of u, v, w yields the required vertex.

Note that C_6 is a partial cube that does not satisfy the quadrangle property.

Exercise 14.5 Hint: Let $u_1, u_2, \ldots, u_6$ be vertices of C_6 in natural order. Set $s_0 = r_0 = u_1$ and consider the request sequence $u_3u_5u_3u_5u_3u_5 \ldots$

Exercise 14.6 Gated sets are convex and clearly Δ-closed. For the converse, argue that if H is convex and Δ-closed, then to every vertex $u \notin H$ there is a unique vertex in H closest to u.

Exercise 14.7 A Hamming graph is quasi-median and it is easy to see that it contains no convex P_3.

Let G be a quasi-median graph with no convex P_3 and assume that it is not a Hamming graph. Then by Theorem 14.11 it is a gated amalgamation of H_1 and H_2. Consider a vertex from $H_1 \setminus H_2$ and a vertex from $H_2 \setminus H_1$ that are as close as possible to find a convex P_3.

Exercise 14.8 Use the three conditions of Mulder (1980a) for the characterization of quasi-median graphs as listed on p. 173.

Exercise 14.9 $x_1 = \mathrm{imp}(u_2, u_3; u_1)$, $x_2 = \mathrm{imp}(u_1, u_3; u_2)$, $x_3 = \mathrm{imp}(u_1, u_2; u_3)$.

Exercise 14.10 Hint: Apply theorems of this chapter.

Chapter 15

Exercise 15.1 Select a diametrical path P of G. Then proceed as in the proof of Theorem 15.1 but disregard the coordinates that correspond to all the vertices of P except one of its endpoints.

Exercise 15.2 Hint: The largest complete subgraph of the strong product of k paths is of order 2^k.

Exercise 15.3 Let $V(P_n) = \{0, 1, \ldots, n-1\}$ and $n = \binom{k}{\lfloor k/2 \rfloor}$. Set

$$X = \{(t_1, \ldots, t_k) \mid t_i \in \{0,1\}, \sum_{i=1}^{k} t_i = \lfloor k/2 \rfloor\},$$

and

$$Y = \{(t_1 + 1, \ldots, t_k + 1) \mid (t_1, \ldots, t_k) \in X\}.$$

Then $|X| = |Y| = \binom{k}{\lfloor k/2 \rfloor} = n$. Consider X and Y as vertex subsets of the strong product of k paths on three vertices and verify that $X \cup Y$ induces an isometric subgraph isomorphic to $K_2 \,\square\, K_n$.

For $n \leq \binom{k}{\lfloor k/2 \rfloor}$, $K_2 \,\square\, K_n$ is an isometric subgraph of graph $K_2 \,\square\, K_{\binom{k}{\lfloor k/2 \rfloor}}$ and hence $\mathrm{sdim}(K_2 \,\square\, K_n) \leq \mathrm{sdim}(K_2 \,\square\, K_{\binom{k}{\lfloor k/2 \rfloor}}) \leq k$.

Exercise 15.4 Hint: In each coordinate of an isometric embedding of G into the strong product of paths, the maximum coordinate and the minimum coordinate differ by at most d. Cut the factors accordingly.

Exercise 15.5 Routine.

Exercise 15.6 Suppose G has finite lattice dimension. Then G is isometric in some $P_n^{\square, r}$. Because $P_n^{\square, r}$ is in turn isometric in some d-cube, G is a partial cube.

The converse follows by the easy fact that Q_d lies isometrically in $P_n^{\square, d}$ for any $n \geq 2$.

Part IV

Chapter 18

Exercise 18.1 Hint: Give an example of arbitrarily large graphs G that are not partial cubes, but where every edge-deleted subgraph is a partial cube.

Exercise 18.2 Hint: Use Proposition 18.2.

Exercise 18.3 Hint: Use the definition of Θ.

Exercise 18.4 Hint: Check Θ for transitivity.

Exercise 18.5 Hint: Use shortest paths from the root of the BFS-ordering to u and v, and the fact that no two edges on shortest paths are in relation Θ.

Exercise 18.6 Hint: Find a counting argument.

Exercise 18.7 Hint: Compute the connected components of the subgraph H of G whose edge set is F and whose vertices are the endpoints of the edges in F. Select a vertex u in every component C. For every edge vw, where v is in C, remove vw and add uw. Then remove all vertices of C except u.

Exercise 18.8 Hint: Use Algorithm 17.1.

Exercise 18.9 Hint: Compute Θ^* in $O(mn)$ time and then use Exercises 18.7 and 18.8.

Exercise 18.10 Routine.

Exercise 18.11 Hint: Use induction on the diameter of G and the labeling method of Procedure 18.4.

Chapter 19

Exercise 19.1 Routine induction.

Exercise 19.2 Hint: Apply Proposition 5.1.

Exercise 19.3 Apply Exercise 19.2 and induction to obtain $W(Q_d) = d2^{2(d-1)}$. For $W(P_n \,\square\, P_m)$, show first that $W(P_n) = \binom{n+1}{3}$.

Exercise 19.4 For the first assertion, note that any pair of nonadjacent vertices contributes at least 2 to the Wiener index. For the second assertion, find a construction that, given a tree T that is not a path, returns another tree with the same number of edges that looks more similar to a path than T and has smaller Wiener index.

Exercise 19.5 $W(L_h) = \frac{1}{3}(16h^3+36h^2+26h+3)$.

Exercise 19.6 Routine.

Chapter 20

Exercise 20.1 Routine.

Exercise 20.2 Hint: Let G be a graph on n vertices and m edges. Then $a(G) \geq m/(n-1)$, by Theorem 20.2. To complete the proof, observe that $\delta n \leq 2m$.

Exercise 20.3 Hint: Apply Theorem 20.2.

Exercise 20.4 Hint: Append a path to K_5.

Exercise 20.5 Hint: Use Corollary 13.6.

Exercise 20.6 Hint: Use Exercise 20.5.

Chapter 21

Exercise 21.1 Hint: To how many sets U_{ab} does a given vertex belong?

Exercise 21.2 Hint: First argue that G is bipartite. Then assume there are three vertices x, y, z with no median, and choose them so that $d(x,y) + d(y,z) + d(z,x)$ is minimized. Use the same set-up to get a contradiction if x, y, z have more than one median.

Exercise 21.3 Hint: Treat the cases where G is connected, resp. disconneted, differently. Observe that the set of central vertices is invariant under automorphisms.

Exercise 21.4 Hint: Consider the case $G = K_2$ separately.

Exercise 21.5 Hint: Use the fact that $G^\triangle$ is a median graph and thus a partial cube.

Chapter 23

Exercise 23.1 Hint: To any two product colors c_1, c_2 there is a chordless square $xyuv$ in G, where xy, uv have color c_1 and yu, xv color c_2. Can the edges yu and uv be in different σ classes of $G - x$?

Exercise 23.2 Hint: Show first that it suffices to prove the relation when i is the product color of vu or vw.

Suppose now that the product color of vw is i. Let $vuzw$ be the unique square containing v,u and w. Suppose next that uz is in the unit-layer $G_i^{v_0}$. Then $p_i(v) = p_i(u), p_i(u) > p_i(z)$ because z is closer to v_0 than u, and $p_i(z) = p_i(w)$. Hence $p_i(v) = p_i(u) > p_i(w)$. Complete the proof by induction with respect to the distance of v from a unit layer.

Exercise 23.3 Hint: Choose two down-edges of v from different σ-classes; use the result of Exercise 23.2 and consider an arbitrary third down-edge (if such an edge exists).

Exercise 23.4 Hint: The unit-layer vertices are coordinatized by their BFS-numbers. Suppose v is not a unit-layer vertex. By Exercise 23.2 its coordinates can be found with k comparisons from the coordinates of two down-neighbors u,w, where vu and vw have different product colors. Show that such a pair of down-neighbors can be found in $d(v)$ time.

Exercise 23.5 Let $G = G_1 \,\square\, G_2$, $I(u,v)$ an interval in G, and R a shortest uv-path. Set $u = (u_1, u_2)$ and $v = (v_1, v_2)$. Clearly $p_1(R)$ is in $I_{G_1}(u_1, v_1)$, and $p_2(R)$ in $I_{G_2}(u_2, v_2)$. (Compare Proposition 5.1.) Hence $I(u,v) \subseteq I_{G_1}(u_1,v_1) \,\square\, I_{G_2}(u_2,v_2)$.

Furthermore, if $(x_1, x_2) \in I(u,v)$, it is easy to construct a shortest uv-path that contains (x_1, x_2) from a shortest u_1, v_1-path that contains x_1, and a shortest u_2, v_2-path that contains x_2. Hence $I(u,v) = I_{G_1}(u_1,v_1) \,\square\, I_{G_2}(u_2,v_2)$. This implies $\langle I(u,v)\rangle = \langle I_{G_1}(u_1,v_1)\rangle \,\square\, \langle I_{G_2}(u_2,v_2)\rangle$.

For arbitrarily many factors, the result follows by induction.

Part V

Chapter 25

Exercise 25.1 If G is not complete, then the neighbors of a vertex of the minimum degree of G form a separating set. Similarly, edges incident to a vertex form a disconnecting set.

Exercise 25.2 Consider a minimal disconnecting set S of G and let the removal of S partitions $V(G)$ into X and Y. First consider the easier case that every vertex of X is adjacent to every vertex of Y. Otherwise, select nonadjacent vertices $x \in X$ and $y \in Y$ and let Z be the set consisting of the neighbors of x in Y and the vertices of $X - \{x\}$ with neighbors in Y. Then Z is a separating set of size at most $\kappa'(G)$.

Exercise 25.3 Use $\delta(G \,\Box\, H) = \delta(G) + \delta(H)$, Exercise 25.1, and the fact that if S is a separating set of G, then $S \times V(H)$ is a is a separating set of $G \,\Box\, H$.

Exercise 25.4 First show that if G and H are connected graphs on at least two vertices, then both $\kappa(G)|V(H)|$ and $\kappa(H)|V(G)|$ are not smaller than $\kappa(G) + \kappa(H)$. Then use Exercise 25.1 and finally apply Theorem 25.1.

Exercise 25.5 Proceed similarly as in the solution to Exercise 25.4. For instance, because $\kappa'(H) \geq 1$ and $|V(G)| - 1 \geq \kappa'(G)$, we get $\kappa'(H)(|V(G)| - 1) \geq \kappa'(G)$ and hence $\kappa'(H)|V(G)| \geq \kappa'(G) + \kappa'(H)$.

Exercise 25.6 Hint: Mimic the example of Figure 25.1.

Exercise 25.7 Hint: Let h be a vertex of H of degree 1, and let h' be its neighbor. Let A and $\overline{A}$ be sets of vertices of G that demonstrate that G has property $\mathcal{P}$. Then consider the (separating) set $(A \times \{h'\}) \cup (\overline{A} \times \{h\})$.

Exercise 25.8 If a subgraph X of $G \circ H$ contains two vertices from different H-layers, then these two vertices are adjacent and any other vertex of X is adjacent to at least one of them. Hence X will be disconnected only if it is a subgraph of an H-layer. Therefore, $\kappa(K_n \circ H) = (n-1)|V(H)| + \kappa(H)$. (In particular, if $H = K_m$ then $K_n \circ K_m = K_{nm}$ and hence $\kappa(K_n \circ K_m) = nm - 1$.)

Exercise 25.9 The smallest n for which $|V(G(n,3))| = n^3 \geq 2\kappa(G(n,3)) + 2$ holds is $n = 5$. Hence the smallest constructed graph has $2 \cdot 5^3 + 2 = 252$ vertices and connectivity $2\kappa(G(5,3)) + 2 = 2 \cdot 60 + 2 = 122$.

Exercise 25.10 Use Proposition 25.7 and the fact that $G \boxtimes K_n = G \circ K_n$. (Alternatively, apply Theorem 25.11 and the fact that complete graphs have no separating sets.)

Exercise 25.11 The value $\delta(G \boxtimes H)$ is (obviously) realized with vertices of minimum degree. The other values are realized with G-towers over $\kappa'(H)$-sets of H and with H-towers over $\kappa'(G)$-sets of G.

Exercise 25.12 Select $X \subset V(G)$ such that $G - X$ is bipartite and $|X| = \kappa_b(G)$ and $Y \subset V(H)$ such that $H - Y$ is bipartite and $|Y| = \kappa_b(H)$. Then show that $(X \times V(H)) \cup (V(G) \times Y)$ is a separating set of $G \times H$.

Chapter 26

Exercise 26.1 Hint: Vertices of a complete subgraph of $G \times H$ project to pairwise different vertices in G and to pairwise different vertices in H.

Exercise 26.2 Suppose K_n is a retract of $G \times K_n$. Then, in particular, K_n is a subgraph of $G \times K_n$, so G must contain a complete subgraph of size at least n. Conversely, if G contains a complete subgraph of size at least n, then $G \times K_n$ contains K_n. Hence $\chi(G \times K_n) = n$, which implies that K_n is a retract of $G \times K_n$.

Exercise 26.3 Let $\chi(G) = n \geq 2$ and suppose that $\chi(G \boxtimes K_2) \leq n + 1$. Let c be an $(n+1)$-coloring of $G \boxtimes K_2$ and $V(K_2) = \{x, y\}$. For $u \in V(G)$, set $f(u) = \min\{c(g,x), c(g,y)\}$ if this minimum is smaller than n, otherwise set $f(u) = n - 1$. Now verify that f is a proper coloring of G, which is a contradiction.

Exercise 26.4 Note $\chi(C_{2k+1} \boxtimes C_{2n+1}) \geq 5$ by Exercise 26.3. To show that $\chi(C_{2k+1} \boxtimes C_{2n+1}) \leq 5$, first find a 5-coloring of $C_5 \boxtimes C_5$ and then extend it to the general case.

Exercise 26.5 Hint: Color $G \circ K_n$ and $G \circ K_m$ with disjoint sets of colors and combine these two colorings into a coloring of $G \circ K_{n+m}$.

Exercise 26.6 If $k = 2$, then $\chi(C_{2k+1} \circ C_{2n+1}) = 8$. If $k > 2$, then $\chi(C_{2k+1} \circ C_{2n+1}) = 7$. The lower bound is from Theorem 26.8. Construction of 7- and 8-colorings is left to the reader.

Exercise 26.7 We have noted (using the duality theorem of linear programming) that $\chi_f(G) = \omega_f(G)$. Assign the weight $1/\alpha(G)$ to any vertex of G. Then $\omega_f(G) \geq |V(G)|/\alpha(G)$ and we are done.

Exercise 26.8 Find a $(2k+1, k)$-coloring of C_{2k+1}. Then apply Theorem 26.19 and Exercise 26.7.

Exercise 26.9 Hint: A (k,d)-coloring of G is just a homomorphism from G to G_k^d.

Exercise 26.10 Hint: Apply Theorems 26.3, 26.17, and 26.19, and the fact that $\chi_f(K_n) = n$.

Exercise 26.11 Hint: As already observed in the text, it suffices to prove the result for $G = C_{2k+1}$. Then apply Exercisess 26.8 and 26.10.

Exercise 26.12 Hint: Let c be a (k,d)-coloring of G. Then verify that $c'(u) = \lfloor \frac{d'}{d} c(u) \rfloor$ defines a (k', d')-coloring of G.

Exercise 26.13 Assume without loss of generality that $k/d = \chi_c(G) \geq \chi_c(H)$. Let f be a (k,d)-coloring of G. By Exercise 26.12 there is also a (k,d)-coloring g of H. Then $h(a,x) = f(a) + g(x) \pmod k$ is a (k,d)-coloring of $G \,\Box\, H$. Thus $\chi_c(G \,\Box\, H) \leq \max\{\chi_c(G), \chi_c(H)\}$. The other inequality follows because $G \,\square\, H$ contains G and H as subgraphs.

Exercise 26.14 By the symmetry of G_k^d we may assume that $u = d$. Let $\alpha > 0$ be the smallest integer such that $\alpha d = 1 \pmod k$. (Such an α exists because by Exercise 26.12 we may assume that $\gcd(k,d) = 1$.) Now verify that $c(i) = i - |\{0 < t \leq \alpha \mid td \leq i\}|$, where $i \in \{0,1,\ldots,k-1\} \setminus \{d\}$ and td is computed in $\mathbb{Z}_k$, is a $(k-\alpha, d-(\alpha d-1)/k)$-coloring of $G_k^d - d$. Verify finally that $\frac{k-\alpha}{d-(\alpha d-1)/k} < \frac{k}{d}$.

Exercise 26.15 Hint: $G \times G$ contains G as a subgraph.

Exercise 26.16 From Lemma 26.29 (or directly) we see that for a 3-chromatic graph G, the 2-coloring graph $\mathcal{C}_2(G)$ has only one edge and is hence 2-chromatic. Then apply Theorem 26.27.

Exercise 26.17 Let $f : \mathcal{I}(G) \to [0,1]$ be a fractional coloring. Then $f' : \mathcal{I}(G \times H) \to [0,1]$ defined with $f'(I \times V(H)) = f(I)$ for all $I \in \mathcal{I}(G)$, and $f'(J) = 0$ for any other independent set of $G \times H$, is a fractional coloring of $G \times H$ with the same weight.

Chapter 27

Exercise 27.1 Let I be an independent set of G, and J an independent set of H. Then $I \times J$ is an independent set of $G \circ H$. Hence $\alpha(G \circ H) \geq \alpha(G) \cdot \alpha(H)$.

For the other direction, let S be an independent set of $G \circ H$. Then the intersection X of S with an H-layer is an independent set of H, so $|X| \leq \alpha(H)$. Moreover, if S intersects two H-layers, then the corresponding vertices of G are not adjacent. Therefore S intersects at most $\alpha(G)$ H-layers and $\alpha(G \circ H) \leq \alpha(G) \cdot \alpha(H)$.

Exercise 27.2 Hint: Only the final inequality is nontrivial. Use Proposition 27.11 and Theorem 27.17 or mimic their proofs.

Exercise 27.3 Hint: Exclude the possibility that a largest independent set has two vertices in some of the layers.

Exercise 27.4 Hint: Verify that $\alpha(K_2 \boxtimes G) = \alpha(G)$ and deduce that $\alpha(C_{2n} \boxtimes G) = n\alpha(G) = \alpha(C_{2n})\alpha(G)$.

Exercise 27.5 Hint: Apply the fact $\alpha(G \boxtimes K_2) = \alpha(G)$.

Exercise 27.6 Hint: There are homomorphisms from $G^{\times,n}$ to G and from G to $G^{\times,n}$.

Exercise 27.7 Let $V(K_n) = \{0,\ldots,n-1\}$, and $U = \{(0,i_2,\ldots,i_k) \mid 0 \leq i_2,\ldots,i_k \leq n-1\}$. The set U contains n^{k-1} different vertices of $K_n^{\times,k}$. To each vertex $u = (0,i_2,\ldots,i_k)$ of U, we assign the set of vertices Q_u defined by $\{(0,i_2,\ldots,i_k),(1,i_2+1,\ldots,i_k+1),\ldots,(n-1,i_2+(n-1),\ldots,i_k+(n-1))\}$, where addition is modulo n. Then $\{Q_u \mid u \in U\}$ is a partition of $V(K_n^{\times,k})$. Moreover, every Q_u induces a complete subgraph on n vertices. Therefore $\alpha(K_n^{\times,k}) \leq n^{k-1}$. On the other hand, the set U is independent, and so $\alpha(K_n^{\times,k}) \geq n^{k-1}$.

Exercise 27.8 Let the state of a given switch be described with an element of $\mathbb{Z}_3$. Then the system can be represented with the graph whose vertices are vectors from $\mathbb{Z}_3^n$ (thus representing the states of the system), two vertices being adjacent if the traffic light changes from one state to the other. Then this graph is isomorphic to $K_3^{\times,n}$. It has chromatic number 3, and as we know from this chapter, any color class in any 3-coloring is a canonical independent set. Hence the traffic light is controlled by the corresponding switch.

Exercise 27.9 Hint: Consider the subproduct of $G \times H$ whose projections are largest complete subgraphs of G and H.

Exercise 27.10 The result for $K_{m,m}^{\times,2}$ follows easily by the structure of the graph. For $K_{m,m}^{\times,n}$, $n \geq 3$, let X be one of the bipartite sets of $K_{m,m}$ and let I to be the set of those vertices $(v_1,v_2,\ldots,v_n)$ of $K_{m,m}^{\times,n}$ for which at least two of v_1,v_2,v_3 belong to X. Verify that I is a maximum independent set. It is clearly non-canonical.

Exercise 27.11 Let I be a maximum independent set of $G \times H$. Summing the degrees of all vertices of I, we get

$$\begin{aligned} |E(G \times H)| &\geq \sum_{v \in I} d_{G \times H}(v) \\ &\geq \delta(G)\,\delta(H)\,\alpha(G \times H)\,. \end{aligned}$$

Because on the other hand $|E(G \times H)| = 2\,|E(G)|\,|E(H)|$ we infer

$$\begin{aligned} |E(G \times H)| &= 2\,|E(G)|\,|E(H)| \\ &\geq \delta(G)\,\delta(H)\,\alpha(G \times H)\,, \end{aligned}$$

which completes the argument.

Exercise 27.12 Apply Exercise 27.11.

Exercise 27.13 Hint: Apply Theorem 27.13.

Exercise 27.14 Partition $V(H)$ into one-element sets.

Exercise 27.15 Hint: Partition $V(G)$ into $\tau(H)$ two-elements sets and $|V(H)|-2\tau(H)$ one-element sets and apply Theorem 27.19.

Exercise 27.16 Because $\chi_f(G) = \chi_c(G) = 2$ holds for a bipartite graph G with at least one edge, the assertion follows from Theorem 27.21.

Alternatively, because G has at least one edge, there is a homomorphism $G \to K_2$. Thus by Theorem 27.20, $I_\Box(G) \geq I_\Box(K_2)$; and because K_2 is a subgraph of G, $I_\Box(G) \leq I_\Box(K_2)$. Clearly, $I_\Box(K_2) = 2$.

Chapter 28

Exercise 28.1 For a dominating set D of $G \Box H$, let a be a vertex of G, and consider $N[a] \Box H$. Then the vertices $\{a\} \times V(H)$ are dominated by the vertices from $D \cap (N[a] \Box H)$. It follows that $|D \cap (N[a] \Box H)| \geq \gamma(H)$.

Exercise 28.2 Let $|V(H)| = n$ and let D be a minimum dominating set of $G \Box H$. Let D_i, $1 \leq i \leq n$, be the intersection of D with the respective G-layers and let S_i be the vertices of the ith G-layer that are not dominated within the G-layer. Then $|D_i| + |S_i| \geq \gamma(G)$. Hence

$$\begin{aligned} |D| &= \sum_{i=1}^{n} |D_i| \geq \sum_{i=1}^{n} (\gamma(G) - |S_i|) \\ &= n\gamma(G) - \sum_{i=1}^{n} |S_i| \geq n\gamma(G) - \Delta(H)|D| \end{aligned}$$

and the conclusion follows immediately.

Exercise 28.3 Hint: Add edges to C_n such that the vertex set of the obtained graph is covered by an according number of K_3's and K_2's.

Exercise 28.4 Let D be a minimum dominating set of G and let $s \in S$. Then $|D \cap N[s]| \geq 1$. Consequently, $|D| = \gamma(G) \geq |S|$. On the other hand, S is a dominating set.

Exercise 28.5 It is not difficult to find a dominating set of size $\lceil (n+1)/2 \rceil$. To show that no smaller dominating set exists, prove first that there exists a minimum dominating set D of $P_2 \Box P_n$ such that D intersects each P_2-layer in at most one vertex. Then every second P_2-layer must contain a vertex of D.

Exercise 28.6 Hint: Construct a dominating set of size $n - \lfloor n/4 \rfloor$ by selecting one and two vertices in every second C_3-layer, respectively. For $n \equiv 2 \pmod 4$, select one additional vertex in the last C_3-layer. To show that this is optimal, prove first that there exists a minimum dominating set D that intersects each C_3-layer in at most two vertices. Let s be the number of C_3-layers with no vertex from D. Because no two such layers can be adjacent, $s \leq \lfloor n/2 \rfloor$. An empty C_3-layer is dominated by two other layers, hence there are at least $\lceil s/2 \rceil$ C_3-layers with two vertices from D. Thus $|D| \geq 2\lceil s/2 \rceil + (n - \lceil s/2 \rceil - s) = n - (s - \lceil s/2 \rceil)$. Now write $n = 4k + t$ and use the fact that $s - \lceil s/2 \rceil$ is maximal when $s = 2k + \lfloor t/2 \rfloor$.

Exercise 28.7 Let the vertices of G_n be the n-subsets of $\mathbb{N}_{n^2} = \{1, 2, \ldots, n^2\}$. Then two such subsets I and J are adjacent if and only if $I \cap J \neq \emptyset$. Hence an independent set of G_n consists of at most n pairwise disjoint subsets. It follows easily that $\gamma^i(G_n) = 1$.

Let $X \subset V(G_n)$, $|X| \leq n-1$. Then $|\mathbb{N}_{n^2} \setminus (\cup_{I \in X} I)| \geq n$. Thus X cannot be a dominating set. Now find (a canonical) dominating set of size n.

Exercise 28.8 Set $V(K_k) = \{1, \ldots, k\}$ and verify that $\{(1,1), (2,2), (3,3)\}$ is a dominating set of $K_n \times K_m$. Moreover, no two vertices (i,j) and (k,l) dominate $K_n \times K_m$. Indeed, assume without loss of generality that $i \neq k$. Then (i,l) is adjacent to neither (i,j) nor (k,l).

Exercise 28.9 Hint: G is symmetric and hence so is $G \times G$. Thus select an arbitrary first vertex to be in a dominating set.

Exercise 28.10 Let $V(K_n) = \{1, \ldots, n\}$. Then $(1,\ldots,1), (2,\ldots,2), \ldots, (r+1, \ldots, r+1)$ form a dominating set of G.

Exercise 28.11 Hint: Consider a couple of cases depending in which layers a vertex of $G \times H$ lies.

Exercise 28.12 Hint: If D is a minimum dominating set of $G \circ H$, then D intersects a given H-layer in at most one vertex.

Chapter 29

Exercise 29.1 Hints: For each edge xy, we note that $\delta(xy) = x + y$ is a sum of two vertices in the same component of G. As vertices cancel in pairs, it follows that $\delta(M) = \sum_{xy \in M} (x + y)$ has an even number of vertices in each component of G, for any element M of the edge space. Conversely, suppose $X \subseteq V(G)$ has an even number of vertices in each component. List the vertices of X that are in a particular component as $x_1, x_2, \ldots, x_{2k}$. For each $1 \leq i \leq 2k-1$, let P_i be a path in G from x_i to x_{2k}. Then $\delta(\sum_{i=1}^{2k-1} P_i) = \{x_1, x_2, \ldots, x_{2k}\}$. Extend this to multiple components in the obvious way.

To prove the statement about the dimension, suppose a given component of G has vertex set $\{x_1, x_2, \ldots, x_\ell\}$. Then the $\ell - 1$ sets $\{x_1, x_2\}$,

$\{x_1, x_3\}$, $\{x_1, x_4\}$, ..., $\{x_1, x_\ell\}$ are clearly independent. Repeat this construction for each component to get a basis for V consisting of $V(G) - c(G)$ elements.

Exercise 29.2 $\mathcal{C}(K_4)$ consists of four triangles, three squares, and the empty set. $\mathcal{C}(K_5)$ contains the empty set and the whole graph, ten triangles, fifteen squares, and twelve 5-cycles. Because $\beta(K_5) = 6$, $|\mathcal{C}(K_5)| = 2^6 = 64$, hence there are 25 more subgraphs in $\mathcal{C}(K_5)$. Find them!

Exercise 29.3 Hint: This is a straightforward application of Equation (29.1). Note that Weichsel's theorem must play a role in the case of $\beta(G\times H)$, as bipartiteness can influence connectedness.

Exercise 29.4 Suppose $\mathcal{B}$ is an MCB, but there is a cycle $C = \sum_{k=1}^{\beta(G)} b_k B_k$ (each b_k is in GL(2)) and $|C| < |B_k|$ for some k with $b_k \neq 0$. Then we can exchange basis element B_k for C and obtain a basis with smaller total length than $\mathcal{B}$, contradicting minimality. Conversely, suppose $\mathcal{B}$ is not an MCB. Assume that the elements B_1, $B_2, \ldots$ are arranged in order of increasing length. Because the Greedy Algorithm cannot terminate with basis $\mathcal{B}$, there must be an element B_p for which the set $\{B_1, B_2, \ldots, B_{p-1}\}$ can be extended to an independent set $\{B_1, B_2, \ldots, B_{p-1}, C\}$ with $|C| < |B_p|$. Necessarily then, $C = \sum_{k=1}^{\beta(G)} b_k B_k$ with $b_i \neq 0$, for some $p \leq i \leq \beta(G)$, and $|C| < |B_i|$.

Exercise 29.5 Hint: For each of the mn layers, take a cycle basis of triangles for that layer, and form the union of all triangles in these bases. Next append to this set the squares in the grid $P_m \,\square\, P_n \subseteq K_m \,\square\, K_n$.

Exercise 29.6 Hint: Note that if a cycle basis consists of a maximal set of independent triangles, plus some squares, then it is an MCB. Show that the graphs in question can be embedded on a torus in such a way that all their triangles bound regions. Use this embedding as an aid in determining the dependencies among cycles.

Exercise 29.7 Let Z be as stated. The homomorphism $\pi^*_{C_p} : \mathcal{C}(C_p \times C_q) \to \mathcal{C}(C_p) = \{0, C_p\}$ must send Z to a subgraph with an even number of edges, so $\pi^*_{C_p}(Z) = 0$. Then for any edge $e = ab \in E(C_p)$, cycle Z must have an even number m_e of edges of form $(a,x)(b,y)$ for which $\pi_{C_p}((a,x)(b,y)) = ab$. Because $2p > |Z| = \sum_{e\in E(C_p)} m_e$, it follows that $m_e = 0$ for some $e \in E(C_p)$. Hence Z is a subgraph of $(C_p - e) \times C_q$. By applying the same argument to the factor C_q (and using $p \leq q$), we see that C_q must have some edge f for which Z is a cycle in the product $(C_p - e) \times (C_q - f)$ of paths. A product of paths has two planar components that can be embedded in the plane so that the boundaries of all interior regions are diamonds. These diamonds span the cycle space, so Z is a sum of diamonds.

Exercise 29.8 Hint: The graph G has no triangles. Hence an MCB will consist of a maximal independent set of squares, plus some longer cycles. Consider embedding G in a torus in such a way that all its squares are faces, and argue as in the proof of Proposition 29.7.

Exercise 29.9 Hint: Use the approach of Exercise 29.8, but note that the graph has two components, so the embedding should be in the disjoint union of two tori.

Chapter 30

Exercise 30.1 Routine.

Exercise 30.2 A bridge of G gives rise to a separating set S in $G \,\square\, C_3$ of size 3. Every one-factor of $G \,\square\, C_3$ must contain at least one edge of S. But $G \,\square\, C_3$ is 5-regular, so there is no one-factorization.

Exercise 30.3 If n and m are both odd, $|V(C_n \circ D_m)|$ is odd, so $C_n \circ D_m$ clearly has no one-factorization.

To show that $C_n \circ D_m$ has a one-factorization if one of n and m is even, we need to find an edge-coloring with $2m$ colors. Let $V(C_n) = \{u_0, u_1, \ldots, u_{n-1}\}$ and $V(D_m) = \{0, 1, \ldots, m-1\}$.

Suppose n is even. Define an edge-coloring c of $C_n \circ D_m$ as follows. To color the edges between the "first" two D_m-layers, set $c((u_0,i)(u_1,j)) = i+j \pmod m$. Next set $c((u_1,i)(u_2,j)) = (i+j \pmod m) + m$ to color the edges between the subsequent two D_m-layers. Because n is even, repeating this double pattern gives a $2m$-edge-coloring.

Suppose m is even. The idea to find a desired coloring goes as follows. Construct four disjoint $m/2\times m/2$ Latin squares. (For this we have used $2m$ disjoint symbols.) Let these Latin squares be blocks of an $m\times m$ matrix A. Then color the edge $(u_0,i)(u_1,j)$ with the (i,j)th element of A. Extend the coloring to the edges between the other D_m-layers by carefully permuting the blocks of the coloring matrix A.

Exercise 30.4 Routine.

Exercise 30.5 Hint: If both m and n are odd, then $P_m \,\square\, P_n$ is a bipartite graph of odd order.

Exercise 30.6 Consider the cases based on the parities of the factors.

Exercise 30.7 If m is even, $C_m \times T$ consists of two connected components. Let m be odd. Then $C_m \times K_2 = C_{2m}$. To show that $C_m \times T$ is not Hamiltonian in all the other cases, consider a C_m-layer corresponding to a pendant vertex of T and note that $C_m \times T$ contains exactly one cycle that contains all vertices of the layer. As this cycle is of length $2m$, the conclusion follows.

Exercise 30.8 Let C be a longest cycle of $C_m \times T$. Then, using the last argument from the solution to Exercise 30.7, C misses at least one vertex of each C_m-layer corresponding to a pendant vertex of T.

Exercise 30.9 The upper bounds follow because $P_m \,\Box\, P_n$ is planar and $P_2 \,\Box\, P_n$ is outerplanar. It remains to find K_4 and K_3 minors.

Exercise 30.10 Routine.

Exercise 30.11 Set $t = \eta(K_m \,\Box\, K_n)$ and use the fact that $\binom{t}{2} \leq |E(K_m \,\Box\, K_n)|$.

Exercise 30.12 For the nontrivial implication, we have the following chain:

$$\begin{aligned} \chi(G \,\Box\, H) &= \chi(G) = \chi(G \,\Box\, K_2) \\ &\leq \eta(G \,\Box\, K_2) \leq \eta(G \,\Box\, H)\,. \end{aligned}$$

Explain each of the (in)equalities above!

Exercise 30.13 Hint: Use Exercise 5.17.

Exercise 30.14 The two-vertex deleted subgraphs are isomorphic to $P_3 \,\Box\, P_3$ with a pendant vertex attached.

Exercise 30.15 Let $G_1 \,\Box\, G_2 \,\Box\, \cdots \,\Box\, G_k$ be the prime factorization of the nontrivial Cartesian product G where $G - v$ contains no product square. This implies that all product squares of G must contain v. Set $G_1^* = G_2 \,\Box\, \cdots \,\Box\, G_k$. If G_1 or G_1^* contains a cycle or a path of length 3, then v cannot meet all product squares of G. Hence both G_1 and G_1^* must be stars.

Part VI

Chapter 31

Exercise 31.1 Hint: Use Zorn's lemma to show that every infinite graph has a spanning forest. For the converse, let S_ι, $\iota \in I$ be an arbitrary collection of sets. Define a multigraph G whose vertex set consists the empty set $\emptyset$ and the sets S_ι, $\iota \in I$, and whose edge set is $\bigcup_{\iota \in I} S_\iota$, where the endpoints of every $e \in S_\iota$ are $\emptyset$ and S_ι. Choose a spanning forest F of G, that is, a maximal acyclic subgraph of G. Observe that it must be a tree. Use it to construct a system of representatives of the sets S_ι, $\iota \in I$.

To avoid multigraphs, subdivide every edge by a single vertex, see Imrich (1977).

Exercise 31.2 Hint: Choose another base point u and use the fact that to every $k \in \mathbb{N}$ there are numbers $m, n \in \mathbb{N}$ such that $N_k(v) \subseteq N_m(u) \subseteq N_n(v)$.

Exercise 31.3 Hint: Prove that every triple of vertices has a unique median.

Exercise 31.4 Routine.

Exercise 31.5 Routine.

Exercise 31.6 Routine.

Exercise 31.7 Routine.

Exercise 31.8 Hint: Find a basis of open sets in this space and use the definition of compact spaces.

Exercise 31.9 Hint: Distinguish the cases where α stabilizes a finite subset of $V(T)$ and the one where this does not hold.

The first case is routine. In the second, consider the orbit of a vertex $v \in T$ under the action of α. Choose a v such that $d(v, \alpha(v))$ is minimal and construct P_Z.

Exercise 31.10 Hint: Think of the construction of $G^\triangle$ in Section 21.3.

Exercise 31.11 Routine.

Exercise 31.12 Hint: Observe that the vertices of G can be considered as 0-1 sequences.

Exercise 31.13 Hint: Show first that all connected components of G are isomorphic. Then use the fact that a connected factor of a Cartesian product must factor every component of the product.

Exercise 31.14 Hint: Make use of the information in the text just before the statement of the exercise.

Exercise 31.15 Hint: Extend the method of the Exercise 31.14.

Exercise 31.16 Hint: Notice that the product must be disconnected and that all components must be isomorphic.

Exercise 31.17 Routine.

Exercise 31.18 Hint: Suppose there is a prime factorization that is different from the given one. Show that there are two decompositions $A \times B$ and $C \times D$ of G with a common refinement that decomposes a factor G_ι into a nontrivial direct product.

Exercise 31.19 Hint: Because any factorization of $S(G)$ is compatible with the given presentation of G as a direct product, $S(G)$ must have infinitely many prime factors.

Exercise 31.20 Hint: Use the fact that $\mathcal{L}(D_2) \times K_3$ and $K_2 \times K_3$ are nonisomorphic graphs in Γ and show that $(\mathcal{L}(D_2) \times K_3) \times G \cong (K_2 \times K_3) \times G$.

Chapter 32

Exercise 32.1 Hint: All components are isomorphic.

Exercise 32.2 Hint: The projections are weak homomorphisms.

Exercise 32.3 Hint: The first projection is a weak homomorphism. Note also that nontriviality is essential.

Exercise 32.4 Suppose $G = G_1 \,\Box \cdots \Box\, G_k$ is unilaterally connected. Because each projection is a weak homomorphism, it follows readily that each factor is unilaterally connected. We now show that at most one factor is not strongly connected. Suppose two factors, say G_1 and G_2, are not strongly connected. Let $x_1, y_1 \in V(G_1)$ be such that G_1 has a directed walk from x_1 to y_1, but not from y_1 to x_1. Choose $x_2, y_2 \in V(G_2)$ similarly. Now consider two vertices of form $(y_1, x_2, \ldots)$ and $(x_1, y_2, \ldots)$ in G. Because G is unilateral there is a directed walk from one to the other. But the projection of this yields a directed walk either from y_1 to x_1 or from y_2 to x_2, a contradiction.

Conversely suppose each factor G_i is unilaterally connected and all but one, say G_1, are strongly connected. Take any two vertices $x = (x_1, x_2, \ldots, x_k)$ and $y = (y_1, y_2, \ldots, y_k)$ in G. Without loss of generality, assume G_1 has a directed walk from x_1 to y_1 (as opposed to y_1 to x_1.) Then each G_i has a directed walk from x_i to y_i, and it is straightforward to assemble these into a directed walk in G from x to y.

The same argument works for the strong product.

Exercise 32.5 By associativity, it suffices to prove this for two factors G and H.

Suppose $G \circ H$ is unilaterally connected but not strongly connected. By Proposition 32.2, G is not strongly connected. That G is unilaterally connected follows immediately from the fact that projection to the first factor is a homomorphism. It follows that G has two vertices x and y for which there is a directed walk from x to y but not from y to x. Thus G has no nontrivial closed directed walk through x. Take $u, v \in V(H)$. Because G is unilaterally connected, there is a directed walk from (x, u) to (x, v) (or the other way round). This walk cannot go outside the layer H^x, for otherwise it projects to a closed directed walk through x in G. Because $H^x \cong H$, we have a directed walk between u and v in H. Thus H is unilaterally connected.

Conversely, suppose both G and H are unilaterally connected and G is not strongly connected. Then $G \circ H$ is not strongly connected by Proposition 32.2. Consider two vertices (x, y) and (u, v) of $G \circ H$. If $x = u$, use the fact that H^x is unilaterally connected to get a directed walk between (x, y) and (u, v). Otherwise, say G has a directed walk from x to u. By definition of the lexicographic product, it is straightforward to extend this to a directed walk from (x, y) to (u, v).

Exercise 32.6 Routine.

Exercise 32.7 Routine.

Exercise 32.8 Suppose $(\lambda)(\mu) \in A(G!)$. It suffices to show that $G^\lambda \cong G^\mu$. Observe that

$$\begin{aligned} xy \in A(G^\mu) &\iff x\mu^{-1}(y) \in A(G) \\ &\iff \lambda(x)\mu\mu^{-1}(y) \in A(G) \\ &\iff \lambda(x)y \in A(G) \\ &\iff \lambda(x)\lambda(y) \in A(G^\lambda). \end{aligned}$$

Thus $\lambda : G^\mu \to G^\lambda$ is an isomorphism.

Exercise 32.9 Hint: A key ingredient is the fact that Theorem 9.11 holds for digraphs.

Exercise 32.10 Hint: Use Theorem 9.11 and Proposition 9.9. Both results hold for digraphs.

Chapter 33

Exercise 33.1 There are three. One is $K_3 \,\Box\, K_3$. In the other two, the nondegenerate edges induce C_9, and the disjoint union $C_6 + C_3$, respectively.

Exercise 33.2 Consider the nontrivial direct bundle with base C_5 and fiber K_2.

Exercise 33.3 Hint: A product coloring will do.

Exercise 33.4 Hint: Consider products of dominating sets.

Exercise 33.5 Hint: See Figure 33.1.

Exercise 33.6 Estimate in how many ways a graph H of distance $\leq k$ from G can be constructed from G. Begin with $H = G$. Add $2k$ isolated vertices to $V(G)$. There are $(n+k)(n+k-1)/2 = O(n^2)$ ways to select a pair of vertices in $V(H)$. If this pair is an edge in G, delete it from H; otherwise we add it to H. Do this k-times.

Bibliography

Abay-Asmerom, G. (1998). Imbeddings of the tensor product of graphs where the second factor is a complete graph. *Discrete Math.*, *182*, 13–19.

Abay-Asmerom, G., Hammack, R. H., Larson, C. E., and Taylor, D. T. (2010). Direct product factorization of bipartite graphs with bipartition-reversing involutions. *SIAM J. Discrete Math.*, *23*, 2042–2052.

Abay-Asmerom, G., Hammack, R. H., Larson, C. E., and Taylor, D. T. (2011). Notes on the independence number in the Cartesian product of graphs. *Discuss. Math. Graph Theory*, *31*, 25–35.

Abay-Asmerom, G., Hammack, R. H., and Taylor, D. T. (2009). Perfect r-codes in strong products of graphs. *Bull. Inst. Combin. Appl.*, *55*, 66–72.

Abello, J., Fellows, M. R., and Stillwell, J. (1991). On the complexity and combinatorics of covering finite complexes. *Australas. J. Combin.*, *4*, 103–112.

Accardi, L., Lenczewski, R., and Sałapata, R. (2007). Decompositions of the free product of graphs. *Infin. Dimens. Anal. Quantum Probab. Relat. Top.*, *10*, 303–334.

Adel′son-Vel′skiĭ, G. M. and Šreĭder, Y. A. (1957). The Banach mean on groups. *Uspehi Mat. Nauk (N.S.)*, *12*, 131–136.

Aharoni, R., Berger, E., and Ziv, R. (2002). A tree version of Kőnig's theorem. *Combinatorica*, *22*, 335–343.

Aharoni, R. and Szabó, T. (2009). Vizing's conjecture for chordal graphs. *Discrete Math.*, *309*, 1766–1768.

Ahlswede, R., Aydinian, H., and Khachatrian, L. H. (1998). The intersection theorem for direct products. *European J. Combin.*, *19*, 649–661.

Aho, A. V., Hopcroft, J. E., and Ullman, J. D. (1974). *The Design and Analysis of Computer Algorithms*. Addison-Wesley, Reading, MA.

Aho, A. V., Hopcroft, J. E., and Ullman, J. D. (1987). *Data Structures and Algorithms*. Addison-Wesley, Reading, MA.

Albertson, M. O. (2005). Distinguishing Cartesian powers of graphs. *Electron. J. Combin.*, *12*, Note 17, 5 pp.

Albertson, M. O. and Collins, K. L. (1985). Homomorphisms of 3-chromatic graphs. *Discrete Math.*, *54*, 127–132.

Albertson, M. O. and Collins, K. L. (1996). Symmetry breaking in graphs. *Electron. J. Combin.*, *3*, Research Paper 18, 17 pp.

Alles, P. (1985). The dimension of sums of graphs. *Discrete Math.*, *54*, 229–233.

Alon, N. (1986). Covering graphs by the minimum number of equivalence relations. *Combinatorica*, *6*, 201–206.

Alon, N., Dinur, I., Friedgut, E., and Sudakov, B. (2004). Graph products, Fourier analysis and spectral techniques. *Geom. Funct. Anal.*, *14*, 913–940.

Alon, N. and Lubetzky, E. (2006). The Shannon capacity of a graph and the independence numbers of its powers. *IEEE Trans. Inform. Theory*, *52*, 2172–2176.

Alon, N., Lubotzky, A., and Wigderson, A. (2001). Semi-direct product in groups and zig-zag product in graphs: Connections and applications (extended abstract). In *42nd IEEE Symposium on Foundations of Computer Science* (pp. 630–637). IEEE Computer Soc., Los Alamitos, CA.

Alon, N., Yuster, R., and Zwick, U. (1997). Finding and counting given length cycles. *Algorithmica*, *17*, 209–223.

Alspach, B., Bermond, J.-C., and Sotteau, D. (1990). Decompositions into cycles I: Hamilton decompositions. In Hahn, G., Sabidussi, G., and Woodrow, R. E. (Eds.), *Cycles and Rays: Basic Structures in Finite and Infinite Graphs*, volume 301 of *NATO ASI Ser., Ser. C*, (pp. 9–18). Kluwer, Dordrecht.

Alspach, B. and George, J. C. (1990). One-factorizations of tensor products of graphs. In Bodendiek, R. and Henn, R. (Eds.), *Topics in Combinatorics and Graph Theory. Essays in Honour of Gerhard Ringel*, (pp. 41–46). Physica-Verlag, Heidelberg.

Alspach, B. and Rosenfeld, M. (1986). On Hamilton decompositions of prisms over simple 3-polytopes. *Graphs and Combinatorics*, *2*, 1–8.

Archdeacon, D., Bonnington, C. P., Pearson, J., and Širáň, J. (1997). The Hadwiger number for the product of two cycles. In *Combinatorics, complexity, & logic (Auckland, 1996)*, Springer Ser. Discrete Math. Theor. Comput. Sci. (pp. 113–120). Springer, Singapore.

Arvind, V., Cheng, C. T., and Devanur, N. R. (2008). On computing the distinguishing numbers of planar graphs and beyond: A counting approach. *SIAM J. Discrete Math.*, *22*, 1297–1324.

Assouad, P. and Deza, M. (1980). Espaces métriques plongeables dans un hypercube: Aspects combinatoires. *Ann. Discrete Math.*, *8*, 197–210. Combinatorics 79 (Proc. Colloq., Univ. Montréal, Montreal, Que., 1979), Part I.

Aubert, J. and Schneider, B. (1981). Décomposition de $K_m + K_n$ en cycles Hamiltoniens. *Discrete Math.*, *37*, 19–27.

Aubert, J. and Schneider, B. (1982). Décomposition de la somme cartesienne d'un cycle et de l'union de deux cycles hamiltoniens en cycles hamiltoniens. *Discrete Math.*, *38*, 7–16.

Aurenhammer, F., Formann, M., Idury, R. M., Schäffer, A. A., and Wagner, F. (1994). Faster isometric embedding in products of complete graphs. *Discrete Appl. Math.*, *52*, 17–28.

Aurenhammer, F. and Hagauer, J. (1995). Recognizing binary Hamming graphs in $O(n^2 \log n)$ time. *Math. Syst. Theory*, *28*, 387–395.

Aurenhammer, F., Hagauer, J., and Imrich, W. (1992). Cartesian graph factorization at logarithmic cost per edge. *Comput. Complexity*, *2*, 331–349.

Avann, S. P. (1961). Metric ternary distributive semi-lattices. *Proc. Amer. Math. Soc.*, *12*, 407–414.

Avis, D. (1981). Hypermetric spaces and the Hamming cone. *Canad. J. Math.*, *33*, 795–802.

Babai, L. (1995). Automorphism groups, isomorphism, reconstruction. In R. L. Graham, M. Grötschel, and L. Lovász (Eds.), *Handbook of Combinatorics* (pp. 1447–1540). Elsevier (North-Holland), Amsterdam.

Balakrishnan, R., Bermond, J.-C., Paulraja, P., and Yu, M.-L. (2003). On Hamilton cycle decompositions of the tensor product of complete graphs. *Discrete Math.*, *268*, 49–58.

Balakrishnan, R. and Paulraja, P. (1998). Hamilton cycles in tensor product of graphs. *Discrete Math.*, *186*, 1–13.

Balakrishnan, R. and Paulraja, P. (2005). Existence of Hamilton cycles in prisms over graphs. In Arumugam, S., Acharya, B. D., and Rao, S. B. (Eds.), *Graphs, Combinatorics, Algorithms and Applications)*, (pp. 17–22). Narosa Publishing House, New Delhi, India.

Bandelt, H.-J. (1982). Characterizing median graphs. Manuscript.

Bandelt, H.-J. (1984). Retracts of hypercubes. *J. Graph Theory*, *8*, 501–510.

Bandelt, H.-J. (2006). Evolutionary networks. In *Encyclopedia of Life Sciences* (pp. DOI: 10.1038/npg.els.0005463). Wiley-Blackwell, New York.

Bandelt, H.-J. and Chepoi, V. (2000). Decomposition and l_1-embedding of weakly median graphs. *European J. Combin.*, *21*, 701–714.

Bandelt, H.-J. and Chepoi, V. (2007). The algebra of metric betweenness. I. Subdirect representation and retraction. *European J. Combin.*, *28*, 1640–1661.

Bandelt, H.-J. and Chepoi, V. (2008a). The algebra of metric betweenness. II. Geometry and equational characterization of weakly median graphs. *European J. Combin.*, *29*, 676–700.

Bandelt, H.-J. and Chepoi, V. (2008b). Metric graph theory and geometry: A survey. In *Surveys on discrete and computational geometry*, volume 453 of *Contemp. Math.* (pp. 49–86). American Mathematical Society, Providence, RI.

Bandelt, H.-J., Dählmann, A., and Schütte, H. (1987). Absolute retracts of bipartite graphs. *Discrete Appl. Math.*, *16*, 191–215.

Bandelt, H.-J., Farber, M., and Hell, P. (1993). Absolute reflexive retracts and absolute bipartite retracts. *Discrete Appl. Math.*, *44*, 9–20.

Bandelt, H.-J., Forster, P., Sykes, B., and Richards, M. (1995). Mitochondrial portraits of human populations using median networks. *Genetics*, *141*, 743–753.

Bandelt, H.-J. and Mulder, H. M. (1983). Infinite median graphs, (0,2)-graphs, and hypercubes. *J. Graph Theory*, *7*, 487–497.

Bandelt, H.-J. and Mulder, H. M. (1988). Regular pseudo-median graphs. *J. Graph Theory*, *12*, 533–549.

Bandelt, H.-J. and Mulder, H. M. (1991). Pseudo-median graphs: Decomposition via amalgamation and Cartesian multiplication. *Discrete Math.*, *94*, 161–180.

Bandelt, H.-J., Mulder, H. M., and Wilkeit, E. (1994). Quasi-median graphs and algebras. *J. Graph Theory*, *18*, 681–703.

Bandelt, H.-J. and Pesch, E. (1989). Dismantling absolute retracts of reflexive graphs. *European J. Combin.*, *10*, 211–220.

Bandelt, H.-J. and Prisner, E. (1991). Clique graphs and Helly graphs. *J. Combin. Theory Ser. B*, *51*, 34–45.

Bandelt, H.-J., Quintana-Murci, L., Salas, A., and Macaulay, V. (2002). The fingerprint of phantom mutations in mitochondrial DNA data. *Amer. J. Human Gen.*, *71*, 1150–1160.

Bandelt, H.-J., Salas, A., and Lutz-Bonengel, S. (2004). Artificial recombination in forensic mtDNA population databases. *Int. J. Legal Med.*, *118*, 267–273.

Bandelt, H.-J. and van de Vel, M. (1987). A fixed cube theorem for median graphs. *Discrete Math.*, *67*, 129–137.

Bandelt, H.-J., Yao, Y.-G., Bravi, C., Salas, A., and Kivisild, T. (2009). Median network analysis of defectively sequenced entire mitochondrial genomes from early and contemporary disease studies. *J. Human Gen.*, *54*, 174–181.

Banič, I., Erveš, R., and Žerovnik, J. (2009). The edge fault-diameter of Cartesian graph bundles. *European J. Combin.*, *30*, 1054–1061.

Banič, I. and Žerovnik, J. (2006). Fault-diameter of Cartesian graph bundles. *Inform. Process. Lett.*, *100*, 47–51.

Barabási, A.-L. and Albert, R. (1999). Emergence of scaling in random networks. *Science*, *286*, 509–512.

Baranyai, Z. and Szasz, G. R. (1981). Hamiltonian decomposition of lexicographic products. *J. Combin. Theory Ser. B*, *31*, 253–261.

Barcalkin, A. M. and German, L. F. (1979). The external stability number of the Cartesian product of graphs (in Russian). *Bul. Akad. Štiince RSS Moldoven.*, *1*, 5–8.

Baron, G. (1968). Über den Baumgraphen eines endlichen ungerichteten Graphen. *Arch. Math.*, *19*, 668–672.

Batagelj, V. and Pisanski, T. (1982). Hamiltonian cycles in the Cartesian product of a tree and a cycle. *Discrete Math.*, *38*, 311–312.

Baumert, L. D., McEliece, R. J., Rodemich, E., Rumsey, H. C., Stanley, R., and Taylor, H. (1971). A combinatorial packing problem. In *Computers in Algebra and Number Theory (proc. SIAM-AMS Sympos. Appl. Math.)*, (pp. 97–108). American Mathematical Society, Providence, RI.

Beaudou, L., Dorbec, P., Gravier, S., and Jha, P. K. (2009). On planarity of direct product of multipartite complete graphs. *Discrete Math. Algorithms Appl.*, *1*, 85–104.

Behzad, M. and Chartrand, G. (1971). *Introduction to the Theory of Graphs*. Allyn and Bacon, Boston.

Behzad, M. and Mahmoodian, E. S. (1969). On topological invariants of the product of graphs. *Canad. Math. Bull.*, *12*, 157–166.

Berger, F. (2004). *Minimum Cycle Bases in Graphs.* PhD dissertation, Technischen Universität München.

Berger, F., Flamm, C., Gleiss, P. M., Leydold, J., and Stadler, P. F. (2004). Counterexamples in chemical ring perception. *J. Chem. Inf. Model.*, *44*, 323–331.

Bermond, J.-C. (1978). Hamiltonian decompositions of graphs, directed graphs and hypergraphs. In *Advances in Graph Theory (Cambridge Combinatorial Conf., Trinity College, Cambridge, 1977)*, volume 3 of *Ann. Discrete Math.*, (pp. 21–28). North-Holland, Amsterdam.

Bermond, J.-C., Germa, A., and Heydemann, M. C. (1979). Hamiltonian cycles in strong products of graphs. *Canad. Math. Bull.*, *22*, 305–309.

Bhat, K. V. S. (1980). On the complexity of testing a graph for n-cube. *Inform. Process. Lett.*, *11*, 16–19.

Bi, Z., Faloutsos, C., and Korn, F. (2001). The DGX distribution for mining massive, skewed data. In *Proceedings of the 6th ACM SIGKDD International Conference on Knowledge Discovery and Data Mining*, (pp. 17–26).

Biebighauser, D. P. and Ellingham, M. N. (2008). Prism-Hamiltonicity of triangulations. *J. Graph Theory*, *57*, 181–197.

Billington, E. J., Cavenagh, N. J., and Smith, B. R. (2010). Path and cycle decompositions of complete equipartite graphs: 3 and 5 parts. *Discrete Math.*, *310*, 241–254.

Birkhoff, G. (1940). *Lattice Theory.* American Mathematical Society, New York.

Bogstad, B. and Cowen, L. J. (2004). The distinguishing number of the hypercube. *Discrete Math.*, *283*, 29–35.

Bollobás, B. (1978). Cycles and semi-topological configurations. In *Theory and Applications of Graphs (Proc. Internat. Conf., Western Mich. Univ., Kalamazoo, 1976)*, volume 642 of *Lecture Notes in Math.*, (pp. 66–74). Springer, Berlin.

Bollobás, B. and Thomason, A. (1979). Set colourings of graphs. *Discrete Math.*, *25*, 21–26.

Bonato, A., Hahn, G., and Tardif, C. (2010). Large classes of infinite k-cop-win graphs. *J. Graph Theory*, *65*, 334–342.

Bondy, J. A. and Hell, P. (1990). A note on the star chromatic number. *J. Graph Theory*, *14*, 479–482.

Bonnington, C. P., Klavžar, S., and Lipovec, A. (2003). On cubic and edge-critical isometric subgraphs of hypercubes. *Australas. J. Combin.*, *28*, 217–224.

Bonnington, C. P. and Pisanski, T. (2004). On the orientable genus of the Cartesian product of a complete regular tripartite graph with an even cycle. *Ars Combin.*, *70*, 301–307.

Borowiecki, M. (1972). Hamiltonian cycles in conjunction of two graphs. *Prace NIMFTP Wrocławskiej Ser. Stud. Materiały*, *6*, 19–26.

Borowiecki, M. and Szelecka, A. (1993). One-factorizations of the generalized Cartesian product and the X-join of regular graphs. *Discuss. Math.*, *13*, 15–19.

Bottreau, A. and Métivier, Y. (1998). Some remarks on the Kronecker product of graphs. *Inform. Process. Lett.*, *68*, 55–61.

Bradshaw, Z. and Hammack, R. H. (2009). Minimum cycle bases of direct products of graphs with cycles. *Ars Math. Contemp.*, *2*, 101–119.

Bradshaw, Z. and Jaradat, M. M. M. (2009). Minimum cycle bases for direct products of K_2 with complete graphs. *Australas. J. Combin.*, *43*, 127–131.

Brandstädt, A., Le, V. B., and Spinrad, J. P. (1999). *Graph Classes*. SIAM Monographs on Discrete Mathematics and Applications. SIAM, Philadelphia.

Brešar, B. (2002). On the natural imprint function of a graph. *European J. Combin.*, *23*, 149–161.

Brešar, B. (2003). Arboreal structure and regular graphs of median-like classes. *Discuss. Math. Graph Theory*, *23*, 215–225.

Brešar, B. (2004). On subgraphs of Cartesian product graphs and S-primeness. *Discrete Math.*, *282*, 43–52.

Brešar, B., Chalopin, J., Chepoi, V., Kovše, M., Labourel, A., and Vaxès, Y. (2010). Retracts of products of chordal graphs. Manuscript.

Brešar, B., Dorbec, P., Goddard, W., Hartnell, B., Henning, M. A., Klavžar, S., and Rall, D. F. (2011). Vizing's conjecture: A survey and recent results. To appear in *J. Graph Theory*, DOI: 10.1002/jgt.20565.

Brešar, B. and Klavžar, S. (2002). On partial cubes and graphs with convex intervals. *Comment. Math. Univ. Carolin.*, *43*, 537–545.

Brešar, B., Klavžar, S., and Rall, D. F. (2007). Dominating direct products of graphs. *Discrete Math.*, *307*, 1636–1642.

Brešar, B. and Rall, D. F. (2009). Fair reception and Vizing's conjecture. *J. Graph Theory*, *61*, 45–54.

Brešar, B. and Špacapan, S. (2007). Edge-connectivity of strong products of graphs. *Discuss. Math. Graph Theory*, *27*, 333–343.

Brešar, B. and Špacapan, S. (2008). On the connectivity of the direct product of graphs. *Australas. J. Combin.*, *41*, 45–56.

Brešar, B. and Tepeh Horvat, A. (2008). Crossing graphs of fiber-complemented graphs. *Discrete Math.*, *308*, 1176–1184.

Brešar, B. and Tepeh Horvat, A. (2009). Cage-amalgamation graphs, a common generalization of chordal and median graphs. *European J. Combin.*, *30*, 1071–1081.

Brešar, B. and Zmazek, B. (2003). On the independence graph of a graph. *Discrete Math.*, *272*, 263–268.

Brown, J. I., Nowakowski, R. J., and Rall, D. (1996). The ultimate categorical independence ratio of a graph. *SIAM J. Discrete Math.*, *9*, 290–300.

Bryant, D. E., El-Zanati, S. I., and Vanden Eynden, C. (2001). Star factorizations of graph products. *J. Graph Theory*, *36*, 59–66.

Bryant, D. E., El-Zanati, S. I., and Vanden Eynden, C. (2007). Erratum to: "Star factorizations of graph products". *J. Graph Theory, 54*, 88–89.

Bueno, L. R. and Horák, P. (2011). On hamiltonian cycles in the prism over the odd graphs. To appear in *J. Graph Theory*, DOI: 10.1002/jgt.20550.

Burr, S. A., Erdős, P., and Lovász, L. (1976). On graphs of Ramsey type. *Ars Combin., 1*, 167–190.

Bussemaker, F. C., Čobeljić, S., Cvetković, D. M., and Seidel, J. J. (1976). Computer investigation of cubic graphs. Technical report, Technological University Einhhoven.

Čada, R., Flandrin, E., and Li, H. (2009). Hamiltonicity and pancyclicity of Cartesian products of graphs. *Discrete Math., 309*, 6337–6343.

Čada, R., Kaiser, T., Rosenfeld, M., and Ryjáček, Z. (2004). Hamiltonian decompositions of prisms over cubic graphs. *Discrete Math., 286*, 45–56.

Catlin, P. A. (1979). Hajós' graph-coloring conjecture: variations and counterexamples. *J. Combin. Theory Ser. B, 26*, 268–274.

Cavenagh, N. J. (1998). Decompositions of complete tripartite graphs into k-cycles. *Australas. J. Combin., 18*, 193–200.

Chae, Y., Kwak, J. H., and Lee, J. (1993). Characteristic polynomials of some graph bundles. *J. Korean Math. Soc., 30*, 229–249.

Chandran, L. S., Kostochka, A., and Raju, J. K. (2008). Hadwiger number and the Cartesian product of graphs. *Graphs Combin., 24*, 291–301.

Chandran, L. S. and Sivadasan, N. (2007). On the Hadwiger's conjecture for graph products. *Discrete Math., 307*, 266–273.

Chang, C. C. (1961). Ordinal factorization of finite relations. *Trans. Amer. Math. Soc., 101*, 259–293.

Chang, C. C. and Morel, A. C. (1960). Some cancellation theorems for ordinal products of relations. *Duke Math. J., 27*, 171–181.

Chang, T. Y. and Clark, W. E. (1993). The domination numbers of the $5 \times n$ and $6 \times n$ grid graphs. *J. Graph Theory, 17*, 81–107.

Chao, C.-Y. (1964). On a theorem of Sabidussi. *Proc. Amer. Math. Soc., 15*, 291–292.

Chastand, M. (1992). Graphes de Hamming invariants dans les graphes quasi-medians. Papier de recherche, no. 40-IEA, Université Jean Moulin.

Chastand, M. (2001). Fiber-complemented graphs. I. Structure and invariant subgraphs. *Discrete Math., 226*, 107–141.

Chastand, M. (2003). Fiber-complemented graphs. II. Retractions and endomorphisms. *Discrete Math., 268*, 81–101.

Chastand, M. and Polat, N. (1996). Invariant Hamming graphs in infinite quasi-median graphs. *Discrete Math., 160*, 93–104.

Chastand, M. and Polat, N. (2006). On geodesic structures of weakly median graphs. I. Decomposition and octahedral graphs. *Discrete Math., 306*, 1272–1284.

Che, Z. and Collins, K. L. (2007). Retracts of box products with odd-angulated factors. *J. Graph Theory*, *54*, 24–40.

Che, Z., Collins, K. L., and Tardif, C. (2008). Odd-angulated graphs and cancelling factors in box products. *J. Graph Theory*, *58*, 221–238.

Chen, G., Piotrowski, W., and Shreve, W. (1996). A partition approach to Vizing's conjecture. *J. Graph Theory*, *21*, 103–111.

Chen, P.-A. (2011). A new coloring theorem of Kneser graphs. *J. Combin. Theory Ser. A*, *118*, 1062–1071.

Cheng, C. T. (2011). A poset-based approach to embedding median graphs in hypercubes and lattices. To appear in *Order*, DOI: 10.1007/s11083-011-9203-7.

Chepoi, V. (1988). *d*-Convexity and isometric subgraphs of Hamming graphs. *Cybernetics*, *1*, 6–9.

Chepoi, V. (1989). Classification of graphs by means of metric triangles. *Metody Diskret. Analiz.*, *49*, 75–93, 96.

Chepoi, V. (1994). Separation of two convex sets in convexity structures. *J. Geom.*, *50*, 30–51.

Chepoi, V. (1996). On distances in benzenoid systems. *J. Chem. Inf. Comp. Sci.*, *36*, 1169–1172.

Chepoi, V., Deza, M., and Grishukhin, V. (1997). Clin d'oeil on L_1-embeddable planar graphs. *Discrete Appl. Math.*, *80*, 3–19.

Chepoi, V. and Klavžar, S. (1997). The Wiener index and the Szeged index of benzenoid systems in linear time. *J. Chem. Inf. Comp. Sci.*, *37*, 752–755.

Chiba, N. and Nishizeki, T. (1985). Arboricity and subgraph listing algorithms. *SIAM J. Comput.*, *14*, 210–223.

Chung, F. R. K. (1997). *Spectral Graph Theory*, volume 92 of *CBMS Regional Conference Series in Mathematics*. Published for the Conference Board of the Mathematical Sciences, Washington, DC.

Chung, F. R. K., Graham, R. L., and Saks, M. E. (1987). Dynamic search in graphs. In Johnson, D. S., Nishizeki, T., Takao, N., and Wilf, H. S. (Eds.), *Discrete Algorithm and Complexity*, volume 15 of *Perspectives in Computing*, (pp. 351–388). Academic Press, San Diego, CA.

Chung, F. R. K., Graham, R. L., and Saks, M. E. (1989). A dynamic location problem for graphs. *Combinatorica*, *9*, 111–132.

Clark, W. E. and Suen, S. (2000). An inequality related to Vizing's conjecture. *Electron. J. Combin.*, *7*, Note 4, 3 pp.

Clarke, N. E. and Nowakowski, R. J. (2005). A tandem version of the cops and robber game played on products of graphs. *Discuss. Math. Graph Theory*, *25*, 241–249.

Cockayne, E. J., Hare, E. O., Hedetniemi, S. T., and Wimer, T. V. (1985). Bounds for the domination number of grid graphs. *Congr. Numer.*, *47*, 217–228.

Coppersmith, D. and Winograd, S. (1990). Matrix multiplication via arithmetic progress-sions. *J. Symbolic Comput.*, *9*, 251–280.

Cormen, T. H., Leierson, C. E., and Rivest, R. L. (1990). *Introduction to Algorithms.* MIT Press, Cambridge, MA.

Craft, D. L. and Schwenk, A. J. (1993). The genus imbedding of the Cartesian product $G \times K_2$ need not include a genus imbedding of its factor G. *Ars Combin.*, *35A*, 29–34.

Culp, L. J. and Hammack, R. H. (2010). An isomorphism theorem for digraphs. *Australas. J. Combin.*, *48*, 205–211.

Dahlhaus, E., Gustedt, J., and McConnell, R. M. (2001). Efficient and practical algorithms for sequential modular decomposition. *J. Algorithms*, *41*, 360–387.

Dakić, T., Nedela, R., and Pisanski, T. (1995). Embeddings of tensor product graphs. In Alavi, Y. and Schwenk, A. (Eds.), *Graph Theory, Combinatorics and Applications: Proc. 7th Quad. Int. Conf. Theor. Appl. Graphs, Vol. 2*, (pp. 893–904). Wiley, New York.

Dewdney, A. K. (1977). Embedding graphs in Euclidean 3-space. *Amer. Math. Monthly*, *84*, 372–373.

Deza, M. and Grishukhin, V. (1993). Hypermetric graphs. *Quart. J. Math. Oxford Ser. (2)*, *44*, 399–433.

Deza, M. M. and Laurent, M. (1997). *Geometry of Cuts and Metrics*, volume 15 of *Algorithms and Combinatorics*. Springer-Verlag, Berlin.

Diestel, R. and Kühn, D. (2003). Graph-theoretical versus topological ends of graphs. *J. Combin. Theory Ser. B*, *87*, 197–206. Dedicated to Crispin St. J. A. Nash-Williams.

Dinur, I., Friedgut, E., and Regev, O. (2008). Independent sets in graph powers are almost contained in juntas. *Geom. Funct. Anal.*, *18*, 77–97.

Dirac, G. A. (1952). A property of 4-chromatic graphs and some remarks on critical graphs. *J. London Math. Soc.*, *27*, 85–92.

Djoković, D. Z. (1973). Distance preserving subgraphs of hypercubes. *J. Combin. Theory Ser. B*, *14*, 263–267.

Dobrynin, A., Entringer, R., and Gutman, I. (2001). Wiener index of trees: Theory and applications. *Acta Appl. Math.*, *66*, 211–249.

Dobrynin, A. A., Gutman, I., Klavžar, S., and Žigert, P. (2002). Wiener index of hexagonal systems. *Acta Appl. Math.*, *72*, 247–294.

Domke, G. S., Hedetniemi, S. T., and Laskar, R. C. (1988). Fractional packings, coverings, and irredundance in graphs. *Congr. Numer.*, *66*, 227–238.

Dörfler, W. (1973). Some results on the reconstruction of graphs. In *Infinite and Finite Sets (Colloq., Keszthely, 1973; dedicated to P. Erdős on his 60th birthday), Vol. I*, volume 10 of *Colloq. Math. Soc. János Bolyai* (pp. 361–383). North-Holland, Amsterdam.

Dörfler, W. (1974). Primfaktorzerlegung und Automorphismen des Kardinalproduktes von Graphen. *Glasnik Mat.*, *9*, 15–27.

Dörfler, W. and Imrich, W. (1970). Über das starke Produkt von endlichen Graphen. *Österreich. Akad. Wiss. Math.-Natur. Kl. S.-B. II*, *178*, 247–262.

Dörfler, W. and Imrich, W. (1972). Das lexikographische Produkt gerichteter Graphen. *Monatsh. Math.*, *76*, 21–30.

Duffus, D., Sands, B., and Woodrow, R. E. (1985). On the chromatic number of the product of graphs. *J. Graph Theory*, *9*, 487–495.

Eaton, N. and Rödl, V. (1996). Graphs of small dimensions. *Combinatorica*, *16*, 59–85.

El-Zahar, M. and Pareek, C. M. (1991). Domination number of products of graphs. *Ars Combin.*, *31*, 223–227.

El-Zahar, M. and Sauer, N. (1985). The chromatic number of the product of two 4-chromatic graphs is 4. *Combinatorica*, *5*, 121–126.

El-Zahar, M. H., Khamis, S. M., and Nazzal, K. M. (2007). On the domination number of the Cartesian product of the cycle of length n and any graph. *Discrete Appl. Math.*, *155*, 515–522.

Elsässer, R., Monien, B., Preis, R., and Frommer, A. (2004). Optimal diffusion schemes and load balancing on product graphs. *Parallel Process. Lett.*, *14*, 61–73.

Eppstein, D. (2004). Algorithms for drawing media. In J. Pach (Ed.), *Lecture Notes in Computer Science*, volume 3383 (pp. 173–183). Springer-Verlag. Proc. 12th Int. Symp. Graph Drawing (GD 2004).

Eppstein, D. (2005). The lattice dimension of a graph. *European J. Combin.*, *26*, 585–592.

Eppstein, D. (2006). Cubic partial cubes from simplicial arrangements. *Electron. J. Combin.*, *13*, Research Paper 79, 14 pp.

Eppstein, D. (2008). Recognizing partial cubes in quadratic time. In *Proceedings of the 19th Annual ACM-SIAM Symposium on Discrete Algorithms*, (pp. 1258–1266). ACM, New York.

Erdős, P., Goodman, A. W., and Pósa, L. (1966). The representation of a graph by set intersections. *Canad. J. Math.*, *18*, 106–112.

Erdős, P. and Rényi, A. (1959). On random graphs. I. *Publ. Math. Debrecen*, *6*, 290–297.

Erdős, P. and Rényi, A. (1960). On the evolution of random graphs. *Magyar Tud. Akad. Mat. Kutató Int. Közl.*, *5*, 17–61.

Fan, C. and Liu, J. (1998). Hamiltonian decompositions of strong products. *J. Graph Theory*, *29*, 45–55.

Farber, M. (1986). An analogue of the Shannon capacity of a graph. *SIAM J. Algebraic Discrete Methods*, *7*, 67–72.

Fáry, I. (1948). On straight line representation of planar graphs. *Acta Univ. Szeged. Sect. Sci. Math.*, *11*, 229–233.

Farzan, M. and Waller, D. A. (1977). Kronecker products and local joins of graphs. *Canad. J. Math.*, *29*, 255–269.

Faudree, R. J., Schelp, R. H., and Shreve, W. E. (1990). The domination number for the product of graphs. *Congr. Numer.*, *79*, 29–33.

Feder, T. (1992). Product graph representations. *J. Graph Theory*, *16*, 467–488.

Feder, T. (1995). Stable networks and product graphs. *Mem. Amer. Math. Soc.*, *116*, no. 555.

Feder, T. (2006). A dichotomy theorem on fixed points of several nonexpansive mappings. *SIAM J. Discrete Math.*, *20*, 291–301.

Feigenbaum, J. (1986). Directed Cartesian-product graphs have unique factorizations that can be computed in polynomial time. *Discrete Appl. Math.*, *15*, 105–110.

Feigenbaum, J. and Haddad, R. W. (1989). On factorable extensions and subgraphs of prime graphs. *SIAM J. Discrete Math.*, *2*, 197–218.

Feigenbaum, J., Hershberger, J., and Schäffer, A. A. (1985). A polynomial time algorithm for finding the prime factors of Cartesian-product graphs. *Discrete Appl. Math.*, *12*, 123–138.

Feigenbaum, J. and Schäffer, A. A. (1986). Recognizing composite graphs is equivalent to testing graph isomorphism. *SIAM J. Comput.*, *15*, 619–627.

Feigenbaum, J. and Schäffer, A. A. (1992). Finding the prime factors of strong direct product graphs in polynomial time. *Discrete Math.*, *109*, 77–102.

Fernández, A., Leighton, T., and López-Presa, J. L. (2007). Containment properties of product and power graphs. *Discrete Appl. Math.*, *155*, 300–311.

Fiduccia, C. M., Scheinerman, E. R., Trenk, A., and Zito, J. S. (1998). Dot product representations of graphs. *Discrete Math.*, *181*, 113–138.

Fiedler, M. (1973). Algebraic connectivity of graphs. *Czechoslovak Math. J.*, *23(98)*, 298–305.

Fink, J. F., Jacobson, M. S., Kinch, L. F., and Roberts, J. (1985). On graphs having domination number half their order. *Period. Math. Hungar.*, *16*, 287–293.

Fischer, M. J. and Meyer, A. R. (1971). Boolean matrix multiplication and transitive closure. In *Conf. Rec. 12th Ann. IEEE Symp. Switching and Automata Theory*, (pp. 129–131). MIT Press, Cambridge.

Fisher, D. C. (1994). Domination, fractional domination, 2-packing, and graph products. *SIAM J. Discrete Math.*, *7*, 493–498.

Fisher, D. C., Ryan, J., Domke, G., and Majumdar, A. (1994). Fractional domination of strong direct products. *Discrete Appl. Math.*, *50*, 89–91.

Fisher, M. J. and Isaak, G. (2008). Distinguishing colorings of Cartesian products of complete graphs. *Discrete Math.*, *308*, 2240–2246.

Fitina, L., Lenard, C., and Mills, T. (2010a). A note on edge-connectivity of the Cartesian product of graphs. Manuscript.

Fitina, L., Lenard, C., and Mills, T. (2010b). A note on the connectivity of the Cartesian product of graphs. *Australas. J. Combin.*, *48*, 281–284.

Fitzpatrick, S. L. and Nowakowski, R. J. (2000). The strong isometric dimension of finite reflexive graphs. *Discuss. Math. Graph Theory*, *20*, 23–38.

Fitzpatrick, S. L. and Nowakowski, R. J. (2001). Copnumber of graphs with strong isometric dimension two. *Ars Combin.*, *59*, 65–73.

Fleischner, H. (1974). The square of every two-connected graph is Hamiltonian. *J. Combin. Theory Ser. B*, *16*, 29–34.

Fleischner, H. (1989). The prism of a 2-connected, planar, cubic graph is Hamiltonian (a proof independent of the four colour theorem). In *Graph Theory in Memory of G. A. Dirac (Sandbjerg, 1985)*, volume 41 of *Ann. Discrete Math.* (pp. 141–170). North-Holland, Amsterdam.

Foldes, S. (1977). A characterization of hypercubes. *Discrete Math.*, *17*, 155–159.

Foregger, M. F. (1978). Hamiltonian decompositions of products of cycles. *Discrete Math.*, *24*, 251–260.

Freudenthal, H. (1931). Über die Enden topologischer Räume und Gruppen. *Math. Z.*, *33*, 692–713.

Fronček, D., Jerebic, J., Klavžar, S., and Kovář, P. (2007). Strong isometric dimension, biclique coverings, and Sperner's theorem. *Combin. Probab. Comput.*, *16*, 271–275.

Frucht, R. (1938). Herstellung von Graphen mit vorgegebener abstrakter Gruppe. *Compositio Math.*, *6*, 239–250.

Gao, G. and Zhu, X. (1996). Star-extremal graphs and the lexicographic product. *Discrete Math.*, *152*, 147–156.

Gedeonova, E. (1990). Constructions of S-lattices. *Order*, *7*, 249–266.

Geller, D. (1976). r-tuple colorings of uniquely colorable graphs. *Discrete Math.*, *16*, 9–12.

Geller, D. and Stahl, S. (1975). The chromatic number and other functions of the lexicographic product. *J. Combin. Theory Ser. B*, *19*, 87–95.

Goddard, W. and Henning, M. A. (2001). Pancyclicity of the prism. *Discrete Math.*, *234*, 139–142.

Godsil, C. and Royle, G. (2001). *Algebraic Graph Theory*, volume 207 of *Graduate Texts in Mathematics*. Springer-Verlag, New York.

Godsil, C. D. (1981). GRRs for nonsolvable groups. In *Algebraic Methods in Graph Theory, Vol. I, II (Szeged, 1978)*, volume 25 of *Colloq. Math. Soc. János Bolyai* (pp. 221–239). North-Holland, Amsterdam.

Goldberg, F. (2009). On the Colin de Verdière numbers of Cartesian graph products. *Linear Algebra Appl.*, *431*, 2285–2290.

Golumbic, M. C. (1980). *Algorithmic Graph Theory and Perfect Graphs*. Academic Press, New York.

Graham, R. L. (1970). On primitive graphs and optimal vertex assignments. In *International Conference on Combinatorial Mathematics (1970)*, volume 175 of *Ann. New York Acad. Sci.*, (pp. 170–186). New York Acad. Sci., New York.

Graham, R. L. (1988). Isometric embeddings of graphs. In Beineke, L. and Wilson, R. (Eds.), *Selected Topics in Graph Theory III*, (pp. 133–150). Academic Press, San Diego, CA.

Graham, R. L. and Pollak, H. O. (1971). On the addressing problem for loop switching. *Bell System Tech. J.*, *50*, 2495–2519.

Graham, R. L. and Pollak, H. O. (1972). On embedding graphs in squashed cubes. In *Graph Theory and Applications (Proc. Conf., Western Michigan Univ., Kalamazoo, 1972; dedicated to the memory of J. W. T. Youngs)*, volume 303 of *Lecture Notes in Math.*, (pp. 99–110). Springer, Berlin.

Graham, R. L. and Winkler, P. M. (1985). On isometric embeddings of graphs. *Trans. Amer. Math. Soc.*, *288*, 527–536.

Graovac, A. and Pisanski, T. (1991). On the Wiener index of a graph. *J. Math. Chem.*, *8*, 53–62.

Gravier, S. (1997). Hamiltonicity of the cross product of two Hamiltonian graphs. *Discrete Math.*, *170*, 253–257.

Gravier, S. and Khelladi, A. (1995). On the domination number of cross products of graphs. *Discrete Math.*, *145*, 273–277.

Gravier, S. and Mollard, M. (1997). On domination numbers of Cartesian products of paths. *Discrete Appl. Math.*, *80*, 247–250.

Greenwell, D. and Lovász, L. (1974). Applications of product colouring. *Acta Math. Acad. Sci. Hungar.*, *25*, 335–340.

Gromov, M. (1981). Groups of polynomial growth and expanding maps. Appendix by Jacques Tits. *Inst. Hautes Etudes Sci. Publ. Math.*, *53*, 53–78.

Gross, J. L. and Yellen, J. (Eds.). (2004). *Handbook of Graph Theory.* Discrete Mathematics and its Applications (Boca Raton). Boca Raton, FL: CRC Press.

Gross, J. L. and Yellen, J. (2006). *Graph Theory and its Applications.* Discrete Mathematics and its Applications (Boca Raton). Chapman & Hall/CRC, Boca Raton, FL.

Grünbaum, B. (1972). *Arrangements and Spreads.* American Mathematical Society Providence, R.I. Conference Board of the Mathematical Sciences Regional Conference Series in Mathematics, No. 10.

Guo, F. and Watanabe, Y. (1990). On graphs in which the Shannon capacity is unachievable by finite product. *IEEE Trans. Inform. Theory*, *36*, 622–623.

Gutman, I. and Cyvin, S. J. (1989). *Introduction to the Theory of Benzenoid Hydrocarbons.* Springer-Verlag, Berlin.

Gutman, I. and Klavžar, S. (1996). A method for calculating Wiener numbers of benzenoid hydrocarbons. *ACH—Models in Chemistry*, *133*, 389–399.

Győri, E. and Plummer, M. D. (1992). The Cartesian product of a k-extendable and an ℓ-extendable graph is $(k + \ell + 1)$-extendable. *Discrete Math.*, *101*, 87–96.

Győri, E. and Imrich, W. (2001). On the strong product of a k-extendable and l-extendable graph. *Graphs Combin.*, *17*, 245–253.

Habib, M. and Maurer, M. C. (1979). On the X-join decomposition for undirected graphs. *Discrete Appl. Math.*, *1*, 201–207.

Haemers, W. (1979). On some problems of Lovász concerning the Shannon capacity of a graph. *IEEE Trans. Inform. Theory*, *25*, 231–232.

Hagauer, J. (1995). Skeletons, recognition algorithm and distance matrix of quasi-median graphs. *Internat. J. Comput. Math.*, *55*, 155–171.

Hagauer, J., Imrich, W., and Klavžar, S. (1999). Recognizing median graphs in subquadratic time. *Theoret. Comput. Sci.*, *215*, 123–136.

Hagauer, J. and Žerovnik, J. (1999). An algorithm for the weak reconstruction of Cartesian-product graphs. *J. Combin. Inform. System Sci.*, *24*, 87–103.

Häggkvist, R., Hell, P., Miller, D. J., and Neumann-Lara, V. (1988). On multiplicative graphs and the product conjecture. *Combinatorica*, *8*, 63–74.

Hahn, G., Hell, P., and Poljak, S. (1995). On the ultimate independence ratio of a graph. *European J. Combin.*, *16*, 253–261.

Hahn, G. and Tardif, C. (1997). Graph homomorphisms: Structure and symmetry. In Hahn, G. and Sabidussi, G. (Eds.), *Graph Symmetry (Montreal, PQ, 1996)*, volume 497 of *NATO Adv. Sci. Inst. Ser. C Math. Phys. Sci.*, (pp. 107–166). Kluwer, Dordrecht.

Hajiabolhassan, H. and Zhu, X. (2003). Circular chromatic number of Kneser graphs. *J. Combin. Theory Ser. B*, *88*, 299–303.

Hajós, G. (1961). Über eine Konstruktion nicht n-färbbarer Graphen. *Wiss. Z. Martin-Luther Univ. Halle-Wittenberg Math. Natur. Reihe*, *10*, 116–117.

Hales, R. S. (1973). Numerical invariants and the strong product of graphs. *J. Combin. Theory Ser. B*, *15*, 146–155.

Halin, R. (1964). Über unendliche Wege in Graphen. *Math. Ann.*, *157*, 125–137.

Hall, Jr., M. (1976). *The Theory of Groups*. Chelsea Publishing Co., New York. Reprinting of the 1968 edition.

Hammack, R. H. (2006). Minimum cycle bases of direct products of bipartite graphs. *Australas. J. Combin.*, *36*, 213–221.

Hammack, R. H. (2007). Minimum cycle bases of direct products of complete graphs. *Inform. Process. Lett.*, *102*, 214–218.

Hammack, R. H. (2008). A cancellation property for the direct product of graphs. *Discuss. Math. Graph Theory*, *28*, 179–184.

Hammack, R. H. (2009). On direct product cancellation of graphs. *Discrete Math.*, *309*, 2538–2543.

Hammack, R. H. and Imrich, W. (2009). On Cartesian skeletons of graphs. *Ars Math. Contemp.*, *2*, 191–205.

Hammack, R. H. and Toman, K. (2010). Cancellation of direct products of digraphs. *Discuss. Math. Graph Theory*, *30*, 575–590.

Harary, F. (1959). On the group of the composition of two graphs. *Duke Math. J.*, *26*, 29–36.

Harary, F. (1969). *Graph Theory*. Addison-Wesley, Reading, MA.

Harary, F. and Trauth, Jr., C. A. (1966). Connectedness of products of two directed graphs. *SIAM J. Appl. Math.*, *14*, 250–254.

Hartnell, B. L. and Rall, D. F. (1991). On Vizing's conjecture. *Congr. Numer.*, *82*, 87–96.

Hartnell, B. L. and Rall, D. F. (1995). Vizing's conjecture and the one-half argument. *Discuss. Math. Graph Theory*, *15*, 205–216.

Hartnell, B. L. and Rall, D. F. (1998). Domination in Cartesian products: Vizing's conjecture. In *Domination in Graphs, Advanced Topics*, volume 209 of *Monogr. Textbooks Pure Appl. Math.*, (pp. 163–189). Dekker, New York.

Hausdorff, F. (1914). *Grundzüge der Mengenlehre.* Veit and Company, Leipzig.

Hedetniemi, S. (1966). Homomorphisms of graphs and automata. Technical Report 03105-44-T, University of Michigan.

Hell, P. (1972). Rétractions de graphes. Ph.D. thesis, Université de Montréal.

Hell, P. and Nešetřil, J. (2004). *Graphs and Homomorphisms*, volume 28 of *Oxford Lecture Series in Mathematics and its Applications.* Oxford University Press, Oxford.

Hell, P. and Rival, I. (1987). Absolute retracts and varieties of reflexive graphs. *Canad. J. Math.*, *39*, 544–567.

Hell, P. and Roberts, F. S. (1982). Analogues of the Shannon capacity of a graph. In *Theory and Practice of Combinatorics*, volume 12 of *Ann. Discrete Math.*, (pp. 155–168). North-Holland, Amsterdam.

Hell, P., Yu, X., and Zhou, H. S. (1994). Independence ratios of graph powers. *Discrete Math.*, *127*, 213–220.

Hell, P., Zhou, H., and Zhu, X. (1994). Multiplicativity of oriented cycles. *J. Combin. Theory Ser. B*, *60*, 239–253.

Hellmuth, M. (2010). *Local Prime Factor Decomposition of Approximate Strong Product Graphs.* PhD thesis, University Leipzig.

Hellmuth, M. (2011). A local prime factor decomposition algorithm. *Discrete Math.*, *311*, 944–965.

Hellmuth, M., Gringmann, L. L., and Stadler, P. F. (2011). Diagonalized Cartesian products of S-prime graphs are S-prime. To appear in *Discrete Math.*, DOI: 10.1016/j.disc.2011.03.033.

Hellmuth, M., Imrich, W., Klöckl, W., and Stadler, P. F. (2009a). Approximate graph products. *European J. Combin.*, *30*, 1119–1133.

Hellmuth, M., Imrich, W., Klöckl, W., and Stadler, P. F. (2009b). Local algorithms for the prime factorization of strong product graphs. *Math. Comput. Sci.*, *2*, 653–682.

Hellmuth, M., Ostermeier, P., and Stadler, P. (2010). Minimum cycle bases of lexicographic products. Manuscript.

Hemminger, R. L. (1968). The group of an X-join of graphs. *J. Combin. Theory*, *5*, 408–418.

Hilton, A. J. W., Rado, R., and Scott, S. H. (1973). A (<5)-colour theorem for planar graphs. *Bull. London Math. Soc.*, *5*, 302–306.

Hilton, A. J. W., Rado, R., and Scott, S. H. (1975). Multicolouring graphs and hypergraphs. *Nanta Math.*, *9*, 152–155.

Hong, S., Kwak, J. H., and Lee, J. (1999). Bipartite graph bundles with connected fibres. *Bull. Austral. Math. Soc.*, *59*, 153–161.

Horák, P., Kaiser, T., Rosenfeld, M., and Ryjáček, Z. (2005). The prism over the middle-levels graph is Hamiltonian. *Order*, *22*, 73–81.

Horton, J. D. and Wallis, W. D. (2002). Factoring the Cartesian product of a cubic graph and a triangle. *Discrete Math.*, *259*, 137–146.

Imrich, W. (1969a). Automorphismen und das kartesische Produkt von Graphen. *Österreich. Akad. Wiss. Math.-Natur. Kl. S.-B. II*, *177*, 203–214.

Imrich, W. (1969b). Graphen mit transitiver Automorphismengruppe. *Monatsh. Math.*, *73*, 341–347.

Imrich, W. (1969c). Über das lexikographische Produkt von Graphen. *Arch. Math. (Basel)*, *20*, 228–234.

Imrich, W. (1970). Graphs with transitive Abelian automorphism group. In *Combinat. Theory Appl.*, volume 4 of *Colloq. Math. Soc. János Bolyai* (pp. 651–656). North Holland, Amsterdam.

Imrich, W. (1971). Über das schwache Kartesische Produkt von Graphen. *J. Combin. Theory Ser. B*, *11*, 1–16.

Imrich, W. (1972a). Assoziative Produkte von Graphen. *Österreich. Akad. Wiss. Math.-Natur. Kl. S.-B. II*, *180*, 203–239.

Imrich, W. (1972b). On products of graphs and regular groups. *Israel J. Math.*, *11*, 258–264.

Imrich, W. (1977). Subgroup theorems and graphs. In *Combinatorial Mathematics, V (Proc. 5th Austral. Conf., Roy. Melbourne Inst. Tech., Melbourne, 1976)*, volume 622 of *Lecture Notes in Math.*, (pp. 1–27). Springer, Berlin.

Imrich, W. (1989). Embedding graphs into Cartesian products. In *Graph Theory and its Applications: East and West (Jinan, 1986)*, volume 576 of *Ann. New York Acad. Sci.*, (pp. 266–274). New York Academy of Sciences, New York.

Imrich, W. (1998). Factoring cardinal product graphs in polynomial time. *Discrete Math.*, *192*, 119–144.

Imrich, W. and Izbicki, H. (1975). Associative products of graphs. *Monatsh. Math.*, *80*, 277–281.

Imrich, W., Jerebic, J., and Klavžar, S. (2008). The distinguishing number of Cartesian products of complete graphs. *European J. Combin.*, *29*, 922–929.

Imrich, W. and Klavžar, S. (1992). Retracts of strong products of graphs. *Discrete Math.*, *109*, 147–154.

Imrich, W. and Klavžar, S. (1993). A simple $O(mn)$ algorithm for recognizing Hamming graphs. *Bull. Inst. Combin. Appl.*, *9*, 45–56.

Imrich, W. and Klavžar, S. (1997). Recognizing Hamming graphs in linear time and space. *Inform. Process. Lett.*, *63*, 91–95.

Imrich, W. and Klavžar, S. (1998). A convexity lemma and expansion procedures for bipartite graphs. *European J. Combin.*, *19*, 677–685.

Imrich, W., Klavžar, S., and Mulder, H. M. (1999). Median graphs and triangle-free graphs. *SIAM J. Discrete Math.*, *12*, 111–118.

Imrich, W. and Klavžar, S. (2006). Distinguishing Cartesian powers of graphs. *J. Graph Theory*, *53*, 250–260.

Imrich, W. and Klavžar, S. (2009). Transitive, locally finite median graphs with finite blocks. *Graphs Combin.*, *25*, 81–90.

Imrich, W. and Klavžar, S. (2011). Two-ended regular median graphs. To appear in *Discrete Math.*, DOI: 10.1016/j.disc.2010.05.002.

Imrich, W., Klavžar, S., and Rall, D. F. (2007). Cancellation properties of products of graphs. *Discrete Appl. Math.*, *155*, 2362–2364.

Imrich, W., Klavžar, S., and Rall, D. F. (2008). *Topics in Graph Theory: Graphs and their Cartesian Product.* A K Peters Ltd., Wellesley, MA.

Imrich, W., Klavžar, S., and Trofimov, V. (2007). Distinguishing infinite graphs. *Electron. J. Combin.*, *14*, Research Paper 36, 12 pp.

Imrich, W. and Klöckl, W. (2007). Factoring directed graphs with respect to the cardinal product in polynomial time. *Discuss. Math. Graph Theory*, *27*, 593–601.

Imrich, W. and Kovše, M. (2009). Lattice embeddings of trees. *European J. Combin.*, *30*, 1142–1148.

Imrich, W. and Peterin, I. (2007). Recognizing Cartesian products in linear time. *Discrete Math.*, *307*, 472–483.

Imrich, W., Peterin, I., Špacapan, S., and Zhang, C.-Q. (2010). NZ-flows in strong products of graphs. *J. Graph Theory*, *64*, 267–276.

Imrich, W., Pisanski, T., and Žerovnik, J. (1997). Recognizing Cartesian graph bundles. *Discrete Math.*, *167/168*, 393–403.

Imrich, W. and Pultr, A. (1991). Classification of tensor products of symmetric graphs. *Comment. Math. Univ. Carolin.*, *32*, 315–322.

Imrich, W. and Seifter, N. (1989). A note on the growth of transitive graphs. In *Proceedings of the Oberwolfach Meeting "Kombinatorik" (1986)*, volume 73, (pp. 111–117).

Imrich, W. and Škrekovski, R. (2003). A theorem on integer flows on Cartesian products of graphs. *J. Graph Theory*, *43*, 93–98.

Imrich, W. and Stadler, P. F. (2002). Minimum cycle bases of product graphs. *Australas. J. Combin.*, *26*, 233–244.

Imrich, W. and Watkins, M. E. (1976). On automorphism groups of Cayley graphs. *Period. Math. Hungar.*, *7*, 243–258.

Imrich, W. and Žerovnik, J. (1994). Factoring Cartesian-product graphs. *J. Graph Theory*, *18*, 557–567.

Imrich, W. and Žerovnik, J. (1996). On the weak reconstruction of Cartesian-product graphs. *Discrete Math.*, *150*, 167–178.

Imrich, W., Zmazek, B., and Žerovnik, J. (2003). Weak k-reconstruction of Cartesian products. *Discuss. Math. Graph Theory*, *23*, 273–285.

Isbell, J. R. (1980). Median algebra. *Trans. Amer. Math. Soc.*, *260*, 319–362.

Ivančo, J. (1988). Some results on the Hadwiger numbers of graphs. *Math. Slovaca*, *38*, 221–226.

Jacobson, M. S. and Kinch, L. F. (1984). On the domination number of products of graphs: I. *Ars Combin.*, *18*, 33–44.

Jacobson, M. S. and Kinch, L. F. (1986). On the domination of the products of graphs II: Trees. *J. Graph Theory*, *10*, 97–106.

Jänicke, S., Heine, C., Hellmuth, M., Stadler, P. F., and Scheuermann, G. (2010). Visualization of graph products. *IEEE Trans. Visualization Comp. Graphics*, *16*, 1082–1089.

Jaradat, M. M. M. (2005). On the edge coloring of graph products. *Int. J. Math. Math. Sci.*, (16), 2669–2676.

Jaradat, M. M. M. (2008). Minimal cycle bases of the lexicographic product of graphs. *Discuss. Math. Graph Theory*, *28*, 229–247.

Jawhari, E. M., Misane, D., and Pouzet, M. (1986). Retracts: Graphs and ordered sets from the metric point of view. In *Combinatorics and Ordered Sets (Arcata, CA, 1985)*, volume 57 of *Contemp. Math.* (pp. 175–226). American Mathematical Society, Providence, RI.

Jensen, T. R. and Toft, B. (1995). *Graph Coloring Problems.* John Wiley & Sons, New York.

Jerebic, J. and Klavžar, S. (2006). On induced and isometric embeddings of graphs into the strong product of paths. *Discrete Math.*, *306*, 1358–1363.

Jerebic, J., Klavžar, S., and Špacapan, S. (2005). Characterizing r-perfect codes in direct products of two and three cycles. *Inform. Process. Lett.*, *94*, 1–6.

Jha, P. K. (1992). Hamiltonian decompositions of products of cycles. *Indian J. Pure Appl. Math.*, *23*, 723–729.

Jha, P. K. (2000). Further results on independence in direct-product graphs. *Ars Combin.*, *56*, 15–24.

Jha, P. K. (2002). Perfect r-domination in the Kronecker product of three cycles. *IEEE Trans. Circuits Systems I Fund. Theory Appl.*, *49*, 89–92.

Jha, P. K. (2003). Perfect r-domination in the Kronecker product of two cycles, with an application to diagonal/toroidal mesh. *Inform. Process. Lett.*, *87*, 163–168.

Jha, P. K., Agnihotri, N., and Kumar, R. (1996). Edge exchanges in Hamiltonian decompositions of Kronecker-product graphs. *Comput. Math. Appl.*, *31*, 11–19.

Jha, P. K., Agnihotri, N., and Kumar, R. (1997). Long cycles and long paths in the Kronecker product of a cycle and a tree. *Discrete Appl. Math.*, *74*, 101–121.

Jha, P. K. and Klavžar, S. (1998). Independence in direct-product graphs. *Ars Combin.*, *50*, 53–63.

Jha, P. K. and Slutzki, G. (1989). An $O(n^2 \log n)$ algorithm for recognizing median graphs. Technical report, Department of Computer Science, Iowa State University.

Jha, P. K. and Slutzki, G. (1992). Convex-expansions algorithms for recognition and isometric embedding of median graphs. *Ars Combin.*, *34*, 75–92.

Jha, P. K. and Slutzki, G. (1993). A note on outerplanarity of product graphs. *Zastos. Mat.*, *21*, 537–544.

Jha, P. K. and Slutzki, G. (1994). Independence numbers of product graphs. *Appl. Math. Lett.*, *7*, 91–94.

Johnson, A., Holroyd, F. C., and Stahl, S. (1997). Multichromatic numbers, star chromatic numbers and Kneser graphs. *J. Graph Theory*, *26*, 137–145.

Jónsson, B. (1982). Arithmetic of ordered sets. In Rival, I. (Ed.), *Ordered Sets (Banff, Alta., 1981)*, volume 83 of *NATO Adv. Study Inst. Ser. C: Math. Phys. Sci.*, (pp. 3–41). Reidel, Dordrecht.

Kaiser, T. and Kriesell, M. (2006). On the pancyclicity of lexicographic products. *Graphs Combin.*, *22*, 51–58.

Kaiser, T., Ryjáček, Z., Kráľ, D., Rosenfeld, M., and Voss, H.-J. (2007). Hamilton cycles in prisms. *J. Graph Theory*, *56*, 249–269.

Kasteleyn, P. W. (1963). Dimer statistics and phase transitions. *J. Mathematical Phys.*, *4*, 287–293.

Kaveh, A. (1995). *Structural Mechanics: Graph and Matrix Methods* (Second ed.)., volume 9 of *Applied and Engineering Mathematics Series*. Research Studies Press Ltd., Taunton.

Kaveh, A. (2006). *Optimal Structural Analysis* (Second ed.). John Wiley & Sons, Chichester.

Kaveh, A. and Mirzaie, R. (2008). Minimal cycle basis of graph products for the force method of frame analysis. *Comm. Numer. Methods Engrg.*, *24*, 653–669.

Kelley, C. A., Sridhara, D., and Rosenthal, J. (2008). Zig-zag and replacement product graphs and LDPC codes. *Adv. Math. Commun.*, *2*, 347–372.

Kel′mans, A. K. (1967). The metric properties of trees. In *Cybernetics and Control (Russian)* (pp. 98–107). Izdat. "Nauka", Moscow.

Kim, D., Kim, H. K., and Lee, J. (2008). Generalized characteristic polynomials of graph bundles. *Linear Algebra Appl.*, *429*, 688–697.

Kim, H. K. and Kim, J. Y. (1996). Characteristic polynomials of graph bundles with productive fibres. *Bull. Korean Math. Soc.*, *33*, 75–86.

Kim, J. Y. (1996). Laplacian spectra of graph bundles. *Commun. Korean Math. Soc.*, *11*, 1159–1174.

Kim, S.-R. (1991). Centers of a tensor composite graph. In *Proceedings of the 22nd Southeastern Conference on Combinatorics, Graph Theory, and Computing (Baton Rouge, LA, 1991)*, volume 81, (pp. 193–203).

Klavžar, S. (1992). Two remarks on retracts of graph products. *Discrete Math.*, *109*, 155–160.

Klavžar, S. (1993). Strong products of χ-critical graphs. *Aequationes Math.*, *45*, 153–162.

Klavžar, S. (2005). Some new bounds and exact results on the independence number of Cartesian product graphs. *Ars Combin.*, *74*, 173–186.

Klavžar, S. (2006). On the canonical metric representation, average distance, and partial Hamming graphs. *European J. Combin.*, *27*, 68–73.

Klavžar, S. (2008). A bird's eye view of the cut method and a survey of its applications in chemical graph theory. *MATCH Commun. Math. Comput. Chem.*, *60*, 255–274.

Klavžar, S. and Gutman, I. (1997). Wiener number of vertex-weighted graphs and a chemical application. *Discrete Appl. Math.*, *80*, 73–81.

Klavžar, S., Gutman, I., and Mohar, B. (1995). Labeling of benzenoid systems which reflects the vertex-distance relations. *J. Chem. Inf. Comp. Sci.*, *35*, 590–593.

Klavžar, S., Lipovec, A., and Petkovšek, M. (2002). On subgraphs of Cartesian product graphs. *Discrete Math.*, *244*, 223–230. Algebraic and topological methods in graph theory (Lake Bled, 1999).

Klavžar, S. and Milutinović, U. (1994). Strong products of Kneser graphs. *Discrete Math.*, *133*, 297–300.

Klavžar, S. and Mohar, B. (1995a). The chromatic numbers of graph bundles over cycles. *Discrete Math.*, *138*, 301–314.

Klavžar, S. and Mohar, B. (1995b). Coloring graph bundles. *J. Graph Theory*, *19*, 145–155.

Klavžar, S. and Mulder, H. M. (1999). Median graphs: Characterizations, location theory and related structures. *J. Combin. Math. Combin. Comp.*, *30*, 103–127.

Klavžar, S., Mulder, H. M., and Škrekovski, R. (1998). An Euler-type formula for median graphs. *Discrete Math.*, *187*, 255–258.

Klavžar, S. and Peterin, I. (2005). Characterizing subgraphs of Hamming graphs. *J. Graph Theory*, *49*, 302–312.

Klavžar, S. and Seifter, N. (1995). Dominating Cartesian products of cycles. *Discrete Appl. Math.*, *59*, 129–136.

Klavžar, S. and Shpectorov, S. (2007). Tribes of cubic partial cubes. *Discrete Math. Theor. Comput. Sci.*, *9*, 273–291.

Klavžar, S. and Shpectorov, S. (2011). Convex excess in partial cubes. To appear in *J. Graph Theory*, DOI: 10.1002/jgt.20589.

Klavžar, S. and Škrekovski, R. (2000). On median graphs and median grid graphs. *Discrete Math.*, *219*, 287–293.

Klavžar, S. and Špacapan, S. (2008). On the edge-connectivity of Cartesian product graphs. *Asian-Eur. J. Math.*, *1*, 93–98.

Klavžar, S., Špacapan, S., and Žerovnik, J. (2006). An almost complete description of perfect codes in direct products of cycles. *Adv. in Appl. Math.*, *37*, 2–18.

Klavžar, S. and Yeh, H.-G. (2002). On the fractional chromatic number, the chromatic number, and graph products. *Discrete Math.*, *247*, 235–242.

Klavžar, S. and Zhu, X. (2007). Cartesian powers of graphs can be distinguished by two labels. *European J. Combin.*, *28*, 303–310.

Klavžar, S. and Zmazek, B. (1996). On a Vizing-like conjecture for direct product graphs. *Discrete Math.*, *156*, 243–246.

Knauer, U. (1987). Unretractive and **S**-unretractive joins and lexicographic products of graphs. *J. Graph Theory*, *11*, 429–440.

Kong, Q.-P., Salas, A., Sun, C., Yao, Y.-G., Fuku, N., Tanaka, M., Zhong, L., Wang, C.-Y., and Bandelt, H.-J. (2008). Distilling artificial recombinants from large sets of complete mtDNA genomes. *PLoS ONE*, *3*, e3016.

König, D. (1936). *Theorie der Endlichen und Unendlichen Graphen.* Akad. Verlag, Leipzig.

Kotlov, A. (2001). Minors and strong products. *European J. Combin.*, *22*, 511–512.

Kotzig, A. (1973). Every Cartesian product of two circuits is decomposable into two Hamiltonian circuits. Rapport 233, Centre de Recherches Mathématiques Montréal.

Kotzig, A. (1979). 1-Factorizations of cartesian products of regular graphs. *J. Graph Theory*, *3*, 23–34.

Kozen, D. (1978). A clique problem equivalent to graph isomorphism. *SIGACT News*, *10*, 50–52.

Kráľ, D., Maxová, J., Podbrdský, P., and Šámal, R. (2004). Pancyclicity of strong products of graphs. *Graphs Combin.*, *20*, 91–104.

Kráľ, D., Maxová, J., Šámal, R., and Podbrdský, P. (2005). Hamilton cycles in strong products of graphs. *J. Graph Theory*, *48*, 299–321.

Kráľ, D. and Stacho, L. (2008). Hamiltonian threshold for strong products of graphs. *J. Graph Theory*, *58*, 314–328.

Kriesell, M. (1997). A note on Hamiltonian cycles in lexicographical products. *J. Autom. Lang. Comb.*, *2*, 135–138.

Kriesell, M. (2006). There exist highly critically connected graphs of diameter three. *Graphs Combin.*, *22*, 481–485.

Křivka, P. (1981a). The dimension of odd cycles and cartesian cubes. In *Algebraic Methods in Graph Theory, Vol. I (Szeged, 1978)*, volume 25 of *Colloq. Math. Soc. János Bolyai* (pp. 435–443). North-Holland, Amsterdam.

Křivka, P. (1981b). Dimension of the sum of two copies of a graph. *Czechoslovak Math. J.*, *31(106)*, 514–520.

Křivka, P. (1985). Dimension of the sum of several copies of a graph. *Czechoslovak Math. J.*, *35*, 347–354.

Křiž, I. (1984). A class of dimension-skipping graphs. *Combinatorica*, *4*, 317–319.

Ku, C. Y. and McMillan, B. B. (2009). Independent sets of maximal size in tensor powers of vertex-transitive graphs. *J. Graph Theory*, *60*, 295–301.

Kuratowski, K. (1930). Sur le problème des courbes gauches en topologie. *Fund. Math.*, *16*, 271–283.

Kwak, J. H. and Kwon, Y. S. (2001). Characteristic polynomials of graph bundles having voltages in a dihedral group. *Linear Algebra Appl.*, *336*, 99–118.

Kwak, J. H. and Lee, J. (1990). Isomorphism classes of graph bundles. *Canad. J. Math.*, *42*, 747–761.

Kwak, J. H. and Lee, J. (1992). Characteristic polynomials of some graph bundles. II. *Linear Multilinear Algebra*, *32*, 61–73.

Kwak, J. H., Lee, J., and Sohn, M. Y. (1996). Isoperimetric numbers of graph bundles. *Graphs Combin.*, *12*, 239–251.

Lamprey, R. H. and Barnes, B. H. (1974). Product graphs and their applications. In *Modeling and simulation (Proc. 5th Annual Pittsburgh Conf., Univ. Pittsburgh, Pittsburgh, Pa., 1974), Vol. 5, Part 2* (pp. 1119–1123). Pittsburgh, Pa.: Instrument Soc. Amer.

Lamprey, R. H. and Barnes, B. H. (1981). A new concept of primeness in graphs. *Networks*, *11*, 279–284.

Lamprey, R. H. and Barnes, B. H. (1995). A characterization of Cartesian-quasiprime graphs. In *Proceedings of the 26th Southeastern International Conference on Combinatorics, Graph Theory and Computing (Boca Raton, FL, 1995)*, volume 109, (pp. 117–121).

Larose, B., Laviolette, F., and Tardif, C. (1998). On normal Cayley graphs and hom-idempotent graphs. *European J. Combin.*, *19*, 867–881.

Larose, B. and Tardif, C. (2000). Hedetniemi's conjecture and the retracts of a product of graphs. *Combinatorica*, *20*, 531–544.

Larose, B. and Tardif, C. (2002). Projectivity and independent sets in powers of graphs. *J. Graph Theory*, *40*, 162–171.

Laskar, R. and Hare, W. (1972). Chromatic numbers for certain graphs. *J. London Math. Soc.*, *4*, 489–492.

Leskovec, J., Chakrabarti, D., Kleinberg, J., Faloutsos, C., and Ghahramani, Z. (2010). Kronecker graphs: An approach to modeling networks. *J. Mach. Learn. Res.*, *11*, 985–1042.

Leskovec, J. and Faloutsos, C. (2007). Scalable modeling of real graphs using kronecker multiplication. In *Proceedings of the 24th International Conference on Machine Learning*, volume 227, (pp. 497–504). ACM, New York.

Leskovec, J., Kleinberg, J., and Faloutsos, C. (2005). Graphs over time: Densification laws, shrinking diameters and possible explanations. In *Proceedings of the 11th ACM SIGKDD International Conference on Knowledge Discovery in Data Mining*, (pp. 177–187). ACM, New York.

Limaye, N. B. and Sarvate, D. G. (1997). On r-extendability of the hypercube Q_n. *Math. Bohemica*, *122*, 249–255.

Linial, N. and Vazirani, U. (1989). Graph products and chromatic numbers. In *Proc. 30th Ann. IEEE Symp. on Found. of Comp. Sci.*, (pp. 124–128). IEEE Comput. Soc. Press, Los Alamos, NM.

Liouville, B. (1978). Sur la connectivité des produits de graphes. *C. R. Acad. Sci. Paris Sér. A-B*, *286*, A363–A365.

Lovász, L. (1967). Operations with structures. *Acta Math. Acad. Sci. Hungar.*, *18*, 321–328.

Lovász, L. (1971). On the cancellation law among finite relational structures. *Period. Math. Hungar.*, *1*, 145–156.

Lovász, L. (1978). Kneser's conjecture, chromatic number, and homotopy. *J. Combin. Theory Ser. A*, *25*, 319–324.

Lovász, L. (1979). On the Shannon capacity of a graph. *IEEE Trans. Inform. Theory*, *25*, 1–7.

Lovász, L., Nešetřil, J., and Pultr, A. (1980). On a product dimension of graphs. *J. Combin. Theory Ser. B*, *29*, 47–67.

Lovász, L. and Plummer, M. D. (1986). *Matching Theory*, volume 121 of *North-Holland Mathematics Studies*. North-Holland, Amsterdam.

Lu, F., Zhang, L., and Lin, F. (2010). Enumeration of perfect matchings of a type of quadratic lattice on the torus. *Electron. J. Combin.*, *17*, Research Paper 36, 14.

Lyndon, R. C. and Schupp, P. E. (2001). *Combinatorial Group Theory*. Classics in Mathematics. Springer-Verlag, Berlin. Reprint of the 1977 edition.

Maamoun, M. and Meyniel, H. (1987). On a game of policemen and robber. *Discrete Appl. Math.*, *17*, 307–309.

Mader, W. (1977). Endlichkeitssätze für k-kritische Graphen. *Math. Ann.*, *229*, 143–153.

Mader, W. (1984). On k-critically n-connected graphs. In *Progress in Graph Theory (Waterloo, 1982)* (pp. 389–398). Academic Press, Toronto, Ont.

Mahmoodian, E. S. (1981). On edge-colorability of Cartesian products of graphs. *Canad. Math. Bull.*, *24*, 107–108.

Manikandan, R. S. and Paulraja, P. (2006). C_p-decompositions of some regular graphs. *Discrete Math.*, *306*, 429–451.

Manikandan, R. S. and Paulraja, P. (2007). C_5-decompositions of the tensor product of complete graphs. *Australas. J. Combin.*, *37*, 285–293.

Manikandan, R. S. and Paulraja, P. (2008). Hamilton cycle decompositions of the tensor product of complete multipartite graphs. *Discrete Math.*, *308*, 3586–3606.

Manikandan, R. S. and Paulraja, P. (2010a). C_7-decompositions of the tensor product of complete graphs. Manuscript.

Manikandan, R. S. and Paulraja, P. (2010b). Hamilton cycle decompositions of the tensor products of complete bipartite graphs and complete multipartite graphs. *Discrete Math.*, *310*, 2776–2789.

McAndrew, M. H. (1963). On the product of directed graphs. *Proc. Amer. Math. Soc.*, *14*, 600–606.

McAndrew, M. H. (1965). On graphs with transitive automorphism groups. *Amer. Math. Soc. Notices*, *12*, 575.

McAvaney, K. L. (1980). A conjecture on two-point deleted subgraphs of Cartesian products. In *Combinatorial Mathematics, VII (Proc. 7th Australian Conf., Univ. Newcastle, Newcastle, 1979)*, volume 829 of *Lecture Notes in Math.* (pp. 172–185). Springer, Berlin.

McEliece, R. J., Rodemich, E. R., and Rumsey, H. C. (1978). The Lovász bound and some generalizations. *J. Combin. Inform. System Sci.*, *3*, 134–152.

McKenzie, R. (1971). Cardinal multiplication of structures with a reflexive relation. *Fund. Math.*, *70*, 59–101.

Meir, A. and Moon, J. W. (1975). Relations between packing and covering numbers of a tree. *Pacific J. Math.*, *61*, 225–233.

Mekiš, G. (2010). Lower bounds for the domination number and the total domination number of direct product graphs. *Discrete Math.*, *310*, 3310–3317.

Mellendorf, S. (1997). Hamilton decompositions of Cartesian products of multicycles. *J. Graph Theory*, *24*, 85–115.

Mendelsohn, E. and Rosa, A. (1985). One-factorizations of the complete graph—a survey. *J. Graph Theory*, *9*, 43–65.

Meng, J. (2003). Connectivity of vertex and edge transitive graphs. *Discrete Appl. Math.*, *127*, 601–613.

Meunier, F. (2005). A topological lower bound for the circular chromatic number of Schrijver graphs. *J. Graph Theory*, *49*, 257–261.

Miklós, D. (1996). On product of association schemes and Shannon capacity. *Discrete Math.*, *150*, 441–447.

Milgram, S. (1967). The small-world problem. *Psych. Today*, *2*, 60–67.

Miller, D. J. (1968). The categorical product of graphs. *Canad. J. Math.*, *20*, 1511–1521.

Miller, D. J. (1970a). The automorphism group of a product of graphs. *Proc. Amer. Math. Soc.*, *25*, 24–28.

Miller, D. J. (1970b). Weak Cartesian product of graphs. *Colloquium Math.*, *21*, 55–74.

Miller, Z. (1978). Contractions of graphs: a theorem of Ore and an extremal problem. *Discrete Math.*, *21*, 261–272.

Moehring, R. H. and Radermacher, F. J. (1984). Substitution decomposition for discrete structures and connections with combinatorial optimization. In *Algebraic and Combinatorial Methods in Operations Research*, volume 19 of *Ann. Discrete Math.*, (pp. 257–355).

Mohar, B. (1984). On edge-colorability of products of graphs. *Publ. Inst. Math. (Beograd) (N.S.)*, *36*, 13–16.

Mohar, B. (1991). The Laplacian spectrum of graphs. In *Graph Theory, Combinatorics, and Applications. Vol. 2 (Kalamazoo, 1988)*, Wiley-Intersci. Publ. (pp. 871–898). Wiley, New York.

Mohar, B. and Pisanski, T. (1983). Edge-coloring of a family of regular graphs. *Publ. Inst. Math. (Beograd) (N.S.)*, *33*, 157–162.

Mohar, B. and Pisanski, T. (1988). How to compute the Wiener index of a graph. *J. Math. Chem.*, *2*, 267–277.

Mohar, B., Pisanski, T., and Shawe-Taylor, J. (1981). Edge-coloring of composite regular graphs. In *Finite and Infinite Sets, Vol. I, II (Eger, 1981)*, volume 25 of *Colloq. Math. Soc. János Bolyai* (pp. 591–600). North-Holland, Amsterdam, New York.

Mohar, B., Pisanski, T., and Škoviera, M. (1988). The maximum genus of graph bundles. *European J. Combin.*, *9*, 215–224.

Mohar, B., Pisanski, T., Škoviera, M., and White, A. T. (1985). The Cartesian product of three triangles can be embedded into a surface of genus 7. *Discrete Math.*, *56*, 87–89.

Mohar, B., Pisanski, T., and White, A. T. (1990). Embeddings of Cartesian products of nearly bipartite graphs. *J. Graph Theory*, *14*, 301–310.

Mulder, H. M. (1978). The structure of median graphs. *Discrete Math.*, *24*, 197–204.

Mulder, H. M. (1980a). *The Interval Function of a Graph.* Mathematical Centre Tracts 132. Mathematisch Centrum, Amsterdam.

Mulder, H. M. (1980b). n-cubes and median graphs. *J. Graph Theory*, *4*, 107–110.

Mulder, H. M. (1990). Triple convexities for graphs. *Rostock. Math. Kolloq.*, *39*, 35–52.

Müller, V. (1979). On colorings of graphs without short cycles. *Discrete Math.*, *26*, 165–176.

Munro, I. (1971). Efficient determination of the transitive closure of a directed graph. *Inform. Process. Lett.*, *1*, 56–58.

Muthusamy, A. and Paulraja, P. (1995). Factorizations of product graphs into cycles of uniform length. *Graphs Combin.*, *11*, 69–90.

Nakayama, T. and Hashimoto, J. (1950). On a problem of G. Birkhoff. *Proc. Amer. Math. Soc.*, *1*, 141–142.

Nash-Williams, C. S. J. A. (1961). Edge-disjoint spanning trees of finite graphs. *J. London Math. Soc.*, *36*, 445–450.

Nash-Williams, C. S. J. A. (1964). Decomposition of finite graphs into forests. *J. London Math. Soc.*, *39*, 12.

Nebeský, L. (1971). Median graphs. *Comment. Math. Univ. Carolin.*, *12*, 317–325.

Nešetřil, J. (1981). Representations of graphs by means of products and their complexity. In *Mathematical Foundations of Computer Science 1981 (Proc. 10th Symp., Štrbske Pleso/Czech. 1981)*, volume 118 of *Lecture Notes Comput. Sci.*, (pp. 94–102). Springer, Berlin.

Nešetřil, J. and Pultr, A. (1978). On classes of relations and graphs determined by subobjects and factorobjects. *Discrete Math.*, *22*, 287–300.

Nešetřil, J. and Rödl, V. (1978). A simple proof of the Galvin-Ramsey property of the class of all finite graphs and a dimension of a graph. *Discrete Math.*, *23*, 49–55.

Nešetřil, J. and Rödl, V. (1985). Three remarks on dimensions of graphs. In *Random Graphs '83, Lect. 1st Semin., Poznan/Pol. 1983*, volume 28 of *Ann. Discrete Math.*, (pp. 199–207). North-Holland, Amsterdam.

Neufeld, S. and Nowakowski, R. J. (1998). A game of cops and robbers played on products of graphs. *Discrete Math.*, *186*, 253–268.

Nickel, C. L. M. (2007). *Random dot product graphs: A model for social networks.* ProQuest LLC, Ann Arbor, MI. Thesis (Ph.D.)–The Johns Hopkins University.

Nieminen, J. (1987). Distance center and centroid of a median graph. *J. Franklin Inst.*, *323*, 89–94.

Niven, I., Zuckerman, H. S., and Montgomery, H. L. (1991). *An Introduction to the Theory of Numbers* (fifth ed.). John Wiley & Sons, New York.

Nowakowski, R. and Rival, I. (1983). The smallest graph variety containing all paths. *Discrete Math.*, *43*, 223–234.

Nowakowski, R. and Rival, I. (1988). Retract rigid Cartesian products of graphs. *Discrete Math.*, *70*, 169–184.

Nowakowski, R. J. and Rall, D. F. (1996). Associative graph products and their independence, domination and coloring numbers. *Discuss. Math. Graph Theory*, *16*, 53–79.

Nowitz, L. A. (1968). On the non-existence of graphs with transitive generalized dicyclic groups. *J. Combinatorial Theory*, *4*, 49–51.

Nowitz, L. A. and Watkins, M. E. (1972a). Graphical regular representations of non-abelian groups, I. *Canad. J. Math.*, *24*, 993–1008.

Nowitz, L. A. and Watkins, M. E. (1972b). Graphical regular representations of non-abelian groups, II. *Canad. J. Math.*, *24*, 1009–1018.

Ovchinnikov, S. (2008). Media theory: Representations and examples. *Discrete Appl. Math.*, *156*, 1197–1219.

Oxley, J. G. (1992). *Matroid Theory.* Oxford Science Publications. The Clarendon Press Oxford University Press, New York.

Ozeki, K. (2009). A degree sum condition for graphs to be prism Hamiltonian. *Discrete Math.*, *309*, 4266–4269.

Parker, E. T. (1973). Edge coloring numbers of some regular graphs. *Proc. Amer. Math. Soc.*, *37*, 423–424.

Pasotti, A. and Pellegrini, M. A. (2010). Symmetric 1-factorizations of the complete graph. *European J. Combin.*, *31*, 1410–1418.

Paul, N. and Tardif, C. (2011). The chromatic Ramsey number of odd wheels. To appear in *J. Graph Theory*, DOI: 10.1002/jgt.20575.

Paulraja, P. (1993). A characterization of Hamiltonian prisms. *J. Graph Theory*, *17*, 161–171.

Paulraja, P. and Sivasankar, S. (2009). Directed Hamilton cycle decompositions of the tensor products of symmetric digraphs. *Graphs Combin.*, *25*, 571–581.

Paulraja, P. and Varadarajan, N. (2004). Independent sets and matchings in tensor product of graphs. *Ars Combin.*, *72*, 263–276.

Payan, C. and Xuong, N. H. (1982). Domination-balanced graphs. *J. Graph Theory, 6*, 23–32.

Pesch, E. (1988). *Retracts of Graphs.* Athenaeum Verlag, Frankfurt.

Pisanski, T. (1980). Genus of Cartesian products of regular bipartite graphs. *J. Graph Theory, 4*, 41–51.

Pisanski, T., Shawe-Taylor, J., and Mohar, B. (1983). 1-Factorization of the composition of regular graphs. *Publ. Inst. Math. (Beograd) (N.S.), 33*, 193–196.

Pisanski, T., Shawe-Taylor, J., and Vrabec, J. (1983). Edge-colorability of graph bundles. *J. Combin. Theory Ser. B, 35*, 12–19.

Pisanski, T. and Tucker, T. W. (2002). Growth in products of graphs. *Australas. J. Combin., 26*, 155–169.

Pisanski, T. and Vrabec, J. (1982). Graph bundles. Preprint Ser. Dep. Math. 20, no. 079, pp. 213–298.

Pisanski, T. and Žerovnik, J. (2009). Hamilton cycles in graph bundles over a cycle with tree as a fibre. *Discrete Math., 309*, 5432–5436.

Pisanski, T., Zmazek, B., and Žerovnik, J. (2001). An algorithm for k-convex closure and an application. *Int. J. Comput. Math., 78*, 1–11.

Polat, N. (1995). Invariant graphs for a family of endomorphisms—a survey. In *Combinatorics and Graph Theory '95, Vol. 1 (Hefei)* (pp. 313–331). World Sci. Publ., River Edge, NJ.

Polat, N. (2001). Convexity and fixed-point properties in Helly graphs. *Discrete Math., 229*, 197–211. Combinatorics, graph theory, algorithms and applications.

Polat, N. (2004). Fixed finite subgraph theorems in infinite weakly modular graphs. *Discrete Math., 285*, 239–256.

Polat, N. and Sabidussi, G. (1994). Fixed elements of infinite trees. *Discrete Math., 130*, 97–102.

Poljak, S. (1991). Coloring digraphs by iterated antichains. *Comment. Math. Univ. Carolin., 32*, 209–212.

Poljak, S. and Pultr, A. (1981). Representing graphs by means of strong and weak products. *Comment. Math. Univ. Carolin., 22*, 449–466.

Poljak, S. and Rödl, V. (1981). On the arc-chromatic number of a digraph. *J. Combin. Theory Ser. B, 31*, 190–198.

Poljak, S., Rödl, V., and Pultr, A. (1983). On a product dimension of bipartite graphs. *J. Graph Theory, 7*, 475–486.

Pultr, A. (1970). Tensor products in the category of graphs. *Comment. Math. Univ. Carolin., 11*, 619–639.

Pultr, A. (1972). Extending tensor products to structures of closed categories. *Comment. Math. Univ. Carolin., 13*, 599–616.

Quilliot, A. (1983). *Homomorphismes, points fixes, rétractions et jeux de poursuite dans les graphes, les ensembles ordonnés et les espaces métriques*. Thèse d'État, Université de Paris VI.

Quilliot, A. (1985a). On the Helly property working as a compactness criterion on graphs. *J. Combin. Theory Ser. A*, *40*, 186–193.

Quilliot, A. (1985b). A retraction problem in graph theory. *Discrete Math.*, *54*, 61–71.

Ramachandran, S. and Parvathy, R. (1996). Pancyclicity and extendability in strong products. *J. Graph Theory*, *22*, 75–82.

Reed, B. A. and Wood, D. R. (2008). Polynomial treewidth forces a large grid-like-minor. arXiv:0809.0724v3 [math.CO].

Reingold, O., Vadhan, S., and Wigderson, A. (2002). Entropy waves, the zig-zag graph product, and new constant-degree expanders. *Ann. of Math. (2)*, *155*, 157–187.

Richards, M., Côrte-Real, H., Forster, P., Macaulay, V., Wilkinson-Herbots, H., Demaine, A., Papiha, S., Hedges, R., Bandelt, H.-J., and Sykes, B. (1996). Paleolithic and Neolithic lineages in the European mitochondrial gene pool. *Amer. J. Human Genetics*, *59*, 185–203.

Ringel, G. (1955). Über drei kombinatorische Probleme am n-dimensionalen Würfel und Würfelgitter. *Abh. Math. Sem. Univ. Hamburg*, *20*, 10–19.

Robertson, N. and Seymour, P. D. (2003). Graph minors. XVI. Excluding a non-planar graph. *J. Combin. Theory Ser. B*, *89*, 43–76.

Robertson, N., Seymour, P. D., and Thomas, R. (1993). Hadwiger's conjecture for K_6-free graphs. *Combinatorica*, *13*, 279–361.

Rollová, E. and Škoviera, M. (2009). Nowhere-zero flows in Cartesian bundles of graphs. Faculty of Mathematics, Physics, and Informatics Comenius University, Bratislava, TR-2009-021.

Romani, F. (1980). Shortest-path problem is not harder than matrix multiplication. *Inform. Process. Lett.*, *11*, 134–136.

Rosenfeld, M. and Barnette, D. (1973). Hamiltonian circuits in certain prisms. *Discrete Math.*, *5*, 389–394.

Roth, R. L. and Winkler, P. M. (1986). Collapse of the metric hierarchy for bipartite graphs. *European J. Combin.*, *7*, 371–375.

Sabidussi, G. (1957). Graphs with given group and given graph-theoretical properties. *Canad. J. Math.*, *9*, 515–525.

Sabidussi, G. (1958). On a class of fixed-point-free graphs. *Proc. Amer. Math. Soc.*, *9*, 800–804.

Sabidussi, G. (1959). The composition of graphs. *Duke Math. J.*, *26*, 693–696.

Sabidussi, G. (1960). Graph multiplication. *Math. Z.*, *72*, 446–457.

Sabidussi, G. (1961). Graph derivatives. *Math. Z.*, *76*, 385–401.

Sabidussi, G. (1964). Vertex-transitive graphs. *Monatsh. Math.*, *68*, 426–438.

Sabidussi, G. (1975). Subdirect representations of graphs. In *Infinite and Finite Sets (Colloq., Keszthely, 1973; dedicated to P. Erdős on his 60th birthday), Vol. III*, volume 10 of *Colloq. Math. Soc. János Bolyai* (pp. 1199–1226). North-Holland, Amsterdam.

Sauer, N. (2001). Hedetniemi's conjecture—a survey. *Discrete Math.*, *229*, 261–292.

Sauer, N. and Zhu, X. (1992). An approach to Hedetniemi's conjecture. *J. Graph Theory*, *16*, 423–436.

Schaar, G., Sonntag, M., and Teichert, H.-M. (1988). *Hamiltonian Properties of Products of Graphs and Digraphs*, volume 108 of *Teubner-Texte zur Mathematik [Teubner Texts in Mathematics]*. BSB B. G. Teubner Verlagsgesellschaft, Leipzig.

Schönberg, I. J. (1938). Metric spaces and positive definite functions. *Trans. Amer. Math. Soc.*, *44*, 522–536.

Schrijver, A. (1979). A comparison of the Delsarte and Lovász bounds. *IEEE Trans. Inform. Theory*, *25*, 425–429.

Seshu, S. and Reed, M. B. (1961). *Linear Graphs and Electrical Networks*. Addison-Wesley, Reading, MA.

Shannon, C. E. (1956). The zero error capacity of a noisy channel. *IRE Trans. Inform. Theory*, *2*, 8–19.

Shearer, J. B. and Watkins, M. E. (1987). Counterexamples to two conjectures about distance sequences. *Discrete Math.*, *66*, 289–298.

Shields, I., Shields, B. J., and Savage, C. D. (2009). An update on the middle levels problem. *Discrete Math.*, *309*, 5271–5277.

Shpectorov, S. V. (1993). On scale embeddings of graphs into hypercubes. *European J. Combin.*, *14*, 117–130.

Shu, J. and Zhang, C.-Q. (2005). Nowhere-zero 3-flows in products of graphs. *J. Graph Theory*, *50*, 79–89.

Simonyi, G. (2006). Asymptotic values of the Hall-ratio for graph powers. *Discrete Math.*, *306*, 2593–2601.

Simonyi, G. and Tardos, G. (2006). Local chromatic number, Ky Fan's theorem and circular colorings. *Combinatorica*, *26*, 587–626.

Sims, J. and Holton, D. A. (1978). Stability of Cartesian products. *J. Combin. Theory Ser. B*, *25*, 258–282.

Škrekovski, R. (2001). Two relations for median graphs. *Discrete Math.*, *226*, 351–353.

Smith, B. R. (2008). Decomposing complete equipartite graphs into cycles of length $2p$. *J. Combin. Des.*, *16*, 244–252.

Smith, B. R. and Cavenagh, N. J. (2010). Decomposing complete equipartite graphs into short odd cycles. *Electron. J. Combin.*, *17*, Research Paper 130, 10.

Sohn, M. Y. and Lee, J. (1995). Characteristic polynomials of complement of graph bundles and their applications. *Linear and Multilinear Algebra*, *39*, 179–190.

Soltan, P. S. and Chepoĭ, V. D. (1987). Solution of the Weber problem for discrete median metric spaces. *Trudy Tbiliss. Mat. Inst. Razmadze Akad. Nauk Gruzin. SSR*, *85*, 52–76.

Sonnemann, E. and Krafft, O. (1974). Independence numbers of product graphs. *J. Combin. Theory Ser. B*, *17*, 133–142.

Špacapan, S. (2008). Connectivity of Cartesian products of graphs. *Appl. Math. Lett.*, *21*, 682–685.

Špacapan, S. (2010). Connectivity of strong products of graphs. *Graphs Combin.*, *26*, 457–467.

Špacapan, S. (2011). The k-independence number of direct products of graphs and Hedetniemi's conjecture. Manuscript.

Stadler, B. M. R. and Stadler, P. F. (2004). The topology of evolutionary biology. In *Modelling in Molecular Biology*, Nat. Comput. Ser. (pp. 267–286). Springer, Berlin.

Stahl, S. (1976). n-tuple colorings and associated graphs. *J. Combin. Theory Ser. B*, *20*, 185–203.

Stillwell, J. (1993). *Classical Topology and Combinatorial Group Theory* (Second ed.)., volume 72 of *Graduate Texts in Mathematics*. Springer-Verlag, New York.

Stong, R. (1991). Hamilton decompositions of Cartesian products of graphs. *Discrete Math.*, *90*, 169–190.

Sun, L. (2004). A result on Vizing's conjecture. *Discrete Math.*, *275*, 363–366.

Tardif, C. (1996). On compact median graphs. *J. Graph Theory*, *23*, 325–336.

Tardif, C. (1997). A fixed box theorem for the Cartesian product of graphs and metric spaces. *Discrete Math.*, *171*, 237–248.

Tardif, C. (1998). Graph products and the chromatic difference sequence of vertex-transitive graphs. *Discrete Math.*, *185*, 193–200.

Tardif, C. (2005a). The fractional chromatic number of the categorical product of graphs. *Combinatorica*, *25*, 625–632.

Tardif, C. (2005b). Multiplicative graphs and semi-lattice endomorphisms in the category of graphs. *J. Combin. Theory Ser. B*, *95*, 338–345.

Tardif, C. (2008). Hedetniemi's conjecture, 40 years later. *Graph Theory Notes N. Y.*, *54*, 46–57.

Taylor, D. T. (2009). Perfect r-codes in lexicographic products of graphs. *Ars Combin.*, *93*, 215–223.

Teh, H. H. and Gan, H. W. (1970). A note on free product of rooted graphs. *Nanyang Univ. J.*, *4*(Part I), 20–23.

Thomason, A. (1989). A disproof of a conjecture of Erdős in Ramsey theory. *J. London Math. Soc.*, *39*, 246–255.

Tits, J. (1970). Sur le groupe des automorphismes d'un arbre. In *Essays on Topology and Related Topics (Memoires dedies a Georges de Rham)*, (pp. 188–211). Springer, New York.

Tóth, Á. (2009). On the ultimate lexicographic Hall-ratio. *Discrete Math.*, *309*, 3992–3997.

Tošić, R. (1986). Search number of the Cartesian product of graphs. *Zb. Rad., Prir.-Mat. Fak., Univ. Novom Sadu, Ser. Mat.*, *16*, 239–243.

Trnková, V. (1976). On products of binary relational structures. *Comment. Math. Univ. Carolin.*, *17*, 513–521.

Trnková, V. (1984). Isomorphisms of products of infinite connected graphs. *Comment. Math. Univ. Carolin.*, *25*, 303–317.

Trnková, V. (1990). Products of metric, uniform and topological spaces. *Comment. Math. Univ. Carolin.*, *31*, 167–180.

Trnková, V. and Koubek, V. (1978). Isomorphisms of products of infinite graphs. *Comment. Math. Univ. Carolin.*, *19*, 639–652.

Trofimov, V. I. (1984). Graphs with polynomial growth. *Mat. Sb. (N.S.)*, *123(165)*, 407–421.

Truszczynski, M. (1983). On a conjecture of Mohar and Pisanski. *Demonstratio Math.*, *16*, 755–759.

Tutte, W. T. (1961). On the problem of decomposing a graph into n connected factors. *J. London Math. Soc.*, *36*, 221–230.

Valencia-Pabon, M. and Vera, J. (2006). Independence and coloring properties of direct products of some vertex-transitive graphs. *Discrete Math.*, *306*, 2275–2281.

van de Woestijne, C. E. (2011). Factors of disconnected graphs and polynomials with nonnegative integer coefficients. arXiv:1103.0709 [math.CO].

van Lint, J. H. and Wilson, R. M. (1992). *A Course in Combinatorics*. Cambridge University Press, Cambridge.

Vesel, A. (1998). The independence number of the strong product of cycles. *Comput. Math. Appl.*, *36*, 9–21.

Vesel, A. and Žerovnik, J. (2002). Improved lower bound on the Shannon capacity of C_7. *Inform. Process. Lett.*, *81*, 277–282.

Vesztergombi, K. (1978). Some remarks on the chromatic number of the strong product of graphs. *Acta Cybernet.*, *4*, 207–212.

Vince, A. (1988). Star chromatic number. *J. Graph Theory*, *12*, 551–559.

Vizing, V. G. (1963). The Cartesian product of graphs (Russian). *Vyčisl. Sistemy*, *9*, 30–43. English translation in *Comp. El. Syst.* 2 (1966) 352–365.

Vizing, V. G. (1968). Some unsolved problems in graph theory. *Russ. Math. Surv.*, *23*, 125–141.

Wagner, K. (1936). Bemerkungen zum Vierfarbenproblem. *Jber. Deutsch. Math. Verein.*, *46*, 21–22.

Wallis, W. D. (1997). *One-Factorizations*. Kluwer, Dordrecht.

Wallis, W. D. and Wang, Z. J. (1985). On one-factorizations of Cartesian products. *Congr. Numer.*, *49*, 237–245.

Wallis, W. D. and Wang, Z. J. (1987). Some further results on one-factorizations of Cartesian products. *J. Combin. Math. Combin. Comput.*, *1*, 221–234.

Watkins, M. E. (2004). Automorphisms. In J. L. Gross and J. Yellen (Eds.), *Handbook of Graph Theory* (pp. 485–504). CRC Press, Boca Raton, FL.

Watkins, M. E. and Zhou, X. (2007). Distinguishability of locally finite trees. *Electron. J. Combin.*, *14*, Research Paper 29, 10 pp.

Weichsel, P. M. (1962). The Kronecker product of graphs. *Proc. Amer. Math. Soc.*, *13*, 47–52.

Welzl, E. (1984). Symmetric graphs and interpretations. *J. Combin. Theory Ser. B*, *37*, 235–244.

West, D. B. (1996). *Introduction to Graph Theory*. Prentice Hall, Upper Saddle River, NJ.

White, A. T. (1970). The genus of repeated Cartesian products of bipartite graphs. *Trans. Amer. Math. Soc.*, *151*, 393–404.

Whitehead, A. N. and Russell, B. (1912). *Principia Mathematica*, volume 2. Cambridge University Press, Cambridge.

Wiener, H. (1947). Structural determination of paraffin boiling points. *J. Amer. Chem. Soc.*, *69*, 17–20.

Wilkeit, E. (1986). *Isometrische Untergraphen von Hamming-Graphen*. Ph.D. thesis, Universität Oldenburg.

Wilkeit, E. (1990). Isometric embeddings in Hamming graphs. *J. Combin. Theory Ser. B*, *50*, 179–197.

Wilkeit, E. (1992). The retracts of Hamming graphs. *Discrete Math.*, *102*, 197–218.

Winkler, P. M. (1983). Proof of the squashed cube conjecture. *Combinatorica*, *3*, 135–139.

Winkler, P. M. (1984). Isometric embedding in products of complete graphs. *Discrete Appl. Math.*, *7*, 221–225.

Winkler, P. M. (1987). Factoring a graph in polynomial time. *European J. Combin.*, *8*, 209–212.

Wood, D. R. (2007). Clique minors in Cartesian products of graphs. arXiv:0711.1189v3 [math.CO].

Xu, J.-M. and Yang, C. (2006). Connectivity of Cartesian product graphs. *Discrete Math.*, *306*, 159–165.

Xu, J.-M. and Yang, C. (2010). Connectivity and super-connectivity of Cartesian product graphs. *Ars Combin.*, *95*, 235–245.

Yan, W. and Zhang, F. (2006). Enumeration of perfect matchings of a type of Cartesian products of graphs. *Discrete Appl. Math.*, *154*, 145–157.

Yeh, Y. N. and Gutman, I. (1994). On the sum of all distances in composite graphs. *Discrete Math.*, *135*, 359–365.

Young, S. J. and Scheinerman, E. (2008). Directed random dot product graphs. *Internet Math.*, *5*, 91–111.

Yu, X. and Zickfeld, F. (2006). Reducing Hajós' 4-coloring conjecture to 4-connected graphs. *J. Combin. Theory Ser. B*, *96*, 482–492.

Zaks, J. (1974). Hamiltonian cycles in products of graphs. *Canad. Math. Bull.*, *17*, 763–765.

Žerovnik, J. (2000). On recognition of strong graph bundles. *Math. Slovaca*, *50*, 289–301.

Zhang, H. (2010). Independent sets in direct products of vertex-transitive graphs. Manuscript, arXiv:1007.0797v1 [math.CO].

Zhang, H. (2011). Primitivity and independent sets in direct products of vertex-transitive graphs. To appear in *J. Graph Theory*, DOI: 10.1002/jgt.20526.

Zhang, Z., Zheng, Y., and Mamut, A. (2007). Nowhere-zero flows in tensor product of graphs. *J. Graph Theory*, *54*, 284–292.

Zheng, Y., Zhang, Z., and Mamut, A. (2009). Nowhere-zero flows in lexicographic product of graphs. *J. Math. Study*, *42*, 30–35.

Zhou, H. (1991a). Multiplicativity. I. Variations, multiplicative graphs, and digraphs. *J. Graph Theory*, *15*, 469–488.

Zhou, H. (1991b). Multiplicativity. II. Nonmultiplicative digraphs and characterization of oriented paths. *J. Graph Theory*, *15*, 489–509.

Zhou, H. and Zhu, X. (1997). Multiplicativity of acyclic local tournaments. *Combinatorica*, *17*, 135–145.

Zhou, M. (1989). Decomposition of some product graphs into 1-factors and Hamiltonian cycles. *Ars Combin.*, *28*, 258–268.

Zhu, X. (1992a). A simple proof of the multiplicativity of directed cycles of prime power length. *Discrete Appl. Math.*, *36*, 313–316.

Zhu, X. (1992b). Star chromatic numbers and products of graphs. *J. Graph Theory*, *16*, 557–569.

Zhu, X. (1996). On the bounds for the ultimate independence ratio of a graph. *Discrete Math.*, *156*, 229–236.

Zhu, X. (1998). A survey on Hedetniemi's conjecture. *Taiwanese J. Math.*, *2*, 1–24.

Zhu, X. (1999). Construction of uniquely H-colorable graphs. *J. Graph Theory*, *30*, 1–6.

Zhu, X. (2001). Circular chromatic number: A survey. *Discrete Math.*, *229*, 371–410. Combinatorics, graph theory, algorithms and applications.

Zhu, X. (2002). The fractional chromatic number of the direct product of graphs. *Glasg. Math. J.*, *44*, 103–115.

Zhu, X. (2011). Fractional Hedetniemi's conjecture is true. To appear in *European J. Combin.*, DOI: 10.1016/j.ejc.2011.03.004.

Ziegler, G. M. (1995). *Lectures on Polytopes*, volume 152 of *Graduate Texts in Mathematics*. Springer-Verlag, New York.

Zmazek, B. and Žerovnik, J. (2002). Algorithm for recognizing Cartesian graph bundles. *Discrete Appl. Math.*, *120*, 275–302.

Zmazek, B. and Žerovnik, J. (2006). On domination numbers of graph bundles. *J. Appl. Math. Comput.*, *22*, 39–48.

Zmazek, B. and Žerovnik, J. (2007). Weak reconstruction of strong product graphs. *Discrete Math.*, *307*, 641–649.

Znoĭko, D. V. (1975). Free products of nets and free symmetrizers of graphs. *Mat. Sb. (N.S.)*, *98(140)*, 518–537, 639.

Author Index

Subject Index

Symbol Index